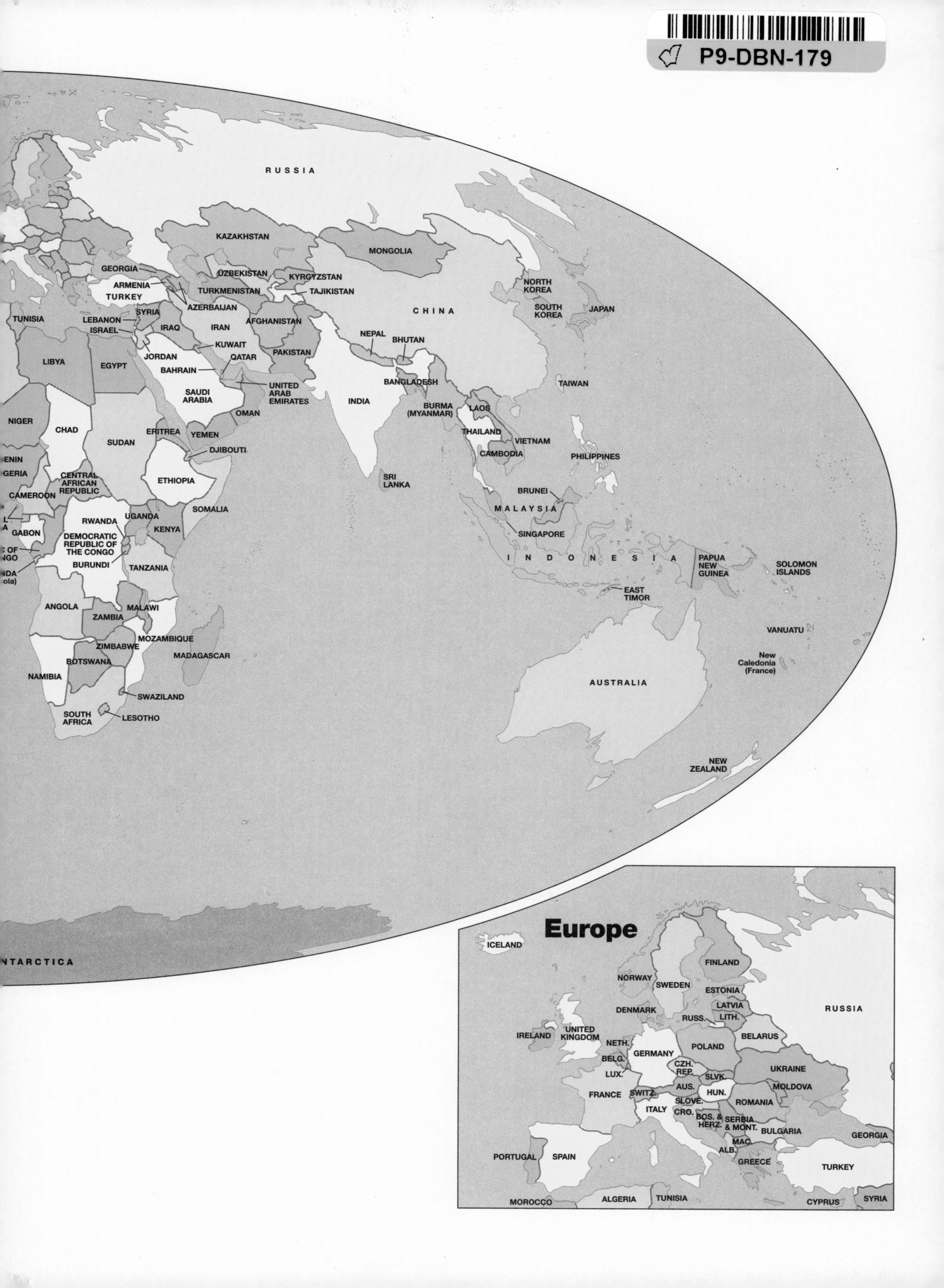
RUSSIA
KAZAKHSTAN
MONGOLIA
GEORGIA
UZBEKISTAN
KYRGYZSTAN
ARMENIA
TURKMENISTAN
TAJIKISTAN
TURKEY
AZERBAIJAN
NORTH KOREA
SOUTH KOREA
JAPAN
CHINA
TUNISIA
LEBANON
SYRIA
ISRAEL
IRAQ
IRAN
AFGHANISTAN
KUWAIT
NEPAL
BHUTAN
JORDAN
QATAR
PAKISTAN
LIBYA
EGYPT
BAHRAIN
SAUDI ARABIA
UNITED ARAB EMIRATES
BANGLADESH
TAIWAN
INDIA
OMAN
BURMA (MYANMAR)
LAOS
NIGER
CHAD
SUDAN
ERITREA
YEMEN
THAILAND
VIETNAM
DJIBOUTI
CAMBODIA
PHILIPPINES
CENTRAL AFRICAN REPUBLIC
ETHIOPIA
SRI LANKA
CAMEROON
BRUNEI
SOMALIA
MALAYSIA
RWANDA
UGANDA
GABON
KENYA
DEMOCRATIC REPUBLIC OF THE CONGO
SINGAPORE
BURUNDI
TANZANIA
INDONESIA
PAPUA NEW GUINEA
SOLOMON ISLANDS
EAST TIMOR
ANGOLA
MALAWI
ZAMBIA
VANUATU
MOZAMBIQUE
ZIMBABWE
MADAGASCAR
New Caledonia (France)
BOTSWANA
NAMIBIA
AUSTRALIA
SWAZILAND
SOUTH AFRICA
LESOTHO
NEW ZEALAND
Europe
ICELAND
FINLAND
NORWAY
SWEDEN
ESTONIA
LATVIA
DENMARK
RUSS.
LITH.
RUSSIA
IRELAND
UNITED KINGDOM
BELARUS
NETH.
GERMANY
POLAND
BELG.
CZH. REP.
UKRAINE
LUX.
SLVK.
MOLDOVA
AUS.
FRANCE
SWITZ.
HUN.
SLOVE.
ROMANIA
ITALY
CRO.
BOS. & HERZ.
SERBIA & MONT.
BULGARIA
GEORGIA
MAC.
PORTUGAL
SPAIN
ALB.
GREECE
TURKEY
MOROCCO
ALGERIA
TUNISIA
CYPRUS
SYRIA

PLACES AND REGIONS IN GLOBAL CONTEXT

Human Geography

swatch
bijoux
NUTUK DEĞİL!.. İŞ - KREŞ İSTİYORUZ
İMECELİ KADINLAR
NUTUK DEĞİL!.. İŞ - KREŞ İSTİYORUZ
İMECELİ KADINLAR
KADERİMİZ
DEĞİLDİR
KESK

FOURTH EDITION

PLACES AND REGIONS IN GLOBAL CONTEXT

Human Geography

Paul L. Knox
Virginia Tech

Sallie A. Marston
University of Arizona

Upper Saddle River, New Jersey 07458

Library of Congress Cataloging-in-Publication Data

Knox, Paul L.
Places and regions in global context : human geography / Paul L. Knox, Sallie A. Marston.—4th ed.
p. cm.
ISBN 0-13-149705-7
1. Human geography. I. Marston, Sallie A. II. Title.
GF41.K56 2007
304.2—dc22

2005036389

Acquisitions Editor: *Jeff Howard*
Editor-in-Chief, Science: *Dan Kaveney*
Associate Editor: *Amanda Griffith*
Production Editor: *Patty Donovan*
Executive Managing Editor: *Kathleen Schiaparelli*
Assistant Managing Editor: *Beth Sweeten*
Managing Editor, Science and Math Media: *Nicole M. Jackson*
Assistant Managing Editor, Science Supplements: *Karen Bosch*
Editor-in-Chief, Development: *Carol Trueheart*
Development Editor: *Ginger Birkeland*
Manufacturing Manager: *Alexis Heydt-Long*
Manufacturing Buyer: *Alan Fischer*
Director of Marketing: *Patrick Lynch*
Media Editor: *Chris Rapp*
Director of Creative Services: *Paul Belfanti*
Creative Director: *Juan R. López*
Art Director: *Maureen Eide*
Interior Design: *Suzanne Behnke*
Cover Design: *Juan R. López/Suzanne Behnke*
Senior Managing Editor, AV Production and Management: *Patricia Burns*
Manager, Production Technologies: *Matthew Haas*
Managing Editor, Art Management: *Abigail Bass*
AV Production Editor: *Jessica Einsig*
Illustrations: *Precision Graphics, MapQuest*
Director, Image Resource Center: *Melinda Patelli-Reo*
Manager, Rights and Permissions: *Zina Arabia*
Interior Image Specialist: *Beth Brenzel*
Cover Image Specialist: *Karen Sanatar*
Photo Researcher: *Teri Stratford*
Image Premission Coordinator: *Frances Toepfer*
Editorial Assistant: *Maragaret Ziegler*
Cover Image: *©Faith Saribas/Reuters/Corbis*

PEARSON
Prentice Hall

Pearson Prentice Hall
Pearson Education, Inc.
Upper Saddle River, New Jersey 07458

Printed in the United States of America
10 9 8 7 6 5 4 3 2 1

ISBN 0-13-149705-7

Pearson Education Ltd., *London*
Pearson Education Australia Pty., Limited, *Sydney*
Pearson Education Singapore, Pte. Ltd
Pearson Education North Asia Ltd., *Hong Kong*
Pearson Education Canada, Ltd., *Toronto*
Pearson Educación de Mexico, S.A. de C.V.
Pearson Education—Japan, *Tokyo*
Pearson Education Malaysia, Pte. Ltd.

Brief Contents

1 Geography Matters 3

2 The Changing Global Context 43

3 Geographies of Population 85

4 Nature and Society 129

5 Cultural Geographies 173

6 Interpreting Places and Landscapes 213

7 The Geography of Economic Development 251

8 Agriculture and Food Production 301

9 The Politics of Territory and Space 343

10 Urbanization 393

11 City Spaces: Urban Structure 427

12 Future Geographies 467

Contents

Lists of Maps xiii

Preface xv

About the Author xxi

Chapter 1 GEOGRAPHY MATTERS 3

Why Places Matter 4

The Influence and Meaning of Places 5 The Interdependence of Places 6 The Interdependence of Geographic Scales 7 Interdependence as a Two-Way Process 8

Interdependence in a Globalizing World 10

Perspectives on Globalization and Interdependence 10

Key Issues in a Globalizing World 13

Geography in a Globalizing World 19

Studying Human Geography 20

Basic Tools 20 Spatial Analysis 22 Regional Analysis 30 Developing a Geographical Imagination 35

Making a Difference: The Power of Geography 36

The Importance of a Geographic Education 37 Geographers at Work 37

1.1 GEOGRAPHY MATTERS—Interdependence Within and Between France and Vietnam 11

1.2 WINDOWS ON THE WORLD—Worlds Apart 14

Chapter 2 THE CHANGING GLOBAL CONTEXT 43

The Pre-Modern World 44

Hearth Areas 44 The Growth of Early Empires 46 The Geography of the Pre-Modern World 47

Mapping a New World Geography 50

Core and Periphery in the New World System 54 The Industrialization of the World's Core Regions 55

Internal Development of the Core Regions 61 Organizing the Periphery 63

Contemporary Globalization 69

The Causes and Consequences of Globalization 72 Globalization and Core-Periphery Differences 75

2.1 GEOGRAPHY MATTERS—Early Geographic Knowledge 49

2.2 GEOGRAPHY MATTERS—Geography and Exploration 52

2.3 GEOGRAPHY MATTERS—The Foundations of Modern Geography 56

2.4 VISUALIZING GEOGRAPHY—Railroads and Geographic Change 64

2.5 GEOGRAPHY MATTERS—Commodity Chains 70

Chapter 3 GEOGRAPHIES OF POPULATION 85

The Demographer's Toolbox 86
Censuses and Vital Records 86 Limitations of the Census 86

Population Distribution and Structure 88
Population Distribution 88 Population Density and Composition 89
Age-Sex Pyramids 92

Population Dynamics and Processes 99
Birth, or Fertility, Rates 99 Death, or Mortality, Rates 103
Demographic Transition Theory 106

Population Movement and Migration 108
Mobility and Migration 108 International Voluntary Migration 109
International Forced Migration 111
Internal Voluntary Migration 115 Internal Forced Migration 118

Population Debates and Policies 119
Population and Resources 119 Population Policies and Programs 120
Sustainable Development, Gender, and Population Issues 124

3.1 GEOGRAPHY MATTERS—**GIS Marketing Applications 94**

3.2 GEOGRAPHY MATTERS—**The Baby Boom and the Aging of the Population 96**

3.3 GEOGRAPHY MATTERS—**Internal Displacement 112**

Chapter 4 NATURE AND SOCIETY 129

Nature as a Concept 130
Nature and Society Defined 131 Nature-Society Interactions 133
U.S. Environmental Philosophies and Political Views of Nature 135
The Concept of Nature and the Rise of Science and Technology 141

The Transformation of Earth by Ancient Humans 142
Paleolithic Impacts 142 Neolithic Peoples and Domestication 143
Early Settlements and Their Environmental Impacts 144

European Expansion and Globalization 145
Disease and Depopulation in the Spanish Colonies 147 Old World Plants and Animals in the New World 148

Human Action and Recent Environmental Change 152
The Impact of Energy Needs on the Environment 152 Impacts of Land-Use Change on the Environment 159

The Globalization of the Environment 166
Global Environmental Politics 166 Environmental Sustainability 167

4.1 GEOGRAPHY MATTERS—**Understanding Cultural Ecology and Political Ecology 136**

4.2 WINDOWS ON THE WORLD—**Uranium Mining and the Impacts on Oceania 150**

4.3 GEOGRAPHY MATTERS—**Global Climate Change and the Kyoto Protocol 162**

Chapter 5 CULTURAL GEOGRAPHIES 173

Culture as a Geographical Process 174

Building Cultural Complexes 176

Cultural Systems 183
Geography and Religion 183 Geography and Language 186
Culture and Society 190

Islamic Cultural Nationalism 193

Culture and Identity 201
Sexual Geographies 201
Ethnicity and the Use of Space 204 Race and Place 205 Gender and Other Identities 205

Globalization and Cultural Change 207
Americanization and Globalization 207 A Global Culture? 208

5.1 GEOGRAPHY MATTERS—**The Culture of Hip-Hop 180**
5.2 GEOGRAPHY MATTERS—**Changing Religious Practices in Latin America and the Caribbean 194**
5.3 GEOGRAPHY MATTERS—**Language and Ethnicity in Africa 200**
5.4 WINDOWS ON THE WORLD—**Separatism in Québec 202**

Chapter 6 INTERPRETING PLACES AND LANDSCAPES 213

Behavior, Knowledge, and Human Environments 214

Place-Making 215
Territoriality 215 People and Places, Insiders and Outsiders 217
Experience and Meaning 219 Images and Behavior 219

Landscape as a Human System 223
Ordinary Landscapes 223 Landscape as Text 226

Coded Spaces 226
Semiotics in the Landscape 226 Sacred Spaces 229

Place and Space in Modern Society 232
Globalization and Place-Making 232
Places as Objects of Consumption 237
Place Marketing 240

6.1 GEOGRAPHY MATTERS—**Jerusalem, the Holy City 234**
6.2 GEOGRAPHY MATTERS—**The Cultural Geography of Cyberspace 237**
6.3 WINDOW ON THE WORLD—**Slow Cities 238**
6.4 VISUALIZING GEOGRAPHY—**Place Marketing and Economic Development 244**

Chapter 7 THE GEOGRAPHY OF ECONOMIC DEVELOPMENT 251

Patterns of Economic Development 252
The Unevenness of Economic Development 252 Resources and Development 257 The Economic Structure of Countries and Regions 260
International Trade, Aid, and Debt 264
Interpretations of Patterns of Development 271

Pathways to Regional Development 273
How Regional Economic Cores Are Created 273

How Core-Periphery Patterns Are Modified 277

Globalization and Economic Development 280

The Global Assembly Line 281

The Global Office 289 The Pleasure Periphery: Tourism and Economic Development 292

7.1 GEOGRAPHY MATTERS—**Sustainable Development 260**

7.2 WINDOW ON THE WORLD—**China's Economic Development 266**

7.3 GEOGRAPHY MATTERS—**Fair Trade 274**

7.4 GEOGRAPHY MATTERS—**The Changing Geography of the Clothing Industry 284**

Chapter 8 AGRICULTURE AND FOOD PRODUCTION 301

Traditional Agricultural Geography 302

Shifting Cultivation 304 Intensive Subsistence Agriculture 307 Pastoralism 308

Agricultural Revolution and Industrialization 309

The First Agricultural Revolution 310 The Second Agricultural Revolution 310 The Third Agricultural Revolution 311 The Industrialization of Agriculture 312

Global Restructuring of Agricultural Systems 313

Forces of Globalization 313

Agricultural Change and Development Policies in Latin America 323 The Organization of the Agro-Food System 324 Food Regimes 325

Social and Technological Change in Global Food Production 327

Two Examples of Social Change 327 Biotechnology Techniques in Agriculture 328

The Environment and Agricultural Industrialization 332

The Impact of the Environment on Agriculture 332 The Impact of Agriculture on the Environment 333

Problems and Prospects in the Global Food System 334

Famine and Undernutrition 335 Genetically Modified Organisms and the Global Food System 335 Urban Agriculture 337

8.1 GEOGRAPHY MATTERS—**The Blue Revolution and Global Shrimp 316**

8.2 GEOGRAPHY MATTERS—**A Look at the Green Revolution 320**

8.3 WINDOW ON THE WORLD—**The New Geography of Food and Agriculture in New Zealand 330**

Chapter 9 THE POLITICS OF TERRITORY AND SPACE 343

The Development of Political Geography 344

The Geopolitical Model of the State 344 Boundaries and Frontiers 346

Geopolitics and the World Order 349

States and Nations 349

Theories and Practices of States 355

International and Supranational Organizations and New Regimes of Global Governance 374
Transnational Political Integration 375 Globalization, Transnational Governance, and the State 376

The Two-Way Street of Politics and Geography 379
The Politics of Geography 379
The Geography of Politics and Geographical Systems of Representation 386

9.1 WINDOW ON THE WORLD—Afghanistan: From the Cold War to the New World Order 368
9.2 GEOGRAPHY MATTERS—State Terrorism in Chechnya 372
9.3 WINDOW ON THE WORLD—The Palestinian-Israeli Conflict 382

Chapter 10 URBANIZATION 393

Urban Geography and Urbanization 394

Urban Origins 396
The Roots of European Urban Expansion 397 Industrialization and Urbanization 403 Imperialism and Peripheral Urbanization 403

Urban Systems 405
City-Size Distributions, Primacy, and Centrality 408 World Cities and the Global Urban System 411 World Urbanization Today 412
Regional Trends and Projections 414 The Periphery and Semiperiphery: Overurbanization and Megacities 415
The Core: Mature Metropolises 418

Globalization and Splintering Urbanism 421

10.1 VISUALIZING GEOGRAPHY—Shock City: Manchester 406
10.2 WINDOW ON THE WORLD—Urban Terrorism 413
10.3 WINDOW ON THE WORLD—The Pearl River Delta: An Extended Metropolis 416

Chapter 11 CITY SPACES: URBAN STRUCTURE 427

Urban Land Use and Spatial Organization 428
Accessibility and Land Use 428 Territoriality, Congregation, and Segregation 428

Traditional Patterns of Urban Structure 429
North American Cities 429 Problems of North American Cities 430 European Cities 435 Islamic Cities 441 Cities of the Periphery: Unintended Metropolises 444

New Patterns: The Polycentric Metropolis 453
Sprawl 454 Packaged Landscapes 455 Globalization and the Quartering of Urban Space 457

11.1 VISUALIZING GEOGRAPHY—"Shock City": Lagos, Nigeria 448
11.2 WINDOW ON THE WORLD—Megapolitan Regions 458
11.3 WINDOW ON THE WORLD—The Globalization of Suburbia 462

Chapter 12 FUTURE GEOGRAPHIES 467

Mapping Our Futures 468

Resources, Technology, and Spatial Change 472
Transportation Technologies 473 Biotechnology 475 Materials Technologies 476 Information Technologies 476

Regional Prospects 478
Uneven Development 478 A New World Order? 479

Critical Issues and Threats 488
Globalizing Culture and Cultural Dissonance 488 Security 489 Sustainability 490

12.1 GEOGRAPHY MATTERS—**Dark Age Ahead? 471**

12.2 GEOGRAPHY MATTERS—**How the World Sees the United States 482**

12.3 GEOGRAPHY MATTERS—**The Asian Brown Cloud 491**

Appendix A: Maps and Geographic Information Systems 495
Map Scales 495
Map Projections 495
Geographic Information Systems 500
Applications of GIS 500
Critiques of GIS 503

Glossary 505

Photo Credits 514

Index 517

List of Maps

GM = Geography Matters; VG = Visualizing Geography; WW = Window on the World

1.5 Chang Jiang (Yangtze) delta region, China, 9
1.6 The human "footprint", 13
1.8 Diffusion of the HIV Virus, 17
1.9 The geography of HIV/AIDS, 17
1.17 Topological space, 26
2.1 Old World hearth areas, 44
2.2 New World hearth areas, 45
2.4 Greek colonies and the extent of the Roman empire, 46
2.5 The precapitalist Old World, circa A.D. 1400, 48
2.6 The Silk Road, 48
2.A GM Ptolemy's map of the world, 50
2.B GM The European Age of Discovery, 52
2.D GM Early topographic map, 54
2.8 The world-system core, semiperiphery, and periphery in 1800, 55
2.10 The spread of industrialization in Europe, 60
2.11 The world-system core, semiperiphery, and periphery in 1900, 60
2.12 The manufacturing belt of the United States, 61
2.4 VG Railroad service in 1860 and 1880, 64
2.15 Major steamship routes in 1920, 66
2.16 The international telegraph network in 1900, 66
2.17 The British Empire in the late 1800s, 67
2.18 The world-system core, semipheriphery, and periphery in 2005, 69
2.20 24-hour trading between major financial markets, 74
2.23 Global internet connectivity, 78
3.2 World population density, 2004, 88
3.3 Population distribution of Egypt, 2004, 90
3.6 Health-care density, 91
3.A GM Walk times to different trade areas, 94
3.B GM Concentrations of Atlanta children ages 3 and 4 without access to Head Start and/or kindergarten, 1998, 95
3.13 World crude birth rates, 2004, 102
3.14 World crude death rates, 2004, 103
3.15 World rates of natural increase, 2004, 104
3.16 World infant mortality rates, 2004, 105
3.17 Adults and children living with HIV/AIDS, 2001, 106
3.20 Global voluntary migration, 1999, 110
3.21 Refugee-sending countries, 1998–2002, 111
3.E GM Internally displaced people worldwide, 2004, 112
3.22 Kurdish and Lebanese diaspora, 1990, 114
3.24 Changing demographic center of the United States, 1790–2000, 116
3.27 Trail of Tears, 1830s, 118
3.30 World population, 2020, 121
4.5 Hurricane Katrina's path, 134
4.B GM St. Vincent and the Grenadines, 137
4.9 The settlement of the world, 142
4.17 European voyages of exploration, 149
4.C WW Oceania, 150
4.19 World population and consumption of energy, 1993–2002, 154
4.24 World distribution of nuclear reactors, 2000, 157
4.25 Global use of woodfuels, 2001, 157
4.26 Percent of hydropower in the electricity supply by country, 2002, 158
4.28 Global acid emissions, 1990s, 160
4.29 Global deforestation, 161
4.32 Desertification in sub-Saharan Africa, 165
5.C GM The sources and diffusion of U.S. rap, 181
5.9 U.S. religious population distribution by county, 2000, 182
5.10 World distribution of major religions, 184
5.11 Origin areas and diffusion of four major religions, 184
5.12 Spread of Buddhism, 185
5.13 Spread of Christianity in Europe, 185
5.14 Pre-Columbian religions in North America, 187
5.18 World distribution of major languages and major language families, 189
5.19 Source area of the Indo-European language family, 190
5.20 Language map of India, 191
5.21 African countries with extinct and threatened languages, 192
5.22 The languages and dialects of France in 1789, 192
5.23 Tongue-tied, 193
5.24 World film production, 2001, 196
5.25 World distribution of TV sets and cable subscribers, 1970 and 2001, 197
5.27 Mulsim world, 199
5.28 Cultural hearth of Islam, 199
5.F GM Language map of Africa, 200
5.G WW Referendum vote on separation from Canada, 203
6.2 Shepherd's map, 216
6.6 Cognitive image of Boston, 220
6.7 Images of Los Angeles, 221
6.8 Preference map of the United States, 222
6.14 Costa's master plan for the new capital of Brasilia, 229
6.19 Sacred sites of Hindu India, 231
6.20 Source areas for pilgrims to Mecca, 232
6.21 Source areas for pilgrims to Lourdes, 233
6.A GM Jerusalem, the Holy City, 234
6.28 London's Docklands, 241
6.4 VG Portsmouth, England, 244
7.1 Gross national income (GNI) per capita, 253
7.3 An index of human development, 2002, 254
7.4 An index of gender empowerment, 2002, 255
7.8 Agricultural land cover, 259

7.10 The geography of primary economic activities, 262
7.11 Emerging growth zones in Pacific Asia, 263
7.14 Index of commodity concentration of exports, 2002, 268
7.15 The debt crisis, 269
7.23 Toyota's global assembly line, 283
7.C GM The changing distribution of clothing manufacturing, 285
7.25 Principle maquiladora centers on the United States-Mexico border, 288
7.28 The clustering of advertising agencies in European cities, 292
7.29 Offshore financial centers, 293
8.3 Global distribution of agriculture, 2005, 304
8.4 Areas of plant and animal domestication, 305
8.11 Mediterranean transhumance routes, 309
8.14 Global distribution of fertilizer use, 1992-1996, 312
8.16 Tractors per 1,000 hectares, 314
8.C GM Leading importers of shrimp, 2004, 318
8.F GM Global distribution of maize production, 321
8.G GM Effects of the green revolution, 322
9.1 The changing map of Europe: 1924, 1989, 2005, 345
9.7 Borders between Egypt and Libya and Sudan, 349
9.8 Township-and-range system, 350
9.9 Nested hierarchy of *de jure* territories, 351
9.10 Territorial growth of the Muscovite/Russian state, 352
9.11 Soviet state expansionism, 1940s and 1950s, 353
9.13 Independent states of the former Soviet Union, 355
9.15 European colonies in Africa, 1946-1912, 357
9.16 Colonization in South America and the Caribbean, 1496-1667, 358
9.19 Participating countries in the League of Nations, 361
9.20 Decolonization of Africa, before and after 1960, 362
9.21 Independent South America, nineteenth century, 363
9.22 Independence in Asia and the South Pacific, before and after 1960, 364
9.23 Territorial divisions of Antartica, 365
9.24 The heartland, 366
9.A WW Afghanistan, 368
9.D GM The Northern Caucasus, 372
9.27 Military deaths in Iraq and Afghanistan, 2005, 374
9.28 U.N. member countries, 375
9.29 Membership in the European Union, 376
9.32 Map of the former Yugoslavia, 380
9.33 The 1860 presidential election, 381
9.F WW Changing of Israel/Palestine, 1923-2005, 383
9.G WW Israeli withdrawal from the Gaza Strip and the West Bank, 384
9.H WW Israeli security fence, 2005, 385
9.35 The vote for president, 1992, 387
9.37 North Carolina's proposed Twelfth Congressional District, 388
10.1 Percentage of each country's population living in urban settlements, 2003, 394
10.2 Rates of growth in urbanization, 2000-2005, 395
10.4 Major cities in A.D. 1000, 398
10.8 The towns and cities of Europe, ca. 1350, 401
10.9 Gateway cities in the world-system periphery, 402
10.10 Growth of Chicago, 404
10.1 VG The opening of the Suez Canal in 1869, 406
10.13 The Spanish urban system, 408
10.14 Functional specialization within an urban system, 409
10.16 Examples of urban centrality, 410
10.B WW Pearl River Delta, 416
11.10 Medieval Arras, 436
11.1 VG Lados Island, 448
11.A WW The Megapolitan regions of the United States, 458
11.B WW The Tokaido megalopolis, 459
11.C WW The I-35 megapolitan corridor, 460
11.D WW The Piedmont megapolitan galaxy, 460
12.2 High-speed rail in Europe, 474
12.3 The geography of technological innovation and achievement, 477
12.5 Index of income inequality, 481
12.6 European growth axes, 484
12.9 Wild zones, 487
12.A GM Asian brown cloud, 491
A.1 Topographic maps, 496
A.2 Isoline maps, 497
A.3 An example of proportional symbols in thematic mapping, 497
A.4 Located charts, 498
A.5 Comparison of map projections, 499
A.6 The Robinson projection, 500
A.7 The Peters projection, 500
A.8 Fuller's Dymaxion projection, 501
A.9 Examples of cartograms, 502
A.11 Map of land cover, 503
A.12 GIS-derived planning map, 504

Preface

Education either functions as an instrument which is used to facilitate integration of the younger generation into the logic of the present system and bring about conformity or it becomes the practice of freedom, the means by which men and women deal critically and creatively with reality and discover how to participate in the transformation of their world.

Paulo Freire

Most people have an understanding of what their own lives are like and a good deal of knowledge of their own neighborhood and perhaps even of the larger city and state in which they live. Yet, even as the countries and regions of the world become more interconnected, most of us still know very little about the lives of people on the other side of our country, or in other societies, or about the ways in which the lives of those people connect to our own. In order to change the world, to make it a better place in which to live for all people, we need to understand not just our little corner of it, but the whole of it, the broad sweep of human geography that constitutes the larger world in which our small corners are just a part.

This book provides an introduction to human geography that will help to provide the means by which young men and women can understand critically the world in which they live. To study human geography, to put it simply, is to study the dynamic and complex relationships between peoples and the worlds they inhabit. Our book gives students the basic geographical tools and concepts needed to understand the complexity of places and regions and to appreciate the interconnections between their own lives and those of people in different parts of the world and to make the world a better place.

Objective and Approach

The objective of the book is to introduce the study of human geography by providing not only a body of knowledge about the creation of places and regions, but also an understanding of the interdependence of places and regions in a globalizing world. The approach is aimed at establishing an intellectual foundation that will enable a life-long and life-sustaining geographical imagination: an essential tool for today's students to confront tomorrow's global, national, regional, and local challenges.

The book takes a fresh approach to human geography, reflecting the major changes that have recently been impressed on global, regional, and local landscapes. These changes include the globalization of industry and the related rapid rise of China and India as economic powerhouses, the upwelling of ethnic regionalisms on the heels of decolonization and the formation of new states, the movement of peoples around the world in search of better lives, the physical restructuring of cities, the transformation of traditional agricultural practices throughout much of the world, global environmental change and the movement for sustainability, the eruptions of war and the struggles for peace, and the emerging trend toward transnational political and economic organizations. The approach used in *Places and Regions in Global Context* provides access not only to the new ideas, concepts, and theories that address these changes but also to the fundamentals of human geography: the principles, concepts, theoretical frameworks, and basic knowledge that are necessary to more specialized studies.

The most distinctive feature of this approach is that it emphasizes the interdependence of both places and processes in different parts of the globe. In overall terms, this approach is designed to provide an understanding of relationships between global processes and the local places in which they unfold. It follows that one of the chief organizing principles is how globalization frames the social and cultural construction of particular places and regions.

This approach has several advantages.

- It captures aspects of human geography that are among the most compelling in the contemporary world—the geographical bases of cultural diversity and their impacts on everyday life, for example.
- It encompasses the salient aspects of new emphases in academic human geography—the new emphasis on sustainability and its role in the social construction of spaces and places, for example.
- It makes for an easier marriage between topical and regional material by emphasizing how processes link them—technological innovation and the varying ways technology is adopted and modified by people in particular places, for example.
- It facilitates meaningful comparisons between places in different parts of the world—how the core-generated industrialization of agriculture shapes gender relations in households both in the core and the periphery, for example.

In short, the textbook is designed to focus on geographical processes and to provide an understanding of the interdependence among places and regions without losing sight of their individuality and uniqueness.

Several important themes are woven into each chapter, integrating them into the overall approach:

- the relationship between global processes and their local manifestations,
- the interdependence of people and places, especially the interactive relationships between core regions and peripheral regions,
- the continuing transformation of the political economy of the world system, and of nations, regions, cities, and localities,
- the social and cultural differences that are embedded in human geographies (especially the differences that relate to race, ethnicity, gender, age, and class).

Chapter Organization

The organization of the book is innovative in several ways. First, the chapters are organized in such a way that the conceptual framework—why geography matters in a globalizing world—is laid out in Chapters 1 and 2 and then deployed in thematic chapters (Chapters 3 through 11). The concluding chapter, Chapter 12, provides a coherent summary of the main points of the text by showing how future geographies may unfold, given what is known about present geographical processes and trends. Second, the conceptual framework of the book requires the inclusion of two introductory chapters rather than the usual one. The first describes the basics of a geographic perspective; the second explains the value of the globalization approach.

Third, the distinctive chapter ordering within the book follows the logic of moving from less complex to more complex systems of human social and economic organization, always highlighting the interaction between people and the world around them. The first thematic chapter (Chapter 3) focuses on human population. Its early placement in the book reflects the central importance of people in understanding geography. Chapter 4 deals with the relationship between people and the environment as it is mediated by technology. This chapter capitalizes on the growing interest in environmental problems and establishes a central theme: that all human geographical issues are about how people negotiate their environment—whether the natural or the built environment. No other introductory human geography textbook includes such a chapter.

The chapter on nature, society, and technology is followed by Chapter 5 on cultural geography. The intention in positioning the cultural chapter here is to signal that culture is the primary medium through which people operate and understand their place in the world. In Chapter 6, the impact of cultural processes on the landscape is explored, together with the ways in which landscape shapes cultural processes.

In Chapter 7, the book begins the move toward more complex concepts and systems of human organization by concentrating on economic development. The focus of Chapter 8 is agriculture. The placement of agriculture after economic development reflects the overall emphasis on globalization. This chapter shows how processes of globalization and economic development have led to the industrialization of agriculture at the expense of more traditional agricultural systems and practices.

The final three thematic chapters cover political geography (Chapter 9), urbanization (Chapter 10), and city structure (Chapter 11). Devoting two chapters to urban geography, rather than a more conventional single chapter, is an important indication of how globalization increasingly leads to urbanization of the world's people and places. The final chapter, on future geographies (Chapter 12), gives a sense of how a geographic perspective might be applied to the problems and opportunities to be faced in the twenty-first century.

Features

To signal the freshness of the approach, the pedagogy of the book employs a unique cartography program, three different boxed features, "Visualizing Geography", "Geography Matters", and "Window on the World", as well as more familiar pedagogical devices such as chapter overviews and end-of-chapter exercises.

Geography Matters: Geography Matters boxes examine one of the key concepts of the chapter, providing an extended example of its meaning and implications through both visual illustration and text. The Geography Matters features demonstrate to students that the focus of human geography is on real world problems.

Visualizing Geography: Visualizing Geography boxes treat key concepts of the chapter by using a photographic essay. This feature helps students recognize that the visual landscape contains readily accessible evidence about the impact of globalization on people and places.

Window on the World: Window on the World boxes take a key concept and explore its application in a particular location. This feature allows students to appreciate the relevance of geographic concepts to world events and brings far-flung places closer to their comprehension.

Pedagogical Structure within Chapters: Each chapter opens with a brief vignette that introduces the theme of the chapter and illustrates why a geographical approach is important. A list of the main points that will be covered in the chapter follows this vignette. Throughout each chapter, key terms are printed in boldface as they are introduced, with capsule definitions of the term in the margin of the same page. These key terms are listed alphabetically at the end of the chapter, together with their location in the text. Figures with extensive captions are provided to integrate illustration with text.

At the end of each chapter, there are five useful devices to help students review. First comes a chapter conclusion that summarizes the overarching themes and concepts of the chapter. Next the main points of the chapter are listed again, but this time they are expanded to include a summary of the text discussion of each main point. Then there is a comprehensive list of key terms for the chapter, followed by a number of suggested additional readings on the topic of the chapter. Each chapter concludes with two sets of exercises, some Internet-based (On the Internet) and some more traditional (Unplugged) Both sets of exercises require students to put into practice several of the key concepts of a chapter.

New to the Fourth Edition

The fourth edition of *Places and Regions in Global Context* represents a thorough revision. Every part of the book was examined carefully with the dual goals of keeping topics and data current and improving the clarity of the text and the graphics. The fourth edition of the book incorporates a comprehensive updating of all of the data, maps, photographs, and illustrative examples. In the text, we have added or expanded upon quite a few topics, including Geographic Information Systems; regional analysis and regionalization; landscape, sense of place, and place-making; globalization and core-periphery differences; the geography of HIV/AIDS; transnational migrants and internally displaced persons; the aging of the global population; the globalization of the environment and the human impacts of global climate change; sustainability; Jihad vs. McWorld; international trade, aid, and debt; flexible production systems; the global food system; genetically modified agricultural products and issues of hunger and malnutrition; urban agriculture; new geographies of war and peace; transnational governance; globalization and new citizenship forms and rights; technology and its impacts on culture; global environmental politics; globalization and uneven urban development; urban terrorism and bioterrorism; and sprawl and packaged landscapes in the polycentric metropolis. These changes are designed to ensure that we offer the most up-to-date coverage of the field of human geography.

The pedagogical structure of the chapters has not been modified in any substantial way over the third edition. The beginning of each chapter continues to feature a section on the main points that will be covered in the chapters. These main points are revisited at the end of each chapter to reinforce the most important points and themes from the chapter. All of the end-of-chapter exercises have been revised, and the web exercises and website have been significantly upgraded and expanded. Lastly, each chapter includes an updated list of suggested further readings.

Supplements

The book includes a complete supplements program for both students and teachers.

For the Student

Study Guide (0-13-154794-1) The study guide includes chapter notes, key terms, 8–10 review questions and 10–15 activities per chapter, and references the text-specific website throughout.

Online Study Guide (http://www.prenhall.com/knox) This innovative online resource center is keyed by chapter to the text. It provides key terms, chapter exercises, thematic exercises, mapping exercises, and additional Internet resources to support and enhance students' study of human geography.

Rand McNally Atlas of World Geography (0-13-959339-X) This atlas includes 126 pages of up-to-date regional maps and 20 pages of illustrated world information tables. It is available FREE when packaged with *Places and Regions in Global Context.* Please contact your local Prentice Hall representative for details.

For the Instructor

Instructor Resource Center on CD-ROM (0-13-154778-X) The Instructor Resource Center on CD-ROM provides high-quality electronic versions of photos and illustrations from the book, as well as customizable PowerPoint lecture presentations, Classroom Response System questions in PowerPoint, and the Instructor's Manual and Test Item File in MS Word format. The CD-ROM includes all of the illustrations and photos from the text in 16-bit low-compression JPEG files. To further guarantee classroom projection quality, all images are manually adjusted for color, brightness, and contrast. For easy reference and identification, the images are organized by chapter.

Instructor's Resource Manual (0-13-154777-1) The Instructor's Resource Manual, intended as a resource for both new and experienced teachers, includes a variety of lecture outlines, additional source materials, teaching tips, advice on how to integrate visual supplements, answers to the end-of-chapter exercises, and various other ideas for the classroom.

Test Bank (0-13154779-8) An extensive array of test questions accompanies the book. These questions are available in hard copy and also on disks formatted for Macintosh and Windows (0-13-154793-3).

Overhead Transparencies (0-13-154776-3) The transparencies feature 225 illustrations from the text, all enlarged for excellent classroom visibility.

Conclusion

The idea for this book evolved from conversations between the authors and colleagues about how to teach human geography in colleges and universities. Our intent was to find a way not only to capture the exciting changes that are rewriting the world's landscapes and reorganizing the spatial relationships between people, but also to demonstrate convincingly why the study of geography matters. Our aim was to show why a geographical imagination is important, how it can lead to an understanding the world and its constituent places and regions, and how it has practical relevance in many spheres of life.

The first edition of this book was written at the culmination of a significant period of reform in geographic education. One important outcome of this reform was the inclusion of geography as a core subject in Goals 2000: Educate America Act (Public Law 103–227). Another was the publication of a set of national geography standards for K–12 education (*Geography for Life*, published by National Geographic Research and Education for the American Geographical Society, the Association of American Geographers, the National Council for Geographic Education, and the National Geographic Society, 1994). This fourth edition builds on these reforms, offering a fresh and compelling approach to college-level geography.

Acknowledgments

We are indebted to many people for their assistance, advice, and constructive criticism in the course of preparing this book. Among those who provided comments on various drafts of the four editions of this book are the following professors:

Christopher A. Airriess *(Ball State University)*
Stuart Aitken *(University of California at San Diego)*
Kevin Archer *(University of South Florida)*
Brian J. L. Berry *(University of Texas at Dallas)*
Brian W. Blouet *(College of William and Mary)*
George O. Brown, Jr. *(Boston College)*
Michael P. Brown *(University of Washington)*
Henry W. Bullamore *(Frostburg State University)*
Edmunds V. Bunske *(University of Delaware)*
Craig Campbell *(Youngstown State University)*
Dylan Clark *(University of Colorado)*
David B. Cole *(University of Northern Colorado)*
Jerry Crampton *(George Mason University)*
Christine Dando *(University of Nebraska, Omaha)*
Fiona M. Davidson *(University of Arkansas)*
Vernon Domingo *(Bridgewater State College)*
Patricia Ehrkamp *(Miami University)*
Nancy Ettlinger *(The Ohio State University)*
Paul B. Frederic *(University of Maine)*
Kurtis G. Fuelhart *(Shippensburg University)*
Gary Fuller *(University of Hawaii at Manoa)*
Wilbert Gesler *(University of North Carolina)*
Melissa Gilbert *(Temple University)*
Jeffrey Allman Gritzner *(University of Montana)*
Douglas Heffington *(Middle Tennessee State University)*
Andrew Herod *(University of Georgia)*
Peter Hugill *(Texas A&M University)*
David Icenogle *(Auburn University)*
Mary Jacob *(Mount Holyoke College)*
Douglas L. Johnson *(Clark University)*
Colleen E. Keen *(Minnesota State University)*
Paul Kelley *(University of Nebraska-Lincoln)*
Thomas Klak *(Miami University)*
James Kus *(California State University, Fresno)*
David Lanegran *(Macalester College)*
James Lindberg *(University of Iowa)*
John C. Lowe *(George Washington University)*
James McCarthy *(Penn State University)*
Byron Miller *(University of Cincinnati)*
Roger Miller *(University of Minnesota)*
John Milbauer *(Northeastern State University)*
Don Mitchell *(Syracuse University)*
Woodrow W. Nichols, Jr. *(North Carolina Central University)*
Richard Pillsbury *(Georgia State University)*
James Proctor *(University of California at Santa Barbara)*
Mark Purcell *(University of Washington)*
Jeffrey Richetto *(University of Alabama)*
Andrew Schoolmaster *(University of North Texas)*
David Schul *(The Ohio State University–Marion)*
Alex Standish *(Rutgers, University)*
Debra Straussfogel *(University of New Hampshire)*
Johnathan Walker *(James Madison University)*
Gerald R. Webster *(University of Alabama)*
Joseph S. Wood *(George Mason University)*
Wilbur Zelinsky *(Penn State University)*

In addition, Michael Wishart (World Bank Photo Library), Earthaline Harried (Photography Division, U.S. Department of Agriculture) and researcher Teri Stratford were especially helpful in our photo gathering. Special thanks go to our development editor, Ginger Birkeland; to our production editor, Patty Donovan; and to Sara Smith at the University of Arizona and Jennifer Woodward at the University of Kentucky for their excellent and insightful research assistance. We want to extend our deepest gratitude, however, to our editor-in-chief, Daniel Kaveney. It is impossible to find the right words to say how fortunate we have been to have worked with Dan. His respect for our ideas and promotion of them, knowledge of and commitment to the discipline of geography, keen intelligence for both the big picture and the little details, and his dedication to quality, have made the writing of this and all the previous editions of this textbook not only possible but an intellectual delight.

Finally, a number of colleagues gave generously of their time and expertise in guiding our thoughts, making valuable suggestions, and providing materials: Alejandro A. Alonso (University of Southern California), Michael Bonine (University of Arizona), Michael Brown (University of Washington), Martin Cadwallader (University of Wisconsin), Judith Carney (University of California, Los Angeles), Sarah Elwood (University of Arizona), Kim Elmore (Centers for Disease Control), Efiong Etuk (Virginia Tech), Antonio Luna Garcia (Universitas de Pompeu Fabra), Rudi Gaudio (University of Arizona), George Henderson (University of Arizona), John Paul Jones, III (University of Arizona), Miranda Joseph (University of

Arizona), Cindi Katz (City University of New York), Diana Liverman (Oxford University), Elaine Mariolle (University of Arizona), Beth Mitchneck (University of Arizona), Natalie Oswin (University of British Columbia), Mark Nichter (University of Arizona), Mimi Nichter (University of Arizona), Asli Ceylan Oner (Virginia Tech), Gearóid Ó Tuathail (Virginia Tech), Mark Patterson (Kennesaw State University), Leland Pederson (University of Arizona), David Plane (University of Arizona), Paul Robbins (University of Arizona), Dereka Rushbrook (University of Arizona), Ralph Saunders (California State University at Dominguez Hills), Neil Smith (City University of New York), Matthew Sparke (University of Washington), Dick Walker (University of California, Berkeley), Marv Waterstone (University of Arizona), Michael Watts (University of California, Berkeley), Clyde Woods (University of Maryland), Keith Woodward (University of Arizona), Emily Young (San Diego Foundation), Chris Uejio (University of Arizona), and Hannes Gerhandt (University of Arizona).

Paul L. Knox
Sallie A. Marston

About the Authors

Paul L. Knox

Paul Knox received his Ph.D. in Geography from the University of Sheffield, England. In 1985, after teaching in the United Kingdom for several years, he moved to the United States to take up a position as professor of urban affairs and planning at Virginia Tech. His teaching centers on urban and regional development, with an emphasis on comparative study. In 1989 he received a university award for teaching excellence. He has written several books on aspects of economic geography, social geography, and urbanization. He serves on the editorial board of several scientific journals and is co-editor on a series of books on World Cities. He is currently a University Distinguished Professor in the College of Architecture and Urban Studies at Virginia Tech.

Sallie A. Marston

Sallie Marston received her Ph.D. in Geography from the University of Colorado, Boulder. She has been a faculty member at the University of Arizona since 1986. Her teaching focuses on the historical, social, and cultural aspects of American urbanization, with particular emphasis on race, class, gender, and ethnicity issues. She received the College of Social and Behavioral Sciences Outstanding Teaching Award in 1989. She is the author of numerous journal articles and book chapters and serves on the editorial board of several scientific journals. In 1994/1995 she served as Interim Director of Women's Studies and the Southwest Institute for Research on Women. She is currently a professor in, and serves as head of, the Department of Geography and Regional Development.

PLACES AND REGIONS IN GLOBAL CONTEXT

Human Geography

1 Geography Matters

MAIN POINTS

■ Geography matters because it is specific places that provide the settings for people's daily lives. In these settings important events happen, and from them significant changes diffuse.

■ Places and regions are highly *interdependent,* each playing specialized roles in complex and ever-changing networks of interaction and change.

■ Some of the most important aspects of the interdependence between geographic scales are provided by the relationships between the global and the local.

■ Human geography provides ways of understanding places, regions, and spatial relationships as the products of a series of interrelated forces that stem from nature, culture, and individual human action.

■ The first law of geography is that "everything is related to everything else, but near things are more related than are distant things."

■ Distance is one aspect of this law, but connectivity is also important, because contact and interaction are dependent on channels of communication and transportation.

In today's world, where places are increasingly interdependent, it is important to know something about human geography and to understand how places affect, and are affected by, one another. Consider, for example, some of the prominent news stories of the first half of 2005. At first glance they were a mixture of achievements, disputes, and disasters that might seem to have little to do with geography, apart from the international flavor of the coverage. The world's richest countries—the G8 group—agreed to write off $40 billion in debts owed by 18 of the world's poorest countries, most of them in Africa; Syria pulled its troops out of Lebanon, allowing the first free elections in Lebanon since1976; a proposed Constitution for the European Union was stalled by referenda in France and the Netherlands; there was ongoing concern over the development of nuclear arms in Iran and North Korea; and, dominating headlines around the world, a continual stream of reports on the war in Iraq. Off the front pages we read of the startling economic growth of China, of concerns over the effects of global warming, controversies over genetically modified foods, and the continuing diffusion of the HIV/AIDS epidemic.

Most of these stories did have important geographical dimensions. The diffusion of HIV/AIDS, for example, is a geographical as well as social and cultural phenomenon. Thus, in the United States as in other countries, the HIV/AIDS epidemic has diffused through the country in a very distinctive geographical pattern. Behind some of the major news stories, geographical processes played a more central role. Stories about economic development, regional territorial disputes, ethnic conflict, and global warming, for example, all have strong geographical elements.

Human geography is about recognizing and understanding the interdependence among places and regions without losing sight of the uniqueness of specific places. *Places* are specific geographic settings with distinctive physical, social, and cultural attributes. *Regions* are territories that encompass many places, all or most of which share attributes different from the attributes of places elsewhere. Basic tools and fundamental concepts enable geographers to study the world in this way. Geographers learn about the world by finding out where things are and why they are there. Maps and mapping, of course, play a key role in how geographers analyze and portray the world. Maps are also important tools for introducing geographers' ideas about the way that places and regions are made and altered.

Business travellers at Amsterdam's Schipol Airport.

WHY PLACES MATTER

An appreciation of the diversity and variety of peoples and places is a theme that runs through all of *human geography*, the study of the spatial organization of human activity and of people's relationships with their environments. This theme is inherently interesting to nearly all of us. *National Geographic* magazine has become a national institution by drawing on the wonder and fascination of the seemingly endless variety of landscapes and communities around the world. Almost 8 million households, representing about 30 million regular readers, subscribe to this magazine for its intriguing descriptions and striking photographs. Millions more read it occasionally in offices, lobbies, or waiting rooms.

Yet many Americans often seem content to confine their interest in geography to the pages of glossy magazines, to television documentaries, or to vacations. It has become part of the conventional wisdom—both in the United States and around the world—that many Americans have little real appreciation or understanding of people and places beyond their own daily routines. This is perhaps putting it too mildly. Surveys have revealed widespread ignorance among a high proportion of Americans, not only of the fundamentals of the world's geography but also of the diversity and variety within the United States itself. One Gallup poll found that 70 percent of young adults in the United States could not find their own country on a map of the world. In a survey of more than 3,000 young adults in Canada, France, Germany, Great Britain, Italy, Japan, Mexico, Sweden, and the United States in 2002, Americans came in next to last in terms of geographic literacy.[1] Top scorers were young adults in Sweden, Germany, and Italy. Despite the daily bombardment of news from the Middle East, Central Asia, and other world trouble spots, roughly 85 percent of young Americans could not find Afghanistan, Iraq, or Israel on a map. The survey found that nearly 30 percent of young Americans could not find the Pacific Ocean, the world's largest body of water; more than half were unable to locate India; and only 19 percent could name four countries that officially acknowledge having nuclear weapons. This lack of geographical knowledge carries over to people's perceptions of the United States and its role in the world. A poll in 2004 found, for example, that most Americans believe that the United States spends more than 20 percent of its budget on aid to poor countries; the actual figure in 2003 was just 0.14 percent of the country's gross national income, putting it last in a table of rich industrial nations. So although most people in the United States are fascinated by different places, relatively few have a systematic knowledge of them. Fewer still understand how different places came to be the way they are, or why places matter in the broader scheme of things. This lack of understanding is important because geographic knowledge can take us far beyond simply glimpsing the inherently interesting variety of peoples and places.

The importance of geography as a subject of study is becoming more widely recognized, however, as people everywhere struggle to understand a world that is increasingly characterized by instant global communications, rapidly changing international relationships, unexpected local changes, and growing evidence of environmental degradation (**Figure 1.1**). Many more schools now require courses in geography than just a decade ago, and the College Board has added the subject to its Advanced Placement program. Meanwhile, many employers are coming to realize the value of employees with expert-

[1]Roper ASW, *2002 Global Geographic Literacy Survey*. Washington, DC: National Geographic—Roper ASW, 2002.

Figure 1.1 Geographic literacy A passer-by looks at maps of the world, Mumbai (Bombay), India. Increasing international interdependency, along with increasing international news coverage, makes people everywhere more engaged in geographic awareness.

ise in geographical analysis and an understanding of the uniqueness, influence, and interdependence of places.

The Influence and Meaning of Places

Places are dynamic, with changing properties and fluid boundaries that are the product of the interplay of a wide variety of environmental and human factors. This dynamism and complexity are what make places so fascinating for readers of *National Geographic*. They are also what make places so important in shaping people's lives and in influencing the pace and direction of change. Places provide the settings for people's daily lives and their social relations (patterns of interaction among family members, at work, in social life, in leisure activities, and in political activity). It is in these settings that people learn who and what they are, how they should think and behave, and what life is likely to hold for them.

Places also exert a strong influence, for better or worse, on people's physical well-being, opportunities, and lifestyle choices. Living in a small town dominated by petrochemical industries, for example, means a higher probability than elsewhere of being exposed to air and water pollution, having a limited range of job opportunities, and having a relatively narrow range of lifestyle options because of a lack of amenities such as theaters, specialized stores and restaurants, and recreational facilities. Living in a central neighborhood of a large metropolitan area, on the other hand, usually means having a wider range of job opportunities and a greater choice of lifestyle options because of the variety of amenities accessible within a short distance. But it also means, among other things, living with a relatively high exposure to crime.

Places also contribute to people's collective memory and become powerful emotional and cultural symbols. Think of the evocative power for most Americans of places like Times Square and the site of the former World Trade Center in New York; the Mall in Washington, D.C.; Hollywood Boulevard in Los Angeles; and Graceland in Memphis. And for many people, ordinary places have special meaning: a childhood neighborhood, a college campus, a baseball stadium, or a family vacation spot. This layering of meanings reflects the way that places are *socially constructed*—given different meanings by different groups for different purposes. Places exist and are constructed by their inhabitants from a subjective point of view. The meanings given to a place may be so strong that they become a central part of the identity of the people experiencing them. People's **identity** is the sense that they make of themselves through their subjective feelings based on their everyday experiences and wider social relations. At the same time, though, the same places will likely be constructed rather differently by outsiders. Your own neighborhood, for example, centered on yourself and your home, is probably heavily laden with personal meaning and sentiment. But your neighborhood may well be viewed very differently, perhaps unsympathetically, by outsiders. This distinction is useful in considering the importance of understanding spaces and places from the viewpoint of the insider—the person who normally lives in and uses a particular place—as well as from the viewpoint of outsiders (including geographers).

Finally, places are the sites of innovation and change, of resistance and conflict. The unique characteristics of specific places can provide the preconditions for new agricultural practices (such as the development of seed agriculture and the use of plow and draft animals that sparked the first agricultural revolution in the Middle East in prehistoric times—see Chapter 8); new modes of economic organization (such as the high-tech revolution that began in Silicon Valley in the late twentieth century); new cultural practices (the punk movement that began in disadvantaged British housing projects, for instance); and new lifestyles (for example, the "hippie" lifestyle that began in San Francisco in the late 1960s). It is in specific locales that important events happen, and it is from them that significant changes spread.

Nevertheless, the influence of places is by no means limited to the occasional innovative change. Because of their distinctive characteristics, places always modify and sometimes resist the imprint of even the broadest economic, cultural, and political trends. Consider, for example, the way that a global cultural trend—rock 'n' roll—was modified in Jamaica to produce reggae, while in Iran and North Korea rock 'n' roll has been resisted by the authorities, with the result that it has acquired an altogether different kind of value and meaning for the citizens of those countries. Similarly, Indian communities in London developed Bhangra—a "world beat" composite of traditional Punjabi music, Mumbai (Bombay) movie scores, and Western disco (**Figure 1.2**). Cross-fertilization with local music cultures in New York and Los Angeles has produced Bhangra rap.

To take a very different illustration, think of the ways some communities have declared themselves "nuclear free" zones: places where nuclear weapons and nuclear reactors are unwelcome or even banned by local laws. By establishing such zones, individual communities are seeking to challenge national trends toward using nuclear energy and maintaining nuclear arms. They are, to borrow a phrase, thinking globally and acting locally. Similarly, some communities have established "GM-free" zones, taking a stance against genetically modified crops and food (**Figure 1.3**). In adopting such strategies, they hope to influence thinking in other communities so that eventually their challenge could result in a reversal of established trends.

In summary, places are settings for social interaction that, among other things:

- structure the daily routines of people's economic and social life;
- provide both opportunities and constraints in terms of people's long-term social well-being;

Figure 1.2 Salsa music spreads to Japan Popular trends in music are easily spread around the world, and in the process they are often modified in innovative ways.

- provide a context in which everyday, commonsense knowledge and experience are gathered;
- provide a setting for processes of socialization; and
- provide an arena for contesting social norms.

The Interdependence of Places

Places, then, have an importance of their own. Yet at the same time most places are *interdependent,* filling specialized roles in complex and ever-changing geographies. The social and economic relations that lend distinctiveness to individual places also operate between places. Some of the social relations that help to shape a particular place stretch beyond it. It is this stretching of social and economic relations across space that connects places and the people who live in them with other places and people.

Consider, for example, the way that Manhattan, New York, operates as a specialized global center of corporate management, business, and financial services while relying on thousands of other places to satisfy its needs. For labor it draws on analysts and managers from the nation's business schools, blue- and pink-collar workers from neighboring boroughs, and skilled professional immigrants from around the world. For food it draws on fruits and vegetables from Florida, dairy produce from upstate New York, and specialty foods from Europe, the Caribbean, and Asia. For energy it draws on coal from southwest Virginia to fuel its power stations. And for consumer goods it draws on specialized manufacturing settings all over the world.

This interdependence means that individual places are tied into wider processes of change that are reflected in

Figure 1.3 Acting locally The town of Überlingen, Germany, has established itself as a "GM-free" zone. Shown here is Cornelia Wiethaler, who initiated the movement to ban genetically-modified crops and food from the town.

broader geographical patterns. New York's attraction for business-school graduates, for example, is reflected in the overall pattern of migration flows that cumulatively affect the size and composition of labor markets around the country: New York's gain is somewhere else's loss. An important goal for human geographers—and a central theme of this book—is to recognize these wider processes and broad geographical patterns without losing sight of the uniqueness of specific places. This means that we have to recognize another kind of interdependence: the interdependence that exists *between different geographic scales.*

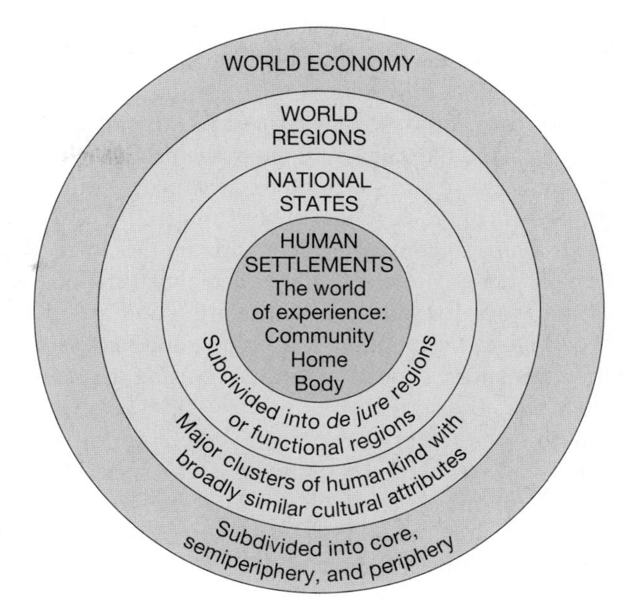

Figure 1.4 Spatial scales There are many scales at which geographic phenomena may be identified, analyzed, and understood. This diagram shows some of the principal scales that are commonly the focus of geographic research. (After S. Marston, P. Knox, and D. Liverman, *World Regions in Global Context.* Upper Saddle River, NJ: Prentice Hall, 2002, p. 18.)

The Interdependence of Geographic Scales

Different aspects of human geography are understood best, and analyzed most effectively, at different spatial scales. At the same time, these different aspects are interrelated and interdependent, so that geographers have to be able to relate things at one scale to things at another. The whole question can be problematic if we do not clarify what "scale" means.

It is useful to think of geographical scales as materializations of real-world processes, not simply different levels of abstraction or convenient devices for zooming in and out from the global context to the detail of local settings. In this sense, scale represents a tangible partitioning of space within which different processes (economic, social, political, etc.) are played out. This partitioning, in turn, often consolidates dominant patterns of geographical organization, at least until some major disruptive change occurs. The Industrial Revolution, for example, changed not only the character of economic development (from agrarian to manufacturing) but also the scales at which industrial production and consumption were organized (from local to national and international). At any particular moment we can identify a sequence of specific scales that represent significant confluences of geographical processes. In today's world the large scale is represented by international, or world regions, large but relatively homogeneous territories with distinctive economic, cultural, and demographic characteristics (**Figure 1.4**). **World regions** are large-scale geographic divisions based on continental and physiographic settings that contain major groupings of peoples with broadly similar cultural attributes. Examples include Europe, Latin America, and South Asia. These regions are constantly evolving as natural resources and technologies create opportunities and constraints to which particular cultures and societies respond.

Superimposed on these regions, sometimes with only approximate fit, are the *de jure* territories of national states. **States** are independent political units with territorial boundaries that are recognized by other states (see the discussion of states, nations, and national states in Chapter 9). *De jure* simply means legally recognized. Territories delimited by formal, legally recognized boundaries—national states, provinces, states, counties, municipalities, special districts, and so on—are known as *de jure* spaces or regions. Because of the inherent power of national governments, especially in relation to the flows of goods, money, and information that underpin reality, national states represent a geographic scale that is often very important. National states tend to be established to fit economic reality as closely as possible at the time of their foundation. Once their boundaries are set, however, principles of national sovereignty mean that these boundaries tend to become regarded by their inhabitants as natural and immutable. However, when economic circumstances change, national states may feel the need to adjust their boundaries or seek other means of accommodating to economic reality, such as joining supranational organizations. **Supranational organizations** are collections of individual states with a common goal that may be economic and/or political in nature and that diminish, to some extent, individual state sovereignty in favor of the group interests of the membership. Examples of supranational organizations include the European Union (EU), the North American Free Trade Association (NAFTA), and the Association of South East Asian Nations (ASEAN).

Within most national states and all international regions are smaller, functional regions. This geographical scale is constructed around specific resources and industries, with their networks of producers, suppliers, distributors, and ancillary activities, and their associated social, cultural and political identities. These are the classic functional regions of traditional regional geography:

the American Corn Belt, the Argentine pampas, the Scottish coalfields, Japan's Pacific Corridor, and the Urals manufacturing region in Russia.

For most people, however, the realm of experience is encompassed by the scale of human settlements. This is the scale that is constructed around the way people's lives are organized through their work, consumption, and recreation. It is also roughly coincident with another important scale of *de jure* territories: local municipalities that provide the framework for public administration and the means for "collective consumption" of certain goods and services (public transport, education, public housing, recreational amenities, and so forth).

Within the realm of experience, there are other significant scales (Figure 1.4). Of these, the scale of community is the most important but also the most difficult to pin down. It is the scale of social interaction—of personal relationships and daily routine. It is a scale that depends a great deal on the economic, social, and cultural attributes of local populations. Much more sharply defined is the scale of the home, which is an important geographic site insofar as it constitutes the physical setting for the structure and dynamics of family and household. It also reflects, in its spatial organization, the differential status accorded to men and women and to the young and the elderly.

Finally, the body and the self represent the most detailed scale that geographers deal with. The body is of interest to geographers because it represents the scale at which differences are ultimately defined—not only through physical attributes (hair, skin, facial features) but also through the socially constructed attributes of the body, such as norms of personal space, preferred bodily styles, and acceptable uses of bodies. Particularly important is the way that, in many cultures, the bodily scale is seen as less relevant to men. That is, men are regarded as able to transcend their body, while women are regarded as limited by bodily functions (for example, menstruation, pregnancy, and childbirth).

The result is that differential geographies are created and experienced by men and women—women's worlds and men's worlds. The self is of interest because it represents the operational scale for cognition, perception, imagination, free will, and behavior. The self has become an important scale of analysis for geographers because of the need to understand the interrelationships between nature, culture, and individual human agency in shaping places and regions.

Perhaps the most important conclusion that we can draw from this examination of scale is that while certain phenomena can be identified and understood best at specific spatial scales, the reality of geography is that social, cultural, political, and economic phenomena are very fluid, constantly being constructed, reinforced, undermined, and rebuilt. Similarly, although certain scales represent manifestations of powerful real-world processes, the real world has to be understood, ultimately, as the product of interdependent phenomena at a variety of spatial scales.

In today's world some of the most important aspects of the interdependence between geographic scales are provided by the relationships between the *global* and the *local* scales. The study of human geography shows not only how global trends influence local outcomes but also how events in particular localities can influence patterns and trends elsewhere.

New York again can illustrate both sides of this relationship. In New York's stock exchanges and financial markets, brokers and clients must, in their own interest, take a global view. Their collective decisions influence stock prices, currency rates, and interest rates around the world, but they also often have very direct outcomes at the local level. Factories in certain localities may be closed and workers laid off because changed currency rates make their products too expensive to export successfully; elsewhere, new jobs may be created because the same change in currency rates puts a local economy at an advantage within the global marketplace. On the other hand, local events can reverberate through New York's stock exchanges and financial markets with global effects. Political instability or natural disasters in a region that produces a key commodity, for instance, can result in changes in global pricing and affect the stock prices of many companies.

Interdependence as a Two-Way Process

One of the most important tenets of human geography is that places are not just distinctive outcomes of geographical processes; they are part of the processes themselves. Think of any city neighborhood, with its distinctive mix of buildings and people. This mix is the product of a combination of processes, including real estate development, the dynamics of the city's housing market, the successive occupancy of residential and commercial buildings by particular groups who move in and then out of the neighborhood, the services and upkeep provided by the city, and so on. Over time these processes result in a distinctive physical environment with an equally distinctive population profile, social atmosphere, image, and reputation. Yet these neighborhood characteristics exert a strong influence in turn on the continuing processes of real estate redevelopment, housing market dynamics, and migration in and out of the neighborhood.

Places, then, are dynamic phenomena. They are created by people responding to the opportunities and constraints presented by their environments. As people live and work in places, they gradually impose themselves on their environment, modifying and adjusting it to suit their needs and to express their values. At the same time, they gradually accommodate both to their physical environment and to the people around them. There is thus a continuous two-way process in which people create and modify places while at the same time being influenced by them (**Figure 1.5**).

Place-making is always incomplete and ongoing, and it occurs simultaneously at different scales. Processes of ge-

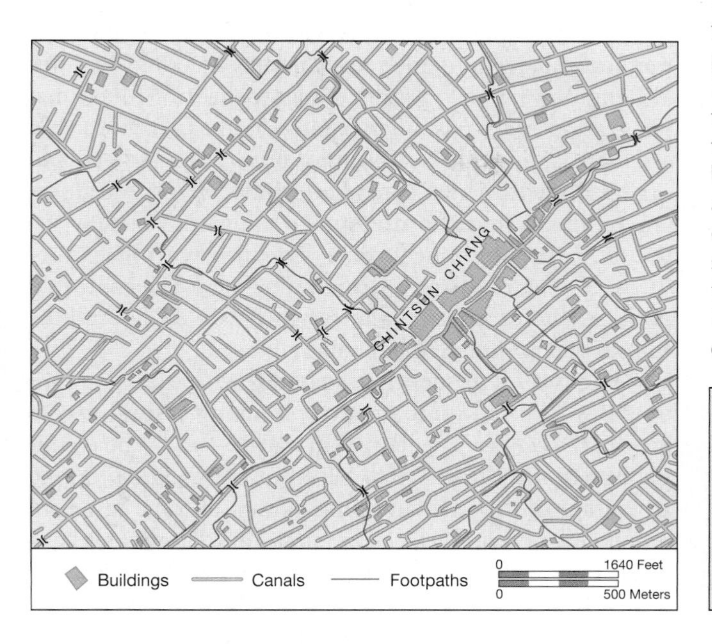

Figure 1.5 Place-making People develop patterns of living that are attuned to the opportunities and constraints of the local physical environment. When this happens, distinctive regional landscapes are produced. These photographs and the map extract show part of the Chang Jiang (Yangtze) delta region of central China. In this watery region there has long been interplay between humans and the natural environment, and the result is a distinctive landscape with tens of thousands of kilometers of canals, irrigation ditches, and drainage channels, and thousands of ponds in rectilinear patterns. This "land of rice and fish" is extremely productive, the humid subtropical climate allowing for a triple-cropping system (two crops of rice, plus one of winter wheat or barley) in many locales.

> It is man who reveals a country's individuality by molding it to his own use. He establishes a connection between unrelated features, substituting for the random effects of local circumstances a systematic cooperation of forces. Only then does a country acquire a specific character, differentiating it from others, till at length it becomes, as it were, a medal struck in the likeness of a people.
>
> Paul Vidal de la Blache *Tableau de la Geographie de la France*, 1903, p. viii

ographic change are constantly modifying and reshaping places, and places are constantly coping with change. It is important for geographers to be sensitive to this kind of interdependence without falling into the trap of overgeneralization, or losing sight of the diversity and variety that constitute the heart of human geography. It is equally important not to fall into the trap of *singularity*, or treating places and regions as separate entities, the focus of study in and of themselves.

INTERDEPENDENCE IN A GLOBALIZING WORLD

As a subject of scientific observation and study, geography has made important contributions both to the understanding of the world and to its development. As we move further into the Information Age, geography continues to contribute to the understanding of a world that is more complex and fast-changing than ever before. With such an understanding, it is possible not only to appreciate the diversity and variety of the world's peoples and places but also to be aware of their relationships to one another and to be able to make positive contributions to local, national, and global development.

Today, in a world that is experiencing rapid changes in economic, cultural, and political life, geographic knowledge is especially important and useful. In a fast-changing world, when our fortunes and ideas are increasingly bound up with those of other peoples in other places, the study of geography provides an understanding of the crucial interdependencies that underpin everyone's lives. One of the central themes throughout this book is the *interdependence* of people and places.

Another central theme of this book is **globalization.** Globalization is the increasing interconnectedness of different parts of the world through common processes of economic, environmental, political, and cultural change. A world economy has been in existence for several centuries, and with it there has developed a comprehensive framework of sovereign national states and an international system of production and exchange. This system has been reorganized several times. Each time it has been reorganized, however, major changes have resulted, not only in world geography but also in the character and fortunes of individual places.

Recently there has been a pronounced change in both the pace and the nature of globalization, leading to a highly interdependent world (see "Box 1.1: Interdependence Within and Between France and Vietnam"). New telecommunication technologies, corporate strategies, and institutional frameworks have combined to create a dynamic new framework for real-world geographies. New information technologies have helped create a frenetic international financial system, while transnational corporations are now able to transfer their production activities from one part of the world to another in response to changing market conditions and changing transportation and communications technologies (see Chapter 7). This locational flexibility has meant that a high degree of functional integration now exists between economic activities that are increasingly dispersed so that products, markets, and organizations are both spread and linked across the globe. Governments, in their attempts to adjust to this new situation, have sought new ways of dealing with the consequences of globalization, including new international political and economic alliances such as NAFTA and the European Union (see Chapter 9).

The interdependence associated with globalization operates in a multitude of ways. In many cases, interdependence seems very unequal in nature—as in the example of a transnational firm based in one country taking advantage of low-cost labor in another. In some cases, interdependence can be seen to be to mutual advantage—as in countries that share the costs and responsibilities of trans-border resource management. In almost every case, however, the outcomes of the increased geographic interdependence associated with globalization are very much open to interpretation. Who "wins" and who "loses" depends very much on one's perspective and the geographic scale.

Perspectives on Globalization and Interdependence

An important aspect of globalization is the widespread perception that the world, through economic and technological forces, is increasingly becoming one shared political and economic space, with events in one region having repercussions for all others, whether near or far. Given that economic and technological forces are breaking down the barriers within and between near and distant places, will the most recent phase of globalization strengthen some regional connections and weaken others, or make regions altogether irrelevant? Alternatively, will globalization enable some regions—core regions, for instance—to create even greater differences of wealth and power than already exist in the world-system? Understanding what the experts believe about globalization will help us to get a better understanding of the complex interdependence between the global and the local.

The number of books published on globalization and its impact on the world's regions has grown at a tremendous rate over the last 15 years. Check out the shelves in any good bookstore and you will find hundreds of them, with whole sections devoted to globalization within political science, sociology, geography, economics, media studies, and business management. It is possible to group the main participants in the contemporary debates about globalization into three general camps: the hyperglobalists, the skeptics, and the transformationalists. While these three viewpoints do not exhaust the range of the debates, they do provide a clear sense of the issues on which the experts on globalization agree and disagree.

1.1

GEOGRAPHY MATTERS

Interdependence Within and Between France and Vietnam

In our increasingly interdependent world, the lives and livelihoods of people separated by vast distances are now linked in many ways. A study by the World Bank[1] has shown how globalization has affected the lives of four very different people in very different places: a Vietnamese peasant, a Vietnamese city dweller, a Vietnamese immigrant to France, and a French garment worker.

Duong is a Vietnamese peasant farmer who struggles to feed his family. He earns the equivalent of $10 a week for 38 hours of work in the rice fields, but he works full-time only six months of the year—during the off-season he can earn very little. His wife and four children work with him in the fields, but the family can afford to send only the two youngest to school. Duong's eleven-year-old daughter stays at home to help with housework, while his thirteen-year-old son works as a street trader in town. By any standard Duong's family is living in poverty. Workers like Duong, laboring in family farms in low- and middle-income countries, account for about 40 percent of the world's labor force.

Hoa is a young Vietnamese city dweller experiencing relative affluence for the first time. In Ho Chi Minh City she earns the equivalent of $30 a week working 48 hours in a garment factory—a joint venture with a French firm. She works hard for her living and spends many hours looking after her three children as well; her husband works as a janitor. But Hoa's family has several times the standard of living of Duong's and, by Vietnamese standards, is relatively well off. There is every expectation that both she and her children will continue to have a vastly better standard of living than her parents had. Wage employees like Hoa, working in the formal sector in low- and middle-income countries, make up about 20 percent of the global labor force.

Françoise is an immigrant in France of Vietnamese origin who works long hours as a waitress to make ends meet. She takes home the equivalent of $220 a week, after taxes and including tips, for 50 hours' work. By French standards she is poor. Legally, Françoise is a casual worker and has no job security, but she is much better off in France than she would have been in Vietnam. Her wage is almost eight times that earned by Hoa in Ho Chi Minh City. Françoise and other service workers in high-income countries account for about 9 percent of the global workforce.

Jean-Paul is a 50-year-old Frenchman whose employment prospects look bleak. For 10 years he has worked in a garment factory in Toulouse, taking home the equivalent of $400 a week—twelve times the average in Vietnam's garment industry. But next month he will lose his job when the factory closes. Unemployment benefits will partly shield him from the shock, but his chances of matching his old salary in a new job are slim. Frenchmen of Jean-Paul's age who lose their jobs are likely to stay unemployed for more than a year, and Jean-Paul is encouraging his son to work hard in school so he can go to college and study computer programming. Workers in industry in high-income countries, like Jean-Paul, make up just 4 percent of the world's labor force.

These four individuals—two living in Vietnam, two in France—have vastly different standards of living and expectations for the future. Employment and wage prospects in Toulouse and Ho Chi Minh City are worlds apart, even when incomes are adjusted for differences in the cost of living. Françoise's poverty wage would clearly buy Hoa a vastly more affluent lifestyle. And many of the world's workers, like Duong, work outside the wage sector, on family farms and in casual jobs, generally earning even lower incomes. The lives of all workers in different parts of the world, however, are increasingly intertwined. French consumers buy the product of Hoa's labor, and Jean-Paul believes it is Hoa's low wages that are taking his job, while immigrant workers like Françoise feel the brunt of Jean-Paul's anger. Meanwhile, Duong struggles to save so that his children can be educated and leave the countryside for the city, where foreign companies advertise new jobs at better wages.

[1] World Bank, *World Development Report.* Washington, DC: The World Bank, 1995, page 1.

The Hyperglobalist View

At one extreme is the view that open markets and free trade and investment across global markets allow more and more people to share in the prosperity of a growing world economy. Economic and political interdependence, meanwhile, creates shared interests that help prevent conflict and foster support for common values. Democracy and human rights, it is asserted, will spread to billions of people in the wake of neoliberal policies that promote open markets and free trade. **Neoliberal policies** are economic policies that are predicated on a minimalist role for the state, assuming the desirability of free markets as the ideal condition not only for economic organization but also for political and social life. Hyperglobalists believe that the current phase of globalization signals the beginning of the end for the nation-state and the "denationalization" of economies. By economic denationalization they mean that national boundaries will become irrelevant with respect to economic processes, and that national governments will not control their once geographically bounded economies but will instead facilitate connections among and between different parts of the world through supranational organizations such as NAFTA and the EU.

The wider implications of the hyperglobalist position is that the world will become borderless as national governments become increasingly meaningless or function merely as facilitators of global capital flows and investments. Hyperglobalizers believe that the nation-state, the primary political and economic unit of contemporary world society, will eventually be replaced by institutions of global governance in which individuals claim transnational allegiances that are founded upon a commitment to neoliberal principles of free trade and economic integration. Politically, the global spread of liberal democracy will reinforce the emergence of a global civilization with its own mechanisms of global governance, replacing the outmoded nation-state with global institutions like the International Monetary Fund (IMF) and the World Trade Organization (WTO).

The Skeptical View

A second broad argument within the globalization literature belongs to the skeptics, those who believe that contemporary levels of global economic integration represent nothing particularly new and that much of the talk about globalization is exaggerated. The skeptics look to the nineteenth century and draw on statistical evidence of world flows of trade, labor, and investment to fortify their position. They argue that contemporary economic integration is actually much less significant than it was in the late nineteenth century, when nearly all countries shared a common monetary system known as the gold standard. The skeptics are also dismissive of the idea that the nation-state is in decline. They argue that national governments are essential to the regulation of international economic activity and that the continuing liberalization of the world economy can only be facilitated by the regulatory power of national governments.

The skeptics assert that their analysis of nineteenth-century economic patterns demonstrates that we are today witnessing not globalization but rather "regionalization," as the world economy is increasingly dominated by three major regional financial and trading blocs: Europe, North America, and East Asia (effectively, Japan). The skeptics understand regionalization and globalization to be contradictory tendencies. They believe that because of the dominance of these three major regional blocs, the world is actually less integrated than it once was because Europe, North America, and Japan control the world economy and limit the participation of other regions in that economy.

The Transformationalist View

According to the transformationalist view, contemporary processes of globalization are historically unprecedented as governments and peoples across the globe confront the absence of any clear distinction between the global and the local, between domestic affairs and international affairs. Like the hyperglobalists, this group understands globalization as a profound transformative force that is changing societies, economies, institutions of government—in short, the world order. In contrast to the hyperglobalists and the skeptics, however, the transformationalists make no claims about the future trajectory of globalization, nor do they see present globalization as a pale version of a more "globalized" nineteenth century. Instead, they see globalization as a long-term historical process that is underlain by crises and contradictions that are likely to shape it in all sorts of unpredictable ways. Moreover, unlike the skeptics, the transformationalists believe that the historically unprecedented contemporary patterns of economic, military, technological, ecological, migratory, political, and cultural flows have functionally linked all parts of the world into a larger global system in which free trade agreements such as NAFTA help to draw regions into a global neoliberal economic framework.

In this book we adhere generally to the transformationalist position. We suggest that we are all heading toward a world where places and regions will experience a wide range of internal changes at the same time that the strength of their connections with other parts of the world will increase. What is perhaps most unsettling about the transformationalist view of globalization is the anticipated increase in disparities in wealth. Transformationalists believe that globalization is leading to increasing social stratification, in which some states and societies are more tightly connected to the global order while others are becoming increasingly marginalized. They contend that there is no evidence to sustain the hyperglobalist claim that the new global social structure is tending toward a global civilization where equality among individuals will eventually prevail. They argue the opposite: that the world will increasingly consist of a three-tiered system—comprising the elites, the embattled, and the marginalized—that cuts across national, regional, and local boundaries (see Chapter 12). Within nations, disparities of wealth—already striking in many countries—will increase, just as they will between nations (see "Box 1.2: Worlds Apart"). Meanwhile, the increasing interdependence among places and regions raises some key issues in relation to the environment, health, and security.

Key Issues in a Globalizing World

Environmental Issues

The sheer scale and capacity of the world economy means that humans are now capable of altering the environment at the global scale. The "footprint" of humankind extends to more than four-fifths of Earth's surface (**Figure 1.6**). Many of the important issues facing modern society are the consequences—intended and unintended—of human modifications of the physical environment. Humans have altered the balance of nature in ways that have brought economic prosperity to some areas and created environmental dilemmas and crises in others. For example, clearing land for settlement, mining, and agriculture provides livelihoods and homes for some but alters physical systems and transforms human populations, wildlife, and vegetation. The inevitable by-products—garbage, air and water pollution, hazardous wastes, and so forth—place enormous demands on the capacity of physical systems to absorb and accommodate them. In addition to the specter of global warming (a result of emissions of gaseous materials into the atmosphere), we also face the reality of serious global environmental degradation through deforestation, desertification, acid rain, loss of genetic diversity, smog, soil erosion, groundwater decline, and the pollution of rivers, lakes, and oceans.

Lake Baykal, in Russia (**Figure 1.7**), provides a salutary example. It is a place of incredible beauty—"the Pearl of Siberia"—that has long been emblematic of the pristine wilderness of the region. But the lake's purity and unique ecosystem have been compromised by environmental mismanagement. The first evidence of this was in the 1950s, when the lake's commercial fish populations nose-dived, partly as a result of overfishing and partly as a result of the construction of the Irkutsk dam (which raised the level of the lake and destroyed many of the shallow-water feeding grounds used by the fish). Then, in the 1960s, increasing levels of pollution were carried into the lake by the Selenga River, which supplies about half of the water that flows into the lake. The

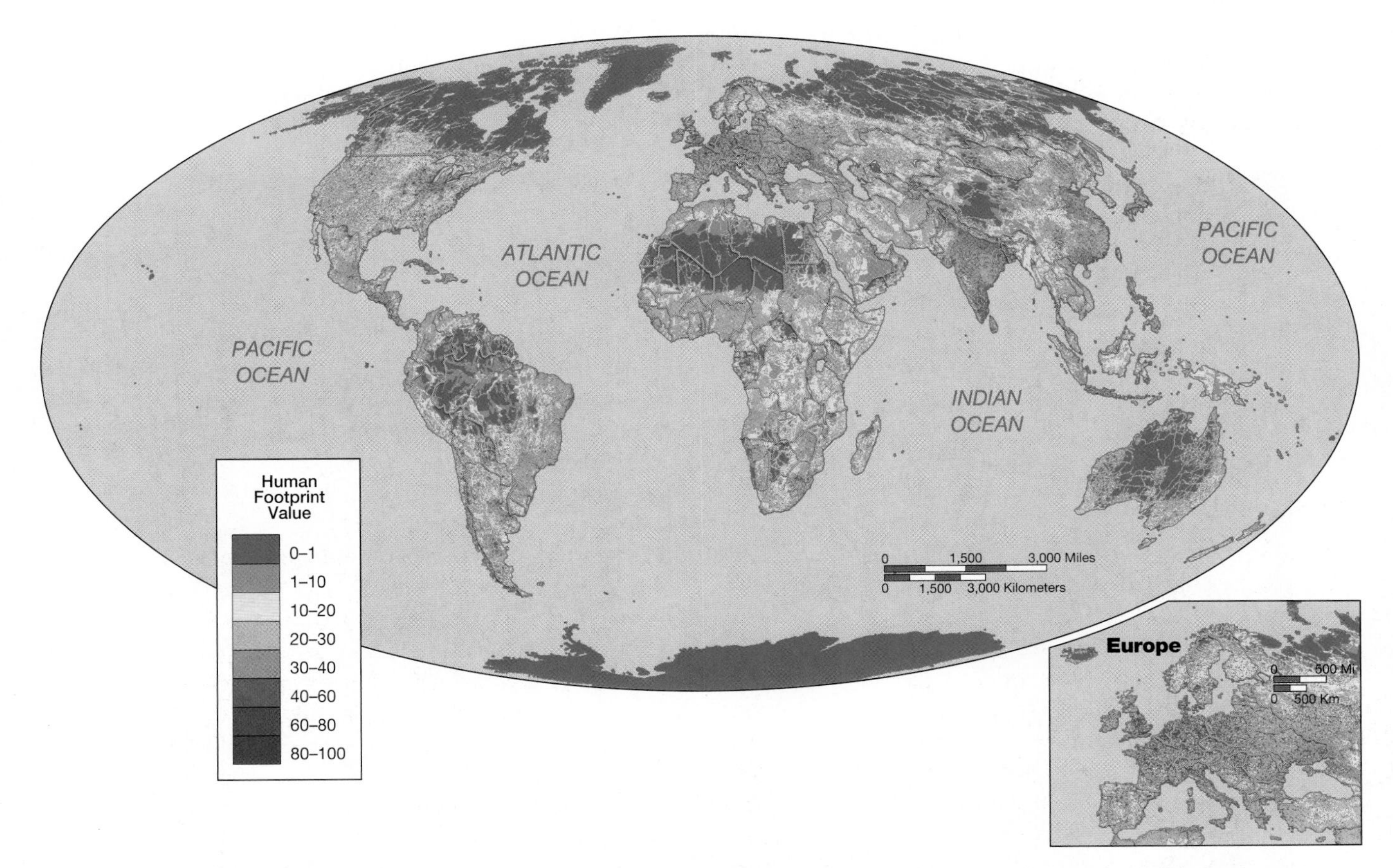

Figure 1.6 The human "footprint" This map, prepared by a team of scientists from the New York–based Wildlife Conservation Society and Columbia University's Center for International Science Information Network (CIESN), shows the extent and intensity of human influence on the land, reflecting population density, agricultural use, access from roads and waterways, electrical power infrastructure, and urbanization. The lower the number, the lesser the overall degree of human influence. (*Source:* www.wcs.org/humanfootprint)

Worlds Apart

Meet Paul Rust and his family, who live and enjoy life in Zug, Switzerland, the richest canton in the world's richest country. And meet Hussein Sormolo and his family, who live in Addis Ababa, capital of the world's poorest country, Ethiopia.

Hussein Sormolo left the village where he was born for the big city in 1978. He left his eight brothers and seven sisters behind, as the land that the family farmed was being forcibly collectivized by a new regime. Hussein, aged 16, traveled the 100 miles north to the city in the back of a truck. A kinsman from the same village took him in until he found a job in a bakery. Paul Rust left his village when he was 17 and also ended up in a bakery. The two men are similar in other ways. Both are friendly, hospitable and generous and love their families. Both work hard. Both like to watch the news. Both are active worshippers, without being religious dogmatists.

Yet their lives are different. Hussein lives with his wife, sons, and daughters in a leaky shack of corrugated asbestos and steel in the Nefas Silk district of Addis Ababa (**Figure 1.A**). Paul lives with his wife in a six-roomed house (not counting the ground-floor apartment where his son Martin lives with his girlfriend) overlooking the lustrous green waters and steep wooded slopes of Lake Aegeri in Zug (**Figure 1.B**).

The income difference is huge, of course. Hussein supports his wife and three younger children on wages of about $280 a year (more than twice the average income in Ethiopia). Paul and his wife, Hedi, draw about $68,000 between them each year from their bakery, though the Rusts are not affluent by Swiss standards (the average income per head in Zug is about $50,000).

It is the rainy season in Addis. Fat raindrops drum against, and often through, the rusting grooves of the corrugated roofs of the houses in Nefas Silk. Nights can be chill and dank. From Debre Zeit road, the busy main north–south street lined with small businesses including the bakery where Hussein works, it's a 10-minute walk to the alley where he lives. Inside the Hussein shack, a single bare light bulb always burns. There is little natural light: There are no glass windows, and the openings punched in the asbestos walls are covered to keep out drafts. Hussein pays his neighbor 18 birr (about $2) a month, almost a tenth of his 200-birr salary, to sublet his electricity supply for the bulb in the shack. The family has no other electrical appliances, apart from a battery-operated radio. Neither Hussein, his wife Rukia, nor his eldest daughter, Fate, 17, who is lucky enough to be at school, have ever used a computer, taken a photograph, or made a phone call. Hussein and Rukia have a pair of shoes

Figure 1.A Hussein Sormolo and his family in Addis Ababa, Ethiopia.

each. They buy new ones every two years. They have no savings and the family doesn't take holidays.

Except for feast days, the family eats the same dish every meal—a grey, spongy, limp bread called injera, spread out like a cloth, and a spicy vegetable stew. Meat, fish, cheese, and eggs are luxuries. They only buy fruit when one of the children is sick. Just under a quarter of the family income goes on cooking charcoal and cans of water. In a country where only a quarter of people in the countryside have access to safe drinking water, Hussein's family is lucky. There is a standpipe around the corner with reasonably clean water. That's about where their luck ends. With their neighbors, they used to have a toilet for 26 people. Now they have no toilet at all.

The Rust house, not counting the apartment, has three toilets, one each in the bathroom, and two shower rooms of the four-story building. On the balconies under its broad, dark, solid eaves are cascades of red flowers. The well-used furnishings inside are not ostentatious, but the building is roomy and comfortable. From the top, there is a loft, four bedrooms, two living rooms, a kitchen, an office, a small wine cellar, a work room, garage parking for three cars (Paul, Hedi, and Martin Rust have one each) with room for another five on the forecourt. The house has its own elevator.

Paul and Hedi are going on vacation for two weeks in Austria this month, and usually take a week at Easter. Each has a mobile phone. The home office has computers and Internet access. They have a TV, a VCR, and a dishwasher. They eat what they want, although their tastes are plain—meat with several vegetables, salad, sometimes a little wine.

Switzerland is a rich country landlocked by other rich countries. Ethiopia is a poor country landlocked by other poor countries. Unlike other African nations, Ethiopia was not a European colony, but its people have endured regular European military incursions, proxy superpower duels, and local wars that have exacerbated the ravages of famine and disease. Famines in the 1970s, 1980s and 1990s killed 1.3 million people. Through the 1970s and 1980s, the country was embroiled in ideological and ethnic civil war. HIV/AIDS has infected 3 million Ethiopians, and kills 300,000 a year.

Hussein knows little about Switzerland. "I heard about Switzerland on the radio but I don't know. I heard it was a rich country, they help poor countries," he said.

Paul thought he could find Ethiopia on the map. Switzerland is not as aloof from the world as it was, he points out: They joined the boycott of apartheid South Africa. He said his brother helped build a dairy in Nepal 20 years ago. His church had adopted a village in Romania, giving it money for a new church and a school. The talk turned to immigration. "The really poor people, they can't come to Switzerland, they need money to get here," said Andrea. "We work, and have our life, we have our own problems," said Andrea. "So we don't think very often of other people's problems. It's a little bit selfish."

Based on an article by James Meek. *The Guardian*, 22 August 2002.

Figure 1.B Paul Rust and his family in Zug, Switzerland.

Figure 1.7 Lake Baykal Lake Baykal is the world's deepest lake, at 1.615 meters (5,300 feet—over a mile), and contains about 20 percent of all the fresh water on Earth—more than North America's five Great Lakes combined. Lake Baykal is also an unusually ancient lake. Most of the world's large lakes are less than 20,000 years old, but evidence from the 7-kilometer-thick sediment at the bottom of the lake shows it to have been in existence for at least 25 million years, perhaps even 50 million. It has a unique ecology, with over 2,500 recorded plant and animal species, 75 percent of which are found nowhere else. These include the nerpa, Baykal's freshwater seal, separated by more than 3,000 kilometers (1,864 miles) from its nearest relative, the Arctic ringed seal. Ecologists have no understanding of how seals ever got to the lake or how they adapted to fresh water. Although Lake Baykal is now on UNESCO's list of World Natural Heritage sites, pollution from numerous paper mills along the lake shore continues to worsen.

Selenga rises in mountain ranges to the south but collects human and industrial waste, much of it untreated, from several large cities before entering Lake Baykal. The Selenga and other rivers also began to carry increasing amounts of agricultural chemicals, such as DDT and PCBs into the lake. Meanwhile, the purity of the lake's waters caught the attention of Soviet economic planners, who saw the lake as a good location for factories that needed plentiful supplies of pure water. The huge Baikalsk Pulp & Paper Mill was opened in the early 1960s to produce high-quality cellulose for the Russian defense industry. The mill pumps 140,000 tons of waste—including deadly dioxins—into the lake every day, along with 23 tons of pollutants into the atmosphere; over the past 40 years, the mill has spewed over a billion tons of waste into the lake. Since 1989 the mill has been partially privatized and now makes pulp for low-quality paper rather than cellulose. When thousands of the lake's freshwater seals began dying in 1997, the lake's fragile ecology came under international scrutiny, and in 1998 the lake was designated a World Heritage Site by UNESCO, the U.N. cultural agency. Nevertheless, it remains to be seen whether Russia can solve its environmental problems at a time when its economy is in disarray.

Health Issues

The increased intensity of international trade and travel has also heightened the risk and speed of the spread of disease. Over the past quarter century, HIV/AIDS (human immunodeficiency virus or acquired immunodeficiency syndrome) quickly spread around the world from a single hearth area. Medical geographers have concluded that the human immunodeficiency virus (HIV), which causes acquired immunodeficiency syndrome, or AIDS, spread in a hierarchical diffusion pattern from a hearth area in Central Africa in the late 1970s (**Figure 1.8**). The virus initially appeared almost simultaneously in the major metropolitan areas of North and South America, the Caribbean, and Europe. These areas then acted as localized diffusion poles for the virus, which next spread to major metropolitan areas in Asia and Oceania and to larger provincial cities in North and South America, the Caribbean, and Europe. Next in this cascading pattern of diffusion were provincial cities in Asia and Oceania and small towns in North and South America, the Caribbean, and Europe. Today Sub-Saharan Africa is more severely affected by HIV/AIDS than any other part of the world, with the United Nations reporting between 25.0 and 28.2 million people infected in 2003—between 50 and 75 percent of the worldwide total (**Figure 1.9**). The infection

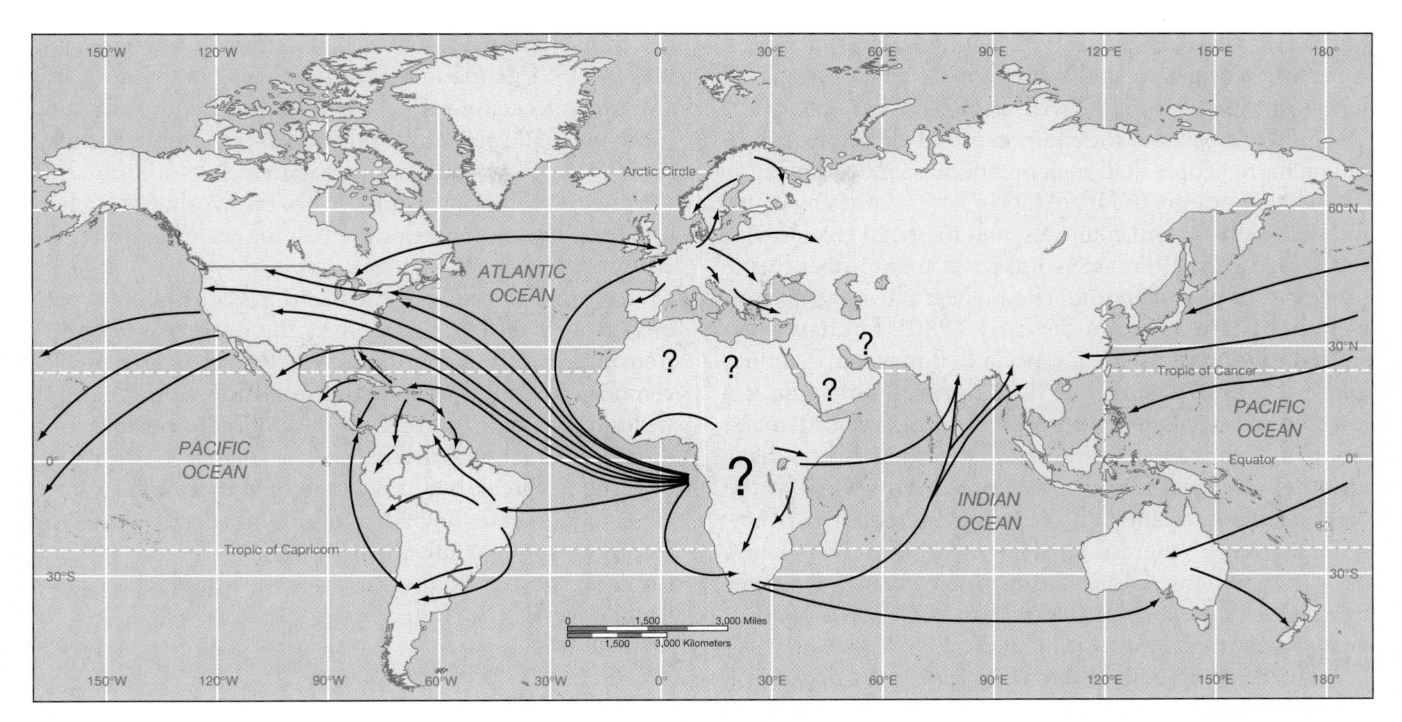

Figure 1.8 Diffusion of the HIV virus The probable early diffusion of HIV/AIDS. (After M. Smallman-Raynor, A. Cliff, and P. Haggett, *London International Atlas of AIDS.* Oxford: Blackwell Reference, 1992, Fig. 4.1(c), p. 146.)

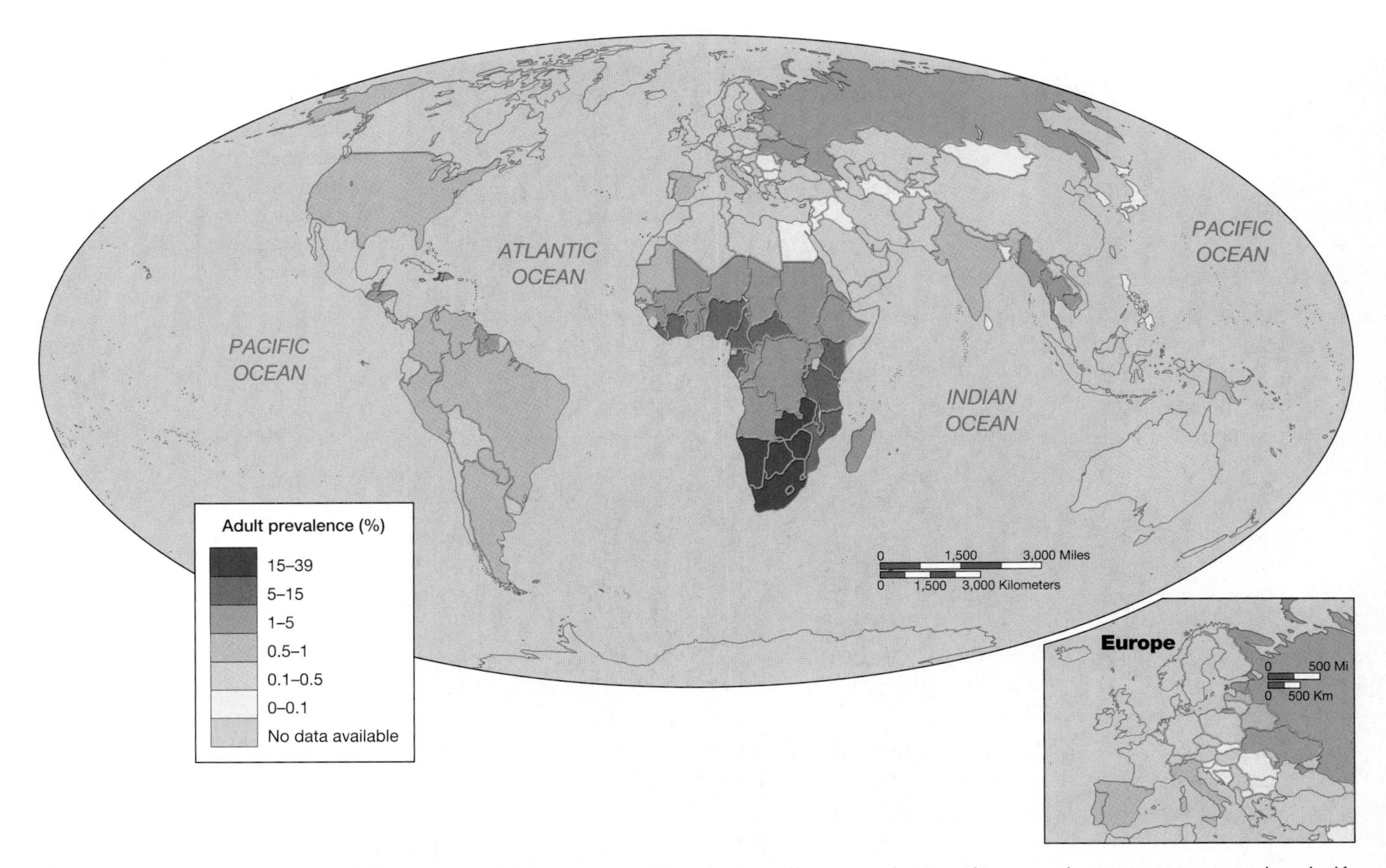

Figure 1.9 The geography of HIV/AIDS In 2003, between 35 and 42 million people were living with HIV/AIDS. More than half of these were in sub-Saharan Africa. (*Source:* http://www.who.int/hiv/facts/hiv2003/en/)

rate is estimated at 8 percent of all adults compared with a 1 percent world rate, and more than 15 million Africans have lost their lives to HIV/AIDS since it was identified in 1981. It has become the main cause of death in Africa, killing more people than malaria and warfare combined.

The geography of HIV/AIDS in Africa varies by country, by regions within countries, and by social groups, and the spread of the disease is linked in many ways to the processes of globalization. The highest rates of infection were in eastern Africa in the early 1980s, but have now shifted to southern Africa, especially Botswana, Zambia, and Zimbabwe, where more than 20 percent of adults are infected. Urban dwellers who have multiple sex partners, including young office workers and migrant workers, have a higher infection rate, as do women who work in the commercial sex trade and the wives and children of migrant workers. Migrant workers have taken the disease back to their home areas. The incidence is generally lower in rural areas, except along major truck routes and in areas where there are a lot of soldiers.

The death of skilled farm laborers has resulted in a decline in agricultural output, and many young professionals critical to the region's future have left their jobs because they have contracted HIV/AIDS. Major industries and companies in southern Africa, such as diamond mines and banks, estimate that absence and loss of employees to HIV/AIDS is costing them more than 5 percent of their profits.

Some countries have had success in combating HIV/AIDS. Uganda and Senegal have promoted aggressive and successful HIV/AIDS education and prevention campaigns and have cut infection rates in half. International agreements with drug companies in combination with new assistance programs from the World Bank, charities, and donor countries are helping to bring down the cost of drugs.

Another example of the health risks associated with increasing interdependency was the outbreak of Severe Acute Respiratory Syndrome (SARS) in China in November 2002 quickly spread around the world, causing widespread panic and serious disruption to business and tourism in East and Southeast Asia (**Figure 1.10**). By the time the outbreak had been contained four months later, almost 3,000 cases (including over 1,400 deaths) had been recorded, some as far afield as Brazil, Canada, Ireland, Romania, Spain, South Africa, Switzerland, the United Kingdom, and the United States.

Security Issues

International terrorism can also be attributed in part to globalization. While terrorism has a long history, it is only recently that terrorist attacks have spilled beyond the sites of local conflict. In targeting the World Trade Center in New York in September 2001 (**Figure 1.11**), Al Quaeda terrorists were striving to tear down a potent

Figure 1.10 The outbreak of Severe Acute Respiratory Syndrome (SARS) in the winter of 2002 caused widespread disruption to international patterns of business and tourism for several months, underlining the increasing vulnerability of populations to disease epidemics as a result of global interdependency.

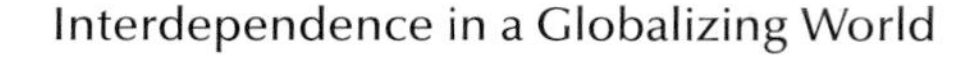

Figure 1.11 International terrorism A shell of what was once part of the facade of one of the twin towers of New York's World Trade Center rises above rubble that remained after both towers were destroyed in the terrorist attack on September 11, 2001. The 110-story towers collapsed after two hijacked airliners carrying scores of passengers slammed into the sides of the towers.

symbol not merely of the economic might of the United States but also of the values of **capitalism** (a form of economic and social organization characterized by the profit motive and the control of the means of production, distribution, and exchange of goods by private ownership) and Western materialism. Provoking an aggressive response from the United States would, it was hoped, prompt an outburst of anti-Americanism around the world. The Al Quaeda terrorists felt that they had a reasonable shot at triggering a new world disorder, pitching the Middle East into anarchy, launching a cultural war against the West, and sowing discord among peoples already uneasy about U.S. cultural and economic hegemony and among governments already unsettled by the U.S. government's increasing tendency to pursue narrow self-interest at the expense of transnational issues of economic inequality, geopolitics, and environmental sustainability.

More generally, as sociologist Ulrich Beck has pointed out, the high degree of interdependence that is now embedded in a globalizing and highly interconnected world has brought about all sorts of security issues. Beck argues that the contemporary world is characterized by the global—or at least transnational—production of risk. In traditional societies the risks faced by individuals and groups were associated mostly with hazards generated by nature (disease, flood, famine, and the like), along with socially determined hazards such as invasion and conquest and regressive forms of thought and culture. The industrial societies of the nineteenth and twentieth centuries, with more powerful technologies and weaponry, faced still more hazards, but they were mostly local and regional in nature. Contemporary society, Beck points out, is characterized by the endemic production of potentially catastrophic risks, so that we now face another set of hazards, many of them uncontrollable and with a global reach. Examples include climate change as a result of human activity; the spread of weapons of mass destruction (i.e., nuclear and biological warfare); the risk of accidents involving ionizing radiation or contamination by radioactivity from nuclear fuel or nuclear waste; the risk of epidemics in the human population resulting from zoonotic diseases (diseases originating with other species—e.g., anthrax, avian flu, ebola, West Nile virus); and the risk of epidemic disease in food animals—as in the devastating outbreak of foot-and-mouth disease, which affects cattle and sheep, in parts of northwestern Europe in 2001. The initial outbreak in England was a result of contaminated animal feed. The spread of the disease was a result of the long-distance trade in animals that characterizes contemporary agriculture.

Overall, Beck argues, we are moving toward a **risk society,** in which the significance of wealth distribution is being eclipsed by the distribution of risk, and in which politics is increasingly about avoiding hazards. As a result, knowledge—especially scientific knowledge—becomes increasingly important as a source of power, while science itself becomes increasingly politicized—as, for example, in the case of global warming.

Geography in a Globalizing World

All this adds up to an intensified global connectedness and the beginnings of the world as an interdependent system. Or, to be more precise, this is how it adds up for the 800 million or so of the world's people who are directly tied to global systems of production and consumption and who have access to global networks of communication and knowledge. All of us in this globalizing world are in the middle of a major reorganization of the world economy and a radical change in our relationships to other people and places.

At first glance it might seem that globalization will render geography obsolete—especially in the more developed parts of the world. High-tech communications and the global marketing of standardized products seem as if they might soon wash away the distinctiveness of people and places, permanently diminishing the importance of differences between places. Far from it. The new mobility of money, labor, products, and ideas actually increases the significance of place in some very important ways:

- ***The more universal the diffusion of material culture and lifestyles, the more valuable regional and ethnic identities become.*** One example of this is the way that the French government has actively resisted the Americanization of the French language and culture by banning the use of English words and phrases and by subsidizing France's domestic movie industry.
- ***The faster the information highway takes people into cyberspace, the more they feel the need for a subjective setting—a specific place or community—they can call their own.*** Examples can be found in the private, master-planned, residential developments that have sprung up around every U.S. metropolitan area. Unlike most previous suburban developments, each of these master-planned projects has been carefully designed to create a sense of community and identity for its residents.
- ***The greater the reach of transnational corporations, the more easily they are able to respond to place-to-place variations in labor markets and consumer markets, and the more often and more radically that economic geography has to be reorganized.*** Global patterns of production are constantly being reorganized as transnational corporations seek to take advantage of geographical differences between places and regions, and as workers and consumers in specific places and regions react to the consequences of globalization.
- ***The greater the integration of transnational governments and institutions, the more sensitive people have become to local cleavages of race, ethnicity, and religion.*** An example is the resurgence of nationalism and regionalism, as in the emergence of the Lega Nord (the Northern League) party in Italy in the 1990s. Lega Nord is a federalist political party whose supporters in northern Lombardy and rural northeastern Italy want to distance themselves from what they view as a distinctively different culture and society in the Italian South.

For some places and regions, globalization is a central reality; for others it is still a marginal influence. While some places and regions have become more closely interconnected and interdependent as a result of globalization, others have been bypassed or excluded. In short, there is no one experience of globalization. All in all, the reality is that globalization is variously embraced, resisted, subverted, and exploited as it makes contact with specific cultures and settings. In the process, places are modified or reconstructed rather than destroyed or homogenized.

STUDYING HUMAN GEOGRAPHY

The study of geography involves the study of Earth as created by natural forces and modified by human action. This, of course, covers an enormous amount of subject matter. There are two main branches of geography: physical and human. **Physical geography** deals with Earth's natural processes and their outcomes. It is concerned, for example, with climate, weather patterns, landforms, soil formation, and plant and animal ecology. **Human geography** deals with the spatial organization of human activities and with people's relationships with their environments. This focus necessarily involves looking at natural physical environments insofar as they influence, and are influenced by, human activity. This means that the study of human geography must cover a wide variety of phenomena. These include, for example, agricultural production and food security, population change, the ecology of human diseases, resource management, environmental pollution, regional planning, and the symbolism of places and landscapes. **Regional geography** combines elements of both physical and human geography. Regional geography is concerned with the way that unique combinations of environmental and human factors produce territories with distinctive landscapes and cultural attributes. The concept of **region** is used by geographers to apply to larger size territories that encompass many places, all or most of which have similar attributes distinct from the attributes of other places.

What is distinctive about the study of human geography is not so much the phenomena that are studied as the way they are approached. The contribution of human geography is to reveal, in relation to a wide spectrum of natural, social, economic, political, and cultural phenomena, *how and why geographical relationships are important.* Thus, for example, human geographers are interested not only in patterns of agricultural production but also in the geographical relationships and interdependencies that are both causes and effects of such patterns. To put it in concrete terms, geographers are interested not only in what specialized agricultural subregions such as the dairy farming area of Jutland, Denmark, are like (just what and how the region produces its agricultural output, what makes its landscapes and culture distinctive, and so on), but also in the role of subregions such as Jutland in national and international agro-food systems (their interdependence with producers, distributors, and consumers in other places and regions—see Chapter 8).

Basic Tools

In general terms, the basic tools employed in geography are similar to those in other disciplines. Like other social scientists, human geographers usually begin with observation. Information must be collected and data recorded. This can involve many different methods and tools. Fieldwork (surveying, asking questions, using scientific instruments to measure and record things), laboratory experiments, and archival searches are all used by human geographers to gather information about geographical relationships. Geographers also use **remote sensing**, the collection of information about parts of Earth's surface by means of aerial photography or satellite imagery designed to record data on visible, infrared, and microwave sensor

Landsat satellite images are digital images captured from spectral bands both visible and invisible to the human eye. Different kinds of vegetation cover, soils, and built environments are reflected by different colors in the processed image. This Landsat image is of part of southern Florida (Miami Beach is to the right of the image).

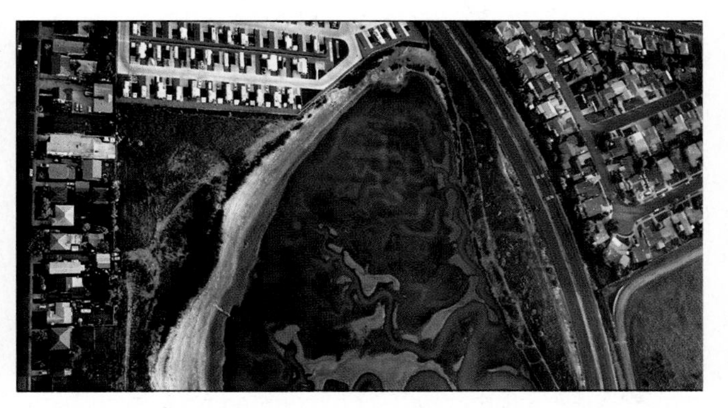

Aerial photographs can be helpful in explaining what would otherwise require expensive surveys and detailed cartography. They are especially useful in working with multidisciplinary teams. This example shows housing and commercial development surrounding a remnant wetland near San Diego, California.

This spectacular image is the most detailed true-color image of the entire Earth to date. Using a collection of satellite-based observations, scientists and visualizers stitched together months of observations of the land surface, oceans, sea ice, and clouds into a seamless, true-color mosaic of every square kilometer (.386 square mile) of the planet. Much of the information contained in this image came from a single remote-sensing device—NASA's Moderate Resolution Imaging Spectroradiometer, or MODIS—flying over 700 km above the Earth on board the Terra satellite.

Figure 1.12 Remotely sensed images Remotely sensed images can provide new ways of seeing the world, as well as unique sources of data on all sorts of environmental conditions. Such images can help explain problems and processes that would otherwise require expensive surveys and detailed cartography. They can be used to identify change in the environment and to monitor and analyze the rate of change. Examples of such applications have included studies of the deforestation of the Amazon rain forest, urban encroachment onto farmland, water pollution, and bottlenecks in highway systems. (*Source:* Top left photo courtesy of Spaceimaging.com; top right photo courtesy Joel Sartore/www.joelsartore.com.)

systems (**Figure 1.12**). For example, agricultural productivity can be monitored by remotely sensed images of crops, and energy efficiency can be monitored by remotely sensed levels of heat loss from buildings.

Once data have been obtained through some form of observation, the next important step is to portray and describe them through *visualization* or *representation*. This can involve a variety of tools, including written descriptions, charts, diagrams, tables, mathematical formulae, and maps. Visualization and representation are important activities because they allow large amounts of information to be explored, summarized, and presented to others. They are nearly always a first step in the analysis of geographical relationships, and they are important in conveying the findings and conclusions of geographic research.

At the heart of geographic research, as with other kinds of research, is the *analysis* of data. The objective of analysis, whether of quantitative or qualitative data, is to discover patterns and establish relationships so that hypotheses can be established and models can be built. Models, in this sense, are abstractions of reality that help explain the real world. They require tools that allow us to generalize about things. Once again we find that geographers are like other social scientists in that they utilize a wide range of analytical tools, including conceptual and linguistic devices, maps, charts, and mathematical equations.

In many ways, therefore, the tools and methods of human geographers are parallel to those used in other sciences, especially the social sciences. In addition, geographers increasingly use some of the tools and methods of the humanities—interpretive analysis and inductive reasoning, for example—together with ethnographic research and textual analysis. The most distinctive tools in the geographer's kit bag are, of course, maps and **geographic information systems (GIS)**. Geographic information systems involve an organized set of computer hardware, software, and spatially coded data that is designed to capture, store, update, manipulate, and display geographically referenced information (see "Appendix 1—Maps and Geographic Information Systems"). As we have seen, maps can be used not only to describe data but also to serve as important sources of data and tools for analysis. Because of their central importance to geographers, they can also be objects of study in their own right.

Spatial Analysis

The study of human geography is easily distinguished by its fundamental concepts. The study of many geographic phenomena can be approached in terms of their arrangement as points, lines, areas, or surfaces on a map. This is known as **spatial analysis.** *Location, distance, space, accessibility,* and *spatial interaction* are five concepts that are key to spatial analysis. Although these concepts may be familiar from everyday language, they require some elaboration.

Location

Location is often nominal, or expressed solely in terms of the names given to regions and places. We speak, for example, of Washington, D.C. or of Georgetown, a location within Washington, D.C. Location can also be used as an absolute concept, whereby locations are fixed mathematically through coordinates of latitude and longitude (**Figure 1.13**). **Latitude** refers to the angular distance of a point on

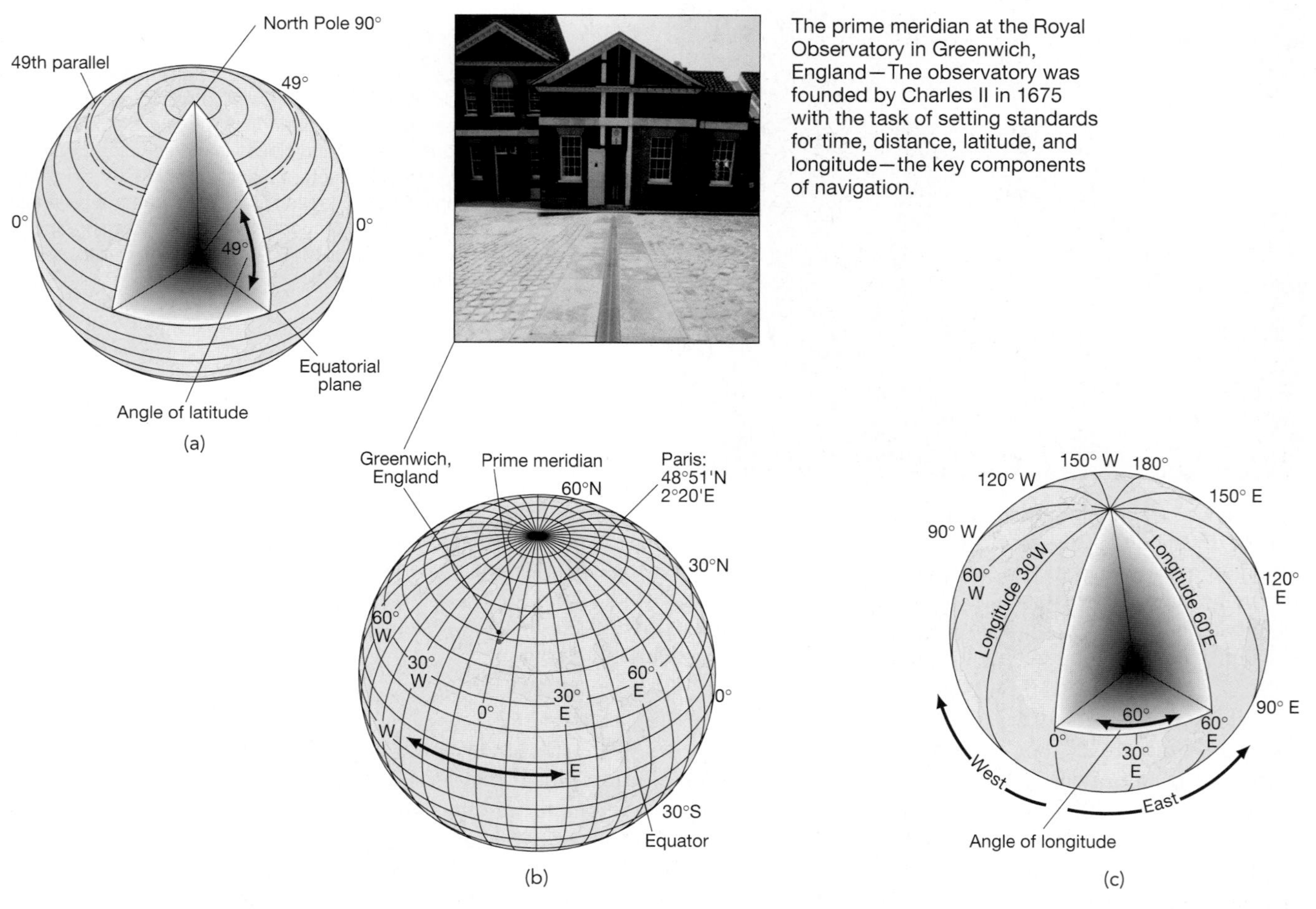

Figure 1.13 Latitude and longitude Lines of latitude and longitude provide a grid that covers Earth, allowing any point on Earth's surface to be accurately referenced. Latitude is measured in angular distance (that is, degrees and minutes) north or south of the equator, as shown in (a). Longitude is measured in the same way, but east and west from the prime meridian, a line around Earth's surface that passes through both poles (North and South) and the Royal Observatory in Greenwich, just to the east of central London, in England. Locations are always stated with latitudinal measurements first. The location of Paris, France, for example, is 48°51′N and 2°20′E as shown in (b). (a) and (c), after R. W. Christopherson, *Geosystems: An Introduction to Physical Geography,* 2nd ed., © 1994, pp. 13 and 15. (b), after E. F. Bergman, *Human Geography: Cultures, Connections, and Landscapes,* © 1995, Figs. 1–10 and 1–13.)

Earth's surface, measured in degrees, minutes, and seconds north or south from the equator, which is assigned a value of 0°. Lines of latitude around the globe run parallel to the equator, which is why they are sometimes referred to as parallels. Longitude refers to the angular distance of a point on Earth's surface, measured in degrees, minutes, and seconds east or west from the *prime meridian* (the line that passes through both poles and through Greenwich, England, which is assigned a value of 0°). Lines of **longitude**, called meridians, run from the North Pole (latitude 90° north) to the South Pole (latitude 90° south). Georgetown's coordinates are precisely 38°55′N, 77°00′E.

Thanks to the **Global Positioning System (GPS)**, it is very easy to determine the latitude and longitude of any given point. The Global Positioning System consists of 21 satellites (plus 3 spares) that orbit Earth on precisely predictable paths, broadcasting highly accurate time and locational information. The GPS is owned by the U.S. government, but the information transmitted by the satellites is freely available to everyone around the world. All that is needed is a GPS receiver. Basic receivers cost less than $50 and can relay latitude, longitude, and height to within 100 meters day or night, in all weather conditions, in any part of the world. Current production models of many automobiles are now equipped with GPS-based navigational systems. The most precise GPS receivers, costing thousands of dollars, are accurate to within a centimeter. The GPS has dramatically increased the accuracy and efficiency of collecting spatial data. In combination with GIS and remote sensing, GPS has revolutionized mapmaking and spatial analysis.

Location can also be *relative*, fixed in terms of site or situation. **Site** refers to the physical attributes of a location: its terrain, its soil, vegetation, and water sources, for example. **Situation** refers to the location of a place relative to other places and human activities: its accessibility to routeways, for example, or its nearness to population centers (**Figure 1.14**). Washington, D.C. has a low-lying riverbank site and is situated at the head of navigation of the Potomac River, on the Eastern Seaboard of the United States.

Finally, location also has a *cognitive* dimension, in that people have cognitive images of places and regions, compiled from their own knowledge, experiences, and impressions. **Cognitive images** (sometimes referred to as mental maps) are psychological representations of locations that spring from people's individual ideas and impressions of these locations. These representations can be

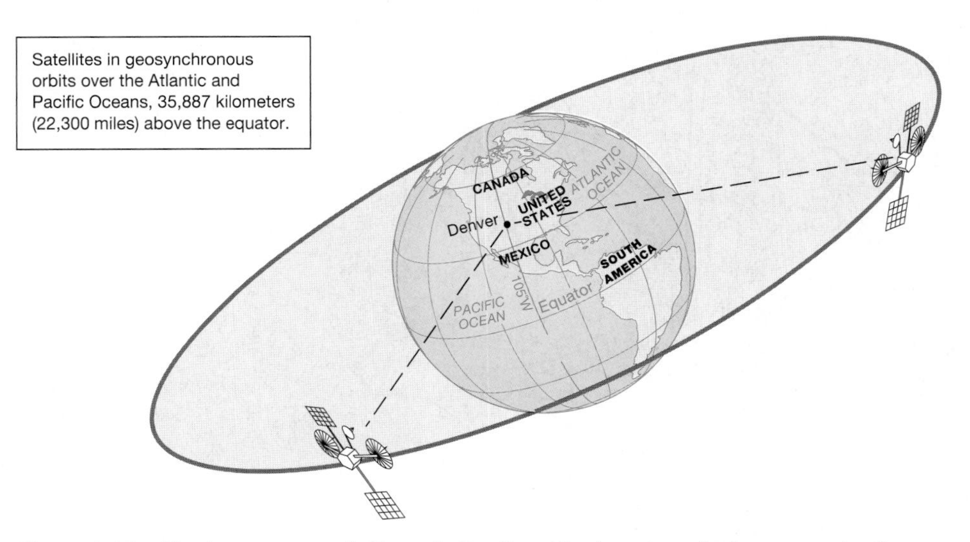

Figure 1.14 The importance of site and situation The location of telecommunications activities in Denver, Colorado, provides a good example of the significance of the geographic concepts of site (the physical attributes of a location) and situation (the location of a place relative to other places and human activities). Denver has become a major center for cable television, with the headquarters of giant cable companies such as Tele-Communications and DirecTV, an industrywide research lab, and a cluster of specialized support companies that together employ over 3,000 people. Denver's site, 1.6 kilometers (1 mile) above sea level, is important because it gives commercial transmitters and receivers a better "view" of communications satellites. Its situation, on the 105th meridian and equidistant between the telecommunications satellites that are in geostationary orbit over the Pacific and Atlantic oceans, allows it to send cable programming directly not just to the whole of the Americas but also to Europe, the Middle East, India, Japan, and Australia—to every continent, in fact, except Antarctica. This is important because it avoids "double-hop" transmission (in which a signal goes up to a satellite, then down, then up and down again), which increases costs and decreases picture quality. Before the location of telecommunications facilities in Denver, places east or west of the 105th meridian would have to double-hop some of their transmissions because satellite dishes would not have a clear "view" of both the Pacific and Atlantic telecommunications satellites.

based on people's direct experiences, on written or visual representations of actual locations, on hearsay, on people's imaginations, or on a combination of these sources. Location in these cognitive images is fluid, depending on people's changing information and perceptions of the principal landmarks in their environment.

Some things may not be located in a person's cognitive image at all. **Figure 1.15** shows one person's cognitive image of Washington, D.C. Georgetown is given a location within this mental map, even though it is some distance from the residence of the person who sketched her image of the city. Less well-known and less distinctive places do not appear on this particular image.

Distance

Distance is also useful as an *absolute* physical measure, whose units we may count in kilometers or miles, and as a *relative* measure, expressed in terms of time, effort, or cost. It can take more or less time, for example, to travel 10 kilometers from point A to point B than it does to travel 10 kilometers from point A to point C. Similarly, it can cost more or less. Geographers also have to recognize that distance can sometimes be in the eye of the beholder. It can seem longer or shorter, more or less pleasant going from A to B as compared to going from A to C. This is **cognitive distance,** the distance that people perceive to exist in a given situation. Cognitive distance is based on people's personal judgments about the degree of spatial separation between points.

The importance of distance as a fundamental factor in determining real-world relationships is a central theme in geography. It was once described as the "first law" of geography: "Everything is related to everything else, but near things are more related than distant things." Waldo Tobler, the geographer from the University of California, Santa Barbara, who put it this way, is one of many who

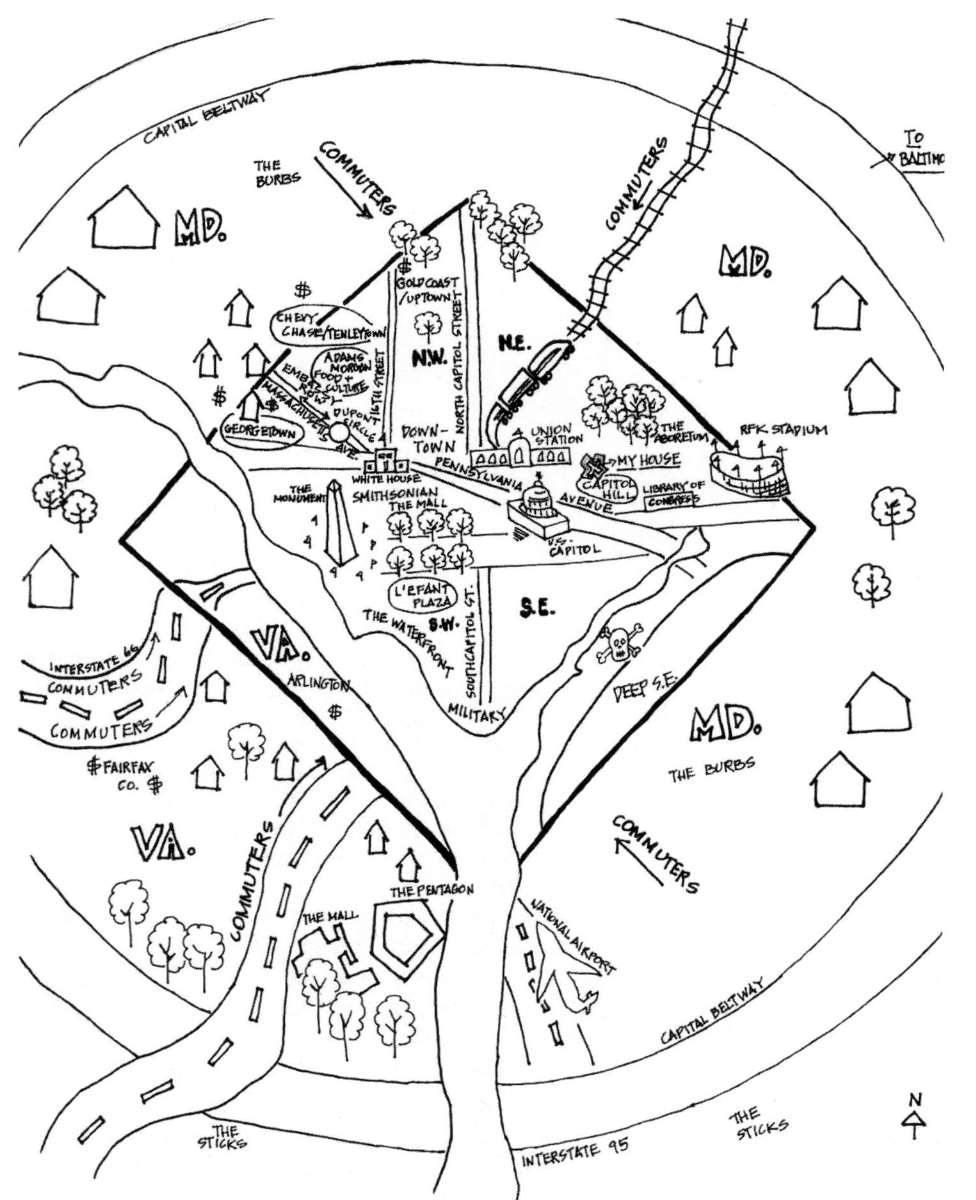

Figure 1.15 One person's cognitive image of Washington, D.C. This sketch was drawn by Rasheda DuPree, an urban affairs major at Virginia Tech, as part of a class exercise in recalling locations within students' hometowns. Rasheda has included many of the District's most prominent landmarks and some of its distinctive districts, including Georgetown. In contrast, there are no recorded locations in the city's southeastern quarter (marked by a skull and crossbones in Rasheda's sketch) or in the eastern outskirts (marked as "the burbs").

have investigated the friction of distance, the deterrent or inhibiting effect of distance on human activity. The **friction of distance** is a reflection of the time and cost of overcoming distance.

What these geographers have established is that these effects are not uniform—that is, they are not directly proportional to distance itself. This is true whether distance is measured in absolute terms (i.e., kilometers) or in relative terms (i.e., time- or cost-based measures). What happens is that the deterrent effects of extra distance tend to lessen as greater distances are involved. Thus, for example, while there is a big deterrent effect in having to travel 2 kilometers rather than 1 to get to a grocery store, the deterrent effect of the same extra distance (1 kilometer) after already traveling 10 kilometers is relatively small.

This sort of relationship creates what geographers call a distance-decay function. A **distance-decay function** describes the rate at which a particular activity or phenomenon diminishes with increasing distance. A typical distance-decay function is described by the graph in **Figure 1.16**, which shows the effects of distance on people's willingness to travel for free medical care.

Distance-decay functions reflect people's behavioral response to opportunities and constraints in time and space. As such, they reflect the **utility** of particular locations to people. The utility of a specific place or location refers to its usefulness to a particular person or group. In practice, utility is thought of in different ways by different people in different situations. The behavior of private firms and their agents or employees, for example, is most often guided by a bottom-line notion of utility that relates to dollar costs or profits. The same individuals will probably use a different notion of utility when it comes to their own lifestyle and the decisions they make in pursuing it. Prestige, convenience, or feelings of personal safety, communality, or happiness may well modify or override financial costs as the measure of utility. The business manager of a supermarket chain, for instance, will almost certainly decide on the utility of potential locations for a new store by weighing criteria based on the projected costs

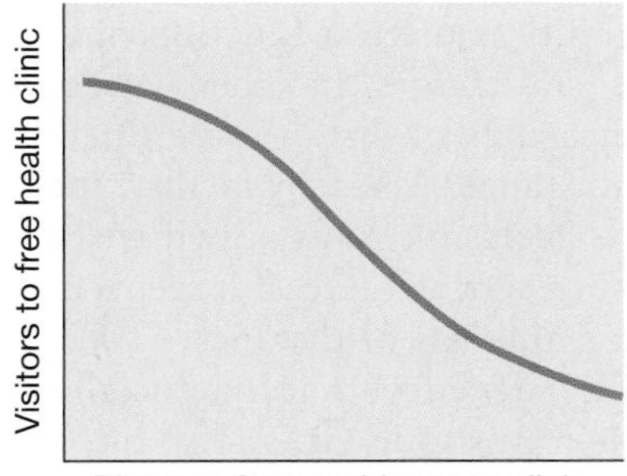

Figure 1.16 The friction of distance The effects of distance on people's behavior can be charted on graphs like this. The farther people have to travel, the less likely they are to do so. In this example, we can see clearly the deterrent effects of distance on people's attendance at a free health clinic.

and revenues for each potential site. In deciding on the utility of potential locations in which to retire, however, that same manager will almost certainly weigh criteria based not only on costs but also on a wide range of quality-of-life aspects of potential retirement places.

The unifying theme here is that however place utility is thought of, people in most circumstances tend to *seek to maximize the net utility of location.* The supermarket chain's business manager, for example, will seek to find the location for the chain's new store that is most likely to yield the greatest profit. Upon retirement he or she will choose to live in the place that represents the best trade-off among housing costs, cost of living, and quality of life. Seeking to maximize the net utility of location means that a great deal of human activity is influenced by what University of Washington geographer Richard Morrill once called the "nearness principle." According to this principle—a more explicit version of Tobler's first law—people will seek to:

- maximize the overall utility of places at minimum effort;
- maximize connections between places at minimum cost; and
- locate related activities as close together as possible.

The result is that patterns of behavior, locational decisions, and interrelations between people and places come to take on fairly predictable, organized patterns.

Space

Like distance, space can be measured in absolute, relative, and cognitive terms. **Table 1.1** lists the concepts human geographers use in talking about space in these various terms. Absolute space is a mathematical space described through points, lines, areas, planes, and configurations whose relationships can be fixed precisely through mathematical reasoning. Several ways of analyzing space mathematically are of use to geographers. The conventional way is to view space as a container, defined by rectangular coordinates and measured in absolute units of distance (kilometers or miles, for example). Other mathematical conceptions of space that geographers sometimes find useful also exist, however. One is **topological space,** defined as the connections between, or connectivity of, particular points in space (**Figure 1.17**). Topological space is measured not in terms of conventional measures of distance but by the nature and degree of connectivity between locations.

Relative measurements of space can take the form of socioeconomic space or of experiential or cultural space (see Table 1.1). Socioeconomic space can be described in terms of sites and situations, routes, regions, and distribution patterns. In these terms spatial relationships have to be fixed through measures of time, cost, profit, and production, as well as through physical distance. Experiential or cultural space is the space of groups of people with common ties, and it is described through the places, territories, and settings whose attributes carry special

TABLE 1.1 Different Kinds of Spaces Analyzed by Human Geographers

Absolute Space: Mathematical Space	Relative Space: Socioeconomic Space	Relative Space: Experiential/Cultural Space	Cognitive Space: Behavioral Space
Points	Sites	Places	Landmarks
Lines	Situations	Ways	Paths
Areas	Routes	Territories	Districts
Planes	Regions	Domains	Environments
Configurations	Distributions	Worlds	Spatial Layouts

Source: Based on H. Couclelis, "Location, Place, Region and Space," in R. Abler et al., *Geography's Inner Worlds.* New Brunswick, NJ: Rutgers University Press, 1992, Table 10.1, p. 231.

meaning for these particular groups. Finally, **cognitive space** is defined and measured in terms of people's values, feelings, beliefs, and perceptions about locations, districts, and regions. Cognitive space can be described, therefore, in terms of behavioral space—landmarks, paths, environments, and spatial layouts.

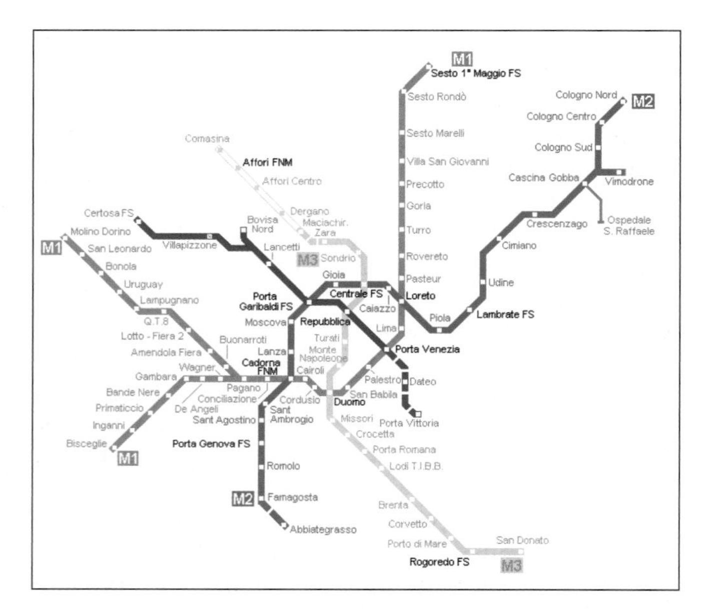

Figure 1.17 Topological space Some dimensions of space and aspects of spatial organization do not lend themselves to description simply in terms of distance. In other words, the connectivity of people and places is often important: whether they are linked, how they are linked, and so on. These attributes of connectivity define topological space. The map of the Metro system in Milan, Italy, is a topological map, showing how specific points are joined within a particular network. The most important aspects of networks of any kind, from the geographer's viewpoint, are their connectivity attributes. These attributes determine the flows of people and things (goods, information) and the centrality of places. As most Milanese know, the Metro system gives Duomo a very high degree of connectivity because trains on both the M1 and M3 lines stop there. Duomo is therefore relatively central within the "space of flows" of passenger traffic in central Milan. Piazza Missori—nearby in absolute terms—is much less central, however, and much less the focus of passenger flows.

Accessibility

Given that people tend to pursue the nearness principle, the concept of accessibility is very important. **Accessibility** is generally defined by geographers in terms of relative location: the opportunity for contact or interaction from a given point or location in relation to other locations. It implies proximity, or nearness, to something. Because it is a fundamental influence on the utility of locations, distance is an important influence on people's behavior. Distance is one aspect of accessibility, but it is by no means the only important aspect.

Connectivity is also an important aspect of accessibility because contact and interaction are dependent on channels of communication and transportation: streets, highways, telephone lines, and wavebands, for example. Effective accessibility is thus a function not only of distance but also of the configuration of networks of communication and transportation. Commercial airline networks provide many striking examples of this. Cities that operate as airline hubs are much more accessible than cities that are served by fewer flights and fewer airlines. Charlotte, N.C. for example (a U.S. Airways hub), is more accessible from Albany, N.Y. than from Richmond, VA even though Richmond is 400 kilometers (248 miles) closer to Albany than Charlotte. To get to Richmond from Albany, travelers must fly to Charlotte or another hub and change—a journey that takes longer and often costs more.

Accessibility is often a function of economic, cultural, and social factors. In other words, relative concepts and measures of distance are often as important as absolute distance. A nearby facility, such as a health-care clinic, is accessible to us only if we can actually afford the cost of getting there, if it seems close according to our own standards of distance, if we can afford to use the facility, if we feel that it is socially and culturally acceptable for us to use it, and so on. To take another example, a day-care center may be located just a few blocks from a single-parent family, but the center is not truly accessible if it opens after the parent has to be at work or if the parent feels that the staff, children, or other parents at the center are from an incompatible social or cultural group.

Spatial Interaction

Interdependence between places and regions can be sustained only through movement and flows. Geographers use the term **spatial interaction** as shorthand for all kinds of movement and flows involving human activity. Freight shipments, commuting, shopping trips, telecommunications, electronic cash transfers, migration, and vacation travel are all examples of spatial interaction. The fundamental principles of spatial interaction can be reduced to four basic concepts: complementarity, transferability, intervening opportunities, and diffusion.

- *Complementarity.* A precondition for interdependence between places is complementarity. For any kind of spatial interaction to occur between two places, there must be a demand in one place and a supply that matches, or complements, it in the other. This complementarity can be the result of several factors. One important factor is the variation in physical environments and resource endowments from place to place. For example, a heavy flow of vacation travel from Swedish cities to Mediterranean resorts is largely a function of climatic complementarity. To take another example, the flow of crude oil from Saudi Arabia (with vast oil reserves) to Japan (with none) is a function of complementarity in natural resource endowments.

  A second factor contributing to complementarity is the international division of labor that derives from the evolution of the world's economic systems. The more developed countries of the world have sought to establish overseas suppliers for their food, raw materials, and exotic produce, allowing the more developed countries to specialize in more profitable manufacturing and knowledge-based industries (see Chapter 2). Through a combination of colonialism, imperialism, and sheer economic dominance on the part of these more developed countries, less powerful countries have found themselves with economies that directly complement the needs of the more developed countries. Among the many flows resulting from this complementarity are shipments of sugar from Barbados to the United Kingdom, bananas from Costa Rica and Honduras to the United States, palm oil from Cameroon to France, automobiles from France to Algeria, school textbooks from the United Kingdom to Kenya, and investment capital from the United States to most of the less developed countries.

  A third contributory factor to complementarity is the operation of principles of specialization and **economies of scale.** Places, regions, and countries can derive economic advantages from the efficiencies created through specialization, which allows for larger-scale operations. Economies of scale are cost advantages to manufacturers that accrue from high-volume production, since the average cost of production falls with increasing output (**Figure 1.18**).

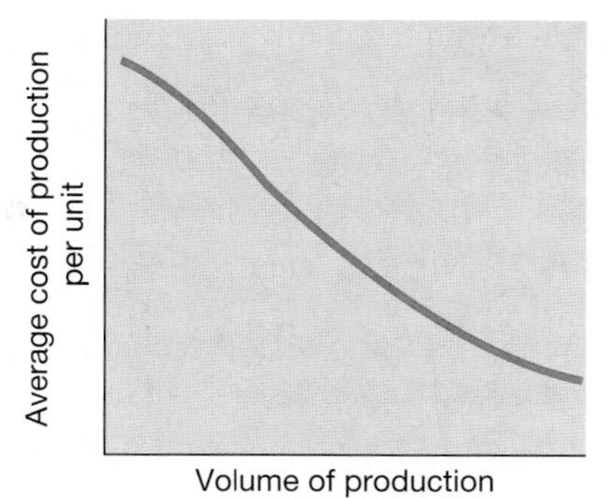

Figure 1.18 Economies of scale In many manufacturing enterprises, the higher the volume of production, the lower the average cost of producing each unit. This is partly because high-volume production allows for specialization and division of labor, which can increase efficiency and hence lower costs. It is also partly because most manufacturing activities have significant fixed costs (such as product design and the cost of renting or buying factory space) that must be paid for irrespective of the volume of production, so that the larger the output, the lower the fixed cost per unit. These savings are known as economies of scale.

  Among other things, fixed costs (for example, the cost of renting or buying factory space, which will be the same—fixed—whatever the level of output from the factory) can be spread over higher levels of output so that the average cost of production falls. Economic specialization results in complementarities, which in turn contribute to patterns of spatial interaction. One example is the specialization of Israeli farmers in high-value fruit and vegetable crops for export to the European Union, which in return exports grains and root crops to Israel.

- *Transferability.* Another precondition for interdependence between places is *transferability,* which depends on the frictional or deterrent effects of distance. Transferability is a function of two things: the costs of moving a particular item, measured in real money and/or time, and the ability of the item to bear these costs. If, for example, the costs of moving a product from one place to another make it too expensive to sell successfully at its destination, then that product does not have transferability between those places.

  Transferability varies between places, between kinds of items, and between modes of transportation and communication. The transferability of coal, for example, is much greater between places that are connected by rail or by navigable waterways than between places connected only by highways. This is because it is much cheaper to move heavy, bulky materials by rail, barge, or ship. The transferability of fruit and salad crops, on the other hand, depends more on the speed of transportation and the availability of specialized refrigerated vehicles so the fruits and vegetables stay fresh. While the transferability of money capital is much greater by telecommunications than it is by surface transportation, it is also higher between places where banks are equipped to deal routinely with electronic transfers. Computer microchips have high transferability because they are easy to handle, and transport costs are a small proportion of their value. Computer monitors, on the other hand, have lower transferability because of their fragility and their relatively lower value by weight and volume.

Transferability also varies over time, as successive innovations in transport and communications technologies and successive waves of **infrastructure** development (canals, railways, harbor installations, roads, bridges, and so on) alter the geography of transport costs. New technologies and new or extended infrastructures have the effect of altering the transferability of particular things between particular places. *As a result, the spatial organization of many different activities is continually changing and readjusting.* The consequent tendency toward a shrinking world gives rise to the concept of **time-space convergence,** the rate at which places move closer together in travel or communication time or costs. Time-space convergence results from a decrease in the friction of distance as new technologies and infrastructure improvements successively reduce travel and communications time between places. Such space-adjusting technologies have, in general, brought places closer together over time (**Figure 1.19**). Overland travel between New York and Boston, for example, has been reduced from 3.5 days (in 1800) to 5 hours (in the 2000s) as the railroad displaced stagecoaches and was in turn displaced by interstate

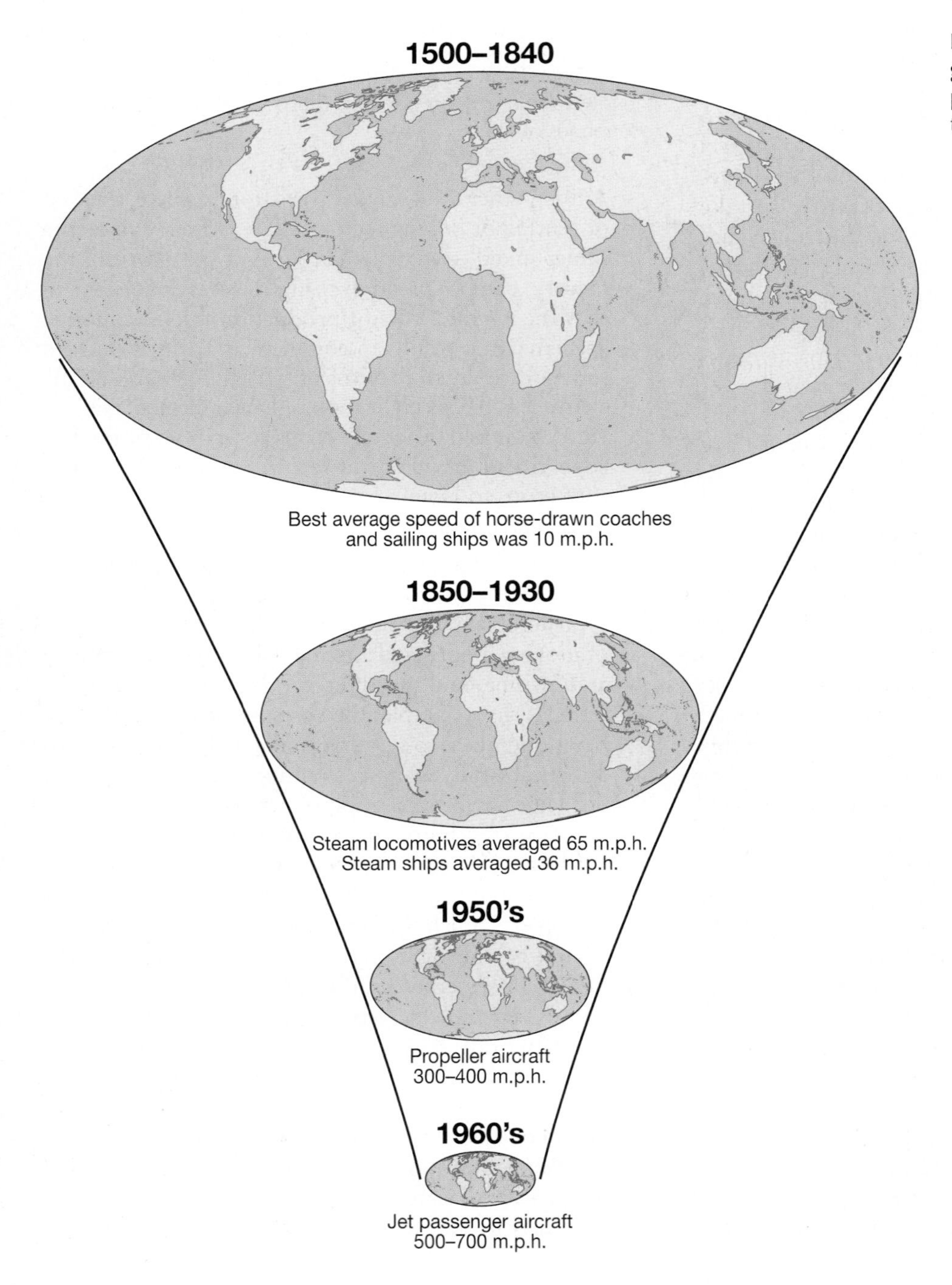

Figure 1.19 Time-space convergence The effects of changing transportation technologies in "shrinking" the world in terms of travel time.

automobile travel. Other important space-adjusting innovations include air travel and air cargo; telegraphic, telephonic, and satellite communications systems; national postal services, package delivery services, and facsimile (fax) machines; and modems, fiber-optic networks, and electronic-mail software.

What is most significant about the latest developments in transport and communication is that they are not only global in scope but also are able to penetrate to local scales. As this penetration occurs, some places that are distant in kilometers are becoming close together, while some that are close in terms of absolute space are becoming more distant in terms of their ability to reach one another electronically (**Figure 1.20**). Much depends on the mode of communication—the extent to which people in different places are "plugged in" to new technologies. Older wire cable can carry only small amounts of information; microwave channels are good for person-to-person communication but depend on line of sight; telecommunications satellites are excellent for reaching remote areas but involve significant capital costs for users, while fiber-optic cable is excellent for areas of high-population density but not feasible for more remote, rural areas. The shrinking of space has important implications for people's everyday conceptions of space and distance and for their level of knowledge about other places.

- *Intervening Opportunity.* While complementarity and transferability are preconditions for spatial interaction, intervening opportunities are more important in determining the *volume* and *pattern* of movements and flows. Intervening opportunities are simply alternative origins and/or destinations. Such opportunities are not necessarily situated directly between two points or even along a route between them. Thus, to take one of our previous examples, for Swedish families considering a Mediterranean vacation in Greece, resorts in Spain, southern France, and Italy are all likely to be intervening opportunities because they can probably be reached more quickly and cheaply than resorts in Greece.

  The size and relative importance of alternative destinations are also important aspects of the concept of intervening opportunity. For our Swedish families, Spanish resorts probably offer the greatest intervening opportunity because they contain the largest aggregate number of hotel rooms and vacation apartments. We can therefore state the principle of intervening opportunity as follows: Spatial interaction between an origin and a destination will be proportional to the number of opportunities at that destination and inversely proportional to the number of opportunities at alternative destinations.

- *Spatial Diffusion.* Disease outbreaks, technological innovations, political movements, and new musical fads all originate in specific places and subsequently spread to other places and regions. The way that things spread through space and over time—**spatial diffusion**—is one of the most important aspects of spatial interaction and is crucial to an understanding of geographic change.

  Diffusion seldom occurs in an apparently random way, jumping unpredictably all over the map. Rather, it occurs as a function of statistical probability, which is often based on fundamental geographic principles of distance and movement. The diffusion of a contagious disease, for example, is a function of the probability of physical contact, modified by variations in individual resistance to the disease. The result

Figure 1.20 The Cybersmith Café, Cambridge, Massachusetts In early 2005 more than 75 percent of American adults had regular access to the Internet. Increasing reliance on the Internet has led to the growth of Internet cafés, catering mostly to people's need for connectivity when they are away from home and/or work.

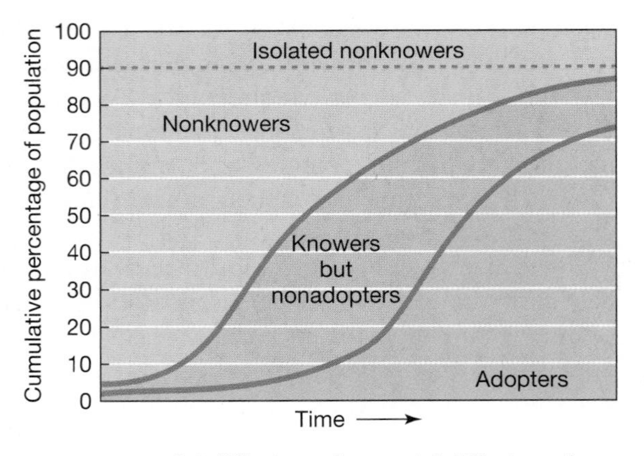

Figure 1.21 Spatial diffusion The spatial diffusion of many phenomena tends to follow an S-curve of slow buildup, rapid spread, and leveling off. In the case of the diffusion of an innovation, for example, it usually takes a while for enough potential adopters to get to know about the innovation, and even longer for a critical mass of them to adopt it. After that the innovation spreads quite rapidly, until most of the potential adopters have been exposed to it. (After D. J. Walmsley and G. J. Lewis, *Human Geography: Behavioral Approaches*. London: Longman, 1984, Fig. 5.3, p. 52.)

is typically a "wave" of diffusion that describes an S-curve, with a slow buildup, rapid spread, and final leveling off (**Figure 1.21**).

It is possible to recognize several spatial tendencies in patterns of diffusion. In *expansion diffusion* (also called *contagion diffusion*—**Figure 1.22a**), a phenomenon spreads because of the proximity of carriers, or agents of change, who are fixed in their location. An example would be the diffusion of an agricultural innovation, such as the use of hybrid seed stock, among members of a local farming community. With *hierarchical diffusion* (also known as cascade diffusion), a phenomenon can be diffused from one location to another without necessarily spreading to people or places in between (**Figure 1.22b**). An example would be the spread of a fashion trend from large metropolitan areas to successively smaller cities, towns, and rural settlements. Many patterns and processes of diffusion reflect both expansion and hierarchical diffusion (**Figure 1.22c**), as different aspects of human interaction come into play in different geographic settings. The diffusion of outbreaks of communicable diseases, for example, usually involves a combination of expansion and hierarchical diffusion, as in the case of HIV/AIDS (see p. 16).

Regional Analysis

Not all geographic phenomena are most effectively understood through spatial analysis. Geographers also seek to understand the complex relationships between peoples and places in terms of the similarities and differences among and between them and the identities and qualities associated with them. Here the key concepts are *regionalization, landscape,* and *sense of place.*

Regionalization

The geographer's equivalent of *scientific classification* is **regionalization,** with individual places or areal units being the objects of classification. The purpose of regionalization is to identify regions of one kind or another. There are several ways in which individual areal units can be assigned to regions. One is that of *logical division,* or "classification from above." This involves partitioning a universal set of areal units into successively larger numbers of regions, using more specific criteria at every stage. Thus, a classification of countries might be achieved by first differentiating between rich and poor countries, then dividing both rich and poor countries into those that have a trade surplus and those that have a deficit, and so on. A second way in which individual areal units can be assigned to regions is that of *grouping,* or "classification from below." This involves searching for regularities or significant relationships among areal units and grouping them in successively smaller numbers of classes, using a broader measure of similarity at each stage.

An implicit assumption in this type of classification of areal units into regions is that each unit is homogeneous with respect to the attribute or attributes under consideration. Where this assumption holds true, the result of regional classification is a set of formal regions. **Formal regions** are groups of areal units that have a high degree of homogeneity in terms of particular distinguishing features (such as religious adherence or household income). Few phenomena, however, exhibit such homogeneity over large areal units. For this reason, geographers also recognize **functional regions** (sometimes referred to as nodal regions)—regions within which, while there may be some variability in certain attributes (again, for example, religious adherence and income), there is an overall coherence to the structure and dynamics of economic, political, and social organization.

The concept of functional regions allows us to recognize that the coherence and distinctive characteristics of a region are often stronger in some places than in others. This point is illustrated by geographer Donald Meinig's *core-domain-sphere* model, which he set out in his classic essay on the Mormon region of the United States (**Figure 1.23**). In the core of a region the distinctive attributes are very clear; in the domain they are dominant but not to the point of exclusivity; in the sphere they are present but not dominant.

In addition to questions of classification, the art and science of regional analysis must consider questions of geographic scale, for we can (and must) see the world as a mosaic of small regions that exist within successively larger frameworks (see Figure 1.4). These frameworks are closely related, as both cause and effect, to the formal boundaries that have evolved (and that are continually challenged and amended) under national and international law.

Finally, people's own conceptions of place, region, and identity may resonate with or against these boundaries to

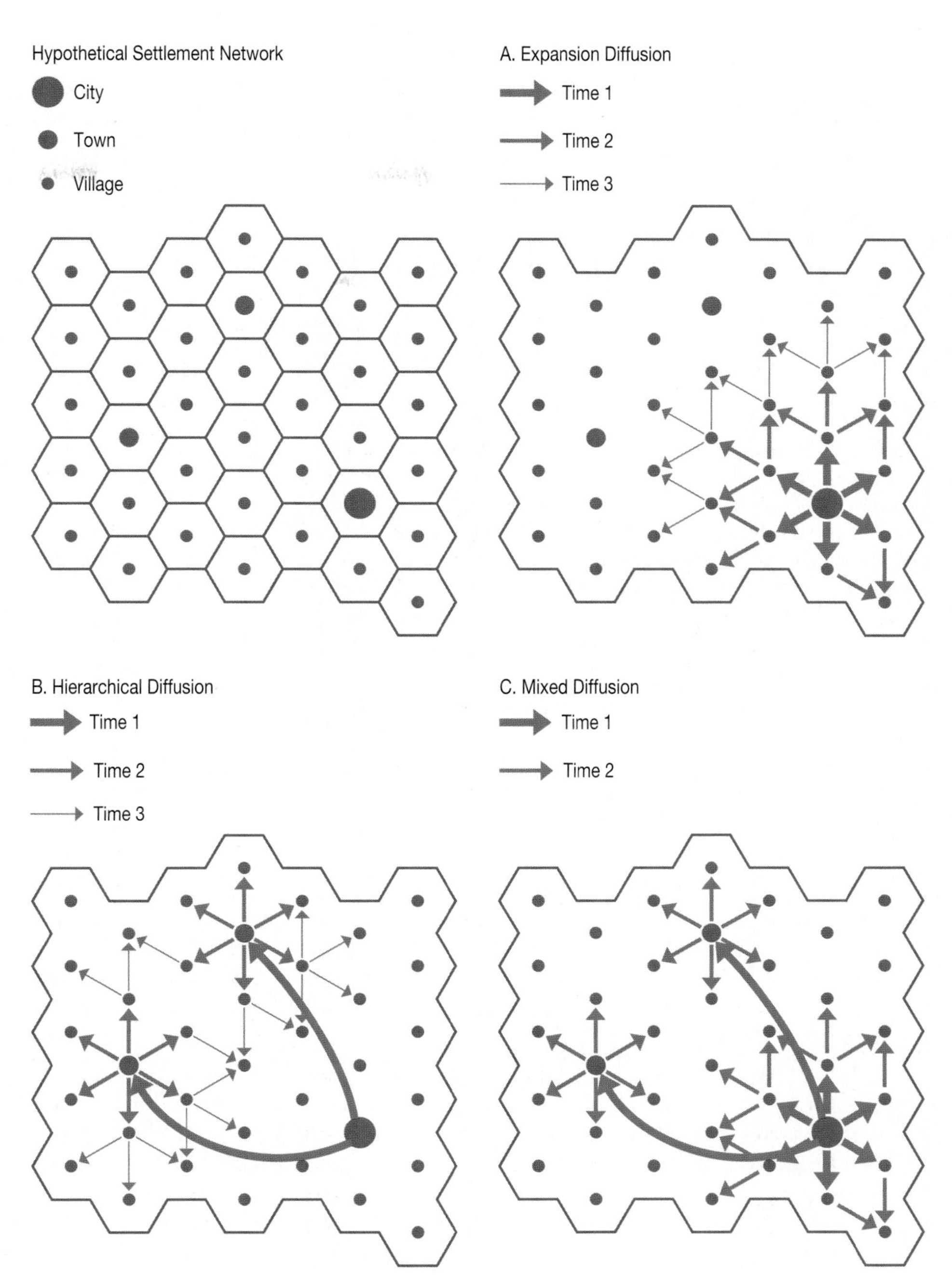

Figure 1.22 Patterns of spatial diffusion (a) Expansion diffusion (for example, the spread of an innovative agricultural practice, such as the use of hybrid seed stock, across a rural region); (b) hierarchical diffusion (the spread of a fashion trend from large metropolitan areas to smaller cities and towns); (c) mixed diffusion (the spread of a contagious disease across a region). (After E. K. Cromley and S. L. McLafferty, *GIS and Public Health.* New York: Guilford Press, p. 193.)

generate strong feelings of regionalism and sectionalism that feed back into the processes of place-making and regional differentiation. **Regionalism** is a term used to describe situations in which different religious or ethnic groups with distinctive identities coexist within the same state boundaries, often concentrated within a particular region and sharing strong feelings of collective identity. If such feelings develop into an extreme devotion to regional interests and customs, the condition is known as **sectionalism.** Regionalism often involves ethnic groups whose aims include autonomy from a national state and the development of their own political power (see Chapter 9). In certain cases, enclaves of ethnic minorities are claimed by the government of a country other than the one in which they reside. Such was the case, for example, of Serbian enclaves in Croatia, claimed by nationalist Serbs. The assertion by the government of a country that a minority living outside its formal borders belongs to it historically and culturally is known as **irredentism.** In some circumstances, as with Serbia's claims on Serbian enclaves in Croatia in the early 1990s, irredentism can lead to war.

Landscape

Geographers think of landscape as a comprehensive product of human action such that every landscape is a complex repository of society. It is a collection of evidence about our character and experience, our struggles and triumphs as humans. To understand better the meaning of landscape, geographers have developed different categories of landscape types based on the elements contained within them. **Ordinary landscapes** (or vernacular landscapes, as they are sometimes called) are the everyday landscapes that people

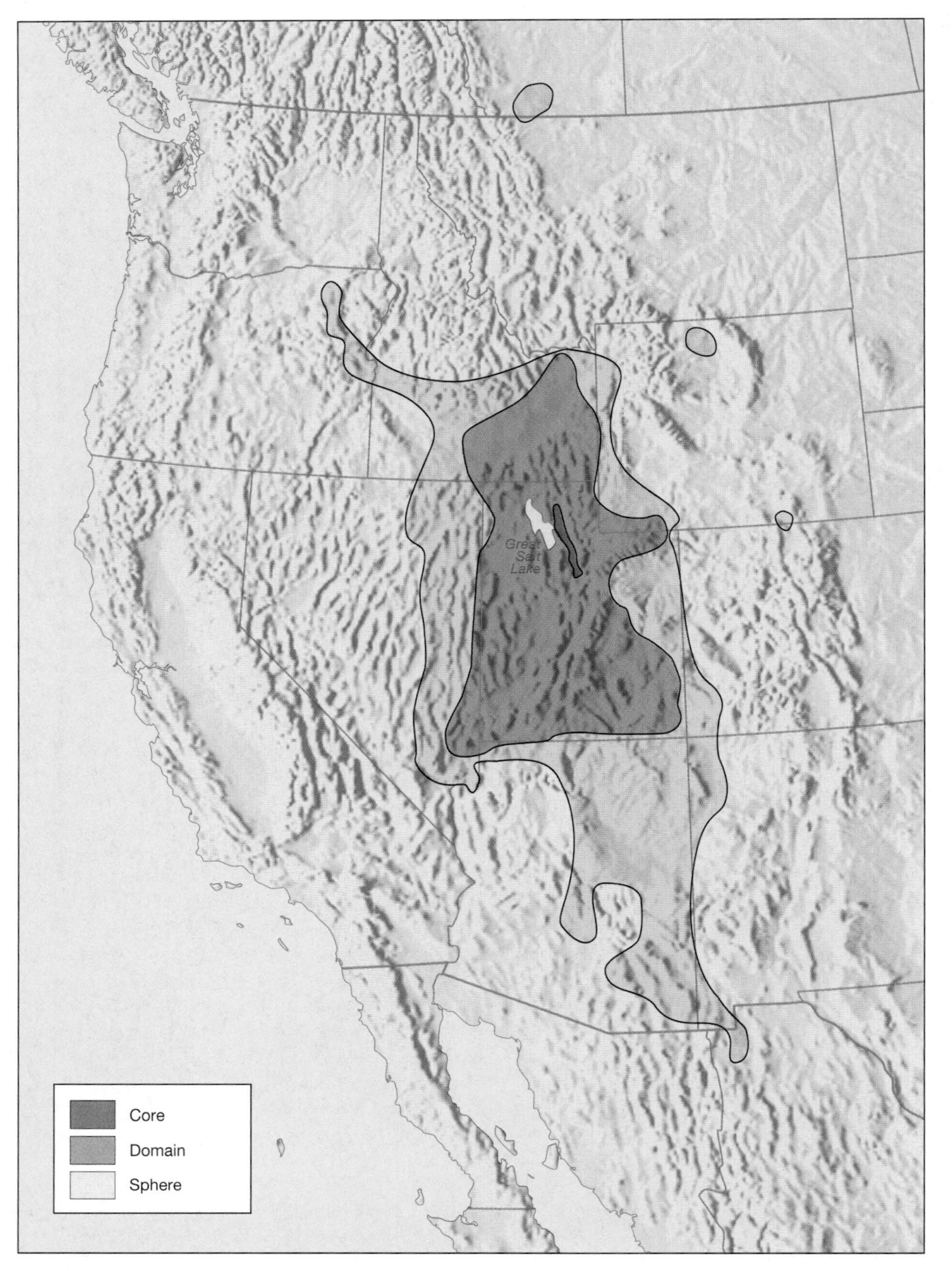

Figure 1.23 The Mormon culture region Cultural attributes often gradually shade from one region to another, rather than having a single, clear-cut boundary. Geographer Donald Meinig's work on the Mormon culture region of the United States identified a "core" region, which exhibits all the attributes of Mormon culture; a "domain" where not all these attributes may be present (or may be less intense); and a "sphere," where some attributes of Mormon culture are present but often as a minority with the attributes of another culture region. (After D. Meinig, "The Mormon Culture Region: Strategies and Patterns in the Geography of the American West," *Annals, Association of American Geographers, 55,* 1965, pp. 191–220.)

create in the course of their lives together. From crowded city centers to leafy suburbs and quiet rural villages, these are landscapes that are lived in and changed and that in turn influence and change the perceptions, values, and behaviors of the people who live and work in them.

Symbolic landscapes, by contrast, represent particular values or aspirations that the builders and financiers of those landscapes want to impart to a larger public. For example, the neoclassical architecture of the buildings of the federal government in Washington, D.C., along with the streets, parks, and monuments of the capital, constitute a symbolic landscape intended to communicate a sense of power, but also of democracy in its imitation of the Greek city-state. Some landscapes become powerfully symbolic of national identity. Nation-building depends heavily on stories of golden ages, enduring traditions, heroic deeds, shared hardships, and dramatic destinies, all located in traditional (or promised) home lands with hallowed sites and scenery. Landscapes thus become a way of picturing a nation. With the creation of modern Italy during the *Risorgimento* ("revival through unification"—1815–1861), for example, the classical Tuscan landscape (**Figure 1.24**) became emblematic of Italy itself and has attracted landscape painters, romantic poets, and novelists ever since. Similarly, the West of Ireland (**Figure 1.25**) came to symbolize the whole of Ireland to Irish nationalists in the early twentieth century—partly because it was seen as

Figure 1.24 The power of place In some countries, particular landscapes have become powerfully symbolic of national identity. In Italy it is the classical landscape of Tuscany, with its scattered farms and villas, elegant cypress trees, silvery-green olive trees, and rolling fields with a rich mixture of cereals, vegetables, fruit trees, and vines.

the region least affected by British colonization, but also because its bare and rugged landscape seemed to contrast so strikingly with the more bucolic rural landscapes (**Figure 1.26**) through which England was popularly imagined.

Geographers now recognize that there are many layers of meaning embedded in the landscape, meanings that can be expressed and understood differently by different social groups at different times. Landscapes reflect the lives of ordinary people as well as the more powerful, and they reflect their dreams and ideas as well as their material lives. The messages embedded in landscapes can be read as signs about values, beliefs, and practices, though not every reader will take the same message from a particular landscape (just as people may differ in their interpretation of a passage from a book). In short, landscapes both produce and communicate meaning, and one of our tasks as geographers is to interpret those meanings.

Sense of Place

The experience of everyday routines in familiar settings allows people to derive a pool of shared meanings. Often this carries over into people's attitudes and feelings about themselves and their locality. When this happens, the result is a self-conscious sense of place. The concept of a **sense of place** refers to the feelings evoked among people as a result of the experiences and memories they associate with a place and to the symbolism they attach to that place. It can also refer to the character of a place as seen by outsiders: its distinctive physical characteristics and/or its inhabitants.

For *insiders,* this sense of place develops through shared dress codes, speech patterns, public comportment, and so

Figure 1.25 The national landscape of Ireland The rugged landscapes of the West of Ireland have come to symbolize the whole country to many people, both within Ireland and beyond.

Figure 1.26 Aylesford, England The well-ordered and picturesque landscape of the southern parts of rural England have long been taken to be emblematic of England as a whole, and of the values and ideals of its people—even though urban and industrial development, together with modern agricultural practices, have brought about significant changes to both landscapes and society.

on. A crucial concept here is that of the **lifeworld,** the taken-for-granted pattern and context for everyday living through which people conduct their day-to-day lives without having to make it an object of conscious attention. People's experience of everyday routines in familiar settings leads to a pool of shared meanings. People become familiar with one another's vocabulary, speech patterns, dress codes, gestures, and humor as a result of routine encounters and shared experiences in bars and pubs, cafes and restaurants, shops and street markets, and parks. This is known as **intersubjectivity:** shared meanings that are derived from the lived experience of everyday practice. Elements of daily rhythms (such as mid-morning grocery shopping with a stop for coffee, the *aperitivo* en route from work to home, and the after-dinner stroll) are all critical to the density of routine encounters and shared experiences that underpin the intersubjectivity that is the basis for a sense of place within a community (**Figure 1.27**). The same is true of elements of weekly rhythms, such as street markets and farmers' markets; and of seasonal rhythms, such as festivals.

These rhythms, in turn, depend on certain kinds of spaces and places: not only streets, squares, and public open spaces but also "third places" (after home, first, and workplace, second): the sidewalk cafés, pubs, post offices, drug stores, corner stores, and family-run *trattoria* that are the loci of routine activities and socio-cultural transactions. Third places accommodate "characters," "regulars," and newcomers, as well as routine patrons and, like public spaces, facilitate casual encounters as well as settings for sustained conversations. The nature and frequency of routine encounters and shared experiences depends a great deal on the attributes of these spaces and places.

A sense of place also develops through familiarity with the history and symbolism of particular elements of the

Figure 1.27 Intersubjectivity Routine encounters such as this, in Chiavenna, Italy, help to develop a sense of community and a sense of place among residents.

Figure 1.28 Community art Community art can provide an important element in the creation of a sense of place for members of local communities. This example is from the Mission district of San Francisco.

physical environment—a local mountain or lake, the birthplace of someone notable, the location of some particularly well-known event, or the expression of community identity through community art (**Figure 1.28**). Sometimes it is deliberately fostered by the construction of symbolic structures such as monuments and statues. Often it is a natural outcome of people's familiarity with one another and their surroundings. Because of this consequent sense of place, insiders feel at home and "in place."

For *outsiders,* a sense of place can be evoked only if local landmarks, ways of life, and so on are distinctive enough to evoke a significant common meaning for people who have no direct experience of them. Central London, for example, is a setting that carries a strong sense of place to outsiders who have a sense of familiarity with the riverside panoramas, busy streets, and distinctive monuments and historic buildings that together symbolize the heart of the city.

Developing a Geographical Imagination

A **geographical imagination** allows us to understand changing patterns, processes, and relationships among people, places, and regions. Developing this capacity is increasingly important as the pace of change around the world increases to unprecedented levels. Whereas much of the world had remained virtually unchanged for decades, even centuries, the Industrial Revolution and long-distance, high-speed transportation and communications brought a rapid series of rearrangements to the countryside and to towns and cities in many parts of the world. Today, with a globalized economy and global telecommunications and transportation networks, places have become much more interdependent, and still more of the world is exposed to increasingly urgent imperatives to change.

It is often useful to think of places and regions as representing the cumulative legacy of successive periods of change. Following this approach, we can look for superimposed layers of development (**Figure 1.29**). We can show how some patterns and relationships last, while others are modified or obliterated. We can show how

Figure 1.29 Places as the cumulative legacy of change To understand places properly, we have to see them as the cumulative legacy, not just of buildings and structures from different periods of the past but also of the laws, institutions, customs, and so on that developed within each of these periods. This photograph of Istanbul, Turkey, is a very striking example, with modern urban development interlayered with surviving fragments of Roman, Ottoman, and nineteenth-century development.

different places bear the imprint of different kinds of change, perhaps in different sequences and with different outcomes. To do so, we must be able to identify the kinds of changes that are most significant.

Recognizing the General and the Unique

We can prepare our geographical imagination to deal with an important aspect of spatial change by making a distinction between the *general* and the *unique*. This distinction helps account for geographical diversity and variety because it provides a way of understanding how and why one kind of change can result in a variety of spatial outcomes: It is because the *general effects* of a particular change always involve some degree of modification as they are played out in different environments, giving rise to *unique outcomes*.

Although we can usually identify some general outcomes of major episodes of change, there are almost always some unique outcomes, too. Let us take two related examples. The Industrial Revolution of nineteenth-century Europe provides a good example of a major period of change. A few of the general spatial outcomes were increased urbanization, regional specialization in production, and increased interregional and international trade. At one level, places could be said to have become increasingly alike: generic coalfield regions, industrial towns, ports, downtowns, worker housing, and suburbs.

It is clear, however, that these general outcomes were mediated by the different physical, economic, cultural, and social attributes of different places. Beneath the dramatic overall changes in the geography of Europe, new layers of diversity and variety also existed. Industrial towns developed their own distinctive character as a result of their manufacturing specialties, their politics, the personalities and objectives of their leaders, and the reactions and responses of their residents. Downtowns were differentiated from one another as the general forces of commerce and land economics played out across different physical sites and within different patterns of land ownership. They also were differentiated as local socioeconomic and political factors gave rise to different expressions of urban design. Meanwhile, some places came to be distinctive because they were almost entirely bypassed by this period of change, their characteristics making them unsuited to the new economic and spatial order (**Figure 1.30**).

Figure 1.30 Hersbrück, Germany Hersbrück was once an important and prosperous regional center on an overland trade route—the "Golden Road"—between Nuremberg and Prague. After 1806, when Napoleon redrew the political map of Europe, the reorganization of the European economy, together with the onset of the Industrial Revolution, left Hersbrück somewhat isolated and economically disadvantaged. Hersbrück was never drawn into the industrial development of Germany and is not well connected to the transportation infrastructure of canals, railways, or major highways.

The second example of general and unique outcomes of change is the introduction of the railroad, one of the specific changes involved in the Industrial Revolution. In general terms, the railroad contributed to time-space convergence, to the reorganization of industry into larger market areas, to an increase in inter-regional and international trade, and to the interconnectedness of urban systems. Other unique outcomes, however, have also contributed to distinctive regional geographies. In Britain the railroad was introduced to an environment that was partially industrialized and densely settled, and one of the main outcomes was that the increased efficiencies provided by the railroad helped to turn Britain's economy into a highly integrated and intensively urbanized national economy. In Spain, however, the railroad was introduced to an environment that was less urbanized and industrialized and less able to afford the costs of railroad construction. The result was that the relatively few Spanish towns connected by the railroads gained a massive comparative advantage. This situation laid the foundation for a modern space-economy that was much less integrated than Britain's, with an urban system dominated by just a few towns and cities.

MAKING A DIFFERENCE: THE POWER OF GEOGRAPHY

The study of geography has become an essential basis for understanding a world that is more complex and faster changing than ever before. Through an appreciation of the diversity and variety of the world's peoples and places, geography provides real opportunities not only for contributing to local, national, and global development but also for understanding and promoting multicultural, international, and feminist perspectives on the world.

The Importance of a Geographic Education

In the United States a decade of debate about geographic education has resulted in a widespread acceptance that being literate in geography is essential in equipping citizens to earn a decent living, to enjoy the richness of life, and to participate responsibly in local, national, and international affairs. In response to the inclusion of geography as a core subject in the *Goals 2000: Educate America Act* (Public Law 103–227), a major report on the goals of geographic education was produced jointly by the American Geographical Society, the Association of American Geographers, the National Council for Geographic Education, and the National Geographic Society.[2] The report emphasizes the importance of being geographically informed; understanding that geography is the study of people, places, and environments from a spatial perspective; and appreciating the interdependent worlds in which we live:

> The power and beauty of geography allow us to see, understand, and appreciate the web of relationships between people, places, and environments.
>
> At the everyday level, for example, a geographically informed person can appreciate the locational dynamics of street vendors and pedestrian traffic or fast-food outlets and automobile traffic; the routing strategies of school buses in urban areas and of backpackers in wilderness areas; the land-use strategies of farmers and of real estate developers.
>
> At a more expanded spatial scale, that same person can appreciate the dynamic links between severe storms and property damage or between summer thunderstorms and flash floods; the use of irrigation systems to compensate for lack of precipitation . . . ; the seasonal movement of migrant laborers in search of work and of vacationers in search of sunshine and warmth.
>
> At a global level, the geographically informed person can appreciate the connections between cyclical drought and human starvation in the Sahel or between the Chernobyl nuclear disaster and the long-term consequences to human health and economic activities throughout eastern and northwestern Europe; the restructuring of human migration and trade patterns as the European Union becomes increasingly integrated or as the Pacific rim nations develop a commonality of economic and political interests; and the uncertainties associated with the possible effects of global warming on human society or the destruction of tropical rain forests on global climate.[3]

[2]Geography Education Standards Project, *Geography for Life: National Geography Standards 1994*. Washington, DC: National Geographic Research and Exploration, 1994.
[3]*Geography for Life*, p. 29.

Most people want to understand the intrinsic nature of the world in which they live. Geography enables them to understand where they are both literally and figuratively. Geography provides knowledge of Earth's physical and human systems and of the interdependency of living things and physical environments. That knowledge, in turn, provides a basis for people to cooperate in the best interests of our planet. Geography also captures the imagination: It stimulates curiosity about the world and the world's diverse inhabitants and places, as well as about local, regional, and global issues. By better understanding the world, people can overcome parochialism and ethnocentrism. Last but not least, geography has utilitarian value in the modern world. As the interconnectedness of the world accelerates, the practical need for geographic knowledge becomes more critical. With a strong grasp of geography, people are better equipped to solve issues not only at the local level but also the global level.[4]

Geographers at Work

Geography, then, is very much an applied discipline as well as a means of understanding the world. Geographers employed in business, industry, and government are able to use geographic theories and techniques to understand and solve a wide variety of specific problems. A great deal of the research undertaken by geography professors also has an applied focus. As a result, geography is able to make a direct and significant contribution to society. Because of the broad nature of the field, these contributions cover every aspect of human activity and every scale from the local to the global. A number of examples reflect this:

- ***International Affairs.*** Geographers' knowledge and understanding of regional histories and geographies, along with their ability to analyze the interdependence of places and regions, enables them to effectively contribute to discussions of international policy. Geographers' work within governmental agencies, corporations, and nonprofit institutions in shaping international strategies is especially important in the context of accelerating globalization.
- ***Location of Public Facilities.*** Geographers use specialized techniques to analyze the location patterns of particular population groups, to analyze transportation networks, and to analyze patterns of geographic accessibility to alternative sites. Such analysis enables geographers to determine the most effective locations for new public facilities, such as clinics, emergency rooms, social centers for the elderly, and shelters for the homeless.
- ***Marketing and Location of Industry.*** Similar techniques are used in determining the most efficient, or most profitable, location for new factories, stores, and offices. Geographical research is also used to analyze

[4]*Geography for Life*, pp. 23–24.

the changing geography of supply and demand, allowing industry to determine whether, and where, to relocate. Basic techniques of geographic analysis are also used in geodemographic research, an important aspect of marketing.

- ***Geography and the Law.*** Geographical analysis is increasingly required in helping to resolve complex social and environmental issues. One important example is the issue of racial segregation and the implications of residential segregation for policies such as school busing. Another example is the issue of property development and the implications of environmental hazards such as flooding, coastal erosion, toxic waste dumps, and earthquake fault zones for policies, codes, and regulations affecting development. A third is the issue of drawing up geographical boundaries for political units in ways that ensure equal representation by population size and racial and political composition—even as populations are changing between one election and another.
- ***Disease Ecology.*** By analyzing social and environmental aspects of human diseases, geographers are able to shed light on the causes of disease, to predict the spread of particular outbreaks, and to suggest ways in which the incidence of disease might be controlled.
- ***Urban and Regional Planning.*** Urban and regional planning adopts a systematic, creative approach to address and resolve physical, social, and economic problems of neighborhoods, cities, suburbs, metropolitan areas, and larger regions. Planners work directly on preserving and enhancing the quality of life in communities, protecting the environment, promoting equitable economic opportunity, and managing growth and change of all kinds. Planning has roots in engineering, law, architecture, social welfare, and government, but geography, because of its focus on the interdependence of peoples and places, offers the best preparation for specialized professional training in urban and regional planning.
- ***Economic Development.*** Geographers' ability to understand the interdependence of places and to analyze the unique economic, environmental, cultural, and political attributes of specific regions enables them to contribute effectively to strategies and policies aimed at economic development. Geographers are involved in applied research and policy formulation concerning economic development all over the world, addressing the problems not only of individual places and regions but also of the entire world economy.
- ***Homeland Security.*** Geographers' knowledge and understanding of geopolitics, political geography, demographics, medical geography, and cultural geography, together with an appreciation of the interdependent relationships among local, regional, and global systems, provides a sound basis for work in many areas of homeland security. Knowing how places and regions "work" also means knowing about their vulnerability to potential security risks. A regional specialization may also be very relevant to certain occupations, while proficiency in using and interpreting GIS applications is fundamental to most aspects of homeland security.

These are just a few examples of how geography is critical in today's world. The career choices for geography majors are diverse, challenging, and exciting. The single most popular choice for geography majors is, in fact, a career in marketing for retailing or industrial companies. Another popular choice is an administrative, managerial, or analytical post in local, state, or federal government. Most geography graduates are able to find careers in which they have the opportunity to make a positive contribution to the world through their analytical skills. These careers include cartography, geographic information systems (GIS), laboratory analysis, private consulting, urban and regional planning, international development, teaching, and management in private industry.

CONCLUSION

Human geography is the systematic study of the location of peoples and human activities across Earth's surface and of their relationships to one another. An understanding of human geography is important both from an intellectual point of view (that is, understanding the world around us) and a practical point of view (for example, contributing to environmental quality, human rights, social justice, business efficiency, political analysis, and government policymaking).

While modern ideas about the study of human geography developed from intellectual roots going back to the classical scholarship of ancient Greece, as the world itself has changed, our ways of thinking about it have also changed. What is distinctive about the study of human geography today is not so much the phenomena that are studied as the approach taken by the investigator. The contribution of human geography is to reveal, in relation to economic, social, cultural, and political phenomena, how and why geographical relationships matter in terms of cause and effect.

Geography matters because it is in specific places that people learn who and what they are and how they should think and behave. Places are also a strong influence, for better or worse, on people's physical well-being, their opportunities, and their lifestyle choices. Places also contribute to peoples' collective memory and become powerful emotional and cultural symbols. Places are the sites of innovation and change, of resistance and conflict.

To investigate specific places, however, we must be able to frame our studies of them within the compass of the entire globe. This is important for two reasons. First, the world consists of a complex mosaic of places and regions that are interrelated and interdependent in many ways. Second, place-making forces—especially economic, cultural, and political forces that influence the distribution of human activities and the character of places—are increasingly operating at global and international scales. The interdependence of places and regions means that individual places are tied into wider processes of change that are reflected in broader geographical patterns. An important issue for human geographers is to recognize these wider processes and broad geographical patterns without losing sight of the uniqueness of specific places.

This global perspective leads to the following principles:

- Each place, each region, is largely the product of forces that are both local and global in origin.
- Each is ultimately linked to many other places and regions through these same forces.
- The individual character of places and regions cannot be accounted for by general processes alone. Some local outcomes are the product of unusual circumstances or special local factors.

MAIN POINTS REVISITED

- **Geography matters because it is specific places that provide the settings for people's daily lives.**

  Places are settings for social interaction that, among other things, structure the daily routines of people's economic and social lives; provide both opportunities for—and constraints on—people's long-term social well-being; establish a context in which everyday commonsense knowledge and experience are gathered; provide a setting for processes of socialization; and provide an arena for contesting social norms.

- **Places and regions are highly *interdependent*, each filling specialized roles in complex and ever-changing networks of interaction and change.**

  Individual places are tied into wider processes of change that are reflected in broader geographical patterns. An important issue for human geographers is to recognize these wider processes and broad geographical patterns without losing sight of the individuality and uniqueness of specific places. Processes of geographic change are constantly modifying and reshaping places, and places are constantly coping with change.

- **In today's world some of the most important aspects of the interdependence between geographic scales are provided by the relationships between the global and the local.**

  The study of human geography shows not only how global trends influence local outcomes but also how events in particular localities can come to influence patterns and trends elsewhere. With an understanding of these trends and outcomes, it is possible not only to appreciate the diversity and variety of the world's peoples and places but also to be aware of their relationships to one another and to be able to make positive contributions to local, national, and global development.

- **Human geography provides ways of understanding places, regions, and spatial relationships as the products of a series of interrelated forces that stem from nature, culture, and individual human action.**

  Places are dynamic phenomena. They are created by people responding to the opportunities and constraints presented by their environments. As people live and work in places, they gradually impose themselves on their environment, modifying and adjusting it to suit their needs and express their values. At the same time, people gradually accommodate both to their physical environment and to the people around them. There is thus a continuous two-way process in which people create and modify places while at the same time being influenced by the settings in which they live and work. Places are not just distinctive outcomes of geographical processes; they are part of the processes themselves.

- **The first law of geography is, "Everything is related to everything else, but near things are more related than are distant things."**

  A great deal of human activity is influenced by the "nearness principle," according to which people tend to seek to maximize the overall utility of places at minimum effort, to maximize connections between places at minimum cost, and to locate related activities as close together as possible. In doing so, people are responding to the friction of distance, the deterrent or inhibiting effect of distance on human activity. A distance-decay function describes the rate at which a particular activity or phenomenon diminishes with increasing distance from a given point.

- **Connectivity is also important, because contact and interaction are dependent on channels of communication and transportation: streets, highways, and telephone lines, for example.**

  Interdependence between places and regions can be sustained only through movement and flows. Accessibility and spatial interaction are two of the fundamental concepts that distinguish the study of human geography. The fundamental principles of spatial interaction can be reduced to four basic concepts: complementarity, transferability, intervening opportunities, and diffusion.

KEY TERMS

accessibility (p. 26)
capitalism (p. 19)
cognitive distance (p. 24)
cognitive image (p. 23)
cognitive space (p. 26)
distance-decay function (p. 25)
economies of scale (p. 27)
formal region (p. 30)
friction of distance (p. 25)
functional region (p. 30)
geodemographic research (p. 38)
geographical imagination (p. 35)
geographic information systems (GIS) (p. 22)
Global Positioning System (GPS) (p. 23)
globalization (p. 10)
human geography (p. 20)
identity (p. 5)
infrastructure (p. 28)
intersubjectivity (p. 34)
irredentism (p. 31)
latitude (p. 22)
lifeworld (p. 33)
longitude (p. 23)
neoliberal policies (p. 12)

ordinary landscapes (p. 31)
physical geography (p. 20)
place (p. 20)
region (p. 20)
regional geography (p. 20)
regionalism (p. 31)
regionalization (p. 30)
remote sensing (p. 20)
risk society (p. 19)
sectionalism (p. 31)
sense of place (p. 33)
site (p. 23)
situation (p. 23)
social relations (p. 5)
spatial analysis (p. 22)
spatial diffusion (p. 29)
spatial interaction (p. 27)
states (p. 7)
supranational organization (p. 7)
symbolic landscapes (p. 32)
time-space convergence (p. 28)
topological space (p. 25)
utility (p. 25)
world region (p. 7)

ADDITIONAL READING

Agnew, J., D. Livingstone, and A. Rogers (eds.), *Human Geography: An Essential Anthology.* Cambridge, MA: Blackwell, 1996.

Beck, U., *Risk Society: Towards a New Modernity.* Thousand Oaks, CA: Sage, 1992.

Boyd, A., *An Atlas of World Affairs,* 11th ed. New York: Routledge, 1998.

Cloke, P., P. Crang, and M. Goodwin, *Envisioning Human Geographies.* London: Edward Arnold, 2004.

Dorling, D., and D. Fairbairn, *Mapping: Ways of Seeing the World.* London: Addison Wesley Longman, 1997.

Giddens, A., *Runaway World: How Globalization Is Shaping Our Lives.* New York: Routledge, 2000.

Gould, P., *Becoming a Geographer.* Syracuse: Syracuse University Press, 1999.

Harvey, D. W., *Explanation in Geography.* London: Edward Arnold, 1969.

Hubbard, P., *Thinking Geographically.* Continuum International Publishing, 2002.

Johnston, R. J. and J. D. Sidaway, *Geography and Geographers: Anglo-American Human Geography Since 1945,* 6th ed. London: Edward Arnold, 2004.

Livingstone, D. N., *The Geographical Tradition: Episodes in the History of a Contested Enterprise.* Oxford: Blackwell, 1993.

Phillips, M., *Contested Worlds: An Introduction to Human Geography.* Aldershot, U.K.: Ashgate, 2005.

Unwin, T., *The Place of Geography.* Upper Saddle River, NJ: Prentice Hall, 1996.

Warf, B. (ed.), *Encyclopedia of Human Geography.* Thousand Oaks, CA: Sage, 2005.

Wilford, J. N., *The Mapmakers.* New York: Vintage Books, 2000.

EXERCISES

At the end of each chapter, you will find exercises and activities based on using the Internet, along with some that do not require access to the Internet. This book has its own home page on the Internet, where you will find additional resources—maps, photographs, data—as well as exercises and activities that relate to each chapter. You will also find an evaluation checklist and suggestion form that you can mail to the authors electronically.

On the Internet

The Internet exercises for this chapter are designed to reinforce the geographic concepts presented. For example, concept-review exercises are provided for the major ideas introduced, and these exercises can be immediately graded electronically. The online Web site also focuses on several map exercises that will help you understand and compare map scale and projection. There is a critical-thinking exercise on the development of geographic thought. Finally, to emphasize the idea that geography matters, we have created an exercise that allows you to explore geographic information systems (GIS) used to create digital maps, scaled from outer space to the smallest plot of land, that focus on such things as site and situation information, and more.

Unplugged

1. Consider geographical interdependence from the point of view of your own life. Take an inventory of your clothes, noting where possible the location each garment was manufactured. Where did you buy the garments? Was the store part of a regional, national, or international chain? What can you find out about the materials used in the garments? Where were they made?
2. Describe, as exactly and concisely as possible, the site (see p. 23) of your campus. Then describe its situation (see p. 23). Can you think of any reasons why the campus is sited and situated where it is? Would there be a better location? If so, why?
3. Choose a local landscape, one with which you are familiar, and write a short essay (500 words, or two double-spaced typed pages) on how the landscape has evolved over time. Note especially any evidence that physical environmental conditions have shaped any of the human elements in the landscape, together with any evidence of people having modified the physical landscape.
4. Figures 1.24 through 1.26 show several examples of landscapes that have acquired a strong symbolic value because of the buildings, events, people, or histories with which they are associated. List five additional landscapes that have strong symbolic value to a large number of citizens of your own state or country, and state in 25 words or less why each setting has acquired such value.

McDonald's
2·3F
客席
喫茶室
ルノアール
BF
マクドナ
ハンバー
McDonald's
1万円
現金プレゼント

2 The Changing Global Context

The story is told of a little Japanese girl who arrives in Los Angeles, sees a McDonald's restaurant, tugs her mother's sleeve, and says, "Look, Mother, they have McDonald's in this country, too."

It has become a cliché about the twenty-first century that everywhere will come to look like everywhere else, with the same McDonalds, Pizza Huts, and Kentucky Fried Chickens, the same television programming with Hollywood movies and TV series, and the same malls selling the same Nike shoes, Philips electronics, and GAP clothing. Another cliché is that instantaneous global telecommunications, satellite television, and the Internet will soon overthrow all but the last vestiges of geographical differentiation in human affairs. Large corporations, according to this view, will no longer have strong ties to their home country, scattering their activities around the world in search of low-cost, low-tax locations. Employees will work as effectively from home, car, or beach as they could in the offices that need no longer exist. Events halfway across the world will be seen, heard, and felt with the same immediacy as events across town. National differences and regional cultures will dissolve, the cliché has it, as a global marketplace brings a uniform dispersion of people, tastes, and ideas.

Such developments are, in fact, highly unlikely. Even in the Information Age, geography will still matter and may well become more important than ever. Places and regions will undoubtedly change as a result of the new global context of the Information Age. But geography will still matter because of several factors: transport costs, different resource endowments, fundamental principles of spatial organization, people's territorial impulses, the resilience of local cultures, and the legacy of the past.

In this chapter we take a long-term, big-picture perspective on changing human geographies, emphasizing the continuing interdependence among places and regions. We show how geographical divisions of labor have evolved with the growth of a worldwide system of trade and politics and with the changing opportunities provided by successive technology systems. As a result of this evolution, the world is now structured around a series of core regions, semiperipheral regions, and peripheral regions; and globalization seems to be intensifying many of the differences among places and regions, rather than diminishing them.

A McDonald's restaurant in Tokyo

MAIN POINTS

- The modern world-system has evolved through several distinctive stages, each of which has left its legacy in different ways on particular places, depending on their changing role within the world-system.
- At the end of the eighteenth century, the new technologies of the Industrial Revolution brought about the emergence of a global economic system that reached into almost every part of the world and into virtually every aspect of people's lives.
- Places and regions are part of a world-system that has been created as a result of processes of private economic competition and political competition between national states.
- Today the world-system is highly structured and is characterized by three tiers: core regions, semiperipheral regions, and peripheral regions.
- The growth and internal colonization of the core regions could take place only with the foodstuffs, raw materials, and markets provided by the colonization of the periphery.
- Within each of the world's major regions, the successive technological innovations have transformed regional geographies.
- Globalization has intensified the differences between the core and the periphery and has contributed to the emergence of a digital divide and an increasing division between a fast world (about 15 percent of the world's population) and a slow world (the remaining 85 percent) with contrasting lifestyles and levels of living.

THE PRE-MODERN WORLD

The essential foundation for an informed human geography is an ability to understand places and regions as components of a constantly changing global system. In this sense, all geography is historical geography. Built into every place and each region is the legacy of a sequence of major changes in world geography. We can best understand the consequences of these changes for specific places and regions by thinking of the world as an evolving, competitive, political-economic system that has developed through successive stages of geographic expansion and integration. This evolution has affected the roles of individual places in different ways. It has also affected the nature of the interdependence among places. It explains why places and regions have come to be distinctive and how this distinctiveness has formed the basis of geographic variability. To understand the sequence of major changes in world geography, we need to begin with the hearth areas of the first agricultural revolution.

Hearth Areas

Systematically differentiated human geographies began with minisystems. A **minisystem** is a society with a single cultural base and a *reciprocal* social economy. That is, each individual specializes in particular tasks (tending animals, cooking, or making pottery, for example), freely giving any excess product to others, who reciprocate by giving up the surplus product of their own specialization. Such societies are found only in subsistence-based economies. Because they do not have (or need) an extensive physical infrastructure, minisystems are limited in geographic scale. Before the first agricultural revolution, in prehistoric times, minisystems had been based on hunting-and-gathering societies that were finely tuned to local physical environments.

The first agricultural revolution was a transition from hunter-gatherer minisystems to agricultural-based minisystems that were both more extensive and more stable. The transition began in the Proto-Neolithic (or early Stone Age) period, between 9000 and 7000 B.C., and was based on a series of technological preconditions: the use of fire to process food, the use of grindstones to mill grains, and the development of improved tools to prepare and store food. One key breakthrough was the evolution and diffusion of a system of slash-and-burn agriculture (also known as "swidden" cultivation—see Chapter 8). **Slash-and-burn** is a system of cultivation in which plants are harvested close to the ground, the stubble left to dry for a period, and then ignited, the burned stubble providing fertilizer for the soil. In Neolithic times (7000 to 5500 B.C.) this system involved planting familiar species of wild cereals or tubers on scorched land. Slash-and-burn did not require special tools or weeding, provided that cultivation was switched to a new plot after a few crops had been taken from the old one. Another key breakthrough was the domestication of cattle and sheep, a technique that had become established in a few regions by Neolithic times.

As cultural geographer Carl Sauer pointed out in his book *Agricultural Origins and Dispersals* (1952), these agricultural breakthroughs could take place only in certain geographic settings: where natural food supplies were plentiful; where the terrain was diversified (thus offering a variety of habitats and species); where soils were rich and relatively easy to till; and where there was no need for large-scale irrigation or drainage. Archaeological evidence suggests that the breakthroughs took place independently in several agricultural hearth areas and that agricultural practices diffused slowly outward from each (**Figures 2.1** and **2.2**). **Hearth areas** are geographic settings where new practices have developed and from which they have

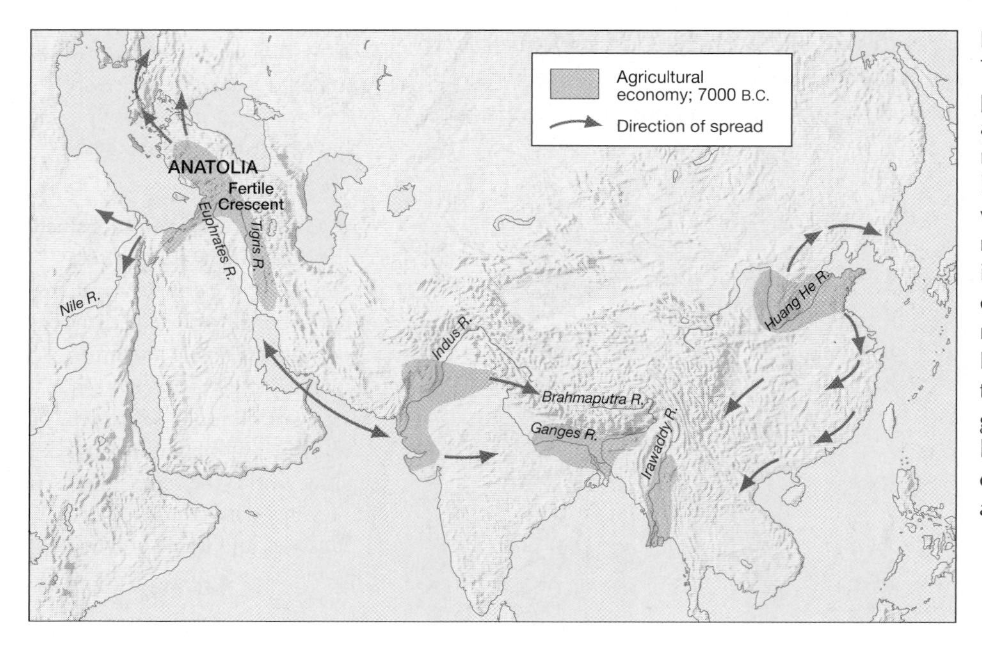

Figure 2.1 Old World hearth areas The first agricultural revolution took place independently in several hearth areas, where hunter-gatherer communities began to experiment with locally available plants and animals in ways that eventually led to their domestication. From these hearth areas, improved strains of crops, domesticated animals, and new farming techniques diffused slowly outward. Farming supported larger populations than were possible with hunting and gathering, and the extra labor allowed the development of other specializations, such as pottery making and jewelry.

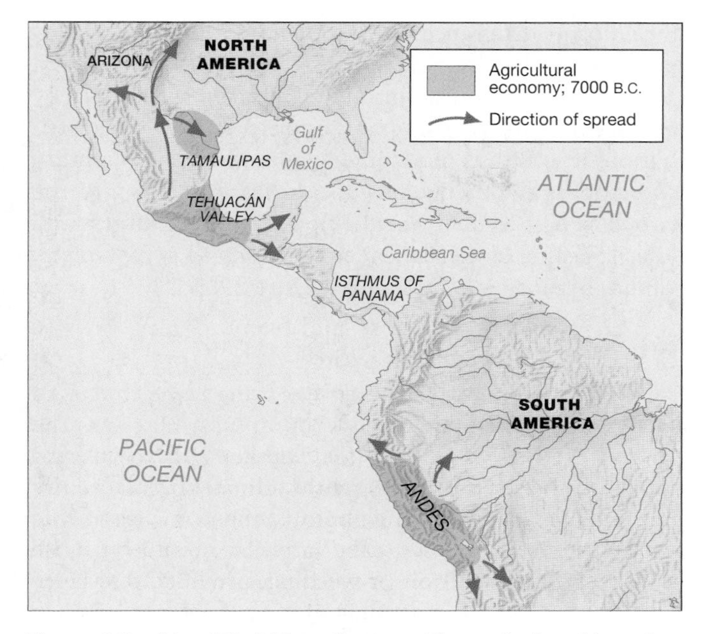

Figure 2.2 New World hearth areas The agricultural hearth area in Central America eventually gave rise to the Aztec and Mayan world-empires, while the hearth areas of the Andes in South America developed into the Inca world-empire.

subsequently spread. The main agricultural hearth areas were situated in four broad regions:

- *In the Middle East:* in the so-called Fertile Crescent around the foothills of the Zagros Mountains (parts of present-day Iran and Iraq), along the floodplains of the Tigris River and Euphrates River, around the Dead Sea Valley (Jordan and Israel), and on the Anatolian Plateau (Turkey).
- *In South Asia:* along the floodplains of the Ganga (Ganges), Brahmaputra, Indus, and Irawaddy rivers (Assam, Bangladesh, Burma, and northern India).
- *In China:* along the floodplain of the Huang He (Yellow) River.
- *In the Americas:* in Mesoamerica (the middle belt of the Americas that extends north to the North American Southwest and south to the Isthmus of Panama) around Tamaulipas and the Tehuacán Valley (Mexico) and in Arizona and New Mexico, and along the western slopes of the Andes in South America.

This transition to food-producing minisystems had several important implications for the long-term evolution of the world's geographies:

1. It allowed much higher population densities and encouraged the proliferation of settled villages.
2. It brought about a change in social organization, from loose communal systems to systems that were more highly organized on the basis of kinship. Kin groups provided a natural way of assigning rights over land and resources and of organizing patterns of land use.
3. It allowed some specialization in nonagricultural crafts, such as pottery, woven textiles, jewelry, and weaponry.
4. This specialization led to a fourth development: the beginnings of barter and trade between communities, sometimes over substantial distances.

Most minisystems vanished a long time ago, although some remnants have survived to provide material for Discovery Channel and National Geographic "Life in a Time Warp" specials (**Figure 2.3**). Examples of these residual and fast-disappearing minisystems are the bushmen of the Kalahari, the hill tribes of Papua New Guinea, and the tribes of the Amazon rain forest. They contribute powerfully to regional differentiation and sense of place in a few enclaves around the world, but their most important contribution to contemporary human geographies is that they

Figure 2.3 A remnant minisystem: Himba tribe, Namibia Minisystems were rooted in subsistence-based social economies that were organized around reciprocity, each person specializing in certain tasks and freely giving any surplus to others, who in turn passed on the surplus from their own work. Today few of the world's remnant minisystems are "pure," unaffected by contact with the rest of the world. Their economies and societies have become more complex, using money, for example, as the basis for exchange. Some of the most striking illustrations of the imprint of globalization on local communities are those in which mass-produced consumer goods have found their way into remnant minisystems.

provide a stark counterpoint to the landscapes and practices of the contemporary world-system. A **world-system** is an *interdependent* system of countries linked by political and economic competition. The term, which was coined by historian Immanuel Wallerstein, is hyphenated to emphasize the interdependence of places and regions around the world.

The Growth of Early Empires

The higher population densities, changes in social organization, craft production, and trade brought about by the first agricultural revolution provided the preconditions for the emergence of several world-empires. A **world-empire** is a group of minisystems that have been absorbed into a common political system while retaining their fundamental cultural differences. The social economy of world-empires can be characterized as redistributive-tributary. That is, wealth is appropriated from producer classes by an elite class in the form of taxes or tribute. This redistribution of wealth is most often achieved through military coercion, religious persuasion, or a combination of the two. The best-known world-empires were the largest and longest lasting of the ancient civilizations—Egypt, Greece, China, Byzantium, and Rome (**Figure 2.4a**).

Urbanization

These world-empires brought important new elements to the evolution of the world's geographies. One was the emergence of *urbanization* (see Chapter 10). Towns and cities became essential as centers of administration, as military garrisons, and as theological centers for the ruling

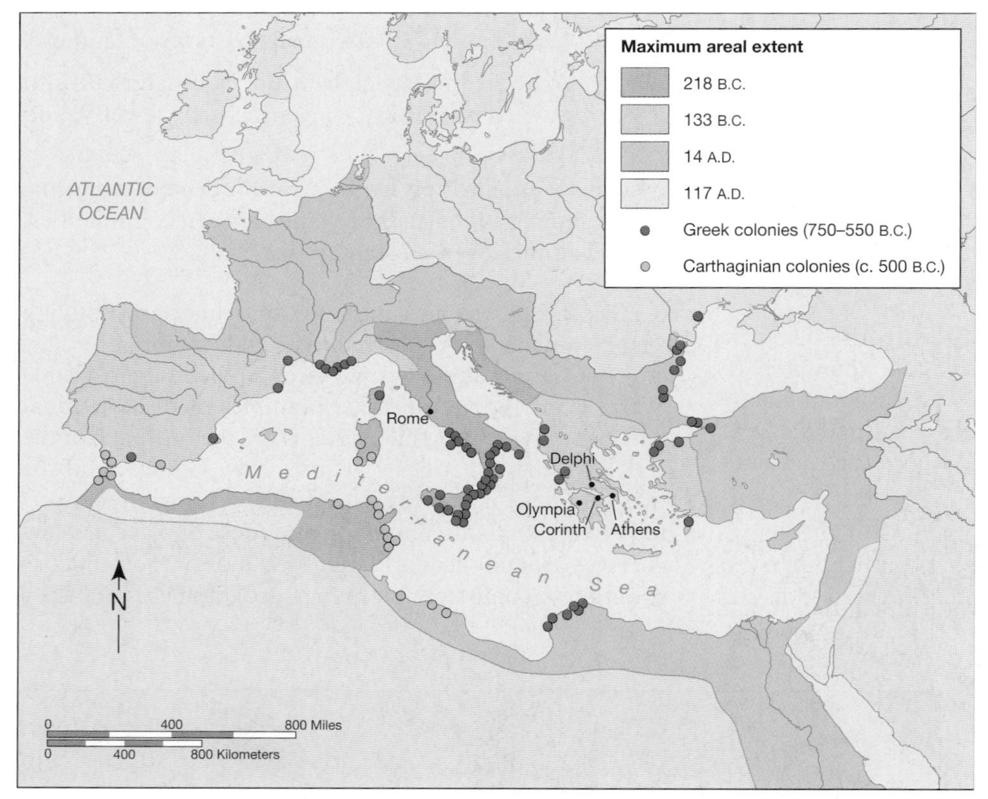

(a)

(b)

(c)

Figure 2.4 Greek colonies and the extent of the Roman empire (a) This map shows the distribution of the Greek poleis (city-states) and Carthaginian colonies and the spread of the Roman empire from 218 B.C. to A.D. 117. (*Source:* After R. King et al., *The Mediterranean.* London: Arnold, 1977, pp. 59 and 64.) (b) The Roman world-empire relied on a highly developed infrastructure of roads, settlements, and utilities. Roman road building was an important aspect of empire building, and many of the roads that were built by Romans became established as major routes throughout Europe. Wherever they could, Roman surveyors laid out roads in straight lines. This photo shows part of a Roman road outside Albe, Italy. (c) The remains of the Claudian aqueduct near Rome. Completed in A.D. 38, the whole project is a testament to the surveying and engineering skills of the Romans.

classes, who were able to use a combination of military and theological authority to hold their empires together. As long as these early world-empires were successful, they gave rise not only to monumental capital cities but also to a whole series of smaller settlements, which acted as intermediate centers in the flow of tribute and taxes from colonized territories.

The most successful world-empires, such as the Greek and Roman, established quite extensive urban systems. In general, the settlements in these urban systems were not very large—typically ranging from a few thousand inhabitants to about 20,000. The seats of empire grew quite large, however. The Mesopotamian city of Ur, in present-day Iraq, for example, has been estimated to have reached a population of around 200,000 by 2100 B.C., while Thebes, the capital of Egypt, is thought to have had more than 200,000 inhabitants in 1600 B.C. Athens and Corinth, the largest cities of ancient Greece, had populations between 50,000 and 100,000 by 400 B.C. Rome at the height of the Roman Empire (around A.D. 200) may have had as many as a million inhabitants. The most impressive thing about these cities, though, was not so much their size as their degree of sophistication: elaborately laid out, with paved streets, piped water, sewage systems, massive monuments, grand public buildings, and impressive city walls.

Colonization

Another important contribution of world-empires to evolving world geographies was **colonization,** the physical settlement in a new territory of people from a colonizing state. In part, this was an indirect consequence of the operation of the **law of diminishing returns.** This law refers to the tendency for productivity to decline after a certain point with the continued addition of capital and/or labor to a given resource base. World-empires could support growing populations only if overall levels of productivity could be increased. While some productivity gains could be achieved through better agricultural practices, harder work, and improvements in farm technology, a fixed resource base meant that as populations grew, overall levels of productivity fell. For each additional person working the land, the gain in production per worker was less. The usual response of empire builders was to enlarge the resource base by colonizing nearby land. This colonization had immediate spatial consequences in terms of establishing dominant/subordinate spatial relationships between hearths and colonies. It was also important in establishing hierarchies of settlements and creating improved transportation networks. The military underpinnings of colonization also meant that new towns and cities came to be carefully sited for strategic and defensive reasons.

The legacy of these important changes is still apparent in today's landscapes. The clearest examples are in Europe, where the Roman world-empire colonized an extensive territory that was controlled through a highly developed system of towns and connecting roads. Most Western European cities that are important today had their origin as Roman settlements. In quite a few it is possible to trace the original street layouts. In some it is possible to glimpse remnants of defensive city walls, paved streets, aqueducts, viaducts, arenas, sewage systems, baths, and public buildings (**Figure 2.4b, c**). In the modern countryside, meanwhile, we can still read the legacy of the Roman world-empire in arrow-straight roads built by Roman engineers and maintained and improved by successive generations. These early world-empires were also significant in developing a base of geographic knowledge (see Box 2.1: "Early Geographic Knowledge").

Some world-empires were exceptional in that they were based on a particularly strong central state, with totalitarian rulers who were able to organize large-scale communal land-improvement schemes using forced labor. These world-empires were found in China, India, the Middle East, Central America, and the Andean region of South America. Their dependency on large-scale land-improvement schemes (particularly irrigation and drainage schemes) as the basis for agricultural productivity has led some scholars to characterize them as *hydraulic societies.* Today their legacy can be seen in the landscapes of terraced fields that have been maintained for generations in places like Sikkim, India; and East Java, Indonesia.

The Geography of the Pre-Modern World

Figure 2.5 shows the generalized framework of human geographies in the Old World as they existed around A.D. 1400. The following characteristics of this period are important:

1. Harsher environments in continental interiors were still characterized by isolated, subsistence-level, kin-ordered hunting-and-gathering minisystems.
2. The dry belt of steppes and desert margins stretching across the Old World from the western Sahara to Mongolia was a continuous zone of kin-ordered pastoral minisystems.
3. The hearths of sedentary agricultural production extended in a discontinuous arc from Morocco to China, with two main outliers, in the central Andes and in Mesoamerica.

The dominant centers of global civilization were China, northern India (both of them hydraulic variants of world-empires), and the Ottoman Empire of the eastern Mediterranean. They were all linked by the Silk Road, a series of overland trade routes between China and Mediterranean Europe (**Figure 2.6**).

By A.D. 1400, other important world-empires had developed in Southeast Asia, in Muslim city-states of coastal North Africa, in the grasslands of West Africa, around the gold and copper mines of East Africa, and in the feudal kingdoms and merchant towns of Europe. Over time all of

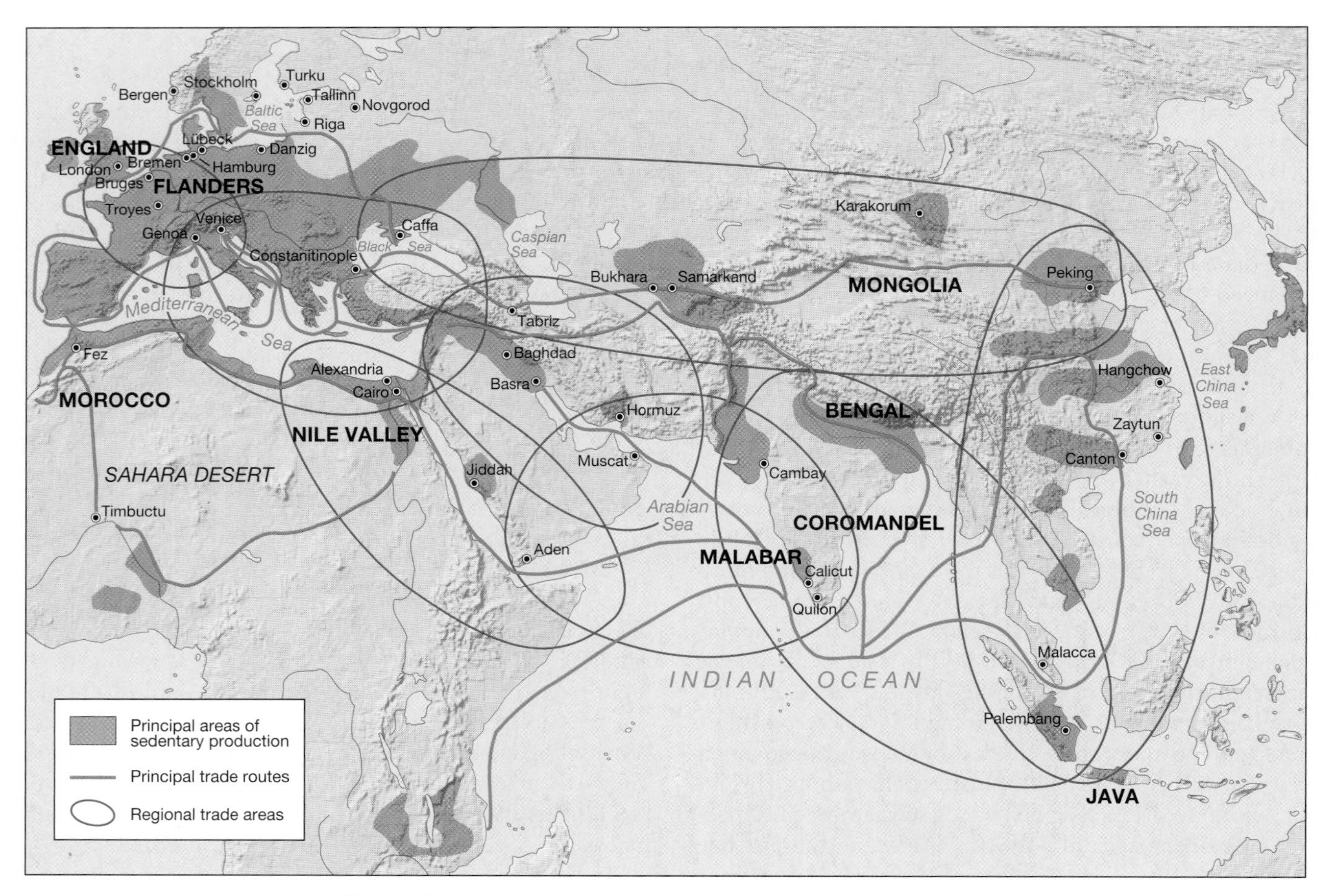

Figure 2.5 The precapitalist Old World, circa A.D. 1400 Principal areas of sedentary agricultural production are shaded. Some long-distance trade took place from one region to another, but for the most part it was limited to a series of overlapping regional circuits of trade. (*Source:* After R. Peet, Global Capitalism: Theories of Societal Development. New York: Routledge, 1991; J. Abu-Lughod, *Before European Hegemony: The World-System A.D. 1200–1350.* New York: Oxford University Press, 1989; and E. R. Wolf, *Europe and the People Without History.* Berkeley: University of California Press, 1983.)

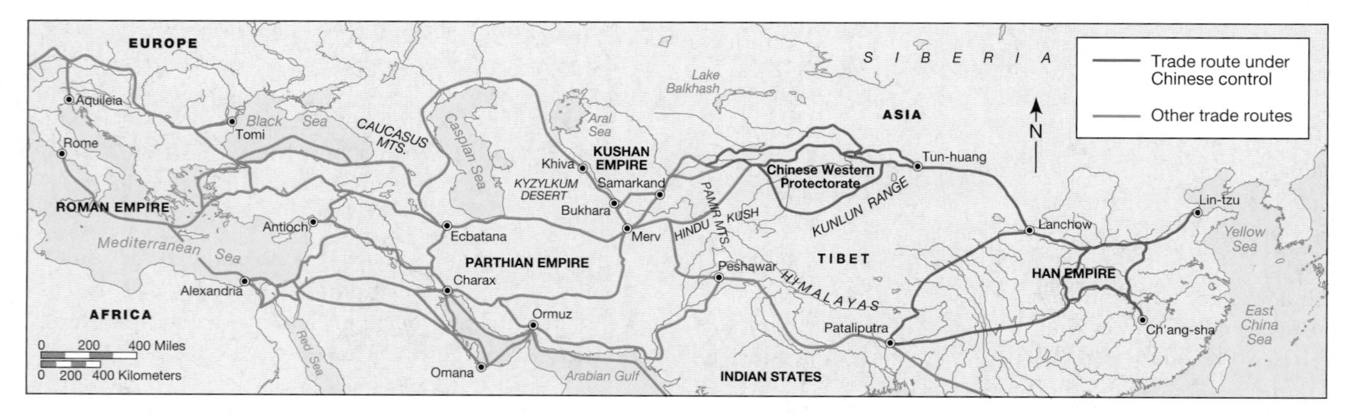

Figure 2.6 The Silk Road This map shows the trade routes of the Silk Road as they existed between 112 B.C. and A.D. 100. From Roman times until Portuguese navigators found their way around Africa and established seaborne trade routes, the Silk Road provided the main East-West trade route between Europe and China. This shifting trail of caravan tracks facilitated the exchange of silk, spices, and porcelain from the East and gold, precious stones, and Venetian glass from the West. The ancient cities of Samarkand, Bukhara, and Khiva stood along the Silk Road, places of glory and wealth that astonished Western travelers such as Marco Polo in the thirteenth century. These cities were East-West meeting places for philosophy, knowledge, and religion, and in their prime they were known for producing scholars in mathematics, music, architecture, and astronomy, such as Al Khoresm (780–847), Al Biruni (973–1048), and Ibn Sind (980–1037). The cities' prosperity was marked by impressive feats of Islamic architecture.

2.1

GEOGRAPHY MATTERS

Early Geographic Knowledge

Several of the early world-empires were important in developing the intellectual importance and utility of geographic knowledge. Greek scholars developed the idea that places embody fundamental relationships between people and the natural environment and that the study of geography provides the best way of addressing the *interdependencies* between places and between people and nature. The Greeks were also among the first to appreciate the practical importance and utility of geographic knowledge, not least in politics, business, and trade. The word "geography" is in fact derived from the Greek language, the literal translation meaning "earth-writing" or "earth-describing." As Greek civilization developed, descriptive geographical writing came to be an essential tool for recording information about sea and land routes and for preparing colonists and merchants for the challenges and opportunities they would encounter in faraway places.

As Greek civilization blossomed, other aspects of geography also came to be valued. One important strand of scholarship, drawing on mathematics and astronomy, was concerned with the measurement of Earth and its accurate representation. Another important strand was philosophical. It was from this philosophical strand that the foundations of human geography were established. Strabo's 17-volume *Geography*, written between 8 B.C. and A.D. 18, was particularly influential. Strabo systematically described places in order to address what he saw as the distinctive local relationships between nature and society. Today this sort of descriptive approach to geographical differentiation is known as *chorology* (or *chorography*) and is often referred to as the *regional approach* to geography. The problem with chorology is that in focusing on the distinctiveness of individual parts of Earth's surface (that is, distinctive places or homogeneous regions), it can lose sight not only of the whole but also of the relationships between places and regions. This problem was recognized by another Greek scholar, Ptolemy (ca. A.D. 90–168), whose eight-volume *Guide to Geography* was concerned with developing a comprehensive view of the world (see **Figure 2.A**).

Both Strabo and Ptolemy were writing at the height of the Roman Empire, and their work was part of the mainstream of classical scholarship. The Romans, however, were less interested than the Greeks in the scholastic and philosophical aspects of geography, though they did appreciate geographical knowledge as an aid to conquest, colonization, and political control. With the decline and fall of the Roman Empire, however, geography and geographical knowledge were neglected.

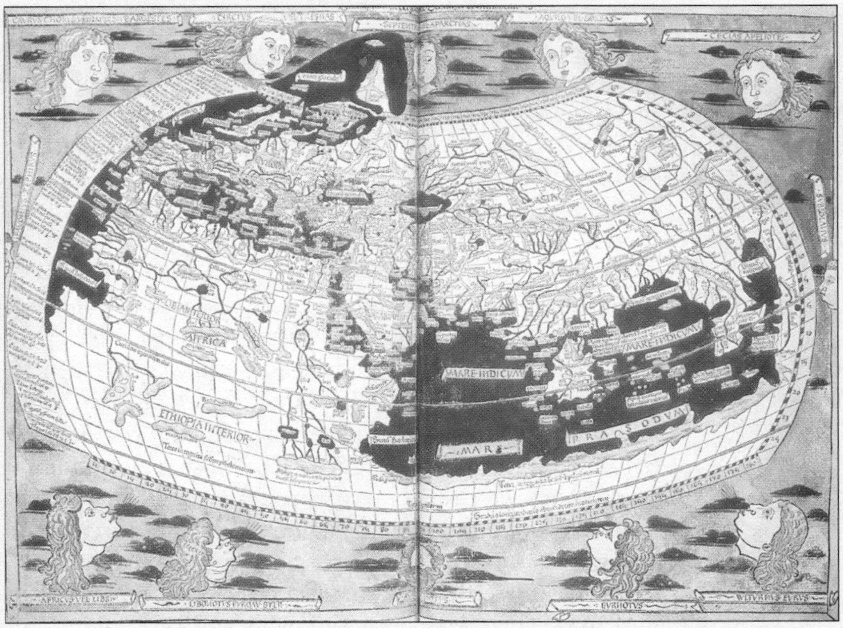

Figure 2.A Ptolemy's map of the world Ptolemy began his *Guide to Geography* with an explanation of how to construct a globe, together with its parallels and meridians, and then showed how to project the world onto a plane surface. His map of the world stood for centuries as the basis for cartography. This example, published in 1482 by Leinhart Holle in Ulm, Germany, was typical of the basic map of the world in use at the time that Cristóbal Colón (Columbus) was considering the feasibility of going to China by sailing west. Unfortunately, Ptolemy had disregarded Eratosthenes' earlier calculations on the circumference of the Earth. As a result, Ptolemy's painstaking work, apparently so precise, contained fundamental inaccuracies that led generations of philosophers, monarchs, and explorers to underestimate the size of the globe. When Columbus did sail west from Spain in search of Asia, it was partly as a result of reading Ptolemy, whose miscalculations had led to an exaggerated conception of the size of Asia.

these more developed realms were interconnected through trade, which meant that several emerging centers of capitalism existed. Port cities were particularly important, and among the leading centers were the city-state of Venice; the Hanseatic League of independent city-states in northwestern Europe (including Bergen, Bremen, Danzig, Hamburg, Lübeck, Riga, Stockholm, and Tallinn; and affiliated trading outposts in other cities, including Antwerp, Bruges, London, Turku, and Novgorod); and Cairo, Calicut, Canton, and Malacca, in North Africa and Asia. Traders in these port cities began to organize the production of agricultural specialties, textiles, and craft products in their respective hinterlands. The **hinterland** of a town or city is its sphere of economic influence—the tributary area from which it collects products to be exported and through which it distributes imports. By the fifteenth century several regions of budding capitalism existed: northern Italy, Flanders, southern England, the Baltic Sea region, the Nile Valley, Malabar, Coromandel, Bengal, northern Java, and southeast coastal China (Figure 2.5).

Between roughly A.D. 500 and A.D. 1400, geographic knowledge was preserved and expanded by Chinese and Islamic scholars. Chinese maps of the world from the same period were more accurate than those of European cartographers because the Chinese were able to draw on information brought back by Imperial China's admirals, who successfully navigated much of the Pacific and Indian oceans. They recognized, for example, that Africa was a southward-pointing triangle, whereas on European and Arabic maps of the time it was always represented as pointing eastward.

With the rise of Islamic power in the Middle East and the Mediterranean in the seventh and eighth centuries A.D., centers of scholarship emerged throughout these regions, including Baghdad, Damascus, Cairo, and Granada, Spain. Here, surviving Greek and Roman texts were translated into Arabic by scholars such as Al-Battani, Al-Farghani, and Al-Khwarazmi. These Islamic scholars were also able to draw on Chinese geographical writing and cartography, brought back by traders along the Silk Road. The requirement that the Islamic religious faithful undertake at least one pilgrimage to Mecca created a demand for travel guidebooks. It also brought scholars from all over the Arab world into contact with one another, stimulating considerable debate over different philosophical views of the world and of people's relationship with nature.

MAPPING A NEW WORLD GEOGRAPHY

The modern world-system had its origins in parts of fifteenth-century Europe, when exploration beyond European shores began to be seen as an important way of opening up new opportunities for trade and economic expansion. By the sixteenth century new techniques of shipbuilding and navigation had begun to bind more and more places and regions together through trade and political competition. As a result, more and more peoples around the world became exposed to one another's technologies and ideas. Their different resources, social structures, and cultural systems resulted in quite different pathways of development, however. Some societies were incorporated into the new, European-based international economic system faster than others; some resisted incorporation; and some sought alternative systems of economic and political organization. Some parts of the world were barely penetrated, if at all, by the European world-system. Australia and New Zealand, for example, were discovered by Europeans only in the late eighteenth century. Regions not yet absorbed into the world-system are called **external arenas.**

With the emergence of this modern world-system at the beginning of the sixteenth century, a whole new geography began to emerge. Although several regions of budding capitalist production existed, and although imperial China could boast of sophisticated achievements in science, technology, and navigation, it was European merchant capitalism that reshaped the world. Several factors motivated European overseas expansion. A relatively high-density population and a limited amount of cultivable land meant that a continuous struggle occurred to provide enough food. Meanwhile, the desire for overseas expansion was intensified both by competition among a large number of small monarchies and by inheritance laws that produced large numbers of impoverished aristocrats with little or no land of their own. Many of these were eager to set out for adventure and profit.

Added to these motivating factors were the enabling factors of innovations in shipbuilding, navigation, and gunnery. In the mid-1400s, for example, the Portuguese developed a cannon-armed ship—the caravel—that could sail anywhere, defend itself against pirates, pose a threat to those who were initially unwilling to trade, and carry enough goods to be profitable. The quadrant (1450) and the astrolabe (1480) enabled accurate navigation and mapping of ocean currents, prevailing winds, and trade routes (see Box 2.2: "Geography and Exploration"). Naval power enabled the Portuguese and the Spanish to enrich their economies with capital from gold and silver plundered from the Americas.

Equipped with better maps and navigation techniques, Europeans not only sent adventurers in search of gold and silver but also to commandeer land, decide on its use, and exploit coerced labor to produce high-value crops (such as sugar, cocoa, and indigo) on **plantations,** large landholdings that usually specialize in the production of one particular crop for market (**Figure 2.7**). Those regions whose populations were resistant to European disease and which also had high population densities, a good resource base, and strong states were able to keep Europeans at arm's length. For the most part, these regions were in South and East Asia. Their dealings with Europeans were conducted through a series of coastal trading stations. Textiles were an important commodity, as reflected by English language:

> The word satin comes from the name of an unknown city in China that Arab traders called Zaitun. Khaki is the Hindi word for dusty. The word "calico" comes from India's southwestern coastal

Figure 2.7 A cotton plantation in the East Indies Plantation agriculture dominated the many European colonies in the nineteenth century.

> city of Calicut; chintz, from the Hindi name for a printed calico; cashmere, from Kashmir. Percale comes from the Farsi word pargalah. Another Farsi derivative is seersucker, whose bands of alternating smooth and puckered fabric prompted a name that literally means milk and sugar. Still another Farsi borrowing is taffeta, which comes from the Farsi for "spun." The coarse cloth we call muslin is named for Mosul—the town in Iraq—while damask is a short form of Damascus. Cotton takes its name from qutun, the Arabic name of the fiber.[1]

Within Europe, meanwhile, innovations in business and finance (banking, loan systems, credit transfers, commercial insurance, and courier services, for example) helped to increase savings, investment, and commercial activity. European merchants and manufacturers also became adept at **import substitution**—copying and making goods previously available only by trading. The result was the emergence of Western Europe as the core region of a world-system that had penetrated and incorporated significant portions of the rest of the world.

For Europe this overseas expansion stimulated still further improvements in technology. These included new developments in nautical mapmaking, naval artillery, shipbuilding, and the use of sail. The whole experience of overseas expansion also provided a great practical school for entrepreneurship and investment. In this way, the self-propelling growth of merchant capitalism was intensified and consolidated.

For the periphery, European overseas expansion meant dependency (as it has ever since for many of the world's peripheral regions). At worst, territory was forcibly occupied and labor systematically exploited. At best, local traders were displaced by Europeans, who imposed their own terms of economic exchange. Europeans soon destroyed most of the Muslim shipping trade in the Indian Ocean, for example, and went on to capture a large share of the oceangoing trade within Asia, selling Japanese copper to China and India, Persian carpets to India, Indian cotton textiles to Japan, and so on.

As revolutionary as these changes were, however, they were constrained by a technology that rested on wind and water power, on wooden ships and structures, and on wood for fuel. Grain mills, for example, were built of wood and powered by water or wind. They could generate only modest amounts of power, and only at sites determined by physical geography, not human choice. Within the relatively small European landmass, wood for structural use and for fuel competed for acreage with food and textile fibers.

More important, however, was the size and strength of timber, which imposed structural limits on the size of buildings, the diameter of waterwheels, the span of bridges, and so on. In particular, it imposed limits on the size and design of ships, which in turn imposed limits on the volume and velocity of world trade. The expense and relative inefficiency of horse- or ox-drawn wagons for overland transportation also meant that for a long time the European world-system could penetrate into continental interiors only along major rivers.

After 300 years of evolution, roughly between 1450 and 1750, the world-system had incorporated only parts of the world. The principal spheres of European influence were Mediterranean North Africa; Portuguese and Spanish colonies in Central and South America; Indian ports and trading colonies; the East Indies; African and Chinese ports; the Greater Caribbean; and British and French ter-

[1]B. Wallach, *Understanding the Cultural Landscape*. New York: Guilford Press, 2005, p. 148.

Geography and Exploration

The Portuguese, under the sponsorship of Dom Henrique (known as Prince Henry the Navigator), established a school of navigation and cartography and began to explore the Atlantic and the coast of Africa. **Cartography** is the name given to the system of practical and theoretical knowledge about making distinctive visual representations of Earth's surface in the form of maps. **Figure 2.B** shows the key voyages of

Captain Cook landing on New Hebrides (present day Vanuatu) in 1774.

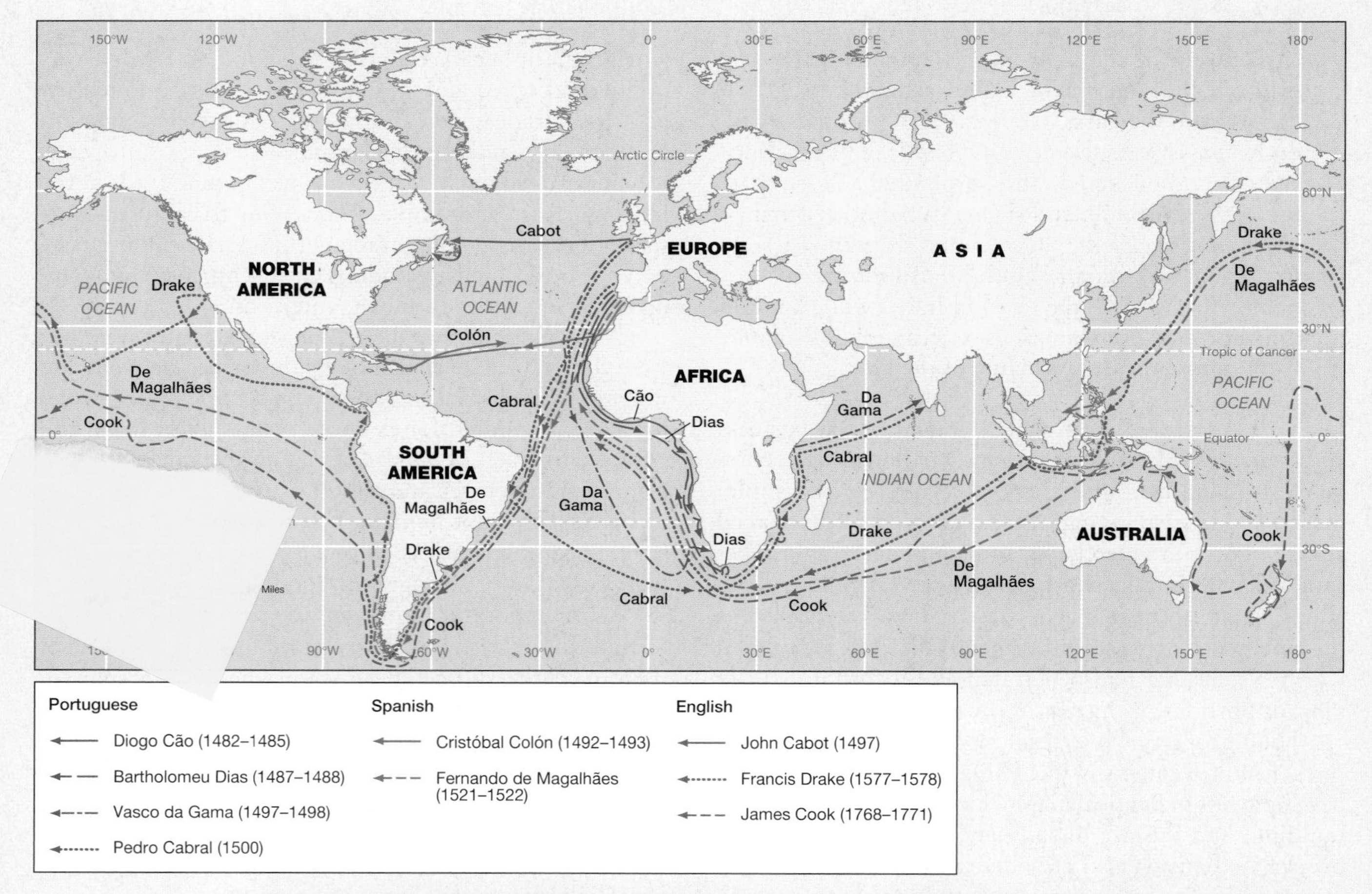

Figure 2.B The European Age of Discovery The European voyages of discovery can be traced to Portugal's Prince Henry the Navigator (1394–1460), who set up a school of navigation and financed numerous expeditions with the objective of circumnavigating Africa in order to establish a profitable sea route for spices from India. The knowledge of winds, ocean currents, natural harbors, and watering places built up by Henry's captains was an essential foundation for the subsequent voyages of Cristóbal Colón (Columbus), da Gama, de Magalhães (Magellan), and others. The end of the European Age of Discovery was marked by Captain James Cook's voyages to the Pacific.

discovery. Portuguese explorer Bartholomeu Dias reached the Cape of Good Hope (the southern tip of Africa) in 1488. In 1492 Cristóbal Colón (Christopher Columbus), from Genoa, Italy, sailed to Hispaniola (the island that is now Haiti and the Dominican Republic) under the sponsorship of the Castillian (Spanish) monarchy. Six years later Vasco da Gama reached India; two years after that, Pedro Cabral crossed the Atlantic from Portugal to Brazil. A small fleet of Portuguese ships reached China in 1513, and the first circumnavigation of the globe was completed in 1522 by Juan Sebastián del Cano, a survivor of the expedition led by Portuguese navigator Fernando de Magalhães (better known to English speakers as Magellan) in 1519. Portuguese successes inspired other countries to attempt their own voyages of discovery, all of them in pursuit of commercial advantage and economic gain. Between them these explorations led to an invaluable body of knowledge about ocean currents, wind patterns, coastlines, peoples, and resources.

Geographical knowledge acquired during this Age of Discovery was crucial to the expansion of European political and economic power in the sixteenth century. In societies that were becoming more commercially oriented and profit conscious, geographical knowledge became a valuable commodity in itself. Information about regions and places was a first step toward controlling and influencing them, and this in turn was an important step toward amassing wealth and power. At the same time, every region began to be opened up to the influence of other regions because of the economic and political competition that was unleashed by geographical discovery. As the New World was being affected by European colonists, missionaries, and adventurers, the countries of the Old World found themselves pitched into competition with one another for overseas resources. Meanwhile, new crops, like corn and potatoes, introduced to Europe from the New World, profoundly affected local economies and ways of life.

The growth of a commercial world economy meant that objectivity in cartography and geographical writing became essential. Navigation, political boundaries, property rights, and rights of movement all depended on accuracy and impartiality. Success in commerce depended on clarity and reliability in describing the opportunities and dangers presented by one region or another. International rivalries required sophisticated understandings of the relationships among nations, regions, and places. Geography became a key area of knowledge. The historical period in Europe known as the Renaissance (from the mid-fourteenth to the mid-seventeenth centuries) saw an explosion of systematic mapmaking and the development of new **map projections** (see Appendix 1) and geographical descriptions (**Figure 2.C**).

Figure 2.C Dutch master Johannes Vermeer's painting *The Geographer* (1668–1669) In Renaissance Europe, the study of geography not only contributed to the growth of scientific knowledge but also helped to support European overseas expansion. Vermeer's geographer is surrounded with accurately rendered cartographic objects, including a wall chart of the seacoasts of Europe, published by Willem Blau in 1658, and a globe made by Jodocus Hondius in 1618.

Throughout the seventeenth and eighteenth centuries, the body of geographic knowledge increased steadily as more and more of the world was explored, using increasingly sophisticated techniques of survey and measurement. Some of the most important advances in scientific cartographic techniques took place in France, where Louis XIV appointed an official cartographer, Nicolas Sanson, to produce a set of accurate maps of French territory. This was achieved through innovative techniques of measurement. Levels, surveying instruments, and telescopes, together with improved mathematical methods, helped to establish an accurate framework of known points throughout France. These points formed the basis of a system of triangulation, allowing French cartographers to produce advanced and innovative maps that became the model for cartographers in other countries. In 1747, under the direction of the Cassini family, the first of a series of 182 sheet maps designed to cover the whole of France was published. Engraved on copper plate and printed at a scale of 1:86,400 (1 inch to 2,400 yards), these maps showed towns in plan view; used a variety of symbols to represent important landmarks such as churches and mills; used a series of short parallel lines (hachures) instead of perspective drawings to show relief; and used tints to show various categories of land cover, such as forest and marsh (**Figure 2.D**).

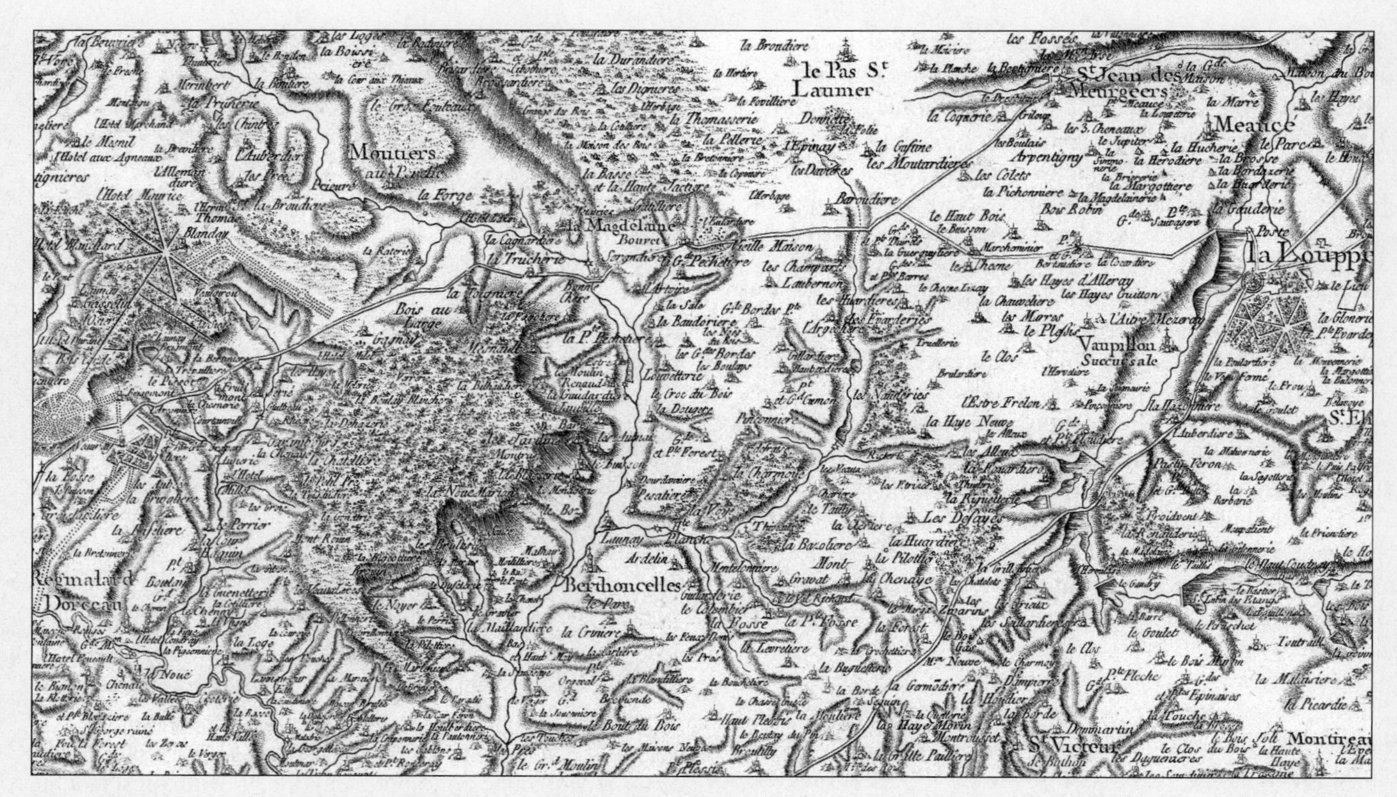

Figure 2.D Early topographic map The famous Cassini triangulation surveys of France were conducted by four generations of the Cassini family in the late 1600s to mid-1700s. In 1672 Jean Dominique Cassini (Cassini I), Royal Astronomer of the Paris Observatory, began to consider new ways to produce more accurate maps through triangulation, similar to the techniques used by astronomers to measure the size of the planets and the Sun. In 1744, he and his son, Jacques Cassini (Cassini II) produced the first accurate survey of an entire nation. In 1747, Louis XV asked Ceasar-Francois Cassini (Cassini III) to create an even more precise national map of France. This was completed by his son Jacques Dominique Cassini (Cassini IV), with 180 maps covering all of France at a scale of 1:86,400. These 180 "cartes de l'Academie" maps of France were a superb product, far more accurate and detailed than any previous maps. This extract shows part of Sheet 27, Chartres, published in 1767.

ritories in North America. The rest of the world functioned more or less as before, with slow-changing geographies based on modified minisystems and world-empires that were only partially and intermittently penetrated by market trading.

Core and Periphery in the New World System

With the new production and transportation technologies of the Industrial Revolution (from the late 1700s), capitalism truly became a global system that reached into virtually every part of the inhabited world and into virtually every aspect of people's lives. Human geographies were recast again, this time with a more interdependent dynamic. New production technologies, based on more efficient energy sources, helped raise levels of productivity and create new and better products that stimulated demand, increased profits, and created a pool of capital for further investment. New transportation technologies triggered successive phases of geographic expansion, allowing for internal development as well as for external colonization and **imperialism** (the deliberate exercise of military power and economic influence by powerful states in order to advance and secure their national interests—see Chapter 7). As a result, human geography came to be re-interpreted by new generations of scholars (see Box 2.3: "The Foundations of Modern Geography").

Since the seventeenth century, the world-system has been consolidated with stronger economic ties between countries. It has also been extended, with all the world's countries eventually becoming involved to some extent in the interdependence of the capitalist system. While there have been some instances of resistance and adaptation, the overall result is that a highly structured relationship between places and regions has emerged. This relationship has been organized around three tiers: *core, semiperipheral,* and *peripheral* regions. These broad geographic divisions have evolved—and are still evolving—through a combination of processes of private economic competition and competition among states.

The **core regions** of the world-system at any given time are those that dominate trade, control the most advanced technologies, and have high levels of productivity within diversified economies. As a result, they enjoy relatively high per capita incomes. The first core regions of the world-system were the trading hubs of Holland and England, joined soon afterward by France (**Figure 2.8**). The continuing success of core regions depends on their dominance and exploitation of other regions. This dominance in turn depends on the participation of these other regions within the world-system. Initially, such participation was achieved by military enforcement, then by European colonialism. **Colonialism** involves the establishment and maintenance of political and legal domination by a state over a separate and alien society. This domination usually involves some colonization (that is, the physical settlement of people from the colonizing state) and always results in economic exploitation by the colonizing state. After World War II the sheer economic and political influence of the core regions was sufficient to maintain their dominance without political and legal control, and colonialism was gradually phased out. Regions that have remained economically and politically unsuccessful throughout this process of incorporation into the world-system are peripheral. **Peripheral regions** are characterized by dependent and disadvantageous trading relationships, by primitive or obsolescent technologies, and by undeveloped or narrowly specialized economies with low levels of productivity.

Transitional between core regions and peripheral regions are semiperipheral regions. **Semiperipheral regions** are able to exploit peripheral regions but are themselves exploited and dominated by core regions. They consist mostly of countries that were once peripheral. This semiperipheral category underlines the fact that neither peripheral status nor core status is necessarily permanent. The United States and Japan both achieved core status after having been peripheral; Spain and Portugal, part of the original core in the sixteenth century, became semiperipheral in the nineteenth century but are now once more part of the core. Quite a few countries, including Brazil, India, Mexico, South Korea, and Taiwan, have become semiperipheral after first having been incorporated into the periphery of the world-system and then developing a successful manufacturing sector that moved them into semiperipheral status.

An important determinant of these changes in status is the effectiveness of states in ensuring the international competitiveness of their domestic producers. They can do this in several ways: by manipulating markets (protecting domestic manufacturers by charging taxes on imports, for example); by regulating their economies (enacting laws that help to establish stable labor markets, for example); and by creating physical and social infrastructures (spending public funds on road systems, ports, educational systems, and so on). Because some states are more successful than others in pursuing these strategies, the hierarchy of three geographical tiers is not rigid. Rather, it is fluid, providing a continually changing framework for geographical transformation within individual places and regions.

The Industrialization of the World's Core Regions

Beginning in the late eighteenth century, a series of technological innovations in power and energy, transportation, and manufacturing processes resulted in some crucial changes in patterns of economic development. Each of these major clusters of technological innovations created new demands for natural resources as well as labor forces and markets. The result was that each major cluster of technological innovations—called technology systems—

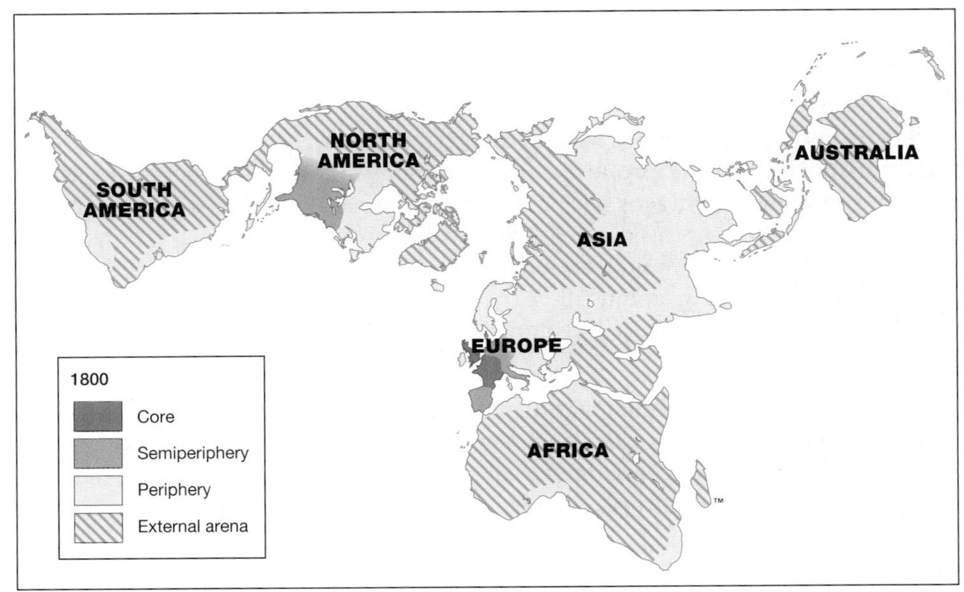

Figure 2.8 The world-system core, semiperiphery, and periphery in 1800 (*Source:* Map projection, Buckminster Fuller Institute and Dymaxion Map Design, Santa Barbara, CA. The word Dymaxion and the Fuller Projection™ Map design are trademarks of the Buckminster Fuller Institute, Santa Barbara, California, © 1938, 1967 & 1992. All rights reserved.)

2.3 GEOGRAPHY MATTERS

The Foundations of Modern Geography

The foundations of geography as a formal academic discipline can be traced to the writing of a few key scholars in the eighteenth and nineteenth centuries. These included Immanuel Kant, Alexander von Humboldt, Karl Ritter, and Friedrich Ratzel. Their contribution, in general terms, was to move geography away from straightforward descriptions of parts of Earth's surface toward explanations and generalizations about the relationships of different phenomena within and among particular places.

Writing in the latter part of the eighteenth century, Immanuel Kant (1724–1804) distinguished between specific fields of knowledge represented by disciplines such as physics and biology, and two general fields of knowledge: geography, which classified things according to space (that is, regions), and history, which classified things according to time (that is, periods or epochs). Kant was an influential philosopher, and his belief in the intellectual importance of geography was an important element in establishing the subject as a formal discipline. His interpretation of the field of geography was also influential. He recognized the importance of commercial geography, theological geography, moral geography (concerning people's customs and ways of life), mathematical geography (concerning the Earth's shape and motion), and political geography—and saw them all as being heavily influenced by the underlying physical geography.

Figure 2.E Alexander von Humboldt German geographer Alexander von Humboldt's statue stands outside the Berlin university that is named after him.

German geographer Alexander von Humboldt (**Figure 2.E**) (1769–1859) set about the task of collecting and analyzing data about the relationships between the spatial distribution of rocks, plants, and animals. He traveled for five years (1799–1804) in South America, collecting data in order to identify relationships and make generalizations. Von Humboldt emphasized the mutual causality among species and their physical environment: the interdependence of people, plants, and animals with one another within specific physical settings. In this way, von Humboldt showed how people, like other species, have to adapt

tended to favor different regions and different kinds of places. **Technology systems** are clusters of interrelated energy, transportation, and production technologies that dominate economic activity for several decades at a time—until a new cluster of improved technologies evolves. What is especially remarkable about technology systems is that they have come along at about 50-year intervals. Since the beginning of the Industrial Revolution, we can identify four of them:

1790–1840: early mechanization based on water power and steam engines; development of cotton textiles and ironworking; development of river transport systems, canals, and turnpike roads.

1840–1890: exploitation of coal-powered steam engines; steel products; railroads; world shipping; and machine tools.

1890–1950: exploitation of the internal combustion engine; oil and plastics; electrical and heavy engineering; aircraft; radio and telecommunications.

1950–1990: exploitation of nuclear power, aerospace, electronics, and petrochemicals; development of limited-access highways and global air routes.

A fifth technology system, still incomplete, began to take shape in the 1980s with a series of innovations that are now being commercially exploited:

to their environment and how their behaviors also affect the environment around them.

German geographer Karl Ritter (1779–1859) was the founder of the tradition of regional geography. Ritter's 20-volume *Erdkunde*, published between 1817 and 1859, was a monumental work of comparison and classification. Like von Humboldt, Ritter saw geography as an integrative science, able to reveal the interdependencies between people and nature. Ritter's approach was to establish a framework for scientific comparisons and generalizations by dividing the continents into broad physical units and then subdividing these into coherent regions with distinctive attributes.

By the mid-nineteenth century, thriving geographical societies had been established in a number of cities, including Berlin, London, Frankfurt, Moscow, New York, and Paris. By 1899 there were 62 geographical societies worldwide, and university chairs of geography had been created in many of the most prestigious universities around the world. It must be said, however, that geography was seen at first in narrow terms, as the discipline of exploration. Because the importance of geography was linked so clearly to European commercial and political ambitions, ways of geographic thinking also changed. Places and regions tended to be portrayed from a distinctly European point of view and from the perspective of particular national, commercial, and religious interests. Geography mattered, but mainly as an instrument of colonialism.

One result of geography's involvement with colonialism and imperialism was that the discipline fostered ethnocentrism and masculinism. **Ethnocentrism** is the attitude that one's own race and culture are superior to those of others. **Masculinism** is the assumption that the world is, and should be, shaped mainly by men for men. These trends became more and more explicit as European dominance increased, reaching a peak in the late nineteenth century at the height of European geopolitical influence.

We should also note that most of the geographic writing in the nineteenth century was strongly influenced by environmental determinism. **Environmental determinism** is a doctrine holding that human activities are controlled by the environment. It rests on a belief that the physical attributes of geographical settings are the root not only of people's physical differences (skin color, stature, and facial features, for example) but also of differences in people's economic vitality, cultural activities, and social structures. One of the most influential of the pioneer academic geographers who propagated this belief was Friedrich Ratzel (1844–1904). Ratzel, a German, was strongly influenced by Charles Darwin's theories about species' adaptation to environmental conditions and competition for living space. (Darwin's *On the Origin of Species* was published in 1859.) Ratzel's own ideas were carried to the United States by Ellen Churchill Semple, a former student of his who went on to teach at the University of Chicago and, later, Clark University. One of the best-known U.S. proponents of Ratzel's ideas was Ellsworth Huntington, who taught geography at Yale University between 1907 and 1917. Ratzel argued that civilization and successful economic development are largely the result of invigorating climates, which he defined as temperate climates with marked seasonal variations and varied weather but without prolonged extremes of heat, humidity, or cold. Huntington used this assertion to provide a rationale for the domination of northwestern Europe and the northeastern United States within the world economy. Environmental determinists thus tend to think in terms of the influence of the physical environment on people rather than the other way around. The idea that peoples' social and economic development and behavior are fundamentally shaped by their physical environment lasted well into the twentieth century, though geographers now regard it as simplistic.

1990 onward: exploitation of solar energy, robotics, microelectronics, biotechnology, advanced materials (fine chemicals and thermoplastics, for example), and information technology (digital telecommunications and geographic information systems, for example).

Each of these technology systems has rewritten the geography of economic development as it has shifted the balance of advantages between regions (**Figure 2.9**). From the mid-1800s, industrial development spread to new regions, whose growth then became interdependent with the fortunes of others through a complex web of production and trade.

The Industrial Revolution began in England toward the end of the eighteenth century and eventually resulted not only in the complete reorganization of the geography of the original European core of the world-system but also in an extension of the world-system core to the United States and Japan.

Europe: Three Waves of Industrialization

In Europe, three distinctive waves of industrialization occurred. The first, between about 1790 and 1850, was based on the initial cluster of industrial technologies (steam engines, cotton textiles, and ironworking) and was very localized (**Figure 2.10**). It was limited to a few regions in

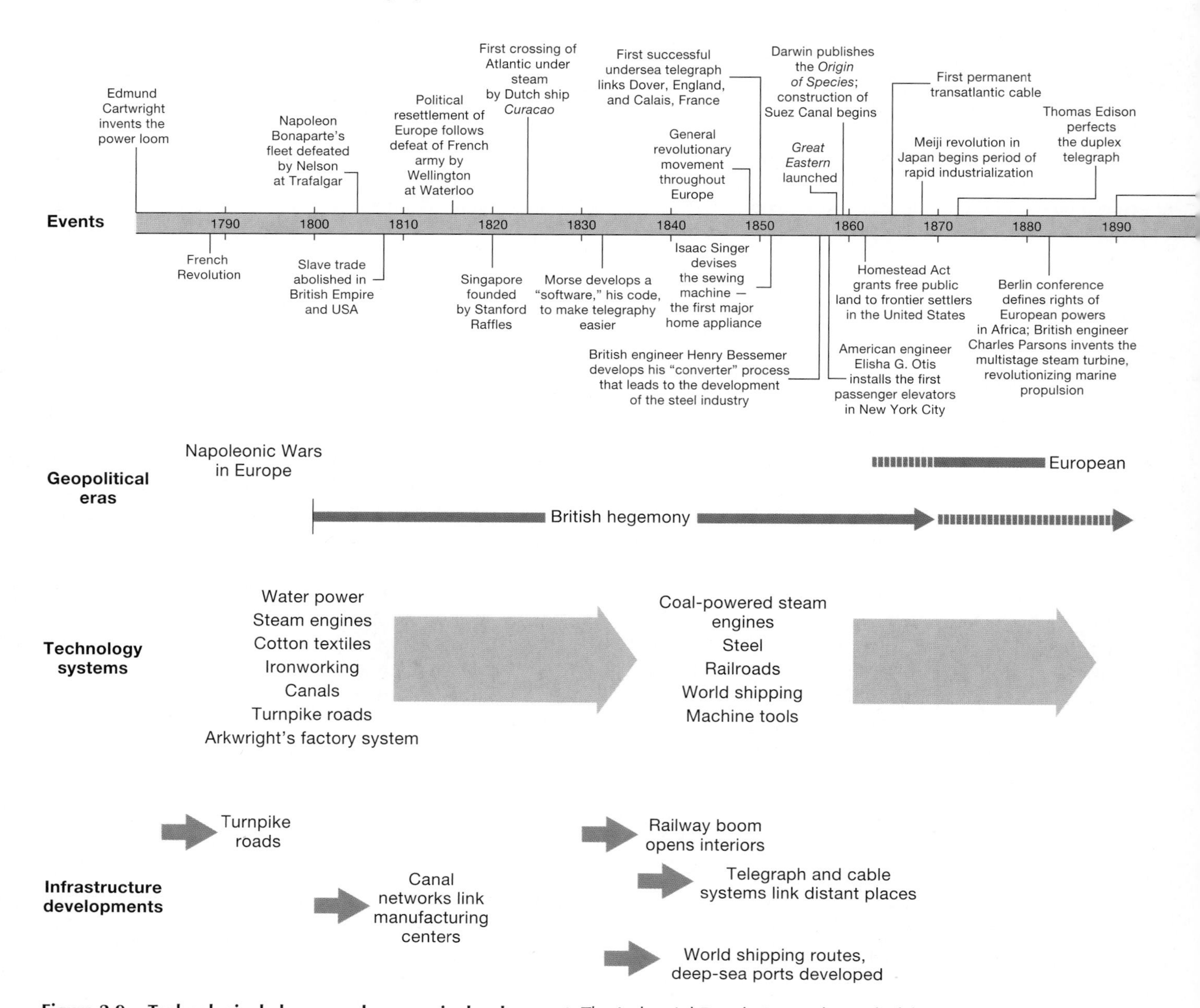

Figure 2.9 Technological change and economic development The Industrial Revolution at the end of the eighteenth century was driven by a technology system based on water power and steam engines, cotton textiles, ironworking, river transport systems, and canals. Since then there have been several more technology systems, each opening new geographic frontiers and rewriting the geography of economic development while shifting the balance of advantages between regions. Overall, the opportunities for development created by each new technology system have been associated with distinctive economic epochs and long-term fluctuations in the overall rate of change of prices in the economy.

Britain where industrial entrepreneurs and workforces had first exploited key innovations and the availability of key resources (coal, iron ore, water). Although these regions shared the common impetus of certain key innovations, each of them retained its own technological traditions and industrial style. From the start, then, industrialization was a regional-scale phenomenon.

The second wave of industrialization, between about 1850 and 1870, involved the diffusion of industrialization to most of the rest of Britain and to parts of northwest Europe, particularly the coalfields of northern France, Belgium, and Germany (Figure 2.10). This second wave also brought a certain amount of reorganization to the first-wave industrial regions as a new cluster of technologies (steel, machine tools, railroads, steamships) brought new opportunities, new locational requirements, new business structures, and new forms of societal organization. New opportunities were created as railroads and steamships made more places accessible, bringing their resources and their markets into the sphere of industrialization. New materials and new technologies (steel, machine tools) created opportunities to manufacture and market new products. These new activities brought some significant changes in the logic of industrial location.

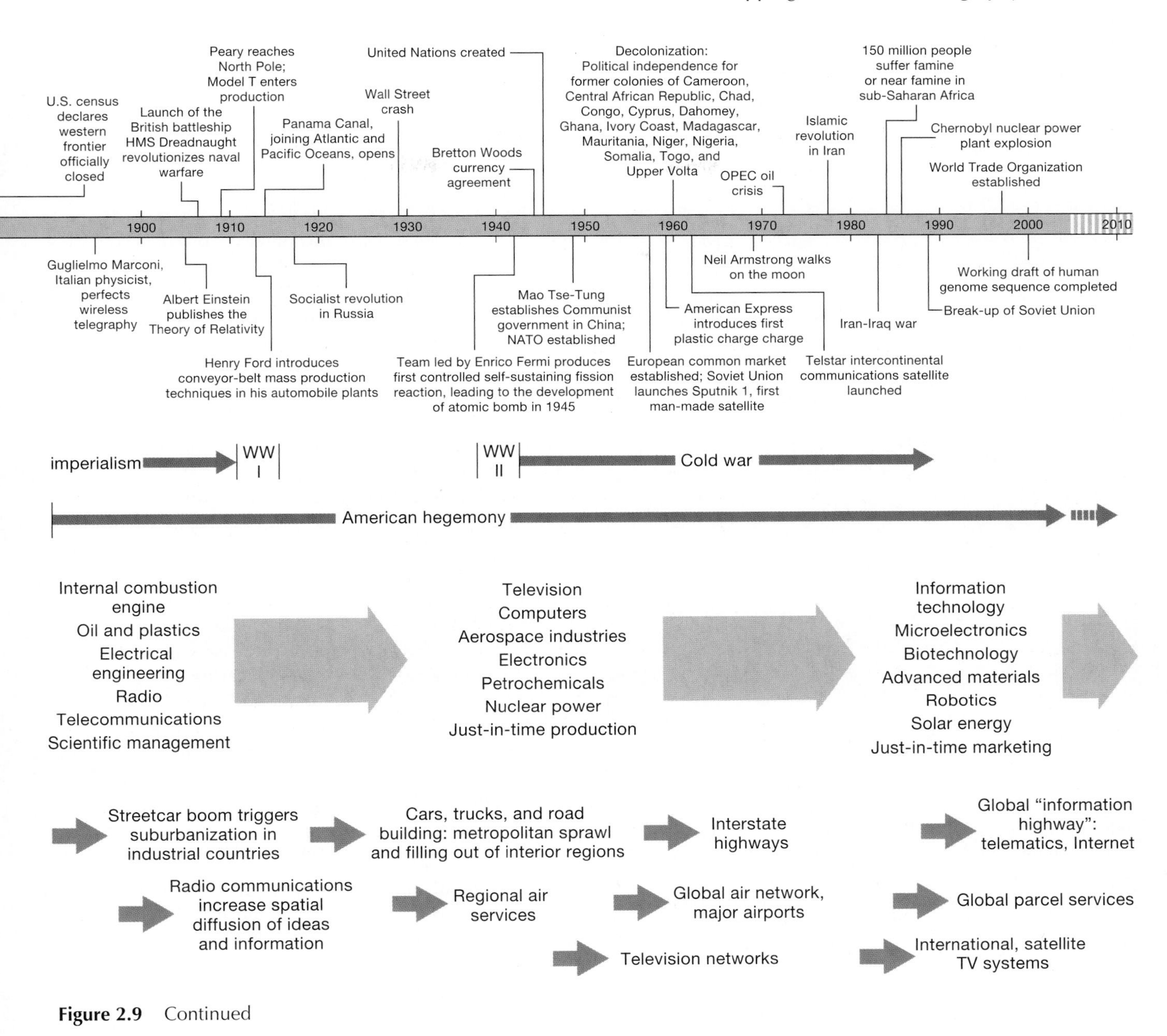

Figure 2.9 Continued

The importance of railway networks, for example, attracted industry away from smaller towns on the canal systems toward larger towns with good rail connections. The importance of steamships for coastal and international trade attracted industry to larger ports. At the same time, the importance of steel produced concentrations of heavy industry in places with nearby supplies of coal, iron ore, and limestone. The scale of industry increased as new technologies and improved transportation made larger markets accessible to firms. Local, family firms became small companies that were regional in scope. Small companies grew to become powerful firms serving national markets, and specialized business, legal, and financial services emerged within larger cities to assist them. The growth of new occupations transformed the structure of social classes, and this transformation in turn came to be reflected in the politics and landscapes of industrial regions.

The third wave of industrialization, between 1870 and 1914, saw a further reorganization of the geography of Europe as yet another cluster of technologies (including electricity, electrical engineering, and telecommunications) imposed different needs and created new opportunities. During this period, industrialization spread for the first time to more remote parts of Britain, France, and Germany and to most of the Netherlands, southern Scandinavia, northern Italy, eastern Austria, and Catalonia, Spain. The overall result was to create a core within a core. Within the world-system core, processes of industrialization, modernization, and urbanization had forged a core of prosperity centered on the "Golden Triangle" stretching between London, Paris, and Berlin.

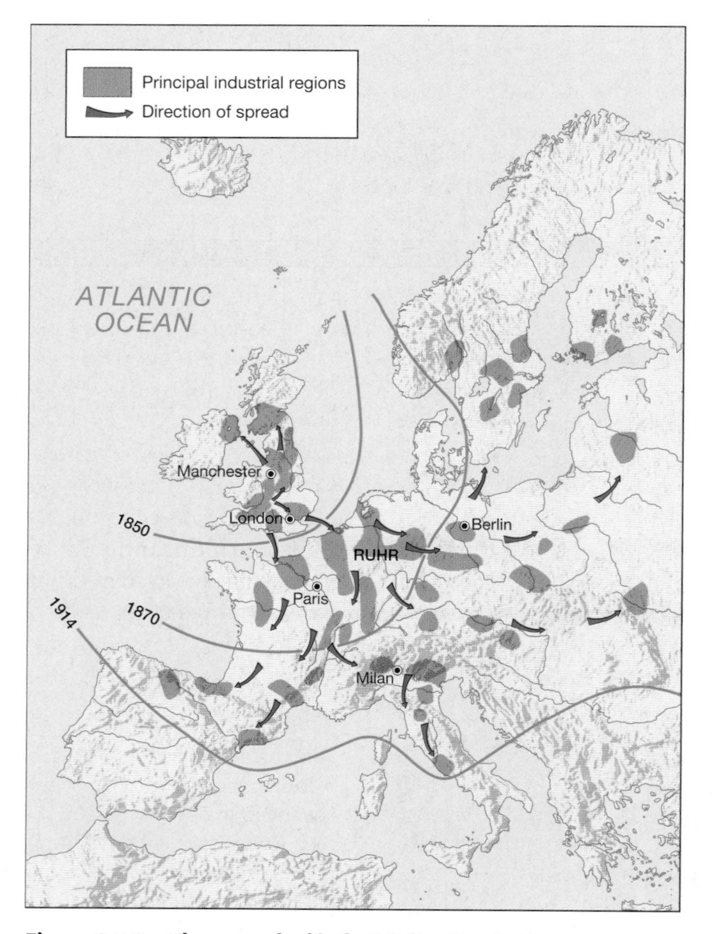

Figure 2.10 The spread of industrialization in Europe European industrialization began with the emergence of small industrial regions in several parts of Britain, where early industrialization drew on local mineral resources, water power, and early industrial technologies. As new rounds of industrial and transportation technologies emerged, industrialization spread to other regions with the right locational attributes: access to raw materials and energy sources, good communications, and large labor markets.

The United States Manufacturing Belt

By the end of the nineteenth century, the core of the world-system had itself extended to include the United States and Japan (**Figure 2.11**). The United States, politically independent just before the onset of the Industrial Revolution, was able to make the transition from the periphery to the core because of several favorable circumstances. Its vast natural resources of land and minerals provided the raw materials for a wide range of industries that could grow and organize without being hemmed in and fragmented by political boundaries. Its population, growing quickly through immigration, provided a large and expanding market and a cheap and industrious labor force. Its cultural and trading links with Europe provided business contacts, technological know-how, and access to capital (especially British capital) for investment in a basic infrastructure of canals, railways, docks, warehouses, and factories.

As in Europe, industrialization developed around pre-existing centers of industrialization and population and was shaped by the resource needs and market opportunities of successive clusters of technology. America's industrial strength was established at the beginning of the twentieth century with the development of a new cluster of technologies that included the internal combustion engine, oil and plastics, electrical engineering, and radio and telecommunications (see Chapter 7). The outcome was a distinctive manufacturing region, known as the Manufacturing Belt, that by 1920 stretched from Boston and Baltimore in the East to Milwaukee and St. Louis in the West (**Figure 2.12**)—another core within a core.

Japan's First Economic Miracle

Japan remained a feudal world-empire for a long time before vaulting into the core of the world-system. In 1868 a revolution deposed the feudal Tokugawa regime and restored an old imperial dynasty—the Meiji—whose backers deliberately set out to modernize the country as quickly

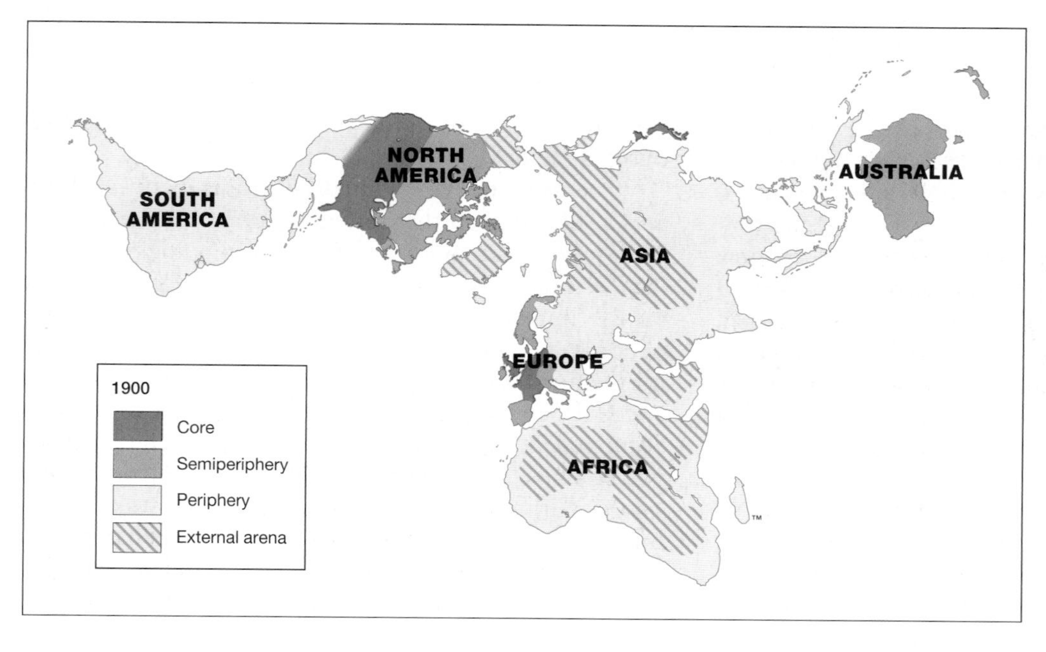

Figure 2.11 The world-system core, semiperiphery, and periphery in 1900 (*Source:* Map projection, Buckminster Fuller Institute and Dymaxion Map Design, Santa Barbara, CA. The word Dymaxion and the Fuller Projection™ Map design are trademarks of the Buckminster Fuller Institute, Santa Barbara, California, © 1938, 1967 & 1992. All rights reserved.)

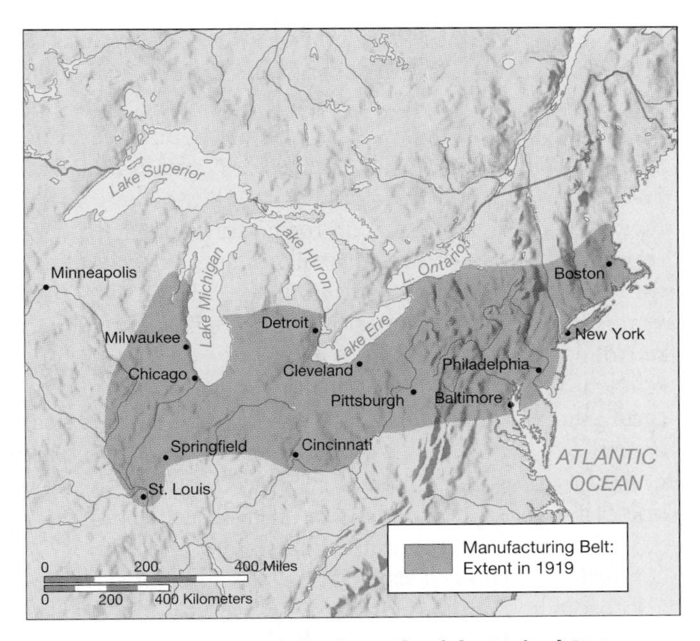

Figure 2.12 The Manufacturing Belt of the United States The cities of this region, already thriving industrial centers that were well connected through the early railroad system, were ideally placed to take advantage of a series of crucial shifts between 1880 and 1920: a general upsurge in demand for consumer goods; the increased efficiency of the telegraph system and postal services; advances in manufacturing technologies; and the opening up of the national market through the extension of the railroad system. Individual cities began to specialize in their industrial products (e.g., grain milling in Minneapolis, brewing in Milwaukee and St. Louis, coachbuilding and furniture in Cincinnati, agricultural machinery in Springfield, Illinois), producing in volume for the national market rather than local ones. This specialization in turn required an increase in commodity flows between the cities of the Manufacturing Belt, which bound the region more closely together.

as possible, whatever the cost. Under the slogan "National Wealth and Military Strength," the government took an unprecedented role, strong and deliberate, in creating capitalist enterprises and establishing a modern infrastructure. The costs were borne chiefly by a domestic workforce (whose rewards for hard work were meager) and by the victims of Japanese military aggression in Korea, Formosa (now Taiwan), and Manchuria.

Just as Japanese industry was becoming established with a base in textiles and shipbuilding, World War I provided a timely opportunity to expand productive capacity. With much European and American industry diverted to supply war materials, Japanese textile manufacturers were able to expand into Asian and Latin American markets. Meanwhile, with a great deal of merchant shipping destroyed by the hostilities, Japanese shipping industries took the opportunity to expand their merchant fleet. The profits from this commercial activity paid for the rebirth of the Japanese navy, which by 1918 had a dozen battleships and battle cruisers, with 16 more under construction. The United States at the time had only 14 battleships, with three under construction. Within 50 years of the Meiji revolution, Japan had joined the core of the world-system.

Internal Development of the Core Regions

Within the world's core regions, the transformation of regional geographies hinged on successive innovations in transport technology. These innovations opened up agrarian interiors and intensified interregional trading networks. Farmers were able to mechanize their equipment, while manufacturing companies were able to take over more resources and more markets.

The first phase of this internal geographic expansion and regional integration was based on an old technology: the canal (**Figure 2.13**). Merchant trade and the beginnings of industrialization in both Britain and France were underpinned by extensive navigation systems that joined one river system to another. By 1790, France had just over 1,000 kilometers (620 miles) of canals and canalized rivers; Britain had nearly 3,600 kilometers (2,230 miles). The Industrial Revolution provided both the need and the capital for a spate of additional canal building

Figure 2.13 Canal systems The canal systems that opened up the interiors of Europe and North America in the eighteenth century were initially dependent on horse power. Early examples carried horse-drawn barges. Later, barges were able to utilize steam- and oil-powered engines, extending the importance of larger canals and navigable rivers. Barges are still an important mode of transport for bulky goods. This photograph shows coal being moved up the River Rhine from coalfields in the Ruhr to power stations in southern Germany.

that began to integrate and extend emerging industrial regions.

In Britain 2,000 more kilometers (1,240 miles) of canals were built between 1790 and 1810. In France, which did not industrialize as early as Britain, 1,600 additional kilometers (990 miles) were built between 1830 and 1850. In the United States, the key event was the opening of the Erie Canal in 1825, which enabled New York, a colonial gateway port, to reorient itself toward the nation's growing interior. The Erie Canal was so profitable that it set off a "canal fever" that resulted in the construction of some 2,000 kilometers (1,240 miles) of navigable waterways in the next 25 years. This canal system helped to bind the emergent Manufacturing Belt together.

The scale of the United States was such that a network of canals was a viable proposition only in more densely settled areas. The effective colonization of the interior could not take place until the development of steam-powered transportation—first riverboats and then railroads. The first steamboats were developed in the early 1800s, offering the possibility of opening up the vast interior by way of the Mississippi and its tributaries (**Figure 2.14**). By 1830 the technology and design of steamboats had been perfected and navigable channels had been established. The heyday of the river steamboat was between 1830 and 1850. During this period, vast acreages of the U.S. interior were opened up to commercial, industrialized agriculture—especially cotton production for export to British textile manufacturers. At the same time, river ports such as New Orleans, St. Louis, Cincinnati, and Louisville grew rapidly, extending the frontier of industrialization and modernization.

Railroads and Regional Economic Integration

By 1860 the railroads had taken over the task of internal development, further extending the frontier of settlement and industrialization and intensifying the use of previously developed regions (see Box 2.4: "Visualizing Geography—Railroads and Geographic Change," page 64). The railroad originated in Britain, where George Stephenson engineered the world's first commercial railroad, a 20-kilometer (12.4-mile) line between Stockton and Darlington that was opened in 1825. The famous *Rocket,* the first locomotive for commercial passenger trains, was designed mainly by Stephenson's son Robert for the Liverpool & Manchester line, which opened four years later. The economic success of this line sparked two railroad-building booms that eventually created a highly efficient transportation network for Britain's manufacturing industry. In other core countries, where sufficient capital existed to license (or copy) the locomotive technology and install the track, railroad systems led to the first full stage of economic and political integration.

While the railroads integrated the economies of entire countries and allowed vast territories to be colonized, they also brought some important regional and local restructuring and differentiation. In the United States, for example, the railroads led to the consolidation of the Manufacturing Belt. They also contributed to the mushrooming of Chicago as the focal point for railroads that extended the Manufacturing Belt's dominance over the West and South. This reorientation of the nation's transportation system effectively ended the role of the cotton regions of the South as outliers of the British trading system. Instead, they became outliers of the U.S. Manufacturing Belt, supplying factories in New England and the Mid-Atlantic Piedmont. This left New Orleans, which had thrived on cotton exports, to cope with an abrupt end to its phenomenal growth.

Tractors, Trucks, Road Building, and Spatial Reorganization

In the twentieth century the internal combustion engine powered further rounds of internal development, integration, and intensification. The replacement of horse-

Figure 2.14 Steamboats The interior of continental North America was first opened up as a result of steamboat trade along the Mississippi and its tributaries. Cotton and other plantation crops were shipped north to manufacturing cities, while grain crops were shipped south and both grain and cotton were exported. This illustration shows another important aspect of riverboat commerce: the migration of African-Americans from the South in search of jobs in the manufacturing cities of the Northeast.

drawn farm implements with lightweight tractors powered by internal combustion engines, beginning in the 1910s, amounted to a major revolution in agriculture (see Chapter 8). Productivity was increased, the frontiers of cultivable land were extended, and vast amounts of labor were released for industrial work in cities. The result was a parallel revolution in the geographies of both rural and urban areas.

The development of trucks in the 1910s and 1920s suddenly released factories from locating near railroads, canals, and waterfronts. Trucking allowed goods to be moved farther, faster, and cheaper than before. As a result, trucking made it feasible to locate factories on inexpensive land on city fringes, and in smaller towns and peripheral regions where labor was cheaper. It also increased the market area of individual factories and reduced the need for large product inventories. This decentralization of industry, in conjunction with the availability of buses, private automobiles, and massive road-building programs, brought about another phase of spatial reorganization. The outcomes of this phase were the specialized and highly integrated regions and urban systems of the modern core of the world-system. This integration was not simply a question of their being interconnected through highway systems. It also involved close economic linkages among manufacturers, suppliers, and distributors—linkages that enabled places and regions to specialize and develop economic advantages (see Chapter 7).

Organizing the Periphery

Parallel with the internal development of core regions were changes in the geographies of the periphery of the world-system. Indeed, the growth and internal development of the core regions simply could not have taken place without the foodstuffs, raw materials, and markets provided by the colonization of the periphery and the incorporation of more and more territory into the sphere of industrial capitalism.

As soon as the Industrial Revolution had gathered momentum in the early nineteenth century, the industrial core nations embarked on the inland penetration of the world's midcontinental grassland zones in order to exploit them for grain or stock production. This led to the settlement, through the emigration of European peoples, of the temperate prairies and pampas of the Americas; the veld in southern Africa; the Murray-Darling Plain in Australia; and the Canterbury Plain in New Zealand. At the same time, as the demand for tropical plantation products increased, most of the tropical world came under the political and economic control—direct or indirect—of one or another of the industrial core nations. In the second half of the nineteenth century, and especially after 1870, there was a vast increase in the number of colonies and the number of people under colonial rule.

The International Division of Labor

The fundamental logic behind all this colonization was economic: the need for an extended arena for trade, an arena that could supply foodstuffs and raw materials in return for the industrial goods of the core. The outcome was an international division of labor, driven by the needs of the core and imposed through its economic and military strength. This **division of labor** involved the specialization of different people, regions, and countries in certain kinds of economic activities. In particular, colonies began to specialize in the production of commodities meeting certain criteria:

- Where an established demand existed in the industrial core (for foodstuffs and industrial raw materials, for example).
- Where colonies held a **comparative advantage** in specializations that did not duplicate or compete with the domestic suppliers within core countries (tropical agricultural products like cocoa and rubber, for example, simply could not be grown in core countries).

The result was that colonial economies were founded on narrow specializations that were oriented to and dependent upon the needs of core countries. Examples of these specializations were many: bananas in Central America; cotton in India; coffee in Brazil, Java, and Kenya; copper in Chile; cocoa in Ghana; jute in East Pakistan (now Bangladesh); palm oil in West Africa; rubber in Malaya (now Malaysia) and Sumatra; sugar in the Caribbean islands; tea in Ceylon (now Sri Lanka); tin in Bolivia; and bauxite in Guyana and Surinam. Most of these specializations persist today. Thus, for example, 48 of the 55 countries in sub-Saharan Africa still depend on just three products—tea, cocoa, and coffee—for more than half of their export earnings.

This new global economic geography took some time to establish, and the details of its pattern and timing were heavily influenced by technological innovations. The incorporation of the temperate grasslands into the commercial orbit of the core countries, for example, involved successive changes in regional landscapes as critical innovations—such as barbed wire, the railroad, and refrigeration—were introduced. The single most important innovation stimulating the international division of labor, however, was the development of metal-hulled, oceangoing steamships. This development was cumulative, with improvements in engines, boilers, transmission systems, fuel systems, and construction materials adding up to produce dramatic improvements in carrying capacity, speed, range, and reliability. The construction of the Suez Canal (opened in 1869) and the Panama Canal (opened in 1914) was also critical, providing shorter and less hazardous routes between core countries and colonial ports of call. By the eve of World War I the world economy was effectively integrated by a sys-

Railroads and Geographic Change

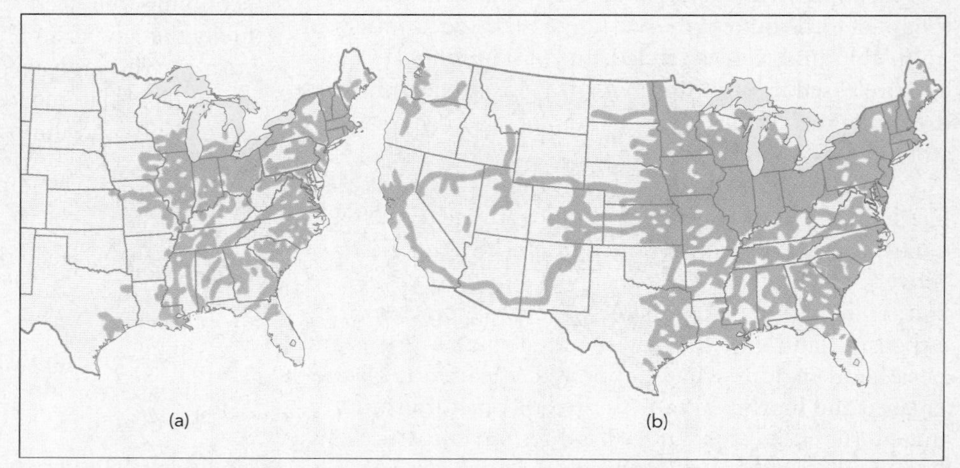

Railroad service in (a) 1860 and (b) 1880 Areas shown in brown are no more than 24 kilometers (15 miles) from a railroad line. (*Source:* P. Hugill, *World Trade Since 1431.* Baltimore: Johns Hopkins University Press, 1993, p. 179.)

Railroad construction was labor intensive but required little more than common farm equipment applied to a graded surface. Most of the labor force consisted of immigrants: European immigrants on railroads driven westward from the East Coast, and Chinese immigrants on those driven eastward from the West Coast.

The first passenger-train services averaged little more than 20 to 35 kilometers per hour (15–20 m.p.h.), but locomotive technology changed rapidly and made it easier, faster, and cheaper to conquer the vast interior distances of the United States. Between 1830 and 1845 the United States created the world's largest rail system—some 5,458 kilometers (3,688 miles) compared to Britain's 3,083 kilometers (2,069 miles), Germany's 2,956 kilometers (1,997 miles), France's 817 kilometers (552 miles), and Belgium's 508 kilometers (343 miles).

The Illinois Central Railroad issued a promotional poster in the 1860s celebrating the role it envisaged—that its trunk line from Chicago to New Orleans would significantly affect the settling of the American interior. Note the deliberate juxtaposition of the high technology of the railroad locomotive and the telegraph line, contrasting with the obsolescent technologies of the stagecoach (inset top right) and barge (inset bottom right). Even the oceangoing ship is shown with masts and sails to diminish its status in comparison to the railroad locomotive.

The poster issued 50 years later, on the other hand, shows the railroad locomotive in the company of several important new technologies: aircraft, automobiles and trucks, and electric streetlights.

The railroads reorganized the geography of America's cities and opened up its interior regions. Railway stations reordered land-use patterns in central business districts, where new hotels and department stores competed for sites near the station's entrance. Farther away, the railroad lines cut swaths through the urban fabric, separating neighborhoods from one another and establishing the basis for a radial framework for the physical development of the city.

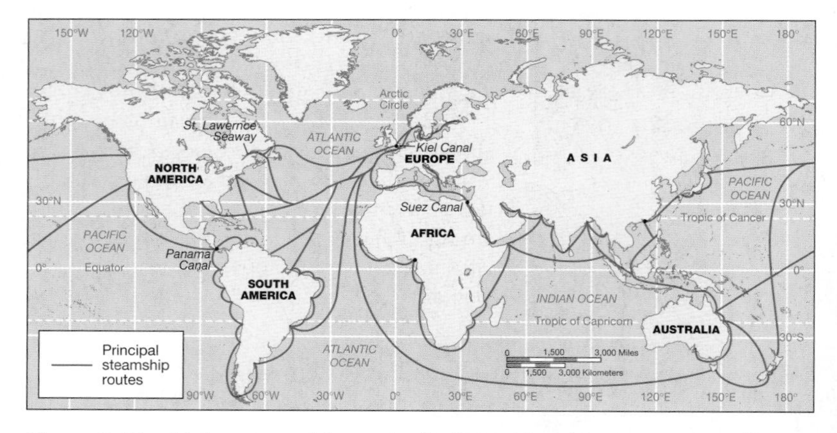

Figure 2.15 Major steamship routes in 1920 The shipping routes reflect (1) the transatlantic trade between the bipolar core regions of the world-system at the time and (2) the colonial and imperial relations between the world's core economies and the periphery. Transoceanic shipping boomed with the development of steam-turbine engines for merchant vessels and with the construction of shipping canals, such as the Kiel Canal, the Suez Canal, the St. Lawrence Seaway, and the Panama Canal. When the 82 kilometers (51 miles) of the Panama Canal opened in 1914, shipping could move between the Atlantic and the Pacific without having to go around South America, saving thousands of kilometers of steaming. The poster dates from 1924.

tem of regularly scheduled steamship trading routes (**Figure 2.15**). This integration, in turn, was supported by the second most important innovation stimulating the international division of labor: a network of telegraph communications (**Figure 2.16**) that enabled businesses to monitor and coordinate supply and demand across vast distances on an hourly basis.

The international division of labor brought about a substantial increase in trade and a huge surge in the overall size of the capitalist world economy. The peripheral regions of the world contributed a great deal to this growth. By 1913 Africa and Asia provided more *exports* to the world economy than either North America or the British Isles. Asia alone was *importing* almost as much, by

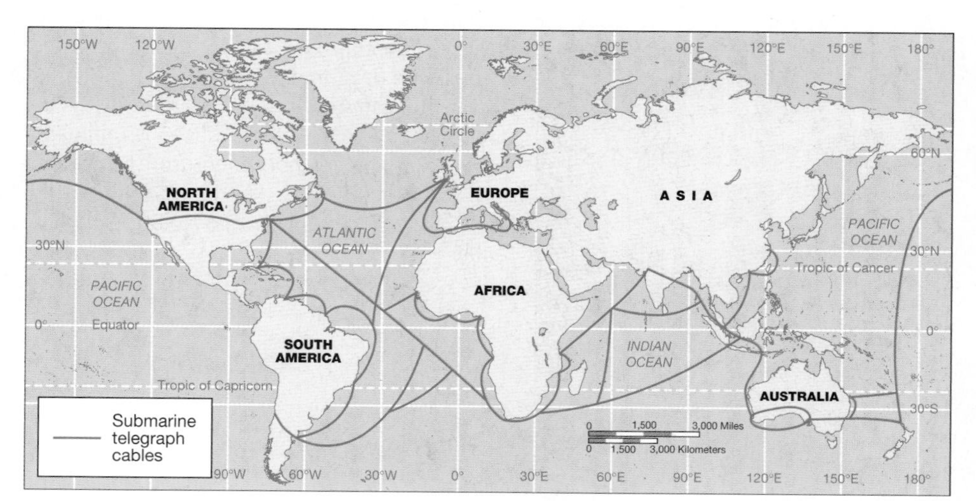

Figure 2.16 The international telegraph network in 1900 For Britain, submarine telegraph cables were the nervous system of its empire. Of the global network of 246,000 kilometers (152,860 miles) of submarine cable, Britain had laid 169,000 kilometers (105,015 miles).

value, as North America. The industrializing countries of the core bought increasing amounts of foodstuffs and raw materials from the periphery, financed by profits from the export of machinery and manufactured goods. Britain, the dominant power of the period, drew on a trading empire that was truly global (**Figure 2.17**).

Patterns of international trade and interdependence became increasingly complex. Britain used its capital to invest not just in peripheral regions but also in profitable industries in other core countries, especially the United States. At the same time, these other core countries were able to export cheap manufactured items to Britain. Britain financed the purchase of these goods, together with imports of food from its dominion states (Canada, South Africa, Australia, and New Zealand) and colonies, through the export of its own manufactured goods to peripheral countries. India and China, with large domestic markets, were especially important. Thus there developed a widening circle of exchange and dependence, with constantly switching patterns of trade and investment.

Imperialism: Imposing New Geographies on the World

The incorporation of the periphery was by no means entirely motivated by this basic logic of free trade and investment. Although Britain was the dominant power in the late nineteenth century, several other European countries (notably Germany, France, and the Netherlands), together with the United States—and later Japan—were competing for global influence. This competition developed into a scramble for territorial and commercial domination. The core countries engaged in preemptive geographic expansionism in order to protect their established interests and to limit the opportunities of others. They also wanted to secure as much of the world as possible—through a combination of military oversight, administrative control, and economic regulation—in order to ensure stable and profitable environments for their traders and investors. This combination of circumstances defined a new era of imperialism.

In the final quarter of the nineteenth century, a surge of imperialism brought a competitive form of colonialism that resulted in a scramble for territory. Africa, more than any other peripheral region, was given an entirely new geography. It was carved up into a patchwork of European colonies and protectorates in just 34 years, between 1880 and 1914, with little regard for either physical geography or the preexisting human geographies of minisystems and world-empires. Whereas European interest had previously focused on coastal trading stations and garrison ports, it now extended to the entire continent.

Within just a few years the whole of Africa became incorporated into the modern world-system, with a geography that consisted of a hierarchy of three kinds of spaces. One consisted of regions and localities organized by European colonial administrators and European investors to produce commodities for the world market. A second consisted of zones of production for local markets, where peasant farmers produced food for consumption by laborers engaged in commercial mining and agriculture. The third consisted of widespread regions of subsistence agriculture whose connection with the world-system was as a source of labor for the commercial regions.

Meanwhile, the major powers jostled and squabbled over small Pacific islands that had suddenly become valuable as strategic coaling stations for their navies and merchant fleets. Resistance from indigenous peoples was quickly brushed aside by imperial navies with iron steamers, high-explosive guns, and troops with rifles and cannons. Euro-

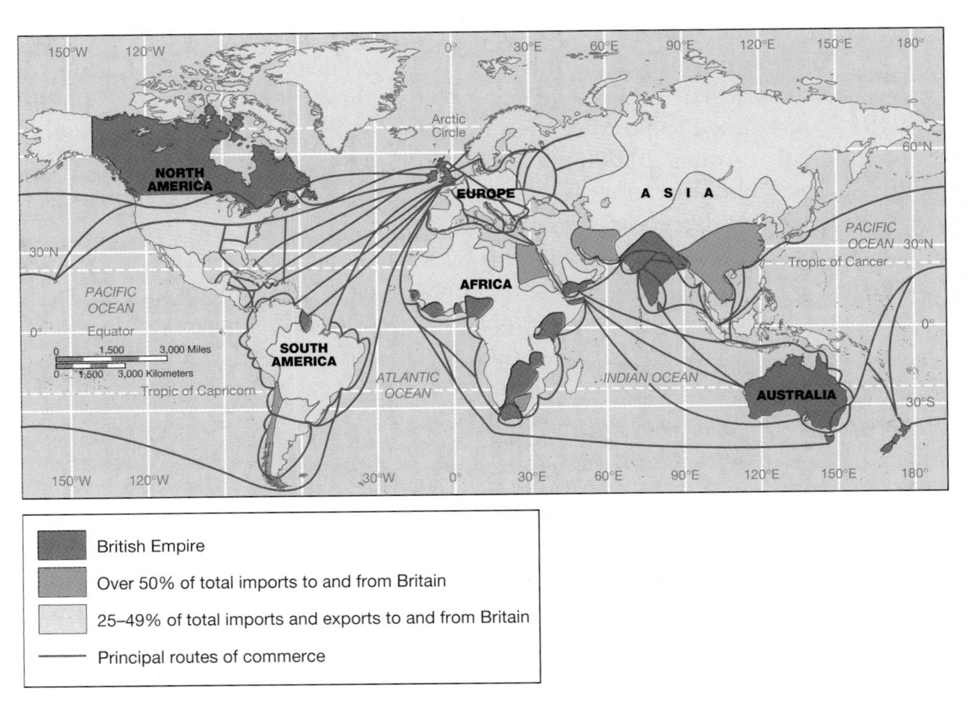

Figure 2.17 The British Empire in the late 1800s Protected by the all-powerful Royal Navy, the British merchant navy established a web of commerce that collected food for British industrial workers and raw materials for its industries, much of it from colonies and dependencies appropriated by imperial might and developed by British capital. So successful was the trading empire that Britain also became the hub of trade for other States. (*Source:* After P. Hugill, *World Trade Since 1431*. Baltimore: Johns Hopkins University Press, 1993, p. 136.)

pean weaponry was so superior that Otto von Bismarck, the founder and first chancellor (1871–1890) of the German empire, referred to these conflicts as "sporting wars." Between 1870 and 1900, European countries added almost 22 million square kilometers (10 million square miles) and 150 million people to their spheres of control—20 percent of the Earth's land surface and 10 percent of its population.

The imprint of imperialism and colonization on the geographies of the newly incorporated peripheries of the world-system was immediate and profound. The periphery was rendered almost entirely dependent on European and North American capital, shipping, managerial expertise, financial services, and news and communications. Consequently, it also became dependent on European cultural products: language, education, science, religion, architecture, and planning. All of these influences were etched into the landscapes of the periphery in a variety of ways as new places were created, old places were remade, and regions were reorganized.

The Third World and Neocolonialism

The imperial world order began to disintegrate shortly after World War II, however. The United States emerged as the new hegemonic power, the dominant state within the world-system core, which came to be called the First World. The Soviet Union and China, opting for alternative paths of development for themselves and their satellite countries, were seen as a Second World, withdrawn from the capitalist world economy. Their pursuit of alternative political economies was based on radically different values.

By the 1950s many of the old European colonies began to seek political independence. Some of the early independence struggles were very bloody because the colonial powers were initially reluctant to withdraw from colonies where strategic resources or large numbers of European settlers were involved. In Kenya, for example, a militant nationalist movement known as the Mau Mau launched a campaign of terrorism, sabotage, and assassination against British colonists in the early 1950s. Their actions killed more than 2,000 white settlers between 1952 and 1956; in return, 11,000 Mau Mau rebels were killed by the colonial army and 20,000 put into detention camps by the colonial administration. By the early 1960s, however, the process of decolonization had become relatively smooth. (In 1962, Jomo Kenyatta, who had been jailed as a Mau Mau leader in 1953, became prime minister of a newly independent Kenya.) The periphery of the world-system now consisted of a Third World of politically independent states, some of which adopted a policy of nonalignment, vis-à-vis the geopolitics of the First and Second Worlds. They were nevertheless still highly dependent, in economic terms, on the world's core countries.

As newly independent peripheral states struggled to be free of their economic dependence through industrialization, modernization, and trade from the 1960s onward, the capitalist world-system became increasingly integrated and interdependent. The old imperial patterns of international trade broke down and were replaced by more complex patterns. Nevertheless, the newly independent states were still influenced by many of the old colonial links and legacies that remained intact. The result was a neocolonial pattern of international development. **Neocolonialism** refers to economic and political strategies by which powerful states in core economies indirectly maintain or extend their influence over other areas or people. Instead of formal, direct rule (colonialism), controls are exerted through such strategies as international financial regulations, commercial relations, and covert intelligence operations. Because of neocolonialism, the human geographies of peripheral countries continued to be heavily shaped by the linguistic, cultural, political, and institutional influence of the former colonial powers and by the investment and trading activities of their firms.

At about the same time, a new form of imperialism was emerging. This was the *commercial imperialism* of giant corporations. These corporations had grown within the core countries through the elimination of smaller firms by mergers and takeovers. By the 1960s quite a few of them had become so big that they were *transnational* in scope, having established overseas subsidiaries, taken over foreign competitors, or simply bought into profitable foreign businesses.

These **transnational corporations** have investments and activities that span international boundaries, with subsidiary companies, factories, offices, or facilities in several countries. By 2004, over 61,000 transnational corporations were operating, 90 percent of which were headquartered in the core states. Between them these corporations control about 900,000 foreign affiliates and account for over $18 trillion in worldwide sales. Transnational corporations have been portrayed as imperialist by some geographers because of their ability and willingness to exercise their considerable power in ways that adversely affect peripheral states. They have certainly been central to a major new phase of geographical restructuring that has been under way for the last 30 years or so. This phase has been distinctive because an unprecedented amount of economic, political, social, and cultural activity has spilled beyond the geographic and institutional boundaries of states. It is a phase of *globalization,* a much fuller integration of the economies of the worldwide system of states and a much greater interdependence of individual places and regions from every part of the world-system (**Figure 2.18**).

The current phase of globalization also involves a distinctive new geopolitical element that has been described as the "new imperialism": the imperialism of the United States, the world's only effective superpower. While Americans do not like to think of their country as territorially aggressive or exploitative, the "war on terror" and invasion of Afghanistan and Iraq following the Al-Quaida attacks in 2001 are widely interpreted elsewhere in the world as comprising an exercise in imperialism, motivated in large part by a desire for military control over global oil resources. This interpretation of the United States as the instigator of a new imperialism has been reinforced by the country's military threats against Iran and North Korea, its gulag-style detention of prisoners without trial, its deployment of spe-

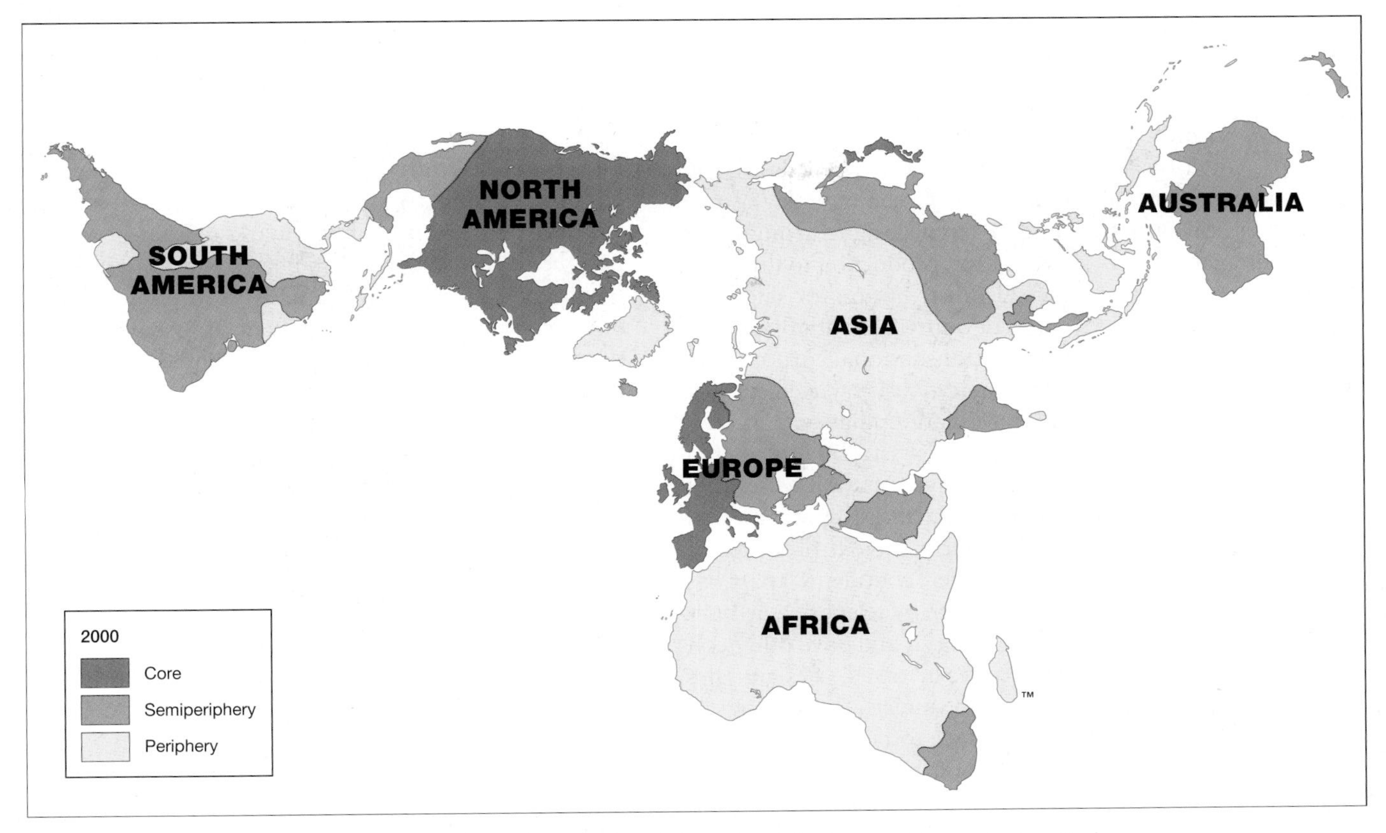

Figure 2.18 The world-system core, semiperiphery, and periphery in 2005 (*Source:* Map projection, Buckminster Fuller Institute and Dymaxion Map Design, Santa Barbara, CA. The word Dymaxion and the Fuller Projection™ Map design are trademarks of the Buckminster Fuller Institute, Santa Barbara, California, © 1938, 1967 & 1992. All rights reserved.)

cial forces around the globe, and its unilateral rejection of international environmental treaties and international aid agreements. What is less widely discussed in the opinion pages of the world's newspapers is that this new imperialism is also seen by some academics as the result of a highly competitive global economic environment in which the United States is no longer able to achieve superiority through innovation, product design, productivity and marketing, and so has had to resort to military intervention.

CONTEMPORARY GLOBALIZATION

Globalization has been under way since the inception of the modern world-system in the sixteenth century. The basic framework for globalization has been in place since the nineteenth century, when the competitive system of states fostered the emergence of international agencies and institutions; global networks of communication; a standardized system of global time; international competitions and prizes; international law; and internationally shared notions of citizenship and human rights. Global connections today, though, differ in at least four important ways from those in the past. First, they function at much greater *speed* than ever before. Second, globalization operates on a much larger *scale*, leaving few people unaffected and making its influence felt in even the most remote places. Third, the *scope* of global connections is much broader and has *multiple dimensions:* economic, technological, political, legal, social, and cultural, among others. Fourth, the interactions and interdependencies among numerous global actors have created a new level of *complexity* for the relationships between places and regions.

The most distinctive feature of globalization over the past 30 years or so is a decisive increase in the proportion of the world's economic and cultural activities that are transnational in scope. This increase is linked to a significant shift in the *nature* of international economic activity. Flows of goods, capital, and information that take place within and between transnational corporations are becoming more important than imports and exports between countries. At the same time, all these flows and activity have helped to spread new values around the world. These new values range from changing consumer lifestyle preferences to increased concern about global resources, global environmental change, and famine relief.

The contemporary world economy is constituted through the myriad commodity chains that crisscross global space. **Commodity chains** are networks of labor and production processes that originate in the extraction or production of raw materials and whose end result is the delivery and consumption of a finished commodity (see Box 2.5: "Commodity Chains"). These networks often

Commodity Chains

A commodity chain is a network of labor and production processes whose end result is a finished commodity. Global commodity chains link the progression of a commodity from design through procurement of raw materials and production to import or export to the point of sale, distribution for sale, marketing, and advertising. They are often entirely internal to the global operations of transnational corporations.

Advances in telecommunications, management techniques, transportation, informatics, finance, and other services to industry have made possible the segmentation of corporate production lines, as well as services, across multiple settings. Manufacturing companies can design a product in one country, have it produced by contractors in various countries continents apart, sell the product with its brand name by telephone or Internet almost anywhere in the world, and have other contractors deliver it. The services involved—design, sales, financing, and delivery—can be undertaken without the various actors ever meeting face to face (except for the carrier services). Advances in technology and management have also permitted the reproduction and standardization of services and products on a global basis. Certain patented services, such as fast-food restaurants (McDonald's, Kentucky Fried Chicken, Burger King), rely on computer-regulated technology to deliver a standard service and product over time and geographic space.

Almost every mass-marketed manufactured product involves a complex commodity chain. Take, for example, the manufacture of Lee Cooper jeans (**Figure 2.F**). Designed in the United States, advertised globally, retailed in stores across Europe, the United States, and the metropolises of semi-peripheral and peripheral countries, their manufacture draws on labor and products from around the world. The final stages of the commodity chain of one particular pair of jeans sold in a large discount store of a provincial British city were "in a van that came up the A12 [road] from Lee Cooper's warehouse at Staples Corner, just at the bottom of the M1 [highway] in North London. . . . Before that they came through the Channel Tunnel in a lorry from similar warehouses in Amiens, France and before that, by boat and train from Tunis in Tunisia . . .[1]"

There are three broad types of global commodity chains. The first is producer-driven, in which large, often transnational, corporations coordinate production networks. A good example of this is the U.S. pharmaceutical industry. Research and development of drugs are conducted in the United States; materials and components are produced in a vertically integrated global production line (i.e., a commodity chain); the drugs are assembled and marketed out of a semiperipheral site (in particular, Puerto Rico and Ireland), where companies enjoy certain competitive advantages not only in the industrial segment of assembly but in marketing, financial, and management services.

A second type of commodity chain is consumer-driven, where large retailers, brand-name merchandisers, and trading companies influence decentralized production networks in a variety of exporting countries, often in the periphery. A good example is the case of Lee Cooper jeans described here. Another good example is the discount chain-store company Wal-Mart, which contracts directly with producers in low-wage countries (China, in particular) for the bulk of its merchandise, which is sold in the United States to the accompaniment of advertising that invokes community-oriented and even patriotic themes.

The third type, the marketing-driven commodity chain, represents a hybrid of the first two types. It involves the production of inexpensive consumer goods—such as colas, beers, breakfast cereals, candies, cigarettes, and infant formula—that are global commodities and carry global brands yet are often manufactured in the periphery and semiperiphery for consumption in those regions. These commodities take their globalized status not only from their recipes and production techniques but, even more important, from their globally contrived cultural identities.

Different forms of commodity chains provide varying opportunities for firms and national economies to enter into and improve their position within the global division of labor. They are an important dimension of the complex transnationalization of economic space. They are also an important dimension of the complex currents of cultural globalization. In addition to facilitating the standardization of products and services (for those who can afford them) around the world, commodity chains also reflect the inequalities that are inscribed into the global economy. If core-country consumers stop to think about the origins of many of the products they consume, they may recognize what is happening "down the line," where poverty wages and grim working and living conditions are a precondition for the beginning of many commodity chains. In contrast, those who work on the farms and plantations and in the workshops and factories at the beginning of commodity chains are acutely aware—thanks to contemporary media—of the dramatically more affluent lifestyles of those who will eventually consume the fruits of their labor.

[1]Abrams, F. and J. Astill. Story of the blues, *The Guardian*. May 29, 2001, p. 2; quoted in L. Crewe, "Unravelling fashion's commodity chains." In A. Hughes and S. Reimer (eds.), *Geographies of Commodity Chains*. London: Routledge, 2004, p. 201.

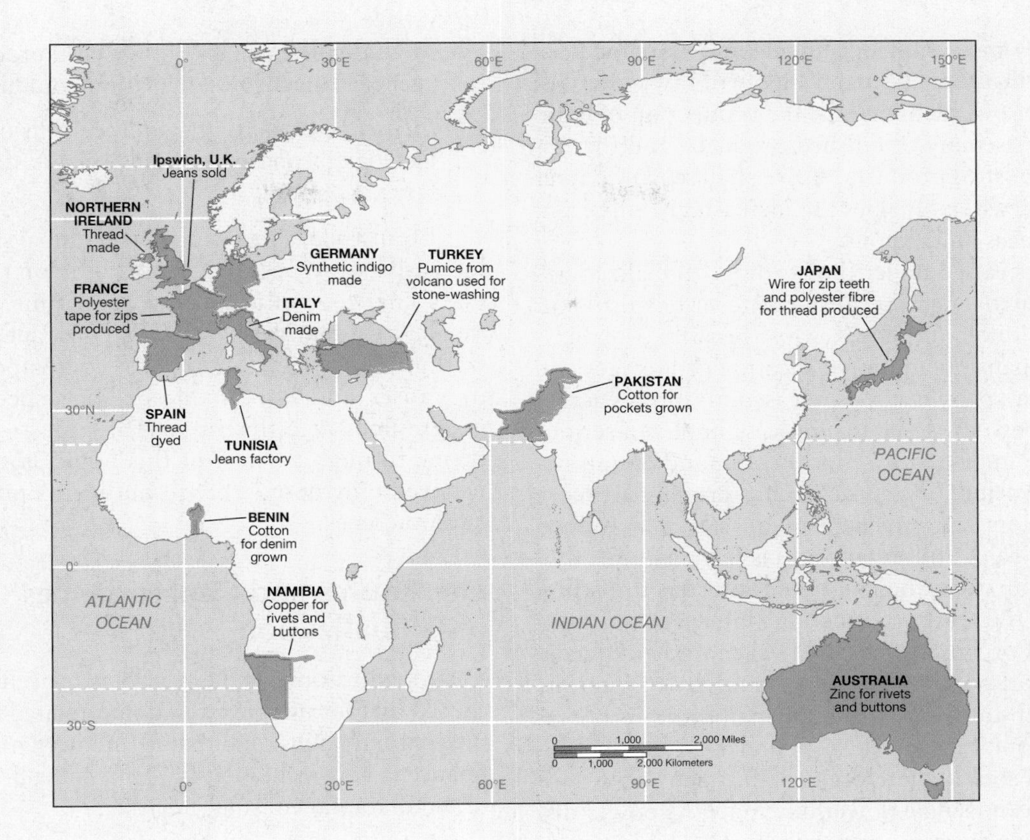

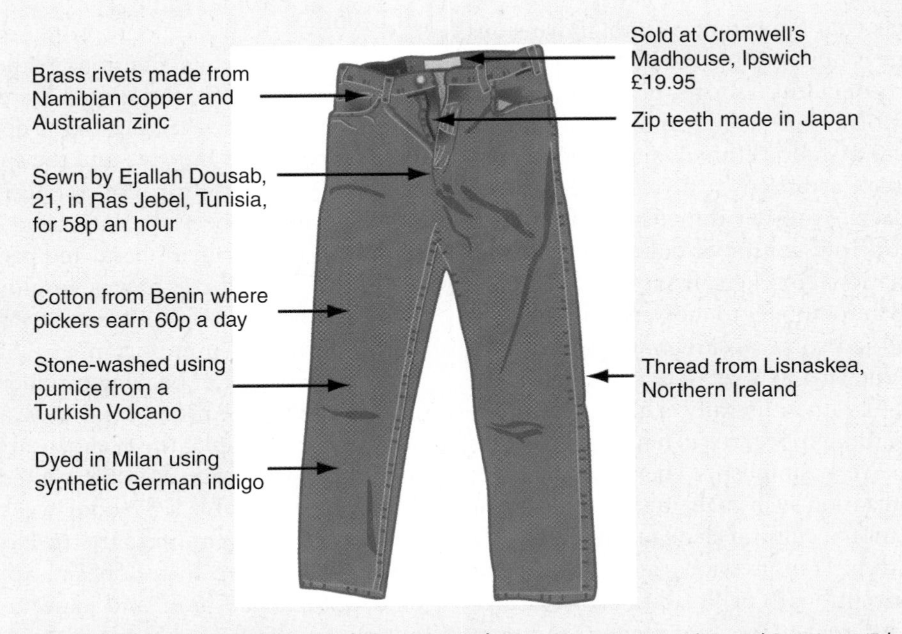

Figure 2.F The making of a pair of Lee Cooper jeans (*Source:* A. Hughes and S. Reimer (Eds.). *Geographics of Commodity Chains*. N.Y.: Routledge, 2004.)

span countries and continents, linking into vast global assembly lines the production and supply of raw materials; the processing of raw materials; the production of components; the assembly of finished products; and the distribution of finished products. As we shall see in Chapter 7, these global assembly lines are increasingly important in shaping places and regions.

The issues raised under the heading "globalization" are often controversial because the term has different meanings for different people. Most broadly, globalization is the expansion and intensification of linkages and flows of capital, people, goods, ideas, and cultures across national borders. To some this process implies a serious decline in the importance of local communities and national governments. Globalization has produced a more complex system of interdependent states in which *trans*national rules and organizations have gained influence. States pursuing their national interests are still a major force, but corporations and international nongovernmental organizations (INGOs) can now critically influence world politics. No clear hierarchy of interests common to all states exists, and the major powers are reluctant to use military force except under extreme provocation. World society therefore contains many centers of power and has no single power hierarchy. As power disperses and goals diverge, a new pattern of complex interdependence is emerging.

Globalization also has important cultural dimensions (see Chapter 5). One is quite simply the diffusion around the world of all sorts of cultural forms, practices, and artifacts that had previously been confined to specific places or regions. Examples include "ethnic" and regional cuisine, "world" music, Caribbean carnivals, and "charismatic" Christian sects. Another dimension of cultural globalization derives from consumer culture: everything that is sold in international markets, from sneakers, replica soccer shirts, and automobiles to movies and rock concert tours. This had led some observers to believe that globalization is producing a new set of universally shared images, practices, and values: literally, a global culture.

Most observers, though, regard such notions of a single global culture as overly simplistic. They point to other dimensions of globalization such as the fusion of different regional and international cultural elements that creates a hybridization of culture. Transgressions, adaptations, and subversions of "conventional" or "traditional" cultures originate in many different places and regions and then spread, resulting in a series of open-ended "global" cultures. Cultural anthropologist Arjun Appadurai has described five kinds of cultural flows that contribute to these global cultures:

Ethnoscapes: produced by flows of people, including tourists, immigrants, refugees, exiles, and guest workers.

Technoscapes: resulting from the diffusion of goods, technologies, and architectural styles.

Finanscapes: produced by rapid flows of money in currency markets and stock exchanges.

Mediascapes: images of the world produced by news agencies, magazines, television, and film.

Ideoscapes: resulting from the diffusion of ideas and ideologies, concepts of human rights, democracy, welfare, and so on.

Meanwhile, a more integrated global system has also increased awareness of a set of common problems—climate change, pollution, disease, crime, poverty, and inequality—that many see as a consequence of globalization. The globalization of the contemporary world—its causes and effects on specific aspects of human geographies at different spatial scales—is a recurring theme through the rest of this book. For the moment, we need only note in broad outline its principal causes and outcomes.

The Causes and Consequences of Globalization

The globalization of the past quarter century has been caused by four important and interrelated factors: a new international division of labor, an internationalization of finance, a new technology system, and a homogenization of international consumer markets.

A New International Division of Labor

The new international division of labor has involved three main changes. First, the United States has declined as an industrial producer, relative to the development of Japan, the resurgence of Europe, and the spectacular growth of China as industrial producers. Second, manufacturing production has been decentralized from these core regions to some semiperipheral and peripheral countries. In 2004, U.S.-based companies employed about 7 million workers overseas, 80 percent of whom were in manufacturing jobs. An important reason for this trend has been the prospect of keeping production costs low by exploiting the huge differential in wage rates around the world. For example, the compensation (including benefits) of apparel workers in the United States in 2002 was $13.06 per hour for a 37-hour week, whereas for a 60-hour week their counterparts in Hong Kong were paid US$4.49 an hour. In Indonesia, apparel workers were paid US$0.15 an hour and expected to work up to 70 hours a week. A third result of the new international division of labor is that new specializations have emerged within the core regions of the world-system: high-tech manufacturing and **producer services** (that is, services such as information services, insurance, and market research that enhance the productivity or efficiency of other firms' activities or that enable them to maintain specialized roles). One significant reflection of this new international division of labor is that global trade has grown much more rapidly over the past 30 years than global production—a clear indication of the increased economic integration of the world-system.

The Internationalization of Finance

The second factor contributing to today's globalization is the internationalization of finance: the emergence of global banking and globally integrated financial markets. These changes are, of course, tied in to the new international division of labor. In particular, they are a consequence of massive increases in levels of international direct investment. Between 1975 and 2004, private capital flows increased from less than 5 percent of world GDP to just under 23 percent. Until the early 1970s, U.S.-based transnational corporations accounted for about two-thirds of the total outflows of foreign direct investment (FDI), and about four-fifths of this was directed toward Canada and the more advanced industrial national economies of Western Europe. By 2004, U.S. transnational corporations' share of the total had dropped to less than half, while direct foreign investment by Japanese, Canadian, and German corporations increased significantly. Meanwhile, another source of foreign direct investment had begun to show up: transnational corporations based in semiperipheral countries. Several of the top 100 corporations in the world are now based in these newly industrializing countries (NICs). Sinopec Corp, a Chinese petroleum and chemical corporation, is bigger than Boeing, while Samsung, the South Korean electronics firm, is bigger than Japan's Toshiba, Germany's BMW, and the U.S. conglomerate of Procter and Gamble.

Along with these changes in the sources of investment have been changes in destination. Core countries now absorb most of the inflows of capital: Between 1998 and 2004 the United States, the European Union, and Japan accounted for three-quarters of global inflows of foreign direct investment. They also accounted for 85 percent of outflows, so that the dominant outcome of globalization is a circular movement of FDI between North America, Western Europe, and Japan, at the expense of peripheral regions. Nevertheless, as we shall see in Chapter 7, the globalization of economic activity has also brought significant flows of capital into NICs such as Argentina, Brazil, China (including Hong Kong), Malaysia, Mexico, Singapore, and South Korea. As noted in Chapter 1, all of these changes have been fostered by neoliberal policies and institutions such as the World Trade Organization and World Economic Forum (**Figure 2.19**).

Meanwhile, the capacity of computers and information systems to deal very quickly with changing international conditions has added a speculative component to the internationalization of finance. All in all, about $1.2 trillion worth of currencies are exchanged every day. The volume of international investment and financial trading has created a need for banks and financial institutions that can handle investments on a large scale, across great distances, quickly and efficiently. The nerve centers of the new system are located in just a few places—London, Frankfurt, New York, and Tokyo, in particular. Their activities are interconnected around the clock for almost 24 hours (**Figure 2.20**), and their networks penetrate into every part of the globe.

A New Technology System

The third factor contributing to globalization is a new technology system (see p. 56) based on a combination of innovations, including solar energy, robotics, microelectronics, biotechnology, and digital telecommunications and information systems. This new technology system has required the geographical reorganization of the core economies. It has also extended the global reach of finance and industry and permitted a more flexible approach to investment and trade. Especially important in this regard have been new and improved technologies in transport and communications—the integration of shipping, railroad, and highway systems through containerization (**Figure 2.21**); the introduction of wide-bodied cargo jets;

Figure 2.19 The World Economic Forum, 2005 Participants in the World Economic Forum walk around in the congress center in Davos, Switzerland. Some 2,000 high-ranking representatives from the world of politics and economics meet annually in the Swiss ski resort of Davos for the World Economic Forum.

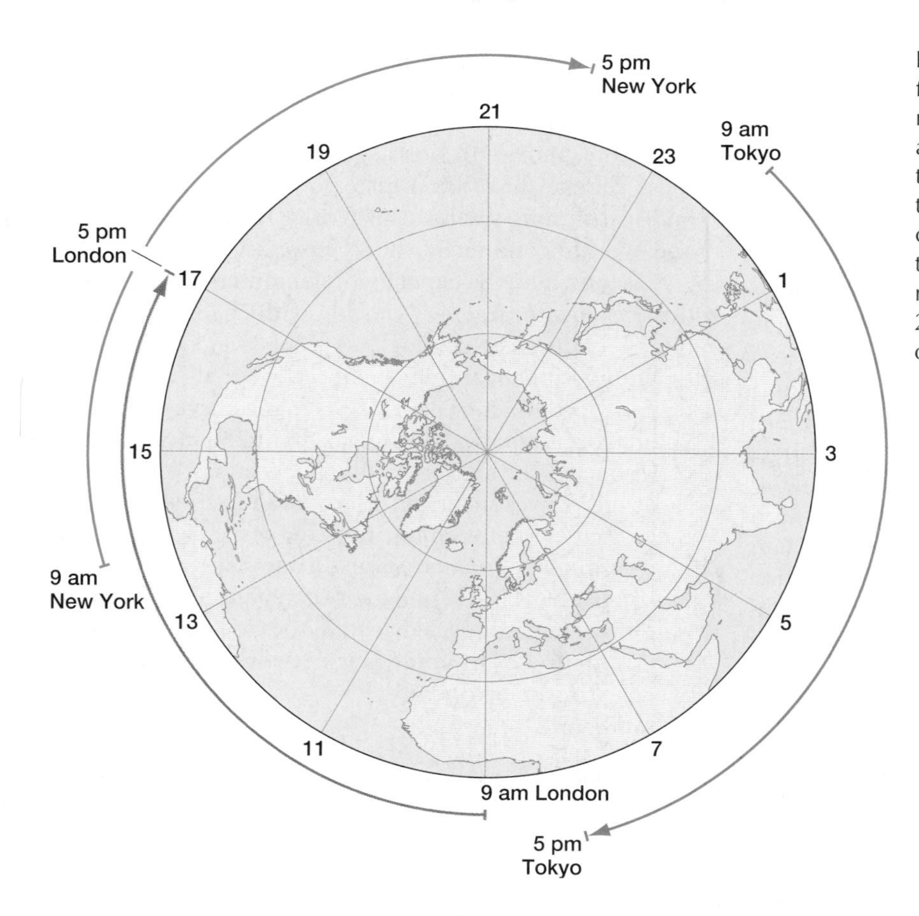

Figure 2.20 24-hour trading between major financial markets Office hours in the two most important financial centers—New York and London—overlap one another because the two cities are situated in broadly separated time zones. While these markets are both closed, Tokyo offices are open. This means that between them the world's three major financial centers span the globe with almost 24-hour trading in currencies, stocks, and other financial instruments.

Figure 2.21 The impact of containerization on world trade Containerization revolutionized long-distance transport because it did away with the slow, expensive, and unreliable business of loading and unloading ships with manual labor. Before containerization, ships spent one day in port for every day at sea; after containerization, they spent a day in port for every 10 days at sea. By 1965 an international standard for containers had been adopted, making it possible to transfer goods directly from ship to rail to road and allowing for a highly integrated global transport infrastructure. The average container ship today holds 4,000 20-foot containers, but some are able to carry 6,000 or 8,000. Containerization requires a heavy investment in both vessels and dockside handling equipment, however. As a result, container traffic has quickly become concentrated in a few ports that handle high-volume transatlantic and trans Pacific trade. This photograph shows a Hyundai container ship leaving the port of Long Beach, California.

and the development of fiber-optic networks, communications satellites, and electronic mail and information retrieval systems. Finally, many of these telecommunications technologies have also introduced a wider geographical scope and faster pace to many aspects of political, social, and cultural change, as we shall see in subsequent chapters.

Global Consumer Markets

The fourth factor in globalization has been the growth of consumer markets. Among the more affluent populations of the world, similar social processes create parallel trends in consumer taste. A new and materialistic international culture has taken root, in which people save less, borrow more, defer parenthood, and indulge in affordable luxuries that are marketed as symbols of style and distinctiveness. This culture is easily transmitted through the new telecommunications media, and it has been an important basis for transnational corporations' global branding and marketing of "world products" (German luxury automobiles, Swiss watches, British raincoats, French wines, American soft drinks, Italian shoes and designer clothes, and Japanese consumer electronics, for example). Nine of the ten most trusted brands in India in 2003, for example, were products of British or American transnational corporations: Colgate, Dettol, Pond's, Lux, Pepsodent, Brittania, Rin, Surf, and Close-Up. Tata Salt was the only Indian brand in the top ten list.

This materialistic international culture is reinforced through other aspects of globalization, including the internationalization of television, especially CNN, MTV, Star Television, and the syndication of TV movies and light entertainment series. The number of television sets per 1,000 people worldwide doubled between 1980 and 2000, while multimedia industries have been booming. The global market for popular cultural products carried by these media is becoming concentrated, however. At the core of the entertainment industry—film, music, and television—there is a growing dominance of U.S. products, and many countries have seen their homegrown industries wither. Hollywood obtains more than 50 percent of its revenues from overseas, up from just 30 percent in 1980. Movies made in the United States account for about 50 percent of the market in Japan, 70 percent in Europe, and 85 percent in Latin America. Similarly, U.S. television series have become increasingly prominent in the programming of other countries.

Globalization and Core-Periphery Differences

Just as globalization has been driven by several interrelated factors, so the outcomes of globalization are manifest in different ways. The single most dramatic outcome of the globalization that has resulted from all these changes is the consolidation of the core of the world-system. The core is now a close-knit triad of the geographic centers of North America, the European Union, and Japan. These three geographic centers are connected through three main circuits, or flows, of investment, trade, and communication: between Europe and North America, between Europe and the Far East, and among the regions of the Pacific Rim. **Figure 2.22**, for example, shows just how dominant North America has become in accounting for flows of international telephonic communication. As we shall see in Chapter 7, this consolidation of the core of the world-system is having some profound effects on economic geography. Within the core regions, for example, a new hierarchy of regional economic specialization has been imposed by the locational strategies of transnational corporations and international financial institutions.

Globalization, although incorporating more of the world more completely into the capitalist world-system, has intensified the differences between the core and the periphery. According to the United Nations Development Program, the gap between the poorest fifth of the world's population and the wealthiest fifth increased more than threefold between 1960 and 2000. Per capita GDP in the world's 20 richest countries is now 40 times higher than in the world's poorest 20 countries. Some parts of the periphery have almost slid off the economic map. In 55 countries per capita income actually fell during the 1990s. In sub-Saharan Africa, economic output fell by one-third during the 1980s and stayed low during the 1990s, so that people's standard of living there is now, on average, lower than it was in the early 1960s. In 2004 the fifth of the world's population living in the highest-income countries had:

- 75 percent of world income (the bottom fifth had just 1 percent);
- 83 percent of world export markets (the bottom fifth had just 1 percent);
- 76 percent of world telephone lines, today's basic means of communication (the bottom fifth had just 1.5 percent).

While 1.3 billion people around the world struggle to live on less than US$1.00 a day, the world's richest 200 individuals have a net worth of more than $1 trillion between them. The world's top three billionaires alone possess more assets than the combined Gross National Product of all of the least developed countries (whose combined population exceeds 600 million people). OECD countries (the Organization for Economic Cooperation and Development, an association of 30 industrialized countries), with 19 percent of global population, control 71 percent of global trade in goods and services, and consume 16 times more than the poorest 20 percent of the world's population. More than 850 million people are still clinically malnourished and permanently hungry, including 142 million in India, 221 million in China, and 204 million in Sub-Saharan Africa.

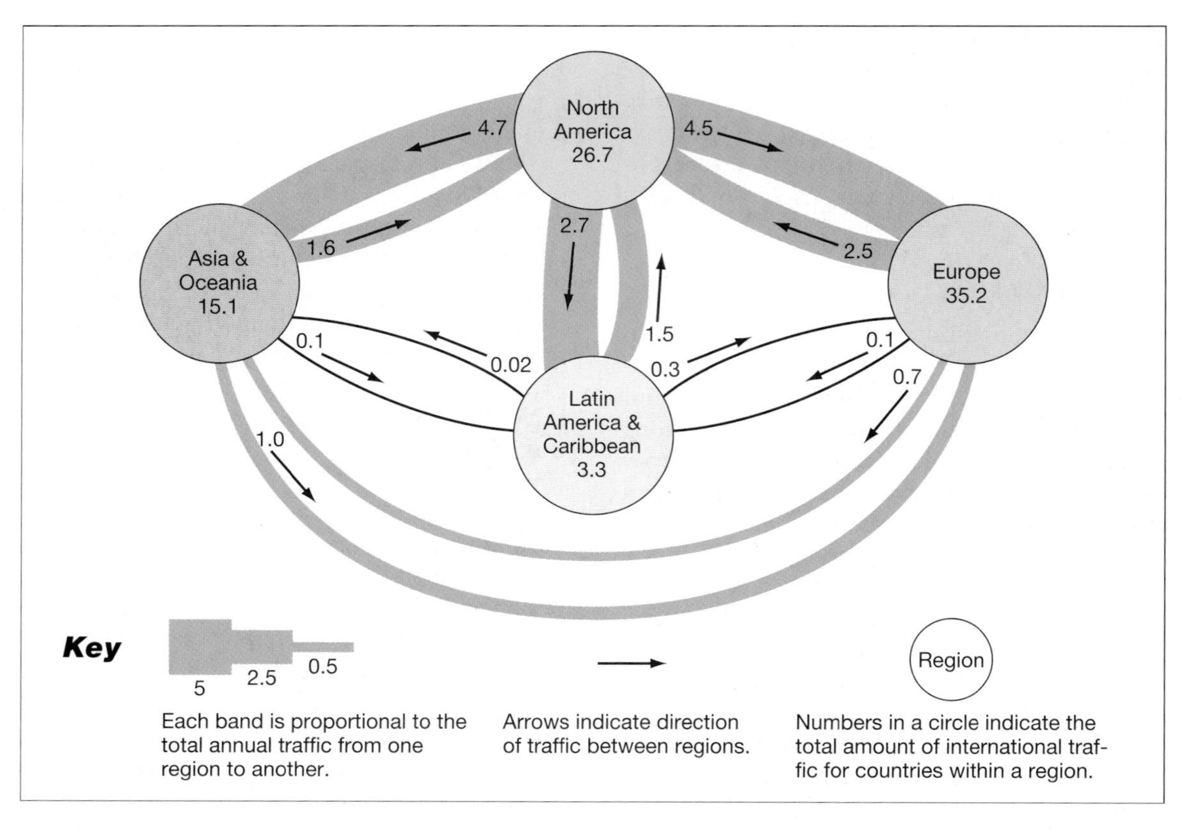

Figure 2.22 Communications flows between major world regions This diagram shows the flows, in billions of minutes of telecommunications traffic over public telephone networks, between major regions. (*Source:* G. C. Staple (ed.), *TeleGeography 1999.* Washington, DC: TeleGeography Inc., 1999, Fig. 4, p. 255.)

Such enormous differences lead many people to question the equity of the geographical consequences of globalization. The concept of **spatial justice** is important here, because it requires us to consider the distribution of society's benefits and burdens at different spatial scales, taking into account both variations in people's need and in their contribution to the production of wealth and social well-being. In particular, many people, nations, and ethnic groups around the world feel marginalized, exploited, and neglected as a result of the quickening pace of transformation that has come with globalization. Across much of the peripheral world, the perception of injustice has been brewing for a long time. Resentment at past colonial and imperial exploitation has been compounded as the more affluent places and regions of the world have become increasingly dependent on the cheap labor and resources of the periphery and as transnational businesses have displaced the traditional economic and social practices of peripheral and semiperipheral regions under the banner of modernization. Thinking about spatial justice is an important aspect of the "geographical imagination" described in Chapter 1 and is a recurring theme in the remainder of this book.

Meanwhile, some other important and controversial issues are tied to economic globalization: the depletion of the ozone layer, for example, together with threats to biodiversity and marine life. Carbon emissions, an important factor in global climate change, tend to reflect core-periphery patterns of economic development. On a per capita basis, the United States generates roughly seven times more carbon emissions than China and nine times more than India. Of equal concern to many are the various local outcomes of the operation of the international economy: resource depletion and environmental degradation in some regions, ecotourism in others, industrialization in still others, and so on. Globalization has also been accompanied by increased international flows of narcotics and laundered money, a more extensive diffusion of communicable diseases like tuberculosis and HIV/AIDS, and greater freedom of movement for terrorists and weapons of mass destruction. These unwelcome consequences of globalization pose huge challenges to an international community more interconnected—and therefore more vulnerable—than ever before.

Nevertheless, it is important to bear in mind that the "shrinking" of the world through processes of globalization has been uneven. There is an increasing division between the "fast world" and the "slow world." The **fast world** consists of people, places, and regions directly involved as producers and consumers in transnational industry, modern telecommunications, materialistic consumption, and international news and entertainment. The **slow world,** which accounts for about 85 percent of the world's population, consists of people, places, and regions whose participation in transnational industry, modern telecommunications, materialistic consumption, and international news and entertainment is limited. The slow world consists chiefly of the impoverished periphery, but it also includes many rural backwaters, declining manufacturing regions, and disadvantaged slums in core countries, all of them bypassed by this latest phase in the evolution of the modern world-system.

The Fast World and the Slow World

The fast world is very much a product of the expansion of capitalism on a global scale. Because capitalism is an inherently competitive system, capitalists find themselves in an endless race to seek out new markets and reduce what is known as the "turnover time" of capital: the amount of time it takes for money invested to fund the costs of new production to be returned with a profit through the sale of goods and services. In the global capitalist system, time costs money, and the inevitable result is a steady acceleration in the pace of life. Not surprisingly, the center of gravity of the fast world is the tri-polar core of the world-system. But the fast world also extends throughout the world to the more affluent regions, neighborhoods, and households that are "plugged in" to the contemporary world economy, whether as producers or consumers of its products and culture.

The pivotal moment in the creation of the fast world was a "system shock" to the international economy that occurred in the mid-1970s. World financial markets, swollen with U.S. dollars by the U.S. government's deficit budgeting and by huge currency reserves held by OPEC (Organization of Petroleum Exporting Countries) after they had orchestrated a four-fold increase in the price of crude oil, quickly evolved into a new and sophisticated system of international finance, with new patterns of investment and disinvestment that led to some radical socioeconomic changes. New social formations emerged as part of new, post-industrial societies in most OECD countries. A new, transnational material culture emerged around the consumption of globally branded products. And, as money accelerated around local, national, and international circuits of capital, so the pace of everyday life quickened.

The leading edge of the fast world is the Internet, the global web of computer networks that began in the United States in the 1970s as a decentralized communication system sponsored by the United States Department of Defense. Today the Internet has become the world's single most important mechanism for the transmission of scientific and academic knowledge. Roughly 50 percent of its traffic is electronic mail; the rest consists of scientific documents, data, bibliographies, electronic journals, bulletin boards, and a user interface to the Internet, the World Wide Web. In 2005 approximately one billion people had access to the Internet. The Internet has been doubling in networks and users every year since 1990, but most Internet users are still in the world's core regions. In mid-2005 about 27 percent were in North America, and another 28 percent were in Europe. The rest were in Japan, Australia, and New Zealand, and in the fragmentary outposts of the fast world that are embedded within the larger metropolitan areas of the periphery and semiperiphery. Overall, more than 65 percent of all Internet traffic originates in, or is destined for, North America. These imbalances between the fast world and the slow world are part of a digital divide that exists at every spatial scale (**Figure 2.23**).

The division between fast and slow worlds is, of course, something of a caricature. In fact, the fast world encompasses almost everywhere but not everybody. As a result, human geography now has to contend with the apparent paradox of people whose everyday lives are lived partly in one world, partly in another. Consider, for example, the shantytown residents of Mexico City. With extremely low incomes, only makeshift housing, and little or no formal education, they are nevertheless knowledgeable about international soccer, music, film, and fashion and are even able to copy fast-world consumption through castoffs and knockoffs. Much the same could be said about the impoverished residents of rural Appalachia (substitute NASCAR racing for international soccer) and, indeed, about most regions of the slow world. Very few regions remain largely untouched by globalization.

At first glance the emergence of the fast world—with its transnational architectural styles, dress codes, retail chains, and popular culture, and its ubiquitous immigrants, business visitors, and tourists—seems as if it might have brought a sense of placelessness and dislocation, a loss of territorial identity, and an erosion of the distinctive sense of place associated with certain localities. Yet the common experiences associated with globalization are still modified by local geographies. The structures and flows of the fast world are variously embraced, resisted, subverted, and exploited as they make contact with specific places and specific communities. In the process, places and regions are reconstructed rather than effaced. Often this involves deliberate attempts by the residents of a particular area to create or re-create territorial identity and a sense of place. Inhabitants of the fast world, in other words, still feel the need for enclaves of familiarity, centeredness, and identity. Human geographies change, but they don't disappear.

Jihad vs. McWorld

Related to core-periphery and fast world/slow world contrasts and interdependencies are broad cultural differences that have been characterized by political scientist Benjamin Barber as "Jihad vs. McWorld." In his characterization, McWorld is shorthand for the pop culture and shallow materialism that is part of Western, capitalist modernization. "Jihad" is shorthand for cultural values that are underpinned by religious fundamentalism, traditional tribal allegiances, and opposition to Western materialism. (Note that the term *jihad* properly refers to a struggle waged as a religious duty on behalf of Islam.) Neither "Jihad" nor "McWorld" is conducive to a healthy

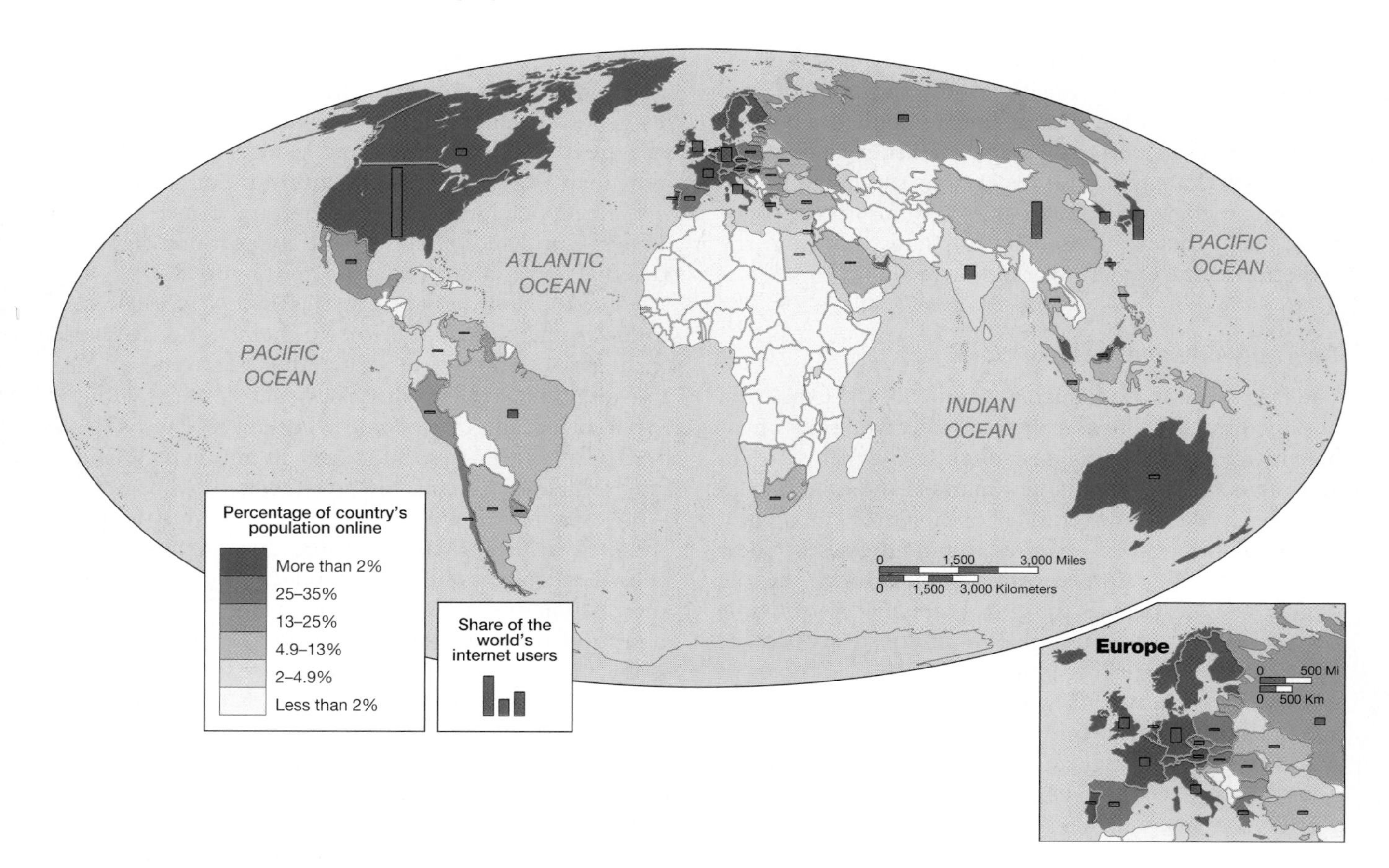

Figure 2.23 Global Internet connectivity Like all previous revolutions in transportation and communications, the Internet is effectively reorganizing space. Although often spoken of as "shrinking" the world and "eliminating" geography, the Internet is highly uneven in its availability and use. About 70 percent of all traffic on the Internet originates from, or is addressed to, North America. In contrast, some peripheral countries have almost no Internet connectivity, while in others the costs are prohibitive. Even in Europe, Japan, and North America, Internet connectivity is decidedly uneven in socioeconomic terms. It is the medium of the "fast world" of big business and affluent consumers. This map shows the percentage of the total population in each country in 2004 with access to the Internet (indicated by the density of shading) and the relative share of the world's Internet users accounted for by each country (indicated by the vertical bars). (*Source:* M. Zook, http://www.zookNIC.com/.)

democracy or civil society, argues Barber, while tensions between the two make for potentially volatile situations.

At the heart of these tensions is a marked disillusionment with the West, especially within traditional Islamic societies. Across much of the world, modernization is now taken to mean Westernization and, more specifically, Americanization (**Figure 2.24**). While most Americans think of modernization as necessary and good, many other people see it as having resulted in their exploitation and humiliation. In most peripheral countries, only a minority can enjoy Western-style consumerism, though the impoverished majority is acutely aware of the affluence of the core countries. While the gap between rich and poor countries has actually been widening for several decades, the U.S. aid budget—already low compared to the aid budgets of other developed countries—has been declining substantially. The United States therefore tends to be seen as a swaggering superpower, rigging the world system in its own interests but doing relatively little by way of economic or humanitarian aid.

In the Arab world, the sense of grievance at such injustice, along with a sense of dismay at the way modernization is undermining Islamic values and traditions, has been intensified by opposition to U.S. policy toward the Middle East, which is seen as being cynically geared to U.S. oil interests and support for Israel. In addition, there is widespread resentment at the injustices experienced by Palestinians, at the suffering of Iraqi civilians resulting from the U.S. invasion of Iraq, at the presence of U.S. troops in Saudi Arabia, and at the repressive and corrupt nature of U.S.-backed Persian Gulf governments.

Meanwhile, the world's core countries, acting on the confident assumption that people everywhere want Western-style modernization, have until recently failed to recognize the cultural resentment, the sense of injustice, and the genuine rejection of modernization that exists in many parts of the world. The terrorist attacks on the Pentagon and on New York's World Trade Center in September 2001 brought an end to such assumptions. In the aftermath of the attacks it became clear that resentment to-

Figure 2.24 Westernization Turkish Moslems chat near an illuminated billboard advertising a Turkish Internet company at Istanbul airport.

ward the United States and its economic and foreign policies has bred widespread hostility and even hatred.

Mobilization Against Globalization

Economic globalization is in many ways still in its beginning stages, but already it has brought a great deal of change to the economic, cultural, and political geography of places and regions throughout the world. A great deal of this change has been progressive, bringing increased overall levels of economic well-being, a strengthening of free enterprise and democracy, and an enriched flow of products, ideas, and culture among and between places and regions. Inevitably, though, as with previous epochs of economic change, economic globalization has also brought problems as some places and regions—the U.S. Manufacturing Belt, northern England, central Scotland, and northeast France, for example—have experienced disinvestment in order that capital could be made available for more profitable investments elsewhere. Economic globalization has also undercut the power of national and local governments to regulate economic affairs and has erased a great deal of local diversity because of the economic success of global products: the "McDonaldization of everywhere."

Fundamental geographic differences—in climate, resources, culture, and so on—mean that economic globalization is variously embraced, modified, or resisted in different parts of the world. Indeed, there has emerged a countermovement, a "mobilization against globalization," that could well affect the whole dynamic of economic globalization as it is played out over the next decade or two. One form of this mobilization is exemplified by the efforts of activists who use the legal system in order to resist what they see as the undesirable local outcomes of transnational business practices. In the United States, activists have resurrected a 1789 alien tort law that was originally designed to provide redress for foreigners against sea pirates and slavers. The law had already been used to pursue individuals such as Ferdinand Marcos, the late dictator of the Philippines. In the late 1990s activists pursued transnational corporations, accusing them of helping to suppress human rights. For example, four U.S. retailers and clothing manufacturers charged with unethical labor practices in a $1 billion alien tort suit that had been filed on behalf of some 50,000 garment workers in Saipan agreed to settle in 1999. While admitting no liability, Nordstrom, Gymboree, Cutter & Buck, and J. Crew agreed to pay $1.25 million into a fund to support the independent monitoring of their overseas suppliers.

A second form of mobilization is old-fashioned popular protest (**Figure 2.25**). French farmers, for example, regularly take to tactics such as blocking streets with tractors, with produce, with farmyard manure, or with farm animals in protest over trade-liberalization policies. A third form of mobilization is to counter the prevailing trend with alternatives, as in the case of Mecca Cola (**Figure 2.26**).

Fourth, and perhaps most significant in terms of future cultural struggles between local interests and transnational business interests, mobilization can be organized by coalitions of nongovernmental organizations (NGOs). This form of mobilization against globalization became much more powerful in the 1990s as a result of the Internet. Groups such as Kenya's Consumers' Information Network, Ecuador's Acción Ecologica, and Trinidad and Tobago's Caribbean Association for Feminist Research and Action are linked through scores of Web sites, listserves, and discussion groups to U.S., European, and Asian counterparts. NGOs set the agenda for the Earth Summit in Rio de Janeiro in 1992 and lobbied govern-

(a) (b)

Figure 2.25 Anti-globalization demonstrations Globalization often leads to the downward convergence of wages and environmental standards, an undermining of democratic governance, and a general recoding of nearly all aspects of life to the language and logic of global markets. (a) A massive anti-globalization march marks the end of the European Social Forum in Paris in 2003. (b) Italian carabinieri guard a McDonald's restaurant in Genoa against possible attacks from demonstrators protesting against genetically-modified foods in 2000.

ments to attend; they publicized the Chiapas rebellion in Mexico in 1994, thereby preventing the Mexican government from suppressing it violently. In 1997 a loose alliance of 350 NGOs from 23 countries set out to ban land mines; they soon persuaded 122 nations to sign on to a treaty. In 1998 another NGO alliance, this time reckoned to number 600 groups in nearly 70 countries, blocked a painstakingly negotiated treaty on international multilateral investment.

In 1999 more than 775 NGOs registered with the World Trade Organization (WTO) and took more than 2,000 observers to the WTO summit in Seattle, Washington. They also helped to organize some 70,000 protesters who took part in the most extensive teach-ins and demonstrations in the United States since the Vietnam War. "No Globalization without Representation" read the placards in the protest march, and "Hey, hey, ho, ho, WTO has got to go." "Whose world? Our world! Whose streets? Our streets!" chanted the crowd. Riot police fired tear gas, pepper spray, and rubber bullets at peaceful protesters blockading the WTO meeting, while roving gangs of anarchists smashed windows, overturned newspaper stands, and attacked cars. When demonstrators refused to disperse, the city declared a state of civil emergency. The WTO meetings themselves, delayed by the protests, ended in collapse. In 2001 a young man was killed by paramilitary po-

Figure 2.26 Mecca Cola A can of Mecca-Cola is displayed in Putrajaya, near Kuala Lumpur. Mecca-Cola was brewed to boycott American brands as a protest against the U.S.-led war on Iraq, according to its founder, Tawfik Mathlouthi. Sales of Mecca-Cola have spread to 48 countries since it was introduced in 2002.

Figure 2.27 Genoa protests In July 2001 leaders of the so-called G-8 group of major industrial powers—Canada, France, Germany, Italy, Japan, Russia, the United Kingdom, and the United States—met behind miles of fence-and-concrete barricades in Genoa, Italy, for their annual meeting. An estimated 100,000 antiglobalization protesters converged on the city, a small minority of them clashing with police in fierce riots. Protesters hurled cobblestones, smashed windows, set fire to cars and trash bins, and looted storefronts. Police responded with water cannons, tear gas and nightsticks, clubbing some protesters into submission and killing a 20-year-old man.

lice during antiglobalization riots in Genoa, Italy (**Figure 2.27**). Since then, economic summit meetings have been rescheduled to meet in hard-to-reach locations in an attempt to curb the effectiveness of demonstrations.

These demonstrations highlight some of the central issues that surround economic globalization. Economic globalization depends on free trade, but should the abolition of economic protectionism be accompanied by the abolition of social and environmental protection? The WTO's mandate is the "harmonization" of safety and environmental standards among member nations as well as the removal of tariffs and other barriers to free trade. Most people support free trade, but not if it harms public health, and not if it is based on child labor.

There have been several examples of how the free trade principles embodied in the WTO can erode national environmental standards. In 1997 Venezuela and Brazil, on behalf of their gasoline producers, challenged U.S. Environmental Protection Agency (EPA) regulations on gasoline quality, which were designed to ensure minimal levels of pollution. The WTO ruled in favor of Venezuela. The EPA subsequently had to change its regulations, leaving it with a weakened ability to enforce federal air-quality standards. When the European Union banned U.S. beef raised with the assistance of hormone injections in 1998, the United States took the case to the WTO. The WTO ruled that the European Union's act was illegal under international trading rules, even if the Europeans were distrustful of the hormone-injected meat. Another WTO ruling in 1998 undermined the U.S. Endangered Species Act. The United States has attempted to protect endangered sea turtles from extinction by requiring that shrimp boats install devices that allow the turtles to escape their nets. The law applied to all shrimp sold in the United States, but the WTO ruled that this was a restraint on other countries' ability to trade freely.

CONCLUSION

Places and regions everywhere carry the legacy of a sequence of major changes in world geography. The evolution of world geography can be traced from the prehistoric hearths of agricultural development and human settlement, through the trading systems of the precapitalist, pre-industrial world, to the foundations of the geography of the modern world. These foundations were cast through industrialization, the colonization of many parts of the world, and the spread of an international market economy. Today these foundations can be seen in the geography of the Information Age, a geography that now provides a global context for places and regions.

Today's world is highly integrated. Places and regions have become increasingly interdependent, linked through complex and rapidly changing commodity chains that are orchestrated by transnational corporations. Using new technology systems that allow for instantaneous global telecommunications and flexible patterns of investment and production, these corporations span the fast world of the core and the slow worlds of the periphery. This integration does tend to blur some national and regional differences as the global marketplace brings a dispersion of people, tastes, and ideas. The overall result, though, has been an intensification of differences between the core and the periphery. Within this new global context, local differences in resource endowments remain, and people's territorial impulses endure. Many local cultures continue to be resilient or adaptive. Fundamental principles of spatial organization also continue to operate. All this ensures that even as the world-system becomes more and more integrated, places and regions continue to be made and remade. The new global context is filled with local variety that is constantly changing, just as the global context itself is constantly responding to local developments.

MAIN POINTS REVISITED

- **The evolution of the modern world-system has exhibited distinctive stages, each of which has left its legacy in different ways on particular places, depending on their changing role within the world-system.**

  The modern world-system was first established over a long period that began in the late fifteenth century. More and more peoples around the world became exposed to one another's technologies and ideas over the next five centuries. Their different resources, social structures, and cultural systems resulted in quite different pathways of development, however. Some societies were incorporated into the new, European-based international economic system faster than others; some resisted incorporation; and some sought alternative systems of economic and political organization.

- **At the end of the eighteenth century, the new technologies of the Industrial Revolution brought about the emergence of a global economic system that reached into almost every part of the world and into virtually every aspect of people's lives.**

  New transportation technologies triggered successive phases of geographic expansion, allowing for an intensive period of external colonization and imperialism. The core of the world-system grew to include the United States and Japan, while most of the rest of the world was systematically incorporated into the capitalist world-system as a dependent periphery.

- **Places and regions are part of a world-system that has been created as a result of processes of private economic competition and political competition between national states.**

  Each place and region carries out its own particular role within the competitive world-system. Because of these different roles, places and regions are dependent on one another. The development of each place affects, and is affected by, the development of many other places.

- **Today the world-system is highly structured and is characterized by three tiers: core regions, semiperipheral regions, and peripheral regions.**

  The core regions of the world-system are those that dominate trade, control the most advanced technologies, and have high levels of productivity within diversified economies. Peripheral regions are characterized by dependent and disadvantageous trading relationships, by primitive or obsolescent technologies, and by undeveloped or narrowly specialized economies with low levels of productivity. Semiperipheral regions are able to exploit peripheral regions but are themselves exploited and dominated by the core regions. This three-tiered system is fluid, providing a continually changing framework for geographical transformation within individual places and regions.

- **The growth and internal colonization of the core regions could only take place with the foodstuffs, raw materials, and markets provided by the colonization of the periphery.**

  In the eighteenth and nineteenth centuries, the industrial core nations embarked on the inland penetration of the world's midcontinental grassland zones in order to exploit them for grain or stock production. At the same time, as the demand for tropical plantation products increased, most of the tropical world came under the political and economic control—direct or indirect—of one or another of the industrial core nations. For these peripheral regions, European overseas expansion meant political and economic dependency.

- **Within each of the world's major regions, successive technological innovations have transformed regional geographies.**

  Each new system of production and transportation technologies has helped raise levels of productivity and create new and better products that have stimulated demand, increased profits, and created a pool of capital for further investment. This investment, however, has taken place in new or restructured geographic settings.

- **Globalization has intensified the differences between the core and the periphery and has contributed to the emergence of a digital divide and an increasing division between a fast world (about 15 percent of the world's population) and a slow world (about 85 percent of the world's population) with contrasting lifestyles and levels of living.**

  The leading edge of the fast world is the Internet, which is now the world's single most important mechanism for the transmission of scientific and academic knowledge. Today flows of goods, capital, and information that take place within and between transnational corporations are becoming more important than imports and exports between countries. At the same time, all these flows have helped to spread new values—from consumer lifestyle preferences to altruistic concerns with global resources, global environmental change, and famine relief—around the "fast" world.

KEY TERMS

cartography (p. 52)
colonialism (p. 55)
colonization (p. 47)
commodity chain (p. 69)
comparative advantage (p. 63)
core regions (p. 55)
division of labor (p. 63)
environmental determinism (p. 57)
ethnocentrism (p. 57)
external arena (p. 50)
fast world (p. 76)
hearth areas (p. 44)
hinterland (p. 50)
imperialism (p. 54)
import substitution (p. 51)
law of diminishing returns (p. 47)
map projections (p. 53)
masculinism (p. 57)
minisystem (p. 44)
neocolonialism (p. 68)
peripheral regions (p. 55)
plantation (p. 50)
producer services (p. 72)
semiperipheral regions (p. 55)
slash-and-burn (p. 44)
slow world (p. 76)
spatial justice (p. 76)
technology systems (p. 56)
transnational corporations (p. 68)
world-empire (p. 46)
world-system (p. 46)

ADDITIONAL READING

Appadurai, A., "Disjuncture and difference in the global cultural economy," in M. Featherstone (ed.). *Global Culture: Nationalism, Globalization, and Modernity.* London: Sage, 1990, pp. 295–310.

Barber, B. R., *Jihad vs. McWorld: How Globalism and Tribalism Are Reshaping the World.* New York: Ballantine Books, 1995.

Daniels, P., and W. F. Lever (eds.), *The Global Economy in Transition.* New York: Addison Wesley Longman, 1996.

Diamond, J., *Guns, Germs, and Steel.* New York: W. W. Norton, 1997.

Diamond, J., *Collapse: How Societies Choose to Fail or Succeed.* New York: Viking, 2005.

Harvey, D., *The New Imperialism.* Oxford: Oxford University Press, 2003.

Held, D., A. McGrew, D. Goldblatt, and J. Perraton, *Global Transformations.* Malden, MA: Blackwell, 1999.

Held, D. and A. McGrew (eds.), *The Global Transformations Reader.* Malden, MA: Blackwell, 2000.

Hopkins, A. G. (ed.). *Globalization in World History.* London: Pimlico, 2002.

Hughes, A. and S. Reimer (eds.), *Geographies of Commodity Chains.* New York: Routledge, 2004.

Hugill, P., *World Trade Since 1431.* Baltimore: Johns Hopkins University Press, 1993.

Hutton, W. and A. Giddens (eds.), *Global Capitalism.* New York: New Press, 2000.

Johnston, R. J., P. J. Taylor, and M. Watts (eds.), *Geographies of Global Change: Remapping the World.* Cambridge, MA: Blackwell, 2002.

Haywood, J., *Historical Atlas of the 19th Century World, 1783–1914.* New York: Barnes and Noble, 1998.

Knox, P. L., J. Agnew, and L. McCarthy, *The Geography of the World Economy,* 4th ed. New York: Routledge, 2003.

Landes, D. S., *The Wealth and Poverty of Nations.* New York: W. W. Norton, 1999.

Lechner, F. J., and J. Boli (eds.), *The Globalization Reader.* Malden: Blackwell, 2000.

Milanovic, B., *Worlds Apart: Measuring International and Global Inequality.* Princeton, NJ: Princeton University Press, 2005.

Mittelman, J. H., *The Globalization Syndrome: Transformation and Resistance.* Princeton, NJ: Princeton University Press, 2000.

O'Loughlin, J., L. Staeheli, and E. Greenburg (eds.), *Globalization and Its Outcomes.* New York: Guilford Press, 2004.

O'Meara, P., H. D. Mehlinger, and M. Krain (eds.), *Globalization and the Challenges of a New Century.* Bloomington: Indiana University Press, 2000.

Royal Geographical Society, *Atlas of Exploration.* New York: Oxford University Press, 1997.

Wallach, B., *Understanding the Cultural Landscape.* New York: Guilford Press, 2005. See especially Part II ("Historical Developments").

EXERCISES

On the Internet

The Internet exercises for this chapter take you back to the time of Alexander the Great and, through a critical-thinking exercise, help you to understand (through the actions of Alexander and the consequences for his empire) key geographic concepts such as core and periphery regions, the development of transportation, and how the Greek culture managed to diffuse throughout the known world. The concept-review exercises will help your understanding of concepts such as world-system, semiperipheral regions, slow world, and commodity chain. There is also a set of online map exercises that focus on industrialization, satellite images, transportation, trade, and more. You also have an opportunity to explore key terms in depth through Internet search engines that provide examples of how these terms are used.

Unplugged

1. The present-day core regions of the world-system, shown in Figure 2.18, are those that dominate trade, control the most advanced technologies, have high levels of productivity within diversified economies, and enjoy relatively high per capita incomes. What statistical evidence can you find in your local library to support this characterization? *Hint:* Look for data in annual compilations of statistics published in annual reviews such as the Encyclopaedia Britannica's *Yearbook* and in the annual reports of organizations such as the United Nations Development Programme (UNDP), the World Bank, and the World Resources Institute.

2. The idea of an international division of labor is based on the observation that different countries tend to specialize in the production or manufacture of particular commodities, goods, or services. In what product or products do the following countries specialize: Bolivia, Ghana, Guinea, Libya, Namibia, Peru, and Zambia? You will find the data you need in a good statistical yearbook, such as the Encyclopaedia Britannica's *Yearbook,* or in a world reference atlas or economic atlas.

GJ-18
T-656

3 Geographies of Population

Over the next five decades, the world's population is likely to reach nearly 10 billion people, if present rates of population growth continue. Of late, while there is some optimism that the rate of world population growth has peaked and is slowing, it is clear that the large absolute numbers of human beings on the planet will have long-term impacts on all aspects of human geography, especially the environment. For instance, in another 20 years, given current growth rates, experts expect that nearly a third of the world's population will be living under water-scarce or water-stressed conditions. Producing enough food from croplands and fisheries for the global population will also be difficult over the next two decades. And the amount of forest—critical to both planetary and human well-being because trees absorb carbon dioxide and produce oxygen, anchor soils, help regulate the water cycle, and provide habitat for millions of species—will also be dangerously limited. There is no doubt that environmental impacts would be less dramatic if population growth leveled off, or better yet, if there was no more growth at all. One of the biggest unknowns of the twenty-first century is how to to achieve a balance between a growing population and a vulnerable environment.

Population growth in the twentieth century was the result of the phenomenal decline in death rates. Birth and death are the two variables that shape overall population growth and change. In the twenty-first century, most experts agree that lowering birth rates will be key to harnessing population growth. For population geographers, however, knowing fertility, or birth figures, and mortality, or death figures, is not enough. They also want to know where different levels of fertility and mortality are occurring, why they are occurring, and what the consequences of these changes—environmental, economic, political or otherwise—are for the remaining population. In this chapter we examine population distribution and structure, as well as the dynamics of population growth and change, with a special focus on spatial variations and implications. In short, we want to know the locations of population clusters, the numbers of men and women and old and young, and the different ways in which these dynamics combine to create overall population growth or decline. We also look at population movements and the models and concepts that population experts have developed to understand better the potential problems posed by human populations.

MAIN POINTS

- Population geographers depend on a wide array of data sources to assess the geography of populations. Chief among these sources are censuses, while other sources include vital records and public health statistics.
- Population geographers are largely concerned with the same sorts of questions that other population experts study, but they also investigate "the why of where." *Why* do particular aspects of population growth and change (and problems) occur *where* they do, and what are the implications of these factors for the future of places?
- Two of the most important factors that make up population dynamics are birth and death. These variables may be examined in simple or complex ways, but in either case the reasons for the behavior of these variables are as important as the numbers themselves.
- A third crucial force in population change is the movement of populations. The forces that push populations from particular locations as well as those that pull them to move to new areas are key to understanding new settlement patterns. Population migration may not always be a matter of choice.
- Perhaps the most pressing question facing scholars, policymakers, and other interested individuals was articulated at the U.N. Millennium Summit in 2000: How can the global economy provide the world's growing population with enough food and safe drinking water, and a sustainable environment, so that all the world's people have the basic necessities for enjoying happy, healthy, and satisfying lives?

THE DEMOGRAPHER'S TOOLBOX

Demography, or the study of the characteristics of human populations, is an interdisciplinary undertaking. Geographers study population to understand the areal distribution of Earth's peoples. They are also interested in the reasons for, and the consequences of, the distribution of populations from the international to the local level. While historians study the evolution of demographic patterns, and sociologists study the social dynamics of human populations, geographers focus on the spatial patterns of human populations, the implications of such patterns, and the reasons for them. Using many of the same tools and methods of analysis as other population experts, geographers think of population in terms of the places that populations inhabit. They also consider populations in terms of the way that places are shaped by populations and in turn shape the populations that occupy them.

Censuses and Vital Records

Population experts rely on a wide array of instruments and institutions to carry out their work. Government entities, schools, and hospitals collect information on births, deaths, marriages, migration, and other aspects of population change. The most widely known instrument for assessing the state of the population is the census, a survey the ancient Romans developed to obtain information for tax collection.

At a simple level, a **census** is a straightforward count of the number of people in a country, region, or city. Censuses, however, are not usually so simple. Most are also directed at gathering other information about people, such as previous residences, marital status, and income.

Many of the developed countries of the Western Hemisphere comprehensively assess the characteristics of their national populations every 10 years. In the United States, for example, the Bureau of the Census has surveyed the population every 10 years since 1790. The information gathered is used to apportion seats in the U.S. House of Representatives, as well as to redistribute federal tax funds and other revenues to states, counties, and cities.

In addition to the census, population experts employ other data sources to assess population characteristics. One such source is **vital records,** which report births, deaths, marriages, divorces, and the incidence of certain infectious diseases (**Figure 3.1**). These data are collected, and records of them are kept by local, county, state, and other levels of government. Schools, hospitals, police departments, prisons, other public agencies such as the Immigration and Naturalization Service in the United States, and international organizations like the World Health Organization also collect demographic statistics that are useful to population experts.

Limitations of the Census

Census enumerations are an extremely expensive and labor-intensive undertaking for any governmental jurisdiction, and as a result, they occur rather infrequently—usually no more than every five or 10 years. Historically, the United States has undertaken a population census every 10 years, but in 1985 it introduced a quinquennial (every five years) census to augment the decennial (every 10 years) one. Most prominent among the reasons for initiating an additional, quinquennial, census was that data collected every 10 years quickly became obsolete.

In many peripheral and semiperipheral countries, governments are not always able to finance a decennial census such as the kinds of comprehensive surveys undertaken in more developed countries like France or Germany. The small country of Andorra in southwestern Europe conducted its first comprehensive census in 1948; the next in 1954; followed by its last one in 1975, over 30 years ago! In southeast Asia, Cambodia conducted its first complete census in 1962 and did not conduct another until 1998. Off the east coast of Africa, the semi-autonomous island of Zanzibar (loosely part of the United Republic of Tanzania) has not conducted a census since 1958. The incompatibility of enumeration dates makes comparisons among, between, and within countries quite difficult. For example, the United States conducted a decennial census on April 1, 2000. China conducted a comparable decennial census on November 1, 2000, its fifth nationwide census. However, the seven-month difference in collection times between the two censuses makes comparisons difficult. While demographers compare the populations of countries all the time, they can do so only in very general terms because processes such as international migration and the difference caused by gaps between census collection dates can have a significant impact on the actual numbers.

Self-identification of ethnicity or race by indicators such as chief language spoken at home or skin color mask any number of other, often more meaningful, cultural attachments. A problematic category in the last several U.S. censuses has been the word "Hispanic," a term used to refer to residents whose cultural heritage derives from a Spanish-speaking country. Hispanics have in common ancestral ties to Latin America or Spain, but this apparent commonality obscures a great deal of cultural differentiation because a self-identified Hispanic can be Mexican American, Puerto Rican, Cuban American, Spanish, or someone from all but a couple of the South or Central American countries.

Possibly even more complicated is the new "multiracial" category available on the 2000 census. The category is intended to allow individuals who feel they do not truly belong to one racial category—because their parents or grandparents were not from the same racial category—not to have to choose between White and Asian, for instance. The appearance of the category on the census form reflects the complex, polyglot world of U.S. de-

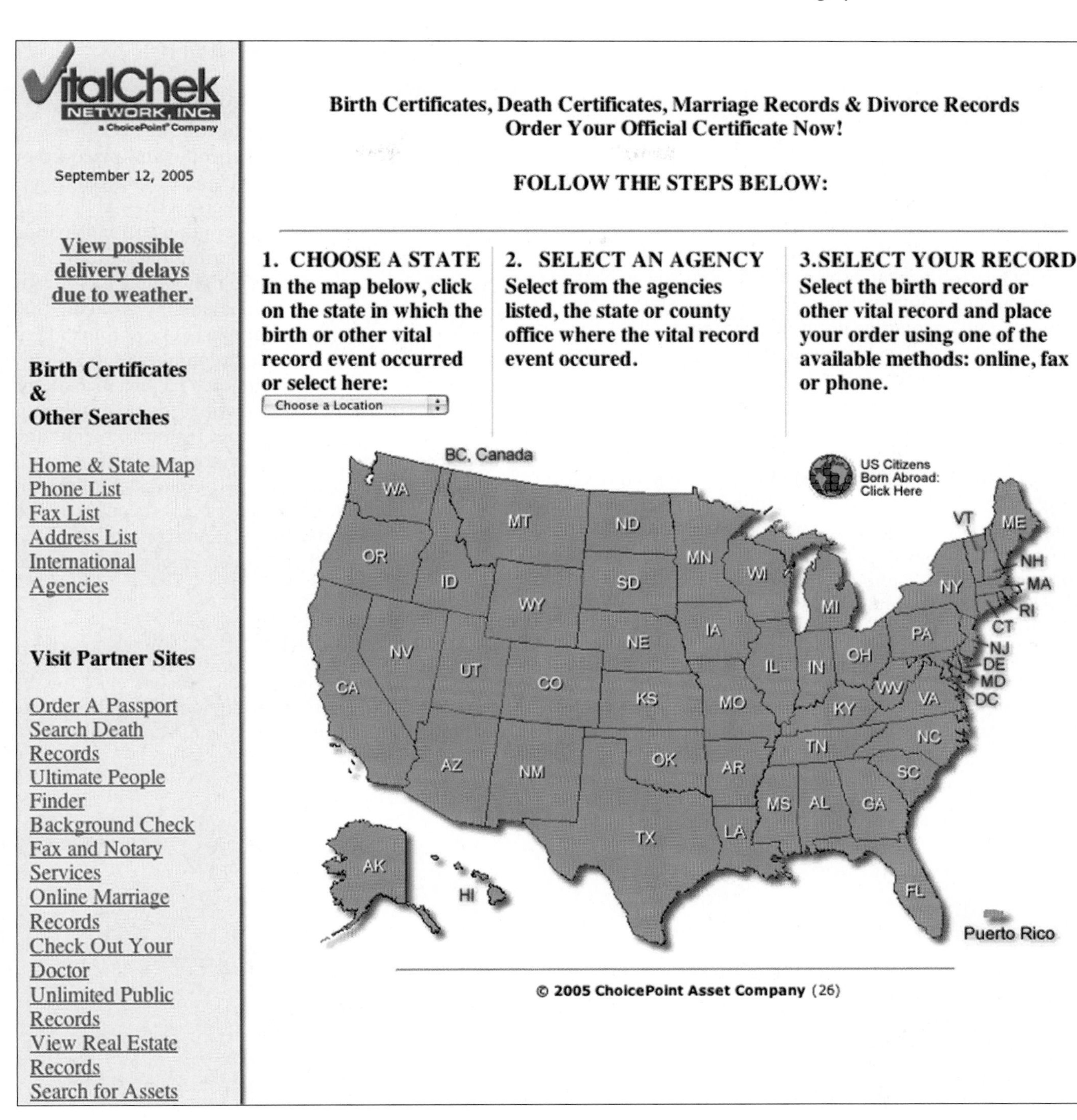

Figure 3.1 Vital Records on the Web Shown here is one of the many Web sites that make U.S. vital records more readily accessible to the public. While some sites are free and may be sponsored by state governments, others charge for the service of providing vital information. The sites are especially popular with genealogists.

mographics. But it is also a confusing category, and few people actually chose it. It is not even clear that those who did choose it really understood the Census Bureau's racial categories.

The bottom line is that no census is entirely comprehensive (or always comprehensible). All censuses tend to underrepresent non-mainstream kinds of households as well as homeless individuals. For instance, more than 8 million people were not counted in the U.S. 1990 census, most of whom were Hispanics or African Americans. As a result of the 1990 undercount, many municipalities complained to the U.S. Bureau of the Census. The city of Inglewood in metropolitan Los Angeles estimated that 13,400 individuals were not found by the enumerators, leading to an undercount of 10.9 percent, the largest of any city in the country. Because federal revenue-sharing formulas as well as the apportionment of U.S. House seats are tied to population numbers, urban political officials from many of America's large and medium-size cities complained extensively about the numbers generated from the

1990 census. The designers of the 2000 census suggested a correction for the 1990 undercounting problem by proposing a sampling plan. The aim of the plan was to sample statistically for overcounted as well as undercounted individuals so that minority residents would be acknowledged and counted. Thus, although every individual would not actually be counted through the submission of a census form, all would be accounted for by statistically manipulating a sample of the counted population. In a case brought before a federal district court by Republican Newt Gingrich, the sampling plan was defeated, and it was believed that undercounting occurred again in the 2000 census. After several reevaluations of the data, however, the Census Bureau determined that, in fact, there was not an undercount of the total population but rather a net overcount of 1.3 million persons. When looking at specific populations the Census Bureau found that not only had Non-Hispanic Whites been overcounted, so had children between the ages of 0–9. It was actually Native Hawaiian and Other Pacific Islanders and American Indians and Alaskan Natives who had been underercounted. The counting of the U.S. population has a real impact on peoples' lives. Since billions of dollars in federal grants for education, transportation, and health care are distributed on the basis of population, undercounted communities can lose significant financial support for local schools, roads, and health-care services, while overcounted communities can reap more grants than warranted.

POPULATION DISTRIBUTION AND STRUCTURE

Because human geographers explore the interrelationships and interdependencies between people and places, they are interested in demography. Population geographers bring to demography a special perspective—the spatial perspective—that emphasizes description and explanation of the "where" of population distribution, patterns, and processes. For instance, the seemingly simple fact that as of early 2005 the world was inhabited by 6 billion, 500 million people is one that geographers like to think about in a more complex way. While this number is undeniably phenomenal and increasing with each passing second, the most important aspect of this number for geographers is its uneven spatial expression from region to region and from place to place. Equally important are the implications and impacts of these differences. Looking at population numbers, geographers ask themselves two questions: Where are these populations concentrated, and what are the causes and consequences of such a population distribution?

Population Distribution

At a basic level, many geographic reasons exist for the distribution of populations throughout the globe. As the world population density map demonstrates (**Figure 3.2**),

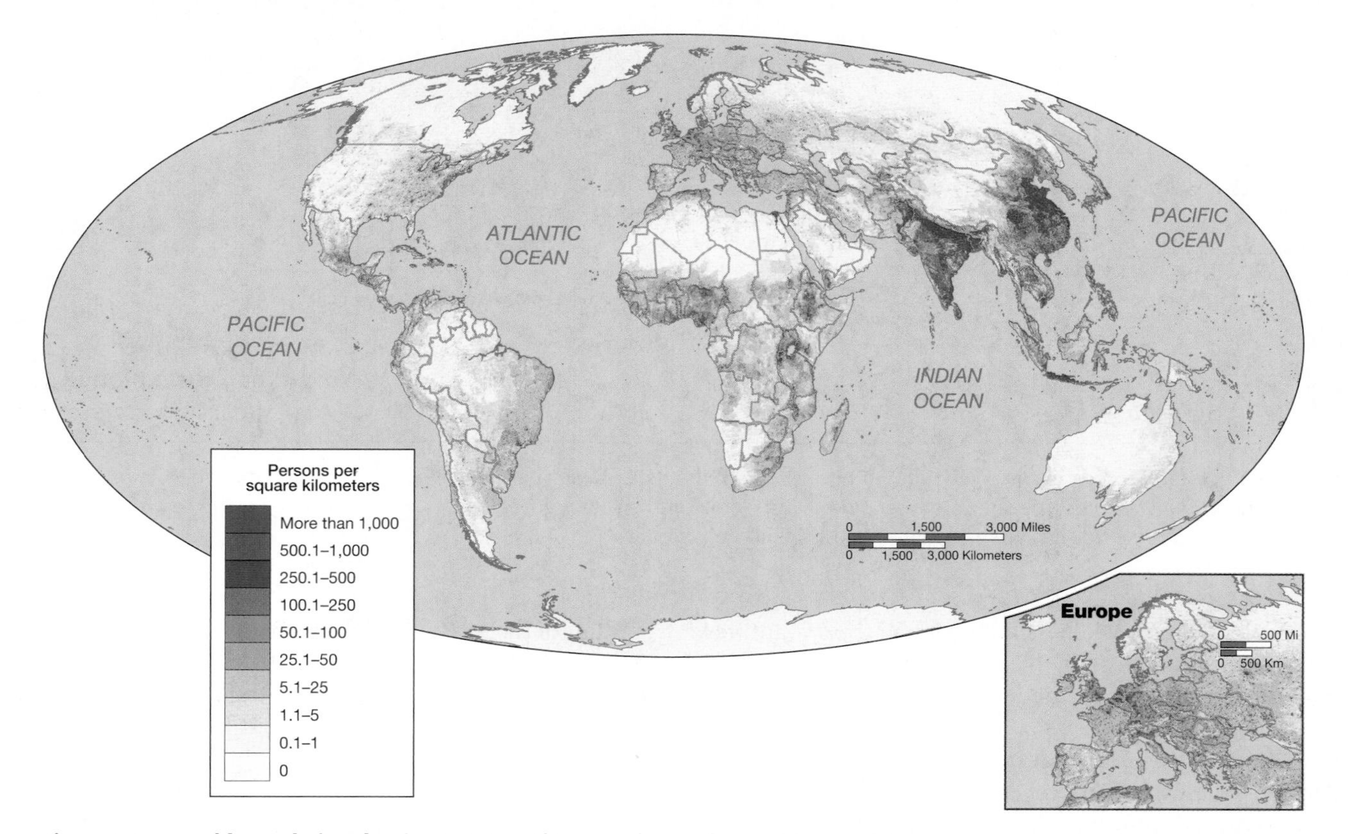

Figure 3.2 World population density, 2004 As this map shows, the world's population is not uniformly distributed across the globe. Such maps are useful in understanding the relationships between population distribution and the national contexts within which they occur. China and India are the largest countries in the world with respect to population, while the core countries have much smaller populations. (After World Bank, *World Development Indicators 2004,* World Bank: Washington, DC.)

some areas of the world are very heavily inhabited, others only sparsely. Some areas contain no people whatsoever. Overall, Bangladesh and the Netherlands, for example, have high population densities. India, however, displays a different pattern, with especially high population concentrations along the coasts and rivers but a relatively low population density in its western sector. The sparsest inland population concentrations in the world occur in the densely forested frontier regions of the Amazon. Thus, environmental and physical factors are important influences on population distributions and concentrations.

Degree of accessibility, topography, soil fertility, climate and weather, water availability and quality, and type and availability of other natural resources are some of the factors that shape population distribution. Other factors are also crucial—first and foremost are a country's political and economic experiences and characteristics. For example, the high population concentrations along Brazil's Atlantic coast date back to the trade patterns set up during Portuguese colonial control in the sixteenth and seventeenth centuries. Another important factor is culture as expressed in religion, tradition, or historical experience. Cities like Medina and Mecca, in the Middle East, comprise important population concentrations in no small part because they are Islamic sacred sites.

Table 3.1 shows population estimates in terms of traditional continental distributions. From the table it is clear that Asia is far and away the most populous continent, including 61 percent of the world's inhabitants in more than 40 countries. In Asia, China and India alone constitute 37 percent of the continental population total and 21 percent of the world total. Running a distant second and third are Africa, with 13 percent of the world's population, and Europe, with 12 percent.

The population clusters that take shape have a number of physical similarities. Almost all of the world's inhabitants live on 10 percent of the land. Most live near the edges of land masses, near the oceans or seas, or along rivers with easy access to a navigable waterway. Approximately 90 percent live north of the equator, where the largest proportion of the total land area (63 percent) is located. Finally, most of the world's population lives in temperate, low-lying areas with fertile soils.

Population numbers are significant not only on a global scale but also at other levels. Population concentrations within countries, regions, and even metropolitan areas are also important. For example, much of the population of North Africa is distributed along the coastal areas where most of the large cities are. Egypt, however, is an exception. While it does possess a significant coastal population, a large part of its population is also distributed narrowly in a line running through the interior of the country along the banks of the Nile River, which marks the essential importance of water to this desert population (**Figure 3.3**).

Population Density and Composition

Another way to explore population is in terms of density, a numerical measure of the relationship between the number of people and some other unit of interest expressed as a ratio. For example, crude density is probably the most common measurement of population density. **Crude density,** also called **arithmetic density,** is the total number of people divided by the total land area. The metropolitan area of Dallas–Fort Worth, a classic low-density urban settlement predicated on the widespread use of automobiles and among the most populous cities in the United States, has a population density of approximately 2,187 persons per square kilometer (3,500 per square mile; **Figure 3.4**). By comparison, the city of Chicago, a classic high-density urban settlement predicated on the use of mass transit, has a population density of 7,969 persons per square kilometer (12,750 persons per square mile). New York, the most densely populated city in the U.S. has a density of 23,671 persons per square kilometer (24,000 persons per square mile; **Figure 3.5**).

The limitation of the crude density ratio—and hence the reason for its "crudeness"—is that it is one-dimensional. It tells us very little about the opportunities and obstacles that the relationship between people and land contains. For that, we need other tools for exploring population density, such as nutritional density or agricultur-

TABLE 3.1 World Population Estimates by Continents

Continent	Number of Inhabitants (in millions)	% of Total Population
Africa	885	14
Asia	3875	60.5
Oceania	33	0.5
Europe	728	11.5
North America	326	5
Latin America and Caribbean	549	8.5
TOTAL	6396	100

Source: Population Reference Bureau Web site, www.prb.org. *World Population Data Sheet, 2004.*

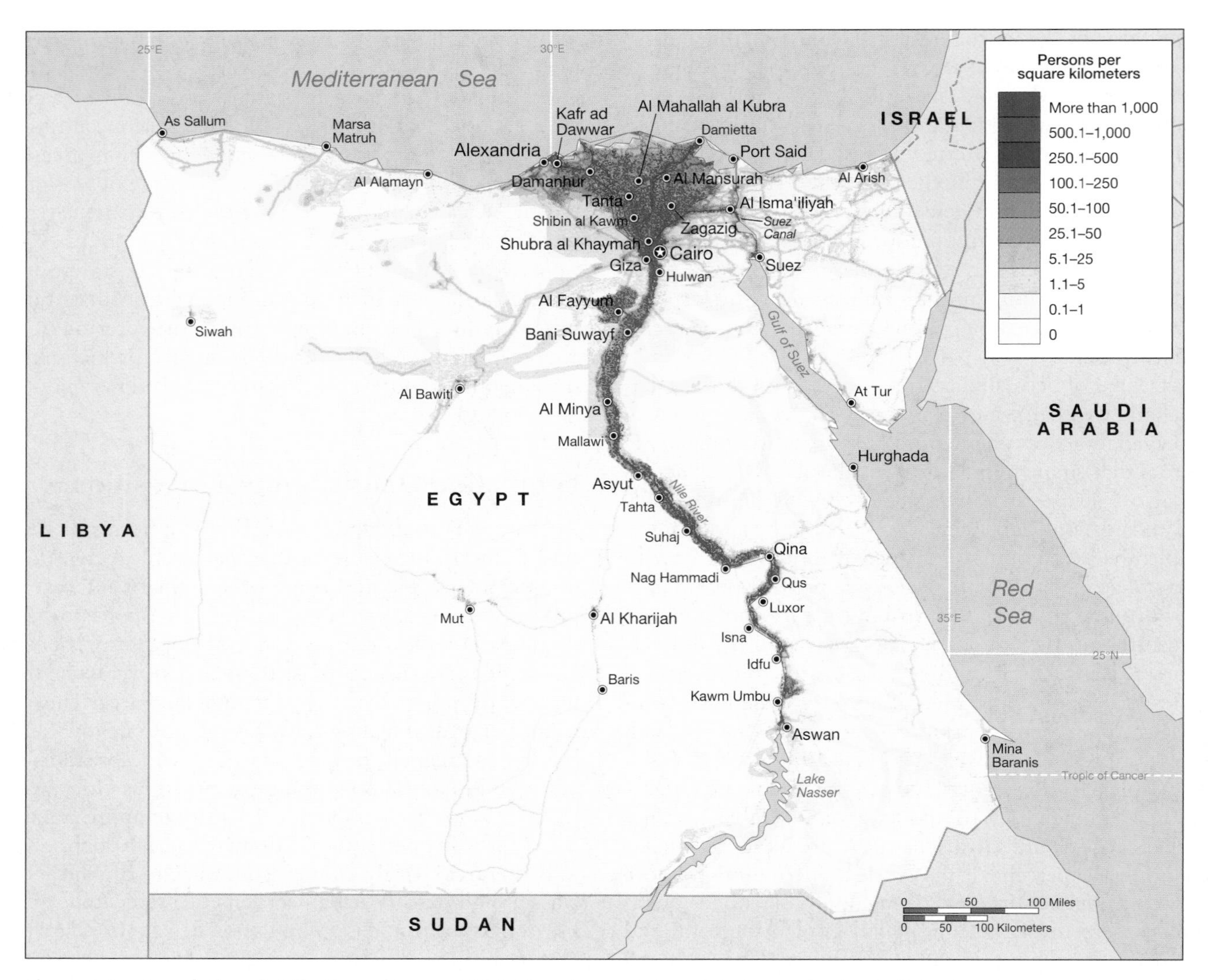

Figure 3.3 Population distribution of Egypt, 2004 Egypt's population distribution is closely linked to the proximity of water. In the north, the population clusters along the Mediterranean and in the interior, along the banks of the Nile River. (*Source:* Gridded Population of the World [GPW], Version 2. Palisades, NY: CIESIN, Columbia University. Available at http://sedac.ciesen.org/plu/gpw/index.html?main.html@2)

Figure 3.4 Population density, Dallas–Fort Worth The urban landscape of metropolitan Dallas–Fort Worth is low profile and widely spread. Few if any physical boundaries exist to inhibit growth in Dallas–Fort Worth. More important, however, is that land is relatively cheap—in part because it is more abundant, but also because it is not as important in the world system as New York City. Dallas–Fort Worth is a regional urban hub, whereas New York City's importance in the world economic system is national and international.

Figure 3.5 Population density, New York City Urban form in New York City is the result of many factors. Most important in terms of its population size is its role as a central—if not *the* central—node in the world system of cities. Such a prominent role is built upon both a varied and extensive population base. The urban economy is complex, requiring a wide range of labor skills from professional and managerial to low-skilled service-oriented workers.

al density. **Nutritional density** is the ratio between the total population and the amount of land under cultivation in a given unit of area. **Agricultural density** is the ratio between the number of agriculturists—people earning their living or subsistence from working the land—per unit of farmable land in a specific area. **Figure 3.6** is a map of the ratio of population to the number of physicians, an indicator of the quality of a country's health care.

In addition to exploring patterns of distribution and density, population geographers also examine population in terms of its composition—that is, the subgroups that constitute it. Understanding population composition enables geographers to gather important information about population dynamics. For example, knowing the composition of a population in terms of the total number of males and females, number and proportion of senior

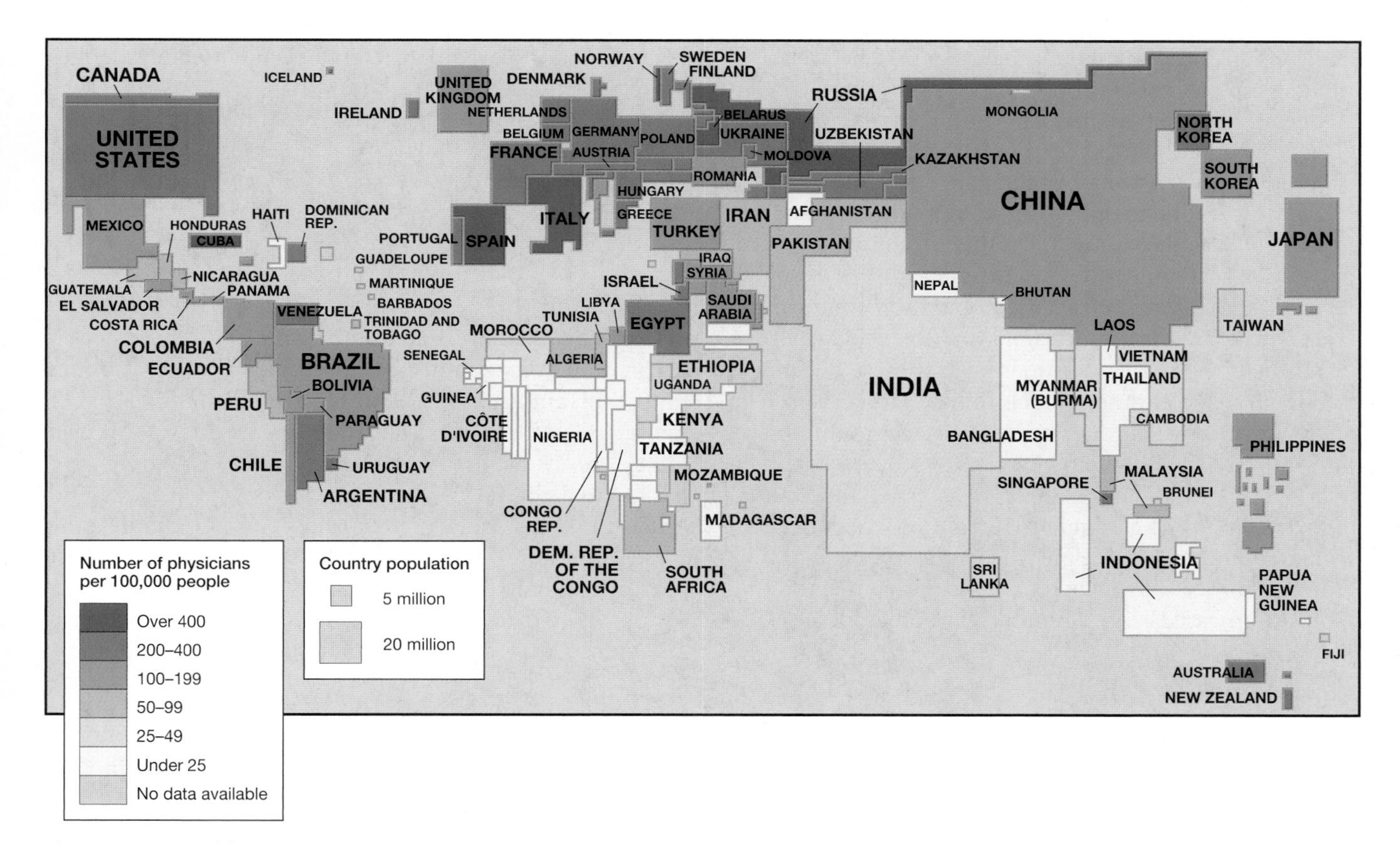

Figure 3.6 Health-care density Another measure of population density is reflected in this map, which shows the number of people per physician in the total population. The core countries and China have the highest ratio of doctors to overall population. Most of the continent of Africa, except South Africa and Egypt, Libya, and Morocco in North Africa, has the lowest ratio, reflecting another dimension of core-periphery inequality. (*Source:* H. Veregin (ed.), *Goode's World Atlas,* 21st ed., Rand McNally & Co, 2005, p. 37.)

citizens and children, and number and proportion of people active in the workforce provides valuable insights into the ways in which the population behaves now and how it might behave in the future.

Facing unique challenges are countries with a population that contains a high proportion of old people—a situation most core countries will soon be facing as their "baby boom" generation ages. The **baby boom** generation includes those individuals born between 1946 and 1964. A considerable amount of a country's resources and energies will be necessary to meet the needs of a large number of people who may no longer be contributing in any significant fashion to the creation of the wealth necessary for their maintenance. There might also be a need to import workers to supplement the relatively small working-age population.

Knowing the number of women of childbearing age in a population, along with other information about their status and opportunities, can provide valuable information about the future growth potential of that population. For example, populations in core countries like Denmark, which has a small number of women of childbearing age relative to the total population size, but women with high levels of education, socioeconomic security, and wide opportunities for work outside the home, will generally grow very slowly if at all. Populations in peripheral countries like Kenya, on the other hand, where a large number of women of childbearing age have low levels of education and socioeconomic security and relatively few employment opportunities, will continue to experience relatively high rates of population growth, barring unforeseen changes. The variety that exists within a country's population very much shapes the opportunities and challenges it must confront nationally, regionally, and locally.

Understanding population composition, then, not only can tell us much about the future demographics of regions but is also quite useful in the present. For example, businesses in core countries use population composition data to make marketing decisions and to decide where to locate their operations. For many years these businesses used laborious computer models to help target their markets. With the recent development of geographic information systems (GIS), however, this process has been greatly simplified. Some examples of assessing the location and composition of particular populations, known as **geodemographic analysis,** are provided in Box 3.1: "GIS Marketing Applications," which discusses private marketing applications as well as community-oriented ones.

Age-Sex Pyramids

Areal distributions are not the only way that demographers portray population distributions. To display variations within particular subgroups of a population or with respect to certain of its descriptive aspects, such as births or deaths, demographers also often use bar graphs displayed both horizontally and vertically. The most common way for demographers to graphically represent the composition of the population is to construct an **age-sex pyramid,** which is a representation of the population based on its composition according to age and sex. An age-sex pyramid is actually a bar graph displayed horizontally. Ordinarily, males are portrayed on the left side of the vertical axis and females on the right. Age categories are ordered sequentially from the youngest at the bottom of the pyramid to the oldest at the top. By moving up or down the pyramid, one can compare the opposing horizontal bars in order to assess differences in frequencies for each age group. Age-sex pyramids also allow demographers to identify changes in the age and sex composition of populations. For example, an age-sex pyramid depicting Germany's 2000 population clearly revealed the impact of the two world wars, especially with regard to the loss of large numbers of males of military age and the deficit of births during those periods (**Figure 3.7**). Demographers call population groups like these cohorts. A **cohort** is a group of individuals who share a common *temporal* demographic experience. A cohort is not necessarily based only on age, however. Cohorts may be defined based on criteria such as time of marriage or time of graduation.

In addition to revealing the demographic implications of war or other significant events, age-sex pyramids can provide information necessary to assess the potential impacts that growing or declining populations might have. As illustrated in **Figure 3.8**, the shape of an age-sex pyramid varies depending on the proportion of people in each age cohort. The pyramid for Mali, for example, shown in Figure 3.8a, reveals that many dependent children, ages 0 to 14, exist relative to the rest of the population. The considerable narrowing of the pyramid toward the top indicates that the population has been growing very rapidly in recent years. The shape of this pyramid is typical of peripheral countries with high birth rates and low death rates.

Serious implications are associated with this type of pyramid, however. First, in the absence of high productivity and wealth, resources are increasingly stretched to their limit to accommodate even elemental schooling, nutrition, and health care for the growing number of dependent children. Furthermore, when these children reach working age, a large number of jobs will have to be created to enable them to support themselves and their families. Also, as they form their own families, the sheer number of women of childbearing age will almost guarantee that the population explosion will continue. This will be true unless strong measures are taken, such as intensive birth-control campaigns, improved education and outside opportunities for women, and modifications of cultural norms that place a high value on large family size.

In contrast, the pyramid for the United States (Figure 3.8b) illustrates the typical shape for a country experiencing a slow rate of growth. In fact, the population would not be growing much at all if it were not for substantial immigration from other countries. Note that a perceptible bulge exists in this pyramid, representing the

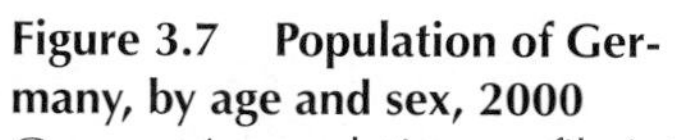

Figure 3.7 Population of Germany, by age and sex, 2000 Germany's population profile is that of a wealthy core country that has passed through the postwar baby boom and currently possesses a low birth rate. It is also the profile of a country whose population has experienced the ravages of two world wars. (*Source:* J. McFalls, Jr., "Population: A Lively Introduction," 4th ed., *Population Bulletin, 58*(4), 2003, p. 27.)

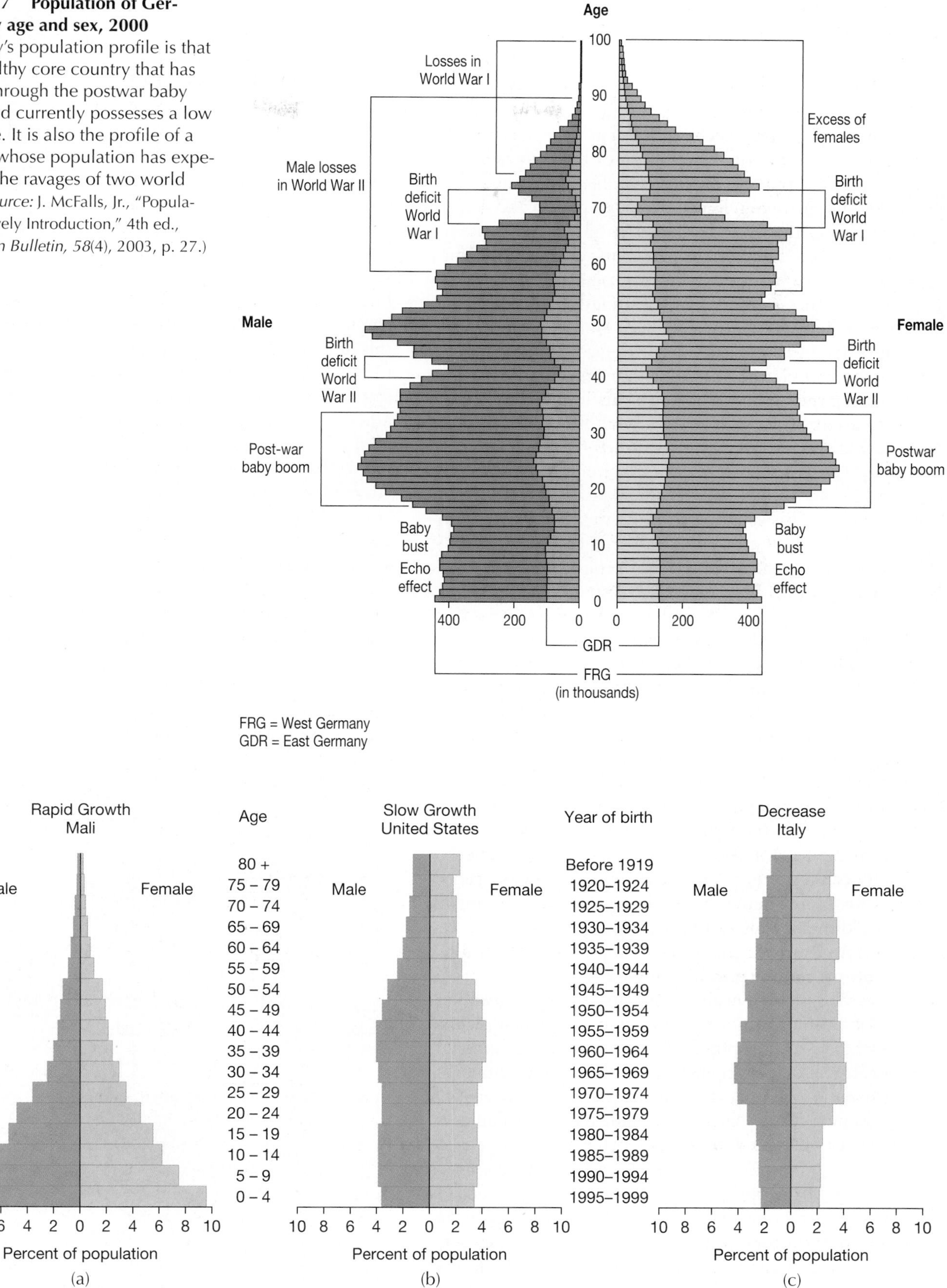

Figure 3.8 Population pyramids of rapid, slow, and zero population growth (ZPG), 2000 Population pyramids vary with the age and gender structure of the population being depicted. We can derive important information about the population growth rates of different countries over time by analyzing changes in the numbers of people in each sex and age category. (After J. McFalls, Jr., "Population: A Lively Introduction," 4th ed., *Population Bulletin, 58*(4), 2003, p. 24.)

GIS Marketing Applications

Geographic information systems, or GIS, possess the powerful ability to reveal spatial patterns in demographic data such as the occurrence of crime and the distribution of air pollution or through economic or political practices such as retail location analysis or land-use planning. Recognizing the potential commercial applications, marketing firms and in-house site-location departments have readily incorporated GIS into their daily operations. Prior to using GIS for marketing, businesses made crude estimates of the location and behavior of their clientele. By using GIS to analyze demographic data, marketing companies are now able to provide businesses with maps identifying areas of potential customers. With this information, businesses can engage in promotional "sales blitzes," targeting specific customers identified from GIS analysis.

The key to using GIS effectively in marketing is the ability to link demographic data to particular locations. This is known as *georeferencing*. Demographic data are linked to ZIP code areas and telephone prefix zones. Then the data are analyzed to determine whether meaningful spatial relationships exist among specified areal units, based on given demographic data. For example, the residences of people with a particular level of education may cluster around a few contiguous ZIP code regions.

Demographic data such as income level, education level, home value, age, and number of household residents provide marketing firms with information that can be linked to certain consumer goods. For example, people who have high income and high education levels tend to own luxury cars. Luxury-car dealerships would want to target such households. A GIS could produce a map that would show the general location of such households, and advertisements could then be distributed throughout this area.

An example of businesses analyzing demographic data with a GIS are shopping-center developers, who use the system to evaluate mall location opportunities. In addition to the demographic data—which include variables such as population density, income, and education—developer databases also often contain information about the characteristics of competing shopping centers, as well as tenant and ownership data.

Another marketing area in which GIS is used is locational analysis—especially determining where to locate a business. The success of businesses such as gasoline stations and fast-food restaurants depends on locational factors such as traffic density. GIS is also aptly suited for modeling traffic patterns, which are frequently determined by studying demographic data that indicate how far, how often, and in which direction people travel to work. Many GIS systems have models that can be used to predict traffic patterns. By identifying areas with high traffic volumes, businesses can choose strategic locations. Other conditions, such as zoning restrictions and proximity to competition, can also be factored into a GIS to reduce the number of potential retail sites.

Figure 3.A shows how Starbucks Coffee Co. uses GIS to assess different geographic scales of the coffee consumption market through a visual analysis of walk times to different trade areas that have high volumes of pedestrian traffic. This kind of analysis—based on sales forecasting data—helps retail outlets like Starbucks to explore the potential of specific sites so that sound site-selection decisions can be made. Maps like this help site selection teams at retail and food outlets to lower the risk of locating a store by improving the chances of catering to a responsive customer base. In addition to incorporating existing government-generated data—such as U.S. census data; 3- and 5-digit ZIP codes; highway data; and state, county, and city street maps—these maps also contain "store intercept studies," which are data generated by the company based on sales and customer characteristics. These sorts of studies have encouraged Starbucks Coffee Co. to move beyond its traditional affluent customer base into more demographically diverse areas, such as Harlem in New York City, and be successful there.

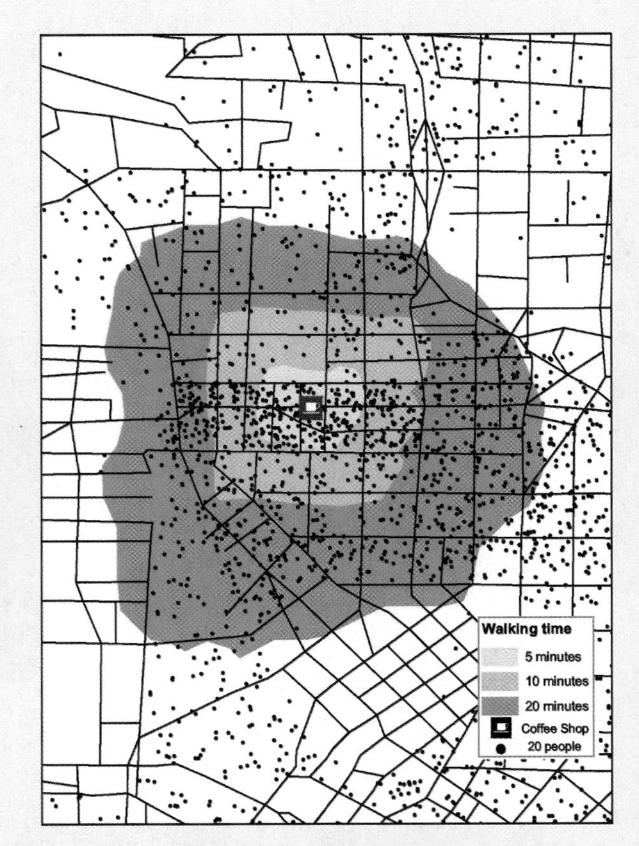

Figure 3.A Walk times to different trade areas Maps like this help Starbucks to understand the catchment area from which an individual outlet is likely to draw its customers.

It should also be pointed out, however, that the potential for invasion of privacy through the coupling of different sorts of data sets is very real. For instance, when you shop in U.S. grocery stores with electronic inventory systems and pay for your groceries by credit card or check, a record of your purchases can now be joined with your name. Such a coupling allows marketers to target you based on information on sales and manufacturers' coupons. It also, potentially, gives them access to information about what prescription drugs you are taking, how much alcohol you consume, what sorts of magazines you buy, or which videos you rent. Thus, while GIS and the massive data sets that support it open up new opportunities for marketing, the possibility also exists for the erosion of privacy.

It is also possible to use GIS, however, for all sorts of non-commercial sectors including government as well as private citizens. For instance, public participation GIS, also known as PPGIS, is a way that communities, especially poor and minority communities, can use the technology to offer their own solutions to local issues and problems from crime to conservation. Two Georgia planners, for example, have employed a PPGIS model to help Atlanta's inner-city poor neighborhoods—among the poorest in the country—by teaching them how to use GIS in order to produce knowledge that will help them make more informed decisions. One of the ways that professional planners helped poor Atlanta neighborhoods was through a project aimed at addressing the decline of housing conditions. Through the efforts of the planners and some volunteers, selected neighborhood residents were trained to identify building code violations. Next, the residents were taught how to enter the code violation data into a GIS program so that maps showing the distribution of the violations could be generated. The maps were used to persuade city officials to take action to improve code enforcement.

A second PPGIS project undertaken in Atlanta, known as the Atlanta Project (TAP), focused on improving access to early childhood development programs. The goal of the project was to lobby state and federal agencies for more classrooms in TAP's poorest neighborhoods. **Figure 3.B** is one of the maps produced by the neighborhood-university partnership in 1998. The map shows where the need for pre-kindergarten education programs were the greatest. The purpose of introducing GIS technology was to help residents and community organizations establish the dimension of their needs as well as to provide them with a tool to approach government agencies for a remedy. The efforts were successful in that the majority of TAP applications for pre-kindergarten programs were funded for some of the most needy neighborhoods in the city.

Source: This feature was adapted from J. Francica, "Location Analysis Tools Help Starbucks Brew Up New Ideas," *Business Geographics*, September/October, pp. 12–15, 2000; and D. S. Sawici, and P. Burke, "The Atlanta Project: Reflections on PPGIS Practice," in *Community Participation and Geographic Information Systems,* W. J. Craig, T. M. Harris and D. Weiner (eds.), pp. 89–100, 2002.

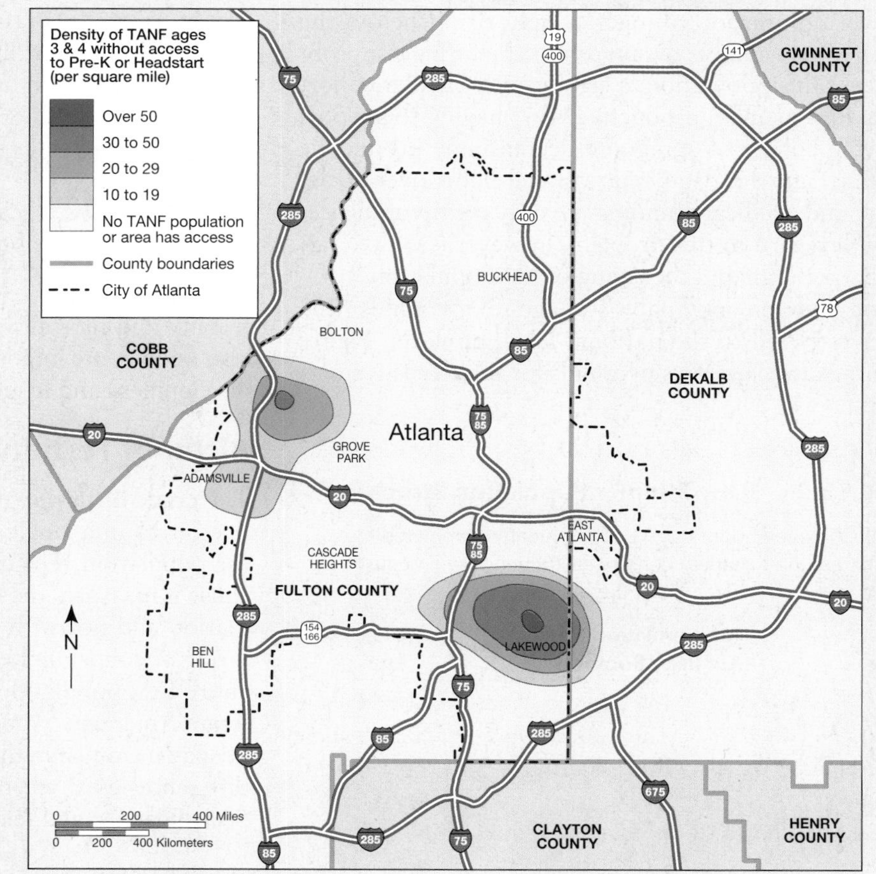

Figure 3.B Concentrations of Atlanta children ages 3 and 4 without access to Head Start and/or pre-kindergarten, 1998 Maps produced by communities can help them make the point to government agencies that much-needed resources are deficient. This map shows the areas of greatest need for educational resources for young children receiving Temporary Assistance for Needy Families (TANF). (*Source:* D. S. Sawicki and P. Burke, "The Atlanta Project: Reflections on PPGIS practice," in W. J. Craig, T. M. Harris & D. Weiner (eds.), *Community Participation and GIS.* London & NY: Taylor and Francis, p. 97.)

baby boom cohort. The period immediately after the Second World War was one of tremendous economic growth as well as political and technological change, as the United States took its place as the leading core country. With a great deal of prosperity and a rosy future, returning Americans reproduced at a faster rate than at any other time in our demographic history.

While Italy has a pyramid somewhat similar to that of the United States (Figure 3.8c), one key difference exists: Where the United States has a slow rate of growth, Italy has a zero population growth rate. Thus, in Italy, where the birth rate is below replacement level, the pyramid is very columnar, hardly a pyramid at all. People are equally distributed among the cohorts, though the base is perceptibly narrower. Italy—as well as many other European countries, such as Denmark, Spain, and Germany—can expect many of the same demographic challenges, including the extremely large elderly population faced by the United States, in not too many years. In all these countries, however, high levels of production and wealth, combined with low birth rates, translate into a generally greater capacity to provide not only high levels of health, education, and nutrition but also jobs, as those children grow up and join the workforce. Whether these opportunities are equitably distributed among individual members of the population remains to be seen.

Table 3.2 and **Figure 3.9** show the present state and potential future impact of the baby boom cohort, the largest population cohort in U.S. demographic history. Figure 3.9 provides a series of pyramids that illustrate how the configuration changes as the boomers age. The narrower column of younger people rising below the boomer cohort in these pyramids reveals the biggest problem facing this population: a significantly smaller cohort moving into its main productive years having to support a growing cohort of aging and decreasingly productive boomers (**Figure 3.10**). Congressional fights over Social Security and Medicare funding are only the tip of the iceberg with regard to this problem, however (see Box 3.2: "The Baby Boom and the Aging of the Population").

Construction and analysis of age-sex pyramids need not be restricted to the national level, however. **Figure 3.11**, for example, shows pyramids for three census tracts within the same city. Here, at the local scale, variations exist in the geography of population composition that are as informative as those constructed at the national scale. The proportional representation of each cohort, by age and sex, raises provocative questions about why such spatialized differences exist and what the implications of the distributions might be for social services, marketing strategies, and a host of other applications. For example, a census tract with a large number of children may lead decision makers to push for provision of greater opportunities for organized sports and perhaps a children's health-care facility. In contrast, a growing concentration of older people might prompt the opening of a senior citizen center and a geriatric-care facility.

A critical aspect of the population pyramid is the **dependency ratio,** which is a measure of the economic impact of the young and old on the more economically productive members of the population. Traditionally, in order to assess this relation of dependency in a particular population, demographers will divide the total population into three age cohorts, sometimes further dividing those cohorts by sex. The **youth cohort** consists of those members of the population who are less than 15 years of age and generally considered to be too young to be fully active in the labor force. The **middle cohort** consists of those members of the population aged 15 to 64, who are considered economically active and productive. Finally, the **old-age cohort** consists of those members of the population aged 65 and older, who are considered beyond their economically active and productive years. By dividing the population into these three groups, it is possible to obtain a measure of the dependence of the young and old upon the economically active, and the impact of the dependent population upon the independent.

TABLE 3.2 Baby Boomer Population Structure

While the baby boom has demographically dominated the last half of the twentieth century, its influence will begin to wane in the early decades of the new one.

Year	% of Population Who Are Baby Boomers	Age
1990	30	25–44
2000	20	35–59
2020	15	55–79
2040	7	75–85+

POPULATION DYNAMICS AND PROCESSES

In order to evaluate a different understanding of population growth and change, experts look at two significant factors: fertility and mortality. Birth and death rates, as they are also known, are important indicators of a region's level of development and its place within the world economy.

Birth, or Fertility, Rates

The **crude birth rate** (**CBR**) is the ratio of the number of live births in a single year for every thousand people in the population. The crude birth rate is indeed crude, because it measures the birth rate in terms of the total population and not with respect to a particular age-specific group or cohort. For example, as of 2004, the CBR of the entire U.S. population was 14.0, while the birth rate for Asian American women aged 20 to 24 was 72.0 and for Hispanic women in the same age category 184.6. Clearly, differences exist when we look at specific groups and especially at age and sex cohorts at their reproductive peak.

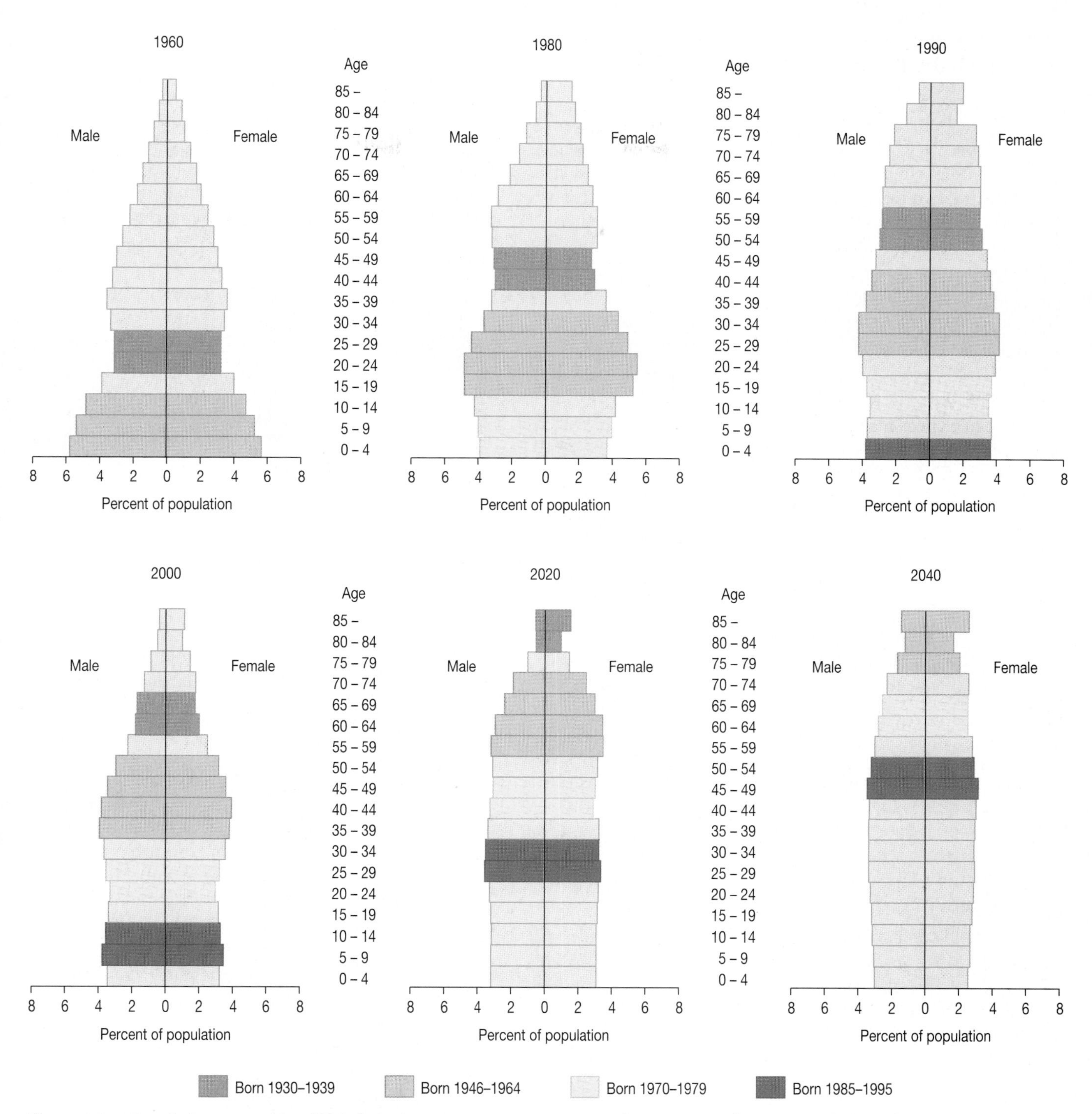

Figure 3.9 Population pyramids of U.S. baby boomers, 1960–2040 Population pyramids are not only based on past or present statistics but may also be constructed on future estimates. In this series of pyramids, we can clearly see the progression of the baby boomers, shown in blue, up the pyramid as their cohort ages. Note how the "pyramid" becomes a column by the year 2040 as birth rates remain below death rates for each cohort. Note also the significantly higher number of women compared with men in the oldest age group for this same year, reflecting the statistical tendency for women to live longer than men. (After L. Bouvier and C. DeVita, "The Baby Boom: Entering MidLife," *Population Bulletin, 46*(3), 1991.)

Although the level of economic development is a very important factor shaping the CBR, other, often equally important influences also affect CBR. In particular, it may be heavily affected by the demographic structure of the population, as graphically suggested by age-sex pyramids. In addition, an area's CBR is influenced by women's educational achievement, religion, social customs, and diet and health, as well as politics and civil unrest. Most demographers also believe that the availability of birth-control methods is critically important to a country's or region's birth rate (**Figure 3.12**). A world map of the CBR (**Figure 3.13**) shows high levels of fertility in most of the

The Baby Boom and the Aging of the Population

The baby boom generation—those individuals born between 1946 and 1964 in the United States—has been described as "one of the most powerful and enduring demographic influences on this nation." It has certainly been one of the most studied, analyzed, vilified, and sanctified. The case of the baby boom generation provides a useful way of understanding the complex of factors that shape demographic change.

Why so many births between 1946 and 1964? Most demographers cannot give a definitive answer to this question. In fact, in the 1940s, population experts were predicting that the American population would stop growing. And although increases in births following a war are expected, the extended surge in births that followed the end of the Second World War came as a surprise to population experts, policymakers, government officials, and most others. Demographic transition theory would have led demographers to anticipate declining births as urbanization and "modernization" accelerated after the war. Sociological theories and predictions based on past trends were not very useful, however, in anticipating the unprecedented phenomenon of the baby boom. To understand it, then, we need to examine a whole host of factors.

Demographic Factors

Demographers insist that the baby boom not be seen as a direct or indirect result of the end of the war. In fact, although marriages (and divorces) did increase dramatically after the war, the accompanying rise in birth rates accounts only for the time from 1946 to 1947. By 1950 birth rates had actually fallen slightly (**Figure 3.C**). In 1951 the birth rates once again began to climb and continued to be high until 1964.

Political and Economic Factors

It is significant that the end of the Second World War witnessed a phenomenal expansion in the U.S. economy. Initially stimulated by the war, it was extended by important transformations in transportation and technology. Additionally, government had expanded and created programs for education and housing that helped returning veterans start married life with property and the opportunity for improving their economic status by attending college. Transportation—especially through the construction of the interstate highway system—helped to fuel suburbanization, which meant growth in the building industry, in automobile manufacturing, and in the manufacturing of durable goods for the home. Not surprisingly, demand for labor was high in a growing economy, and young people, most of whom had attained a higher level of education than the previous generation, were able to obtain good jobs with relatively high wages, decent benefits, and reasonable prospects for promotion.

The Aging of the Population

It is important to recognize that something like the baby boom occurred not only in the United States but also in most of the core countries of the world and even in parts of the periphery. The result is that there is a very large cohort of individuals, currently in their late thirties/early forties to their mid- to late fifties, that has had and will continue to have tremendous impacts on the rest of the population, especially as they enter their sixties, seventies, and eighties. In fact, the most fundamental demographic transformation of the twenty-first century is the aging of the population (**Figure 3.D**). In the U.S., this is a result not only of the aging of the baby boom cohort but also of longer life expectancies and falling fertility rates all across the globe. In the year 2000 the median age of the world's populations was 26.6, but by 2050 it is expected to increase to 37.8. The aging of the population is currently most pronounced in the core countries, where in 1998 the number of older persons (60 and over) actually exceeded the number of children (under 15) for the first time. As a result, for the core countries the median age of the pop-

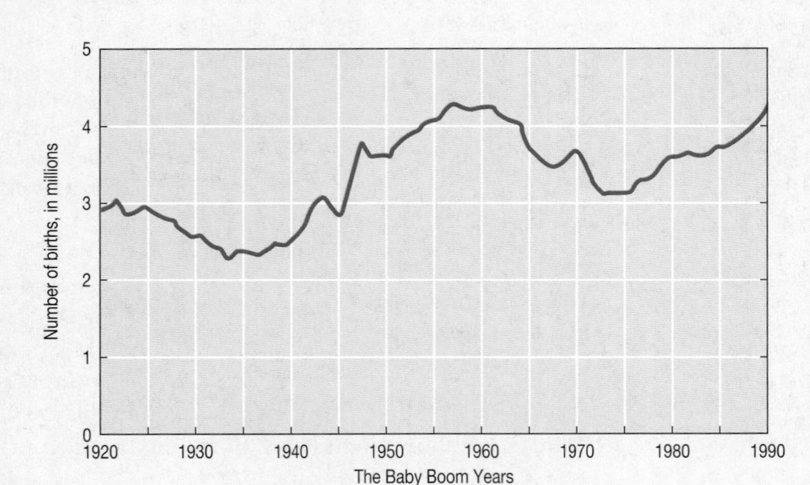

Figure 3.C U.S. baby boom crude birth rate The chart very clearly shows the big increase in births following the Second World War. (*Source:* L. Bouvier and C. DeVita, "The Baby Boom: Entering MidLife," *Population Bulletin, 46*(3), 1991.)

Figure 3.D Elderly people in United States In the core countries, men and women are living longer through lifestyle changes and improved health care.

ulation will rise from 37.5 in 2000 to 45.6 in 2050, in many cases resulting in decreasing national populations more generally. **Table 3.A** illustrates some of the core countries and regions that are expected to experience increasing numbers of elderly as well as overall population decline. In most cases, populations that are getting both older and smaller are also likely to experience a decrease in the ratio of persons of working age (15–64) to older persons (65 and older); that is, there will be fewer working age people to support the needs of the aging population. What this means is that for countries like Spain, Italy, Japan, and Germany, just to name a few, the prospect of smaller and older populations raises some very serious concerns about employment, economic growth, health care, pensions and social support services. One possible "solution" to the problem of the aging of the core's population that could have immediate effects is increasing immigration opportunities for young migrants from countries and regions that are experiencing rapid population growth.

And yet, although the aging of the population has been slower in peripheral countries because of high twentieth-century birth rates, the next 50 years will see the median age rise there as well. By 2050 the median age in countries of the periphery is expected to be 36.7 years, up from 24.4 years in 2000. This is actually a bigger jump than in the core. East and Southeast Asia are the regions whose population is aging most rapidly, with Africa aging the least.

In the United States the aging of the population is expected to have increasing impacts on the economy, politics, society, and culture. A significant difference of opinion exists among population experts on the future impacts of aging baby boomers. Some see all these highly educated workers as continuing to stimulate the economy in terms of improved productivity and increasing innovations. Others worry that this cohort will eventually place unfulfillable demands on the nation's social service system. Certainly the debates about health-care and social security reform are related to the increasing demands that this group is beginning to place on the health-care and social service system. In addition, the pressures generated by its sheer size will certainly continue well into the first half of the twenty-first century. For example, by 2030 the oldest surviving boomers will be age 84 and the youngest will be 65. At that point a projected 65 million people in the United States will be over the age of 65, compared with roughly 30 million today.

The Impacts on Younger Americans

Demographers, in addition to identifying the existence of the baby boom generation, have also demonstrated the presence of a "baby bust" generation. Two ways exist to determine the extent of this cohort. One way is through annual birth rates; if we use this measure, this generation includes those individuals born between 1965 and 1980, when births were below the baby boom average of 4 million births per year. A second way to determine the extent of the baby bust cohort is in terms of the total fertility rate. Between 1947 and 1964 the TFR ranged between 3.1 and 3.7, well above replacement level. In 1965, however, the TFR dropped below replacement level for the first time in U.S. history, and in 1976 it hit an all-time low of 1.7. Between 1977 and 1987 the TFR remained pretty stable at 1.8, and in 1990 it climbed to 2.09—close to replacement level, marking the end of the baby bust generation.

While the term baby bust refers to the demographic features of a cohort, cultural terms have also been offered to describe this group as well as groups that have followed it. Those persons born between 1965 and 1975 have also been called Generation X. The cultural label placed on Generation Xers is "slackers" because some of the most famous of this birth cohort, such as Kurt Cobain of the band Nirvana, were associated with grunge music and the anti-establishment attitude it stood for. Those individuals born after the Xers—between 1982 and 1994—are sometimes called Generation Y, the Echo Boomers or the NetGeneration. As their dates make clear, this group is not part of the baby bust cohort. Importantly, they represent a small rise or a "boomlet" in birth rates which "echoes" the baby boom, though on a much smaller scale than what happened in the late 1940s, 1950s and early 1960s. They also represent the first generation born "digital," that is, into a society wired into the Internet.

Generations X and Y will, of course, be the ones that have to "mop up" after the flood of boomers moves up their career ladders toward retirement and beyond. The tremendous size of the baby boom cohort is likely to affect the career and job mobility of the busters, as well as labor costs, which will affect taxes, health care, and other benefit costs. These older workers will be more costly for their employers, as their salaries and benefits increase in line with their seniority.

The actual impact that the baby boom generation will have into the first half of the twenty-first century remains to be seen. What is clear, however, is that it carries the future of the nation's economic and social security along with it. Because of its sheer size, the cohort will continue to influence public policy through its political participation at the ballot box and as national, regional, and local leaders and officials. Most important for the years to come, however, is that boomers will force the issues of generational equity and the allocation of scarce resources across all cohorts to the center of political and ethical debate.

Source: Adapted from "The Baby Boom: Entering Midlife," by Leon Bouvier and Carol De Vita in *Population Bulletin* 46(3), 1991; and *An Aging World: 2001* by Kevin Kinsella and Victoria A. Velkoff, U.S. Census Bureau, Series P95/01-1, U.S. Government Printing Office, Washington, DC, 2001.

TABLE 3.A Countries Where Population is Expected to Decline, 2000 and 2050

	Population (thousands)		Population Change		Percent 65 years or older		Change in proportion 65 years or older
Country or area*	**2000**	**2050**	**(thousands)**	**(per cent)**	**2000**	**2050**	**(per cent)**
Austria	8,211	7,094	−1,117	−14	15	30	106
Belarus	10,236	8,330	−1,907	−19	14	25	86
Belgium	10,161	8,918	−1,243	−12	17	28	65
Bosnia and Herzegovina	3,972	3,767	−205	−5	10	27	171
Bulgaria	8,225	5,673	−2,552	−31	16	30	88
China, Hong Kong SAR[a]	6,927	6,664	−263	−4	11	33	217
Croatia	4,473	3,673	−800	−18	15	26	77
Cuba	11,201	11,095	−105	−1	10	27	176
Czech Republic	10,244	7,829	−2,415	−24	14	33	144
Denmark	5,293	4,793	−500	−9	15	24	59
Estonia	1,396	927	−469	−34	14	29	107
Finland	5,176	4,898	−278	−5	15	26	72
Germany	82,220	73,303	−8,917	−11	16	28	73
Greece	10,645	8,233	−2,412	−23	18	34	92
Hungary	10,036	7,488	−2,548	−25	15	28	92
Italy	57,298	41,197	−16,101	−28	18	35	92
Japan	126,714	104,921	−21,793	−17	17	32	86
Latvia	2,357	1,628	−728	−31	14	27	86
Lithuania	3,670	2,967	−704	−19	13	27	102
Luxembourg	431	430	−1	0	14	27	84
Netherlands	15,786	14,156	−1,629	−10	14	28	104
Poland	38,765	36,256	−2,509	−6	12	26	118
Portugal	9,875	8,137	−1,738	−18	16	31	99
Romania	22,327	16,419	−5,908	−26	13	31	131
Russian Federation	146,934	121,256	−25,678	−17	13	25	100
Slovakia	5,387	4,836	−551	−10	11	27	139
Slovenia	1,986	1,487	−499	−25	14	32	131
Spain	39,630	30,226	−9,404	−24	17	37	117
Sweden	8,910	8,661	−249	−3	17	27	53
Switzerland	7,386	6,745	−641	−9	15	30	104
Ukraine	50,456	39,302	−11,154	−22	14	27	91
United Kingdom	58,830	56,667	−2,163	−4	16	25	56
Yugoslavia	10,640	10,548	−92	−1	13	23	73

[a]As of 1 July 1997, Hong Kong became a Special Administrative Region (SAR) or China.

*Countries or areas with 150,000 persons or more in 1995.

Source: U.N. Population Division, *World Population Prospects: The 1998 Revision.*

Figure 3.10 The Net Generation Members of the Net Generation, people who are currently in their mid-twenties, are faced with the awesome burden of having to help support a huge, aging baby boom population. In addition to the economic burden, many commentators have observed that they also face a cultural burden—the difficulty of establishing a distinct identity. Because there are so many baby boomers compared with later generations, their preferences and styles tend to dominate consumption patterns. Now in possession of a disproportionate share of the national wealth, baby boomers have a great deal of buying power, which translates into a significant impact on the music industry, film industry, and other forms of contemporary cultural expression.

periphery of the world economy and low levels of fertility in the core. The highest birth rates occur in Africa, the poorest region in the world.

The crude birth rate is only one indicator of fertility and in fact is somewhat limited in its usefulness, telling very little about the potential for future fertility levels. Two other indicators formulated by population experts—the total fertility rate and the doubling time—provide more insight into the potential of a population. The **total fertility rate** (TFR) is a measure of the average number of children a woman will have throughout the years that demographers have identified as her childbearing years, approximately ages 15 through 49 (**Table 3.3**). Whereas the CBR indicates the number of births in a given year, the TFR is a more predictive measure that attempts to portray what birth rates will be among a particular cohort of women over time. A population with a TFR of slightly higher than two has achieved replacement-level fertility. This means that birth rates and death rates are approximately balanced and there is stability in the population.

Closely related to the TFR is the doubling time of the population. The **doubling time,** as the name suggests, is a measure of how long it will take the population of an area to grow to twice its current size. For example, a country whose population increases at 1.8 percent per year will have doubled in about 40 years. In fact, world population is currently increasing at this rate. By contrast, a country whose population is increasing 3.18 percent annually will double in only 22 years—the doubling time for Kenya. Birth rates and the population dynamics we can project

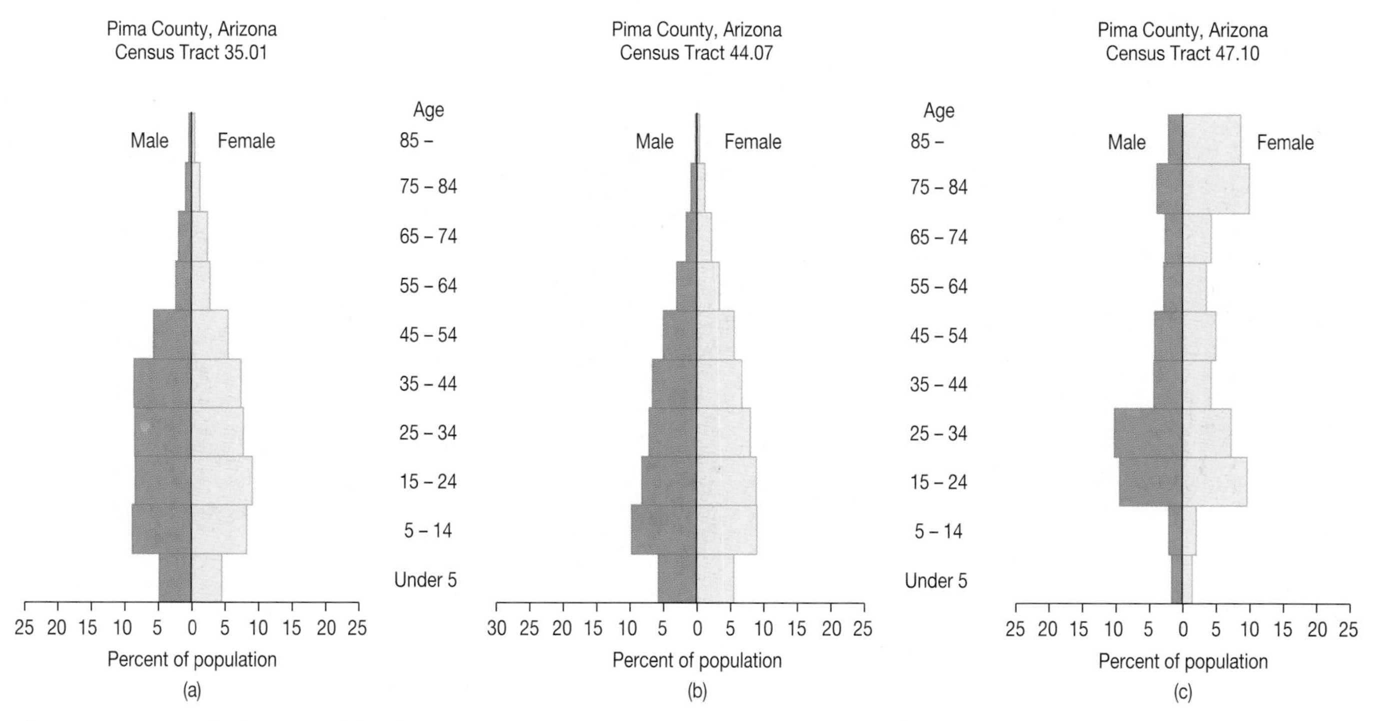

Figure 3.11 Population pyramids of Tucson census tracts, 2000 Population pyramids are as useful for analyzing populations within a single region as for an entire country. Here we see how age-sex profiles can also vary within different census tracts of the same city. The tracts show that even within a city, variation in populations can be substantial. Information like this can be very valuable in decision-making and policymaking at state, regional, and local levels as well as for marketers who can target different populations through direct mailings. (After U.S. Department of Commerce, *Bureau of Census, 1990.*)

Figure 3.12 Birth-control promotion in India Controlling fertility has been an objective of population policy at the international level for several decades. For many years, however, birth-control programs were rejected by peripheral countries because having children was regarded as an economically sound decision. Birth-control programs coupled with improved educational and economic opportunities for women have proved to be far more effective than birth-control policies alone. But in India, issues of ethnicity complicate things because one ethnic group is fearful that if it limits its births, it will soon be outnumbered by another ethnic group.

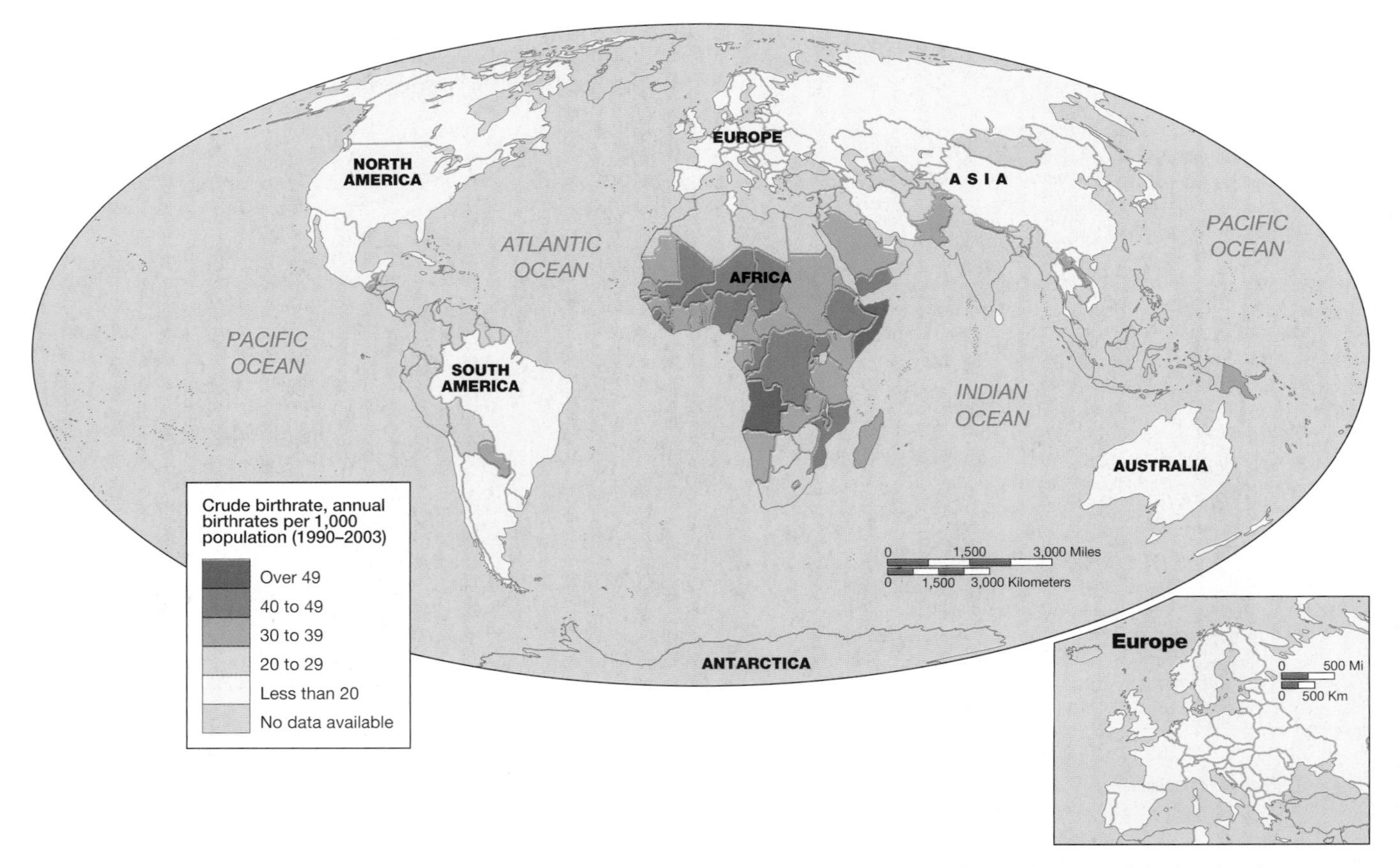

Figure 3.13 World crude birth rates, 2004 Crude birth rates and crude death rates are often indicators of the levels of economic development in individual countries. Compare, for example, Australia, a core country, which offers a stark contrast to statistics for Ethiopia, a very poor and underdeveloped peripheral country. (After World Bank, *World Development Indicators, 2004,* World Bank, Washington, DC.)

TABLE 3.3 Total Fertility Rates for Selected Countries, 2000

Country	Total Fertility Rate (TFR)
Afghanistan	6.8
Argentina	2.4
China	1.7
India	3.1
Angola	6.8
Mauritius	1.9
Spain	1.3
Russia	1.4
United States	2.0

from them, however, tell us only part of the story of the potential of the population for growth. We must also know the death, or mortality, rates.

Death, or Mortality, Rates

Countering birth rates and shaping overall population numbers and composition is the **crude death rate** (**CDR**), the ratio of the number of deaths in one year to every thousand people in the population. As with crude birth rates, crude death rates often roughly reflect levels of economic development—countries with low birth rates generally have low death rates (**Figure 3.14**).

Although often associated with economic development, CDR is also significantly influenced by other factors. A demographic structure with more men and elderly people, for example, usually means higher death rates. Other important influences on mortality include health-care availability, social class, occupation, and even place of residence. Poorer groups in the population have higher death rates than the middle class. In the United States, coal miners have higher death rates than schoolteachers. Also in the United States, urban areas often have higher death rates than rural areas. The difference between the CBR and CDR is the rate of **natural increase**—the surplus of births over deaths—or the rate of **natural decrease**—the deficit of births relative to deaths (**Figure 3.15**).

Death rates can be measured for both sex and age cohorts, and one of the most common measures is the **infant mortality rate.** This figure reflects the annual number of deaths of infants less than one year of age compared to the total number of live births for that same year. The figure is usually expressed as number of deaths during the first year of life per 1,000 live births.

The infant mortality rate has been used by researchers as an important indicator both of a country's health-care

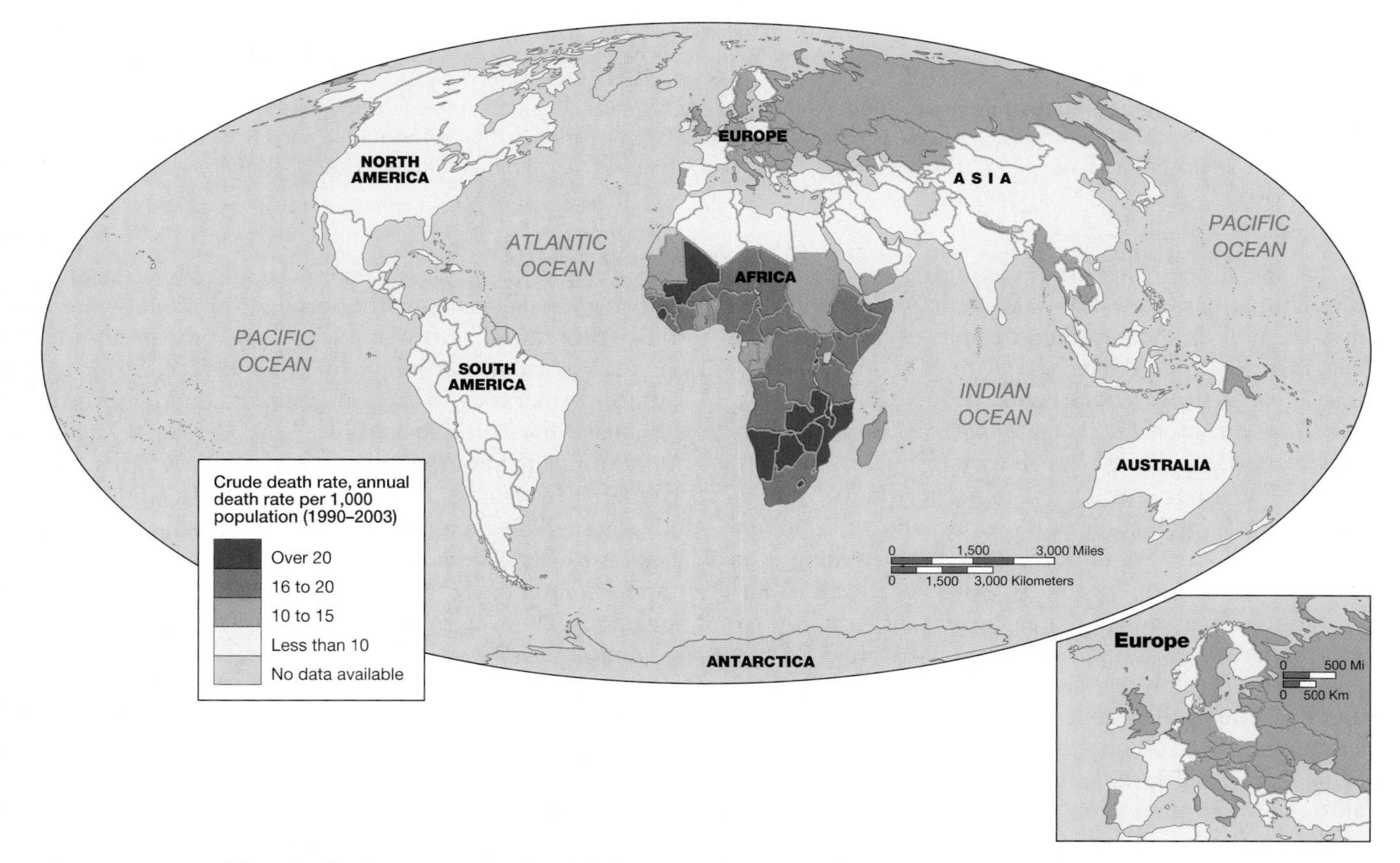

Figure 3.14 World crude death rates, 2004 The global pattern of crude death rates varies from crude birth rates. Most apparent is that the difference between highest and lowest crude death rates is relatively smaller than is the case for crude birth rates, reflecting the impact of factors related to the middle phases of the demographic transition. (After World Bank, *World Development Indicators, 2004,* World Bank, Washington, DC.)

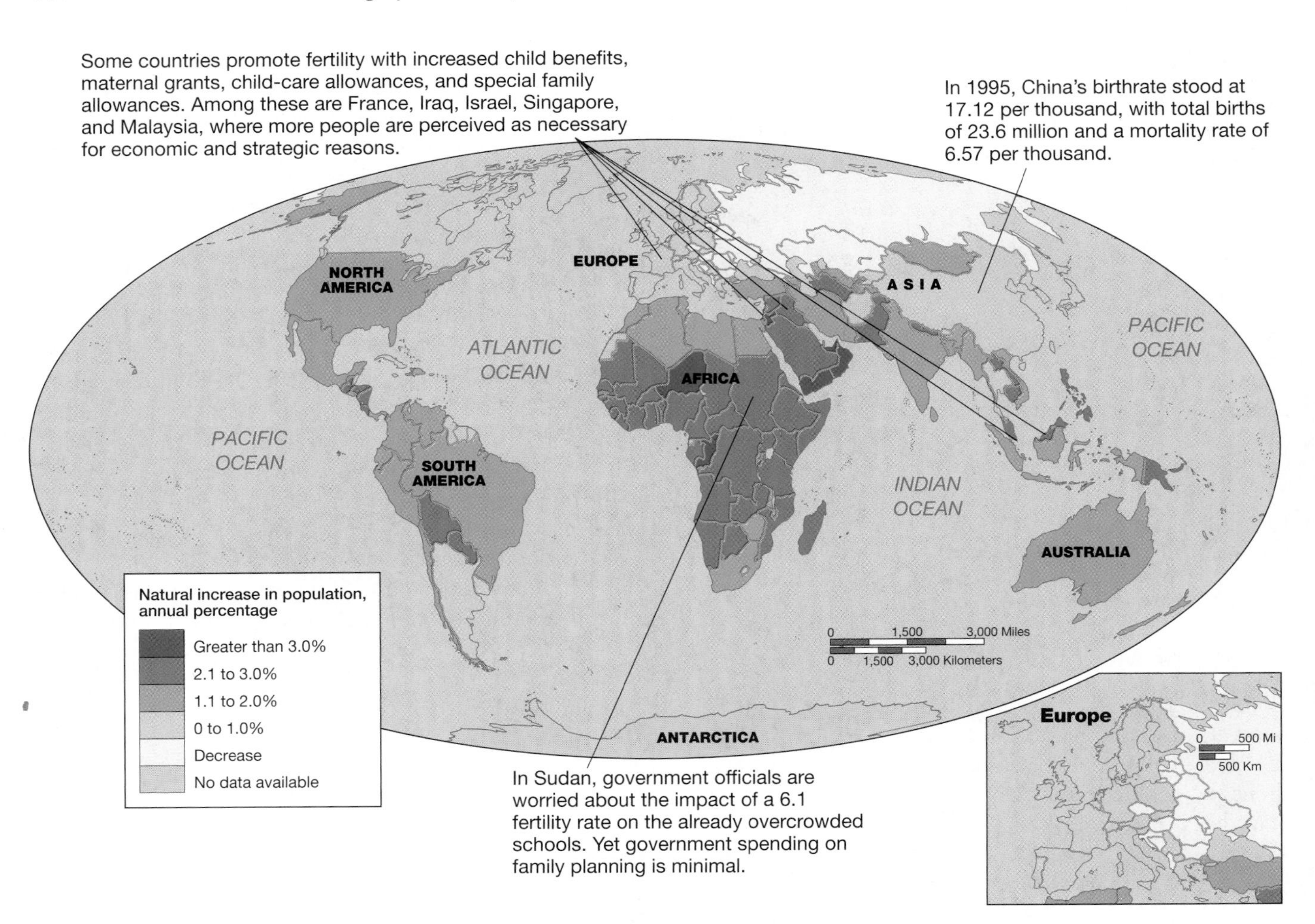

Figure 3.15 World rates of natural increase, 2004 As the map shows, rates of natural increase are highest in sub-Saharan Africa, the Near East, and parts of Asia, as well as parts of South and Central America. Europe and the United States and Canada, as well as Australia and parts of central Asia and Russia, have slow to stable rates of natural increase. (After World Bank, *World Development Indicators 2004,* World Bank: Washington, DC.)

system and of the general population's access to health care. Global patterns show that while infant mortality rates are high in the peripheral countries of Africa and Asia, they are low in the more developed countries of Europe and North America (**Figure 3.16**). Generally, the core's low rates reflect adequate maternal nutrition and the wider availability of health-care resources and personnel. When patterns are examined below the global scale, at the level of countries, regions, and cities, infant mortality rates are not uniform. In the United States, for example, African Americans, as well as other ethnic minorities in urban and rural areas, suffer infant mortality rates that are twice as high as the national average. In east central Europe, Bulgaria has a 12.3-per-thousand infant mortality rate, while the Czech Republic has a rate of 3.9 per thousand. The point is that global patterns often mask regional and local variation in mortality rates for both infants and other population cohorts.

Related to infant mortality and the crude death rate is **life expectancy,** the average number of years an infant newborn can expect to live. Not surprisingly, life expectancy varies considerably from country to country, region to region, and even from place to place within cities and among different classes and racial and ethnic groups. In the United States an infant born in 2000 could expect to live more than 77.0 years. If we begin to specify the characteristics of that infant by sex and race, however, variation emerges. An African American male born in 2000 has a life expectancy of 71.8 years, while an Anglo-American male born the same year can expect to live, on average, 77.8 years.

Another key factor influencing life expectancy is epidemics, which can quickly and radically alter population numbers and composition. In our times, epidemics can spread rapidly over great distances, largely because people and other disease carriers can now travel from one place to another very rapidly. Epidemics can have profound effects at various scales, from the international to the local, and reflect the increasing interdependence of a shrinking globe. They may affect different population groups in different ways and, depending on the quantity and quality of health and nutritional care available, may have a greater or lesser impact on different localities.

One of the most widespread epidemics of modern times is HIV/AIDS (human immunodeficiency virus/acquired immunodeficiency syndrome). The disease is now

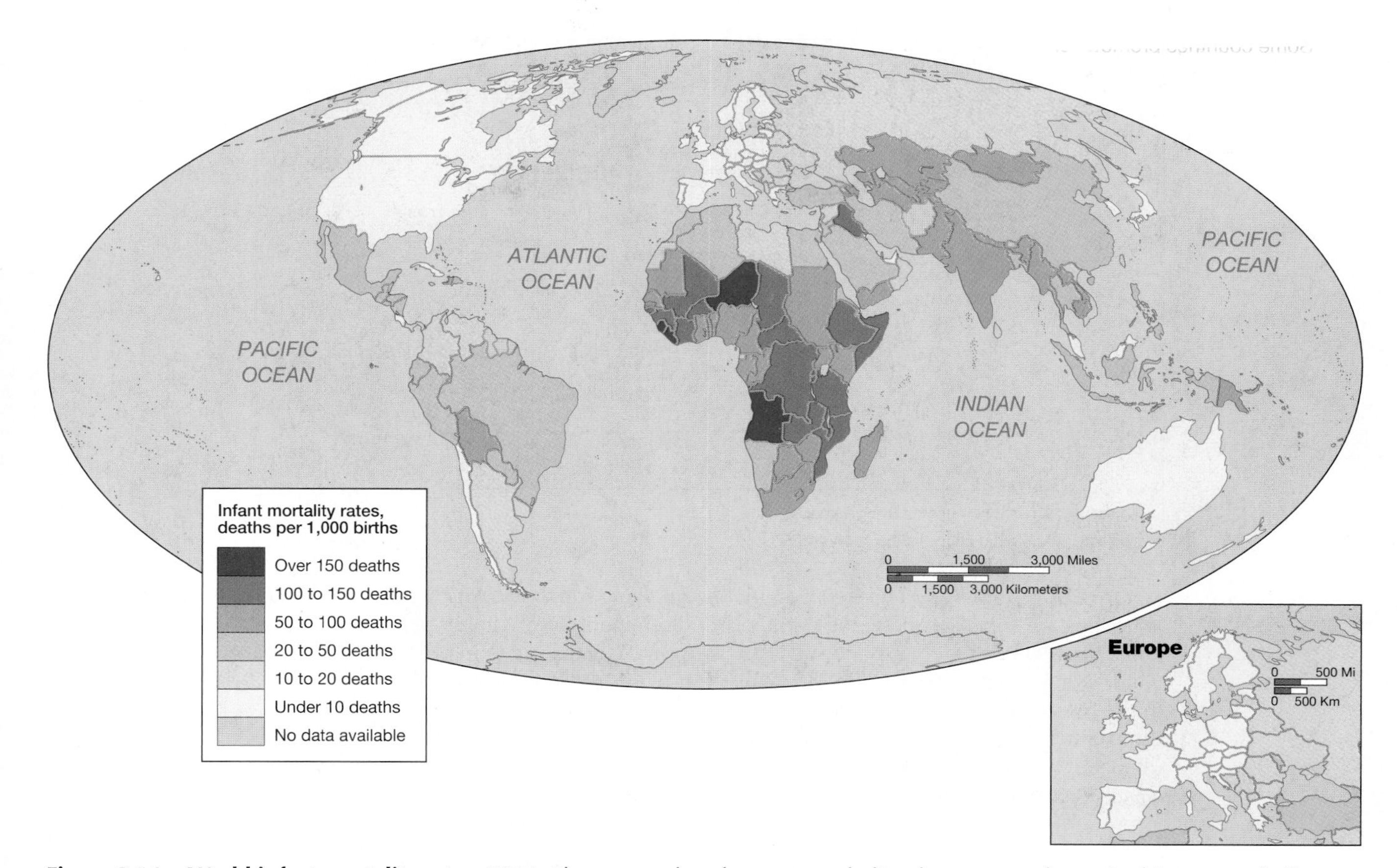

Figure 3.16 World infant mortality rates, 2004 The geography of poverty underlies the patterns shown in this map and allows us to analyze the linkages between population variables and social conditions. Infant mortality rates seem generally to parallel crude death rates, with sub-Saharan Africa generally reporting the highest rates. These rates reflect a number of factors, including inadequate or completely absent maternal health care as well as poor nutrition for infants. (After *Hammond Atlas of the World.* New York: Oxford University Press, 1993. Update from the World Bank, *World Development Indicators, 2004,* World Bank: Washington, DC.)

a serious problem in regions ranging from Southeast Asia to sub-Saharan Africa, and also affects certain populations in many core countries of Europe as well as the United States. To date, the Islamic countries of the Middle East and North Africa, as well as South America, seem to be least affected, though in the last several years there has been a marked increase in the spread of HIV/AIDS in the Middle East and North Africa, especially in Djibouti, Somalia, and the Sudan. Eastern Europe—especially the Russian Federation—is experiencing the fastest-growing epidemic in the world, with the number of new HIV infections rising steeply. In 2001 there were 1 million people living with HIV/AIDS there. Transmission in this region appears to occur largely through the high rates of injecting drug use among young people.

Medical geographers have made important contributions to the study of the diffusion of HIV/AIDS. **Medical geography** is a subarea of the discipline that specializes in understanding the spatial aspects of health and illness. This spatial perspective includes disease mapping as well as the distribution and diffusion of health and illness. The importance of medical geography lies in the fact that the landscape of disease that we are seeing today is changing: Diseases we thought were eradicated, such as tuberculosis, are coming back; diseases that were previously unknown are emerging, such as ebola; and diseases that have maintained geographic limitations, such as dengue hemorrhagic fever, are spreading. Perhaps the most popular health issue addressed by medical geographers is HIV/AIDS. With respect to this disease, medical geographers ask: Where did AIDS originate? Why did it originate there? How did it spread to surrounding areas? How is the HIV/AIDS crisis related to other things? How has the epidemiology of HIV/AIDS changed over time?

In the United States, for example, HIV/AIDS first arose largely among male homosexuals and intravenous drug users who shared needles. Geographically, early concentrations of AIDS occurred in places with high concentrations of these two subpopulations, such as San Francisco, New York, Los Angeles, and southern Florida. Since first appearing, the disease has hierarchically diffused to other groups, often manifesting itself differently in each group, including hemophiliacs and other recipients of blood transfusions. It has had perhaps the most severe impact in inner city areas but has cropped up in every region of the United States, increasingly appearing in the male and female heterosexual population. Especially alarming is the increasing incidence of HIV/AIDS among heterosexual teenagers and twenty-year-olds in this country, a group largely untouched by HIV/AIDS previously.

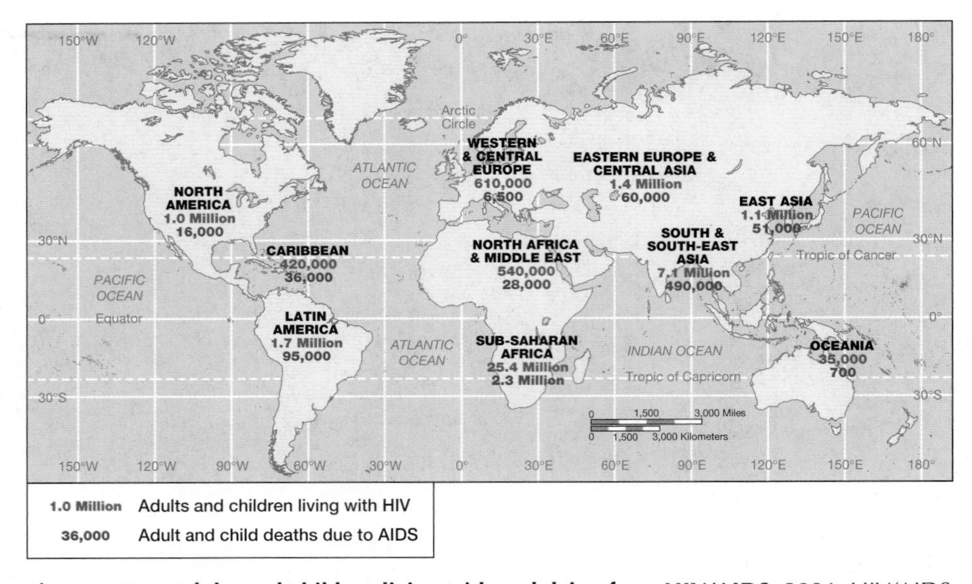

Figure 3.17 Adults and children living with and dying from HIV/AIDS, 2001 HIV/AIDS infections are concentrated in the periphery and semiperiphery, with 89 percent of people with HIV/AIDS living in Asia and sub-Saharan Africa. Compare the number of adults and children living with HIV/AIDS in Africa with those in North America or Europe. Deaths from HIV/AIDS have also been highest in Africa. Campaigns in some African countries, especially in Zimbabwe and Uganda, have helped to stem the number of new cases and ultimately the number of deaths. (After U.N. AIDS Organization Web site: http:www/unaids.org/worldaids/2001/EPIgraphics2001/EPIgraphic6_en.gif and EPIgraphic4_en.gif)

The pattern of the disease is different, however, in central Africa, where it is overwhelmingly associated with heterosexual, non-drug users and affects both sexes equally. The geographical diffusion of HIV/AIDS in Africa has occurred along roads, rivers, and coastlines, all major transportation routes associated with regional marketing systems. The central African nations, including Congo (formerly Zaire), Zambia, Uganda, Rwanda, and the Central African Republic, have been especially hard hit. For sub-Saharan Africa it is estimated that approximately 28 million people are infected with the virus (**Figure 3.17a**). The impact is worst in urban areas, though no area has been immune to the disease's spread. High death rates in these countries reflect in no small part the ravages of HIV/AIDS (**Figure 3.17b**). That these countries continue to experience high growth rates, however, is a reflection of the huge proportion of the population that is in the peak years of biological reproduction.

Although no cure yet exists, some countries have successfully slowed the spread of HIV/AIDS. Finland, for example, has significantly inhibited the diffusion of HIV through an intensive public relations campaign and the provision of top-notch health services to all its citizens. Few of the sub-Saharan African countries, mired in poverty, inadequate health-care systems, and political inefficiency and often civil unrest, are able to approach these levels of activity, although it appears as if Uganda, which has extremely high rates of HIV/AIDS, may be experiencing a decline in new cases. Additionally, in the summer of 2000, activists in Durban, South Africa, were effective in forcing several large pharmaceutical companies to slash the price of HIV/AIDS drugs for victims of the disease there and in other parts of the African continent. Many Southeast Asian countries also face difficult economic constraints in attempting to slow the diffusion of the disease. In Cambodia and Thailand, however, prompt large-scale prevention programs have begun to stem the spread of the epidemic. After peaking at around 140,000 cases in 1991, the number of new HIV cases dropped to an estimated 21,000 in 2003, due in large part to lower drug prices and very aggressive and more widespread anti-retroviral treatment.

Demographic Transition Theory

Many demographers believe that fertility and mortality rates are directly tied to the level of economic development of a country, region, or place. Pointing to the history of demographic change in core countries, they contend that many of the economic, political, social, and technological transformations associated with industrialization and urbanization lead to a demographic transition. A **demographic transition** is a model of population change in which high birth and death rates are replaced by low birth and death rates. Once a society moves from a preindustrial economic base to an industrial one, population growth slows. According to the demographic transition model, the slowing of population growth is attributable to

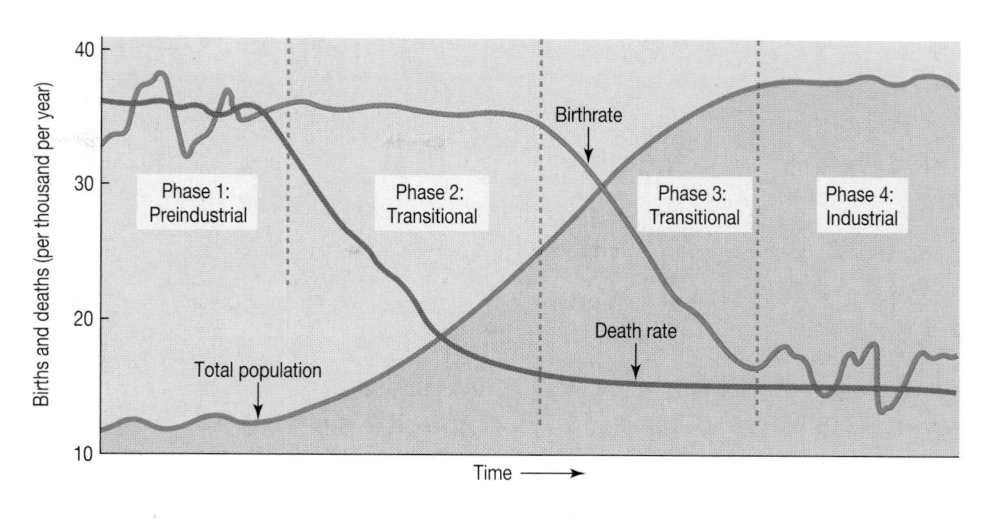

Figure 3.18 Demographic transition model The transition from a stable population based on high birth and death rates to one based on low birth and death rates progresses in clearly defined stages, as illustrated by this graph. With basic information about a country's birth and death rates and total population, it is possible to identify that country's position within the demographic transition process. Population experts disagree about the usefulness of the model, however. Many insist it is applicable only to the demographic history of core countries.

improved economic production and higher standards of living brought about by better health care, education, and sanitation.

As **Figure 3.18** illustrates, the high birth and death rates of the preindustrial phase (Phase 1) are replaced by the low birth and death rates of the industrial phase (Phase 4) only after passing through the critical transitional phase of steady birth rates and falling death rates (Phase 2) and then more moderate rates (Phase 3) of natural increase (increase through birth, not migration) and growth. This transitional phase of rapid population growth is the direct result of early and steep declines in mortality at the same time that fertility remains at levels characteristic of a place that has not yet been industrialized.

Some demographers have observed that many peripheral and semiperipheral countries appear to be stalled in the transitional phase, or caught in a "demographic trap." **Figure 3.19** illustrates the disparity between birth and death rates for core and peripheral countries. Despite a sharp decline in mortality rates, most peripheral countries retain relatively high fertility rates.

The reason for this lag in declining fertility rates relative to mortality rates is that while societies have developed new and more effective methods for fighting infectious diseases, social attitudes about the desirability of large families are only recently changing. Historical trends in birth and death rates and natural increase are shown in **Table 3.4** for Scotland and England. Over a roughly 50-year period (1870–1920), both places were able to reduce their rates of natural increase by nearly one-third. In the fourth phase of postindustrial development (not shown in the table), birth and death rates have both stabilized at a low level, which means that population growth rates are very slow and birth rates are more likely to oscillate than death rates. Whereas England and Scotland, early industrializers, passed through the demographic transition over a period of about 150 years beginning in the mid-nineteenth century, most peripheral and semiperipheral countries have yet to complete the transition.

Although the demographic transition model is based on actual birth and death statistics, many population geographers and other population experts increasingly question its generalizability to all places and all times. In fact, while the model adequately describes the history of population change in the core countries, it appears less useful for explaining the demographic history of countries

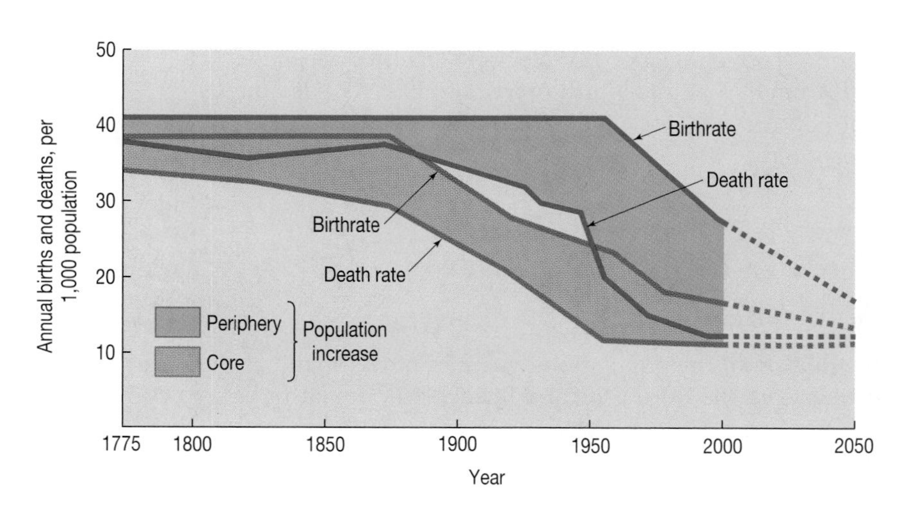

Figure 3.19 World trends in birth and death rates, 1775–2050 The graph is an illustration of the impact of affluence on reproductive choices. It also portrays an optimistic view of the future, where for peripheral regions, birth rates will continue to fall as we move closer to the midpoint of the next century. For the core, the projection is that birth and death rates will stay low. (After T. Allen and A. Thomas, *Poverty and Development in the 1990s*. London: Oxford University Press, 1992.)

TABLE 3.4 Birth and Death Rates for England and Scotland, 1870–1920

The CBRs and CDRs for England and Scotland for the period between 1870 and 1920 illustrate countries moving between Phase 2 and Phase 3 of the demographic transition, in which death rates are lower than birthrates. England and Scotland are clear examples of the way in which the demographic transition has been theorized to operate in the core countries. During the 50-year period covered in the table, both countries were completing their transformation as key industrial regions.

	1870	1880	1890	1900	1910	1920
			Crude Birthrate			
England	35.2	35.4	32.5	29.9	27.3	22.7
Scotland	35	34.8	32.3	30.2	28.4	24
			Crude Death Rate			
England	22.5	21.4	19.2	18.2	15.4	14.6
Scotland	22.1	21.6	19.2	18.5	16.6	15.3

Source: M. Anderson and D. Morse, *Scottish Demography,* vol. 42, 1993.

and regions in the periphery. Such criticism has led to its contested significance for understanding population geography. Among other criticisms is that industrialization—which, according to the theory, is central to moving from Phase 2 to Phases 3 and 4—is seldom domestically generated in the peripheral countries. Instead, foreign investment seems to drive peripheral industrialization. As a result, the features of demographic change witnessed in the core countries, where industrialization was largely a result of internal capital investment, such as higher living standards, have not been as dramatic in many peripheral countries. Other critics of the demographic transition model point to several factors undermining a demographic transition fueled by economic growth: the shortages of skilled laborers, the absence of advanced educational opportunities for all members of the population (especially women), and limits on technological advances. In other words, while a demographic transition may be a characteristic experience of the core regions of the globe, its applicability to the periphery needs to be premised on a different set of factors. Whether or not the demographic transition model, derived from core countries' experiences, is directly relevant to the periphery, it is becoming increasingly clear that some sort of demographic transformation is occurring globally, resulting in a decrease in the TFR from Europe and North America to East Asia and the Caribbean, with Africa and parts of the Middle East being exceptions.

POPULATION MOVEMENT AND MIGRATION

In addition to the population dynamics of death and reproduction, the third critical influence on population is the movement of people from place to place. Individuals may make far-reaching, international or intraregional moves, or they may simply move from one part of a city to another. For the most part, mobility and migration reflect the interdependence of the world-system. For example, global shifts in industrial investment result in local adjustments to those shifts as populations move or remain in place in response to the creation or disappearance of employment opportunities.

Mobility and Migration

One way to describe such movement is by the term **mobility,** the ability to move from one place to another, either permanently or temporarily. Mobility may be used to describe a wide array of human movement, ranging from a journey to work (for example, a daily commute from suburb to city or suburb to suburb) to an ocean-spanning permanent move.

The second way to describe population movement is in terms of **migration,** a long-distance move to a new location. Migration involves a permanent or temporary change of residence from one neighborhood or settlement (administrative unit) to another. Moving from a particular location is defined as **emigration,** also known as out-migration. Moving to a particular location is defined as **immigration** or in-migration. Thus, a Russian who moves to Israel emigrates from Russia and immigrates to Israel. This type of move, from one country to another, is termed **international migration.** Moves may also occur within a particular country or region, in which case they are called **internal migration.** Both permanent and temporary changes of residence may occur for many reasons but most often involve a desire for economic betterment or an escape from adverse political conditions, such as war or oppression.

Governments are concerned about keeping track of migration numbers and rates, as well as the characteristics of the migrant populations, because these factors can have profound consequences for political, economic, and cultural conditions at national, regional, and local scales. For example, a peripheral country that has experienced substantial out-migration of highly trained professionals, such as Cuba, may find it difficult to provide needed services such as health care. Benefiting from lower labor costs are countries that have received large numbers of low-skilled in-migrants willing to work for extremely low wages, such as the United States, Germany, and France. These countries may also face considerable social stress in times of economic recession, when unemployed citizens begin to blame the immigrants for "stealing" their jobs or receiving welfare benefits.

Demographers have developed several calculations of migration rates. Calculation of the in-migration and out-migration rates provides the foundation for discovering gross and net migration rates for an area of study. **Gross migration** refers to the total number of migrants moving into and out of a place, region, or country. **Net migration** refers to the gain or loss in the total population of that area as a result of the migration. Migration is a particularly important concept because the total population of a country, region, or locality is dependent on migration activity as well as on birth and death rates.

Migration rates, however, provide only a small portion of the information needed to understand the dynamics of migration and its effects at all scales of resolution. In general terms, migrants make their decisions to move based on push factors and pull factors. **Push factors** are events and conditions that impel an individual to move from a location. They include a wide variety of possible motives, from the idiosyncratic, such as an individual migrant's dissatisfaction with the amenities offered at home, to the dramatic, such as war, economic dislocation, or ecological deterioration. **Pull factors** are forces of attraction that influence migrants to move to a particular location. Again, factors drawing individual migrants to chosen destinations may range from the highly personal (such as a strong desire to live near the sea) to the very structural (such as strong economic growth and thus relatively lucrative job opportunities).

Usually the decision to migrate is a combination of both push and pull factors, and most migrations are voluntary. In **voluntary migration** an individual chooses to move. Where migration occurs against the individual's will, push factors can produce **forced migration.** Oftentimes the decision to migrate is a mixed one reflecting both forced and voluntary factors. Forced migration (both internal and international) remains a critical problem in the contemporary world. The United Nations reported that a total of 17,084,100 million people were forced migrants or refugees in 2004. These migrants may be fleeing a region or country for many reasons, but some of the most common are war, famine (often war-induced), life-threatening environmental degradation or disaster, or governmental coercion or oppression. The terrorist attacks on the World Trade Center and the Pentagon on September 11 caused refugee movements not only in Central Asia, especially Afghanistan, but also globally. And while **refugees**—individuals who cross national boundaries to seek safety and asylum—are a significant global problem, **internally displaced persons (IDPs)**—individuals who are uprooted within the boundaries of their own country because of conflict or human rights abuse—are also growing globally (see Box 3.3: "Internal Displacement)."

International Voluntary Migration

Migration does not always involve force or even involve a permanent change of residence. Voluntary migration can occur for any number of reasons such as high wage differentials between places, better experience and job opportunities elsewhere, family links abroad, or local underemployment of unemployment conditions (**Figure 3.20**). Temporary labor migration has long been an indispensable part of the world economic order and has at times been actively pursued by governments and companies alike. Individuals who migrate temporarily to take up jobs in other countries are known as **guest workers.** Sending workers abroad is an important economic strategy for many peripheral and semiperipheral countries; it not only lessens local unemployment but also enables workers to send substantial amounts of money to their families at home. This arrangement helps to supplement the workers' family income and supports the dominance of the core in global economic activities. It is not always an ideal situation, however, because economic downturns in the guest workers' host country may result in a large decrease in remittances received by the home country, thus further aggravating that country's economic situation.

It is also important to know the gender of temporary workers and the gender-based differences in the types of work performed. The Philippines, for example, has an Overseas Contract Worker (OCW) program that links foreign demand for workers to Philippine labor supply. The proportion of men and women OCWs is approximately equal, but the jobs they hold are gender-biased. Men receive most of the higher level positions, while women are largely confined to the service and entertainment sectors. Some interesting geographical variations exist as well. Although constituting a small percentage of the total number of women working abroad, a large proportion of female OCWs who have professional positions (such as doctors and nurses) work in the United States and Canada. More typical of Filipina OCW experience, however, are the patterns in Hong Kong, Singapore, and Japan. In Hong Kong and Singapore, Filipinas are almost exclusively employed as domestic servants; in Japan most work in the "entertainment industry," a term often synonymous with prostitution. These women in many ways have been transformed from individuals to commodities—many are even chosen by catalog. The OCW program has been criticized by feminist groups as well as human rights organizations. Many regard the OCW treatment of women as a contemporary form of slavery.

An increasingly important category of international migrants are **transnational migrants,** so called because they set up homes and/or work in more than one nation-

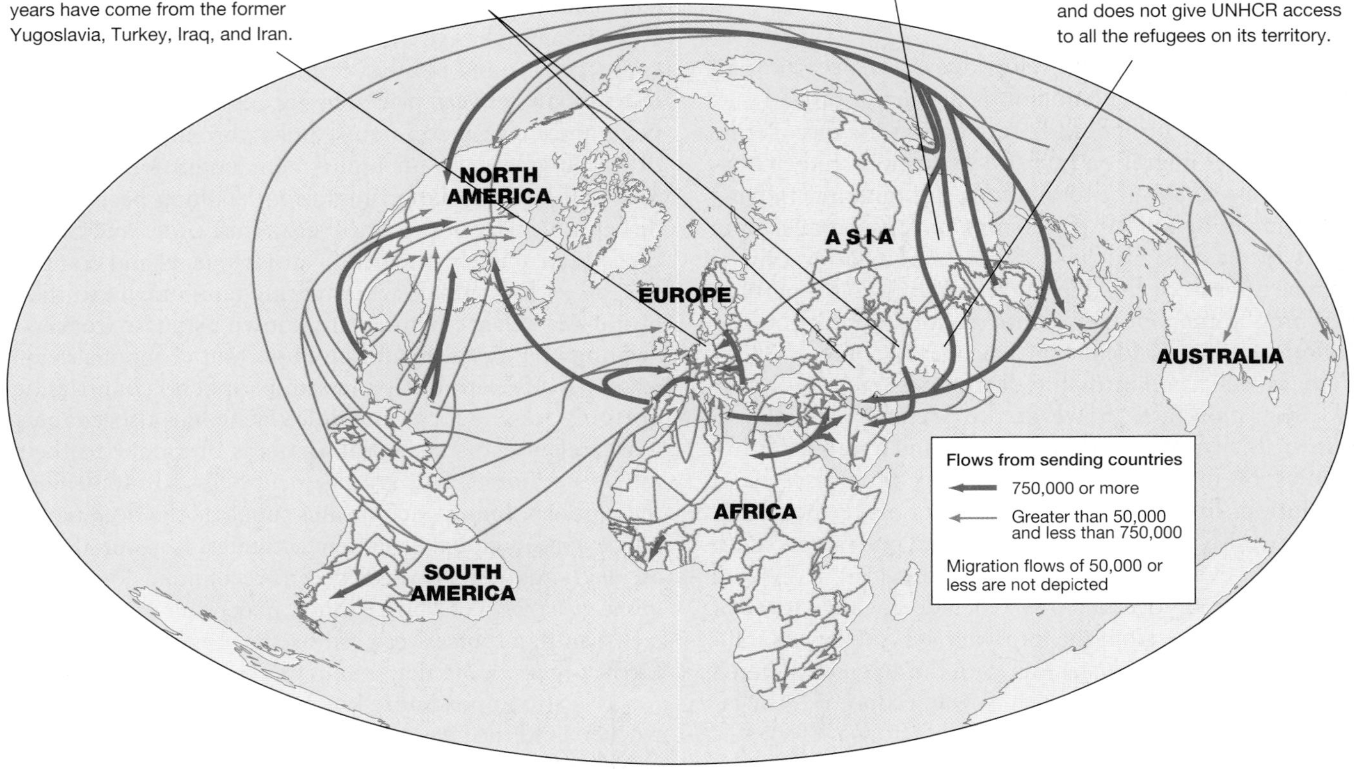

Figure 3.20 Global voluntary migration, 1999 This map illustrates very complex flows of people across borders who have migrated by choice. While each of the flows represents a cluster of individual decisions, it is noteworthy that, generally speaking, the flows emanate from the periphery and are moving toward the core. Intracontinental migration, such as that in South America and Africa, represents the apparent pull of economic opportunity for residents of relatively poorer countries.

state. Sometimes these migrants are low-paid workers, as in the case of many Mexican migrants to the United States who take up jobs in the unskilled sectors of manufacturing, agriculture, and the service economy. These migrants characteristically possess domiciles in Mexican cities, towns, or villages as well as in the United States, maintaining regular contact between the two places through visits and phone calls and sending substantial sums of money to support family members in Mexico to improve their residences there. Other transnational migrants occupy the higher end of the socioeconomic spectrum, such as the Hong Kong Chinese, who have established residences in Vancouver, Canada. Geographer Katharyne Mitchell has written extensively about this group of transnational migrants who moved to Canada in the 1980s and 1990s as a result of Hong Kong's pending transfer from British to Chinese (PRC) control in 1997. A business immigration program, intended to encourage Canadian economic development, enabled prospective migrants with a personal net worth of C$500,000 (US$405,433) and a promise to invest C$350,000 (US$283,803) in a Canadian venture to jump to the head of a long immigration waiting list and receive a visa. The Hong Kong Chinese who have migrated to Vancouver maintain extensive and substantial property and business investments in both places and after three years are eligible to become Canadian citizens, while maintaining their citizenship status in Hong Kong as well. Transnational migrants seek destinations other areas of the world than North America, with substantial migrant labor communities also occuring in France, German, Saudi Arabia and South Africa, to name a few.

International Forced Migration

Forced migration is a world-wide phenomenon (**Figure 3.21**). A recent example of international forced migration is of Lebanese and Kurdish people in the Middle East. As **Figure 3.22** shows, these two groups have been scattered widely throughout the globe because of war and civil strife. Over the last century, and because of the tensions existing in this multiethnic, multireligious society, Lebanon has lost a great deal of its population to forced migration. The 1975–1990 civil war in Lebanon resulted in a particularly large number of migrations. The failure of the Kurds to establish an autonomous state in the early twentieth century led to their being split among Iran, Iraq, and Turkey, with a small minority in Syria. Many Kurds have moved to different parts of the region or left it altogether because of military aggression, persecution, and the failure of repeated attempts to establish a Kurdish state. Other prominent examples of international forced migration include the migration of Jews from Germany and Eastern Europe preceding the Second World War and the deportation of Armenians from the Ottoman Empire after the First World War. **Figure 3.23** shows the scattering of Palestinians from their homeland since the establishment of the state of Israel in 1948. The refugees have been fleeing ongoing fighting as well as discrimination. It is still too soon to tell what the new political climate of renewed negotitations between Palestine and Israel might bring, but the death of Yasser Arafat and the election of Mahmoud Abbas as the new Prime Minister of the Palestinian people provides some optimism that the map of Palestinian refugees and internally displaced persons could change in the next few years. This new map will hinge not only on the return of formerly occupied Israeli land to Palestine, but also to the granting of the right of return to refugees who have lived away from Palestine for generations in camps in Lebanon, Syria, and Jordan as well as in many other countries of the world like the United States, Great Britain and Canada.

Recently—and especially since September 11—many European countries, faced with rising numbers of refugees seeking political asylum, have tightened their previously liberal asylum policies. Since September 11, the immigration issue has frequently been reconfigured as a security issue and national governments across Europe have voted to restrict the conditions under which asylum would be granted. This is despite the fact that many European countries are in need of immigrants to help counter labor shortages in skilled and unskilled jobs. In short, in the last several years national elections in which xenophobic politicians were either elected or garnered a sizable portion of the vote in normally liberal countries, such as the

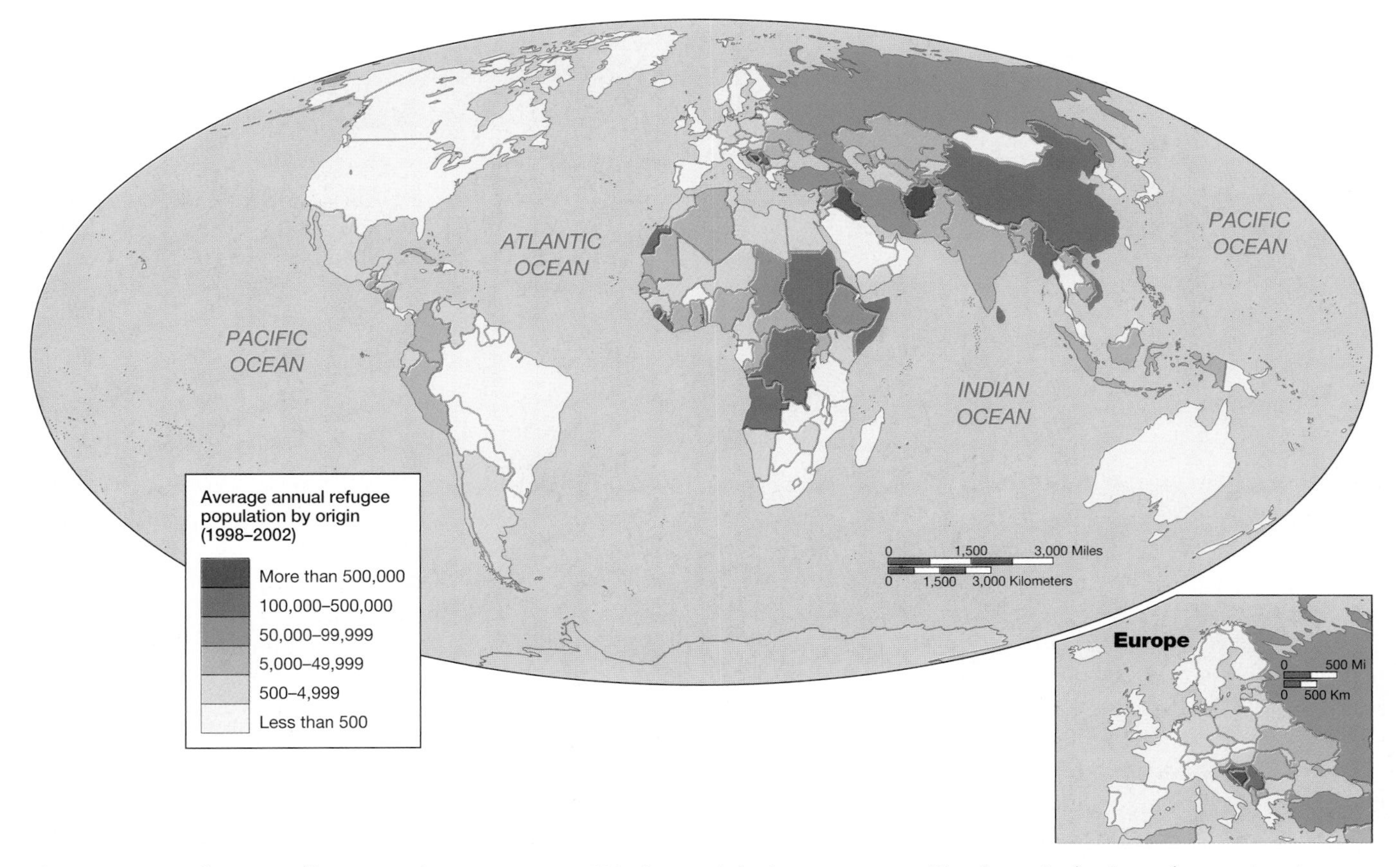

Figure 3.21 Refugee-sending countries, 1998–2002 War is certainly the most compelling factor in forcing refugee migration. Shown are the sending countries, those whose internal situations propelled people to leave. What is perhaps most distressing about this graphic is that refugee populations have increased over the last decade almost exclusively in the periphery. (After http://www.unhcr.ch/static/statistical_yearbook/2002/maps04.pdf)

3.3 GEOGRAPHY MATTERS

Internal Displacement

While the global humanitarian community has paid close attention to refugee populations for decades, the problem of internally displaced persons has only recently begun to draw international attention. In 2004, around the globe, there were an estimated 25 million internally displaced persons (IDPs)—individuals who were uprooted within their own countries due to civil conflict or human rights violations, sometimes by their own governments (**Figure 3.E**). This number is actually twice as high as the global refugee population and it is frequently the case that the plight of IDPs is actually worse than refugees. This is because their own governments are either unable or unwilling to provide them with the protection or assistance they have a right to expect. It is also the case that the international community is either unaware of them or has not secured the resources needed to help them (**Table 3.B**).

TABLE 3.B Regional Distribution of IDPs (2003) and Refugees (2002)

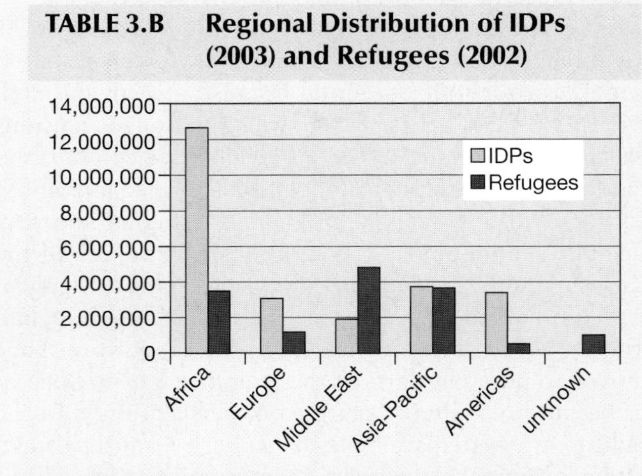

Source: Global IDP Database, UNHCR, USCR, UNWRA.

Two of the world's regions with the most difficult IDP challenges are Africa and Latin America. Africa possesses nearly 13 million IDPs caused by rebel activities and inter-communal violence as well as some direct abuse by national governments. National security forces and government-backed militias have deliberately displaced large numbers of people in Zimbabwe, the Democratic Republic of Congo and Côte d'Ivoire. In Liberia and Somalia, IDPs had virtually nowhere to go to escape attacks and find safe shelter; many were killed or died of hunger and disease. In Sudan, more than half a million people fled the western state of Darfur in 2003, as

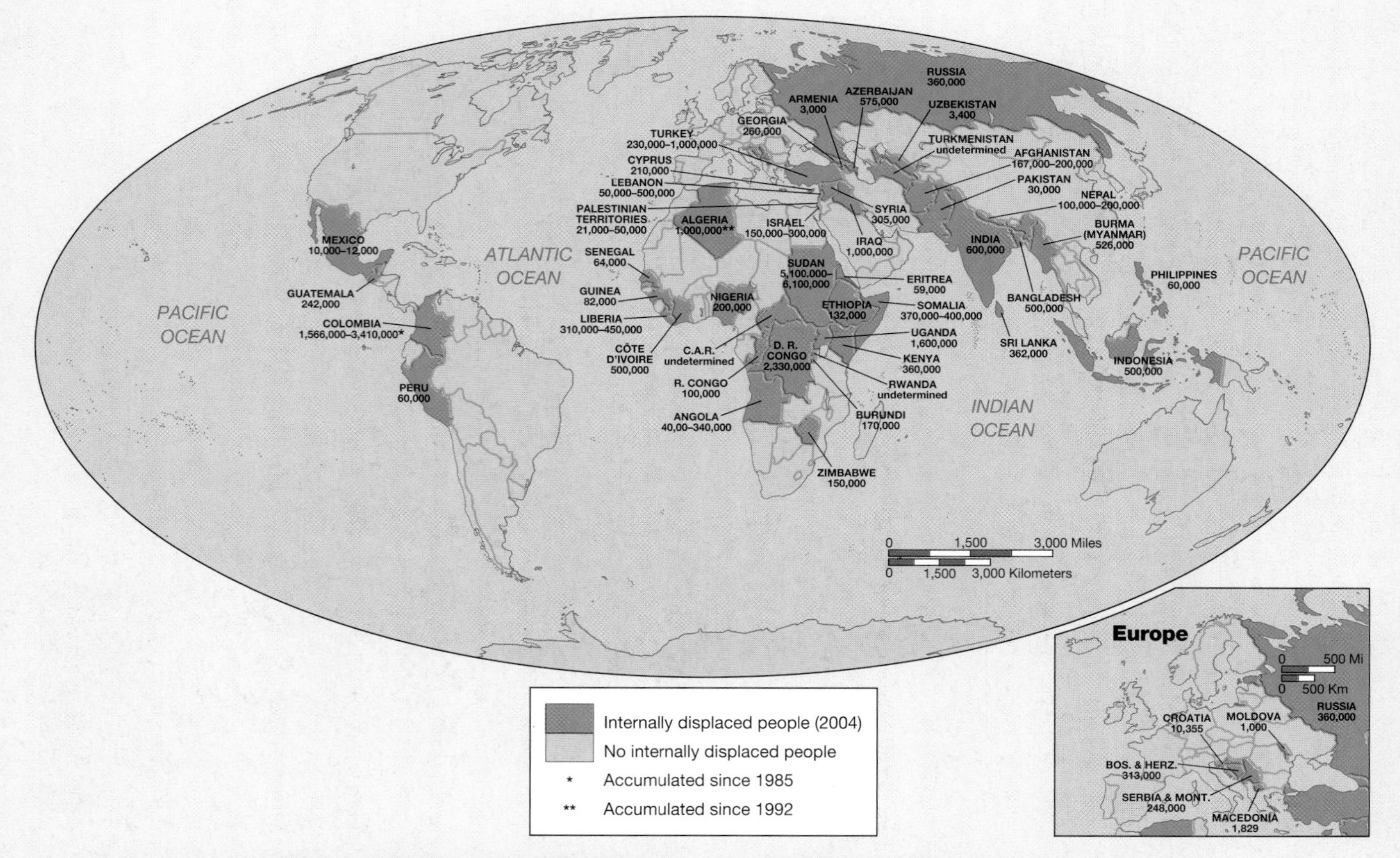

Figure 3.E Internally Displaced People Worldwide, 2004 This map provides an indication of the geographical extent and numbers of people displaced from their livelihoods and homes but still living within their own countries, oftentimes in temporary camps provided by humanitarian organizations or the U.N. (*Source:* Norwegian Refugee Council, Global Internally Displaced Persons Project, http://www.idpproject.org/index.htm)

Figure 3.F Darfur, Sudan internally displaced persons Shown here are Sudanese women and children who have fled the village of Darfur in order to escape the attacks of the Janjaweed tribal militia, sponsored by the Sudanese government, that have been occuring throughout the southern region of the country since early 2003. In mid-2005 there were nearly 2 million internally displaced Sudanese in camps like this one.

a response to attacks by government troops and raids by militias reportedly backed by the government (**Figure 3.F**). In the South American country of Colombia, both guerrillas and paramilitaries continue to target civilian populations through arbitrary killings, looting, and destruction of property in order to depopulate rural areas for political and economic gains and to control or regain strategic territories. Often such acts have been accompanied by other serious human rights violations such as torture and rape. Following the end of civil war in Peru (2000) and Guatemala (1986), where military campaigns were launched by the governments against the population—and where similar abuses had occurred—the process of returning IDPs to their homes has been only partially implemented.

The U.S.-led "war on terror" has had a decidedly negative effect on the worldwide displacement crisis by encouraging governments to seek military solutions to conflicts. As a result, international humanitarian and human rights standards, including those relating to the protection of IDPs, have been undermined or ignored altogether in numerous places around the globe (**Table 3.C**). In fact, there is little doubt that the international anti-terrorism campaign has enabled some non-democratic governments to characterize armed opposition movements as "terrorists" and to present their counter-insurgency operations as part of the international "war on terror." This

TABLE 3.C Countries Where the War on Terror Has Affected the Protection of IDPs

Afghanistan
Colombia
Indonesia
Iraq
Nepal
Israel/Palestinian Territories
Philippines
Russian Federation
Uganda

Sources: Global IDP Database, UNHCR, USCR, UNWRA

Figure 3.G Peace and return in the Balkans Pictured here are Muslims who have returned to Republika Srpska in Bosnia to rebuild their homes and recuperate their lives. Since the 1990s wars have ended about 1.8 million people have returned to the Balkans.

posture has earned these governments—many with a long history of instability, military coups and human rights violations—substantial military support, mainly from the United States. In Indonesia and the Philippines, for example, tens of thousands of people have been displaced because of counter-insurgency operations conducted under the banner of the "war on terror." Importantly, these military campaigns were ongoing before the "war on terror" was launched in 2001. And yet, by invoking the rhetoric of terror, they have, ironically, undermined the protection of civilians causing them to flee their homes for safe havens elsewhere in the country.

The IDP story, however, is not an entirely pessimistic one. On the positive side, almost two million people in Angola and a half a million people in Indonesia have been able to return to their homes. There is also some reason to be optimistic that peace processes underway in Liberia, Burundi, the Balkans, and Sri Lanka might result in the full return of displaced persons (**Figure 3.G**). This is already beginning to occur in Sri Lanka and the Balkans.

Source: Adapted from *Internal Displacement: A Global Overview of Trends and Developments in 2003,* by the Global IDP Project of the Norwegian Refugee Council, Geneva, 2003.

Netherlands, Switzerland and France, make it quite clear that Europeans are becoming increasingly resistant to absorbing refugee populations.

In other parts of the world, though, refugee populations are pouring across borders and creating very difficult conditions for national governments. Sudan, for example, is in the difficult position of having displaced several million of its own citizens as a result of civil war, while at the same time hosting over 750,000 refugees from neighboring countries. The situation provides a clear example of the openness of the borders that define many African states in contrast to the tight border security in most core

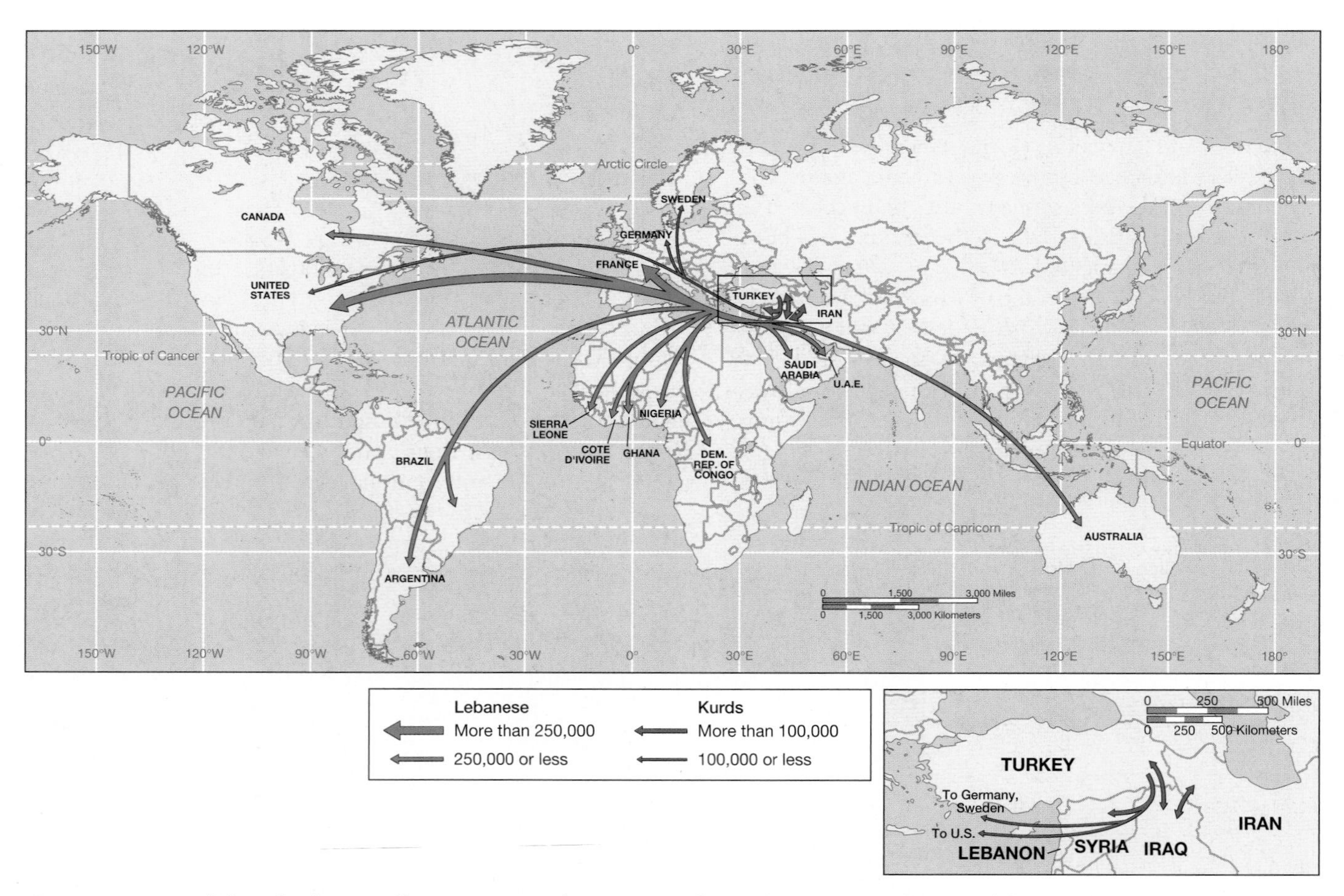

Figure 3.22 Kurdish and Lebanese diaspora, 1990 The most significant diasporic populations of the Middle East and North Africa during modern times are the Palestinians, Lebanese, and Kurdish peoples. This map shows the scattering of the Lebanese and the Kurds (the dispersal of the Palestinians is shown in Figure 3.23). (After Aaron Segal, *An Atlas of International Migration.* London; Hans Zell, 1993, pp. 95 and 103.)

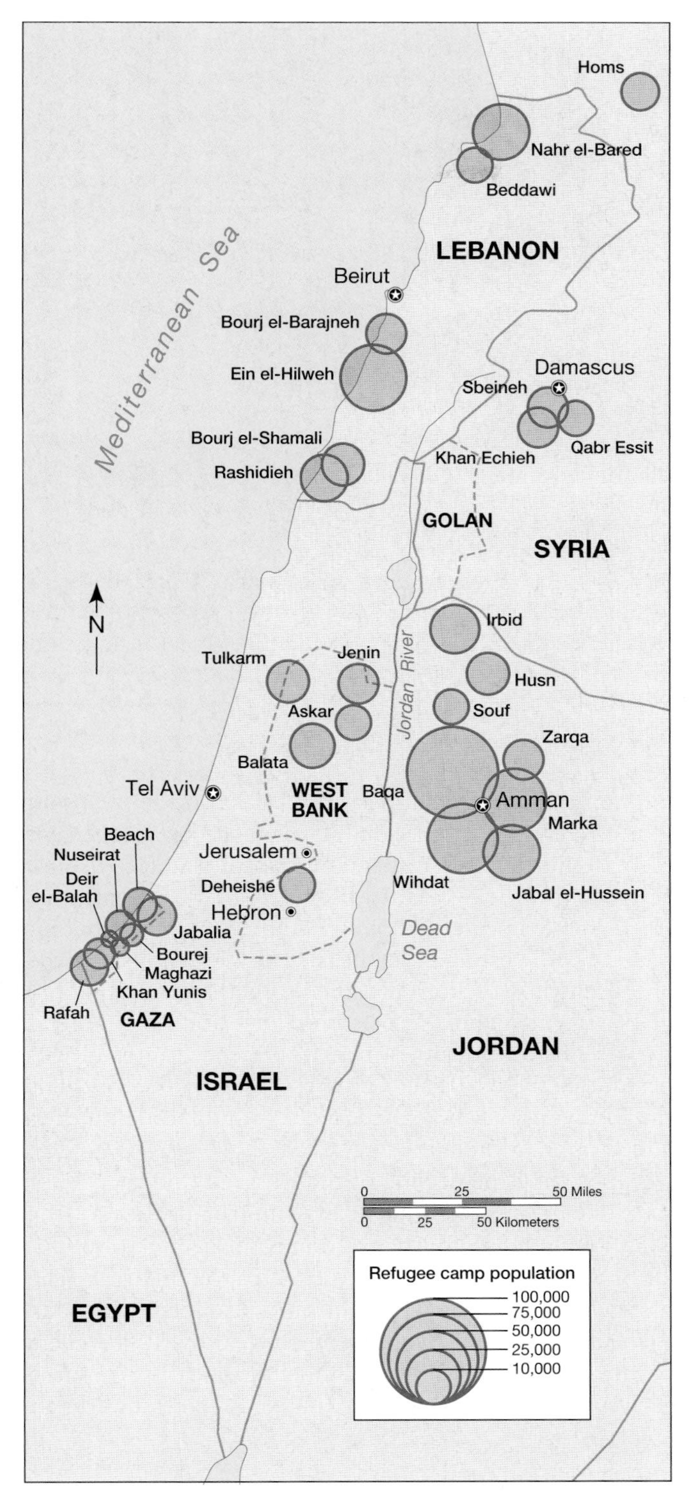

Figure 3.23 Palestinian refugees in the Middle East, 2000 This map shows the dispersion of Palestinian refugees—in camps and elsewhere—in the states around Israel. One of the biggest sticking points in negotiations between Israelis and Palestinians is the question of whether refugees will be allowed to return and, if so, where they will be allowed to settle, given that most of their land has been occupied by Israeli settlers. More than 200,000 Jews have settled on Palestinian land. The settlers include some of the most radical Jewish nationalists in the region, and although the most recent round of removal resulted in protests, the settlers were ultimately forced to move as sections of the West Bank were returned to Palestinian control. (After *The Guardian,* 14 October 2000, p. 5.)

countries. In some cases, refugee populations remain in a host country for long periods of time. In Jordan and Lebanon, for example, some of the Palestinian refugee camps are multigenerational, dating back to the 1948 Arab–Israeli war. High fertility rates in some of these refugee camps put considerable strain on the humanitarian aid resources of both international organizations and the host countries.

Internal Voluntary Migration

One way to understand the geographical patterns of migration is to think in terms of waves of migration. In the United States, for example, three important and overlapping waves of internal migration over the past two centuries altered the population geography of the country. As with a great deal of migration activity, these three major migrations were tied to broad-based political, economic, and social changes.

The first internal migration wave began with colonization and increased steadily through the twentieth century. This wave had two parts. The first was characterized by the large movement of people from the settled Eastern Seaboard into the interior of the country. This westward expansion, which began spontaneously during the British colonial period—in blatant disregard for British restrictions on such expansion—became official settlement policy after the American Revolution. The federal government encouraged migration over the course of more than a century as part of the country's expansionist strategies. **Figure 3.24** illustrates how the demographic center of the United States moved westward in the years between 1790 and 2000 as more and more areas were settled, largely through transportation and communications innovations, capital investment, and large internal population movements. A second part consisted of massive rural-to-urban migration associated with industrialization, especially occurring during the mid-nineteenth to the early twentieth centuries. **Table 3.5** illustrates how, between 1860 and 1920, the United States was transformed from a rural to an urban society as industrialization created new jobs and as increasingly redundant numbers of agricultural workers (along with foreign immigrants) moved to urban areas to work in the manufacturing sector.

The second great migration wave, which began early in the 1940s and continued through the 1970s, was the massive and very rapid movement of mostly African Americans out of the rural South to cities in the South, North, and West. Although African Americans formed considerable populations in cities such as Chicago and New York before the onset of this wave, the mechanization of cotton picking pushed additional large numbers of these people out of the rural areas. Tenant cotton picking was a major source of livelihood for African Americans in the Deep South until mechanization reduced the number of jobs available. At the same time, pull factors attracted them to the large cities. In the early 1940s, large numbers of jobs

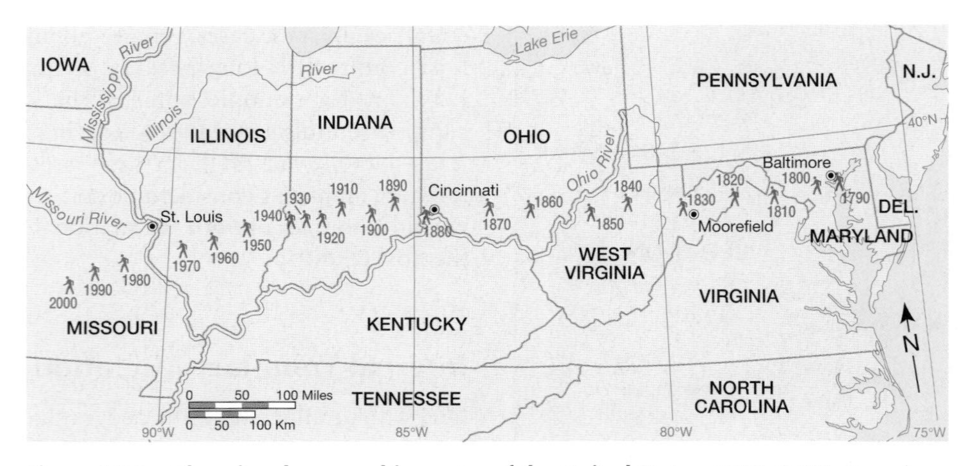

Figure 3.24 Changing demographic center of the United States, 1790–2000 Knowing where the demographic center of a country's population is located at given times allows us to track that population's movement, growth, distribution, and concentration. Here we can clearly perceive the expansion of the American frontier and see that although the East continued to be the most populated part of the country early on, the demographic center moved as more and more of the population dispersed to the West and the population of the West grew through natural increase. (Reprinted with permission of Prentice Hall, adapted from J. M. Rubenstein, *The Cultural Landscape: An Introduction to Human Geography,* 5th ed., © 1996, p. 122, and updated with data from the U.S. 2000 census.)

in the defense-oriented manufacturing sector became available as other urban workers joined the war effort. This second wave of migration can be seen as part of a wider pattern of rural-to-urban migration among agricultural workers as industrialization spread globally.

After World War II a more important catalyst drove this migration: the increasing emphasis on high levels of mass consumption, which reoriented industry toward production of consumer goods and which in turn stimulated large increases in the demand for unskilled and semiskilled labor. The impact on the geography of racial distribution in the United States was profound.

The third internal migration wave began shortly after World War II and continues into the present. Following the end of the war and directly related to the impact of governmental defense policies and activities on the country's politics and economy, the region of the United States lying below the thirty-seventh parallel, also known as the Sunbelt, emerged. Between 1950 and 1990, this region, which includes fifteen states and extends from North Carolina in the east to southern California in the west and Florida and Texas to the south, experienced a 97.9 percent increase in population. Beginning in the late 1970s and early 1980s, the West also began to grow dramatically. At the same time that the West and the South were booming, the Midwest and Northeast, known variously as the Snowbelt, Frostbelt, or Rustbelt, together grew by only 33.3 percent. **Figure 3.25** is one illustration of the relative decline and disinvestment experienced in the Rustbelt during the 1970s and early 1980s. Since the late 1980s the Midwest and the Northeast have stabilized and even added population in most cases.

Preceding the decline of the Rustbelt and the rise of the Sunbelt has been the steady suburbanization of the U.S. population. **Suburbanization** is the growth of population along the fringes of large metropolitan areas. The first evidence of U.S. suburbanization can be traced back to the late eighteenth century, when wealthy city-dwellers began seeking more scenic residential locations. By the early twentieth century, residents fled to the suburbs to

TABLE 3.5 United States Rural to Urban Population Change, 1860–1920

Between 1860 and 1920, the United States went from a primarily agricultural country to a primarily industrial one. In the United States industrialization was consistently associated with urbanization: As the country industrialized, its population urbanized, with machines replacing human labor on the farm and former farmhands and others operating machines in factories.

	1860	1870	1880	1890	1900	1910	1920
Total U.S. population (in millions)	31.4	38.5	50.1	62.9	75.9	91.9	105.7
Total urban population (in millions)	6.2	9.9	14.1	22.1	30.1	41.9	54.1
% Urban	19.8	25.7	28.2	35.1	39.7	45.7	51.2
% Increase in total population	—	22.7	30.1	25.5	20.7	21.0	14.9
% Increase in urban population	—	59.3	42.7	56.4	36.4	39.2	29.0

Figure 3.25 U.S. rust belt In the mid-twentieth century, the most important automobile manufacturers in the United States located their factories in the Midwest, where skilled labor and raw materials were readily available. By the late 1970s and 1980s, cities like Flint, Michigan, began losing their automobile employment base as companies, such as Ford, moved to suburban locations in the Midwest or out of the region either to the South or to foreign locations, such as Mexico, where labor is far cheaper and environmental laws less stringent.

get away from the new immigrants and their increasing hold over urban political machines (**Figure 3.26**).

The process really took on its own life, however, with the introduction of new transportation technologies—first with horse-drawn streetcars, then intra-urban rail services, and finally automobiles. Each innovation in transportation enabled people to travel longer distances to and from work within the same or shorter time period. They chose to move to the suburbs in massive numbers, not in the least because the suburbs were—and are—considered by many to be healthier and more desirable places to raise a family. It is important, however, to understand that deeper political and economic factors, beyond individual lifestyle choices, were also at work.

The most compelling explanation for the large-scale population shift characteristic of the third migration wave was the pull of economic opportunity. Rather than being invested in upgrading the aged and obsolescent urban industrial areas of the Northeast and Midwest, venture capital was invested in Sunbelt locations, where cheaper land, lower labor costs, and the absence of labor union organizations made the introduction of manufacturing and service-sector activity more profitable. By the 1990 census there was a decrease in the amount of in-migration to the Sunbelt, but the geography of the U.S. population at the end of the twentieth century is nevertheless almost diametrically opposed to that of 150 years ago. This new population distribution—largely suburban and more evenly spread across the country—illustrates the way in which political and economic transformations play an especially significant role in shaping individual choice and decision-making.

In other parts of the world, rural-to-urban migration trends have changed population geographies in equally dramatic ways, although local manifestations have often been quite different from those experienced in the United States. Shantytowns ring all major cities of the periphery countries, posing almost insurmountable problems for city managers attempting to provide the most basic health, sanitation, educational, and occupational

Figure 3.26 Suburbanization, 2005 Suburban growth includes not only large tracts of residential development, as pictured here, but large retailers and strip malls, mostly oriented to automobile dependency. As energy costs continue to climb in the first decade of the twentieth century, some urbanists argue that the suburbs will experience problems of mobility and access.

services. In some places, such as Brazil, an effort is ongoing to redirect these migrants to other locations. The Brazilian government constructed highways into the Amazon region, which has drawn some migrants into the frontier area. Nevertheless, pressures on major cities like Rio de Janeiro remain overwhelming.

Another approach aimed at redressing a severe population imbalance has been taken by Indonesia. In 2000, Indonesia's population surpassed 200 million people, more than 60 percent of whom live on the islands of Java and Bali (these two islands comprise just 7 percent of the nation's land base). Because of tremendous population growth, 2 million new jobseekers enter the market annually, causing the government to create programs aimed at inducing the people to move to less densely populated locations.

Efforts to redistribute the population to the less densely settled islands of Indonesia date back to the turn of the nineteenth century, when the area was still under Dutch colonial rule. For many years these efforts involved offering incentives to selected landless and jobless people to move to new agriculturally based settlements in the Outer Islands. Since 1984, however, facing a limited development budget, the government has turned to sponsoring investment in labor-intensive enterprises in established settlement areas. To date, over 1.5 million people have migrated, more than 60 percent of them to one large island: Sumatra.

Internal Forced Migration

One of the best-known forced migrations in the United States was the "Trail of Tears," a tragic episode in which nearly the entire Cherokee Nation was forced to leave their once treaty-protected Georgia homelands for Oklahoma. **Figure 3.27** illustrates this forced migration as well as the removal of four other eastern tribes to territory west of the Mississippi. Despite sustained legal resistance, approximately 16,000 Cherokees were forced to march across the continent in the early 1830s, suffering from drought, food scarcity, bitterly cold weather, and sickness along the way. By some estimates at least a quarter of the Cherokees died as a result of the removal.

Placed within the national and international context, the movement of Native American populations during the nineteenth century can be seen as a response to larger political and economic forces. European populations were participating in a massive migration to the United States, and the national economy was on the threshold of an urban-industrial revolution. The eastern Native American populations posed an obstacle to economic expansion, which itself was dependent upon geographic expansion. Growing Anglo-American prosperity, it was thought, had to be secured by taking Indian land.

Other recent examples of internal forced migration are provided by China and South Africa. In the late 1960s and 1970s, the government of China forcibly relocated 10 to 17 million of its citizens to rural communes in order to enforce Chinese Communist dogma and to ease pressures arising from high urban unemployment. The policy has since been disavowed, but the effects on an entire generation of young people were profound. Another example took place in South Africa between 1960 and 1980, when apartheid policies forced some 3.6 million blacks to relocate to government-created homelands, causing much suffering and dislocation. Indeed, civil war, ethnic conflict, famine, deteriorating economic conditions, and political repression have produced an extraordinary series of internal forced migrations in several sub-Saharan coun-

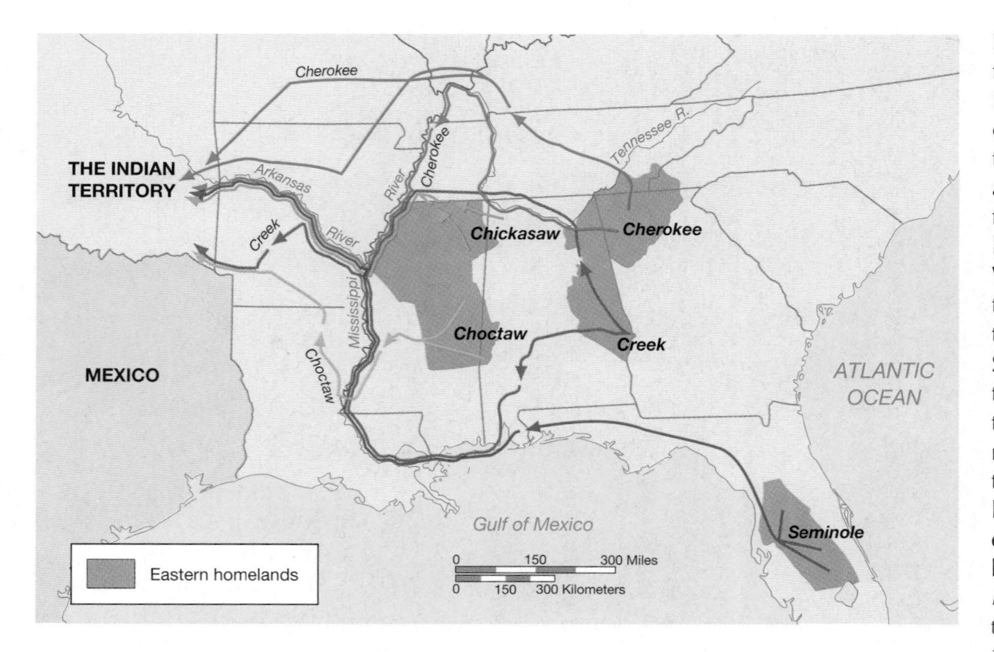

Figure 3.27 Trail of Tears, 1830s In the Indian Removal Act of 1830, lands held by eastern Native American tribes were exchanged for territory that the federal government had acquired west of the Mississippi, effectively eliminating Indian title to land previously protected by treaties. Without government protection of their rights to their land, members of the Choctaw, Chickasaw, Cherokee, Seminole, Creek, and other eastern tribes had to move west to the lands that had been acquired for them. The map shows the routes taken by these tribes as they vacated their ancestral lands for the federally designated Indian Territory in what is now Oklahoma. (After *Atlas of the North American Indian,* by Carl Waldman; maps and illustrations by Molly Braun. Copyright © 1985 by Carl Waldman. Reprinted by permission of Facts on File, Inc.)

tries. These forced migrations, both internal and international in scope, become particularly significant in light of changing population. Forecasts predict that 80 percent of the world population increase in the next decade will take place within the poorest countries of the world. Many of these countries have some of the highest rates of forced migration. The combination bodes ill for improving these countries' prospects for economic and political improvement.

Today forced **eco-migration**—population movement caused by the degradation of land and essential natural resources—has created a new category of migrants and refugees. In Bangladesh, for example, floodplain settlement that began in the 1960s has led to severe losses of life and property and forced temporary relocation of huge numbers of people whenever severe flooding occurred. In Ethiopia the 1984–1985 famine was attributed not only to drought and pest infestation but also to governmental policies and actions that favored urban populations and rendered rural populations especially vulnerable to environmental stress. In the summer of 2002, 10 million Africans in Malawi, Zimbabwe, Lesotho, and Angola faced starvation, with hundreds of thousands having left their villages in hopes of food and employment opportunities in urban areas in the region and beyond (**Figure 3.28**). The tsunami that hit South Asia and parts of southern Africa on December 26, 2004 and killed anywhere from 228,000 to 310,000 people is an example of how natural disasters and tenuous environmental conditions have their greatest impact on poor people, in both the core and the periphery, whose only coping mechanism in the face of impending disaster or devastation is migration. A recent U.N. report indicates that as a result of the tsunami, 111,073 fishing vessels were destroyed or damaged; 36,235 engines were lost or damaged beyond repair; and 1.7 million units of fishing gear—such as nets, tackle, and similar equipment—were destroyed. Without the tools of their livelihood, the poor fishers along the southern coasts of peripheral countries like India, Indonesia, Sri Lanka, Maldives, Burma, and Thailand must find the resources to purchase or build new boats and fishing gear—which is unlikely, as most have no savings to fall back on—or move elsewhere in search of work.

Figure 3.28 Famine migrants Famine is gripping parts of Africa once again as the twenty-first century begins. Drought is the main cause of famine in Zimbabwe. Pictured here are a man and his son who are waiting outside their home for food aid.

POPULATION DEBATES AND POLICIES

One big question occupies the agenda of population experts studying world population trends today: How many people can Earth sustain without depleting or critically straining its resource base? The relationship between population and resources, which lies at the heart of this question, has been a point of debate among experts since the early nineteenth century.

Population and Resources

The debate about population and resources originated in the work of an English clergyman named Thomas Robert Malthus (1766–1834), whose theory of population relative to food supply established resources as the critical limiting condition upon population growth. Malthus's theory was published in 1798 in a famous pamphlet called *An Essay on the Principle of Population*. In this tract Malthus sets up two important postulates:

- Food is necessary to the existence of human beings.
- The passion between the sexes is necessary and constant.

It is important to put the work of Malthus into the historical context within which it was written. Revolutionary changes—prompted in large part by technological innovations—had occurred in English agriculture and industry and were eliminating traditional forms of employment faster than new ones could be created. This condition led to a fairly widespread belief among wealthy members of English society that a "surplus" of unnecessary workers existed in the population. The displaced agriculturists began to be a heavy burden on charity, and so-called poor laws were introduced to regulate begging and public behavior.

In his treatise, Malthus insisted that "the power of the population is indefinitely greater than the power of the earth to produce subsistence." He also believed that if one accepted this premise, a natural law would follow; that is, the population would inevitably exhaust food supplies. Malthus's response to this imbalance was to advocate for the creation of laws to limit human reproduction, especially among poor people.

Malthus was not without his critics, and influential thinkers such as William Godwin, Karl Marx, and Frederich Engels disputed his premises and propositions. Godwin argued that "there is no evil under which the human species cannot labor, that man is not competent to cure." Marx and Engels were in general agreement and insisted that technological development and an equitable distribution of resources would solve what they saw as a fictitious imbalance between people and food.

The debate about the relationship between population and resources continues to this day, with provocative and compelling arguments for both positions being continually advanced. The geographer David Harvey, for example, has explored the population-resources issue in great detail. He has shown that by adopting Malthus's approach, only one outcome is possible—a doomsday conclusion about the limiting effect of resources on population growth. By following Marx's approach, however, quite different perspectives on, and solutions to, the population-resources issue can be generated. These solutions are based on human creativity and socially generated innovation, which allow people to overcome the limitations of their environment.

Neo-Malthusians—people today who share Malthus's perspective—predict a population doomsday. They believe that growing human populations the world over, with their potential to exhaust Earth's resources, pose the most dangerous threat to the environment. Although they acknowledge that the people of the core countries consume the vast majority of resources, they and others argue that only strict demographic controls everywhere will solve the problem, even if they require severely coercive tactics.

A more moderate approach argues that people's behaviors and governmental policies have a much greater impact on the condition of the environment and the state of natural resources than population size in and of itself. Proponents of this approach reject casting the population issue as a biological one in which an ever growing population will inevitably create ecological catastrophe. They also reject framing it as an economic issue in which technological innovation and the sensitivities of the market will regulate population increases before a catastrophe can occur. Rather, they see the issue as a political one—one that governments have tended to avoid dealing with because they lack the will to redistribute wealth or the resources to reduce poverty, a condition strongly correlated with high fertility.

The question of whether too many people exist for Earth to sustain has bedeviled population policymakers and political leaders for most of the second half of the twentieth century. This concern has led to the formation of international agencies that monitor and often attempt to influence population change. It has also led to the organizing of a series of international conferences that have attempted to establish globally applicable population policies. The underlying assumption of much of this policymaking is that countries and regions have a better chance of achieving improvement in their level of development if they can keep their population from outstripping the supply of resources and jobs.

Population Policies and Programs

Contemporary concerns about population—especially whether too many people exist for Earth to sustain—have led to the development of international and national policies and programs. A **population policy** is an official government policy designed to affect any or all of several objectives, including the size, composition, and distribution of population. The implementation of a population policy takes the form of a population program. Whereas a policy identifies goals and objectives, a program is an instrument for meeting those goals and objectives.

Most of the international population policies of the last two decades have attempted to reduce the number of births worldwide. The main instruments developed to address rising fertility took the form of family-planning programs. The desire to limit fertility rates by the international population-planning community is a response to concerns about rapidly increasing global population—an increase that is more significant in the periphery and semiperiphery than in the core countries. Accompanying this situation of imbalanced population growth between the core and the periphery is gross social and economic inequality, as well as overall environmental degradation and destruction.

Figure 3.29 provides a picture of the recent history of world population growth by region and a reasonable projection of future growth. The difference between core and periphery is dramatically illustrated. In mid-2000 the world contained 6 billion people. This means that by the year 2050 the world is projected to contain nearly 9 billion people. In comparison, over the course of the entire nineteenth century, fewer than a billion people were added to the population. **Figure 3.30** shows a projection of world population in the year 2020.

The geography of projected population growth is noteworthy. Over the next century, population growth is predicted to occur almost exclusively in Africa, Asia, and Latin America, while Europe and North America will experience very low, and in some cases zero, population

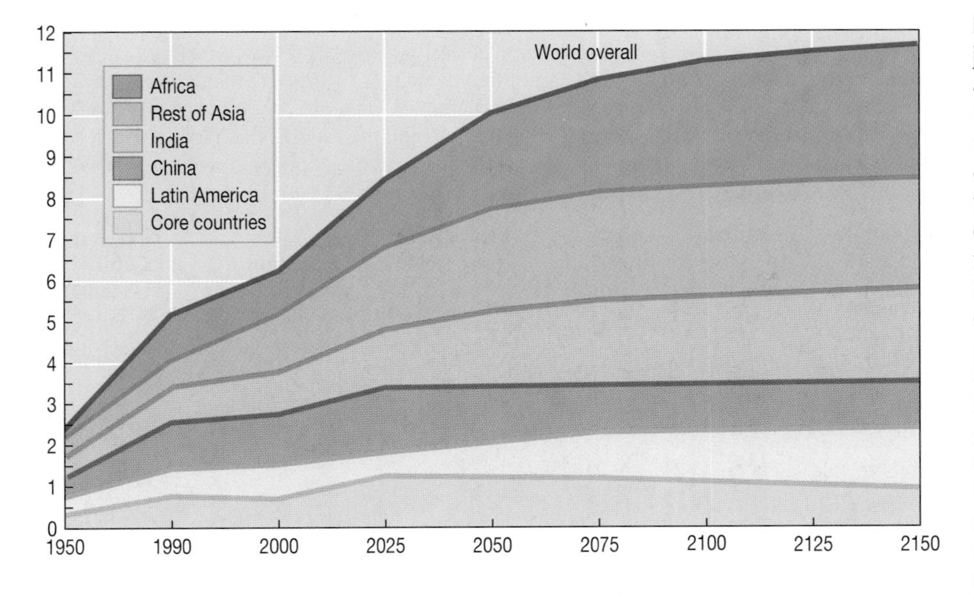

Figure 3.29 World population projection by region This graph represents a medium-variant projection, one that is in the middle of three possible scenarios. In this projection, population continues to expand in the periphery, though in some regions more than others. Africa is projected to experience the greatest growth, followed by Asia (not including China), where growth is expected to level off by 2150. Less dramatic growth is expected to occur in Latin America, while in the core population, numbers remain constant or drop slightly. Though the total number of people in the world will be dramatically greater by 2150, the medium-variant forecast indicates a gradual leveling off of world population. (After I. Hauchler and P. Kennedy (eds.), *Global Trends: The World Almanac of Development and Peace.* New York: Continuum, 1994, p. 109.)

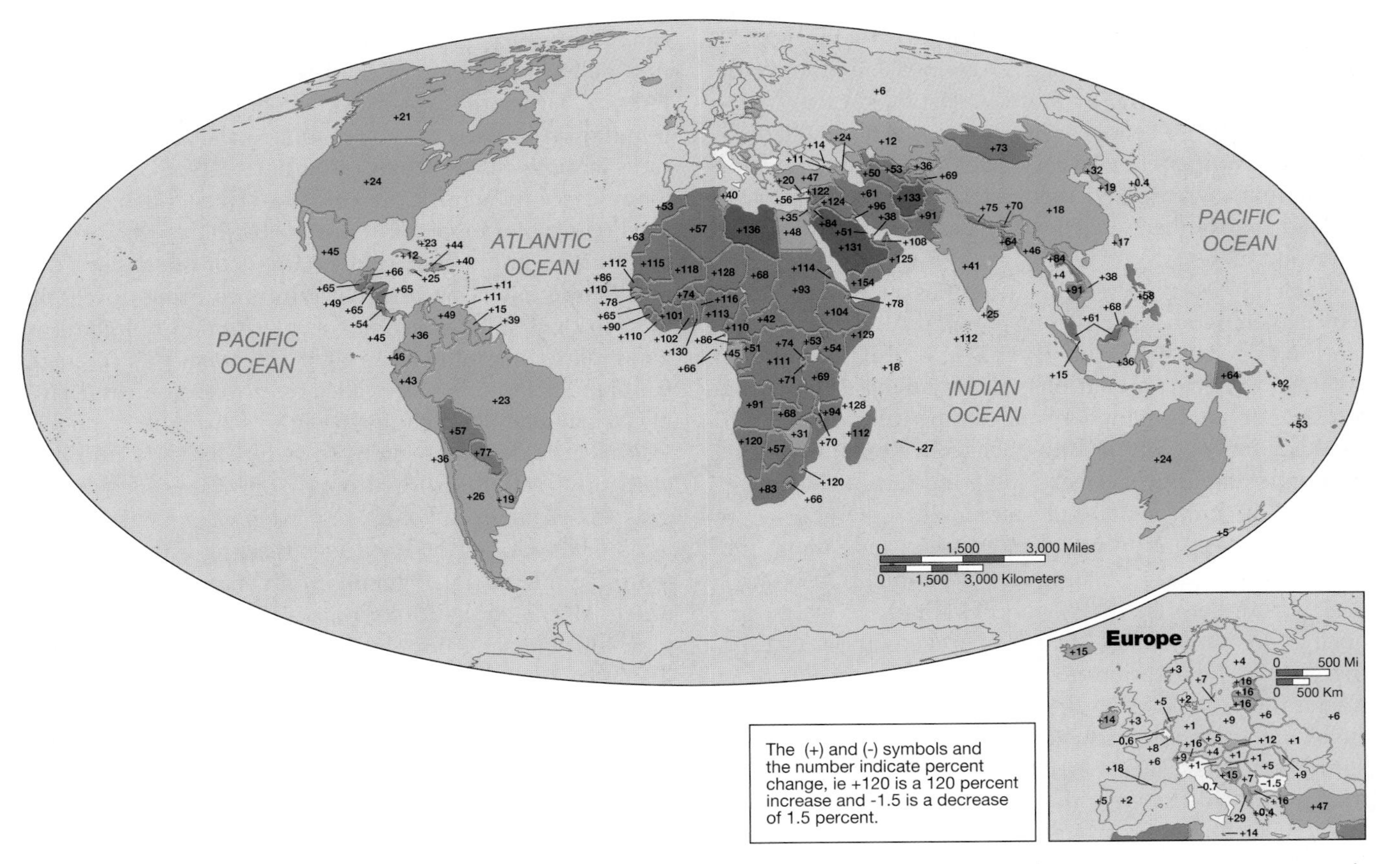

Figure 3.30 World population, 2020 This map provides a sense of how much the populations of various countries are expected to change by the year 2020. Although the populations of nearly all countries are expected to increase, it is clear that some populations will grow far more dramatically than others. Notice the substantial growth expected in Saudi Arabia and Afghanistan in contrast to the United States and Europe, where little if any population growth will occur. Italy, for example, is expected to lose population, while the Netherlands will grow by only 5 percent.

TABLE 3.6 Global Demographic Indicators, 2004

This table illustrates the substantial population pressures emanating from the global periphery which, as of 1998, contained roughly four times the population of the core with a rate of natural increase also nearly four times as great. While Asian countries possess the largest proportion of the world's population, their rate of natural increase is considered moderate and their level of economic development—measured by their GNP—higher than the average overall for the periphery. Most troubling to demographers, however, are the population dynamics of Africa, with a rate of natural increase of 2.4 percent per year and a GNP of only $2,100 per capita.

Region	Population Mid-2004 (millions)	Natural Increase (annual %)	Birthrate (per 1,000 pop.)	Death Rate (per 1,000 pop.)	Life Expectancy (years at birth)	GNP Per Capita ($US 2004)
World	6,396	1.3	21	9	67	7,590
More developed	1,206	0.1	11	10	76	23,690
Less developed	5,190	1.5	24	8	65	3,850
Africa	885	2.4	38	14	52	2,100
Asia	3,875	1.3	20	7	67	4,610
Latin America and Caribbean	549	1.6	22	6	72	6,820
Europe	728	−0.2	10	12	74	17,730
North America	326	0.5	14	8	78	35,390
Australia, New Zealand, and the South Pacific Islands	33	1.0	17	7	75	19,960

Source: 2004 World Population Data Sheet. Washington, DC: Population Reference Bureau.

growth. The differing rates of natural increase listed in **Table 3.6** illustrate this point.

At the turn of the twentieth century the core contained 32 countries with zero population growth. By contrast, the periphery contained 39 countries with rates of natural increase of 3.0 or more. A sustained rate of natural increase of 3.0 per year means a population will double in approximately 24 years. Such remarkable rates of population increase and their relationship to quality of life and environment have led to a concerted international response.

Since 1974, the United Nations has sponsored international conferences at 10-year intervals to develop population policy at the global level. Out of each conference emerged explicit population policies aimed at lowering fertility rates in the periphery and semiperiphery. The 1974 world conference was held in Bucharest, Romania, where two opposing approaches to lowering fertility were expressed. In fact, general disagreement emerged as to whether a population problem even existed. While the core countries were warning that a "population bomb" was ticking because the peripheral countries were producing too many babies, the peripheral countries accused the core of advocating genocide. They argued that it wasn't the periphery but the core that was degrading the environment with its industry and heavy consumption of natural resources, such as petroleum. The peripheral countries promoted the view that the most effective path to lowering fertility was through economic development, not the family-planning policies that the core was advocating. Ten years later, in Mexico City, the positions were reversed and peripheral countries were pressing for increased family-planning programs and the distribution of contraceptives, while core countries had adopted the perspective that "development is the best contraceptive."

The 1994 world population conference took place in Cairo, Egypt. Called the International Conference on Population and Development, the name itself reflects the changes in ideas and practices that 20 years of population policy and programs had wrought. The Cairo conference, rather than focusing on the increase in global population growth rates, pointed to the fact that birth rates in almost every country on Earth are dropping—in many cases by significant amounts. Recognizing that a leveling off of the population was possible in the foreseeable future, conference participants from both the core and the periphery agreed that efforts to bring down the growth rates must continue so that human populations will peak sooner rather than later. The policy that emerged from the Cairo conference called for governments not simply to make family-planning programs available to all but also to take deliberate steps to reduce poverty and disease; improve educational opportunities, especially for girls and women; and work toward environmentally sustainable development. The Cairo meeting was the first of the three international population conferences in which nearly all the core and peripheral countries agreed on a plan for achieving population stabilization, a plan that would encourage freedom of choice in the matter of family size, population policies and programs, and development policies and practices. The goals established at the Cairo meeting have been embraced and extended in the most recent population conference held at the United Nations in 2000, the U.N. Millenium Summit (**Figure 3.31**).We discuss the summit in greater detail below.

All the world population conferences have recognized that the history, social and cultural practices, development level and goals, and political structures for countries and even regions within countries are highly variable, and that one rigid and overarching policy to limit fertility will not

Figure 3.31 U.N. Millennium Summit, 2000 U2 star Bono, Nigerian President Olusegin Obasanjo, and U.N. Secretary General Koffi Anan at the World Summit in New York where Bono presented a petition signed by 21 million people urging world leaders of the G8 nations to cancel peripheral countries' debt. Debt forgiveness is believed to be critical to enabling peripheral countries to develop economically and thus lower birth rates through an improving quality of life.

work for all. Whereas some programs and approaches will be effective for some countries seeking to cut population numbers, they will be fruitless for others.

For instance, China's family-planning policy of one child per household appears to be effective in driving down the birth rate and damping the country's overall population growth. Because so little data is available on China, however, it is not clear whether the policy is operating in the same fashion throughout the country. For instance, some population experts believe an urban-rural bias exists. That is, while the policy is adhered to in the cities, it is disregarded in the countryside.

Family-planning regulations include authorizing the age of marriage, offering incentives to couples to have only one child, and mandating increasingly severe disincentives to couples who have larger families. In India, family-planning policies offering free contraceptives and family-planning counseling have resulted in lowering the birth rate. Other countries, such as Sri Lanka, Thailand, and Cuba, as well as the Indian state of Kerala, have lowered their birth rates not through regulation of family size but by increasing access to social resources such as health care and education, particularly for women.

Indeed, it is now a widely accepted belief among demographers and policymakers that a close relationship exists between women's status and fertility. Women who have access to education and employment tend to have fewer children because they have less of a need for the economic security and social recognition that children are thought to provide (**Figure 3.32**). In Botswana, for instance, women with no formal education have, on average, 5.9 children, while those with four to six years of school have just 3.1 children. In Senegal, women with no education give birth to an average of seven children. In contrast, the average number of children born to a woman with 10 years of education drops to 3.6. The numbers are comparable for Asia and South America.

More equality between men and women inside and outside the household is also believed to have a significant impact on reducing fertility. Giving both men and women choices about birth control, and educating them about the implications of such choices, appears to be especially successful in small island populations—such as those of Bali, Barbados, and Mauritius—with historically high population growth. In Mauritius in just 24 years (between 1962 and 1986) the introduction of voluntary constraints lowered the total fertility rate from 5.8 to 1.9. It is hardly any wonder, then, that the 1994 population conference placed such a clear and well-received empha-

Figure 3.32 Educating girls in Eritrea Improving the economic status of women is central to the success of controlling population growth. Access to education and employment security are seen as critical factors shaping a woman's decisions about how many children to have and when to have them.

sis on (1) rejecting coercive measures, including government sterilization quotas, that force people to violate their personal moral codes and (2) improving the rights, opportunities, and economic status of girls and women as the most effective way of easing global population growth.

Sustainable Development, Gender, and Population Issues

Globalization appears to lead to increasing levels of participation by women in the formal labor force, which is expected to lead to an increase in their economic and social value overall and a decline in birth rates globally. Large firms producing for export tend to employ women in assembly-line jobs because they can be hired for lower wages than men. And yet, while increasing labor-force participation has its benefits for women, it can also be based on implicit and explicit gender discrimination. Women constitute a large share of workers in informal subcontracting—often in the garment industry—at low wages and under poor conditions. Globalization is also associated with increasing levels of home work, telework and part-time work. In the United Kingdom the share of workers in such positions rose from 17 percent in 1965 to 40 percent in 1991. Similar changes have taken place in many other countries, and in most of them women constitute 70 or 80 percent of home-, tele-, and part-time workers. This is a mixed blessing. Informal work arrangements can accommodate women's care obligations in the family, but such jobs are typically precarious and underpaid.

Recognizing the central importance of economic development to improving the lives of women, men, and children throughout the world, international agencies are increasingly turning to sustainable economic development as a way of limiting births and ensuring an improved quality of life through the reduction of poverty. A host of related conferences have been held to address this theme, the most central one being the U.N. Millennium Summit. The goals, listed in **Table 3.7**, are aimed at reducing poverty by improving economic development through aid, trade, and debt relief and through the enhancement of democratic governance institutions and careful attention to the impact of development on the environment. The U.N.D.P. Millennium Development Goals are based on a partnership between core and peripheral countries. With respect to governance transformations, for instance, U.N.D.P. is working with an oil company and Amnesty International in Venezuela to provide the country's judges with a comprehensive understanding of human rights laws, regulations, and issues. With respect to the environment, U.N.D.P. is working with farmers in Ethiopia by supporting the planting and marketing of traditional crops and, in the process, strengthening the country's Biodiversity Research Institute and encouraging farmers to create biodiversity banks, while the crops make their incomes more secure.

TABLE 3.7 Millennium Development Goals (MDGs)

The goals and targets are based on the U.N. Millennium Declaration and the U.N. General Assembly has approved them as part of the Secretary General's road map toward implementing the declaration. U.N.D.P. worked with other U.N. departments, funds, and programs, the World Bank, the International Monetary Fund and the Organization for Economic Cooperation Development to identify over 40 quantifiable indicators to assess progress.

Goals and Targets
Goal 1: Eradicate extreme poverty and hunger
Goal 2: Achieve universal primary education
Goal 3: Promote gender equality and empower women
Goal 4: Reduce child mortality
Goal 5: Improve maternal health
Goal 6: Combat HIV/AIDS, malaria and other diseases
Goal 7: Ensure environmental sustainability*
Goal 8: Develop a Global Partnership for Development*

*The selection of indicators for Goals 7 and 8 is subject to further refinement.

The eight major Millennium Development Goals reflect the neoliberal turn in international development. As discussed in Chapter 1, neoliberalism promotes a reduction in the role and budget of government, including reduced subsidies and the privatization of formerly publicly owned and operated concerns such as utilities. The goal of neoliberal development policies—such as the one being widely advanced by the U.N.D.P.—is to enable peripheral countries to achieve core economic standards of wealth and prosperity while recognizing that preexisting conditions will have to be taken into account to construct a place-specific development path. As the goals imply, enabling more sustainable economic development worldwide is seen as a way of also shaping population growth and the quality of life for population in the periphery. It is also a way of opening up new markets for core products and services and extending the capitalist world-system.

CONCLUSION

The geography of population is directly connected to the complex forces that drive globalization. And since the fifteenth century, the distribution of the world's population has changed dramatically as the capitalist economy has expanded, bringing new and different peoples into contact with one another and setting into motion additional patterns of national and regional migrations. When capitalism emerged in Europe in the fifteenth century, the world's population was experiencing high birth rates, high death rates, and relatively low levels of migration or mobility. Four hundred years later, birth, death, and migration rates vary—sometimes quite dramatically—from region to region, with core countries experiencing low death and birth rates and peripheral and semiperipheral countries generally experiencing high birth rates and fairly low death rates. Migration rates vary within and outside the core. These variations may be seen to reflect the level and intensity of political, economic, and cultural connectedness between core and periphery.

The example of formerly colonized peoples migrating to their ruling countries in search of work provides insights into the dynamic nature of the world economy. The same can be said of U.S. migrants who in the 1970s and 1980s steadily left their homes in the Northeast and Midwest to take advantage of the employment opportunities that were emerging in the Sunbelt: the Southeast, Southwest and West. Both examples show the important role that people play in acting out the dynamics of geographic variety.

In the final analysis, death rates, birth rates, and migration rates are the central variables of population growth and change. These indicators tell us much about transforming regions and places as elements in a larger world-system. Globalization has created many new maps as it has unfolded; the changing geography of population is just one of them.

MAIN POINTS REVISITED

- **Population geographers depend on a wide array of data sources to assess the geography of populations. Chief among these are censuses, while other sources include vital records and public health statistics.**

  Census collection methods vary from country to country and are often conducted in different years, making it difficult to compare different places.

- **Population geographers are largely concerned with the same sorts of questions that other population experts address, but they also investigate "the why of where."** ***Why*** **do particular aspects of population growth and change (and problems) occur** ***where*** **they do, and what are the implications of these factors for the future of places?**

  A geographic perspective is sensitive to the important influences of place and sees geographic factors as helping to explain population growth and change.

- **Two of the most important factors that make up population dynamics are birth and death. These variables may be examined in simple or complex ways, but in either case the reasons for their behavior are as important as the numbers themselves.**

  Birth and death rates are fairly crude measures of population change, so population geographers also look at factors such as the particular experiences of certain age cohorts or race cohorts and how those factors influence birth and death rates.

- **A third crucial force in population change is the movement of populations. The forces that push populations from particular locations as well as those that pull them to move to new areas are key to understanding new settlement patterns. Population migration may not always be a matter of choice.**

  Migration is one of the most important factors affecting the distribution of world population today. For some countries whose birth rates are especially low, migration is one way of reversing that trend.

- **Perhaps the most pressing question facing scholars, policymakers, and other interested individuals was articulated at the U.N. Millennium Summit in 2000: How can the global economy provide the world's growing population with enough food and safe drinking water and a sustainable environment so that all the world's people have the basic necessities for enjoying happy, healthy, and satisfying lives?**

  The most recent approach to limiting population growth is to look toward sustainable economic development as a way of limiting births and ensuring an improved quality of life.

KEY TERMS

age-sex pyramid (p. 92)
agricultural density (p. 91)
arithmetic density (p. 89)
baby boom (p. 92)
census (p. 86)
cohort (p. 92)
crude birth rate (CBR) (p. 99)
crude death rate (CDR) (p. 103)
crude density (p. 89)
demographic transition (p. 106)
demography (p. 84)
dependency ratio (p. 99)
doubling time (p. 101)
eco-migration (p. 119)
emigration (p. 108)
forced migration (p. 109)
geodemographic analysis (p. 92)
gross migration (p. 109)
guest workers (p. 109)
immigration (p. 108)
infant mortality rate (p. 103)
internally displaced persons (IDPs) (p. 109)
internal migration (p. 108)
international migration (p. 108)
life expectancy (p. 104)
medical geography (p. 105)
middle cohort (p. 99)
migration (p. 108)
mobility (p. 108)
natural decrease (p. 103)
natural increase (p. 103)
net migration (p. 109)
nutritional density (p. 91)
old-age cohort (p. 99)
population policy (p. 120)
pull factors (p. 109)
push factors (p. 109)
refugees (p. 109)
suburbanization (p. 116)
total fertility rate (TFR) (p. 101)
transnational migrant (p. 109)
vital records (p. 86)
voluntary migration (p. 109)
youth cohort (p. 99)

ADDITIONAL READING

Applegard, R. T. (ed.), *Emigration Dynamics in Developing Countries*. Brookfield VT: Ashgate, 1998.

Bean, F. D., *At the Crossroads: Mexican Migration and U.S. Policy*. Lanham, MD, and London: Rowman & Littlefield, 1997.

Bernard, R. M., and B. R. Rice, *Sunbelt Cities: Politics and Growth Since World War II*. Austin: University of Texas Press, 1983.

Birdsall, N., A. C. Kelley, and S. A. W. Sinding, *Population Matters: Demographic Change, Economic Growth and Poverty in the Developing World.* New York: Oxford University Press, 2001.

Bouvier, L., and C. DeVita, "The Baby Boom—Entering Midlife," *Population Bulletin 46*(3): 2–33, 1991.

Castles, S., and M. J. Miller, *The Age of Migration: International Population Movements in the Modern World* (3rd ed.), New York: Guilford Press, 2003.

Cernea, M. M. and C. McDowell, *Risks and Reconstruction: Experiences of Resettlers and Refugees,* 2000.

Kinsella, K. and V. A. Velkoff, *An Aging World: 2001.* U.S. Census Bureau, Series P95/01-1, U.S. Government Printing Office, Washington, DC, 2001.

Klein, Herbert S., *A Population History of the United States.* Cambridge: Cambridge University Press, 2004.

Kunstler, J. H., *The Long Emergency: Surviving the Converging Catastrophes of the Twenty-First Century.* New York: Atlantic Monthly Press, 2005.

Lemann, N., *The Promised Land: The Great Black Migration and How It Changed America.* New York: Alfred Knopf, 1991.

Mann, C. C., "How Many Is Too Many?" *Atlantic Monthly,* February 1993, pp. 47–67.

McFalls, J. Jr., *Population: A Lively Introduction,* 4th ed., *Population Bulletin, 58*(4), 2003, p. 27.

McIntosh, A. C. and J. Finkle, "The Cairo Conference on Population and Development: A New Paradigm?" *Population and Development Review, 21*(2), 223–260, 1995.

Macunovich, D. J., *Birth Quake: The Baby Boom and Its Aftershocks.* Chicago: University of Chicago Press, 2002.

Portes, A. and R. Rumbaut, *Immigrant America: A Portrait.* Berkeley: University of California Press, 1996.

Suárez-Orozco, M. M., *Crossings: Mexican Immigration in Interdisciplinary Perspectives.* Cambridge, MA, and London: Harvard University, David Rockefeller Center for Latin American Studies; distributed by Harvard University Press, 1998.

U.N. Development Programme, *Partnerships to Fight Poverty: Annual Report.* New York and Oxford: Oxford University Press, 2001.

U.N. Population Division, *World Population Ageing: 1950–2050.* New York: United Nations, 2002.

U.N. Population Division, "Population Ageing and Living Arrangements of Older Persons: Critical Issues and Policy Response," United Nations *Population Bulletin,* Special Issue Nos. 42/43. New York: United Nations, 2001.

U.N. Population Fund, *The State of World Population, 2000: Lives Together, Worlds Apart: Men and Women in a Time of Change.* New York: The Fund, 2000.

U.N. Population Fund, *The State of World Population.* New York: The Fund, 2001.

World Bank, *World Development Indicators.* Washington, DC: World Bank, 2000.

EXERCISES

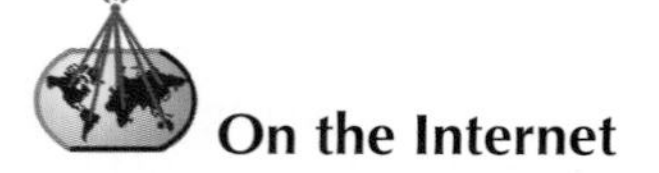

On the Internet

The Internet exercises for this chapter use Internet Web sites to help you understand its key concepts. For example, through an interactive map and socioeconomic database, you view country demographics, economics, geography, and social indicators. Using an animated program, you can illustrate and explain various age-sex pyramids among different countries. There are a number of review questions regarding population movements and migration, along with several map exercises based on population density and composition. A critical-thinking exercise focuses on the impact on our society of the baby boom generation as it reaches maturity. Other exercises help you examine the geography of Latinos in the United States and the policies of population and resources.

Unplugged

1. The distribution of population is a result of many factors, such as employment opportunities, culture, water supply, climate, and other physical environment characteristics. Look at the distribution of population in your state or province. Is it evenly distributed, or is the majority of people found in only a few cities? What role do you think these various factors have played in influencing where people live in your state or province? Can you think of other reasons for this distribution?
2. Immigration is an important factor contributing to the increase in the population of the United States. Chances are your great-grandparents, grandparents, parents, or even you immigrated to, or migrated within, your country of residence. Construct your family's immigration or migration history. What were some of the push and pull factors influencing your family's decision to immigrate to or migrate within your country?
3. During the 1990s an interesting example of return migration occurred. Some areas of the United States that had previously lost large numbers of African Americans began to regain them. Compare data from the 1980 census with data from the 2000 census for the southeastern region of the United States. Which states experienced an increase of African-American migrants? What are some of the characteristics of this migrant stream, and why might it have occurred?
4. Every few years U.N.E.S.C.O. publishes a data book on global refugee statistics indicating both sending and receiving countries, among other variables. Use your library's most recent data book and identify one country that has been a large sender of refugees and the country that has been the largest receiver of those refugees. The data book provides information not only on the numbers of refugees but on their age, gender, and other variables. Discuss some of the demographic implications for both countries if the refugee population was not to be allowed to return to the sending country.

S.W.A.T.

Spray DEET Repellent • Wear Long Sleeves • Avoid Standing Water • Tell Your Commu

ELIMINATE WEST NILE VIRUS

4 Nature and Society

MAIN POINTS

- Nature and society constitute a complex relationship. It is our view in this text that nature is both a physical realm and a social creation.
- Because in this text we regard nature as a social creation, it is important to understand the many views of nature in society today as well as the history of those views. The most prominent view of nature in Western culture is derived from the Judeo-Christian tradition, which is founded on a belief that humans should dominate nature.
- Humankind's relationships with nature have developed over the course of human history, beginning with the early Stone Age. The early human history included people who revered nature, as well as those who abused it. Urbanization and industrialization have had extremely degrading impacts on the environment.
- The globalization of the political economy has meant that environmental problems are also global in scope. Deforestation, acid rain, and nuclear fallout affect us all. Many new ways of understanding nature have emerged in the last several decades in response to global environmental crises.
- Sustainability has recently become a predominant way of approaching global economic development and environmental change. In addition, new institutional frameworks, including conventions, protocols, and organizations, are rapidly emerging to promote global sustainability.

In addition to the spread of people, ideas, money and goods, globalization, from its earliest beginnings to the present, is also about the spread of disease. Within the Old World, diseases carried by insects, rodents and people spread from one region to another fairly regularly. An outbreak of the bubonic plague occurred in China in the 1330s and spread to western Asia and Europe and finally into Britain through the rats in the hulls of ships that carried the traders who travelled across these regions. After killing over a third of Europe's population, the disease recurred regularly in epidemic proportions until the seventeenth century. The contact between the Old World and the New, in the sixteenth and seventeenth centuries, geographically extended the spread of diseases that wiped out entire, previously insulated native populations in the New World. The Arawac, an indigenous group in the Carribbean, was completely wiped out by a flu carried by pigs on Columbus's ships.

During the twentieth century, developments in international public health enabled a wide range of diseases affecting people around the globe either to be eradicated completely or largely controlled, at least in some parts of the world. For example, malaria and typhoid have disappeared from western Europe, North America, Australia and Japan but still affect populations throughout the periphery. Tuberculosis still kills about two million people a year and sickens another eight million; malaria causes one million deaths a year with more than 41 percent of the world's population at risk of acquiring it. Both of these diseases affect populations in the periphery because of deteriorating health systems, growing drug and insecticide resistance, and war. But it is also the case that global climate change is increasingly being seen as playing a role in the resurgence of diseases, once thought to be eradicated, not only in the periphery but also in the core.

Climate change is increasingly drawing the attention of scientists, environmental groups, and health professionals world wide because it points to changing environmental conditions as the cause of the spread of a range of infectious diseases once restricted to the periphery. One of the diseases of the twentieth century thought to have been wiped out but recurring recently in the United States is West Nile Virus (WNV). The disease was first isolated in the West Nile District of Uganda in 1937 and began to appear in Europe, the Middle East, west and central Asia, Oceania, and most recently North America in the 1990s. Transmitted to humans by mosquitoes and birds, the most serious manifestation of WNV infection is encephalitis (inflammation of the brain) which is fatal in humans as well as in certain domestic and wild birds. WNV cases in human populations have now occurred in all fifty states of the United States except for Alaska, Hawai'i, and Oregon.

Climate scientists and epidemiologists in the United States and elsewhere have begun to collaborate in order to better understand the conditions that produce the disease. They believe the climate factors that accelerate the disease's life cycle—mild winters coupled with prolonged droughts and heat waves—are the result of global climate change, which is a long-term transformation in Earth's atmosphere. The conditions producing climate change are thought to be caused by human activity such as

West Nile Virus Arizona Health Department Campaign

deforestation and fossil fuel burning which lead to alterations in the atmosphere resulting in warmer temperatures, a decline in sea ice and snow cover in the Northern Hemisphere, a retreat of mountain glaciers, and an increase in precipitation in high latitudes. Climate change is one of the results of the interaction of nature and society, a change that is likely to have long term consequences that may be difficult to reverse.

The relationship between people and the environment is perhaps the most central of all relationships within the discipline of geography. Indeed, the discipline consists of those who study natural systems and those who study human systems. In this chapter we explore the ways that society has used technology to transform and adapt to nature, together with the impact of those technological adaptations on humans and the environment.

NATURE AS A CONCEPT

As discussed in Chapter 2, a simple model of the nature-society relation is that nature, through its awesome power and subtle expressions, limits or shapes society. This model is known as environmental determinism. A second model posits that society also shapes and controls nature, largely through technology and social institutions. This second model, explored in this chapter, emphasizes the complexity of nature-society interactions.

Interest in the relationship between nature and society has experienced a resurgence over the last two or three decades largely because the scope of environmental problems seems to have increased from those that are locally or regionally defined to those that have implications for the whole planet. The single most dramatic manifestation of this interest occurred in the summer of 1992, when more than 100 world leaders and 30,000 other participants attended the second Earth Summit in Rio de Janeiro (the first Earth Summit was held in Stockholm in 1972). The central focus of the agenda was to ensure a sustainable future for Earth by establishing treaties on global environmental issues, such as climate change and biodiversity. The signatories to the 1992 Earth Summit conventions created the Commission on Sustainable Development to monitor and report on implementation. A five-year review of Earth Summit progress was made in 1997 through a U.N. General Assembly meeting in special session. In July 2002, world leaders and other environmentally concerned individuals met for another Earth Summit, this time in Johannesburg, South Africa (**Figure 4.1**).

Some very important and significant, though not necessarily dramatic, changes have occurred since Rio. One change has been the emergence of international institutions to facilitate and monitor environmental improvements. Another has been real progress on the global phase-out of leaded gasoline. A third has been rising scientific and popular interest in global environmental issues.

Renewed interest in the nature-society relationship is the result of the persistence, increasing number, and wider impact of environmental crises. This interest has led to attempts to rethink the relationship. For example, at the beginning of the twentieth century, Gifford Pinchot, as an influential citizen and later as director of the National Conservation Commission, advocated environmental conservation as the nation's forests and wild lands were increasingly given over to development. In 1962 Rachel L. Carson, in a groundbreaking book, *Silent Spring*, warned of the dangers of agricultural pesticides to ecosystems. Yet as the new century begins, the pesticide problem persists in both peripheral and core countries. In the past, technology emerged as the apparent solution to most environmental problems, but today's technological progress seems often to aggravate rather than to solve such problems (**Figure 4.2**). As a result, researchers and activists have begun to ask different questions and abandon the assumption that technology is the *only* solution. For instance, 2004 Nobel Laureate and scientist Wangari Maathai has led a movement advocating tree-planting as a way of addressing the linkage between environmental conservation and economic development (**Figure 4.3**). The Green Belt Movement, as it is known, works to conserve local biodiversity, prevent soil erosion, and increase the forest cover of Kenya. At present Kenya's forest cover is less than 2 percent. The GBM is therefore fully engaged in a reforestation campaign aimed at preserving and improving local biological diversity. The purchase of seedlings

Figure 4.1 2002 Earth Summit, Johannesburg, South Africa Global concern about environmental issues persists as the new century unfolds. Pictured here are attendees at the 2002 Earth Summit, with U.N. Secretary General Kofi Annan at the podium.

for the campaign helps to generate income for local groups at the same time that reforestation is advanced.

The GBM and others like it have influenced environmental experts, including a number of geographers, to conceptualize nature not as something apart from humans but as inseparable from us. These experts believe that nature and questions about the environment need to be considered in conjunction with society, which shapes our attitudes toward nature and how we identify sources of and solutions to environmental problems. Such an approach—uniting nature and society as interactive components of a complex system—enables us to ask new questions and consider new alternatives to current practices with respect to nature.

But before asking new questions about nature, this text examines the nature-society relationship by looking first at different approaches to it. We then examine how changing conceptions of nature have translated into very different uses of, and adaptations to, it. We conclude with an examination of sustainable development as a way of addressing global environmental problems and the new institutional frameworks and activist organizations that are emerging to promote sustainability.

Figure 4.3 Wangari Maathai In 2004, environmental activist, scientist, and Assistant Minister for Environment and Natural Resources in Kenya became the first African woman to receive a Nobel Prize. Dr. Maathai received the prize for her work in the Green Belt Movement (GBM) in Kenya and the pan-GBM. The GBM is a grass-roots organization involved in tree planting/biodiversity conservation, civic and environmental education, advocacy and networking, food security, and capacity building for women and girls. The aim of the organization is to address issues of sustainable development through environmental alternatives to technology.

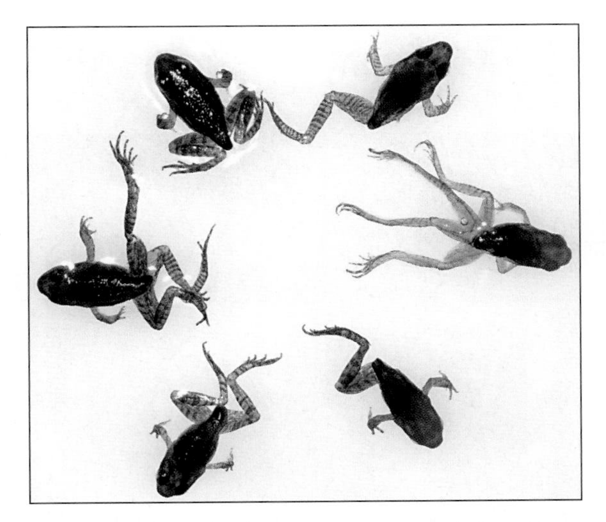

Figure 4.2 Deformities in frogs A 2005 study by a Pennsylvania State University scientist and published in the *Proceedings of the National Academy of Sciences* found that the combination of pesticide contamination and parasite infection has caused missing legs and extra legs in wood frogs in 43 states in the United States and five provinces in Canada. Both factors acting together are thought to be responsible for the mutations. It is believed that exposure to agricultural chemicals may weaken amphibian immune systems, making the frogs more vulnerable to parasitic infection leading to limb deformities. World-wide, amphibians have been decreasing at an alarming rate and scientists like this one believe that global climate change and other human induced factors are responsible for this decline.

Nature and Society Defined

The central concepts of this chapter—nature and society—have very specific meanings. Although we discuss the changing conceptions and understandings of nature in some detail, we hold to one basic conception here, that **nature** is a social creation as much as it is the physical universe that includes human beings. Therefore, understandings of nature are the product of different times and different needs. Nature is not only an object, it is a reflection of society in that philosophies, belief systems, and ideologies shape the way people think about and use nature. The relationship between nature and society is two-way: society shapes people's understandings and uses of nature at the same time that nature shapes society. The amount of shaping by society is dependent to a large extent on the state of technology and the constraints on its use at any given time.

Society is the sum of the inventions, institutions, and relationships created and reproduced by human beings across particular places and times. Society's relationship with nature is just one of its relationships, and the social relationship to nature varies from place to place among different social groups.

The relationship between society and nature is usually mediated through technology. Knowledge, implements, arts, skills, and the sociocultural context all are components of technology. If we accept that all of these components are relevant to technology, then we can provide a definition that has three distinguishable though equally important aspects. **Technology** is defined as:

- physical objects or artifacts (for example, the plow);
- activities or processes (for example, steelmaking);
- knowledge or know-how (for example, biological engineering).

This definition recognizes tools, applications, and understandings as critical components of the human production of technology. The manifestations and impacts of technology can be measured in terms of concepts, such as level of industrialization and per capita energy consumption.

The definitions provided in this section reflect current thinking on the relationship between society and nature. For centuries humankind, in response to the constraints of the physical environment, has been as much influenced by prevailing ideas about nature as by its realities. In fact, prevailing ideas about nature have changed over time, as evidence from literature, art, religion, legal systems, and technological innovations makes abundantly clear.

A recent attempt to conceptualize the relationship between social and environmental changes has emerged from concern with global environmental changes. Based on the premise that individual societal changes can be both subtly and dramatically related to environmental changes, a formula for distinguishing the sources of social impacts on the environment has been advanced and is now widely used. The formula, known as $I = PAT$, relates human population pressures on environmental resources to the level of affluence and access to technology in a society. More specifically, the formula states that $I = PAT$, where I (impact on Earth's resources) is equal to P (population) times A (affluence, as measured by per capita income) times T (a technology factor). For example, the differential impact on the environment of two households' energy use in two different countries would equal the number of people per household times the per capita income of the household times the type of technology used to provide energy for that household (**Figure 4.4**).

Each of the variables in the formula—population, affluence, and technology—is complex. For example, with regard to population numbers, it is generally believed that fewer people on the planet will result in fewer direct pressures on resources. Some argue, however, that increased world population is quite desirable, since more people

Figure 4.4 Affluence differences in Iceland and Guatemala As the $I = PAT$ formula suggests, the level of affluence of households plays an important role in their impact on the global environment. Pictured here are two families, one from the core and the other from the periphery. Both families are pictured outside their homes with all of their possessions arrayed around them. The extensive range of possessions shows the Icelandic family, composed of two parents and four children, to be far more affluent than the Guatemalan family, composed of two parents and three children. Iceland ranks eighth in affluence among the 183 members of the United Nations; Guatemala's rank is 114. The Icelandic family possesses two radios, one stereo, two televisions, one VCR, one home computer, two automobiles, and a private airplane. The Guatemalan family possesses one battery-operated radio and no telephone, but would like to acquire a television set.

means more labor coupled with more potential for the emergence of innovation to solve present and future resource problems. Clearly there is no simple answer to the question of how many people are too many.

Affluence also cannot simply be assessed in terms of "less is better." Certainly, increasing affluence—a measure of per capita consumption multiplied by the number of consumers and the environmental impacts of their technologies—is a drain on Earth's resources and a burden on Earth's ability to absorb waste. Yet how much affluence is too much is difficult to determine. Furthermore, evidence shows that the core countries, with high levels of affluence, are more effective than the poor countries of the periphery at protecting their environments. Unfortunately, core countries often do so by exporting their noxious industrial processes and waste products to peripheral countries. By exporting polluting industries and the jobs that go with them, however, core countries may also be contributing to increased affluence in the receiving countries. Given what we know about core countries, such a rise fosters a set of social values that ultimately leads to better protection of the environment in a new place. It is difficult to identify just when environmental consciousness goes from being a luxury to a necessity. The role of affluence in terms of environmental impacts is, in short, like population, difficult to assess.

Not surprisingly, the technology variable is no less complicated. Technologies affect the environment in three ways:

- through the harvesting of resources;
- through the emission of wastes in the manufacture of goods and services;
- through the emission of waste in the consumption of goods and services.

A technological innovation can shift demand from an existing resource to a newly discovered, more plentiful one. In addition, technology can sometimes be a solution and sometimes a problem. Both principles can be seen in the case of nuclear energy, widely regarded as cleaner and more efficient than coal or oil as energy sources. Producing this energy creates hazards, however, which scientists are still unable to prevent.

It is therefore clear that increases in human numbers, in levels of wealth, and in technological capacity are key components of social and economic progress whose impact on the environment has been extremely complex. In the last 100 years this complexity has increasingly come to be seen as a triple-barreled threat to the quality of the natural world and the availability and quality of environmental resources. Before we look more carefully at the specific impacts of populations, affluence, and technology on nature, we need to look first at how differing social attitudes toward nature shape the human behaviors that are a basis for $I = PAT$.

Nature-Society Interactions

The concept of *adaptation* to the natural environment is part of the geographical subfield of cultural ecology most closely associated with the work of Carl Sauer and his students. **Cultural ecology** is the study of how human society has adapted to environmental challenges such as aridity and steep slopes through technologies such as irrigation and terracing and the organization of people to construct and maintain these systems. These adaptations can be seen clearly in the landscape, such as the rice terraces of Southeast Asia or the canals and reservoirs of the southwestern United States. More recent adaptations include the use of biotechnology and agricultural chemicals to increase agricultural production and the development of new pharmaceuticals to cope with diseases.

Human adaptation has gone far beyond responses to natural constraints to produce widespread modification of environment and landscapes. In some cases, the human use of nature has resulted in environmental degradation or pollution. For example, overcultivation of steep slopes can result in erosion of the soil needed for subsequent agricultural production, and the use of agricultural chemicals has caused the contamination of adjacent rivers and lakes by chemicals that are toxic to fish and humans. The Industrial Revolution produced a dramatic growth in the emissions of waste material to land, water, and atmosphere, and resulted in serious air pollution and health problems in many areas.

The massive transformation of nature by human activity led geographers such as Neil Smith and Margaret Fitzsimmons to claim that we can no longer talk about "natural" environments or untouched wilderness. They use the phrase *social production of nature* to describe the refashioning of landscapes and species by human activity, especially capitalist production and labor processes.

Geographers have played a major role in highlighting the global scope of this transformation in their discussions of the human dimensions of global environmental change, in which they explore the social causes and consequences of changes in global environmental conditions. Of particular concern are global patterns of fossil-fuel use and land-use change that are producing serious changes in climate and biodiversity through carbon dioxide–induced global warming or deforestation.

Global climate change is causing sea levels to rise as polar ice caps melt and has increased the frequency of violent storms. Warmer oceans surrender greater quantities of water as evaporation. Warmer surface temperatures and more humid air masses intensify weather systems, resulting in fiercer cyclones and hurricanes. In summer 2005 the twin disasters associated with global warming—violent storms and flooding—came together in the United States as Hurricane Katrina bore down on a wide swath of the Gulf Coast the extended from Pensacola, Florida to New Orleans, Louisiana (**Figure 4.5**). The hurricane destroyed extensive sections of the built environment and caused the flooding of low lying areas, especially Greater

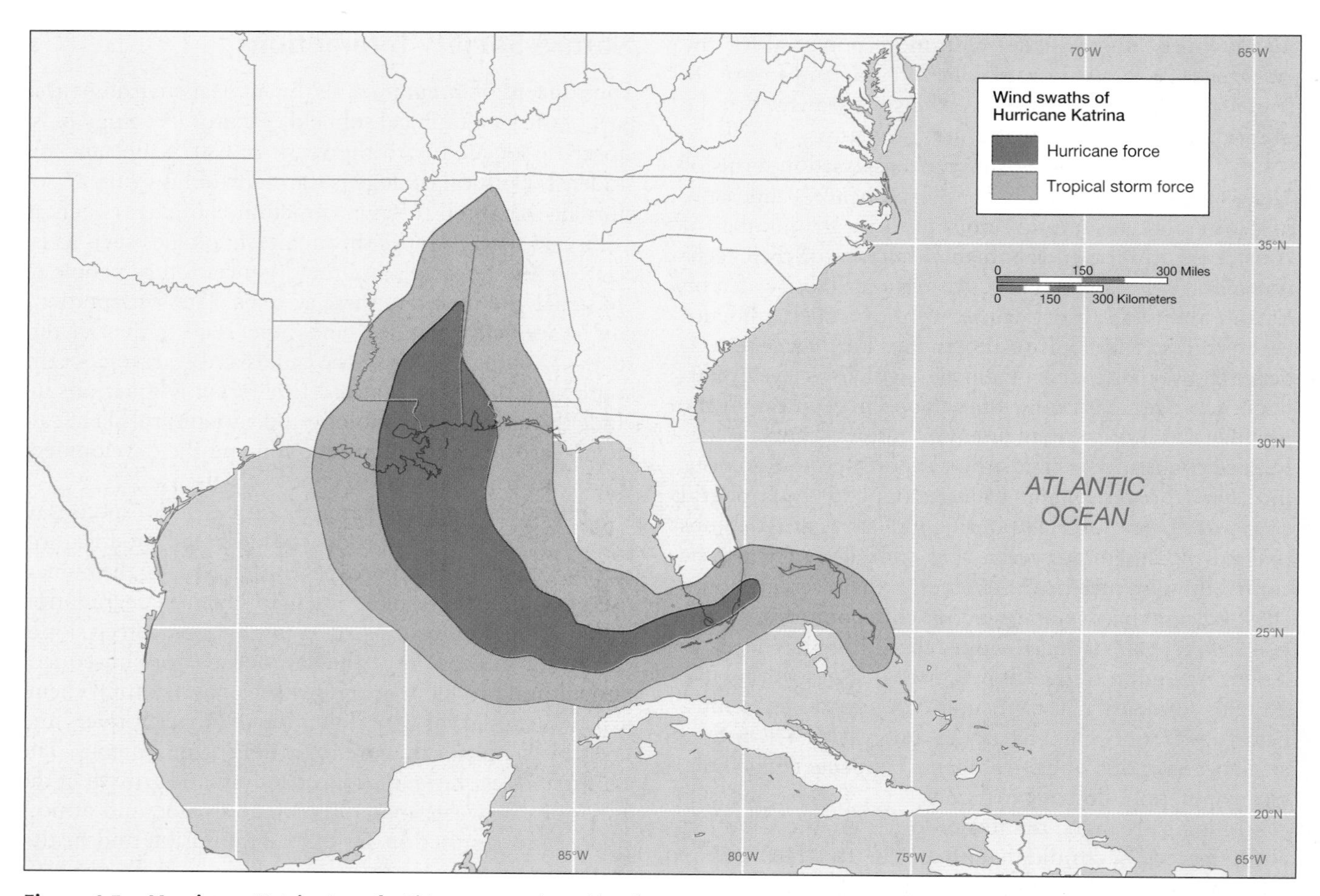

Figure 4.5 Hurricane Katrina's path This map produced by the National Oceanographic and Atmospheric Administration and the National Weather Service provides a graphic sense of the trail of Hurricane Katrina as it swept across southern Florida and then up the Gulf of Mexico making landfall a second time across the tip of the Florida panhandle, Alabama and Louisiana. (*Source:* http://www.nhc.noaa.gov/refresh/graphics_at2+shtml/084543.shtml?swath?large)

New Orleans—with a population of 1.4 million—where thousands died or were injured and more than one million people were displaced. The media, politicians, government officials, planners, and engineers labeled Hurricane Katrina the worst natural disaster the United States has ever experienced.

It is still too soon to identify definitively all the root causes of the calamity, but many scientists and public figures are arguing that Katrina was far from a "natural" disaster. While there is no doubt that the force of the winds slamming the Gulf Coast were extreme, the fact that districts and parishes in and around New Orleans (where the most dramatic impacts occurred) flooded and so many people (who had not evacuated) died was not because of the hurricane but was the result of political and social factors that could have been avoided. As with Hurricane Andrew, the violent storm that hit the Miami-Dade area in 1992, weak building codes and poor building code enforcement in New Orleans, made the wood-frame housing stock especially susceptible to damage from high winds. And critically needed improvements to the sinking levees—that hold back the Mississippi River and Lake Ponchartrain and enable various districts of the city of New Orleans to sit nearly five feet below sea level—were repeatedly postponed by the federal government. The strong winds mostly brought down power and communication lines leaving the region without electricity or phone service but the housing stock fairly in tact. But the storm surge that accompanied Katrina breached the levees and caused flooding to occur in over 80 percent of the city. Add to these technological vulnerabilities the fact that although the city's evacuation plan worked well for many, tens of thousands of people who were too poor or disabled or fearful to find their way out of the city before the storm hit and were thus put directly in harm's way. Blaming nature for the flooding and the tremendous loss of life in New Orleans, if not also in the rest of the affected Gulf region, is difficult to maintain. This is especially so when comparing Katrina's impact on the Gulf Coast of the United States in 2005 to Hurricane Ivan's impact on Cuba just a year before. Ivan was a Category 5 storm with 160-mile per hour winds when it hit the west coast of Cuba; Katrina a Category 4 storm with winds of 125–140 miles per hour when it hit the Gulf Coast. Yet 1.5 million Cubans successfully evacuated to higher ground ahead of the storm. They also evacuated pets, TVs, and refrigerators so people weren't reluctant to leave because their belongings would be unprotected. And though

20,000 houses were destroyed, no one died, no curfews were imposed, and no looting or violence took place. After Hurricane Ivan, the United Nations International Secretariat for Disaster Reduction cited Cuba as a model for hurricane preparation. Atmospheric scientists have been arguing for over a decade that global warming is putting Atlantic and Gulf coastal cities in the United States in a very vulnerable position as higher than normal ocean temperatures help to intensify hurricanes and tropical storms. Many believe that Hurricane Katrina is not an example of a "perfect storm" where essential meteorological elements combine to produce an extreme and devastating event, but the perfect example of how global climate change in combination with human practices, is transforming Earth's environment in dramatic and devastating ways.

During the twentieth century, global sea level rose by 20 centimeters (7.9 inches), and a recent report by Britain's Meteorological Office warned that flooding will increase more than ninefold over the twenty-first century, with four-fifths of the increase coming in South and Southeast Asia. A rising sea level would be disastrous for some countries. About 70 percent of Bangladesh, for example, is at sea level, as is much of Egypt's most fertile land in the Nile delta. In contrast, farmers in much of Europe and North America would welcome a local rise in mean temperatures, since it would extend their options for the kinds of crops that they could profitably raise.

The causes and consequences of these global climate changes vary considerably by world region. For example, the industrial countries have higher carbon dioxide emissions. Increased carbon dioxide emissions are contributing to rising temperatures through the trapping of heat in Earth's atmosphere. In order to survive in many of the world's peripheral regions, the rural poor are often impelled to degrade and destroy their immediate environment by cutting down forests for fuelwood, leading to the destruction of forests, which help to cool Earth's surface. Thus, both the core and the periphery are contributing to the problem of global change in different but equally significant ways.

Population growth patterns and the changing geography of economic development allow us to predict with some confidence that the air and water pollution generated by low-income countries will more than double in the next 15 years, as they become more industrialized. Thus, environmental problems are becoming inseparable from processes of demographic change, economic development, and human welfare. In addition, regional environmental problems are becoming increasingly enmeshed in matters of national security and regional conflict. Since the principles of cultural ecology did not explain the political dimensions of ecological questions, cultural ecologists in the 1980s began moving away from a strict focus on particular cultural groups' relationship with the environment, instead placing that relationship within a wider context. The result is political ecology, the merging of political economy with cultural ecology. **Political ecology** stresses that human-environment relations can be adequately understood only by reference to the relationship of patterns of resource use to political and economic forces (see Box 4.1: "Understanding Cultural Ecology and Political Ecology").

U.S. Environmental Philosophies and Political Views of Nature

As mentioned, nature is a construct that is very much shaped by social ideas and values. As a result, different societies and different cultures have very different views of nature. In the contemporary world, views of nature are dominated by the Judeo-Christian tradition, which understood Man to be superior to nature, such that nature is something to be tamed or dominated. But other views of nature have emerged and departed dramatically from the Judeo-Christian tradition, especially the environmental philosophies that became popular in the nineteenth and early twentieth centuries and the more radical political views of nature that gained prominence in the late twentieth century. We examine both of these approaches in this section in terms of changing U.S. philosophies and views of nature, which are largely representative of the core of the world economy.

Henry David Thoreau (1817–1862), an American naturalist and activist, perhaps best illustrates the Western incorporation of North American Indian conceptions of nature into ecological approaches that began emerging in the mid-nineteenth century in the United States. Thoreau lived and studied the natural world around the town of his birth, Concord, Massachusetts. He is most famous for his book *Walden,* which chronicles the two years he spent living and observing nature in solitude in a house he built on Walden Pond, a mile and a half from the village green of Concord. Thoreau represents a significant alternative to the "Man-over-nature" approach that characterized his times. Many people credit him as the originator of a U.S. ecological philosophy.

Thoreau was impressed with the power of nature. He often described its unrestrained and sometimes explosive capacity, which he thought had the potential to overthrow Man's dominion if left unchecked. He also emphasized the interrelatedness of the natural world, where birds depended upon worms, fish depended upon flies, and so on, along the food chain (**Figure 4.6**). Most notably, however, Thoreau regarded the natural world as an antidote to the negative effects of technology on the landscape and the American character. Concord was just 20 miles west of Boston and an equal distance south of the booming mill towns of Lowell and Lawrence. Although he spent his life in a more or less rural setting, the Industrial Revolution was in full force all around Thoreau, and he was keenly aware of its impacts. In fact, Thoreau's approach to the natural world was very much a response to the impacts on nature of the early forces of globalization. His research into the animals and plants that surrounded Concord was an attempt to reconstruct the landscape as it had existed before colonization and massive European immigration.

4.1

GEOGRAPHY MATTERS

Understanding Cultural Ecology and Political Ecology

The impact of Spanish agricultural innovations on the culture of the indigenous people of the Central Andes region of South America (an area encompassing the mountainous portions of Peru, Bolivia, and Ecuador) presents an excellent case study in cultural ecology. The transformation of Andean culture began when Pizarro arrived in Peru from Spain in 1531 and set about vanquishing the politically, technologically, and culturally sophisticated Incas. The Spaniards brought with them not only domestic plants and animals (mainly by way of Nicaragua and Mexico) but also knowledge about how to fabricate the tools they needed and a strong sense of what was necessary for a "civilized" life.

By the 1590s a bundle of Spanish cultural traits had been integrated into the Central Andean rural culture complex, creating a hybridized rural culture. The hybridized culture—and cultural landscape—combined a much simplified version of Spanish material life with important (though altered) Incan practices of crop growing, herding, agricultural technology, and settlement patterns. That this hybrid culture complex remains identifiable today, even after four centuries and in the face of contemporary globalizing forces, is due to the peasants' strong adherence to custom, geographic isolation, and poverty.

By 1620 the indigenous Andean people had lost 90 percent of their population and had been forced to make significant changes in their subsistence lifestyles (an illustration of demographic collapse as discussed on pages 144–148). The Inca empire, with its large population base, had once engaged in intensive agriculture practices, including building and maintaining irrigation systems, terracing fields, and furrowing hillsides. With the severe drop in population and consequent loss of labor power, the survivors turned to pastoralism because herding requires less labor than intensive agriculture. Ultimately, it was the introduction of Old World domesticated animals that had the greatest impact on the Central Andes (**Figure 4.A**).

Cultural ecologists study the material practices (food production, shelter provision, levels of biological reproduction) as well as the nonmaterial practices (belief systems, traditions, social institutions) of cultural groups. Their aim is to understand how cultural processes affect groups' adaptation to the environment. Whereas the traditional approach to the cultural landscape focuses on human impacts on the landscape or its form or history, cultural ecologists seek to explain how cultural processes affect adaptation to the environment. Cultural adaptation involves the complex strategies human groups employ to live successfully as part of a natural system. Cultural ecologists recognize that people are components of complex ecosystems and that the way they manage and consume resources is shaped by cultural beliefs, practices, values, and traditions, as well as by larger institutions and power relationships.

The cultural ecology approach incorporates three key points:

- Cultural groups and the environment are interconnected by systemic interrelationships. Cultural ecologists examine how people manage resources

Figure 4.A Andean woman weaving Though sheep are not indigenous to the Andes, they have been widely adopted in this region since the colonial period. Sheep are well adapted to high altitudes and provide wool and meat. Shown here is a woman weaving accompanied by another woman.

through a range of strategies to comprehend how the environment shapes culture, and vice versa.

- Cultural behavior is examined as a function of the cultural group's relationship to the environment through both material and nonmaterial cultural elements. Such examinations are conducted through intensive fieldwork.
- Most studies in cultural ecology investigate food production in rural and agricultural settings in the periphery in order to understand how change affects the relationship between cultural groups and the environment.[1]

Cultural ecologists look at food production, demographic change and its impacts on ecosystems, and ecological sustainability. The scale of analysis is not on cultural areas or cultural regions, but on small groups' adaptive strategies to a particular place or setting.

In the Andean example above, cultural ecologists have been able to understand complex relationships between two cultural groups and their environment, showing how the groups' choices were shaped by both culture and environmental conditions. Some critics have argued, however, that cultural ecology leaves out other intervening influences of the relationship between culture and the environment: the impact of political and economic institutions and practices.

During the 1980s cultural ecologists began moving away from a strict focus on a particular group's interactions with the environment, instead placing that relationship within a wider context. The result is political ecology, the merging of political economy with cultural ecology. Political ecology stresses that human-environment relations can be adequately understood only by reference to the relationship of patterns of resource use to political and economic forces. Just as with the study of agriculture, industrialization, urbanization, and comparable geographical phenomena, this perspective requires an examination of the impact of the state and the market on the ways in which particular groups utilize their resource base.

Political ecology incorporates the same human-environment components analyzed by cultural ecologists. However, because political ecologists frame cultural ecology within the context of political and economic relationships, they go beyond what cultural ecologists seek to understand.

A case study of the banana industry on St. Vincent and the Grenadines, an island nation in the Caribbean, illustrates the difference (**Figure 4.B**).[2] Beginning in the 1980s, agriculturalists in the main island of St. Vincent shifted to banana production for export at the same time that local food production began to decline. Without recognizing the impacts of politics and the wider economy, it would be impossible to understand why these two processes have been occurring simultaneously. Disincentives and incentives have both played a role. Disincentives to maintain local food production include marketing constraints, crop theft, competition from inexpensive food exports, and inadequate government assistance. Incentives to produce for export include state subsidies to export-oriented agriculture and access to credit for banana producers, as well as a strong British market for Caribbean bananas. As a result, local food production, although faced with the same environmental conditions as banana production, does not enjoy the same political and economic benefits. Because production for export is potentially more lucrative and an economically safer option, and to some extent because of changing dietary practices, local food production is a less attractive option for agriculturalists.

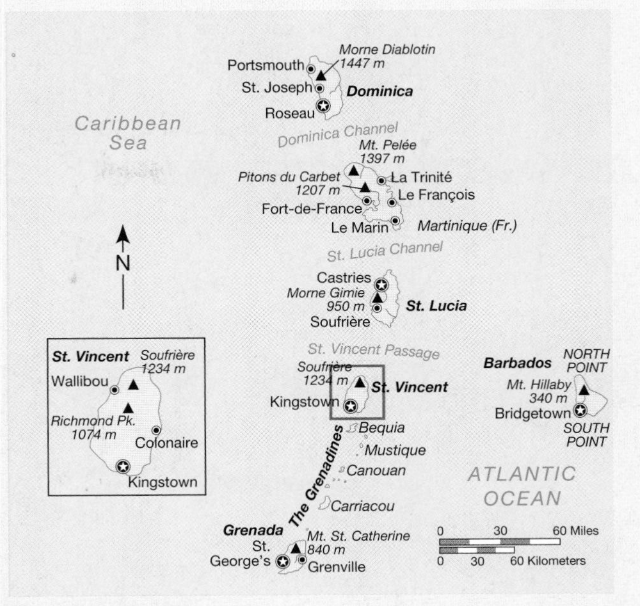

Figure 4.B St. Vincent and the Grenadines These island and island groupings are part of the chain of the Lesser Antilles in the Caribbean Sea. The total population is about 115,000, occupying about 390 square kilometers (150 square miles).

As the St. Vincent case illustrates, the political ecology approach provides a framework for understanding how the processes of the world economy affect local cultures and practices. It also indicates how state policies and practices and economic demand in the global economy shape local decision making. Furthermore, local cultural practices (especially dietary) are being abandoned as people develop a taste for low-cost and convenient imported agricultural commodities such as flour and rice. Unfortunately, however, production for export also opens up the local economy to the fluctuations of the wider global economy. Recent changes in European Union policy on banana imports, for example, are having negative effects on banana production in St. Vincent.

[1]K. Butzer, "Cultural Ecology," in G. L. Gaile and C. J. Wilmot (eds.), *Geography in America.* Columbus, OH: Merrill Publishing Co., 1989, p. 192.

[2]Adapted from L. Grossman, "The Political Ecology of Bananas: Contract Farming, Peasants, and Agrarian Change in Eastern Caribbean." Chapel Hill, NC: University of North Carolina Press, 1998.

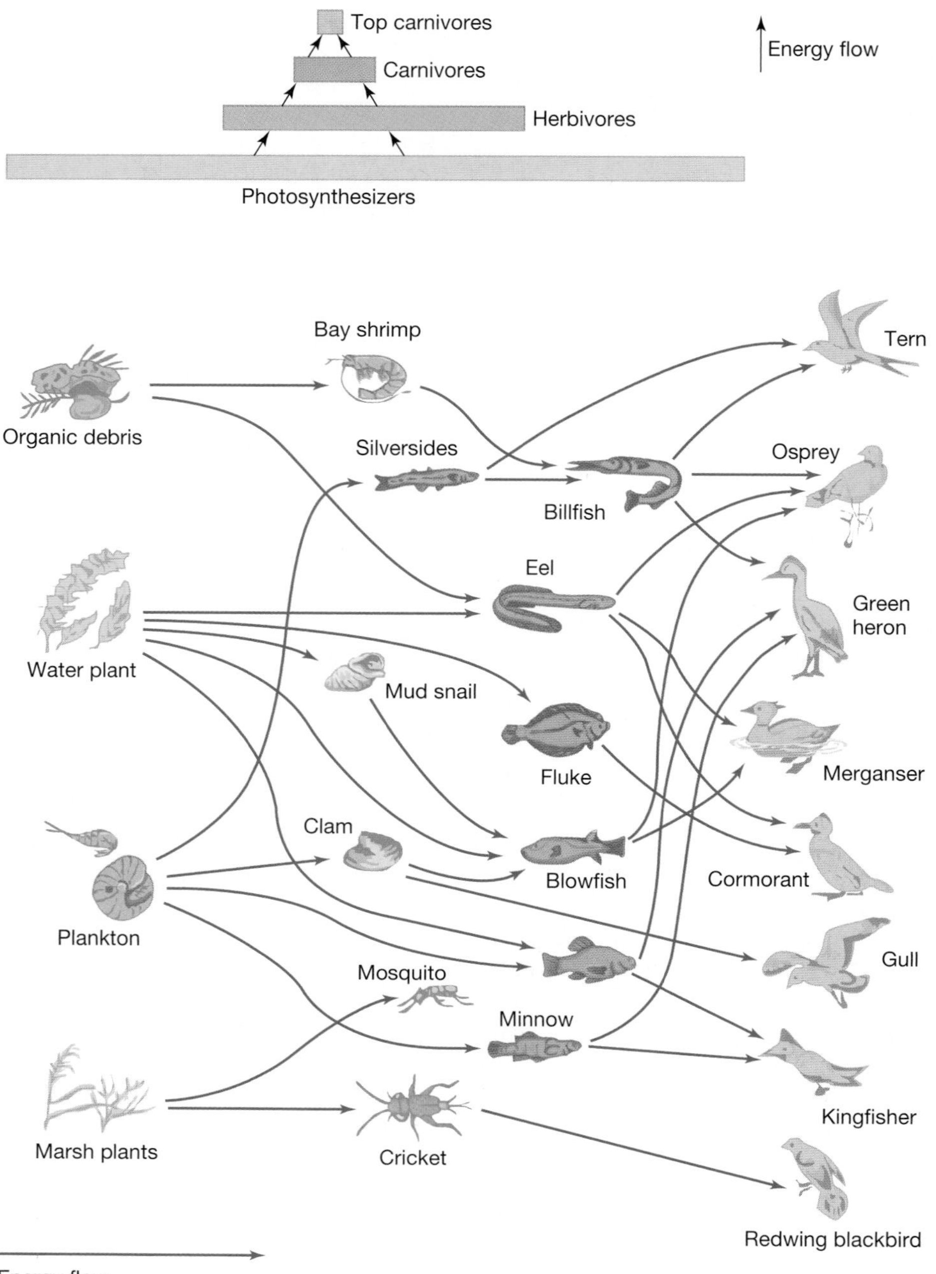

Figure 4.6 Generalized food chain Although a contemporary diagram, this illustration of the food chain in a Long Island estuary demonstrates the nineteenth-century naturalist's view. Plants, animals, and insects are all seen to be part of a complex whole, such that elimination or injury to one element affects the entire system. Although most ecosystems have complex food chains containing numerous relationships among the different parts, one rule holds for all: The higher the animal is in a food chain, the fewer there are of that animal. This rule is illustrated in the bar graph at the top of the illustration, which shows that the photosynthesizers (those at the bottom) are numerous whereas the carnivores (at the top) are much less so. The photosynthesizers are the most numerous because they are the most energy efficient. The farther you go up the chain from the photosynthesizers—who are also known as primary producers—the less energy efficient the animal is. (After C. Ponting, *A Green History of the World.* London: Sinclair-Stevenson, 1991, p. 13.)

Thoreau was also a primary force behind romanticism, a movement that originated in Europe. **Romanticism** is a philosophy that emphasizes the interdependence of humans and nature. In direct revolt against those who espoused a Judeo-Christian understanding of nature, the romantics believed that *all* creatures—human and otherwise—were infused with a divine presence that commanded respect and that humans were not exceptional in this scheme. Rather, their divinity issued from humble participation in the natural community.

A branch of American romanticism known as **transcendentalism** has also influenced contemporary understandings of nature. Transcendentalism was espoused most eloquently by Unitarian minister turned poet and philosopher Ralph Waldo Emerson, a neighbor and contemporary of Thoreau's. It encourages people to attempt to rise above nature and the limitations of the body to the point where the spirit dominates the flesh, where a mystical and spiritual life replaces a primitive and savage one. Thoreau and Emerson are two of the most important influences on contemporary ideas about the human-nature relationship.

Another major influence on U.S. environmentalism derives from the writings of George Perkins Marsh, a native Vermonter who in 1864 wrote a treatise entitled *Man and Nature, or Physical Geography as Modified by Human Action* (heavily revised and republished in 1874 as *The Earth as Modified by Human Action*). As the first work to suggest that human beings are significant agents of environmental change, it is considered one of the most important advances in geography, ecology, and resource management in the nineteenth century. Marsh's ideas served as the foundation of the U.S. environmental movement in the twentieth century. Early in that century, writers like Gifford Pinchot and politicians like Theodore Roosevelt drew on the ideas of Thoreau, Emerson, and Marsh to advocate the wise use of natural resources and the conservation of natural environments. Their view that nature should be conserved has persisted to the present. **Conservation** holds that natural resources should be used wisely and that humans should serve as stewards, not exploiters, of the natural world. Conservation implies responsibility to future generations as well as to the natural world itself in the utilization of resources.

The writings of all these individuals eventually helped to inspire a wide range of environmental organizations, including the Environmental Defense Fund, World Watch Institute, Nature Conservancy, and the Sierra Club. The Environmental Defense Fund brings together experts in science, law, and economics to address complex environmental issues and to educate governments and the public about them. World Watch is a nonprofit public-policy research organization whose goal is also to inform policymakers and the public about emerging global problems and trends, and the complex links between the world economy and its environmental support systems. The mission of the Nature Conservancy is: "To preserve the plants, animals and natural communities that represent the diversity of life on Earth by protecting the lands and waters they need to survive." The Sierra Club is an organization founded in the late nineteenth century, which has a similar mission of promoting and protecting Earth's wild places, diverse ecosystems, and resources. All of these organizations are considered mainstream in their approaches in that they work within the political and economic system to achieve their ends.

For those who espouse a more radical approach to nature, the conservation approach described previously is seen as too passive to be truly effective in protecting the environment. These individuals believe that conservation leaves intact the political and economic system that drives the exploitation of nature. They believe that nature is sacred and should be preserved, not used at all. This more extreme position, **preservation,** advocates that certain habitats, species, and resources should remain off limits to human use, regardless of whether the use maintains or depletes the resource in question. The philosophy of groups such as Earth First! is closely aligned with the preservationist perspective. Earth First!, unlike the Sierra Club, operates outside the bounds of mainstream institutional frameworks. Whereas the Sierra Club takes its opponents into the courtroom, Earth First! employs extralegal tactics—often called *ecoterrorist* tactics—such as driving spikes into trees to discourage logging. These "quick strike" actions are intended to halt what are regarded as government or corporate environmental abuses (which may, in fact, be perfectly legal though counter to the Earth First! philosophy).

Greenpeace is yet another environmental organization different from either Earth First! or the Sierra Club. Greenpeace is global in its reach, meaning that both its membership and its areas of emphasis are international (**Figure 4.7**). (We talk more about global environmental organizations later in this chapter.) Focusing on environmental polluters and combining the strategies of both the Sierra Club and Earth First!, Greenpeace utilizes oppositional tactics as well as formal international legal actions. In its membership—with the world headquarters in

Figure 4.7 Greenpeace protest in Europe The sign attached to the Brandenburg Gate in Berlin is displayed by Greenpeace, Europe, one of the many regional branches of the organization. Activists like these and in other environmental organizations throughout Europe and North America were among those who opposed the war in Iraq for humanitarian as well as environmental reasons. The Persian Gulf War (1991) was an environmental disaster as Iraqi troops released oil into the Persian Gulf and ignited 732 oil wells in Kuwait as they fled the country. These protesters and others like them would like to see a different energy source than oil fueling the global economy.

Amsterdam and regional offices in most major industrial countries—as well as its objectives—halting environmental pollution worldwide—Greenpeace articulates the belief that places are interdependent, and what happens in one part of the globe affects us all.

Organizations such as Earth First! and Greenpeace are practical illustrations of new approaches to understanding human interactions with nature that have developed since the publication of *Silent Spring* over 40 years ago. These, as well as other new approaches—environmental ethics, ecofeminism, deep ecology, and environmental justice—take the view that nature is as much a physical universe as it is a product of social thought. All provide different ways of understanding how society shapes our ideas about nature.

Environmental ethics is a philosophical perspective on nature that prescribes moral principles as guidance for our treatment of it. Environmental ethics insists that society has a moral obligation to treat nature according to the rules of moral behavior that exist for treating human beings. An aspect of environmental ethics that has caused a great deal of controversy is the perspective that animals, trees, rocks, and other elements of nature have rights in the same way that humans do. If the moral system of our society insists that humans are to have the right to a safe and happy life, then it is argued that the same rights should be extended to nonhuman nature.

Ecofeminism shares much of this philosophical perspective. Ecofeminists hold that patriarchy—a system of social ideas that values men more highly than women—is at the center of our present environmental malaise. Because patriarchy has equated women with nature, it has promoted the subordination and exploitation of both. The many varieties of ecofeminism range from nature-based spirituality oriented toward a goddess to more political approaches that emphasize resistance and opposition to the dominant masculine models that devalue what is not male. Some ecofeminists are also environmental ethicists. Not only a movement of the core, ecofeminism has also been widely embraced in the periphery, where women are primarily responsible for the health and welfare of their families in environments that are being rapidly degraded. The unifying objective in all of ecofeminism is to dismantle the patriarchal biases in Western culture and replace them with a perspective that values social, cultural, and biological diversity.

Deep ecology, which shares many points with ecofeminism, is an approach to nature revolving around two key components: self-realization and biospherical egalitarianism. Self-realization embraces the view that humans must learn to recognize that they are part of the nonhuman world, whereas biospherical egalitarianism insists that Earth, or the biosphere, is the central focus of all life and that all components of nature, human and nonhuman, deserve the same respect and treatment. Deep ecologists, like environmental ethicists, believe that there is no absolute divide between humanity and everything else, and that a complex and diverse set of relations constitutes the universe. The belief that all things are internally related could enable society to treat the nonhuman world with respect and not simply as a source of raw materials for human use.

Activists in the **environmental justice** movement consider the pollution of their neighborhoods through such elements as nearby factories and hazardous-waste dumps to be the result of a structured and institutionalized inequality that is pervasive in both the capitalist core and periphery. They see their struggles as distinct from the more middle-class and mainstream struggles of groups such as the Sierra Club or even Earth First! and Greenpeace. These activists view their struggles as rooted in their economic status. Thus, these struggles are not about quality-of-life issues, such as whether any forests will be left for recreation, but about sheer economic and physical survival. As a result, the questions raised by environmental justice activists involve the distribution of economic and political resources. Such questions are not easily resolved in courts of law, but speak to more complex issues such as the nature of racism and sexism, and of capitalism as a class-based economic system.

The environmental justice movement is not restricted to the core. Indeed, poor people throughout the world are concerned that the negative impacts of economic development consistently affect them more than the rich. Furthermore, although most people support improving the lives of people throughout the globe who are currently living in poverty, many have serious reservations about trying to raise all living standards to the level that core regions and the elite enjoy. In fact, some globalization experts argue that some of the highest standards of living around the globe will have to be lowered in order to raise others up, because the widespread economic development that universally high standards would require is not environmentally sustainable. In short, because of the limitations set by Earth's resources—both in terms of the amount and quality of key resources—improvements in social well-being at the level now enjoyed by the core cannot be sustained over the long term and may not even be possible to begin with. As a result, discussions of the future of economic globalization are increasingly being accompanied by discussions about sustainability and the ways economic development might proceed without continuing to damage the environment. (We discuss sustainability in greater detail later in this chapter.)

All of these approaches to the environment attest to a growing concern regarding the effects of globalization. Acid rain, deforestation, the disappearance of species, nuclear accidents, and toxic waste have all been important stimuli for newly emerging philosophies about the preferred relationships between society and nature within a globalizing world. While none of these philosophies is a panacea, each has an important critique to offer. More than anything, however, each serves to remind us that environmental crises are not simple, and simple solutions will no longer suffice.

The Concept of Nature and the Rise of Science and Technology

Let us return to the point made earlier: that the most widespread ideas of nature current in Western thought—and ones that have persisted under different labels for thousands of years—are that humans are the center of all creation and that nature in all its wildness was meant to be dominated by humans. This Judeo-Christian belief insists that Man, made in the image of God, was set apart from nature and must be encouraged to control it.

While Christianity held that nature was to be dominated, that idea existed more in the religious and spiritual realm than in the political and social realm. In terms of the conduct of everyday life, it was not until the sixteenth century that Christian theology was conscripted to aid the goals of science. Before 1500 in Europe there existed a widely held view of Earth as a living entity such that human beings conducted their daily lives in an intimate relationship with the natural order of things. The prevailing image was that of an organism, which emphasized interdependence among human beings and between them and Earth. Yet even within this organic idea of nature, we can find two opposing conceptions. One was of a nurturing Earth that provided for human needs in a beneficent way; the other was of a violent and uncontrollable nature that could threaten human lives. In both views Earth and nature were regarded as female.

Francis Bacon (1561–1626) and Thomas Hobbes (1588–1679) were English philosophers who, as prominent promoters of science and technology, were influential in changing the prevailing organic view of nature. Borrowing from Christian theology, they advanced a view of nature as something subordinate to Man. Bacon and Hobbes sought to rationalize benevolent nature as well as to dominate disorderly and chaotic nature.

As feminist environmental historian Carolyn Merchant writes:

> The change in controlling imagery was directly related to changes in human attitudes and behavior toward the earth. Whereas the nurturing earth image can be viewed as a cultural constraint restricting the types of socially and morally sanctioned human actions allowable with respect to the earth, the new images of mastery and domination functioned as cultural sanctions for the denudation of nature. Society needed these new images as it continued the process of commercialization and industrialization, which depended on activities directly altering the earth—mining, drainage, deforestation, and assarting [grubbing up stumps to clear fields]. The new activities used new technologies—lift and force pumps, cranes, windmills, geared wheels, flap valves, chains, pistons, treadmills, under- and overshot watermills, fulling mills, flywheels, bellows, excavators, bucket chains, rollers, geared and wheeled bridges, cranks, elaborate block and tackle systems, worm spur, crown, and lantern gears, cams and eccentrics, ratchets, wrenches, presses, and screws in magnificent variation and combination.[1]

Figure 4.8 illustrates Merchant's point, that by the sixteenth and seventeenth centuries the power of science was too great for the organic idea of nature. Subsequently, a view that nature was the instrument of Man became dominant in Western culture.

Figure 4.8 Commodification of nature This sixteenth-century woodcut illustrates the various uses of wood and water with respect to the mining industry of the late Middle Ages and Renaissance. This scene illustrates the exploitation of nature, in contrast to the view of it as a bountiful mother. (After Georg Agricola woodcut. Georgius Agricola, *De Re Metallica,* translated from the first Latin edition of 1556 by Herbert Clark Hoover/Lou Henry Hoover, p. 337.)

[1]C. Merchant, *The Death of Nature.* San Francisco: Harper and Row, 1979, pp. 2–3.

THE TRANSFORMATION OF EARTH BY ANCIENT HUMANS

Although the previous discussion might suggest that Earth remained relatively unaffected by human action until well into the early modern period, the Paleolithic and Neolithic peoples altered the environment even without machines or elaborate tools. Considerable evidence exists that early humans were very active agents of change. People's perceptions of nature were usually quite influential in shaping their environmental behaviors, although many examples exist of contradictions between attitudes and actions. In this section we see that contemporary humans have inherited an environment that was significantly affected even by the practices of our very earliest ancestors.

Paleolithic Impacts

Although humans are thought to have first inhabited Earth approximately 6 million years ago, almost no evidence exists of how the very earliest hominids, as they are called, used the natural world around them to survive. What is known is that their numbers were not large and that they left little behind in the way of technology or art to help us understand their relationship to nature. The earliest evidence about the environmental relationships of our ancestors comes from the **Paleolithic period** (about 1.5 million years ago), a cultural period also known as the early Stone Age, because this was the period when chipped-stone tools first began to be used.

Hunters and gatherers living on the land in small groups, the early Stone Age people mainly foraged for wild food and killed animals and fish for their survival. Hunting under early Stone Age conditions could not support a growing population, however. It is estimated that on the African grassland, where humans are believed to have first evolved, only two people could survive on the vegetation and wildlife available within about a 2.5-square-kilometer (1-square-mile) area. To help ensure survival, early Stone Age people constantly moved over great distances, which ultimately made them a dispersed species, creating the foundation for the world's population distribution of today (**Figure 4.9**).

Evidence also exists of early Stone Age tools, as well as of the importance of hunting to human existence. The cave paintings of Vallon-pont-d'arc, France, for example, illustrate that hunting was the primary preoccupation of the Stone Age mind (**Figure 4.10**). Because these early peoples lived in small bands and moved frequently in wider and wider ranges, it is tempting to conclude that they had very little impact on their environment. It does appear, however, that Stone Age people frequently used the powerful tool of fire. They used it to attract game, to herd and hunt game, to deflect predators, to provide warmth, and to encourage the growth of vegetation that would attract grazing animals like antelope and deer.

The impact of frequent and widespread fire on the environment can be dramatic. Fire alters or destroys vegetation—from entire forests to vast grasslands (**Figure 4.11**). Fire can encourage the growth and survival of some species, while eliminating others. When fire destroys the vegetation that anchors the soil, however, and is followed by heavy rains, it can lead to soil erosion and, in areas with a steep slope, to the total denudation of the landscape. The use of fire by early Stone Age peoples certainly had all these impacts on the environment.

Archaeologists believe that many large North American animal species disappeared around 11,000 years ago. At the end of the Pleistocene—the geologic and climatic age immediately preceding the one in which we now live—large, slow animals such as the mastodon, mammoth, cave bear, woolly rhino, and giant deer became extinct. These species constituted over two-thirds of the megafauna, or large animals, of the region. A great deal of controversy exists about why these megafauna became extinct and others did not. Climate change or large-scale natural dis-

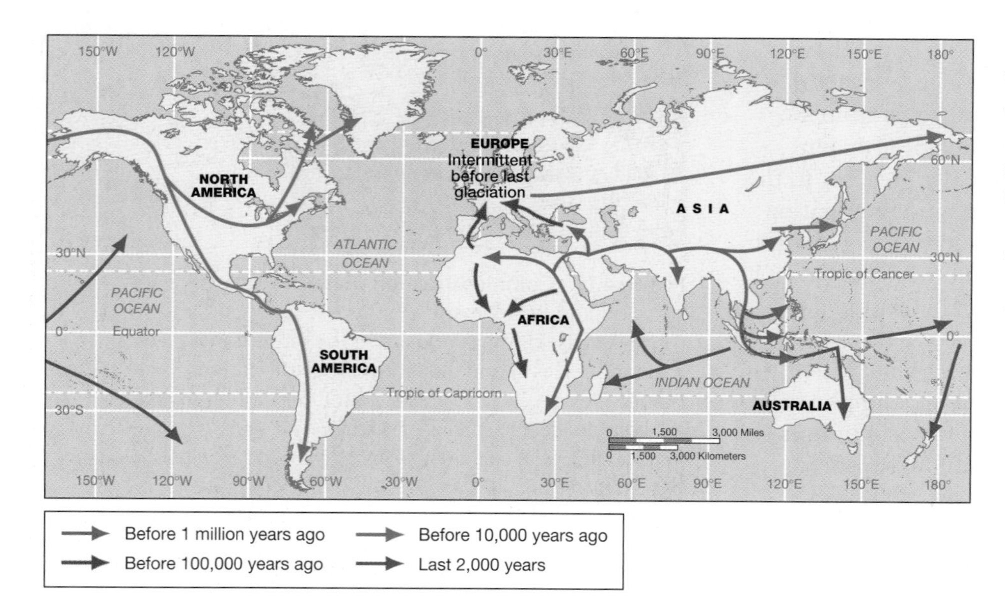

Figure 4.9 The settlement of the world This map shows one theory of the direction and timing of movement of early humans. The constant search for food promoted such movement. The map represents over 1 million years of migration. (After C. Ponting, *A Green History of the World.* London: Sinclair-Stevenson, 1991, p. 25.)

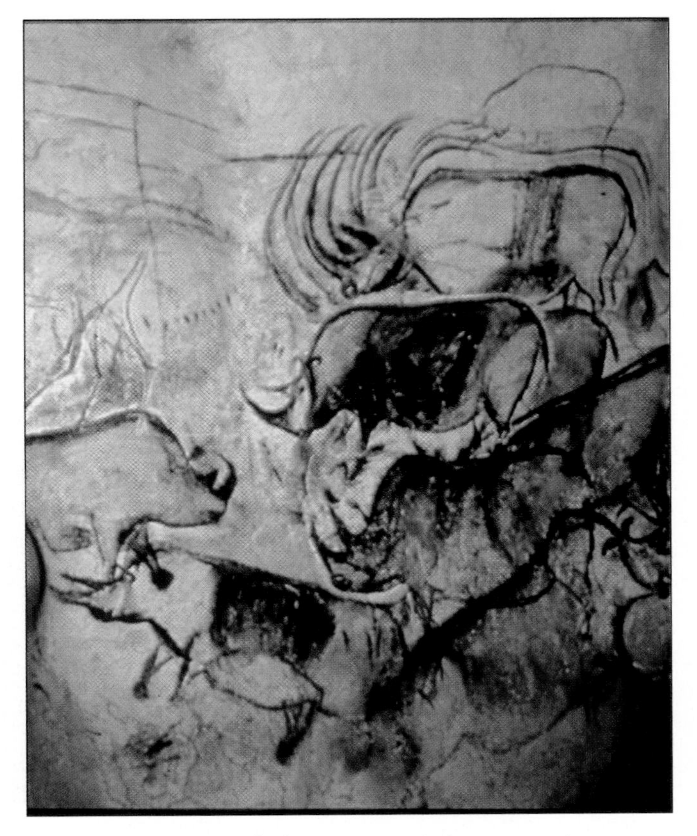

Figure 4.10 Cave paintings Cave paintings are an important record of the imagination of early hunter-gatherers. Some regard these paintings as clear evidence of the development of forward thinking or anticipation among humans. This cave painting from Vallon-pont-d'Arc in southern France portrays large animals hunted for their meat and skins.

asters are not satisfying explanations because neither is particularly selective such that megafauna would be eliminated but not other animals.

Another explanation for the extinction of the larger, slower megafauna is that early Stone Age peoples were directly responsible. While it might seem implausible that primitive hunting techniques could have such a devastating impact, consider the point made earlier that 2.5 square kilometers (one square mile) of vegetation and wildlife were adequate for the survival of only two people. As population sizes increased, more pressure was placed on animal populations to satisfy human food requirements. Those animals easiest to bring down were certainly the slowest and largest—the ones that could not escape quickly.

It is also the case that early Stone Age peoples had, over time, refined their killing technologies, particularly stone blades and spearheads. The double-edged **Clovis point,** for example, increased the likelihood of a kill rather than an injury to an animal. The Clovis point (**Figure 4.12**) is a flaked, bifaced projectile whose length is more than twice its width. Used to kill large animals such as bison, the point is so named because archeologists found the first projectiles in conjunction with kill sites in Clovis, New Mexico. Some Paleolithic hunters used the natural landscape to trap or kill large numbers of animals. Driving animals into canyons where they could be contained, or over cliffs where they would be killed en masse, ensured that huge numbers could be eliminated at one time (**Figure 4.13**). Other Paleolithic groups hunted small game, such as rabbits, using traps and small projectiles.

Neolithic Peoples and Domestication

The credit for the development of agriculture—a technological triumph with respect to nature—goes to the Neolithic peoples, also known as the late Stone Age peoples. While the divide between them and the Paleolithic peoples occurred about 10,000 years ago, it is not known exactly when Neolithic peoples shifted from hunting and gathering to cultivating certain plants and taming and herding wild animals. We have termed that period the First Agricultural Revolution (described in greater detail in Chapter 8). Climatically we know that for many regions of the globe, this period coincided with the end of the last Ice Age, which means that spring slowly began to occur in

Figure 4.11 Fire and its impacts on the landscape Fires can have a devastating impact on the landscape. Especially hot fires can burn not only the surface vegetation but also the organic material in the soil. The loss of these organic materials and other nutrients hinders the regeneration of vegetation. Without vegetation to anchor the soil, heavy rains can cause massive soil erosion. Shown in the photograph is eastern Arizona, the site of the convergence of the Rodeo and Chediski fires in 2002.

Figure 4.12 The Clovis point Clovis points are named after Clovis, New Mexico, where the points were first discovered. The Clovis points in this photo were found in Southern Arizona, and they show the typical shape of the points, which were attached to spears and used to hunt game. (*Source:* Photo courtesy Arizona State Museum © Jerry Jacka Photography.)

places that had not experienced it for thousands upon thousands of years.

At this time, environmental conditions made possible the domestication of plants and animals, which also requires a sedentary lifestyle based in permanent settlements. The first domestication successes of the late Stone Age peoples were with the most docile animals (herbivores) and the hardiest plants (those with large seeds and a tolerance to drought). Once early domestication was established, it became possible for small groups of Neolithic peoples to cease to be nomads. As Chapter 2 discusses, permanent settlement enabled further refinements in domestication. Eventually Neolithic people achieved a truly dramatic innovation—the breeding of plants and animals to produce desired genetic characteristics, such as disease resistance in plants.

The emergence of agriculture changed the course of human history and had important environmental impacts—both negative and positive. One negative impact was the simplification of ecosystems as the multiplicity of wild species began to be replaced by fewer cultivable crops. An **ecosystem** is a community of different species interacting with each other and with the larger physical environment that surrounds it. Along with the vast number of wild species lost has gone the opportunity to understand their benefit to humans and the wider ecosystem. More positively, however, increased crop yields through greater control over available foodstuffs helped to improve human health and eventually increased population growth.

It is from this early period of plant domestication that we also find widespread evidence of a growing appreciation of nature through ritual, religion, and art. Human beings depended upon rain, soil fertility, and an abundance of sunlight to produce a successful harvest, and reverential attitudes toward nature appear to have been pervasive. In many places in both the Old and the New World, people worshiped Earth, Sun, and Rain, and they sacrificed deer and bear to them in an attempt to ensure survival.

Early Settlements and Their Environmental Impacts

What is perhaps the most significant aspect of plant and animal domestication is that it eventually enabled a food surplus to be produced. It also permitted the formation of human settlements in which small groups—probably craftspeople and political and religious elites—were able to live off the surplus without being responsible for its

Figure 4.13 Massive animal kills Paleolithic hunters appear to have used features of the landscape to aid them in hunting large game. Archaeologists believe that the mounds of skeletal remains of large animals found at various sites are evidence of this. It is not clear whether hunters and their kin were even able to consume all the animal flesh made available through such killing methods. It has been speculated that such gross killing methods may have led to the extinction of some species. (After Arthur Lidov/National Geographic Society Image Collection.)

production. Eventually growing numbers of people, bolstered by increasing surpluses, were able to settle in places where water was available and the land cultivable.

The invention of agricultural tools helped to further the domestication of plants and animals as well as multiply the early agriculturalists' impact on the landscape. Among the early tools that enabled humans a greater measure of control over nature were the sickle for harvesting wheat (**Figure 4.14**); the plow for preparing the soil; the yoke for harnessing draft animals, such as oxen, to pull the plow; and the wheel for grinding wheat, creating pottery, and later enabling transportation. The wheel was also used as a pulley to draw water. In Sumer and Assyria, for example, the wheel enabled the development of large-scale irrigation systems.

Irrigation is one of the most significant ways that humans have been able to alter the limits of their environment. Throughout much of the world, in fact, agriculture could not occur without irrigation. And as agriculture has spread, irrigation has increased (**Table 4.1**). Following the success of the Fertile Crescent, agriculture diffused and new settlements emerged. The food-producing minisystems of China, the Mediterranean, Mesoamerica, the Middle East, and Africa were sustained largely through irrigated agriculture. Yet despite the existence of a vast irrigation network and a whole social structure bound up with agricultural production and attendant activities, the cities of the Mesopotamian region collapsed around 4,000 years ago. While there is no undisputed explanation for why this occurred, many researchers believe—based on archaeological evidence—that it was due to environmental mismanagement. The irrigation works became clogged with salt, resulting in increasingly saline soils. To counteract the effect of salt on production, agriculturalists switched to barley, which is more salt-resistant than wheat, but the ultimate result was a significant drop in yields. **Siltation** (the buildup of sand and clay in a natural or artificial waterway) associated with **deforestation** (the removal of trees from a forested area without adequate replanting) also occurred, filling up the deltas for nearly 200 miles (322 kilometers). Eventually the canals filled with salt and the soils became too saline to support cultivation.

Figure 4.14 Wheat and flint sickle blade Perhaps the most significant factor in the spread of agriculture from Mesopotamia was the occurrence of hybrid forms of wheat, one of many wild grasses found in the area. Even before a fertile hybrid emerged, however, people were harvesting the wild forms. Sickle blades made from flint and set into a horn handle were the most common harvesting tools.

TABLE 4.1 World Irrigated Area Since 1700

Date (A.D.)	Area (in thousands of square kilometers/ square miles)
1700	50/19
1800	80/31
1900	480/185
1949	920/355
1959	1490/575
1980	2000/772
1981	2130/822
1984	2200/849
2000	2740/1057

Source: W. Meyer, *Human Impact on the Earth,* 1990. Cambridge: Cambridge University Press, p. 59. *Original source:* B. G. Rozanov, V. Targulian, and D. S. Orlov, "Soils" in *The Earth Transformed by Human Action,* B. L. Turner II, W. L. Clark, R. W. Kates, J. F. Richards, J. T. Matthews, and W. B. Meyer (eds.), 1990. Cambridge: Cambridge University Press. Updated from FAOSTAT Agricultural Database. 2001. Irrigation. 10 July 2001. Web site: http://apps.fao.org/page/form? collection=Irrigation&Domain=Land&servlet=1&language=EN&hostname=apps.fao.org&version=default

While it may seem that poorly informed management led to the demise of Mesopotamian cities, increasingly saline soils currently plague agriculture in California and southwestern Arizona (**Figure 4.15**). And it was not only the Mesopotamians who made environmental mistakes. Other early urban civilizations, such as the Mayans in Central America and the Anasazi of Canyon de Chelly in Arizona, are also thought to have collapsed due to environmental mismanagement of water.

In the following section we examine the period of European expansion and globalization. Although many other important cultures and civilizations affected the environment in the intervening periods, the impacts of their technological developments were much the same as those we have already described. The period of European colonialism, however, had a profoundly different impact from preceding periods in extent, magnitude, and kind. Furthermore, it set the stage for the kinds of environmental problems contemporary society has inherited, perpetuated, and magnified.

EUROPEAN EXPANSION AND GLOBALIZATION

The history of European expansion provides a powerful example of how a society with new environmental attitudes was able to transform nature in radically new ways.

Figure 4.15 Irrigation system near El Centro, Southern California This photo illustrates a large irrigation system in one of the driest areas of the United States. Much of the remaining cultivable land in the United States lies in dry areas such as in the Southwest. Large-scale irrigation is required there to sustain agricultural productivity. Under these circumstances, the application of water to crops is an expensive undertaking—it has to be pumped some distance and is subject to rapid evaporation during the dry, hot season of summer. Irrigation in the Southwest also contributes to the depletion of groundwater supplies. The systems that supply such irrigation are expensive to build and maintain. Much of the water delivered to the agricultural sector in the U.S. Southwest is heavily subsidized by the federal government in order to protect agricultural producers from the negative impact that high water prices would have on the number of farms as well as overall productivity. The agricultural sector is often the largest water user among all sectors (including residential, commercial, and government) in the Southwest.

These new attitudes drew from a newly emerging science and its contribution to technological innovation; the consolidation of the population around Judeo-Christian religious beliefs; and, most important, the development of a capitalist political and economic system.

Initially European expansion was internal—largely contained within its continental boundaries. The most obvious reason for this expansion was population increase: from 36 million in 1000 to over 44 million in 1100, nearly 60 million in 1200, and about 80 million by 1300 (**Figure 4.16**). As population continued to increase, more land was brought into cultivation. In addition, more forest land was cleared for agriculture, animals killed for food, and minerals and other resources exploited for a variety of needs. Forests originally covered upward of 90 percent of Western and Central Europe. At the end of the period of internal expansion, however, around 1300, the forested area was only 20 percent.

The bubonic plague, also known as the Black Death, had temporarily slowed population growth by wiping out over a third of Europe's population in the mid-fourteenth century. By then agricultural settlement had been extended to take up all readily available land and then some. In England, Italy, France, and the Netherlands, for example, marshes and fens had been drained and the sea pushed back or the water table lowered to reclaim and create new land for agriculture and settlement.

In the fifteenth century, Europe initiated its second phase of expansion—external—which not only changed the global political map but launched a period of environmental change that continues to this day. European external expansion—colonialism—was the response to several impulses, ranging from self-interest to altruism. Europeans were fast running out of land, and as we saw in Chapter 2, explorers were being dispatched by monarchs to conquer new territories and enlarge their empires while collecting tax revenues from the monarch's new subjects. Many of these adventurous individuals were also searching for fame and fortune or avoiding religious persecution. Behind European external expansion was also the Christian impulse to bring new souls into the kingdom of God. Other forces behind European colonialism included the need to expand the emerging system of trade, which ultimately meant increased wealth and power for a new class of people—the merchants—as well as for the aristocracy.

Over the centuries, Europe came to control increasing areas of the globe. Two cases illustrate how the introduction of European people, ideologies, technologies, plant species, pathogens, and animals changed not only the environments into which they were introduced but also the societies they encountered.

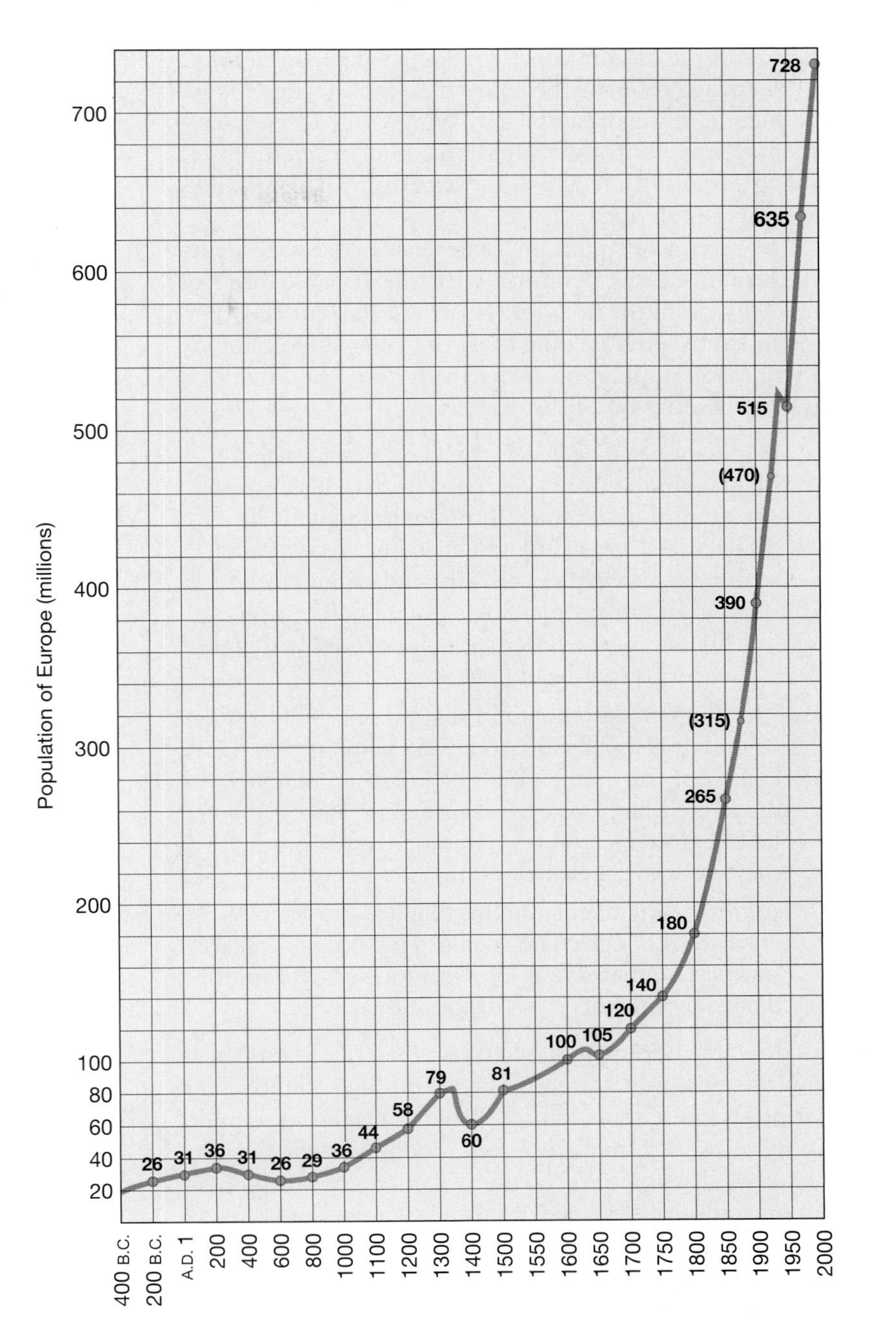

Figure 4.16 Population growth in Europe This graph shows the growth in the European population from 400 B.C. to A.D. 2000. As is apparent from the graph, the growth in European population has been especially dramatic in the last 500 years as a result of capitalist globalization. The increase in human numbers at the beginning of the 1500s was an important push to exploration and colonization beyond the confines of the continent. The dip in the graph from 1300 to 1500 is partially explained by the bubonic plague epidemic called the Black Death, but food shortages also played a significant role in this population decline. Another dip during the middle of the twentieth century shows the effects of two world wars. (After C. McEvedy and R. Jones, *Atlas of World Population History,* London: Allen Lane, 1978, Fig. 1.2, p. 18.)

Disease and Depopulation in the Spanish Colonies

Little disagreement exists among historians that the European colonization of the New World was eventually responsible for the greatest loss of human life in history. Moreover, very little disagreement exists that the primary factor responsible for that loss was disease. New World populations, isolated for millennia from the Old World, possessed immune systems that had never encountered some of the most common European diseases. **Virgin soil epidemics**—where the population at risk has no natural immunity or previous exposure to the disease within the lifetime of the oldest member of the group—were common in the so-called Columbian Exchange, though the exchange in this case was mostly one-way. The **Columbian Exchange** was the interaction between the Old World (Europe) and the New World (the Americas) initiated by the voyages of Columbus. For example, diseases such as smallpox, measles, chicken pox, whooping cough, typhus, typhoid fever, bubonic plague, cholera, scarlet fever, malaria, yellow fever, diphtheria, influenza, and others were unknown in the pre-Columbian New World.

Geographer W. George Lovell has examined the role disease played in the depopulation of some of Spain's New World colonies from the point of initial contact until the early seventeenth century, using several cases to illustrate his point.[2] The first case is Hispaniola (present-day

[2]W. G. Lovell, "Heavy Shadows and Black Night": Disease and Depopulation in Colonial Spanish America. *Annals, Association of American Geographers,* 82:426–443, 1992.

Haiti and the Dominican Republic), where Columbus's 1493 voyage probably brought influenza through the introduction of European pigs carrying swine fever. Subsequent voyages brought smallpox and other diseases, which eventually led to the extinction of the island's Arawak population.

In a second example, in Central Mexico, Lovell writes of Hernán Cortés's contact with the Aztec capital of Tenochtitlán in the first decades of the sixteenth century, which led to a devastating outbreak of smallpox among a virgin soil population. A native Aztec text provides a graphic description of the disease:

> While the Spaniards were in Tlaxcala, a great plague broke out here in Tenochtitlán. It began to spread during the thirteenth month [September 30–October 19, 1520] and lasted for seventy days, striking everywhere in the city and killing a vast number of our people. Sores erupted on our faces, our breasts, our bellies; we were covered with agonizing sores from head to foot.
>
> The illness was so dreadful that no one could walk or move. The sick were so utterly helpless that they could only lie on their beds like corpses, unable to move their limbs or even their heads. They could not lie face down or roll from one side to the other. If they did move their bodies, they screamed with pain.
>
> A great many died from this plague, and many others died of hunger. They could not get up and search for food, and everyone else was too sick to care for them, so they starved to death in their beds.[3]

In a third example, Lovell described the Jesuits' missionizing efforts in northern Mexico during a slightly later period. Because these efforts gathered dispersed population groups into single locations, conditions for the outbreak of disease were created. Contact with Spanish conquistadors in advance of the missionaries had already reduced native populations by perhaps 30 to 50 percent. When groups were confined to smaller areas organized around a mission, mortality rates climbed to 90 percent. Eventually the disease was diffused beyond the initial area of contact as traders carried it across long-distance trade routes to the periphery of the Mayan empire in advance of the Spanish armies and missionaries. The Mayans were not defeated by European technological superiority, but by the ravages of a new disease against which they possessed no natural defenses.

Lovell provides similar descriptions of disease impacts in Mayan Guatemala and the Central Andes of South America that led to devastating depopulation. Scholars refer to the phenomenon of near genocide of native populations as **demographic collapse.** The ecological effect of the population decline caused by the high rates of mortality was the transformation of many regions from productive agriculture to abandoned land. Many of the Andean terraces, for example, were abandoned, and dramatic soil erosion ensued. In contrast, large expanses of cleared land eventually returned to forests in areas such as the Yucatán in present-day Mexico.

Old World Plants and Animals in the New World

A second case study of the environmental effects of European colonization involves the introduction of Old World plants and animals in the New World, and vice versa. The introduction of exotic plants and animals into new ecosystems is called **ecological imperialism,** a term now widely used by geographers, ecologists, and other scholars of the environment. The interaction between the Old and the New World resulted in both the intentional and unintentional introduction of new crops and animals on both sides of the Atlantic. Europeans brought from their homelands many plants and animals that were exotics, that is, unknown to American ecosystems. For example, the Spanish introduced wheat and sugarcane, as well as horses, cattle, and pigs.

These introductions altered the environment, particularly as the emphasis on select species led to a reduction in the variety of plants and animals that constituted local ecosystems. Inadvertent introductions of hardy exotic species included rats, weeds such as the dandelion and thistle, and birds such as starlings, which crowded out the less hardy indigenous species. As with the human population, the indigenous populations of plants, birds, and animals had few defenses against European plant and animal diseases and were sometimes seriously reduced or made extinct through contact.

Contact between the Old and the New Worlds was, however, an exchange—a two-way process—and New World crops and animals as well as pathogens were likewise introduced into the Old World, sometimes with devastating implications. Corn, potatoes, tobacco, cocoa, tomatoes, and cotton were all brought back to Europe; so was syphilis, which spread rapidly throughout the European population.

Contacts between Europe and the rest of the world, though frequently violent and exploitative, were not uniformly disastrous. There are certainly examples of beneficial contacts for both sides. The largely beneficial impacts of the Columbian Exchange were mostly knowledge-based or nutritional. Columbus's voyages (**Figure 4.17**) added dramatically to global knowledge, expanding understanding of geography, botany, zoology, and other rapidly growing sciences. It has been argued that the availability of American gold bullion and silver enlarged Europe's capacity for trade and may even have been indirectly responsible for creating the conditions that launched the Industrial Revolution.

The encounter also had significant nutritional impacts for both sides by bringing new plants to each. European

[3]W. G. Lovell, 1992, p. 429, quoting from M. León-Portilla, *The Broken Spears: The Aztec Account of the Conquest of Mexico*. Boston: Beacon Press, 1962, pp. 92–93.

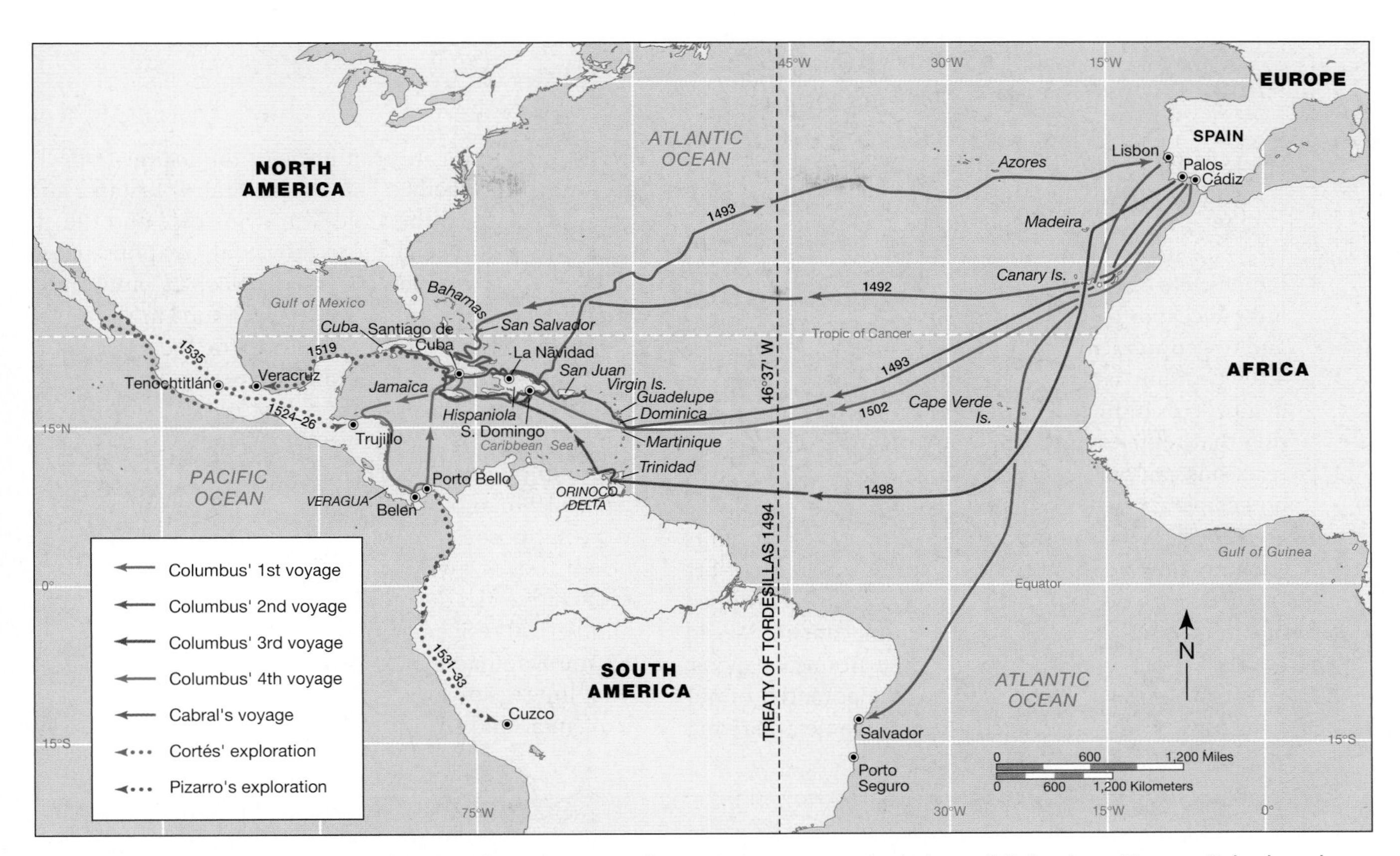

Figure 4.17 European voyages of exploration This maps shows the voyages and missions of Columbus, Pizarro, Cabral, and Cortés. Departing from Portugal and Spain, Columbus encountered several of the islands of the Caribbean, as well as the coastal area of present-day Honduras and Venezuela. (After *The Penguin Atlas of the Diasporas,* by Gerard Chaliand and Jean-Pierre Rageau, translated by A. M. Berrett. Translation copyright © 1995 by Gerard Chaliand and Jean-Pierre Rageau.)

colonization, although responsible for the extermination of hundreds of plant and animal species, was also responsible for increasing the types and amounts of foods available worldwide. It is estimated that the Columbian Exchange may have tripled the number of cultivable food plants in the New World. It certainly enabled new types of food to grow in abundance where they never had grown before, and it introduced animals as an important source of dietary protein. The advantages of having a large variety of food plants are several. For instance, if one crop fails, another more than likely will succeed because not all plants are subject to failure from the same set of environmental conditions.

The introduction of animals provided the New World not only with additional sources of protein but also with additional animal power. Before the Columbian Exchange, the only important sources of animal energy were the llama and the dog. The introduction of the horse, the ox, and the ass created a virtual power revolution in the New World. These animals also provided fibers and, after death or slaughter, hides and bones to make various tools, utensils, and coverings. Most significant in its environmental impact, however, was the ox.

Land that had escaped cultivation because the indigenous digging sticks and tools were unable to penetrate the heavy soil and matted root surface became available to an ox-drawn plow. The result was that the indigenous form of intensive agricultural production (small area, many laborers) was replaced by extensive production (large area, fewer laborers). This transformation, however, was not entirely without negative impacts, such as soil destabilization and erosion.

It is also important to explore the impact of native New World peoples on their environment. The popular image of indigenous peoples living in harmony with nature, having only a minimal impact on their environment, is flawed. In reality, different groups had very different impacts, and it is erroneous to conflate the thousands of groups into one romanticized caricature.

In New England, for example, prior to European contact, groups existed who hunted for wild game and gathered wild foods. More sedentary types also existed, living in permanent and semipermanent villages, clearing and planting small areas of land. Hunter-gatherers were mobile, moving with the seasons to obtain fish, migrating birds, deer, wild berries, and plants. Agriculturalists planted corn, squash, beans, and tobacco and used a wide range of other natural resources. The economy was a fairly simple one based on personal use or on barter (trading corn for fish, for example). The idea of a surplus was foreign here: People cultivated or exploited only as much land and resources as they needed to survive. Land and

Uranium Mining and the Impacts on Oceania

Oceania is a region encompassing Australia, New Zealand, and more than 20,000 other islands—including 11 independent nation islands or island clusters—located in the southern Pacific Ocean. Like oil in the Middle East, the extraction and use of uranium links Oceania to the global hunger for cheap energy and to geopolitical conflicts beyond the region (**Figure 4.C**). Uranium is a radioactive element that can be split in a process of nuclear fission to produce a chain reaction that releases large amounts of energy. Controlled reactions can be used to generate electricity in nuclear power plants, whereas uncontrolled reactions can be used in atomic bombs that release enormous amounts of thermal energy and radioactivity. Uranium first became a desirable commodity after the bombs dropped on Japan at Nagasaki and Hiroshima during World War II demonstrated the power of the atomic weapons, and its desirability increased after the potential of nuclear-powered electrical generation became apparent.

This interest in uranium had important impacts on the South Pacific. The United States, Britain, and France all joined the Cold War arms race and the effort to develop even more powerful weapons based on uranium and related elements, such as plutonium. They chose to test many of the weapons in the Pacific, with devastating implications for local residents and environments. The United States tested its bombs in the Marshall Islands after first relocating the residents of Bikini and Enewetak atolls to other islands, and exploded several types of nuclear weapons between 1946 and 1958. Although the prevailing winds were supposed to carry the radioactive fallout from bomb testing away from inhabited islands, in 1954 radioactive ash dusted the island of Rongelap and its almost 100 residents, including several relocated from Bikini. Radioactive exposure can have serious short- and long-term effects, including acute poisoning, leukemia, and birth defects, so the U.S. government

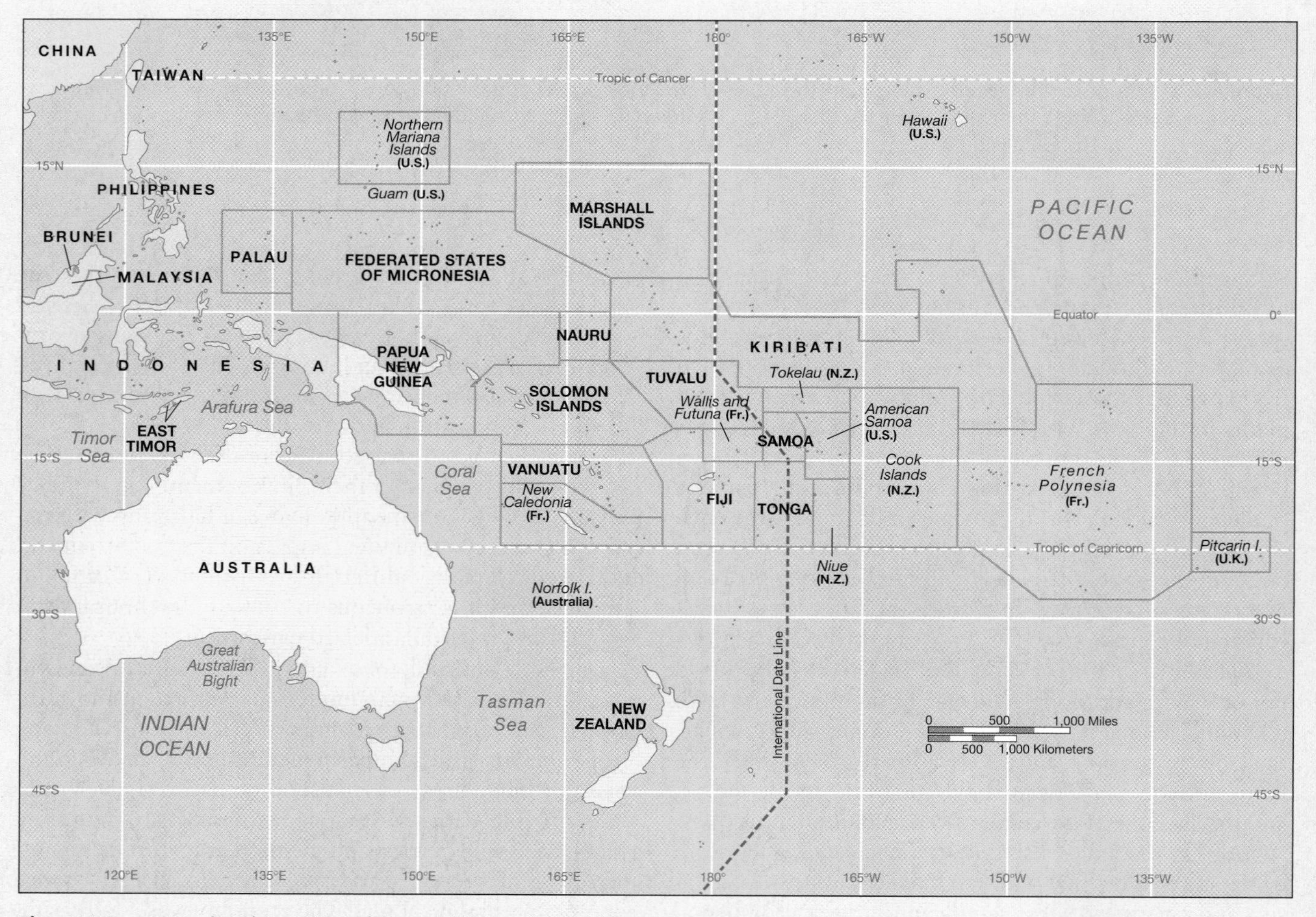

Figure 4.C Oceania This region of the world is made up of more than 20,000 islands and a continent located in the Pacific Ocean. The majority of the islands are located in the southeastern Pacific, including the large countries of Papua New Guinea and New Zealand, as well as the continent of Australia.

evacuated the residents of Rongelap on short notice with little information about the hazard they had been exposed to or warning that they would not be able to return to their homelands. In 1968 the residents of Rongelap and Bikini were told it was safe to return, but those on Bikini later had to be re-evacuated when scientists discovered that dangerous levels of radioactivity persisted in food gathered on the islands. Although the United States has monitored the health of the islanders and established a $90 million trust fund, many residents of the islands remain angry about the experiments that disrupted their lives.

France conducted more than 150 bomb tests on the tiny atolls of Moruroa and Fangataufa in French Polynesia beginning in 1966. The first bombs showered the surrounding regions with radioactivity, reaching as far as Samoa and Tonga hundreds of miles to the west. Opposition from other Pacific islands, including New Zealand and Australia, culminated in boycotts of French products, including wine and cheese, during the 1970s. France moved to underground testing and refused to release information about accidents and monitoring of radioactive pollution or health in French Polynesia. While locals have used the bomb tests as a reason to seek independence from France, international activists have tried to stop the tests. In 1985 the environmental group Greenpeace planned to protest tests by sailing its ship *Rainbow Warrior* to Moruroa, but French intelligence agents scuttled the ship while it was moored in the harbor of Auckland, New Zealand.

The resulting international scandal prompted New Zealand to take a strong stand against nuclear proliferation, banning all nuclear-powered and nuclear-armed vessels from its harbors, breaking off diplomatic relations with France, and taking a leadership role in the antinuclear movement in the Pacific. This created a long-term strain on relations between New Zealand and the United States because U.S. military vessels, which will not admit or deny nuclear capability, were therefore banned from New Zealand. However, New Zealand's actions contributed to the announcement by France in 1996 that it would end nuclear testing after riots in Tahiti and declines in tourism.

The British tested their bombs on Christmas Atoll, now within the nation of Kiribati, and also at several locations in Australia, including the Monte Bello islands off the coast of Western Australia and Maralinga in Southern Australia. Critics now claim that neither the Australian government nor its people were made fully aware of the risks of these tests and that the local Aborigines were heavily exposed to radiation and continue to wander into the contaminated test sites.

The consumption of uranium has also threatened Australian Aborigines through the mining of uranium on or near their lands in northern Australia. The Ranger mine commenced operations in the Northern Territory in 1980 within the boundary of Kakadu National Park, a region of great natural beauty listed as a World Heritage site for both natural and cultural values (**Figure 4.D**). The mine has produced more than 16 million tons (35 billion pounds) of radioactive mine waste and has created serious water-pollution problems in the area. Australia produces 27 percent of the world's uranium, exported to fuel nuclear power stations in the United States, Japan, Europe, Canada, and South Korea, even though Australia itself does not produce electricity from uranium. Great controversy has arisen over proposals to open another mine at Jabiluka on land belonging to the Mirrar Aboriginal Group. Activists have blockaded the mine road, and protests have occurred around Australia.

Thus, uranium links the countries of the Oceania to the global geography of energy consumption and to the desire for geopolitical supremacy in a multitude of ways. Though nuclear testing has been halted in the Pacific, radioactivity persists for thousands of years and will continue to pose risks to people and ecosystems. However, since uranium prices are currently low because few new nuclear power stations are under construction in the aftermath of the Chernobyl accident in Russia in 1986, many countries are seeking to purchase uranium as other supplies become scarce or create different sorts of environmental problems.

Figure 4.D The Ranger uranium mine This mine is located in Kakadu National Park in Australia's Northern Territory. The area is sacred to the Aborigine population and has striking landscapes and ecosystems. Moves to expand the mine have resulted in protests by Aboriginal groups and environmental activists.

resources were shared, without concepts such as private property or land ownership. Fire was used to clear land for planting as well as for hunting. Although vegetation change did occur, it was minimal and not irreversible.

The relationship between some of the the Indians of South and Central America altered their environment as well, though in more dramatic ways. The Aztecs of Mexico and the Incas of Peru had developed complex urban civilizations dependent on dense populations employing intensive agricultural techniques (**Figure 4.18**). These groups were responsible for dramatic environmental modifications through cultivation techniques that included the irrigation of dry regions and the terracing of steep slopes. As we have seen, irrigation over several centuries results in the salinization of soils. In the lowland tropics, intensive agricultural practices resulted in widespread deforestation as people cut and set fire to patches of forest, planted crops, and then moved on when soil fertility declined. A surplus was key to the operations of both societies, as tribute by ordinary people to the political and religious elite was required in the form of food, animals, labor, or precious metals. The construction of the sizable Inca and Aztec empires required the production of large amounts of building materials in the form of wood and mortar. Concentrated populations and the demands of urbanization meant that widespread environmental degradation existed prior to European contact.

Figure 4.18 Machu Picchu Pictured is one of the most important sites of ancient Incan civilization, located in highland Peru on the so-called *ceja de selva,* or "eyebrow of the jungle." Machu Picchu was probably an important ceremonial center but was also a large-scale residential center, as evidenced by the extensive agricultural terraces. Archaeologists, geographers, and other scholars believe Machu Picchu was probably one of the last sites of fierce Incan resistance to the Spanish conquistadors. As the photograph suggests, natural terrain helped in the fortification of the city. Machu Picchu is one of the best-preserved ruins of the Incan empire.

HUMAN ACTION AND RECENT ENVIRONMENTAL CHANGE

No other transition in human history has had the impact on the natural world that industrialization has. When we couple industrialization with its frequent companion, urbanization, we have the two processes that, more than any others, have revolutionized human life and affected far-reaching ecological changes. The changes wrought by industrialization and urbanization have moved beyond a local or a regional scale to affect the entire globe. In this section we explore some of the dramatic contemporary environmental impacts that industrial technology and urbanization have produced. In doing so, we highlight the two issues most central to environmental geography today: energy-use and land-use change.

The Impact of Energy Needs on the Environment

Certainly the most central and significant technological breakthrough of the Industrial Revolution was the discovery and utilization of fossil fuels: coal, oil, and natural gas. Although the very first factories in Europe and the United States relied on waterpower to drive the machinery, hydrocarbon fuels provided a more constant, dependable, and effective source of power. A steady increase in power production and demand since the beginning of the Industrial Revolution has been paralleled, not surprisingly, by an increase in resource extraction and conversion.

At present, the world's population relies most heavily for its energy needs on nonrenewable energy resources that include fossil fuels and nuclear ones, as well as renewable resources such as solar, hydroelectric, wind, and geothermal power. Fossil fuels are derived from organic materials and are burned directly to produce heat. Nuclear energy originates with isotopes, which emit radiation. Most commercial nuclear energy is produced in reactors fueled by uranium (see Box 4.2: "Uranium Mining and the Impacts on Oceania"). Renewable sources of energy, such as the sun, wind, water, and steam, are captured in various ways and used to drive pumps, machines, and electricity generators.

The largest proportion of the world's current consumption of energy resources, 35 percent, is from oil; 24 percent is from coal; 18 percent from gas; 6 percent from hydropower (largely from dams); 5 percent from nuclear power; and 12 percent from biomass (which includes wood, charcoal, crop waste, and dung). The production and consumption of these available resources, however, are geographically uneven, as **Figure 4.19** shows. Fifty percent of the world's oil supplies are from the Middle East, and most of the coal is from the Northern Hemisphere, mainly from the United States, China, and Russia. Nuclear reactors are a phenomenon of the core regions of the world. For example, France generates 90 percent of its electricity from nuclear sources.

The consumption side of energy also varies geographically. It has been estimated that current annual world energy consumption is equal to what it took about one million years to produce naturally. In one year, global energy consumption is equal to about 1.3 billion tons of coal. What is most remarkable is that this is four times what the global population consumed in 1950 and 20 times what it consumed in 1850. And as the $I = PAT$ formula suggests, the affluent core regions of the world far outstrip the peripheral regions in energy consumption. With nearly four times the population of the core regions, the peripheral regions account for less than one-third of global energy expenditures. Yet consumption of energy in the peripheral regions is rising quite rapidly as globalization spreads industries, energy-intensive consumer products such as automobiles, and energy-intensive agricultural practices into regions of the world where they were previously unknown. It is projected that within the next decade or so, the peripheral regions will become the dominant consumers of energy (**Figure 4.20**).

Most important for our discussion, however, is that every stage of the energy conversion process—from discovery to extraction, processing, and utilization—has an impact on the physical landscape. In the coal fields of the world, from the U.S. Appalachian Mountains to western Siberia, mining results in a loss of vegetation and topsoil, in erosion and water pollution, and in acid and toxic drainage. It also contributes to cancer and lung disease in coal miners. The burning of coal is associated with relatively high emissions of environmentally harmful gases, such as carbon dioxide and sulfur dioxide (**Figure 4.21**).

The burning of home heating oil, along with the use of petroleum products for fuel in internal combustion engines, launches harmful chemicals into Earth's atmosphere—causing air pollution and related health problems. The production and transport of oil have resulted in oil spills and substantial pollution to water and ecosystems. Media images of damage to seabirds and mammals after tankers have run aground and spilled oil have shown how immediate the environmental damage can be. Indeed, the oceans are acutely affected by the widespread use of oil for energy purposes. Thousands of tons of oil are spilled into the world's oceans each year from leaking ships, oil drilling, transporting oil, and from natural seeps. Oil drilling can also have other profound environmental consequences in the form of well explosions and fires (**Figure 4.22**).

Natural gas is one of the least noxious of the hydrocarbon-based energy resources because it is converted relatively cleanly. Now supplying nearly one-quarter of global commercial energy, natural gas is predicted to be the fastest-growing energy source in the new century. Reserves are still being discovered, with Russia holding the largest amount—about one-third of the world's total (**Figure 4.23**). While regarded as a preferred alternative to oil and coal, natural gas is not produced or consumed without environmental impacts. The risk of explosions at natural-gas conversion facilities is significant; leakages and losses of gas from distribution systems contribute to the deterioration of Earth's atmosphere.

At the midpoint of the twentieth century, nuclear energy for civilian use was widely promoted as a clearly preferable alternative to fossil fuels. It was seen by many as the answer to the expanding energy needs of core countries, especially as the supply of uranium worldwide was thought to be more than adequate for centuries of use. Nuclear energy was also regarded as cleaner and more efficient than fossil fuels. Although nuclear war was a pervasive threat, and there were certainly critics of nuclear energy even in the early years of its development, the civilian "atomic age" was widely seen as a triumphant technological solution to the energy needs of an expanding global economic system. It was not until serious accidents at nuclear power plants began to occur—such as at Windscale in Britain and Three Mile Island in the United States—that the voices of concerned scientists and citizens began to be heard. These voices described, with incontestable evidence, the problems associated with nuclear energy production, such as ensuring nuclear reactor safety and safely disposing of nuclear waste (which remains radioactive for tens of thousands of years). Since these accidents and the meltdown of the Chernobyl nuclear power plant in Russia in 1986, many core countries have drastically reduced or eliminated their reliance on nuclear energy. Sweden, for example, has committed to eliminate entirely its reliance on nuclear power—from which a large portion of its current electricity use is derived—by the year 2010.

Interestingly, while the majority of core countries have begun to move away from nuclear energy because of the possibility of environmental disaster in the absence of fail-safe nuclear reactors, a few semiperipheral—and especially populous—countries are moving in the opposite direction (**Figure 4.24**). India, South Korea, and China have fledgling nuclear energy programs. So far, no accidents have been associated with nuclear energy production in the periphery. And, because of the rising price of oil, many core countries, such as the United States, who had once abandoned nuclear energy as an acceptable alternative, are now reconsidering it.

While nuclear power problems are largely confined to the core, the periphery is not without its energy-related environmental problems. Because a large proportion of populations in the periphery rely on wood for their

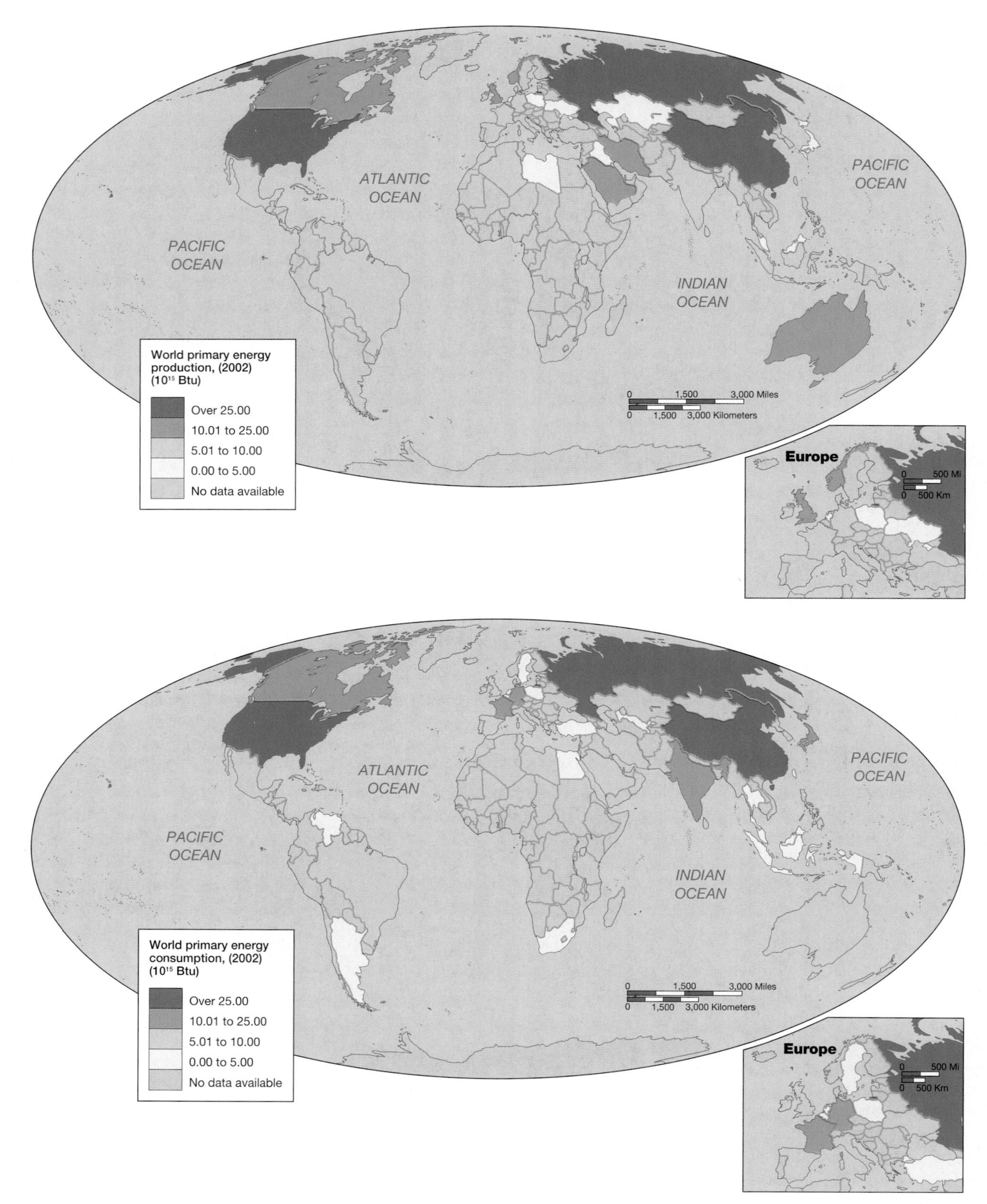

Figure 4.19 World production and consumption of energy, 1993–2002 These paired maps provide a picture of the uneven distribution of the production and consumption of energy resources around the world. The United States is the largest producer and consumer of a range of energy resources. Notice that although the Middle East and North African countries as well as Nigeria are important producers of energy resources, their consumption (as well as that of the rest of the African continent, excluding South Africa) is very low. Japan produces a negligible amount of the total of world energy resources but consumes a relatively high share. (Top map after *International Energy Annual,* 1999, Web site: http://www.eia.doe.gov/iea. Data tables "World Primary Energy Production (Btu)" and "World Primary Energy Consumption (Btu)." Bottom map reprinted with permission of Prentice Hall, from E. F. Bergman, *Human Geography: Cultures, Connections, and Landscapes* © 1995, p. 395. Data from the World Resources Institute, World Resources 1994–95. New York: Oxford University Press, 1994, pp. 334–335.)

Figure 4.20 Pottery furnace, China Black smoke billows from the chimneys of an unlicensed pottery furnace on the outskirts of Wuhan in China's central province of Hubei. The workers use material like old tires and asphalt as fuel, which produces excessive pollutants. Chinese authorities are fighting an uphill battle to eliminate unlicensed factories throughout China which continue to contaminate the environment.

Figure 4.21 Coal mining This coal-mining operation in western Germany is typical of the surface-mining technologies used to exploit shallow deposits in most parts of the world. After deposits are located, the vegetation and overburden (rocks and dirt overlaying the coal seam) are removed by bulldozers and discarded as spoil (or waste material). With the coal seam exposed, heavy equipment (such as the type pictured) is used to mine the deposit. Some countries, such as the United States, require restoration of newly mined landscapes. Sites exploited before these laws were introduced remain unrestored, however. Successful restoration makes it difficult to tell that the area was once a mining site. Unfortunately, many such sites are in arid or semiarid areas where soil and climate prevent full restoration. In addition to substantial land disturbance, the mining and processing of coal resources often cause soil erosion as well as water and air pollution.

Figure 4.22 A tanker oil spill Although the international press is quick to report the disastrous oil spills into the world's oceans, most of the petroleum released into the environment every year comes from nonaccidental sources. The largest contributor to oil pollution of the oceans comes from the extraction process. A very small amount of oil released in oceans is due to major tanker accidents. Tanker spills, like the one shown here, are dramatic because huge amounts of oil end up affecting fairly small areas.

energy needs, as the populations have grown, so has the demand for fuelwood. One of the most immediate environmental impacts of wood burning is air pollution, but the most alarming environmental problem is the rapid depletion of forest resources. With the other conventional sources of energy (coal, oil, and gas) being too costly or unavailable to most peripheral households, wood or other forms of biomass—any form of material that can be used as fuel such as animal wastes, livestock operation residues, and aquatic plants—is the only alternative. The demand for fuelwood has been so great in many peripheral regions that forest reserves are being rapidly used up (**Figure 4.25**).

Figure 4.23 Natural gas processing in Russia Pictured here is a natural gas processing facility in Siberia, where the largest natural gas deposits in the world are located.

Fuelwood depletion is extreme in the highland areas of Nepal, as well as in Andean Bolivia and Peru. The clearing of forests for fuelwood in these regions has led to serious steep-slope soil erosion. In sub-Saharan Africa, where 90 percent of the region's energy needs are met with energy supplied by wood, overcutting of the forests has resulted in denuded areas, especially around rapidly growing cities. And although wood gathering is usually associated with rural life, it is not uncommon for city dwellers to use wood to satisfy their household energy needs as well. For example, in Niamey, the capital of Niger, the zone of overcutting is gradually expanding as the city itself expands. It is now estimated that city dwellers in Niamey must travel from 50 to 100 kilometers (31 to 62 miles) to gather wood. The same goes for inhabitants of Ouagadougou in Burkina Faso, where the average haul for wood is also over 50 kilometers.

Hydroelectric power was also once seen as a preferred alternative to the more obviously environmentally polluting fossil-fuel sources. The wave of dam building that occurred throughout the world over the course of the twentieth century improved the overall availability, quality, cost, and dependability of energy (**Figure 4.26**). Unfortunately, however, dams built to provide hydroelectric power (as well as water for irrigation, navigation, and drinking) for the burgeoning cities of the core and to encourage economic development in the periphery and semiperiphery have also had profound negative environmental impacts. Among the most significant of these impacts are changes in downstream flow, evaporation, sediment

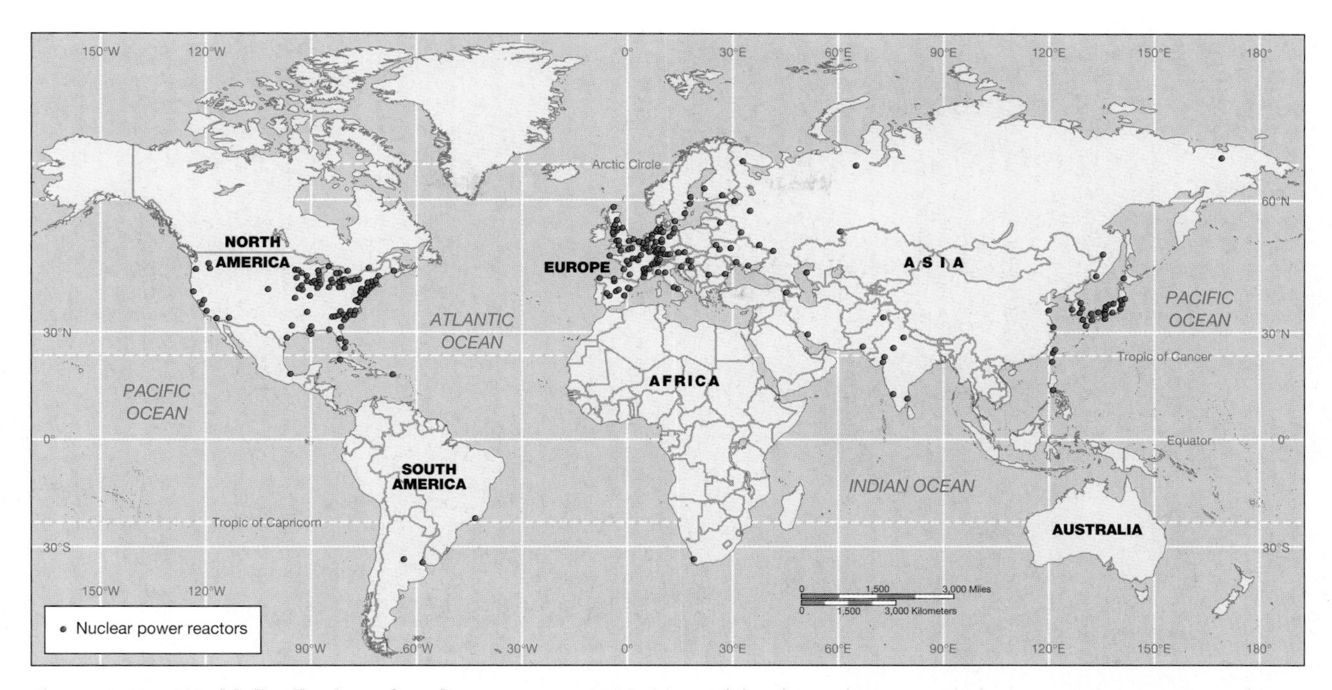

Figure 4.24 World distribution of nuclear reactors, 2000 Most of the dependence on nuclear power is concentrated in core countries. South America and Africa together contain only four nuclear reactors. Whereas some peripheral countries, such as India, are enthusiastic about increasing their nuclear energy production, core countries such as Sweden are phasing out dependence on nuclear power. Australia, where there is a very strong antinuclear movement, is one of the few core countries to have rejected nuclear power altogether. (After International Nuclear Safety Center Web site, http://www.insc.anl.gov/pwrmaps/map/world_map.html, "Maps of Nuclear Power Reactors: World Map" 2002, p.1, retrieved June 15, 2005).

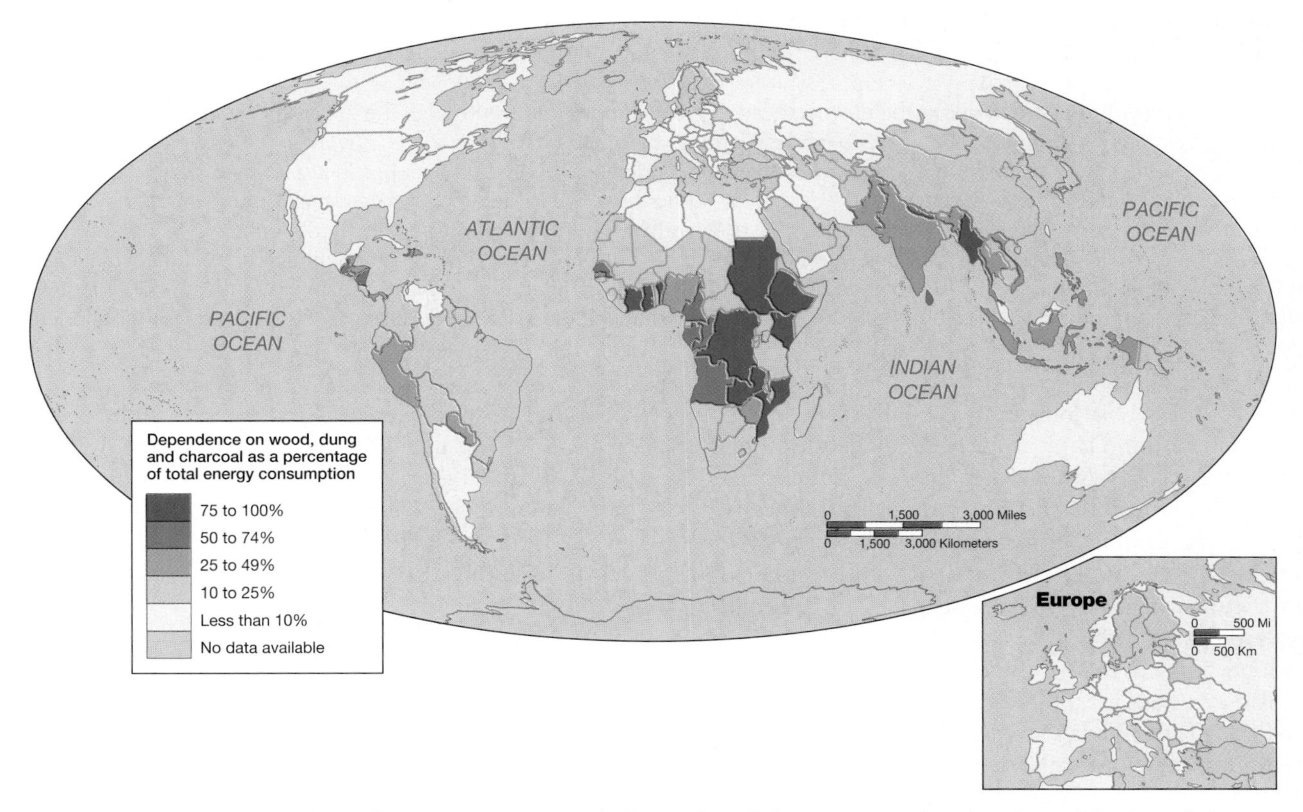

Figure 4.25 Global use of woodfuels, 2001 Firewood, charcoal, and dung are considered traditional fuels, and although their availability is decreasing, dependence upon them is increasing. Dependence on traditional sources of fuel is especially high in the periphery where, in Africa, for example, they are the most important energy source for cooking and heating. Wood and charcoal, although renewable sources, are replenished very slowly. Acute scarcity will be a certainty for most African households in the twenty-first century. (After United Nations Development Programme, 2001. *World Resources 2000–2001, People and Ecosystems: The Fraying Web of Life,* Washington, DC: World Resources Institute, p. 98.)

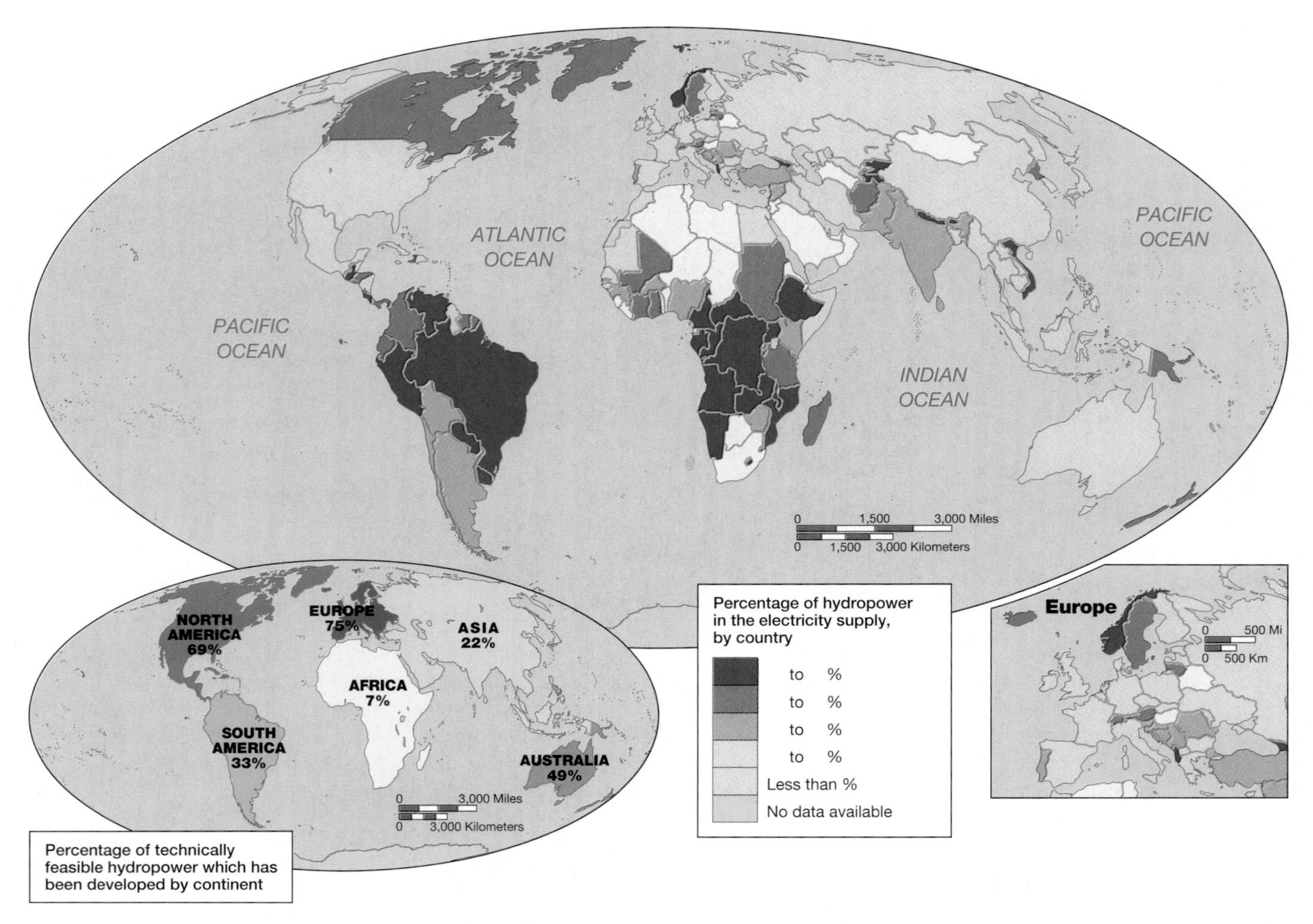

Figure 4.26 Percent of hydropower in the electricity supply by country, 2002 Although the great dam-building era for core countries is now largely completed, many peripheral countries, in a bid to participate more actively in the world economy, are building dams. Only a few countries are almost exclusively dependent on the hydropower produced from dams. These include Norway, Nepal, Zambia, Ghana, Paraguay, and Costa Rica. While the power produced by dams is environmentally benign, the construction of large dams can be extremely destructive of the environment and can dislocate large numbers of people. Still, given the increasing need for electricity by rapidly eveloping peripheral countries, hydropower is becoming a more attractive energy option for many of these countries because of the uncertain supply of oil in the future. The larger map shows the amount of hydropower that is currently available; the smaller one shows the potential for hydropower development, especially for peripheral regions.

transport and deposition, mineral quality and soil moisture, channeling and bank scouring, and aquatic biota and flora, as well as conditions threatening to human health. Furthermore, the construction of dams dramatically alters the surrounding terrain, often with serious consequences. For example, clearance of the forest for dam construction often leads to large-scale flooding. The felled trees are usually left to decay in the impounded waters, which become increasingly acidic. The impounded waters can also incubate mosquitoes, which carry diseases such as malaria. The remedies for such problems are difficult to determine, and many argue that new dam projects should not be undertaken without a clear sense of the complex of indirect social and environmental costs (**Figure 4.27**).

One reason hydroelectric power continues to be appealing, however, is that it produces few atmospheric pollutants as compared to fossil fuels. Indeed, coal and gas power stations as well as factories, automobiles, and other forms of transportation are largely responsible for the increasingly acidic quality of Earth's atmosphere. While people as well as other organisms naturally produce many gases, including oxygen and carbon dioxide, increasing levels of industrialization and motor vehicle use have destabilized the natural balance of such gases, leading to serious atmospheric pollution (see Box 4.3: "Global Climate Change and the Kyoto Protocol"). Increasing the level of acids in the atmosphere are sulfur dioxide, nitrogen oxides, and hydrocarbons, among other gases, which are released into the atmosphere from motor vehicle exhaust, industrial processes, and power generation (based on fossil fuels). If these gases reach sufficient concentrations and are not effectively dispersed in the atmosphere, acid rain can result.

Acid rain is the wet deposition of acids upon Earth through the natural cleansing properties of the atmosphere. Acid rain occurs as the water droplets in clouds absorb cer-

Figure 4.27 Aswan High Dam, Egypt Completed in 1970 at a cost of $1 million, the Aswan High Dam was a significant engineering feat, as well as an important symbol of Egypt's bid for economic independence. The dam is of rock-filled construction and is 111 meters (364 feet) high. The impoundment of water caused by the dam flooded out numerous settlements along the Nile River, requiring the resettlement of tens of thousands of people in both Egypt and Sudan. In addition to its human impacts, the dam affected the natural fertilizing processes of the Nile and flooded out the site of one important ancient temple while restoring another to the open air.

tain gases that later fall back to Earth as acid precipitation. Also included under the term *acid rain* are acid mists, acid fogs, and smog. The effects of acid rain are widespread. Throughout much of the Northern Hemisphere, for example, forests are being poisoned and killed, and soils are becoming too acidic to support plant life. Lakes are becoming acidic in North America and Scandinavia. In urban areas, acid rain is corroding marble and limestone buildings, such as the Parthenon in Athens and St. Paul's Cathedral in London, as well as other historic structures in Europe. **Figure 4.28** illustrates the global problem of acid emissions that come back to Earth as acid rain.

Before giving up all hope that the use of energy can ever be anything but detrimental to the environment, it is important to realize that alternatives exist to fossil fuels, hydroelectric power, and nuclear energy. Energy derived from the sun, the wind, Earth's interior (geothermal sources), and the tides has been found to be clean, profitable, and dependable. Japan, the United States, and Germany all have solar energy production facilities that have proved to be cheap and nonpolluting. Although contributing only small amounts to the overall energy supply, the production of energy from geothermal and wind sources has also been successful in a few locations around the globe. Italy, Germany, the United States, Mexico, and the Philippines all derive some of their energy production from geothermal or wind sources.

Nonrenewable alternative sources of energy, such as fuel cells and cogeneration, are also beginning to appear and become more widely adopted. A fuel cell converts chemical energy directly into electricity by combining hydrogen and oxygen in a controlled reaction. Fuel cells emit no pollution, as the waste exhaust is water vapor and heat. One of the most promising uses of fuel cells is for road transportation. Cogeneration, also known as Combined Heat and Power or CHP, is the production of energy and heat in one single process for dual output streams. In conventional electricity generation, only about 35 percent of the fuel is actually converted to electricity, while the rest is lost as waste heat. Since cogeneration produces both heat and electricity, it is able to achieve efficiency of upwards of 90 percent. It is the most efficient way to use fuel.

Monies to support the development of geothermal, wind, and tidal energy, as well as fuel cells and cogeneration, have been scarce, however, due to the opposition of oil and gas companies, as well as other political factors, such as powerful oil and gas lobbies. While viable alternatives exist to traditional energy sources, the further development of these alternatives is likely to hinge on future political and economic factors.

Impacts of Land-Use Change on the Environment

In addition to industrial pollution and steadily increasing demands for energy, the environment is also being dramatically affected by pressures on the land. The clearing of land for fuel, farming, grazing, resource extraction, highway building, energy generation, and war all have significant impacts. Land may be classified into five categories: forest, cultivated land, grassland, wetland, and areas of settlement. Geographers understand land-use change as occurring in either of two ways: conversion or modification. *Conversion* is the wholesale transformation of land from one use to another (for example, the conversion of forest to settlement). *Modification* is an alteration of existing cover (for example, when a grassland is overlaid with railroad line or when a forest is thinned and not clear-cut). As human populations have increased and the need for land for settlement and cultivation has also increased, changes to the land have followed.

One of the most dramatic impacts of humans upon the environment is loss or alteration of forest cover as it has been cleared for millennia to make way for cultivation and settlement. Forests are cleared not only to obtain land to accommodate increases in human numbers but also to extract the vast timber resources they contain. The approximate chronology and estimated extent of the clearing of the world's forests since preagricultural times are shown in **Table 4.2**. The table shows that the forested area of the world has been reduced by about 8 million square kilometers (about 3 million square miles) since preagricultural times. Rapid clearance of the world's forests has occurred either through logging, settlement, and agricultural clearing or through fuelwood cutting around urban areas. **Figure 4.29** shows the global extent of deforestation in recent years.

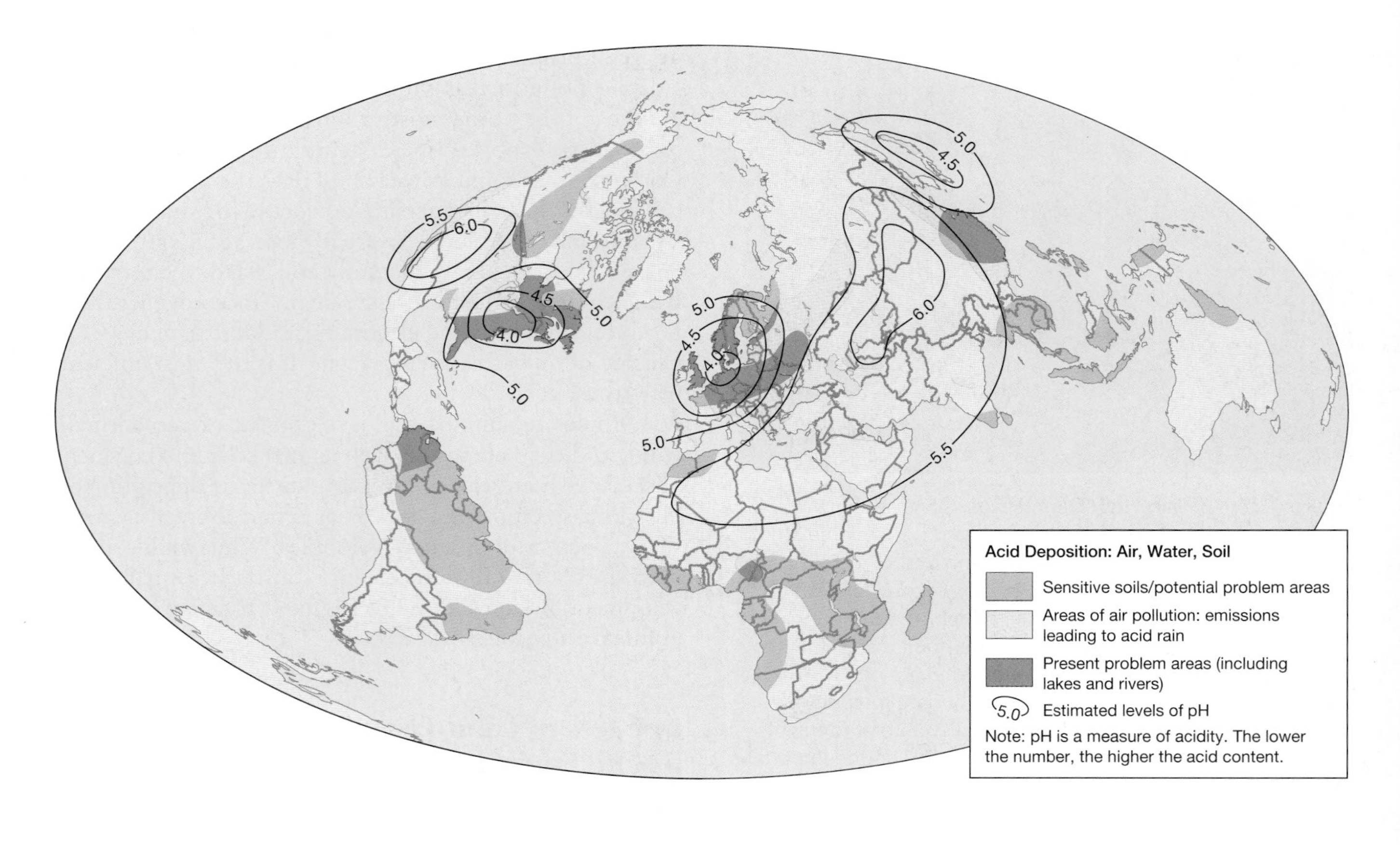

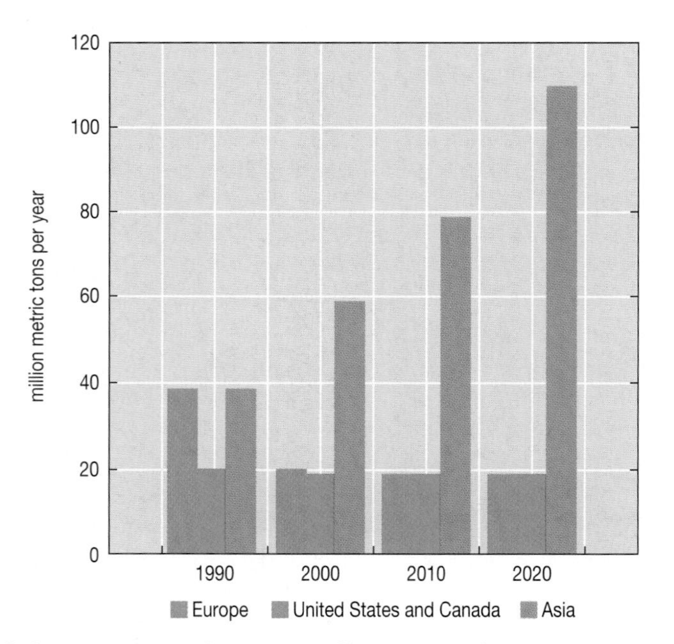

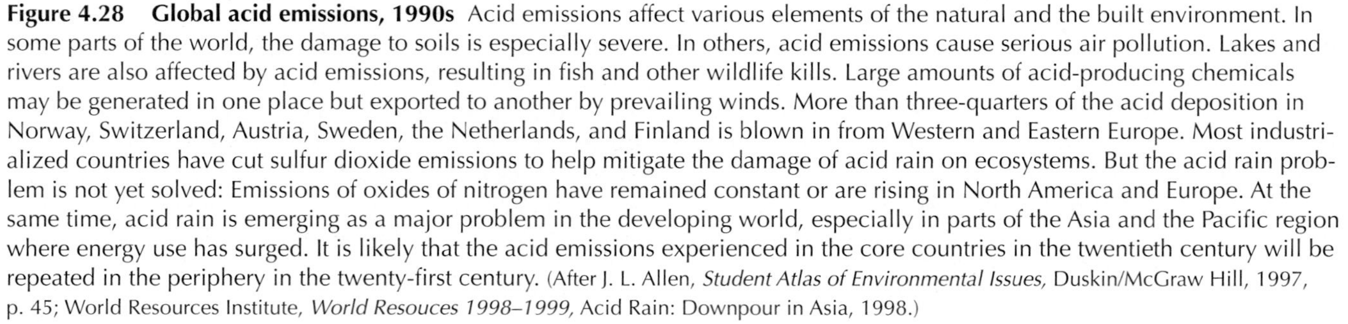
Figure 4.28 Global acid emissions, 1990s Acid emissions affect various elements of the natural and the built environment. In some parts of the world, the damage to soils is especially severe. In others, acid emissions cause serious air pollution. Lakes and rivers are also affected by acid emissions, resulting in fish and other wildlife kills. Large amounts of acid-producing chemicals may be generated in one place but exported to another by prevailing winds. More than three-quarters of the acid deposition in Norway, Switzerland, Austria, Sweden, the Netherlands, and Finland is blown in from Western and Eastern Europe. Most industrialized countries have cut sulfur dioxide emissions to help mitigate the damage of acid rain on ecosystems. But the acid rain problem is not yet solved: Emissions of oxides of nitrogen have remained constant or are rising in North America and Europe. At the same time, acid rain is emerging as a major problem in the developing world, especially in parts of the Asia and the Pacific region where energy use has surged. It is likely that the acid emissions experienced in the core countries in the twentieth century will be repeated in the periphery in the twenty-first century. (After J. L. Allen, *Student Atlas of Environmental Issues,* Duskin/McGraw Hill, 1997, p. 45; World Resources Institute, *World Resouces 1998–1999,* Acid Rain: Downpour in Asia, 1998.)

TABLE 4.2 Estimated Area Cleared (× 1000 km^2)

Region or Country		Pre-1650	1650–1749	1750–1849	1850–1978	Total High Estimate	Total Low Estimate
North America		6	80	380	641	1,107	1,107
	H	18				288	
Central America	L	12	30	40	200	—	282
	H	18					
Latin America	L	12	100	170	637	925	919
Australia, New Zealand, and the South Pacific	H	6	6	6	362	380	
	L	2	4	6	362	—	374
	H	70	180	270	575	1,095	
Former USSR	L	42	130	250	575	—	997
	H	204	66	146	81	497	
Europe	L	176	54	186	81	—	497
	H	974	216	596	1,220	3,006	
Asia	L	640	176	606	1,220	—	2,642
	H	226	80	216	469	759	
Africa	L	96	24	42	469	—	631
Total highest		**1,5**	**758**	**1,592**	**4,185**	**8,057**	
		22					
Total lowest		**986**	**598**	**1,680**	**4,185**		**7,449**

Source: B. L. Turner II, W. C. Clark, R. W. Kates, J. F. Richards, J. T. Mathews, and William B. Meyer, *The Earth as Transformed by Human Action: Global and Regional Changes in the Biosphere over the Past 300 Years.* Cambridge: Cambridge University Press, 1990, p. 180.

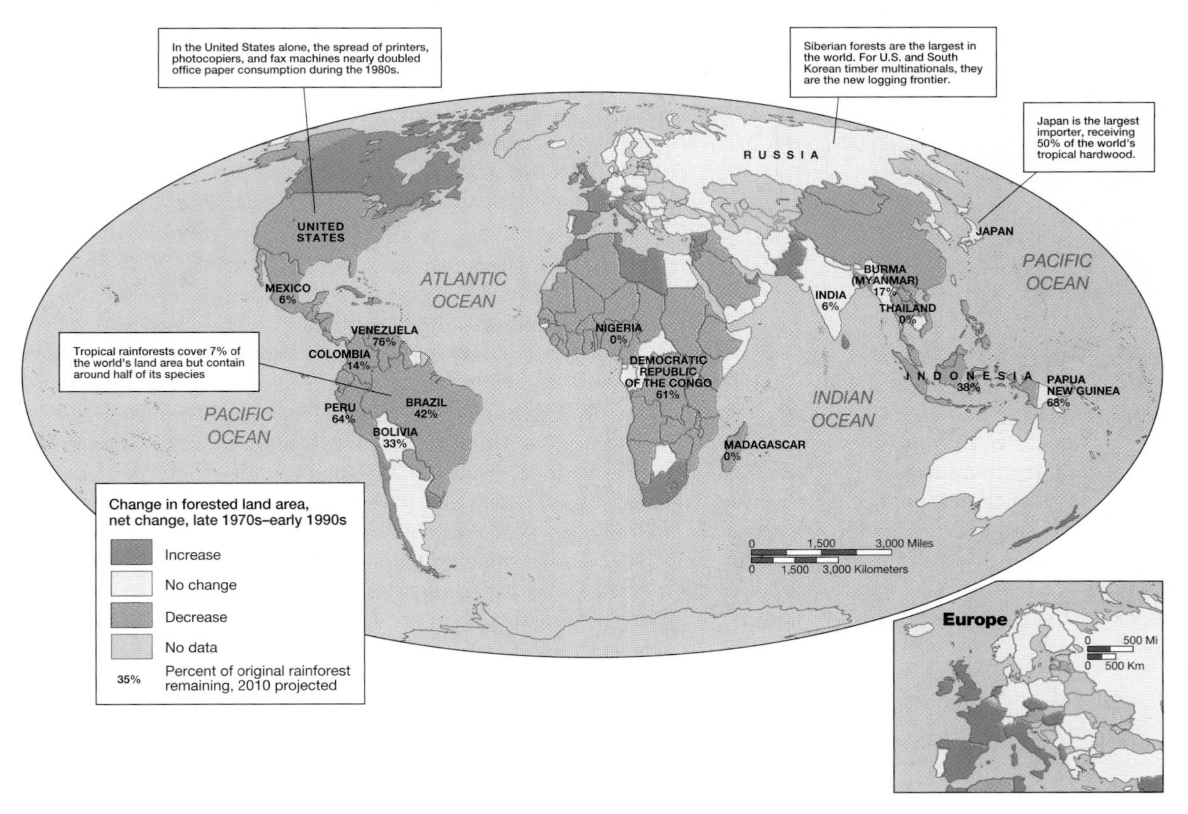

Figure 4.29 Global deforestation The world's forests are disappearing or being reduced or degraded everywhere, but especially in tropical countries. Since agriculture emerged about 10,000 years ago, human activities have diminished the world's forest resources by about twenty-five percent. Whereas forests once occupied about one-third of Earth's surface, they now take up about one-quarter. Playing an important role in the global ecosystem, they filter air and noise pollution, provide a habitat for wildlife, and slow down water runoff, helping to recharge streams and groundwater. They also influence climate at local, regional, and global levels. (After J. Seager, *The New State of the Earth Atlas,* 2nd ed. New York: Simon & Schuster, 1995, pp. 72–73.)

4.3

GEOGRAPHY MATTERS

Global Climate Change and the Kyoto Protocol

Since the 1992 Earth Summit in Rio de Janiero, the international community has been seeking a way to strike a balance between increasing the pace of economic development without further threatening the global environment. The biggest potential threat to the global environment is the impact that increased energy use will have on global climate. At the Rio Earth Summit, 167 nations ratified the Framework Convention on Climate Change with the aim of solving the problem of how to reduce the amount of greenhouse gases—gases that are leading to the warming of the Earth's atmosphere—that are generated by energy use (**Figure 4.E**). An equally critical aim is to ensure that the burden of protecting the environment is shared equitably across all nations.

In December 1997 these nations began to address the problem of balancing global economic development and environmental protection more substantively by forging the Kyoto Protocol. The protocol marks the first time that an attempt was made to limit the amount of greenhouse gas emissions generated by core countries. The aim of the protocol is to cut the combined emissions of greenhouse gases from core countries by roughly 5 percent from their 1990 levels by 2012. (Core countries account for a disproportionate amount of CO_2 emissions, as **Figure 4.F** makes clear.) It also specifies the amount each core nation must contribute toward meeting that reduction goal. Nations with the highest CO_2 emissions—the United States, Japan, and most European nations—are expected to reduce emissions by a range of 6 to 8 percent.

Although the Kyoto Protocol represents a real advance on the 1992 agreement reached in Rio, there are still important issues that have yet to be completely worked out among the 167 nations involved in the protocol. One of the most controversial is whether core countries will be allowed to participate in "emis-

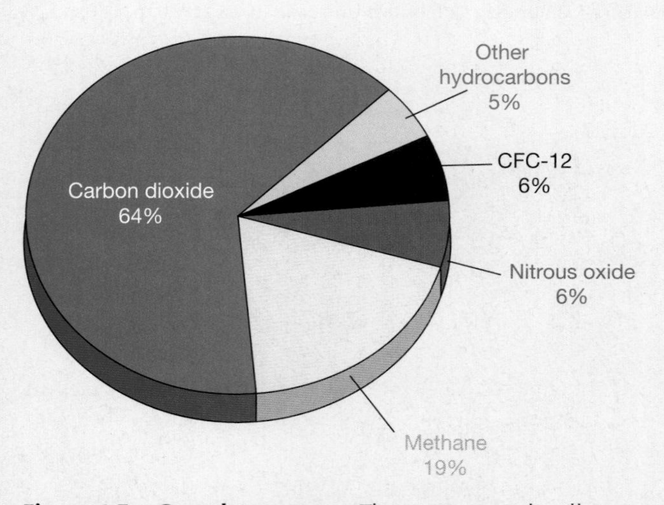

Figure 4.E Greenhouse gases The most central pollutant involved in global climate change is CO_2, carbon dioxide. In addition to CO_2, the Kyoto Protocol focuses on five other greenhouse gases: methane (CH_4), nitrous oxide (N_2O), chlorofluorocarbons (CFCs), and a number of hydrofluorocarbons (HFCs).

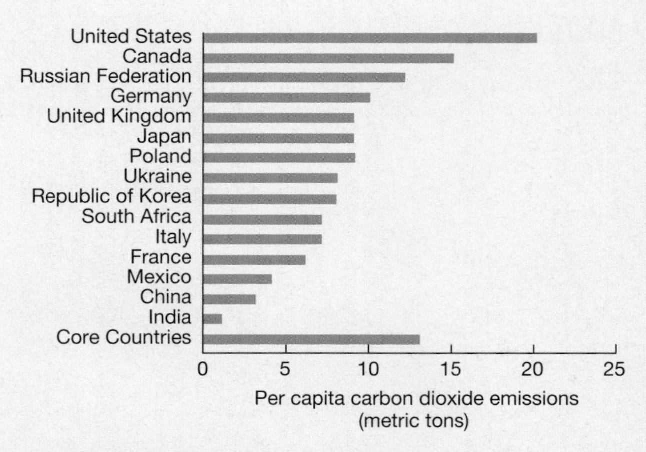

Figure 4.F Per capita CO_2 emissions CO_2 emissions are a good proxy for all greenhouse gases. The graph shows core countries as a whole, plus several core countries with especially high per capita levels of CO_2 emissions as well as a few rapidly developing semiperipheral countries. India and China, with very high populations, have relatively low per capita CO_2 emissions, while the United States and Canada, with populations far lower than India or China, have a massive contribution to CO_2 emissions. This difference is not surprising, given that energy use is highly correlated with level of wealth.

sions trading." In this scenario, a nation whose emissions fall below its treaty limit will be allowed to sell credit for its remaining emissions allotment to another nation, which in turn can use the credit to meet its own treaty obligations. Those who advocate the emissions trading approach to pollution control believe such a program will help curb the cost of controlling greenhouse gases by allowing emissions cuts to occur where they are least expensive.

A second important, and as yet unresolved, issue is the extent to which peripheral nations will be involved in limiting global emissions. While the original 1992 climate treaty placed the burden of reducing global climate change on core countries, which are unquestionably most responsible for the current buildup of greenhouse gases, peripheral countries are also expected to play a role. The Kyoto Protocol, however, does not set any binding limits on peripheral country emissions, nor does it establish a mechanism or timetable for these countries to take on such limits voluntarily. One interesting way of encouraging environmentally sensitive development in peripheral countries is the so-called Clean Development Mechanism. This would allow core countries to invest in projects in peripheral countries that reduce greenhouse gas emissions and in return receive credit for the reductions. The aim is to help peripheral countries develop their economies without increasing the overall contribution to greenhouse gas emissions.

Although it is unlikely that the Kyoto accord will bring about deep emissions cuts, climate negotiators would like to see a new treaty developed that will enable progress to continue well into the twenty-first century. The hope is to stimulate energy policy reform at the same time as new research and development investments bring low-emission technologies to market. It is also possible that the Kyoto Protocol itself could be expanded to include more comprehensive emission cuts designed eventually to stabilize greenhouse gas concentrations at a safe level. **Figure 4.G** shows the projected levels of CO_2 emissions, which are considered a reliable proxy for all greenhouse gases, under various scenarios. Unfortunately, after nine years of international negotiations, in mid-2001 President George Bush announced that the United States would no longer honor its commitment

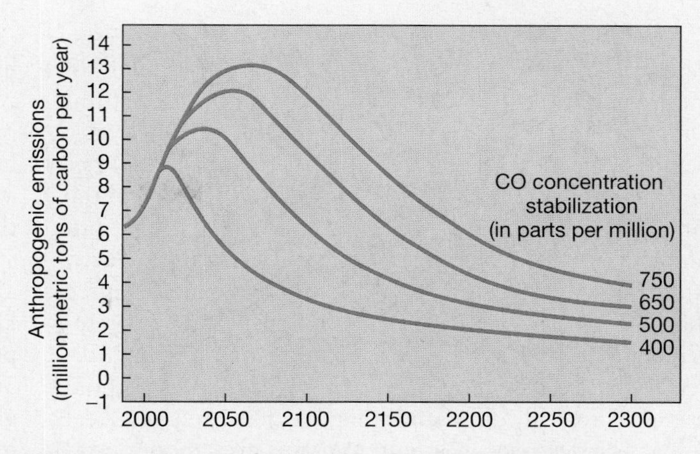

Figure 4.G Impact of proposed reductions in future emissions To stabilize CO_2 emissions, very deep cuts will be necessary for core countries at the same time that peripheral countries must be allowed to pursue economic development. One important way to make this possible is through the development of low-emission technologies.

to the 1997 Kyoto agreement because he feared the ramifications of the Protocol would negatively affect U.S. energy companies and diminish economic growth in the United States and the rest of the globalizing world. The implications of this decision are that the United States will continue to be the world's largest single generator (over 25 percent) of greenhouse gases—emissions that are leading to worldwide rises in temperature. Such temperature increases—known as global warming—have more potential to damage Earth's web of life than any other factor outside of nuclear war or a collision with an asteroid. In addition to causing rising sea levels throughout the world (which could result in widespread loss of property and livelihoods), global warming is also likely to contribute to increases in heat-related deaths (especially respiratory illnesses) and a widening of the range of disease-carrying rodents and bugs (which would cause increases in malaria, dengue fever, and Lyme disease, among other afflictions).

Adapted from World Resources Institute, "Negotiating Climate: Kyoto Protocol Marks a Step Forward," 1999. Web site: http://www.igc.org/trends/kyoto.html

The permanent clearing and destruction of forests, *deforestation,* is currently occurring most alarmingly in the world's rain forests. The U.N. Food and Agricultural Organization has estimated that rain forests globally are being destroyed at the rate of 0.40 hectare (1 acre) per second. Today rain forests cover less than 7 percent of the land surface, half of what they covered only a few thousand years ago. Destruction of the rain forests, however, is not just about the loss of trees, a renewable resource that is being eliminated more quickly than it can be regenerated. It is also about the loss of the biological diversity of an ecosystem, which translates into the potential loss of biological compounds that may have great medical value. The destruction of rain forests is also about destabilizing the oxygen and carbon dioxide cycles of the forests, which may have long-term effects on global climate. Much of the destruction of the South American rain forests is the result of peripheral countries' attempts at economic development. **Figure 4.30** illustrates this point with reference to the Bolivian Amazon rain forest. The introduction of coca production has become an important source of revenue for farmers in the region and has led to the removal of small tracts of forest. Still, removal of the rain forest for agricultural production in Bolivia is minimal when compared to other South American countries, such as Brazil and Colombia.

Figure 4.30 Coca growing in Bolivia The Bolivian portion of the Amazon remains relatively intact as compared with nearby Brazil. Two factors, however, may diminish its relatively pristine state. The first is the increased logging of hardwoods. The second is the production of coca for export as cocaine. Shown in this photograph are coca leaves drying on white cloths in a cleared-out area of the rain forest. Coca is Bolivia's chief export and accounts for over half a million jobs in a largely subsistence economy. In the Chapare region, over 100,000 hectares (247,100 acres) of coca are under production.

Great geographical variability exists with respect to human impacts on the world's forests. For most of the core regions, net clearance of the forests has been replaced by regeneration. Yet for most of the periphery, clearance has accelerated to such an extent that one estimate shows a 50 percent reduction in the amount of forest cover since the early 1900s.

Cultivation is another important component of global land use, which we deal with extensively in Chapter 8. However, one or two points about the environmental impacts of cultivation are pertinent here. During the past 300 years the land devoted to cultivation has expanded globally by 450 percent. In 1700 the global stock of land in cultivation took up an area about the size of Argentina. Today it occupies an area roughly the size of the entire continent of South America. While the most rapid expansion of cropland since the mid-twentieth century has occurred in the peripheral regions, the amount of cropland has either held steady or been reduced in core regions. The expansion of cropland in peripheral regions is partly a response to growing populations and rising levels of consumption worldwide. It is also due to the globalization of agriculture (see Chapter 8), with some core-region production having been moved to peripheral regions. The reduction of cropland in some core regions is a result both of this globalization and of a more intensive use of cropland—utilizing more fertilizers, pesticides, and farm machinery—and new crop strains.

Grasslands are also used productively the world over, either as rangeland or pasture for livestock grazing. Most grasslands are found in arid and semiarid regions that are unsuitable for farming because of lack of water or poor soils (**Figure 4.31**). Some grasslands, however, occur in more rainy regions where tropical rain forests have been removed and replaced by grasslands. Other grasslands occur at the mid-latitudes, such as the tall- and short-grass prairies of the central United States and Canada. Approximately 68 million square kilometers (26 million square miles) of the land surface is currently taken up by grasslands.

Human impacts on grasslands are largely of two sorts. The first is the clearing of grasslands for other uses, most frequently settlement. As the global demands on beef production have increased, so has the intensity of use of the world's grasslands. Widespread overgrazing of grasslands has led to their acute degradation. In its most severe form, overgrazing has led to desertification. **Desertification** is the spread of desert conditions resulting from deforestation, overgrazing, and poor agricultural practices, as well as reduced rainfall associated with climatic change. One of the most severe examples of desertification has been occurring in the Sahel region of Africa since the 1970s. The degradation of the grasslands bordering the Sahara Desert, however, has not been a simple case of careless overgrazing by thoughtless herders. Severe drought, land decline, recurrent famine, and the breakdown of traditional systems for coping with disaster have all combined to create increased pressure on fragile resources, result-

Figure 4.31 African grasslands Also known as savannas, grasslands include scattered shrubs and isolated small trees and are normally found in areas with high-average temperatures and low-to-moderate precipitation. They occur in an extensive belt on both sides of the equator. African tropical savannas, such as the one pictured here in Kenya, contain extensive herds of hoofed animals, including gazelles, giraffes, zebras, wildebeests, antelope and the elephants. Wild species are essential to the safari tourism industry in Africa and would die off without the nourishment that savannas provide.

ing in a loss of grass cover and extreme soil degradation. While the factors behind the human impacts on the Sahelian grasslands are complex, the fact remains that the grasslands have been severely degraded, and the potential for their recovery is still unknown (**Figure 4.32**).

Land included in the wetland category covers swamps, marshes, bogs, peatlands, and the shore areas of lakes, rivers, oceans, and other water bodies. Wetlands can be associated either with salt water or fresh water (**Figure 4.33**). Most of Earth's wetlands are associated with the latter.

Human impacts on wetland environments are numerous. The most widespread has been the draining or filling of wetlands and their conversion to other land uses, such as settlement or cultivation. One reliable estimate places the total area of the world's wetlands at about 8.5 million square kilometers (3.3 million square miles), with about 1.5 million square kilometers (0.6 million square miles) lost to drainage or filling. For example, Australia has lost all of the original 20,000 square kilometers (7,740 square miles) of wetlands to conversion. For the last 400 years or so, people have regarded wetlands as nuisances, if not sources of disease. In core countries, technological innovation made modification and conversion of wetlands possible and profitable. In San Francisco, California, for example, the conversion of wetlands in the mid-nineteenth century allowed speculators and real estate developers to extend significantly the central downtown area into the once marshy edges of San Francisco Bay. The Gold Rush in the Sierra Nevada sent millions of tons of sediment down the rivers into the bay, filling in marshes and reducing its nearshore depth. It is estimated that in 1850 the San Francisco Bay system, which includes San Pablo Bay as well as Suisun Bay, covered approximately 315 square kilometers (about 120 square miles). One hundred years later only about 125 square kilometers (about 50 square miles) remained. By the 1960s the conversion and modification of the wetlands (as well as the effects of

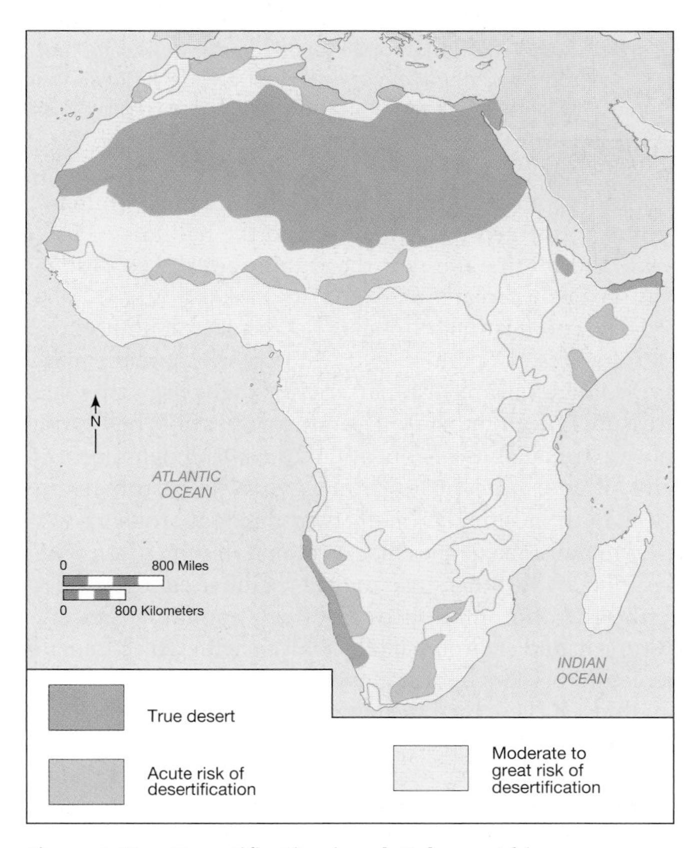

Figure 4.32 Desertification in sub-Saharan Africa Desertification is a mounting problem in many parts of the world, but especially in sub-Saharan Africa (the portion of Africa between North Africa's Sahara Desert and the five countries that make up southern Africa). Overgrazing on fragile arid and semiarid rangelands and deforestation without reforestation are thought to be the chief causes of desertification in this part of Africa.

Figure 4.33 Wetlands Throughout the world, freshwater and saltwater wetlands help to control floods, preserve the water supply, and provide a natural environment for a wide variety of wildlife and a source of recreation for residents of nearby population centers. Shown here is the Yolo Basin Wildlife Area, with the skyline of Sacramento, California, in the background.

pollution pouring directly into the bay) had so dramatically transformed water quality and the habitats of fish, fowl, and marine life that the viability of the ecosystem was seriously threatened.

The combustion of fossil fuels, the destruction of forest resources, the damming of watercourses, and the massive change in land-use patterns brought about by the pressures of globalization—industrialization being the most extreme phase—contribute to environmental problems of enormous proportions. It is now customary to speak of the accumulation of environmental problems we, as a human race, experience as global in dimension. Geographers and others use the term **global change** to describe the combination of political, economic, social, historical, and environmental problems with which human beings across Earth must currently contend. Very little, if anything, has escaped the embrace of globalization, least of all the environment.

In fact, no other period in human history has transformed the natural world as profoundly as the last 500 years. While we reap the benefits of a modern way of life, it is critical to recognize that these benefits have not been without cost. Fortunately, the costs have not been accepted uncritically. Over the last two to three decades, responses to global environmental problems have been on the increase as local groups have mobilized internationally. In the next section we examine popular and institutional responses to global environmental problems.

THE GLOBALIZATION OF THE ENVIRONMENT

Global Environmental Politics

The increasing importance of flows and connections—economic, political, social, and cultural—means that contemporary globalization has resulted in an increasingly shrinking world. In addition to allowing people and goods to travel farther faster and to receive and send information more quickly—the smaller world that globalization has made possible—means that political action has also become global. It can now move beyond the confines of the state into the global political arena, where rapid communications enable complex supporting networks to be developed and deployed, facilitating interaction and decision making. A good example can be seen in the protests that occurred in Seattle, Washington, in 1999 over the World Trade Organization (WTO) meeting, in Genoa, Italy, in July 2001 over the Group of 8 (G8) summit, and in Bangkok in 2003 to protest the meeting of the Asian Pacific Economic Cooperation summit meeting. Telecommunications, and especially the Internet, enabled protest leaders to organize and deploy demonstrators from interested groups all over the world. Such protests reflect a truly global politics that matches the global politics of institutions like the WTO, the International Monetary Fund (IMF), the World Bank, and the G8.

One indication of the increasingly global nature of politics outside of formal political institutions is the increase in environmental organizations whose purview and membership are global (**Figure 4.34**). These organizations have emerged in response to the global impact of contemporary environmental problems, such as fisheries depletion, global warming, the increasing use of genetically modified seeds, and the widespread decline in global biodiversity. Since the 1990s these groups, ranging from lobbying organizations to nongovernmental organizations (NGOs) to direct-action organizations to political parties like the Green Party in Europe and drawing on distinctive traditions and varying levels of resources, have become an important international force. Although states tend to exclude global environmental organizations from formal political decision making, these groups have had important impacts on the creation of international environmental institutions and laws, such as the regulation of international waters and the control of marine pollution (the London Dumping Convention in 1972 and the United Nations Law of the Sea in 1982), as well as specific agreements on protection of such wildlife as polar bears and seals, trade in internationally determined rare species, and on the Antarctic ecosystem.

Most dramatic have been the major conventions signed on the international transport of hazardous waste materials (the Basel convention in 1989), air pollution controls on chloroflurocarbon (CFC) emissions (the Vienna and Montreal Protocols in 1985 and 1987), as well as a range of treaties regulating transboundary acid rain in North America and Europe. Increasingly, agreements and conventions protecting biodiversity are being created, and not a moment too soon. The decline in the diversity of simple foodstuffs, like lettuce, potatoes, tomatoes, and squash, occurred most dramatically over the course of the twentieth century. For instance, in 1903 there were 13 known varieties of asparagus; by 1983 there was just one, or a decline of 97.8 percent. There were 287 known varieties of carrots in 1903, but this has fallen to just 21, a fall of 92.7 percent. A decline in the diversity of foodstuffs means that different resistances to pests inherent in these different varieties have also declined, as have their different nutritional values and tastes.

Moreover, new sources of medicine may be lost not only because of deforestation in tropical forests but also because of the decline in indigenous languages, cultures, and traditions. The Convention on Biological Diversity that emerged from the Rio Summit in 1992 is attempting to protect global biodiversity by preserving and protecting indigenous cultures and traditions, recognizing that many indigenous people have extensive knowledge of local plants and animals and their medicinal uses. As globalization homogenizes languages and draws more and more people into a capitalist market system, traditional knowledge and practices are being lost. The U.N. Environment Program devotes a great deal of its energies to biological and cultural diversity. Even the WTO has begun to recognize the value of indigenous knowledge and the promise of biodiversity

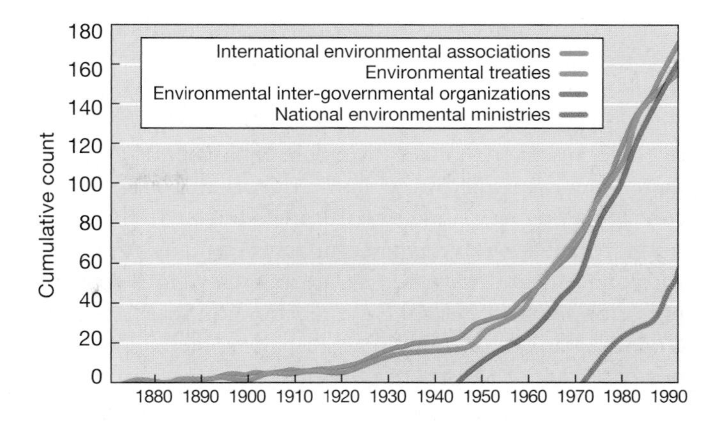

Figure 4.34 Growth of international environmental organizations Since the Second World War, there has been a dramatic increase in global environmental problems, as well as national, international, and voluntary organizations to assess and address those problems. (After data from D. Held, A. McGrew, D. Goldblatt, and J. Perraton, Global Transformations. Cambridge: Polity, 1999, p. 388.)

through its advocacy of intellectual property rights of both corporations and indigenous peoples. For the latter, this means protection against "bioprospecting," the practice in which companies attempt to exploit indigenous knowledge of plants for commercial purposes without compensating those individuals who have developed detailed knowledge of their medicinal value.

Clearly, global environmental awareness is on the rise from both the conservative (such as the WTO) and progressive (such as programs devoted to preserving genetic diversity in seed strains) ends of the political spectrum. This increasing awareness is directly responsible for the staging of global environmental conferences like Rio in 1992, Kyoto in 1997, and Johannesburg in 2002, which have not only affected international laws but continue to shape the debates about and responses to environmental problems. Most recently, the terms of these debates have centered on the concept of sustainability.

Environmental Sustainability

The interdependence of economic, environmental, and social problems, often located within widely different political contexts, means that some parts of the world are ecological time bombs. The world currently faces a daunting list of environmental threats, including the destruction of tropical rain forests and consequent loss of biodiversity; widespread health-threatening pollution; the degradation of soil, water, and marine resources essential to food production; stratospheric ozone depletion; and acid rain. Most of these threats are greatest in the world's periphery, where daily environmental pollution and degradation amount to a catastrophe that will continue to unfold in slow motion in the coming years.

In the peripheral regions there is simply less money to cope with environmental threats. The poverty endemic to peripheral regions also adds to environmental stress. In order to survive, the rural poor are constantly impelled to degrade and destroy their immediate environment, cutting down forests for fuelwood and exhausting soils with overuse. In order to meet their debt repayments, governments feel compelled to generate export earnings by encouraging the harvesting of natural resources. In the cities of the periphery, poverty encompasses so many people in such concentrations as to generate its own vicious cycle of pollution, environmental degradation, and disease. Even climate change, an inherently global problem, seems to pose its greatest threats to poorer, peripheral regions.

A more benign relationship between nature and society has been proposed under the principle of *sustainable development,* a term that incorporates the ethic of intergenerational equity, with its obligation to preserve resources and landscapes for future generations (see Box 7.1; "Sustainable Development"). Geographers such as William Adams and Timothy O'Riordan conceive of sustainable development as including ecological, economic, and social measures to prevent environmental degradation while promoting economic growth and social equality. Sustainable development means that economic growth and change should occur only when the impacts on the environment are benign or manageable and the impacts (both costs and benefits) on society are fairly distributed across classes and regions. This means finding less polluting technologies that use resources more efficiently, and managing renewable resources (those that replenish themselves, such as water, fish, and forests) to ensure replacement and continued yield. In practice, sustainable development policies of major international institutions, such as the World Bank, have promoted reforestation, energy efficiency and conservation, and birth control and poverty programs to reduce the environmental impact of rural populations. At the same time, however, the expansion and globalization of the world economy has resulted in increases in resource use and inequality that contradict many of the goals of sustainable development.

CONCLUSION

The relationship between society and nature is very much mediated by institutions and practices, from technology to religious beliefs. In this chapter we have seen how the nature-society relationship has changed over time and how the globalization of the capitalist world economy has had a more widespread impact on attitudes and practices than any cultural or economic system that preceded it.

Ancient humans apparently displayed a reverential attitude toward the natural world, an attitude still evident among native populations in many parts of the New World, as well as Africa and Asia. With the emergence of Judaism and, later, Christianity, humans adopted a more dominant attitude about nature. The expansion of European trade, followed by colonization and eventually industrialization, broadcast worldwide the belief that humans should take their place at the apex of the natural world. It is the Judeo-Christian attitude toward nature as it was taken up by the emergence of the capitalist economic system that is the most pervasive shaper of nature-society interactions today.

Besides exploring the history of ideas about nature and contemporary environmental philosophies and organizations in the United States, this chapter has also shown that society and nature are interdependent and that events in one part of the global environmental system affect conditions in the system elsewhere. Finally, this chapter has shown that events that have occurred in the past shape the contemporary state of society and nature.

In short, as economies have globalized so has the environment. We can now speak of a global environment in which not only the people but also the physical environments where they live and work are linked in complex and essential ways. Along with the recognition of a globalized environment have come new ways of thinking about global economic development. Sustainable development, one of these new ways of thinking, has come to dominate the agenda of international institutions as well as environmental organizations as the new century begins.

MAIN POINTS REVISITED

- **Nature and society constitute a complex relationship. In our view, nature is both a physical realm and a social creation.**

  Recognizing that nature and society are interactive requires us also to acknowledge that humans are not separate from nature but an integral part of it.

- **Because in this text we regard nature as a social creation, it is important to understand views of nature in society today as well as the history of those views. The most prominent view of nature in Western culture is derived from the Judeo-Christian tradition, one core belief of which is that nature exists to be dominated by humans.**

  Environmental philosophies have emerged and spread in an attempt to counter this view of nature. There is a wide range of philosophies and approaches circulating, most prominently in the United States.

- **Humankind's relationships with nature have developed over the course of human history, beginning with the early Stone Age. The early history of humankind included people who**

revered nature as well as those who abused it. Urbanization and industrialization have had extremely degrading impacts on the environment.

Although other societies have had substantial impacts on nature, the extent of the contemporary core society's impact on the environment is unprecedented. As peripheral countries aim to achieve the level of prosperity enjoyed in the core, their economic practices have similar environmental impacts. The result is that while core countries have begun to limit their negative environmental impacts, peripheral countries in many ways are just beginning to produce their own significant environmental problems.

- **The globalization of the political economy has meant that environmental problems are also global in scope. Deforestation, acid rain, and nuclear fallout affect us all. Many new ways of understanding nature have emerged in the last several decades in response to global environmental crises.**

Some of the most disturbing problems have to do with extensive land-use changes, such as deforestation, as well as with widespread air pollution from the burning of fossil fuels, which many scientists believe is leading to global climate change. In response to these global crises, many new ways of understanding nature have emerged in the last several decades, offering insight into our world as a complexly integrated natural system.

- **Sustainability is fast becoming the most significant approach to addressing global economic and environmental challenges. In addition, new institutional frameworks, including conventions, protocols, and organizations, are rapidly emerging to promote global sustainability.**

New institutions and organizations continue to emerge to demand accountability for a changing global economy that is creating new, more widespread, and oftentimes more disastrous environmental problems.

KEY TERMS

acid rain (p. 158)
Clovis point (p. 143)
Columbian Exchange (p. 147)
conservation (p. 139)
cultural ecology (p. 133)
deep ecology (p. 140)
deforestation (p. 145)
demographic collapse (p. 148)
desertification (p. 164)
ecofeminism (p. 140)
ecological imperialism (p. 148)
ecosystem (p. 144)
environmental ethics (p. 140)
environmental justice (p. 140)
global change (p. 166)
nature (p. 131)
Paleolithic period (p. 142)
political ecology (p. 135)
preservation (p. 139)
romanticism (p. 138)
siltation (p. 145)
society (p. 132)
technology (p. 132)
transcendentalism (p. 138)
virgin soil epidemics (p. 147)

ADDITIONAL READING

Attfield, R., *The Ethics of Environmental Concern.* Athens, GA: University of Georgia Press, 1991.

Bartsch, U., B. Muller, and A. Aaheim, *Fossil Fuels in a Changing Climate: Impacts of the Kyoto Protocol and Developing Country Participation.* New York: Oxford, 2000.

Bebbington, A. "Movements, Modernizations, and Markets: Indigenous Organizations and Agrarian Struggle in Ecuador." In R. Peet and M. Watts (eds.), *Liberation Ecologies: Environment, Development, Social Movements.* London: Routledge, 1996.

Chasek, P. S. *The Global Environment in the 21st Century: Prospects for International Cooperation.* New York: United Nations, 1999.

Cantrill, J.G., and C. L. Oravec (eds.), *The Symbolic Earth: Discourse and Our Creation of the Environment.* Lexington, KY: University Press of Kentucky, 1996.

Centers for Disease Control and Prevention, 2005. West Nile Virus, Division of Vector Borne Infectious Diseases. http://www.cdc.gov/ncidod/dvbid/westnile/background.htm, retrieved June 15, 2005.

Collingwood, R., *The Idea of Nature.* London: Oxford University Press, 1960.

Crosby, A. W., *The Columbian Exchange: Biological and Cultural Consequences of 1492.* Westport, CT: Greenwood Press, 1972.

Committee to Review the U.S. Climate Change Science Program Stratgeic Plan, *Implementing Climate and Global Change Research: A Review of the Draft U.S. Climate Change Science.* Washington, DC: National Research Council, 2004.

Crosby, A. W., *Germs, Seeds, and Animals: Studies in Ecological History.* Armonk, NY: M. E. Sharpe, 1994.

Diamond, I., and G. F. Orenstein, *Reweaving the World: the Emergence of Ecofeminism.* San Francisco: Sierra Club Books, 1990.

Fagen, B., *The Long Summer: How Climate Changed Civilization.* New York: Basic Books, 2004.

Glacken, C., *Traces on the Rhodian Shore.* Berkeley: University of California Press, 1967.

Goudie, A. and D. J. Cuff. *Encyclopedia of Global Change, Environmental Change and Human Society.* New York: Oxford University Press, 2002.

Litfin, K., *The Greening of Sovereignty in World Politics.* Cambridge, MA: MIT Press, 1998.

Lovell, W. G. "Heavy Shadow and Black Night: Disease and Depopulation in Colonial Spanish America," *Annals of the Association of American Geographers 82*(3): 426–443, 1992.

Maathai, W., *The Green Belt Movement: Sharing the Approach and the Experience.* New York: Lantern Books, 2004.

Marsh, G. P., *Man and Nature.* New York: Scribner, 1864.

MacNaughten, P. and J. Urry, *Contested Natures*. Thousand Oaks, CA: Sage, 1998.

Meade, M. S. and R. J. *Erickson, Medical Geography*, 2nd ed. New York: Guildford, 2005.

Merchant, C., *The Death of Nature: Women*. San Francisco: Harper & Row, 1979.

Oelschlager, M., *The Idea of Wilderness: From Prehistory to the Age of Ecology*. New Haven: Yale University Press, 1991.

Peet, R. and M. Watts, *Liberation Ecologies: Environment, Development, Social Movements*. New York: Routledge, 2004.

Peters, R. L., and T. E. Lovejoy, *Global Warming and Biological Diversity*. New Haven: Yale University Press, 1992.

Sayer, J. and B. Campbell, *The Science of Sustainable Development: Local Livelihoods and Global Environment*. Cambridge, England: Cambridge University Press, 2003.

Raven, P. H., *Nature and Human Society: The Quest for a Sustainable World: Proceedings of the '97 Forum on Biodiversity*. Washington, DC: National Academy Press, 2000.

Robbins, P., *Political Ecology: A Critical Introduction*. Oxford: Blackwell, 2004.

Simmons, I. G., *Environmental History: A Concise Introduction*. Oxford: Blackwell, 1993.

Smith, M., and L. Marx, *Does Technology Drive History? The Dilemma of Technological Determinism*. Cambridge, MA: MIT Press, 1995.

Thomas, W. L. (ed.), *Man's Role in Changing the Face of the Earth*. Chicago: University of Chicago Press, 1956.

Turner, B.L. II, et al., *The Earth Transformed by Human Action: Global and Regional Changes in the Biosphere over the Past 300 Years*. New York: Cambridge University Press, 1990.

Wood, D., *Five Billion Years of Global Change: A History of the Land*. New York: Guildford, 2004.

Worster, D., *Nature's Economy: A History of Ecological Ideas*. Cambridge, England: Cambridge University Press, 1977.

Worldwatch Institute, *State of the World 2002: A Worldwatch Institute Report on Progress Toward a Sustainable Society*. New York: W.W. Norton & Co., 2002.

Zimmerer, K., "Human Geography and 'New Ecology': The Prospect and Promise of Integration," *Annals of the Association of American Geography, 84*(1): 108–125, 1994.

EXERCISES

On the Internet

The theme of nature, society, and technology featured in this chapter is explored online by our critical-thinking essay. It directs you to a Web site that explores the *nature, society,* and *technology* of the Incas, a pre-Columbian civilization. You will investigate the Incas' burial customs, deities, economy, transportation system, religion, and origin myth. In a thinking-spatially exercise, you will view and comment on maps demonstrating European expansion, environmental change, and introduction of new diseases, plants, and animals in the New World. A series of multiple-choice questions, with electronic feedback, will test your understanding of chapter concepts.

Unplugged

1. Many communities have begun to produce an index of stress, which is a map of the toxic sites of a city or region. One way to plot a rudimentary map is to use the local phone book as a data source. Use the Yellow Pages to identify the addresses of environmentally harmful and potentially harmful businesses, such as dry-cleaning businesses, gas stations, automotive repair and car-care businesses, aerospace and electronic manufacturing companies, agricultural supply stores, and other such commercial enterprises where noxious chemicals may be produced, sold, or applied. Compile a map of these activities to begin to get a picture of your locale's geography of environmental stress.
2. Locate and read a natural history of the place where your college or university is located. What sorts of plants and animals dominated the landscape there during the Paleolithic period? Do any plants or animals continue to survive in altered or unaltered form from that period?
3. Colleges and universities are large generators of waste, from plain-paper waste to biomedical and other sorts of wastes that can have significant environmental impacts. Identify how your college or university handles this waste stream and how you, as a member of the academic community, contribute to it. Where does the waste go when it leaves the university? Is it locally deposited? Does it go out of state? Remember to trace the stream of all the types of waste, not just the paper.

"YOU DON'T HAVE TO GIVE ME YOUR ANSWER NOW....
YOU CAN TEXT ME LATER."

5 Cultural Geographies

Wel it"s 1 4 th mun e, 2 4 th sho /
3 2 gt rd e now go cat go /
Bt dnt u, txt"n my bl%-scrn fone /
U cn do NEthng bt lay ofv my bl%-scrn fone
(http://books.guardian.co.uk/textpoetry/story/0,12586,854297,00.html)

A shared understanding of words or word clusters is the foundation of language allowing for individuals to communicate with each other. This shared understanding is based on rules—about word sounds and structure—that are understood by a community and enable them to communicate with each other. Communication through language works precisely because we comprehend someone's pronunciation or can read their handwriting, agree what words are referring to and understand how the words in a sentence can, if ordered one way, create a question or in another way a declarative statement. Written and spoken language is full of rules that dictate spelling, punctation, grammar, syntax, and formal practices such as how to write a letter to a friend versus how to write a letter to a prospective employer. At least, that is what linguists, grammarians, editors, teachers and other guardians of language have believed: that there are rules to language and when we disobey those rules all sorts of misunderstandings, some benign and some downright dangerous, will ensue.

But what has been happening over the last 10 to 15 years is that a new generation of American English speakers (and speakers of national languages in other places around the world) act rather differently: They don't follow the rules and inconsistency in language use abounds. Linguist Naomi Baron, who studies computer-mediated communication (CMC), believes there are two key forces, among others, that are contributing to this slow but steady change in language use and practice. They are the mediating influences of technology (computers and cellular phones especially) and a faster pace in everyday life at work, at home, and even at play. E-mail, instant messaging, and cellular phone text messaging are part and parcel of a faster paced life and they are at the heart of a transformation in language, especially written language.

Change, of course, is a natural part of language development such that the words we use are likely to be slightly different from the ones our grandparents used and vastly different from the language used centuries ago. The text poem above illustrates how the written lyrics to a song produced just fifty years ago can be altered dramatically such that even Elvis Presley would be hard pressed to understand them. What is different about today compared to language change in the past is that CMC, through mechanisms like instant messaging, is accelerating changes in language faster than ever before (see NetLingo, 2005 for a dictionary of Internet terms and text messaging acronyms). What is also happening is that the more we use CMC the more it reshapes our lives in terms of how we communicate and interact with each other. That is the point behind the cartoon on the opposite page.

Behind this most recent and rapid transformation of language are young people, the teenagers and twenty-somethings of the new century; that is, most of you reading

Computer-mediated courting

MAIN POINTS

- Though culture is a central, complex concept in geography, it may be thought of as a way of life involving a particular set of skills, values, and meanings.
- Geographers are particularly concerned about how place and space shape culture and, conversely, how culture shapes place and space. They recognize that culture is dynamic and is contested and altered within larger social, political, and economic contexts.
- Like other fields of contemporary life, culture has been profoundly affected by globalization. Globalization, however, globalization has not produced a homogenized culture so much as distinctive impacts in different societies and geographical areas as global forces come to be modified by local cultures.
- Contemporary approaches in cultural geography seek to understand the role played by politics and the economy in establishing and perpetuating cultures, cultural landscapes, and global patterns of cultural traits and complexes.
- Cultural geography has been broadened to include analysis of gender, class, sexuality, race, ethnicity, and life-cycle stage, in recognition that important differences can exist within, as well as between, cultures.
- Globalization does not necessarily mean that the world is becoming more homogeneous. In some ways globalization has made the local even more important than before. In other ways it has meant that local cultures have indeed suffered, as in the case of the disappearance of indigenous languages. In still other cases, globalization has helped to foster hybrid forms that blend distant cultures in unique ways.

this sentence right now! And through your use of CMC, you are not only changing the rules of language use, you are—through your use of new language—also at the forefront of other significant cultural transformations with respect to the relationships and social practices that are enabled and sustained by language. One example is something called continuous partial attention (CPA) where e-mail, cell phones, and text messaging are simultaneously active, with a correspondent partially monitoring all of them. With respect to personal relationships, CMP means that when more than one of your technologies are engaged, no one gets all of your attention but every one gets some of it, and some more than others. Text messaging is a byproduct of CPA, as there isn't enough time—nor does the technology allow—for the kind of formal language and full mental engagement required to produce more traditional written communication. But we still are not certain what the implications of CPA might be for our everyday lives and the relationships that enable them. Will CPA reduce the intensity of our relationships because of divided attention? Will CMC multiply them because the number of contact points are increased? Will both of them lead to changes in the way we communicate in our daily spaces such as classrooms or boardrooms? More than ever before, language—the cultural medium through which we express meaning—and the spaces within which those meanings are produced and comprehended are changing. But the changes are not uniform across the globe nor even across neighborhoods and within households. What we do know is that signficant cultural change around language is already occuring; what we don't know is what that change will mean.

In this chapter, we examine the many ways geographers have explored the concept of culture—including the globalization of culture—and the insights they have gained from these explorations. We ask: What counts as culture? How do geographers study it? How do we come to grips with the fact that U.S. cultural practices are being exported to the far reaches of the globe at the same time that our own cultural practices are being shaped by forces beyond our national borders?

CULTURE AS A GEOGRAPHICAL PROCESS

Geographers seek to understand the manifestations and impacts of culture on geography and of geography on culture. While anthropologists are concerned with the ways in which culture is created and maintained by human groups, geographers are interested not only in how place and space shape culture but also the reverse—how culture shapes place and space.

Anthropologists, geographers, and other scholars who study culture, such as historians, sociologists, and political scientists, agree that culture is a complex concept. Over time our understanding of culture has been changed and enriched. A simple understanding of culture is that it is a particular way of life, such as a set of skilled activities, values, and meanings surrounding a particular type of economic practice. Scholars also describe culture in terms of classical standards and aesthetic excellence in opera, ballet, or literature, for example.

The term "culture" is also used to describe the range of activities that characterize a particular group, such as working-class culture, corporate culture, or teenage culture. Although all of these understandings of culture are accurate, for our purposes they are incomplete. Broadly speaking, **culture** is a shared set of meanings that are lived through the material and symbolic practices of everyday life. Our understanding in this book is that culture is not something that is necessarily tied to a place and thus a fact to be discovered. Rather we regard the connections among people, places, and cultures to be social creations that can be altered and therefore must be explored in order to be understood (**Figure 5.1**). The "shared set of meanings" can include values, beliefs, practices, and ideas about religion, language, family, gender, sexuality, and other important identities. These values, beliefs, ideas, and practices are, because of globalization, increasingly subject to reevaluation and redefinition and can be, and very frequently are, transformed from both within and outside a particular group.

In short, culture is a dynamic concept that revolves around and intersects with complex social, political, economic, and even historical factors. This understanding of culture is part of a longer, evolving tradition within geography and other disciplines, such as anthropology and sociology. We will look more closely at the development of the cultural tradition in geography in the following section, in which we discuss the debates surrounding culture within the discipline.

Figure 5.1 Cultural practices The term *culture* has been used to describe a range of practices characterizing a group. Pictured is a cultural practice known as gothic. Hairstyle, dress, and body adornment, as well as a distinctive philosophy and music, characterize gothic culture. Yet culture is more than just the physically distinguishing aspects of a group. It is also a way groups derive meaning from and attempt to shape the world around them.

For much of the twentieth century, geographers, like anthropologists, have focused most of their attention on material culture, as opposed to its less tangible symbolic or spiritual manifestations. Thus, while geographers have been interested in religion as an object of study, for a long time they have largely confined their work to examining its material basis. For example, they have explored the spatial extent of particular religious practices (the global distribution of Buddhism) and the physical expression of religiosity, such as the appearance of crosses along roadways where fatal traffic accidents have occurred (**Figure 5.2**). In the last 20 or so years, the near-exclusive focus on material cultural practices has changed—driven by the larger changes that are occurring in the world around us.

As with agriculture, politics, and urbanization, globalization has also had complex effects on culture. Terms such as *world music* and *international television* are a reflection of the sense that the world has become a very small place, indeed, and people everywhere are sharing aspects of the same culture through the widespread influence of television and other media, such as radio. Yet, as pointed out in Chapter 2, while powerful homogenizing global forces are certainly at work, the world has not become so uniform that place no longer matters. With respect to culture, just the opposite is true. Place matters more than ever in the negotiation of global forces, as local forces confront globalization and translate it into unique place-specific forms. Box 5.1: "The Culture of Hip-Hop" illustrates the global nature of hip-hop music. The place-based interactions occurring between culture and global political and economic forces are at the heart of cultural geography today. **Cultural geography** focuses on the way in which space, place, and landscape shape culture at the same time that culture shapes space, place, and landscape. As such, cultural geography demarcates two important and interrelated parts. Culture is the ongoing process of producing a shared set of meanings and practices, while geography is the dynamic setting that groups operate in to shape those meanings and practices and in the process to form an identity and act. Geography in this definition can be as small as the body and as large as the globe.

Before we proceed any further, it is important to discuss a significant difference between our view of culture and that of more conventional cultural geographers. Many introductory human geography texts divide culture into two major categories: folk and popular culture. **Folk culture** is seen by specialists as the traditional practices of small groups, especially rural people with a simple lifestyle (compared with modern, urban people), such as the Amish in Pennsylvania or the Gypsies in Europe, who are seen to be homogeneous in their belief systems and practices. **Popular culture,** by contrast, is viewed by some cultural geographers as the practices and meaning systems produced by large groups of people whose norms and tastes are often heterogeneous and change frequently, often in response to commercial products. Hip-hop would be seen by these theorists as an example of popular culture, as would "soccer Moms."

In this text we do not divide culture into categories. We see culture as an overarching process that is shaped by and shapes politics, the economy, and society and cannot be neatly demarcated by reference to the number of characteristics or degree of homogeneity of its practitioners. We see culture as something that can be enduring as well as newly created, but always influenced by a whole range of complex interactions as groups maintain, change, or even create traditions from the material of their everyday lives. For us, there is no purpose served in categorically differentiating between hip-hop and Hinduism, as both are significant expressions of culture and both are of interest to geographers.

Figure 5.2 Road accident memorial Known in Mexico and Spain as *descansos,* these memorials are artistic expressions that mark the place where the soul has left the body in a fatal accident.

BUILDING CULTURAL COMPLEXES

Geographers focus on the interactions between people and culture, and among space, place, and landscape. One of the most influential geographers was Carl Sauer, who taught at the University of California, Berkeley. Sauer was largely responsible for creating the "Berkeley school" of cultural geography. He was particularly interested in trying to understand the material expressions of culture by focusing on their manifestations in the landscape (**Figure 5.3**). This interest came to be embodied in the concept of the **cultural landscape,** a characteristic and tangible outcome of the complex interactions between a human group—with its own practices, preferences, values, and aspirations—and its natural environment. Sauer differentiated the cultural landscape from the natural landscape. He emphasized that the former was a "humanized" version of the latter, such that the activities of humans resulted in an identifiable and understandable alteration of the natural environment. **Figure 5.4** lists the differences between a natural and a cultural landscape.

Figure 5.3 Masai village, Kenya The cultural landscape, as defined by Carl Sauer, reflected the way that cultural and environmental processes came together to create a unique product. Pictured here is a small village where herding is the major occupation. The village is enclosed by thorny brambles and branches harvested from the surrounding area. Within the enclosure, the dwellings are arranged in a unique circular pattern, with the animal pens in the middle of the settlement for easy observation by the residents.

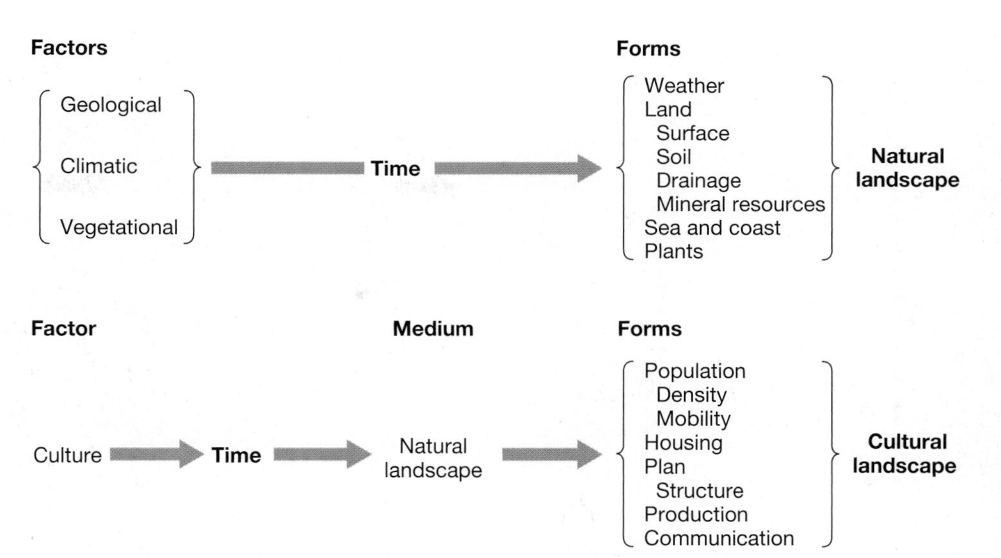

Figure 5.4 Sauer's cultural landscape This figure summarizes the ways the natural and cultural landscapes are transformed. Physical and climatic factors shape the natural landscape. Cultural practices also have an important impact upon it. The results of cultural factors are cultural forms, such as population distributions and patterns and housing. Over time, people—through culture—reshape the natural landscape to meet their needs. (After C. Sauer, "The Morphology of Landscape" in J. Leighly (ed.), *Land and Life: Selections from the Writings of Carl Ortwin Sauer*, pp. 315–350.)

For roughly five decades, interest in culture within geography largely followed Sauer's important work. His approach to the cultural landscape was ecological, and his many published works reflect his attempts to understand the myriad ways that humans transform the surface of Earth. In his own words:

> The cultural landscape is fashioned from a natural landscape by a cultural group. Culture is the agent, the natural area is the medium, the cultural landscape is the result. Under the influence of a given culture, itself changing through time, the landscape undergoes development, passing through phases, and probably reaching ultimately the end of its cycle of development. With the introduction of a different—that is an alien—culture, a rejuvenation of the cultural landscape sets in, or a new landscape is superimposed on remnants of an older one.[1]

In Europe, geographers interested in human interactions with the landscape took slightly different approaches. For example, in Great Britain the approach to understanding the human imprint on the landscape was termed *historical geography*, while in France it was conceptualized as *genre de vie*. **Historical geography,** very simply defined, is the geography of the past. Its most famous practitioner was H. C. Darby, who attempted to understand "cross sections" or sequences of evolution, especially of rural landscapes. **Genre de vie,** a key concept in Vidal de la Blache's approach to cultural geography in France, referred to a functionally organized way of life that was seen to be characteristic of a particular culture group. *Genre de vie* centered on the livelihood practices of a group that were seen to shape physical, social, and psychological bonds (**Figure 5.5**). Although emphasizing some landscape components over others or giving a larger or smaller role to the physical environment, all of these approaches placed the cultural landscape at the heart of their study of human-environment interactions.

H. C. Darby most successfully implemented his historical approach to cultural geography and landscape by developing a geography based on the Domesday Book. William the Conqueror ordered the Domesday compiled in 1085 so he could have a list of his spoils of war. The book provides a rich catalog of the ownership of every tract of land in England and of the conditions and contents of the lands at that time (**Figure 5.6**). For geographers like Darby, such data were invaluable for reconstructing past landscapes.

Vidal de la Blache emphasized the need to study small, homogeneous areas in order to uncover the close relationships that exist between people and their immediate surroundings. He constructed complex descriptions of preindustrial France that demonstrated how the various *genres de vie* emerged from the possibilities and constraints posed by local physical environments. Subsequently, he wrote about the changes in French regions brought on by industrialization, observing that regional homogeneity was no longer the unifying element. Instead, the increased mobility of people and goods produced new, more complex geographies where previously isolated *genres de vie* were integrated into a competitive industrial economic framework. Anticipating the widespread impacts of globalization, de la Blache also recognized how people in various places struggled to cope with the big changes that were transforming their lives.

Geographers also examine specific aspects of culture, ranging from single attributes to complex systems. One simple aspect of culture of interest to geographers is the idea of special traits, which include such things as distinctive styles of dress, dietary habits, and styles of architecture (**Figure 5.7**). A **cultural trait** is a single aspect of the complex of routine practices that constitute a particular cultural group. For example, dietary law for Muslims prohibits the consumption of pork. This avoidance may be said to be a cultural trait of Muslim people. Additionally, in their religious iconography, Muslims have prohibitions against displaying human faces. This, too, is considered a

[1]C. Sauer, "The Morphology of Landscape," in J. Leighly (ed.), *Land and Life: Selections from the Writings of Carl Ortwin Sauer.* Berkeley, CA: University of California Press, 1964, pp. 315–350.

Figure 5.5 Market gardens in Corsica This image shows a rural setting in Corsica, an island nation in the Mediterranean, where commercial agriculture is being undertaken. Farming is a way of life—a *genre de vie*—that can read from the landscape where extensive cultivated fields and isolated farmhouses constitute key elements.

Figure 5.6 Domesday Book Pictured here is the wood and metalwork cover of the Domesday Book, which was written as a report of a survey of land and wealth in 1086. The information contained in the book has been invaluable for reconstructing past landscapes.

cultural trait. Geographers are also interested in learning how cultural traits come together to form larger frameworks for living in the world. Ultimately, cultural traits are not necessarily unique to one group, and understanding them is only one aspect of the complexity of culture. For instance, there are certainly other cultural groups (such as Hindus and Jews) that avoid pork in their diet.

Many cultures also recognize the passage from childhood into adulthood with a celebration or ceremony. Called **rites of passage,** these are ceremonial acts, customs, practices, or procedures that recognize key transitions in human life—birth, menstruation, and other markers of adulthood such as sexual awakening and marriage. Such rites of passage are not uncommon among many of the world's cultures. Some non-Western cultures, for example, send adolescent boys away from the village to experience an ordeal—ritual scarring or circumcision, for example—or to meditate in extended isolation on the new roles they must assume as adults. After an extended absence from the social group, these youths return transformed and ready to lay down their previous childish occupations (**Figure 5.8**).

In Roman Catholicism, the passage of boys and girls into adulthood, traditionally around the age of twelve, is celebrated by confirmation. In this religious ceremony, the confirmed chooses a new name to mark this important spiritual transition. Jews mark the passage of adolescent boys and girls into adulthood with separate religious ceremonies: a bar mitzvah for boys and a bat mitzvah for girls. Although marking the passage into adulthood is a trait of both religious groups, they do not exhibit the trait in exactly the same way. In fact, this and other traits always occur in combination with others. The combination of traits characteristic of a particular group is known as a **cultural complex.** The avoidance of pork, the celebration of bar and bat mitzvahs, and other dietary, reli-

Figure 5.7 Iroquois Longhouse A longhouse is a long, narrow single room building built by peoples in various parts of the world including Asia, Europe and North America. Pictured here is an Iroquois longhouse. On average a typical longhouse was about 60 feet long by 18 feet wide by 18 feet high with doors at each end and no windows. The longhouses were meant to house up to 20 or more families at once.

Figure 5.8 A coming-of-age ceremony, Apache reservation The Apache Indians recognize the passage of female children into adulthood during adolescence. A ceremony initiates a young girl into womanhood, signaling to other members of the community that an important transition has occurred. The photograph shows a young Apache girl about to begin four rotations in recognition of the four stages of her life: infancy, adolescence, adulthood, and old age. Eventually she will be blessed with a shower of pollen, candy, coins, and corn signifying a future of happiness and plenty. Such rites of passage are not uncommon among many of the world's cultures.

The Culture of Hip-Hop

Hip-hop is the popular street culture of U.S. big-city and especially inner city youth. Characterized to some extent by graffiti art, and earlier on by break dancing, hip-hop is understood globally through rap music and a distinctive idiomatic vocabulary. Like most nations, hip-hop has its forebears. These include boxer Muhammad Ali, Jamaican Rastafarian and reggae musician Bob Marley, Black Panther Huey Newton, and funksters James Brown and George Clinton. Also like most nations, the hip-hop nation has its origins in the Bronx, in New York City. But it also has much older roots in the West African storytelling culture known as *griot*. The hip-hop nation has enlarged upon those origins, and now hip-hop is both appreciated and produced on six continents. Hip-hop is a nation that has truly globalized, not because its citizens have migrated far and wide, but because its culture has migrated via telecommunications. Hip-hop has become a nation that exists beyond geography in the music, the clothes, and the language of its citizens. Touré, writing in *The New York Times*, describes the hip-hop nation this way:

> We are a nation with no precise date of origin, no physical land, no single chief. But if you live in the hip-hop nation, if you are not merely a fan of the music but a daily imbiber of the culture, if you sprinkle your conversation with phrases like off the meter (for something that's great) or got me open (for something that gives an explosive positive emotional release), if you know why Dutch Masters make better blunts than Phillies (they're thinner), if you know at a glance why Allen Iverson is hip-hop and Grant Hill is not, if you feel the murders of Tupac Shakur and the Notorious B.I.G. in the 1997–98 civil wars were assassinations (no other words fit), if you can say yes to all of these questions (and a yes to some doesn't count), then you know the hip-hop nation is as real as America on a pre-Columbian atlas.[1]

Although the hip-hop leadership is exclusively black and male, the hip-hop nation crosses all color lines and includes women and gays, though the latter two have been the targets of entrenched sexism and homophobia by a wide range of rappers (**Figures 5.A** and **5.B**). Its pioneers include white graffiti artists and Latinos who influenced break dancing as well as hip-hop DJ (disk jockey) and MC (rapper) styles.

Music is the heart and soul of the hip-hop nation and the geography of U.S. hip-hop—where the most important music has come from—can be crudely divided into East Coast, West Coast, and South Coast, and a newer region in and around Detroit where white rap-metal groups have become popular. The East Coast includes the five boroughs of New York, Long Island, Westchester County, New Jersey, and Philadelphia. The West Coast includes Los Angeles, Compton, Long Beach, Vallejo, and Oakland. The South Coast region is made up of Atlanta, New Orleans, Miami, and Memphis (**Figure 5.C**).

But just as hip-hop has broken out of its regional boundaries, it has transcended national boundaries as well. Hip-hop graffiti art can be found in urban areas as distant as Australia and South Africa. Rap music is as popular in the Philippines as it is in Paris. And individual DJs have had significant influences well beyond their old neighborhoods. For example, Afrika Bambaataa, a former gang member, organized the Universal Zulu Nation over 25 years ago. Bambaattaa, or "Bam," incorporated former gang members into a community-building group that has become a household name in hip-hop circles all around the world. One Web site claims that there are now over 10,000 members of the Zulu Nation worldwide and chapters in every major city in the world.

Figure 5.A Notorious B.I.G. "Biggie" burst onto the hip-hop scene with his platinum 1994 album "Ready to Die?" In 1995 B.I.G. was named Rapper of the Year at the Billboard Awards. Born Christopher Wallace in New Jersey and raised in Brooklyn's Bedford-Stuyvesant neighborhood, the rapper was killed in March 1997. Speculation abounds as to why the murder occurred (as well as the murder of rapper Tupac Shakur). Some believe that B.I.G.'s death was the result of an East Coast–West Coast rap rivalry and payback for Tupac Shakur's murder in Las Vegas months before. Others believe the murder was carried out by a gang upset by B.I.G.'s growing prominence on the West Coast. The murder was characterized in the mainstream press as a drive-by shooting. No suspects have been arrested, and the case remains unsolved to this day. Both murders point to gang culture as an important source of identity in hip-hop.

[1]Touré, "In the End, Black Men Must Lead." *New York Times*, 22 August 1999, Arts and Leisure, p. 1.

Figure 5.B Rapper Eve While rap is unquestionably dominated by men and has historically produced songs that are blatantly misogynist, female rappers are also part of the hip-hop culture and have even become rich and famous entertainers. Pictured here is Eve (Jihan Jeffers) a Philadelphia rap artist with the all-male RuffRyders Records. Other notable female rappers include Bahamadia, Missy Elliot, Lil' Kim, and Lauryn Hill.

But hip-hop is also very clearly about the more local space of the neighborhood or "the 'hood"—rap's dominant spatial trope—which is portrayed in all its complexity in songs, music videos, and hip-hop films like *Breakin'* (1984), *Beat Street* (1985), *Wild Style* (1982) and *8 Mile* (2002). While rap and dancing are central to hip-hop, the local context in which the story, music, or dance unfold is also critical. As hip-hop cultural theorist Murray Forman argues: "Virtually all of the early descriptions of hip-hop practices identify territory and the public sphere as significant factors, whether in visible artistic expression and appropriation of public space via graffiti or b-boying [break dancing], the sonic impact of a pounding bass line, or the discursive articulation of urban geography in rap lyrics [and films]."[2] Hip-hop is effectively about how space and place shape the identities of rappers in particular but also African-Americans more generally, showing how race, space, and place come together to produce the contradiction of "home" as a locus of roots and the foundation of personal history, but also as a site of devaluation vis-à-vis the dominant white society.

Hip-hop, as a youth-oriented cultural product commercialized by multinational corporations but originally homegrown, is to the end of the twentieth century what rock was to the middle of the century. Its influence is enormous, and its practices are likely to persist well into the twenty-first century as its appeal continues to spread globally.

[2]M. Forman, "Ain't No Love in the Heart of the City: Hip-Hop, Space, and Place," in M. Forman and M. A. Neal (eds), *That's the Joint! The Hip Hop Studies Reader.* New York: Routledge, 2004, p. 155.

Sources: Davey D's Hip-Hop Corner at http://www.daveyd.com/index.html; D. Toop, *Rap Attack 2: African Rap to Global Hip Hop*, New York: Serpent's Tail Press, 1991; N. George, *Hip hop America*, New York: Viking Penguin Group, 1998.

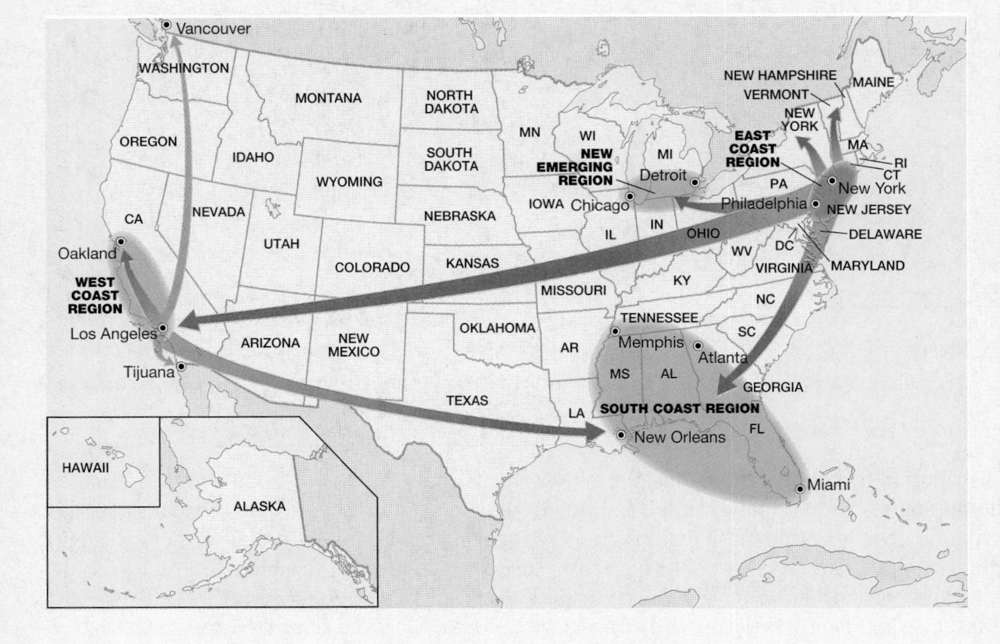

Figure 5.C The sources and diffusion of U.S. rap This map portrays the centers of rap music in the United States today, showing how rap, which began in African American inner city neighborhoods in New York City in the late 1970s, moved westward and then southward. Most recently a hybrid form of rap metal has emerged in the U.S. Midwest urban center of Detroit. The Detroit metropolitan area contains large numbers of African Americans and working-class whites who lost their jobs in the restructuring of the automobile industry in the 1970s and early 1980s. What the rap-metal genre confirms is that although hip-hop culture has its roots in the African American experience, it derives much of its power from issues of poverty and class.

gious, and social practices constitute the cultural complex of Judaism, although it is important to note that even within the cultural complex of Judaism, variation exists among regions and sects.

Another concept key to traditional approaches in cultural geography is the cultural region. Although a cultural region may be quite extensive or very narrowly described and even discontinuous in its extension, it is the area within which a particular cultural system prevails. A **cultural region** is an area where certain cultural practices, beliefs, or values are more or less practiced by the majority of the inhabitants.

For example, the state of Utah is considered to be a Mormon cultural region because the population of the state is dominated by people who practice the Mormon religion and presumably adhere to its beliefs and values. **Figure 5.9** illustrates the overall religious geography of the United States. The map shows that the southeastern part of the United States has a large Baptist population, whereas the West and Southwest, particularly California, Nevada, Arizona, and New Mexico, are dominated by Roman Catholics. As with the large concentration of Roman Catholics in the northeastern and Middle Atlantic states, these concentrations reflect the immigration of people from Catholic countries. In the case of the western/southwestern United States, these are largely populations of fairly recent immigrants from Mexico and other parts of Central America as well as those who occupied the region well before the Anglo population arrived. With respect to the Eastern Seaboard, the dominance of Roman Catholicism reflects older immigration from Catholic Europe, as well as more recent immigration from the Caribbean Basin. In the upper central part of the United States, the scattered concentrations of Lutherans reflect previous immigration from the Scandinavian countries.

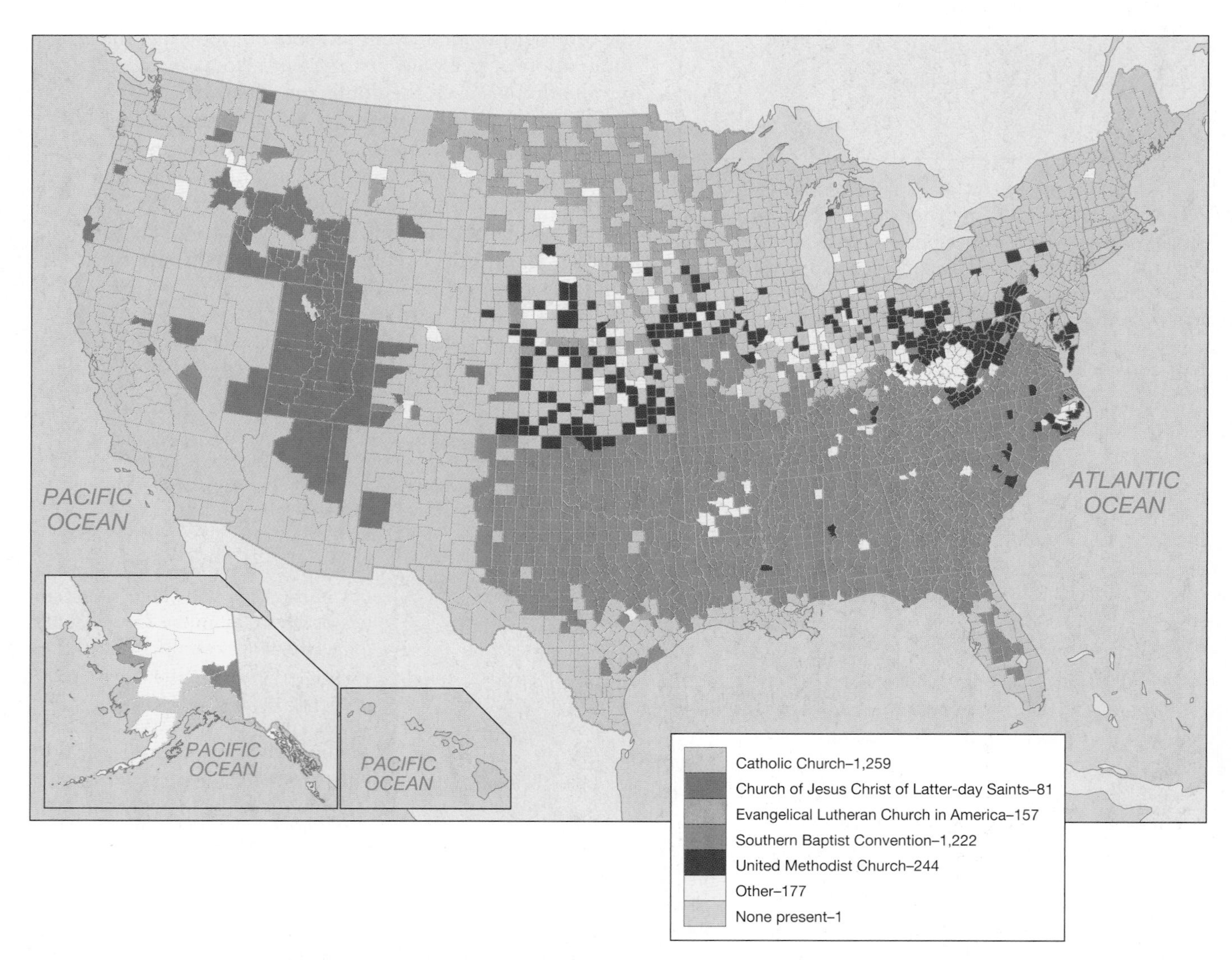

Figure 5.9 U.S. religious population distribution by county, 2000 This map shows the majority religion by county for the United States and illustrates the concept of cultural regions based on religion. It is important to remember, however, that at such a scale, it would be erroneous to assume too much homogeneity within these regions. In each aggregation of counties or region there is likely to be substantial variations in belief systems and practices at the local level. (*Source:* Jones, D. E., S. Doty, C. Grammich, J. E. Horsch, R. Houseal, M. Lynn, J. P. Marcum, K. M. Sanchagrin, and R. H. Taylor, *Religious Congregations and Membership in the United States 2000: An enumeration by region, state and county based on data reported for 149 religious bodies.* Nashville, Tenessee: Glenmary Research Center, 2000, 562).

CULTURAL SYSTEMS

Broader than the cultural complex concept is the cultural system, a collection of interacting components that, taken together, shape a group's collective identity. A **cultural system** includes traits, territorial affiliation, and shared history, as well as other, more complex elements, such as language. In a cultural system it is possible for internal variations to exist in particular elements at the same time that broader similarities lend coherence. For example, Christianity unites all Protestant religions (as well as Catholic ones), yet the practices of particular denominations—Lutherans, Episcopalians, and Quakers—vary. And while Mexicans, Bolivians, Cubans, and Chileans exhibit variations in pronunciation, pitch, stress, and other aspects of vocal expression, they all speak Spanish. This means they share a key element of a cultural system (which, for these nationalities, also includes Roman Catholicism and a Spanish colonial heritage).

Geography and Religion

Two key components of a cultural system for most of the world's people are religion and language. **Religion** is a belief system and a set of practices that recognizes the existence of a power higher than humans. Although religious affiliation is on the decline in some parts of the world's core regions, it still acts as a powerful shaper of daily life, from eating habits and dress codes to coming-of-age rituals and death ceremonies in both the core and the periphery. And, like language, religious beliefs and practices change as new interpretations are advanced or new spiritual influences are adopted. The most important influence on religious change has been conversion from one set of beliefs to another. From the onset of globalization in the fifteenth century, religious missionizing—propagandizing and persuasion—and the conversion of non-Christian souls were key elements in religious change. In the 500 years since the onset of the Columbian Exchange, conversion of all sorts has escalated throughout the globe. In fact, since 1492, traditional religions have become dramatically dislocated from their sites of origin through missionizing and conversion, as well as diaspora and emigration. Whereas missionizing and conversion are deliberate efforts to change the religious views of a person or peoples, diaspora and emigration involve the involuntary and voluntary movement of peoples who bring their religious beliefs and practices to their new locales.

Diaspora is a spatial dispersion of a previously homogeneous group. The processes of global political and economic change that led to the massive movement of the world's populations over the last five centuries have also meant the dislodging and spread of the world's many religions from their traditional sites of practice. Religious practices have become so spatially mixed that it is a challenge to present a map of the contemporary global distribution of religion that reveals more than it obscures. This is because the global scale is too gross a level of resolution to portray the wide variation that exists among and within religious practices. **Figure 5.10** identifies the contemporary distribution of what religious scholars consider to be the world's major religions because they contain the largest number of practitioners globally. As with other global representations, the map is useful in that it helps to present a generalized picture.

Figure 5.11 identifies the source areas of four of the world's major religions and their diffusion from those sites over time. The map illustrates that the world's major religions originated and diffused from two fairly small areas of the globe. The first, where Hinduism and Buddhism (as well as Sikhism) originated, is an area of lowlands in the subcontinent of India drained by the Indus and Ganges rivers (Punjab on the map). The second area, where Christianity and Islam (as well as Judaism) originated, is in the deserts of the Middle East.

Hinduism was the first religion to emerge, among the peoples of the Indo-Gangetic Plain, about 4,000 years ago. Buddhism and Sikhism evolved from Hinduism as reform religions, with Buddhism appearing around 500 B.C. and Sikhism developing in the fifteenth century. It is not surprising that Hinduism helped to produce new religions, because India has long been an important cultural crossroads. As a result, ideas and practices originating in India spread rapidly, at the same time that other ideas and practices were being brought to India from far-flung places and then absorbed and translated to reflect Indian needs and values. For example, Buddhism emerged as a branch of Hinduism in an area not far from the Punjab. At first a very small group of practitioners surrounding Prince Gautama, the founder of the religion, were confined to northern India. Slowly and steadily, however, Buddhism dispersed to other parts of India and was carried by missionaries and traders to China (100 B.C.–A.D. 200, Korea and Japan (A.D. 300–500), Southeast Asia (A.D. 400–600), Tibet (A.D. 700), and Mongolia (A.D. 1500) (see **Figure 5.12**). Not surprisingly, as Buddhism spread, it developed many regional forms, so that Tibetan Buddhism is distinct from Japanese Buddhism.

Christianity, Islam, and Judaism all developed among the Semitic-speaking people of the deserts of the Middle East. Like the Indo-Gangetic religions, these three religions are related. Although Judaism is the oldest, it is the least widespread. Judaism originated about 4,000 years ago, Christianity about 2,000 years ago, and Islam about 1,300 years ago. Judaism developed out of the cultures and beliefs of Bronze Age peoples and was the first monotheistic (belief in one God) religion. Although Judaism is the oldest monotheistic religion, and one that spread widely and rapidly, it is numerically small because it does not seek converts. Christianity developed in Jerusalem among the disciples of Jesus; they proclaimed that he was the Messiah expected by the Jews. As it moved east and south from its hearth area, Christianity's diffusion was helped by missionizing and by imperial sponsorship. The diffusion of Christianity in Europe is illustrated in **Figure 5.13**. Although we discuss Islam and

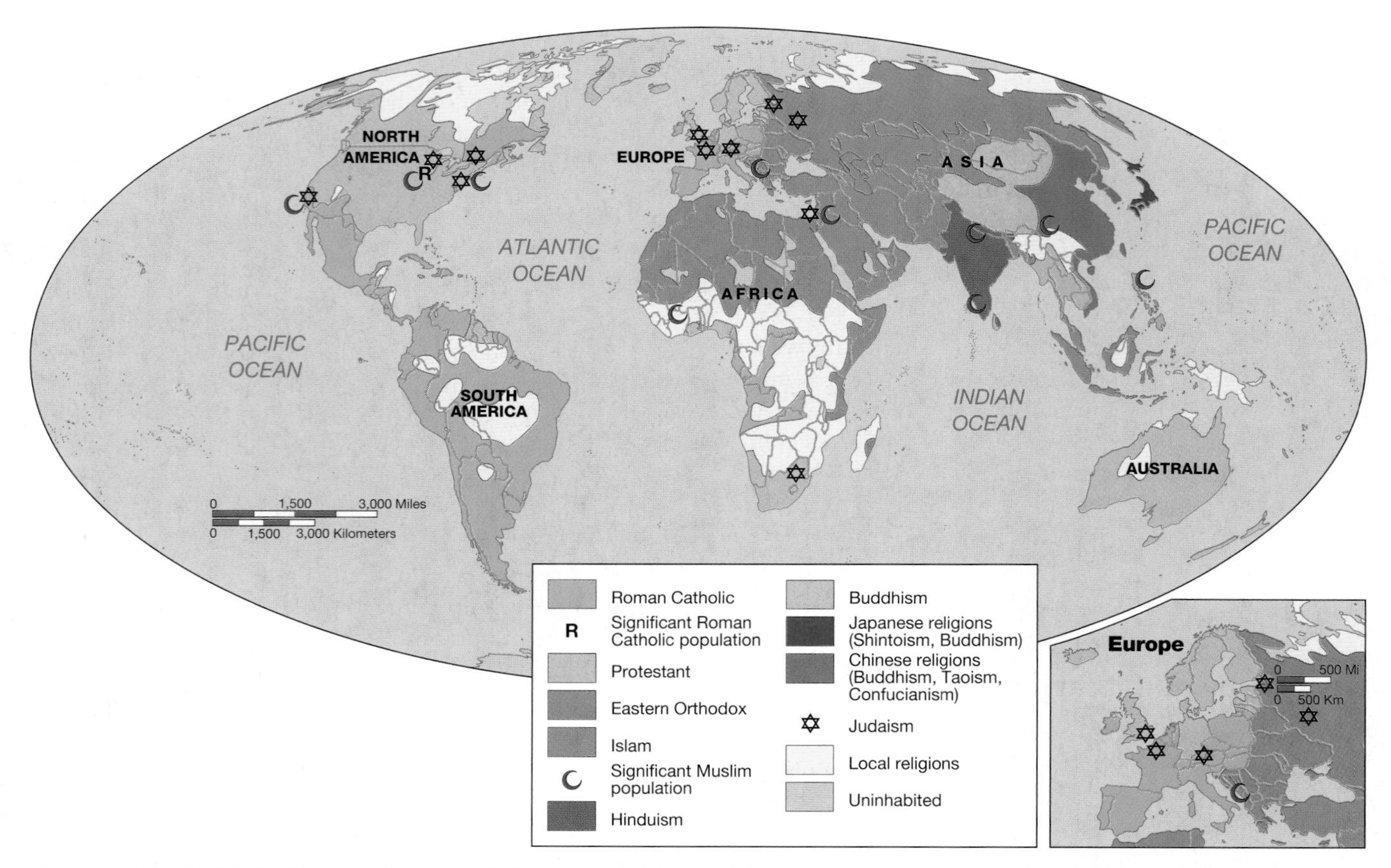

Figure 5.10 World distribution of major religions Most of the world's peoples are members of these religions. Not evident on this map are the local variations in practices, as well as the many smaller religions that are practiced worldwide.

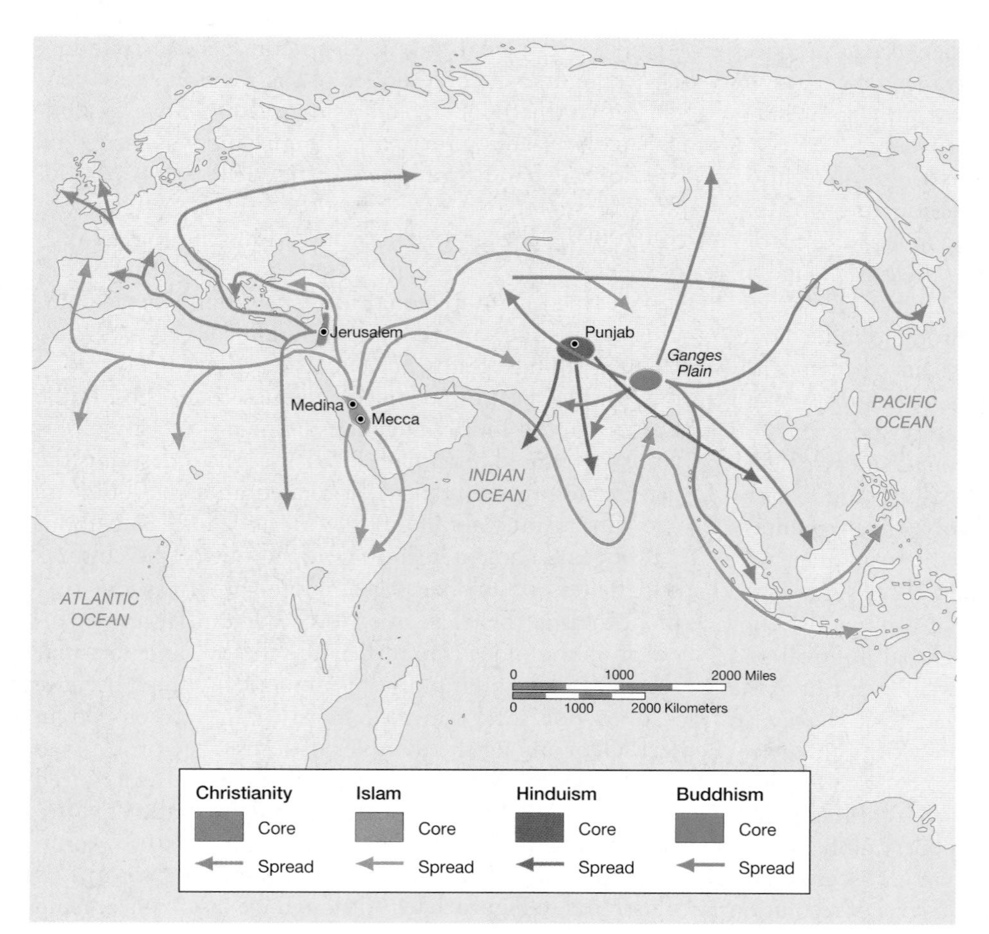

Figure 5.11 Origin areas and diffusion of four major religions The world's major religions originated in a fairly small region of the world. Judaism and Christianity began in present-day Israel and Jordan. Islam emerged from western Arabia. Buddhism originated in India, and Hinduism in the Indus region of Pakistan. The source areas of the world's major religions are also the cultural hearth areas of agriculture, urbanization, and other key aspects of human development.

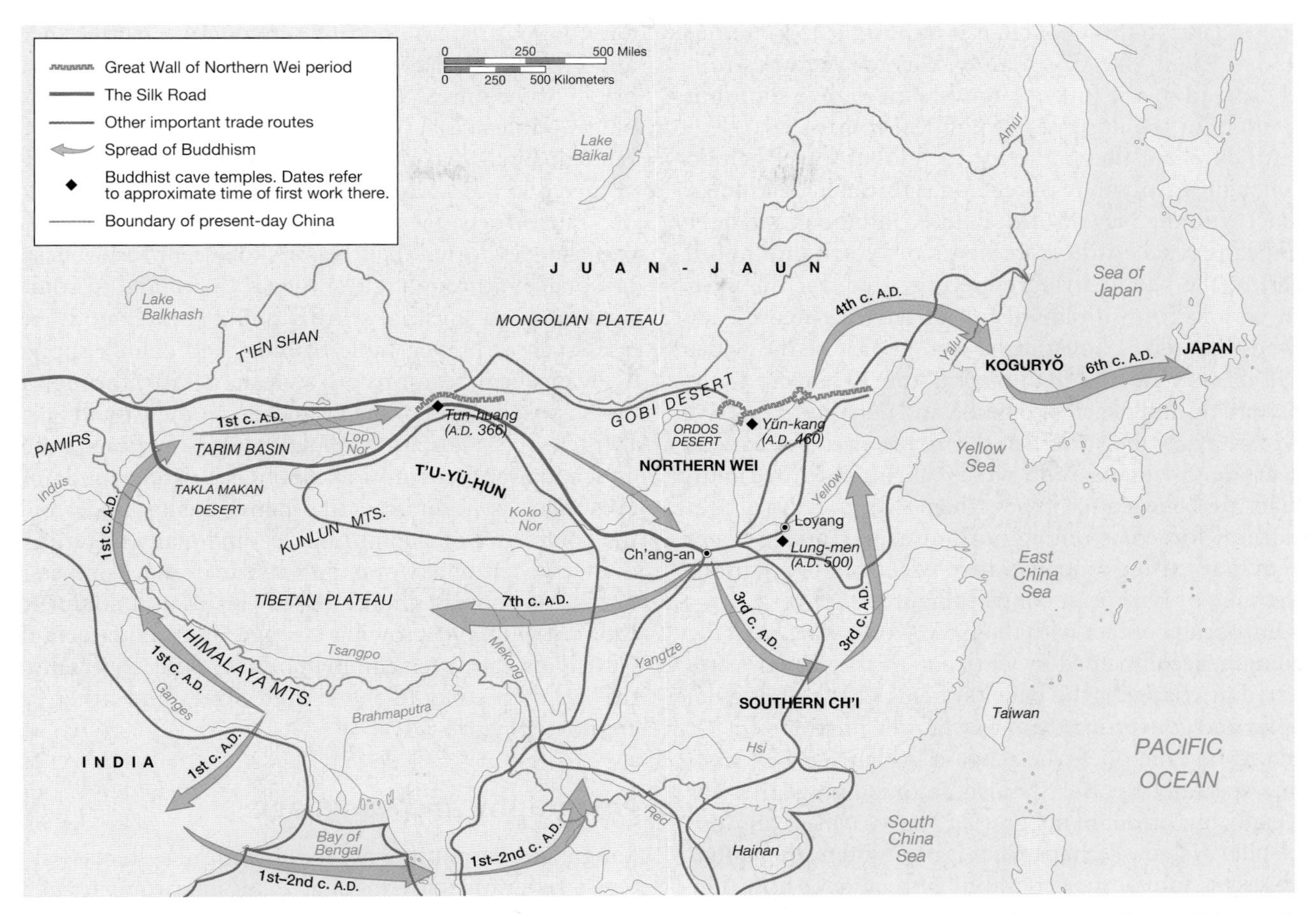

Figure 5.12 Spread of Buddhism This map illustrates the diffusion of Buddhism from its source area in India to China and then from China on to Korea and Japan. Commercial routes, like the Silk Road, were important vectors for the spread of the religion from India to China. Missionaries were responsible for the spread of Buddhism from China to Korea and Japan. (After C. Schirokauer, *A Brief History of Chinese and Japanese Civilizations* 2005, second edition, Florence, KY: Wadsworth.)

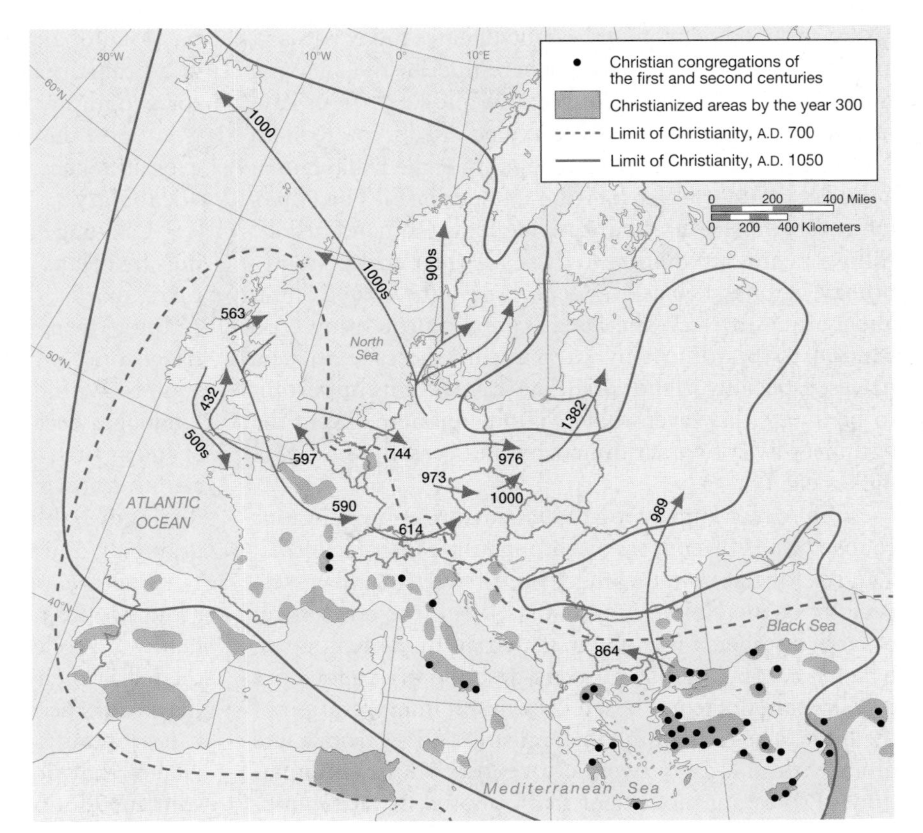

Figure 5.13 Spread of Christianity in Europe Christianity diffused through Europe largely through missionary efforts. Monks and monasteries were especially important as hubs of diffusion in the larger network. Monasteries and other sorts of religious communities are indicated by dots on the map. The shaded areas indicate places where Christian converts dominated by the year 300. (After C. Park, *Sacred Worlds* 1994. New York: Routledge.)

Islamism later in this chapter, it is important to point out here that Islam, like Christianity, was for centuries routinely spread by the force of conversion to save souls but also often for the purposes of political control.

An excellent illustration of the global forces behind the changing geography of religion is through Columbian contact with the New World. Before Columbus and later Europeans reached the continents of North and South America, the people living there practiced, for the most part, various forms of animism and related rituals. They viewed themselves holistically, as one part of the wider world of animate and inanimate nature. They used religious rituals and charms to guide and enhance the activities of everyday life, as well as the more extreme situations of warfare. Shamanism, in which spiritually gifted individuals are believed to possess the power to control preternatural forces, is one important aspect of the belief system that existed among Native American populations at the time of European contact (**Figure 5.14**).

European contact with the New World was, from the beginning, accompanied by Christian missionizing efforts directed at changing the belief systems of the aboriginal peoples and converting them to what the missionizers believed to be "the one, true religion" (**Figure 5.15**). Religion, especially for the Spanish colonizing agents, was especially important in integrating Native Americans into the feudal system. Perhaps what is most interesting about the present state of the geography of religion is how during the colonial period religious missionizing and conversion flowed from the core to the periphery. In the postcolonial period, however, an opposite trend has occurred (see Box 5.2: "Changing Religious Practices in Latin America and the Caribbean"). For example, the fastest-growing religion in the United States today is Islam, and it is in the core countries where Buddhism is making the greatest numbers of converts. While Pope John Paul has been the most widely traveled leader in Roman Catholic history, the same can be said for the Dalai Lama, a tireless world traveler for Tibetan Buddhism. The Pope's efforts are mostly directed at maintaining Roman Catholic followers and attempting to dissuade their conversion to other religions, such as evangelicalism in the United States and Latin America. The Dalai Lama is promoting conversion to Buddhism by carrying its message to new places, especially in the core (**Figure 5.16**). It is important to be aware, however, that religious missionizing in the periphery by core practitioners has not ceased, as Box 5.2 makes clear.

One other impact of globalization upon religious change occurs by conversion through the electronic media. The rise of television evangelism, or *televangelism*—especially in the United States—has led to the conversion of large numbers of people to Christian fundamentalism, which is a term popularly used to describe strict adherence to Christian doctrines based on a literal interpretation of the Bible. The term "fundamentalism" derives from a late nineteenth and early twentieth century transdenominational Protestant movement that opposed the accommodation of Christian doctrine to modern scientific theory and philosophy, especially Darwinian theories about the origin of the universe. It has its origin in a series of pamphlets published between 1910 and 1915 entitled "The Fundamentals: A Testimony to the Truth." Televangelism occurred in the wake of the decline of traditional fundamentalist revivalism—the practice whereby itinerant preachers exhorted their hearers to accept forgiveness of personal sin through faith in Jesus Christ and to commit themselves to spiritual self-discipline and religious exercises such as prayer, Bible reading, and church support. Televangelism began to grow in the 1950s, through the mass consumption of television sets in the United States. By the early 1990s, however, televangelism was in decline as scandal and corruption among Christian fundamentalist television ministers like James Baaker undermined the confidence of the faithful, leading many to return to more conventional worship. While televangelism persists but is less popular, church-based Christian fundamentalism is strong and growing stronger and fundamentalist Christians have been gaining political clout, especially on the national stage in the United States and other core countries (**Figure 5.17**).

Geography and Language

Languages are another aspect of cultural systems that interest geographers. Language is an important focus for study because it is a central aspect of cultural identity. Without language, cultural accomplishments could not be transmitted from one generation to the next. The distribution and diffusion of languages tell much about the changing history of human geography and the impact of globalization on culture (see Box 5.3: "Language and Ethnicity in Africa"). Before looking more closely at the geography of language and the impacts of globalization upon the changing distribution of languages, however, it is necessary to become familiar with some basic vocabulary.

Language is a means of communicating ideas or feelings by means of a conventionalized system of signs, gestures, marks, or articulate vocal sounds. In short, as the chapter opener illustrates, communication is symbolic, based on commonly understood meanings of signs or sounds. Within standard languages (also known as official languages because they are maintained by offices of government such as education and the courts), regional variations, known as **dialects**, exist. Dialects emerge and are distinguishable through differences in pronunciation, grammar, and vocabulary that are place-based in nature.

For the purposes of classification, languages are grouped into families, branches, and groups. A **language family** is a collection of individual languages believed to be related in their prehistorical origin. About 50 percent of the world's people speak a language that is in the Indo-European family. A **language branch** is a collection of languages that possess a definite common origin but have split into individual languages. A **language group** is a

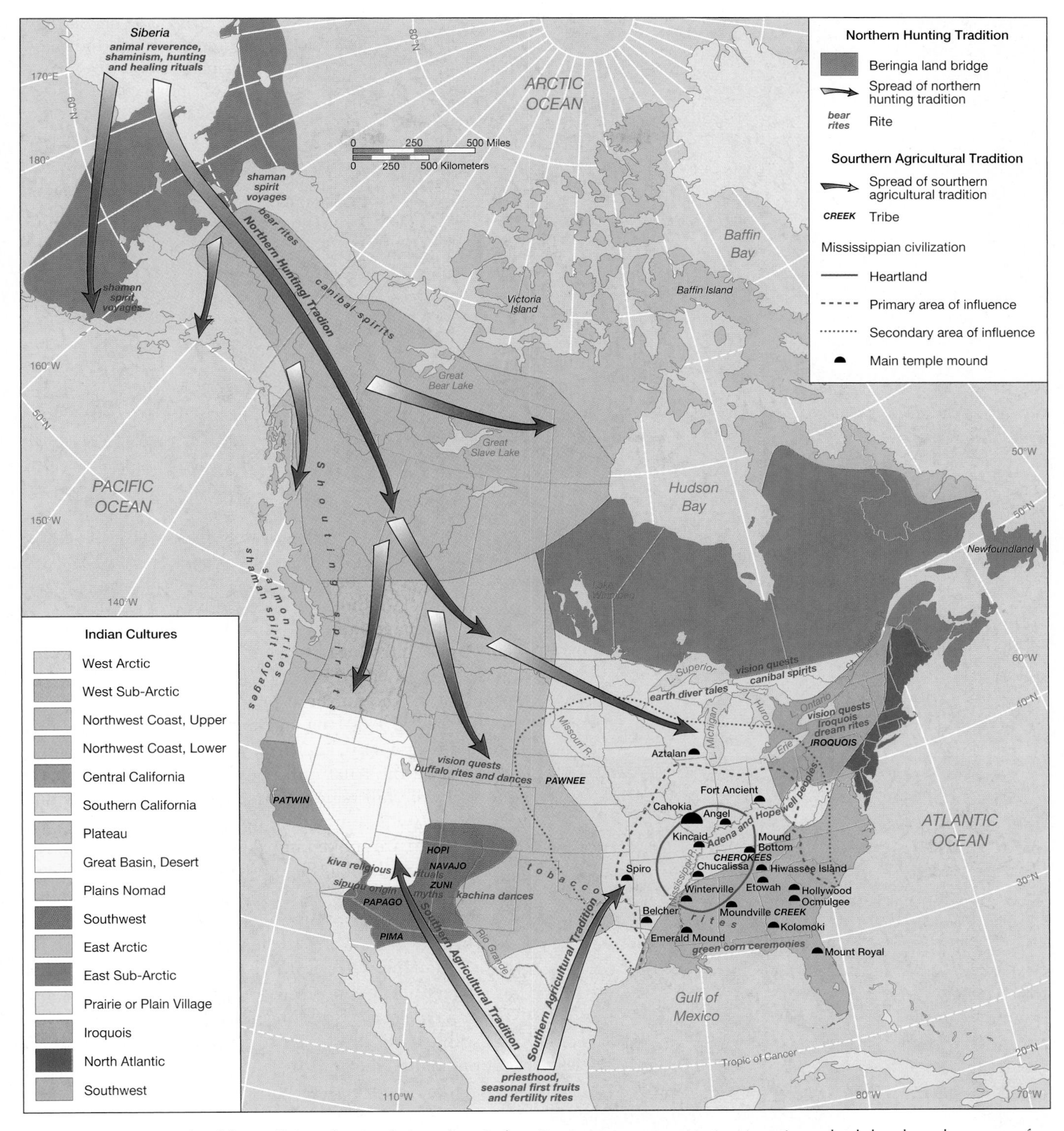

Figure 5.14 Pre-Columbian religions in North America Before European contact, Native Americans had developed a range of religious practices. Religious traditions based on agrarian practices diffused from south to north, while those based on hunting diffused from north to south. (*Source:* Carroll, B. E., *The Routledge Historical Atlas of Religion in America* 2000. New York: Routledge, pp. 15–16).

collection of several individual languages that are part of a language branch, share a common origin in the recent past, and have relatively similar grammar and vocabulary. Spanish, French, Portuguese, Italian, Romanian, and Catalan are a language *group*, classified under the Romance *branch* as part of the Indo-European language *family*.

Traditional approaches in cultural geography have identified the source areas of the world's languages and the paths of diffusion of those languages from their places of origin. Carl Sauer identified the origins of certain cultural practices with the label "cultural hearth." **Cultural hearths** are the geographic origins or sources

Figure 5.15 San Xavier del Bac mission near Tucson, Arizona The influence of missionizers on native populations was profound, as indigenous rituals became mixed with Christian ones. Even today in missions such as San Xavier, native peoples practice Christianity along with components from their native belief system.

of innovations, ideas, or ideologies. Language hearths are a subset of cultural hearths; they are the source areas of languages. **Figures 5.18** and **5.19** show the location of the world's indigenous languages and the hearth area of the Indo-European language. Compare them to see the spread of the latter throughout Europe and into north and South Asia.

Figure 5.16 The Dalai Lama China's refusal to acknowledge an independent Tibet has sent the Dalai Lama on numerous international tours to broadcast the plight of Tibetans, who are experiencing extreme persecution. The practices of Buddhism are increasingly being accepted by Westerners over the teachings of Christianity and Judaism. High-profile practitioners such as the actor Richard Gere have helped to give Buddhism cachet.

In India for example, in the most recent population census, people reported the use of over 1600 languages, which the government collapsed into 200 related ones. These translate into a broad regional grouping of four major language families. The Indo-European family of languages, introduced by the Aryan herdsmen who migrated from Central Asia between 1500 and 500 B.C., is prevalent in the northern plains region, Sri Lanka, and the Maldives. This language family includes Hindi, Bengali, Punjabi, Bihari, and Urdu. Munda languages are spoken among the tribal hill peoples who still inhabit the more remote hill regions of peninsular India. Dravidian languages (which include Tamil, Telegu, Kanarese, and Malayalam) are spoken in southern India and the northern part of Sri Lanka. Finally, Tibeto-Burmese languages are scattered across the Himalayan region (**Figure 5.20**).

In India, the boundaries of many of the country's constituent states were established after partition on the basis of language. Overall, no single language is spoken or understood by more than 40 percent of the people. There have been efforts since India became independent to establish Hindi, the most prevalent language, as the national language, but this has been resisted by many of the states within India, whose political identity is now closely aligned with a different language. In terms of popular media and literature, there is a thriving Hindi and regional language press, while film and television are dominated by Hindi and Tamil, with some Telegu programming.

English, spoken by fewer than 6 percent of the people, serves as the link language between India's states and regions. As in other former British colonies in South Asia, English is the language of higher education, the professions, and national business and government. Without English, there is little opportunity for economic or social mobility. Most children who do attend school are taught only their local language, and so are inevitably restricted

Figure 5.17 Religious right politics The religious right is a strong political force in the United States, and its strength is growing in other countries as well. Pictured here are a mother and daughter celebrating the victory of Repent America defendants in a Philadelphia court in early 2005. A city judge dismissed charges against four members of the conservative Christian group who were arrested in fall 2004 while picketing a street festival for gays and lesbians. The gay marriage amendment was an important issue in the 2004 presidential election.

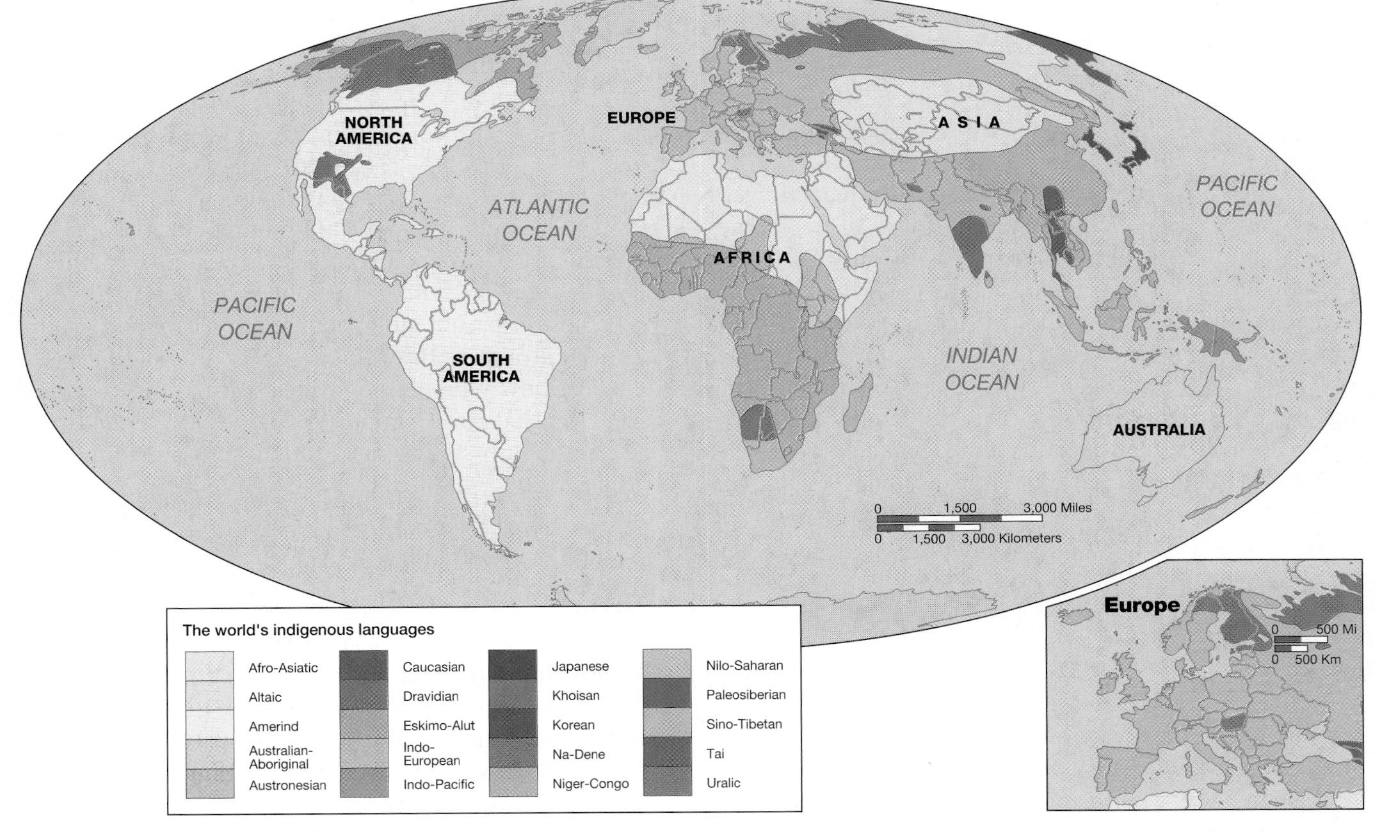

Figure 5.18 World distribution of major languages and major language families Classifying languages by family and mapping their occurrence across the globe provide insights about human geography. For example, we may discover interesting cultural linkages between seemingly disparate cultures widely separated in space and time. We may also begin to understand something about the nature of population movements across broad expanses of time and space. (After E. F. Bergman, *Human Geography: Cultures, Connections, and Landscapes*. Western Hemisphere after *Language in the Americas* by Joseph H. Greenberg; Eastern Hemisphere after David Crystal, *Encyclopedia of Language*.)

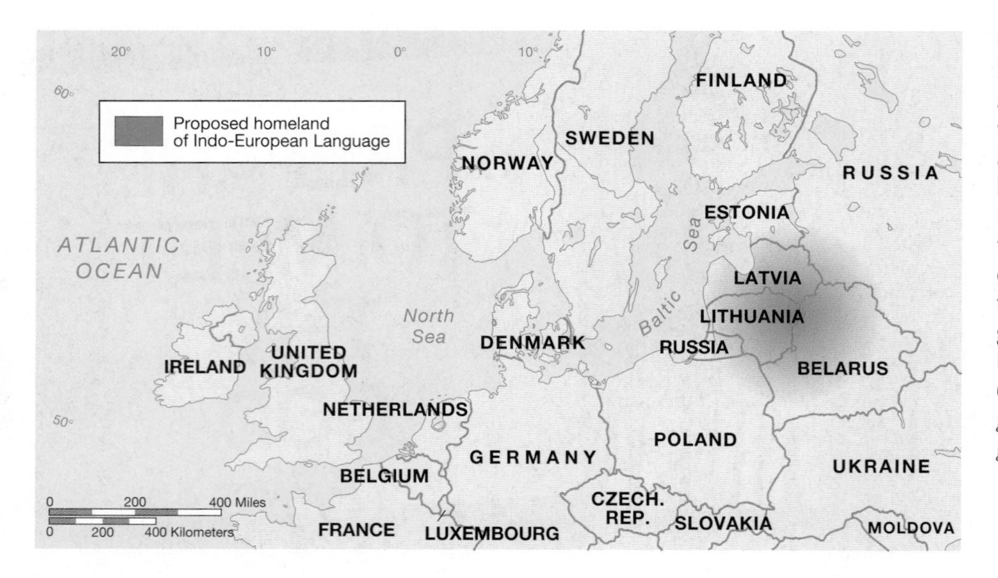

Figure 5.19 Source area of the Indo-European language family Although scholars are not certain where the Indo-European language arose, new evidence suggests that it blossomed in northeast central Europe, north of the Black Sea, in the fifth millennium B.C. As the language diffused west, east, and south, it mutated into the multiplicity of languages that today make up the Indo-European language family. (After C. Renfrew, *Archaeology and Language: The Puzzle of Indo-European Origins,* 1987. London: Cape.

in their prospects. A guard, sweeper, cook, or driver who speaks only Hindi or Urdu will likely do the same work all his life. In contrast, those who can speak English—by definition the upper middle classes—are able to practice their profession or do business in any region of their country or in most parts of the world.

With globalization, the geography of language has become even more dynamic. As mentioned, the plethora of languages and dialects in a region like South Asia makes communication and commerce among different language speakers difficult. It may, furthermore, create problems for governing a population. That is why developing states created one national language to facilitate communication and enable the efficient conduct of state business. In general, where official languages are put into place, indigenous languages may eventually be lost (**Figure 5.21**). Yet the actual unfolding of globalizing forces—such as official languages—works differently in different places and in different times (**Figure 5.22**). The overall trend appears to be toward the loss of indigenous language (and other forms of culture), as **Figure 5.23** shows. It is also important to recognize, however, that religion, language, and other forms of cultural identity can also be a means of challenging the political, economic, cultural, and social forces of globalization as they occur in France, Spain (the Basque separatist movement), Canada (the Québecois movement), and other countries (see Box 5.4: "Separatism in Québec").

Culture and Society

Culture has an important impact on society through its influence on social organization. Social categories like kinship, gang, or generation, or some combination of categories, may figure more or less prominently in a group, depending upon geography. Moreover, the salience of these social categories may change over time as the group interacts with people and forces outside of its boundaries. For example, countries in the Middle East and North Africa have a complex social organization—as complex as that of any other region of the globe—shaped by cultural ties and meaning systems that highlight gender, tribe, nationality, kinship, and family. Global media technologies, such as satellite television and the Internet, are increasingly penetrating the region, however, and the potential for new social forms to emerge and old ones to be reconfigured is increasing. Generally, the predominant forms of social organization in the region have persisted for hundreds of years. It would be incorrect, however, to assume that both subtle and dramatic changes within these forms have not already occurred.

Kinship is a form of social organization that is particularly central to the culture system of the Middle East and North Africa. **Kinship** is normally thought of as a relationship based on blood, marriage, or adoption. This definition needs to be expanded however, to include a *shared notion of relationship* among members of a group. The point is that not all kinship relations are understood by social groups to be exclusively based on biological or marriage ties. While biological ties, usually determined through the father, are important in the Middle East, they are not the only important ties that link individuals and families. In fact, though kinship is often expressed as a "blood" tie among social groups throughout the Middle East, neighbors, friends, even individuals with common economic or political interests are often considered kin. Kinship is such a valued relationship for expressing solidarity and connection that it is often used to assert a feeling of group closeness and as a basis for identity even where no "natural" or "blood" ties are present. This notion of kinship might also be seen displayed among fraternity "brothers," sorority "sisters," gang members, and even police officers and firefighters who feel a strong familial bond with coworkers.

In the Middle East and North Africa, kinship is even an important factor in shaping the spatial relationships of the home, as well as outside the home, determining who can interact with whom and under what circumstances. This is especially the case for the interaction of

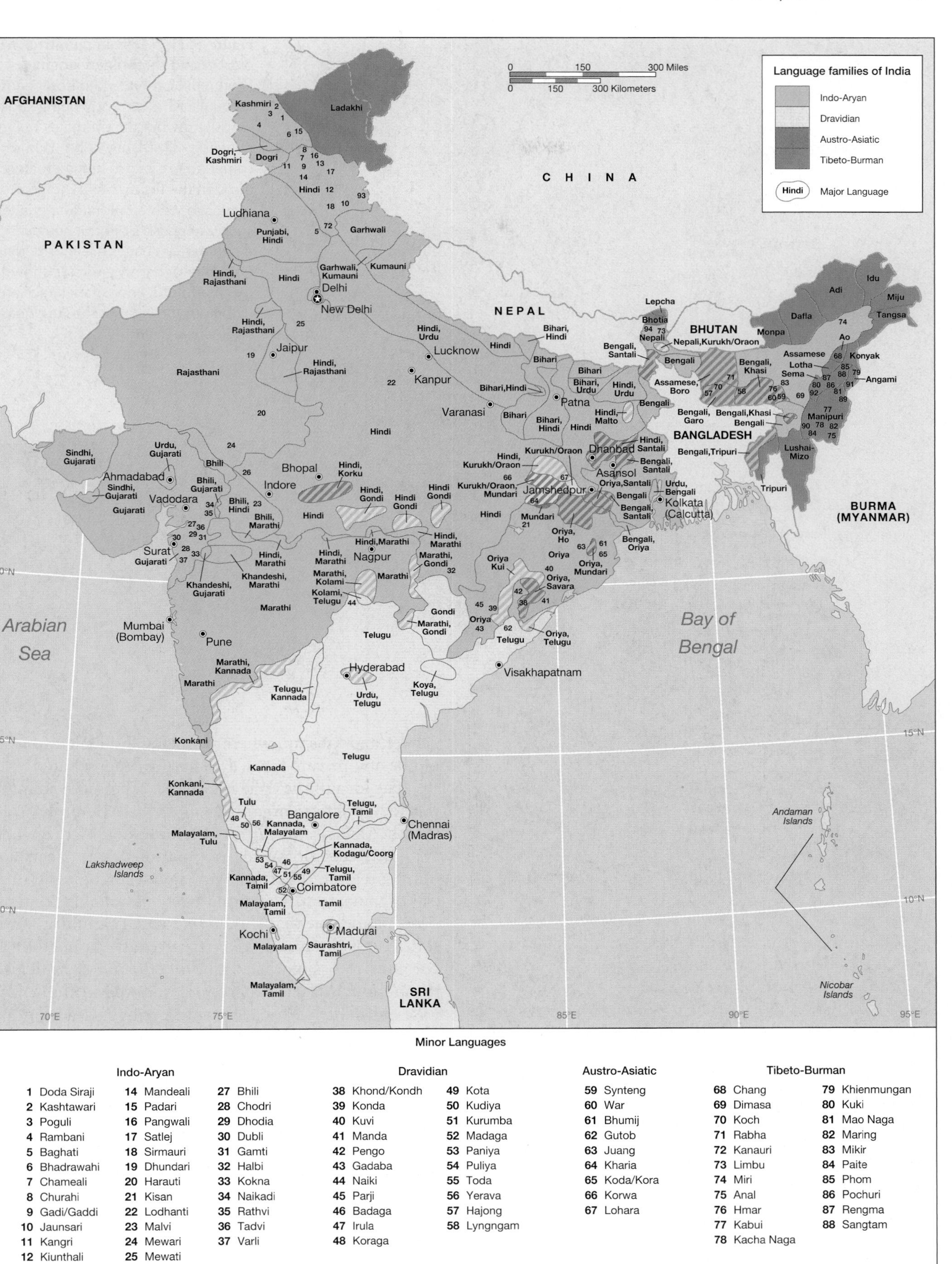

Figure 5.20 Language map of India India's linguistic landscape is complex with hundreds of distinct languages in use. This map provides an illustration of the intricate geography of language on the subcontinent. Included here are the most ? languages . . .

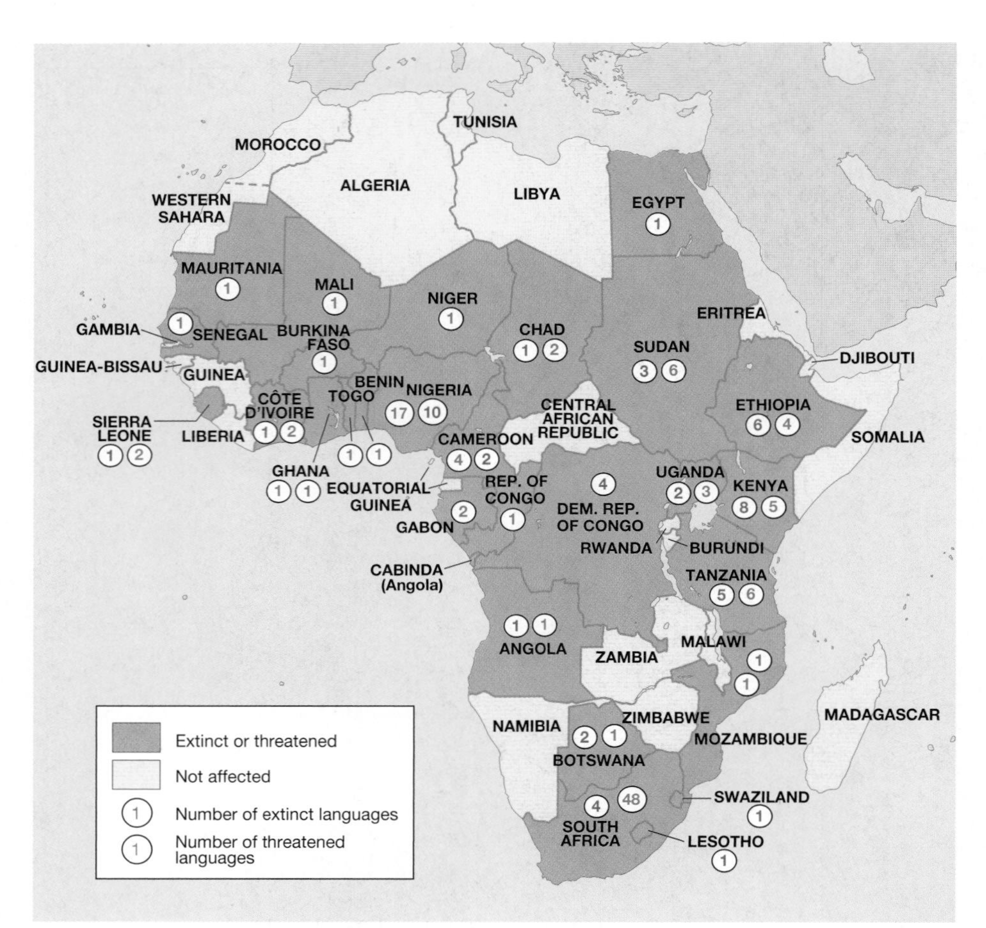

Figure 5.21 African countries with extinct and threatened languages It is not absolutely certain how many languages are currently being spoken worldwide, but the estimates range between 4,200 and 5,600. While some languages are being created through the fusion of an indigenous language with a colonial language, such as English or Portuguese, indigenous languages are mostly dying out. Although only Africa is shown in this map, indigenous languages are dying out throughout the Americas and Asia as well.

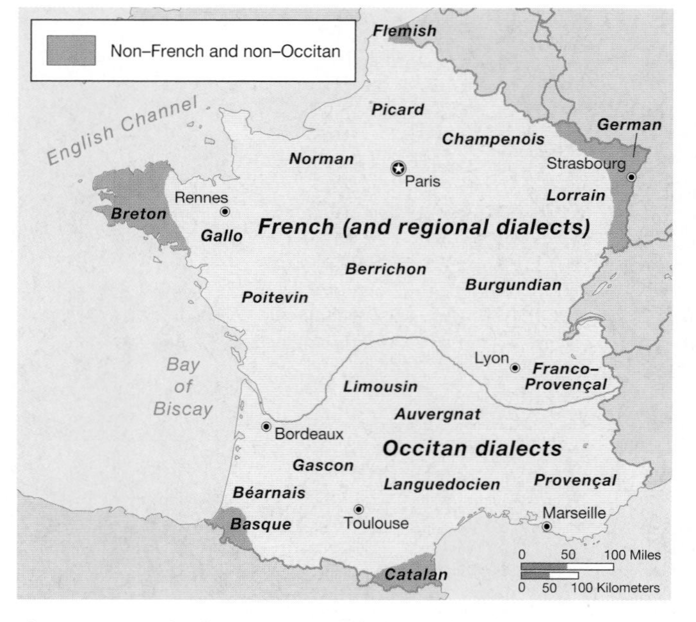

Figure 5.22 The languages and dialects of France in 1789 On the eve of the French Revolution, language diversity in France was not so dissimilar from other European regions that were consolidating into states. Whereas a multiplicity of local languages and dialects prevailed before the emergence of a strong central state, many governments created policies to eliminate them. Local languages made it difficult for states to collect taxes, enforce laws, and teach new citizens. (After D. Bell, "Lingua Populi, Lingua Dei," *American Historical Review,* 1995, p. 1406.)

gender and kinship, where women's and men's access to public and private space are sharply differentiated.

The idea of the tribe is also central to understanding the sociopolitical organization of the Middle East and North Africa, as well as other regions of the world. While tribally organized populations appear throughout the region, the tribe is not a widespread form of social organization. The term "tribe" is a highly contested concept and one that should be treated carefully. For instance, it is often seen as a negative label applied by the colonizers to suggest a primitive social organization throughout Africa. Where it is adopted in the Middle East and North Africa, however, tribe is seen not as a primitive form of social organization, but rather as a valuable element in sustaining modern national identity. Generally speaking, a **tribe** is a form of social identity created by groups who share a set of ideas about collective loyalty and political action. Tribes are grounded in one or more expressions of social, political, and cultural identities created by individuals who share those identities.

The result of shared tribal identity is the formation of collective loyalties that result in primary allegiance to the tribe. External groups may recognize the existence of these self-defined tribal groups and seek to undermine or encourage their persistence. For instance, in early twentieth-century Iran, the state ruthlessly and systematically

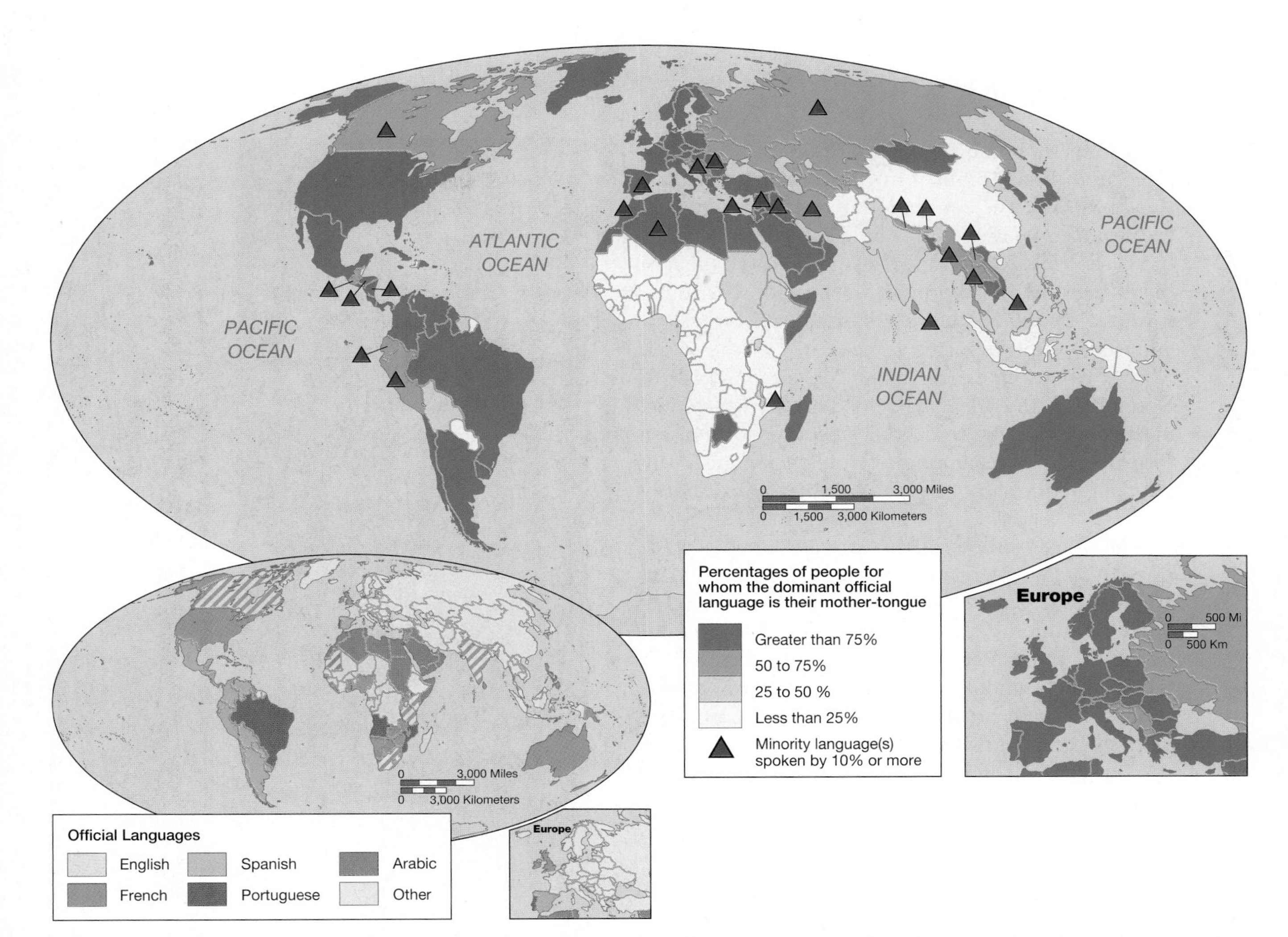

Figure 5.23 Tongue-tied Another way to examine the geography of language is to consider the proportion of people around the world whose mother tongue is not their country's official language. In places like China, where the dominant language is often not the mother tongue, problems of simple communication may pose considerable challenges. While Mandarin is the official language, imposed by the imperial Mandarin bureaucracy, there are six major regional dialects of Mandarin as well as another 41 distinct languages spoken by ethnic minorities in China. In other cases, the disparity between the official language and the mother tongue is the result of colonization. For instance, in South Africa before apartheid was dismantled, there were two official languages, English and Afrikaans—both languages of the colonizers—but over 50 indigenous languages. Since the election of Nelson Mandela in the early 1990s, the number of official languages has grown to 11. The increase gives residents more access to public information and is intended to expand participation in civil society as well as preserve the cultural diversity of the country. (*Source:* After *The New State of World Atlas,* 4th ed., by M. Kidron and R. Segal. Map projection, Buckminster Fuller Institute and Dymaxion Map Design, Santa Barbara, CA. The word *Dymaxion* and the Fuller Projection Dymaxion™ Map design are trademarks of the Buckminster Fuller Institute, Santa Barbara, CA, © 1938, 1967 & 1992. All rights reserved.)

attempted to eliminate tribal affiliations. In contrast, during the European colonial period in Sudan and Morocco, tribes were seen as forms of social organization that might inhibit independence movements, so they were largely promoted and supported by the colonial state.

Throughout the Middle East and North Africa and beyond, globalization is contributing to the transformation of existing culture systems. While some groups accept these changes, others attempt to resist and actively protest them in a whole host of ways, from cultural protection legislation to acts of vandalism and even terrorism.

ISLAMIC CULTURAL NATIONALISM

The protection of regional languages as a way of resisting globalization is just one part of a larger movement that interests geographers and other scholars. The movement, known as **cultural nationalism,** is an effort to protect regional and national cultures from the homogenizing impact of globalization, especially from the penetrating influence of U.S. culture. Figures **5.24** and **5.25** provide a picture of two widespread aspects of U.S. culture—films and television. Many other U.S. products also travel widely outside of U.S. borders (**Figure 5.26**). While many

5.2 GEOGRAPHY MATTERS

Changing Religious Practices in Latin America and the Caribbean

One of the main objectives of Spanish and Portuguese colonialism in Latin America and the Caribbean was the conversion of indigenous peoples to Catholicism. While some indigenous people fiercely resisted missionary efforts, others found ways to blend their own traditions with those of the Roman Catholic Church. The process of conversion was facilitated by the reported appearance of the Virgin Mary of Guadalupe to an Indian convert in Mexico on December 9, 1531, leaving behind her brown-skinned image on his mantle, and by the efforts of some priests to protect local communities from the Spanish government's efforts to obtain land, tribute, and labor by force.

The slave trade brought African religious traditions to Latin America and the Caribbean, and these eventually merged with indigenous and Catholic beliefs to form Candomble and Umbanda in Brazil, Voodoo in Haiti, and Santería in Cuba and other islands (**Figure 5.D**). Candomble and Umbanda are both sects of Santería, with rituals that involve dances, offerings of candles and flowers, sacrifice of animals such as chickens, and mediums and priests who use trances to communicate with spirits that include several Catholic saints. Voodoo (also spelled Voudou, Voudun, and Voudoun) rituals include drumming, prayer, and animal sacrifice to important spirits based on traditional African gods and Catholic saints and are led by priests who act as healers and protectors against witchcraft. Santería is closely connected to the Yoruba religion of West Africa and blends saints with African spirits associated with nature, using rituals similar to other Latin American religions.

Figure 5.D Santería altar This altar in Brazil illustrates how Santería combines animistic elements with Catholic religious symbols.

In the 1970s liberation theology emerged, a new form of Catholic practice that focused on the poor and disadvantaged. It was informed by the perceived preference of Jesus for the poor and helpless and by the writings of Karl Marx and other revolutionaries on

products of U.S. culture are welcomed abroad, many others are not. France, for example, has been fighting for years against the "Americanization" of its language.

Nations can respond to the homogenizing forces of globalization and the spread of U.S. culture in any number of ways. Some groups attempt to seal themselves off from undesirable influences. Other groups attempt to legislate the flow of foreign ideas and values, as in some Muslim countries.

After Christianity, Islam possesses the next largest number of adherents worldwide—about 1 billion. The map in **Figure 5.27** shows the relative distribution of Muslims throughout Europe, Africa, and Asia; **Figure 5.28** shows the heartland of Islamic religious practice.

The Islamic world includes very different societies and regions, from Southeast Asia to Africa. Muslims comprise over 85 percent of the populations of Afghanistan, Algeria, Bangladesh, Egypt, Indonesia, Iran, Iraq, Jordan, Pakistan, Saudi Arabia, Senegal, Tunisia, Turkey, and most of the newly independent republics of Central Asia and the Caucasus (including Azerbaijan, Turkmenistan, Uzbekistan, and Tajikistan). In Albania, Chad, Ethiopia, and inequality and oppression. This new orientation to the poor was espoused by the Second Vatican Council, called by Pope John XXIII in 1962. Priests preached grassroots self-help to organized Christian-based communities and often spoke out against repression and authoritarianism. Some were murdered by powerful interests who saw liberation theology as revolutionary and communistic. For example, Archbishop Oscar Romero in El Salvador was shot to death while saying Mass on March 24, 1980.

In recent decades evangelical Protestant groups with fundamentalist Christian beliefs have grown and spread rapidly in Latin America and the Caribbean. **Figure 5.E** shows the current distribution of these groups. Their message of literacy, education, sobriety, frugality, and personal salvation has become very popular in many rural areas. The conversion of Latin Americans to Protestant faiths grows stronger each year. Today over 15 percent of Latin Americans belong to evangelical churches, with the Latin American Catholic Bishops Conference claiming that 8,000 Latin Americans convert to evangelical Christianity every day. In Brazil, where Evangelical Protestantism is second only to Catholicism in terms of number of adherents, half a million Catholics leave the church per year. In Mexico, there has been a 10 percent decline in the Catholic population since the mid-twentieth century. Observers of evangelical missionizing in Latin America believe that fundamentalist missionaries and ministers are younger and more numerous than Catholic priests and therefore provide a higher minister-to-worshiper ratio than the Catholic priest-to-worshiper ratio. For example, in Mexico, each evangelical pastor serves 230, while each priest serves 8,600. Evangelical churches may also be more involved in indigenous communities than the Catholic churches.

There is also a strong thread of anti-clericalism in many Latin American countries. This partly originated as a reaction to the authority given by the Pope to the Spanish crown, the control of land and labor by the missions, and the Catholic Church's alliance with the landowners and political leaders in colonial and post-independence conservative administrations. After the Mexican Revolution, for example, there was for many years a strict separation of Church and state, with priests forbidden to wear their clerical garb on the street.

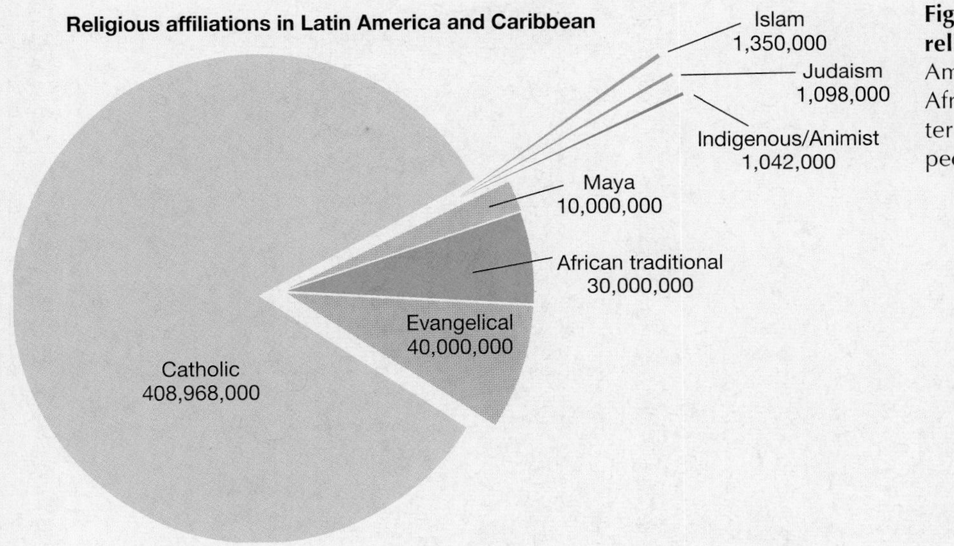

Figure 5.E Latin America and Caribbean religions Although the majority of Latin Americans are Catholic, evangelical and African traditional religions, such as Santería, are also important to millions of people.

Nigeria, Muslims make up 50 to 85 percent of the population. In India, Burma (Myanmar), Cambodia, China, Greece, Slovenia, Thailand, and the Philippines, significant Muslim minorities exist.

Islam is an Arabic term that means "submission," specifically submission to God's will. A **Muslim** is a member of the community of believers whose duty is obedience and submission to the will of God. As a revealed religion, Islam recognizes the prophets of the Old and New Testaments of the Bible, but Muhammad is considered the last prophet and God's messenger on Earth. The Qur'an, the principal holy book of the Muslims, is considered the word of God as revealed to Muhammad by the Angel Gabriel beginning in about A.D. 610. There are two fundamental sources of Islamic doctrine and practice: the Qur'an and the Sunna. Muslims regard the Qur'an as directly spoken by God to Muhammad. The Sunna is not a written document, but a set of practical guidelines to behavior. It is effectively the body of traditions that are derived from the words and actions of the prophet Muhammad.

While Islam holds that God has four fundamental functions—creation, sustenance, guidance, and judgment—

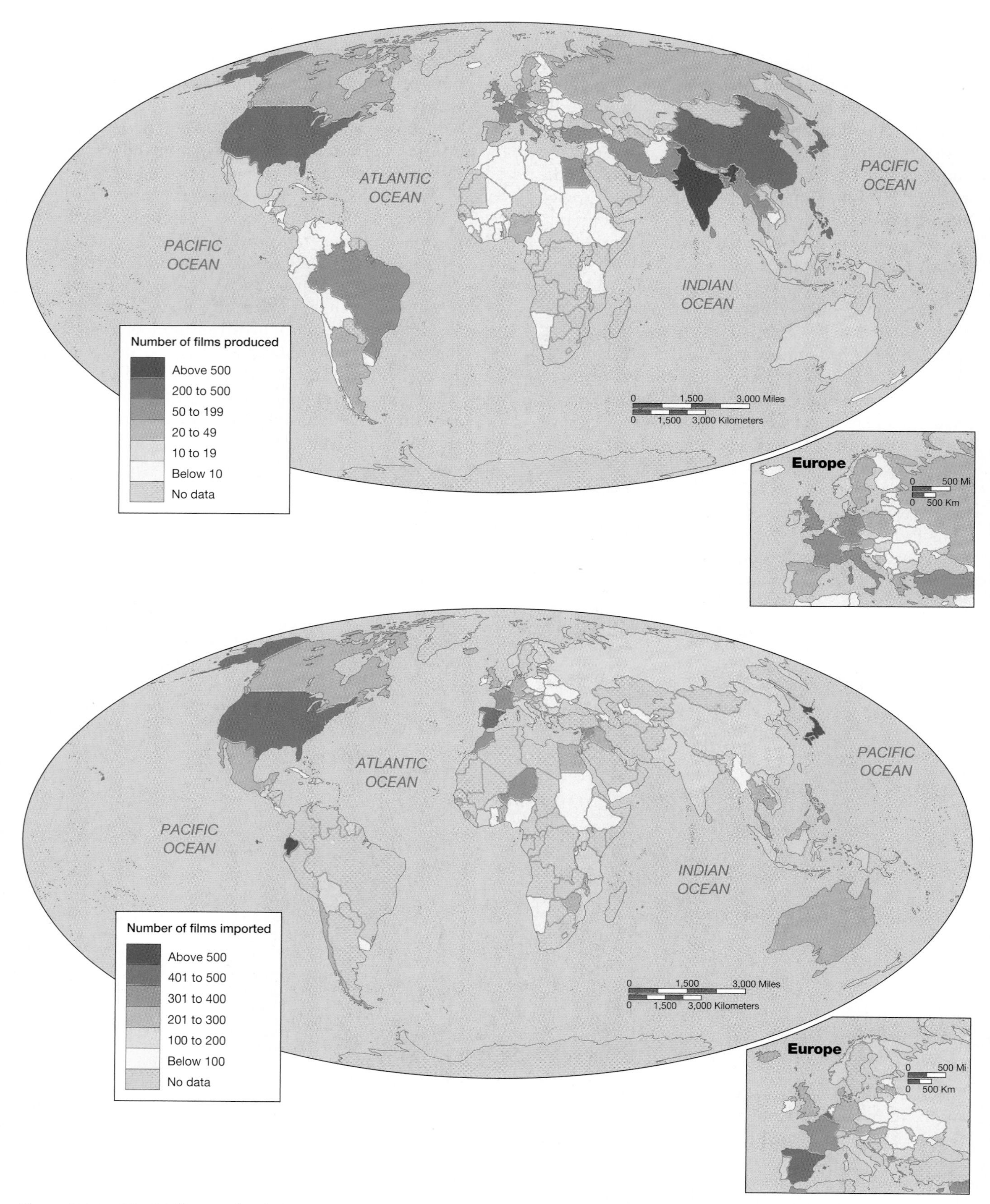

Figure 5.24 World film production, 2001 This map shows the number of films produced by different countries around the world. As expected, core countries are much more likely to be film producers. They also import fewer films as a result. For instance, the United States produces four times more films than it imports. In the semiperiphery, India, a country with an enormously popular film industry, produces eight times more films than it imports. Despite popular beliefs, the United States is not the largest film producer in the world. India far exceeds U.S. production and China and Hong Kong as well as the Philippines are also film production giants (After UNESCO, *Survey on National Cinematography,* 2001 available at http://www.unesco.org/culture/industries/cinema/html_eng/prod.shtml. *World Culture Report,* 2000, table 4: http://www.unesco.org/culture/worldreport/html_eng/stat2/table4.pdf. All rights reserved.)

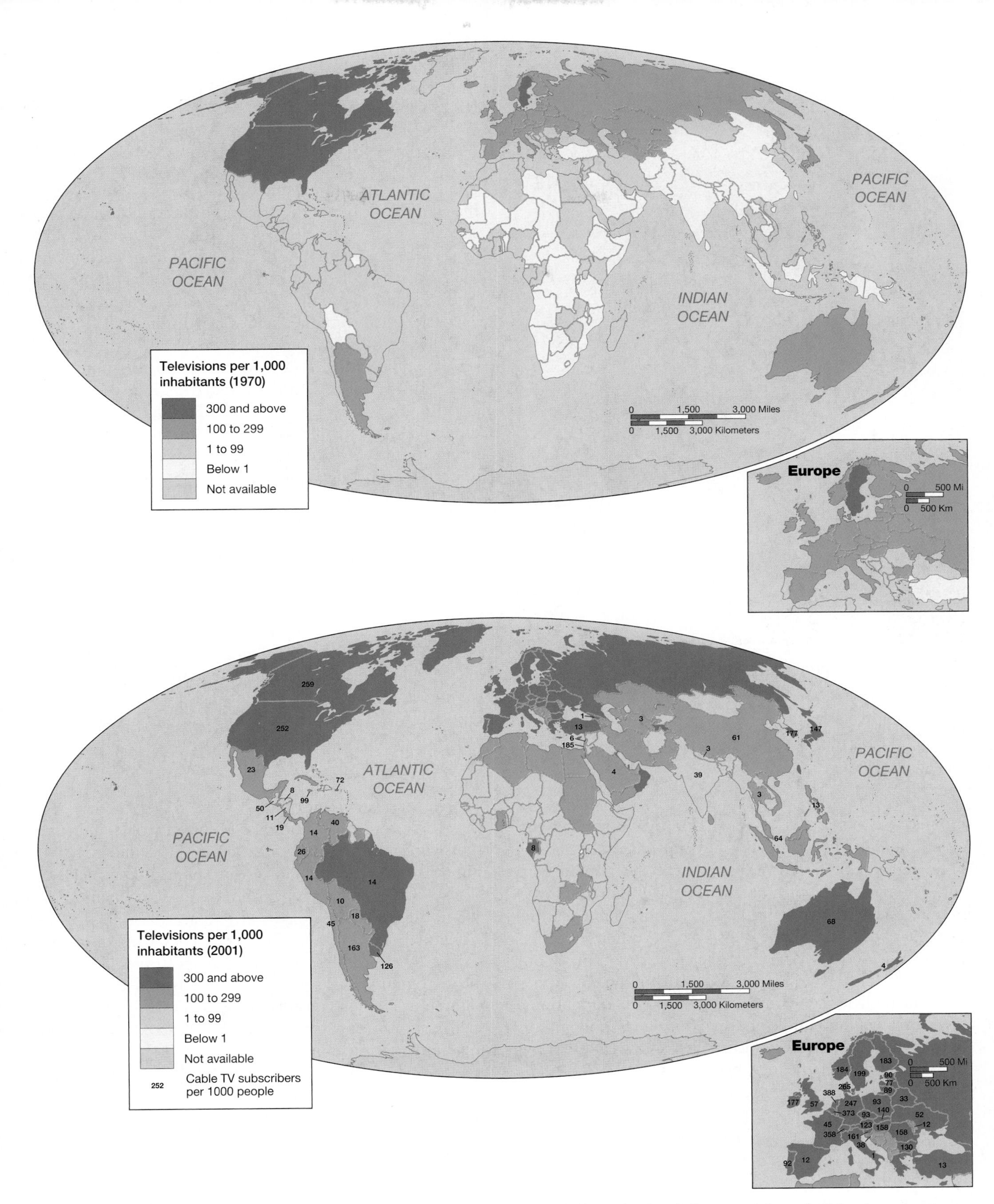

Figure 5.25 World distribution of TV sets and cable subscribers, 1970 and 2001 The ability to receive television broadcasts no matter who is exporting or producing them depends, of course, on access to a television set. In some areas of the periphery—for example, in some of the wealthier South American countries—there are over 300 television sets per 1,000 people. In much of sub-Saharan Africa, however, there are fewer than 10 sets for every 1,000 people. In the core, in countries such as the United States and Japan, there is on average one color television set per household, as well as a high level of ownership of videocassette recorders. Cable subscription opens up a market to even more commercial television products. With respect to television-programming exports, the core vastly overwhelms the periphery in the export of television programs, with the leaders being the United States and the United Kingdom. (After J. M. Rubenstein, *The Cultural Landscape: An Introduction to Human Geography,* 6th ed., © 1999. Update from World Bank, *World Development Indicators 2002,* World Bank: Washington, DC.)

Figure 5.26 U.S. products in Africa Western products like Nike athletic shoes, cell phones, and Coke have increasingly penetrated the markets of peripheral countries, oftentimes displacing native products. Coke is so keen to encourage the wide distribution of its product that it works to tailor the formula to local tastes.

the purpose of people is to serve God by worshiping him alone and adhering to an ethical social order. The actions of the individual, moreover, should be to the ultimate benefit of humanity, not the immediate pleasures or ambitions of the self. There are five primary obligations, known as the five pillars of Islam, that a Muslim must fulfill: repeating the profession of the faith ("There is no god but God; Muhammad is the messenger of God"); praying five times a day facing Mecca; giving alms or charitable donations; fasting from sunup until sundown during the holy month of Ramadan; and making at least one pilgrimmage, or **hajj,** to Mecca if financially and physically able.

The emergence and spread of Islam are linked to the commercial history of the Middle East and North Africa. The geographical origin of Islam is Mecca, in present-day Saudi Arabia. When Islam first emerged, Mecca, where Muhammad was born in A.D. 570, was an important node in the trade routes that at first connected Yemen and Syria and eventually linked the region to Europe and all of Asia. Today Mecca is the most important sacred city in the Islamic world, as well as an important commercial center. Eventually Medina also became an important sacred city because it was the place to which Muhammad fled when he was driven out of Mecca by angry merchants who felt his religious beliefs threatened their commercial practices.

Disagreement over the line of succession from the prophet Muhammad occurred shortly after his death in 632 and resulted in the split of Islam into two main sects, the Sunni and the Shi'i. The Sunni faction argued that the clergy should succeed Muhammad, while the Shi'a argued that Muhammad's son Ali should succeed his father. Ali was killed, and the Sunnis became dominant. They remain the mainstream branch of Islam, but the pattern varies from one country to another. The majority of Iran's 60 million people follow Shi'i, the official state religion of the Islamic Republic of Iran, founded in 1979. The majority of Iraq's population is also Shi'i. It is also important to keep in mind that Islam is practiced differently in many different locales throughout the Middle East and North Africa and that Muslims who have migrated from the region—to Europe and the United States, for instance—are shaped by, and shape the practice of, Islam in the Middle East.

Perhaps one of the most widespread cultural counterforces to globalization has been the rise of Islamism, more popularly, although incorrectly, known as Islamic fundamentalism. Whereas fundamentalism is a general term that describes the desire to return to strict adherence to the fundamentals of a religious system, **Islamism** is an anticolonial, anti-imperial, and generally anticore political movement. In Muslim countries, Islamists resist core, especially Western, forces of globalization—namely modernization and secularization. Not all Muslims are Islamists, although Islamism is the most militant movement within Islam today.

The basic intent of Islamism is to create a model of society that protects the purity and centrality of Islamic precepts through the return to a universal Islamic state—a state that would be religiously and politically unified. Islamists object to modernization because they believe the corrupting influences of the core place the rights of the individual over the common good. They view the popularity of Western ideas as a move away from religion to a more secular (nonreligious) society. Islamists desire to maintain religious precepts at the center of state actions, such as introducing principles from the sacred law of Islam into state constitutions.

Another important aspect of the Islamist movement is the concept of **jihad,** which is a sacred struggle. When this struggle is violently directed against the enemies of Islam, jihad is understood to be a holy war. But jihad can also be a more peaceful struggle to establish Islam as a universal religion through the conversion of nonbelievers. One example of jihad today is the struggle of Shi'ite Muslims for social, political, and economic rights within Sunni-dominated Islamic states.

As popular media reports make clear, no other movement emanating from the periphery is as widespread and has had more of an impact politically, militarily, economically, and culturally than Islamism. Yet Islamism—

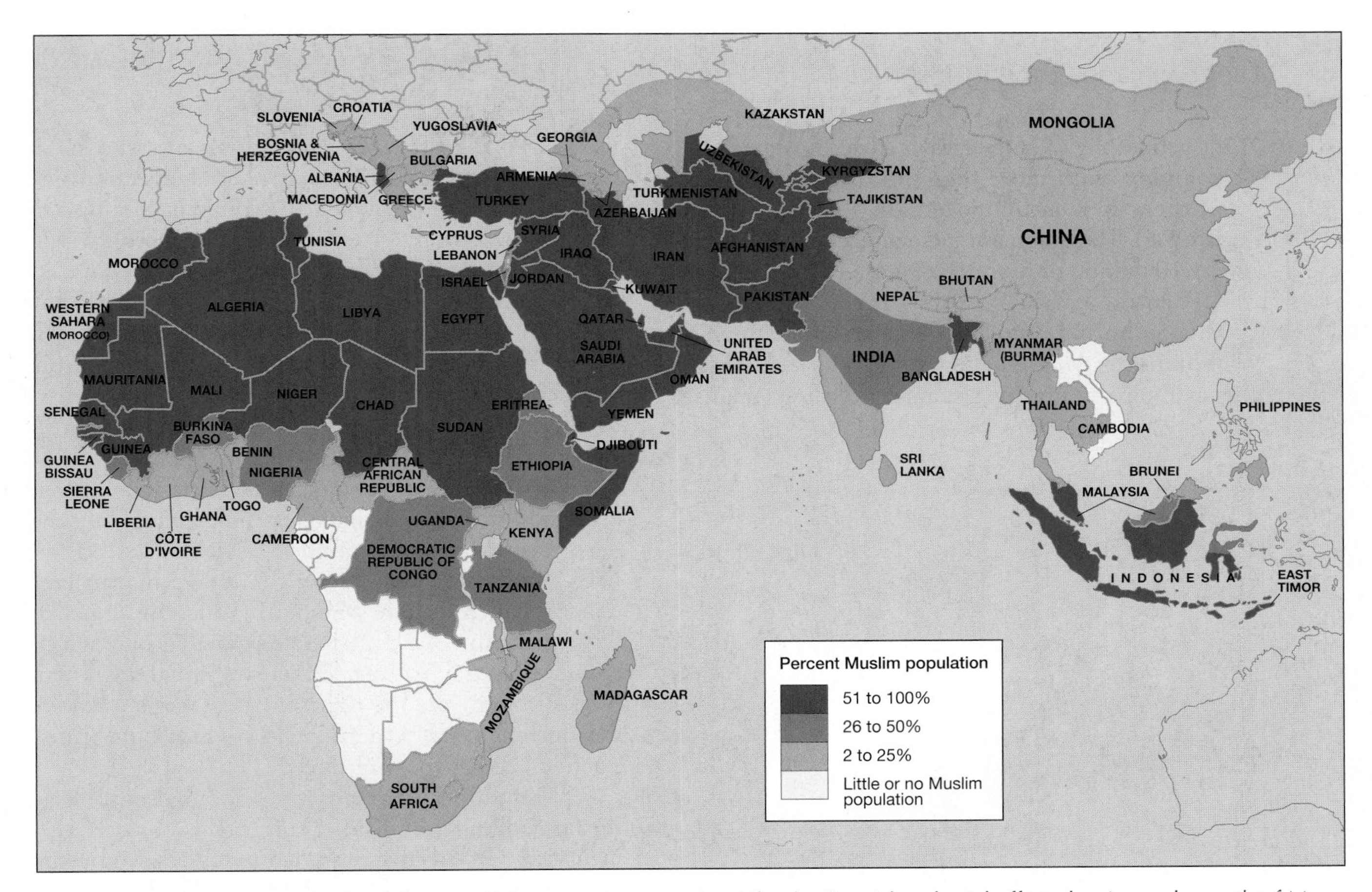

Figure 5.27 Muslim world The diffusion of Islam is quite extensive. Like the Spanish colonial effort, the rise and growth of Muslim colonization were accompanied by the diffusion of the colonizers' religion. The distribution of Islam in Africa, Southeast Asia, and South Asia that we see today testifies to the broad reach of Muslim cultural, colonial, and trade activities. (After D. Hiro, *Holy Wars*. London: Routledge, 1989.)

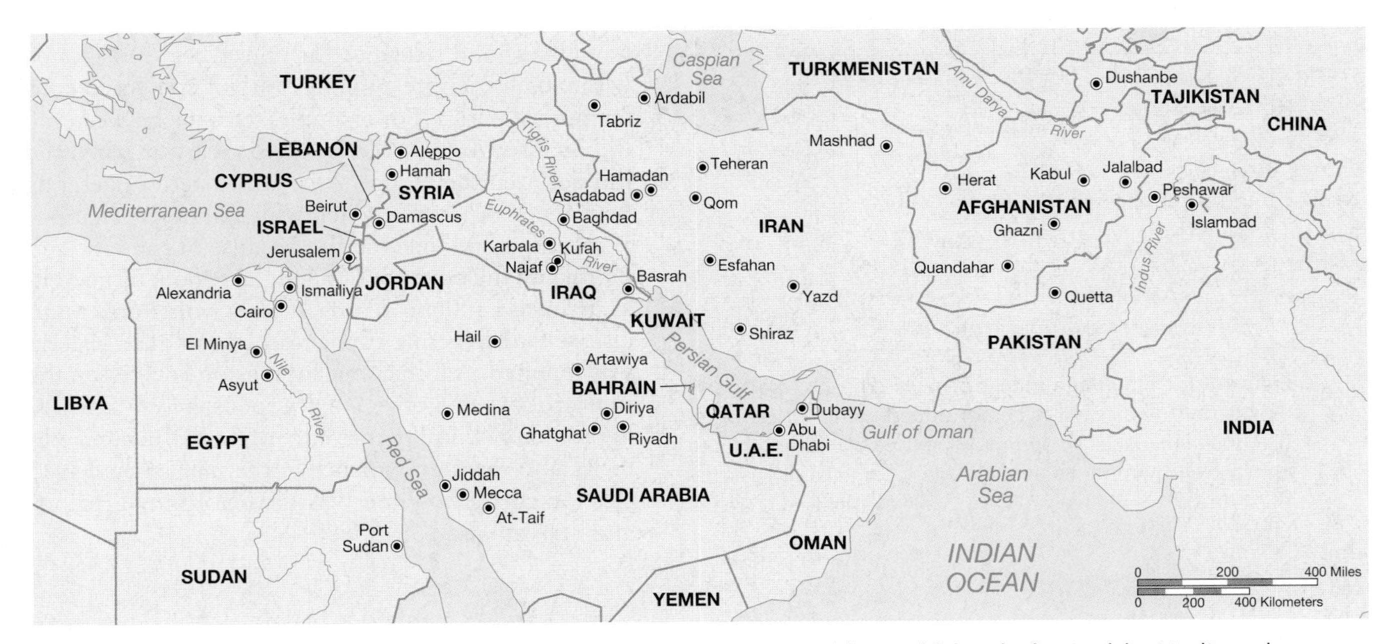

Figure 5.28 Cultural hearth of Islam Islam has reached into most regions of the world, but the heart of the Muslim culture remains the Middle East, the original cultural hearth. It is also in this area that Islamism is most militant. (After D. Hiro, *Holy Wars*. London: Routledge, 1989.)

GEOGRAPHY MATTERS 5.3

Language and Ethnicity in Africa

The geography of languages in Africa is remarkably complex, with more than 800 living languages, 40 of them spoken by more than 1 million people (**Figure 5.F**). The dominant indigenous languages, spoken by 10 million or more, are Hausa (the Sahel), Lingala (Central Africa), Swahili (East Africa), and Tswana-Sotho and Zulu (southern Africa). Hausa and Swahili are trade languages, spoken as second languages by many groups to facilitate trade. English, French, Portuguese, and Afrikaans are also spoken in regions of recent colonial control and education systems or white settlement, and Arabic is common in north Africa. Arabic has strongly influenced Swahili along the east coast of Africa. Because most countries have no dominant indigenous African language, they have often chosen a European language for official business and school systems. The countries with the most coherent overlap between their territory and a dominant African language are Somalia (Somali), Botswana (Tswana), and Ethiopia (Amharic).

The indigenous languages of Africa have been grouped into larger language families, including the Afro-Asiatic languages of north Africa, including Somali, Amharic, and Tuareg; the Nilo-Saharan languages, which include Dinka, Turkana, and Nuer in East Africa; and the largest Niger-Congo group, which includes Hausa, Yoruba, Zulu, Swahili, and Kikuyu. A small family are the Khoisan languages spoken by the Bushmen of southern Africa, which have a distinctive "click" vocalization.

The multiplicity of languages and dialects reflects the large number of distinct cultural or ethnic groups in Africa. Groups are often led by chiefs, and some are monarchies with kings or queens. The largest ethnic groups in Africa are associated with certain dominant languages, such as Hausa, Yoruba, and Zulu, but almost all groups, however large or small, were either split geographically by colonial national boundaries or grouped with neighbors with whom they may have had no affinity.

Attempts to consolidate ethnic groups across boundaries and struggles for power between groups within countries are a major cause of conflict in contemporary Africa. For example, tensions between the Ibo and Yoruba in Nigeria led to civil war when the Ibo declared the independence of eastern Nigeria as Biafra in 1967. The conflict, which drew international attention and intervention because of starvation in Biafra and the presence of oil in the region, resulted in as many as a million deaths (mainly from hunger and disease) and lingering ethnic resentments after Nigeria was reunited. Another evident tension is between the ruling Kikuyu and other ethnic groups in Kenya, where opposition political parties have organized around ethnicity and threaten violence in the face of perceived election corruption and bias toward Kikuyu regions and individuals.

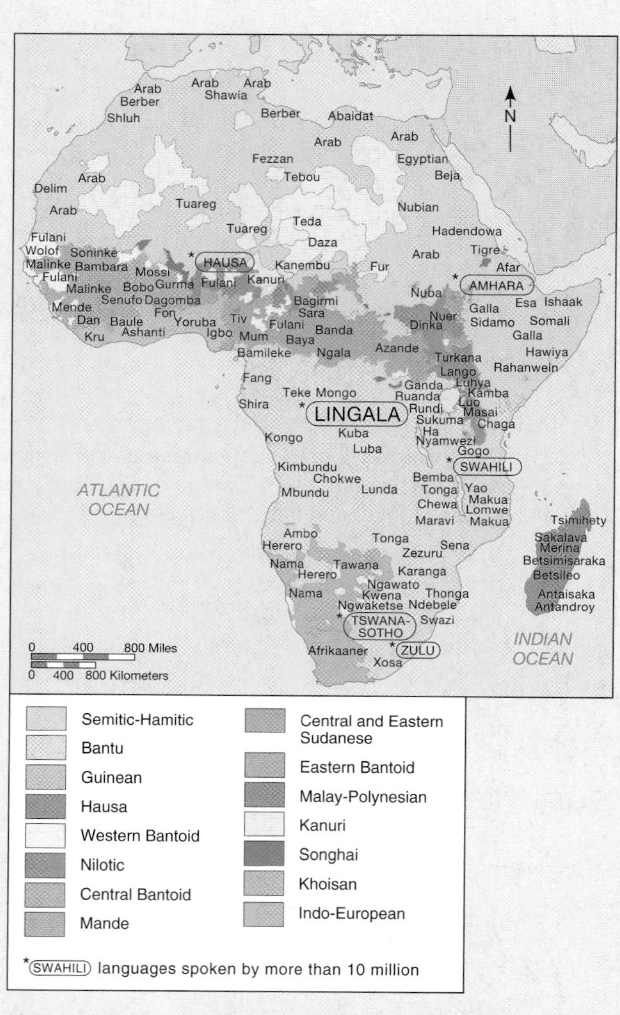

Figure 5.F Language map of Africa The cultural complexity of Africa is clearly demonstrated in the variety of indigenous languages shown in this map. Swahili and Hausa are spoken by millions as the trade languages of East and West Africa, respectively. (After S. Aryeetey–Attoh (ed.), *The Geography of Sub-Saharan Africa.*)

(a)

(b)

Figure 5.29 Muslim women in Afghanistan and Turkey (a) The fundamentalist, theocratic Taliban regime required women to be extremely conservatively dressed when appearing in public. These strictures were dramatically relaxed when the new government took power in early 2002. Women may now exercise a great deal of choice with respect to how to dress in public. Some wear simple head scarves, while others, such as those shown here shopping at a street market in Mazar-e-Sharif, in the northern part of the country, still wear burquas. (b) Three young women stroll down a street in Istanbul. Although Turkey is 99 percent Muslim, it has a secular democratic government and a population with generally liberal attitudes toward the West, especially in urban areas. While some older women and women in the countryside prefer to appear in public with head coverings, much of the younger generation, especially middle- and upper-class urban women, have adopted Western styles of dress and overall appearance.

a radical and sometimes militant movement—should not be regarded as synonymous with the practices of Islam, any more generally than Christian fundamentalism is with Christianity. Islam is not a monolithic religion, and even though all adherents accept the basic pillars, specific practices vary according to the different histories of countries, nations, and tribes. Some expressions of Islam allow for the existence and integration of Western styles of dress, food, music, and other aspects of culture, while others call for the complete elimination of Western influences (**Figure 5.29**).

CULTURE AND IDENTITY

In addition to exploring cultural forms, such as religion and language, and movements, such as cultural nationalism, geographers have increasingly begun to ask questions about other forms of identity. This interest largely has to do with the fact that certain long-established and some more recently self-conscious cultural groups have begun to use their identities to assert political, economic, social, and cultural claims.

Sexual Geographies

One identity that has captured the attention of cultural geographers is sexuality. **Sexuality** is a set of practices and identities that a given culture considers related to each other and to those things it considers sexual acts and desires. One of the earliest and most effective examples of the geographic study of sexuality is the examination of the spatial expression of prostitution. Research on prostitution in California found that sex work—as it is now more commonly known—is spatially differentiated based upon the target clientele as well as systems of surveillance. Female sex workers who service an upper income clientele generally perform their work in the homes, hotels, and other private areas of upper and middle-class society. Female sex workers serving a lower class clientele tend to operate within more public spaces or bring clients to their homes or hotel rooms in the locales where they are "streetwalking."

Not surprisingly, female sex workers at the lower end of the economic spectrum are more likely to be subject to police surveillance and other forms of public scrutiny because their advertising and work activities are largely carried out in public places. In contrast, those oriented toward a wealthier clientele have, in most cases, little or no obvious public identity. The distinction should be clear with respect to geography. The "red-light districts" of most cities are widely known to residents. But sex workers who serve a higher income clientele do not restrict themselves to segregated workspaces.

More typical of contemporary work on sexuality in geography is that which explores the spatial constraints on homosexuality and the ways in which lesbian, gay, bisexual, and transgendered (LGBT) people respond to and reshape them. Two particular areas of research have emerged: gay and lesbian consumerism, and the importance of the body as a site of performance of sexuality and gender identities.

5.4

WINDOW ON THE WORLD

Separatism in Québec

The political movement in the province of Québec that seeks to separate Québec from the rest of Canada is premised on a deep desire to preserve and enrich Québecois culture. The movement to separate from the rest of Canada in order to maintain a distinct French-Canadian identity can be traced back to the early history of colonial Canada, when the British defeated the French and secured possession of the entire Canadian territory in 1763. As the only "vanquished" European settlers in North America, the Québecois are undeniably unique.

Until Québec experienced its so-called "Quiet Revolution" in the 1960s, when a separatist political party—the Parti Québecois—emerged, French-Canadian nationalism was a largely conservative movement. With the French-Canadian family and Roman Catholic religion at its core, Québec oriented itself around clerical, rural, and agricultural values. Its nationalism was backward-looking and traditional. While numerous scholars have advanced any number of explanations for the rise and persistence of French nationalism in Québec, the most compelling one from our perspective is based on its peripheral status in the national and the global economy.

As the Canadian economy became increasingly integrated into the global economy and, more particularly, into the U.S. economy, Québec found itself less able than previously to benefit from U.S. hegemony. Largely because its identity was based on traditional ideas and practices, Québec was ill-equipped to participate in the revolutionary changes transpiring around it. As western Canada's economic power increased with the increasing importance of the energy sector, and as Toronto transformed itself based on the growing service sector, enabling it to become the financial and industrial capital of Canada, Québec was left behind.

Recognizing that these changes posed a major threat to the social, economic, and political survival of Québec's cultural identity, Québec's elites saw a way out of peripheralization by opting out of the Canadian national project altogether. Seizing control of the provincial government of Québec, the elites were able to translate their vision for a separate Québec by creating a polity that wished not simply to try to hold on to the old ways, but to achieve complete independence in order to determine their own path through the new economy.

The desire of the Québec separatists to opt out of their regional/provincial status and become an independent nation with a distinct identity became translated into provincial referenda on independence. In both 1980 and 1985 the separatists forced votes that, if they had passed, would have signaled the desire of residents of the province to separate politically from Canada. Both times the referenda failed. In 1995 another referendum was on the ballot asking voters if they wished to stay as part of Canada; in this referendum, "yes" votes to remain in Canada exceeded "no" votes by a very small margin (**Figure 5.G**).

It is too soon to tell whether Québec will push once more for independence from Canada. There are signs that the province is actually becoming less interested in separation, given its increasing prosperity though incorporation into the global economy. Globalization and Québec's growing independent relation-

The open expression of sexual preference in consumption has become fully developed in a phenomenon called "pink spending." Use of the word *pink* is a direct reference to the pink triangle Hitler required homosexuals to wear in Nazi concentration camps. Consumer support by LGBT consumers of openly gay businesses has led to the establishment of identifiably gay spaces in the form of shopping districts and neighborhoods. These openly gay spaces have enabled gay communities in many core countries to gain significant political power in cities, regions, and even nationwide. The gay-pride parades that occur in major U.S. cities and elsewhere around the world each year are also a reminder of the demographic and political power of gays and lesbians. The parades exemplify attempts by homosexuals to resist the dominant ideology about sexuality and to occupy the spaces of cities on an equal footing with heterosexuals (**Figure 5.30**).

Given the emphasis that the forces of globalization have placed on commodities and consumption in the contemporary world, it is not surprising that gays and lesbians constitute a market for advertisers, retailers, clubs, and restaurants. Rejecting tourist destinations that may emphasize nuclear-family fun or romantic heterosexual vacation spots, gays and lesbians can consume different and specially marketed alternative destinations. Global gay tourism guides explicitly market places like Amsterdam in Europe, San Francisco in the United States, and Rio de Janeiro in Brazil as "gay capitals," offering alternative places of tourism for gay consumers.

A second approach to understanding homosexuality, particularly as it is represented in contrast to hetereosex-

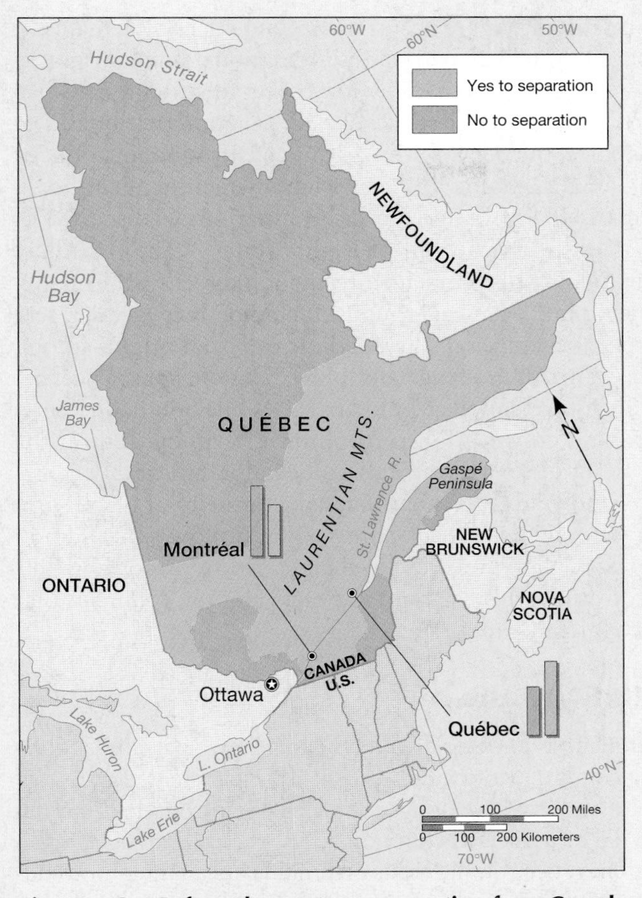

Figure 5.G Referendum vote on separation from Canada The distribution of the vote reflects the ethnic division of upper and lower Québec. While lower Québec is urbanized, with overwhelming numbers of French-speaking Canadians, upper Québec is largely rural and contains mostly non-Francophone communities.

ship with the United States have begun to influence its desire for nation-state status. Some scholars believe that rather than undermining Québecois national identity, NAFTA and Canada's increasing connections to Europe have enhanced it. In short, there is no clear answer at this point to the question of whether Québec will remain a distinct society within Canada or become a new nation-state. What is clear is that the context for making that decision is rapidly changing, as changes in the Canadian state brought about by globalization have begun to open up ways for Québec to maintain its distinct identity, to prosper economically, and to control more of its own destiny without having to secede from Canada.

In fact, a recent poll shows that for the first time since the 1990s, 50 percent of the residents of Québec support sovereignty for the province. At the same time, however, two-thirds of them indicated they were proud to be Canadian and most still wanted the province to remain part of Canada. In short, it is sovereignty but not separatism that increasingly characterizes the political desire of more and more Québec residents. It would appear that the question of Québec's relationship to Canada hinges on the attitudes of different generations and time may change which of those attitudes will prevail. Those residents of Québec who opposed separation from Canada, or the "traditional federalists," were born before World War II and they are dying off. The baby boomers of the Quiet Revolution support sovereignty but wish to remain with Canada. It appears that today's young adults—the children of Bill 101, the landmark language law from 1977 that put French at the forefront of education—are likely to support independence from Canada. Time will tell whether the right conditions prevail for separation, when this generation holds the reins of power.

uality, is research and writing on the body and the ways in which individuals literally perform their sexual identities through clothing, attitude, body language, and grooming, such as hairstyles and body adornments. Research on the body has also been used to decode the social and cultural practices that express other kinds of social identity, such as gender and class. But the work has been most fully developed in its application to sexuality.

The central theoretical position of this research is that sexual identities are learned and performed through standard taken-for-granted practices that most people tend to take as "normal" or "natural." Using gender as a starting point and recognizing it as a key feature of all human identity that marks and defines us as "sexed" beings, researchers in this area contend that gender is not something we essentially are (because of a given set of physical characteristics), but something we do (something we enact by the way we present our bodily selves to the world). We may be born as biological females, but we learn how to act as women. An illustration of the argument that sexuality is something that we perform through gendered acts is the gay skinhead or the lipstick lesbian. Generally, the skinhead is socially understood to be a hypermasculine, heterosexual male. But the gay skinhead complicates this construction by challenging the idea that all hypermasculine men are straight.

Research on sexuality and the body in geography recognizes that space is central to our understanding. Where gay or straight identities can be performed plays a central role in who occupies those spaces, how they are occupied, and even what sexual identity you might be performing if you occupy one space or another. But issues of sexuality and space go beyond the performance of sexual identity to

Figure 5.30 Gay pride parade, São Paulo, Brazil Almost 2 million LGBT people and their supporters—many in lavish Carnival costumes and waving rainbow-colored flags—paraded in Brazil's biggest city in May 2005 to celebrate gay pride and call for the legalization of civil unions between homosexuals. In comparison, San Francisco's gay pride parade typically attracts tens of thousands of people, and the 2004 world gay pride day celebrations in Berlin attracted between 200,000 and 500,000 participants.

the emergence of new political cultures that are being constructed to protect the rights of lesbian, gay, and transgendered people, in particular national and international spaces.

With the increasing attention being paid to human rights issues around the globe, sexual rights have also been put onto the agenda. For example, the Department of Reproductive Health and Research of the World Health Organization provides a Web site on sexual health that includes a set of working definitions intended to provide a foundation for discussion about sexual health around the world. These working definitions include the right of all persons, free of coercion, discrimination and violence, to, among other things, choose their sexual partners, receive sexuality education, and pursue a satisfying, safe, and pleasurable sexual life. While these definitions are provisional and not the official policy of the World Health Organization, they provide some insight into the transformations that are occurring around issues of sexuality—especially as they pertain to international human rights.

"Rights discourses" with respect to sexuality are also extending to other spaces, besides the international. For instance, the new South African constitution, introduced in 1996, contains a Bill of Rights which explicitly protects LGBT people from discrimination. Clause 9(3) of the South African Bill of Rights states: "The state may not unfairly discriminate directly or indirectly against anyone on one or more grounds, including race, gender, sex, pregnancy, marital status, ethnic or social origin, colour, *sexual orientation,* age, disability, religion, conscience, belief, culture, language and birth." While South Africa's is the first constitution to prohibit discrimination on the grounds of sexual orientation, Ecuador, Fiji, and Portugal have now added similar language to theirs. It is too soon to tell whether other countries will follow suit and add sexual rights protection to their constitutions or how this protection will work out in practice in the countries that have already adopted it.

Ethnicity and the Use of Space

Ethnicity is another area in which geographers are exploring cultural identity. **Ethnicity** is a socially created system of rules about who belongs to a particular group based upon actual or perceived commonalities, such as language or religion. A geographic focus on ethnicity is an attempt to understand how it shapes and is shaped by space, and how ethnic groups use space with respect to mainstream culture. For cultural geographers, territory is also a basis for ethnic group cohesion (see Chapter 9 for more on territory). For example, cultural groups—ethnically identified or otherwise—may be spatially segregated from the wider society in ghettos or ethnic enclaves (see Chapter 11). Or they may use space to declare their subjective interpretations about the world they live in and their place in it.

The use of the city streets by many different cultural groups demonstrates this point. Nineteenth-century immigrants to U.S. cities used the streets to broadcast their ideas about life in their adopted country. Ethnic parades in the nineteenth century—such as St. Patrick's Day for the Irish and Columbus Day for the Italians—were often very public declarations about the stresses that existed between and among classes, cultures, and generations.

An endless array of spaces besides city streets are also important to the making and remaking of ethnic identity, whether for a few hours or for generations. Ethnic restaurants are often places where individuals and groups sharing a common place of origin and cuisine come together to enjoy those foods as well as confirm their connections to each other and to a distant place. Even immigrant Web sites and blogs, where people far from their homeland can communicate with others from that same place, are virtual spaces where ethnic identity is made and remade through the words and ideas that are exchanged.

Geographers Maria Price and Courtney Whitworth have explored how Latino immigrants in Washington,

Figure 5.31 Chinatown, San Francisco Chinatowns have been portrayed as voluntary places chosen by immigrant Chinese. But they can also be seen as places of exclusion and racism created by the dominant society that prefers to segregate people of difference from the wider society.

D.C. create vibrant, though temporary, ethnic spaces through immigrant soccer leagues. They write: "Leagues create a cultural space that is familiar, entertaining, practical, inexpensive, transnational, and ephemeral, where immigrants gather to reaffirm their sense of identity and belonging."[2] They also make it clear that it is not just that the soccer field itself constitutes a vibrant ethnic space, but it is more so that the social networks that enable the leagues to function are themselves a kind of "in between" space that links the lived space of Washington, D.C. with the remembered space of their home country. Like Web sites that link distant immigrants through discussions about the home country and the pleasures and difficulties of living abroad, the social networks that sustain soccer leagues also create a transnational space of ethnic identity.

Race and Place

Geographers also use prevailing ideas and practices with respect to race to understand the shaping of places.[3] **Race** is a problematic classification of human beings based on skin color and other physical characteristics. **Racialization** is the practice of creating unequal castes where whiteness is considered the norm or unexceptional. Biologically speaking, however, no such thing as race exists within the human species. Yet consider the categories of race and place that correspond to "Chinese" and "Chinatown." Powerful Western ideas about Chinese as a racial category enabled the emergence and perpetuation of Chinatowns in many North American cities (**Figure 5.31**). In this and other cases, the visible characteristics of hair, skin, and bone structure made race into a category of difference that was (and still is) widely accepted and often spatially expressed.

The mainstream approach to neighborhood is to see it as a spatial setting for systems of affiliation more or less chosen by people with similar skin color. Yet cultural geographers have begun to overturn this approach and to see neighborhoods as spaces that affirm the dominant society's sense of identity. For example, from the perspective of white society, nineteenth-century Chinatowns were the physical expression of what set the Chinese apart from whites. The distinguishing characteristics revolved around the way the Chinese looked, what they ate, their non-Christian religion, opium consumption, gambling habits, and other "strange" practices. Place—Chinatown—maintained and manifested differences between white and Chinese society. Furthermore, place continues to be a mechanism for creating and preserving local systems of racial classification and for containing geographical difference within defined geographical confines. The homelands of South Africa and the dismantling of apartheid there also illustrate the interaction of race and place, though at a much larger scale.

What remains unexamined in the kinds of perceptions and practices that create racialized places as different as Chinatown and the homelands of South Africa is the taken for grantedness of whiteness. Whiteness is seen to be the norm or the standard against which all other visible differences are compared. But whiteness, as geographers Owen Dwyer and John Paul Jones, III, have shown, is itself a category of difference that depends on visible distinctions and not biological ones and is always constructed in relation to other categories. Recently, researchers in the humanities and social sciences have begun to "denaturalize" whiteness by investigating the way it has been constructed as a social category in different periods and places as well as the ways that whiteness operates in particular sites such as classrooms, on the street, and in boardrooms. It is only when we begin to challenge the naturalness of whiteness and the spaces that enable it, that we can truly begin to undo racist practices.

Gender and Other Identities

Gender is an identity that has received a great deal of attention by cultural geographers within the last two to three decades. **Gender** is a category reflecting the social differences between men and women. As with other forms of identity, it implies a socially created difference in power

[2]M. Price and C. Whitworth, "Soccer and Latino Cultural Space: Metropolitan Washington *Fútbol* Leagues," in *Hispanic Spaces, Latino Places: Community and Cultural Diversity in Contemporary America,* D. Arreola (ed.), pp. 167–186. Austin: University of Texas Press, 2004.

[3]Adapted from K. Anderson, "The Idea of Chinatown: The Power of Place and Institutional Practice in the Making of a Racial Category." *Annals of the Association of American Geographers* 77(4), 1987, pp. 580–598.

between groups. In the case of gender, the power difference gives an advantage to males over females and is not biologically determined, but socially and culturally created. Gender interacts with other forms of identity and can intensify power differences among and between groups. Furthermore, the implications of these differences are played out differently in different parts of the world.

For example, although gender differences play an important part in shaping social life for men and women in the Middle East, as elsewhere around the globe, there is no single Islamic, Christian, or Jewish notion of gender that operates exclusively in the region. Many of us have formed stereotypes about the restricted lives of Middle Eastern women because of the operation of rigid Islamic traditions. It is important to understand, however, that these are indeed stereotypes and do not capture the great variety in gender relations that exist in the Middle East and North Africa across lines of class, generation, level of education, and geography (urban versus rural origins), among other factors. What is pervasive throughout the Middle East and North Africa, as well as in many other societies throughout the world, is an ideological assumption that women should be subordinate to men. This view is largely held by both men and women in Middle Eastern and North African societies. Interestingly, it seems that men regard women's subordination as something natural, something that is effectively determined by biology. In contrast most women in the Middle East tend to regard their subordination as something social, something that is the product of the society in which they live and operate, and therefore something that can be negotiated and manipulated. The gender systems that operate in a wide variety of contexts in the Middle East and North Africa are derived in large part from some of the same notions about men and women that inform gender systems in Western societies.

Another way to explore the implications of gender systems is to complicate them by introducing other indentity elements such as class. Among South Asia's poor, for example, women bear the greatest burden and the most suffering. South Asian societies are intensely patriarchal, though the form that patriarchy takes varies by region and class. The common denominator among the poor throughout South Asia is that women not only have the constant responsibilities of motherhood and domestic chores but also have to work long hours in informal-sector occupations (**Figure 5.32**). In many poor communities, 90 percent of all production occurs outside of formal employment, more than half of which is the result of women's efforts. In addition, women's property rights are curtailed, their public behavior is restricted, and their opportunities for education and participation in the waged labor force are severely limited. Women's subservience to men is deeply ingrained within South Asian cultures, and it is manifest most clearly in the cultural practices attached to family life, such as the custom of providing a dowry to daughters at marriage. The preference for male children is reflected in the widespread (but illegal) practice of selective abortion and female infanticide. Within marriages, many (but by no means all) poor women are routinely neglected and maltreated. More extreme are the cases—usually reported only when they involve middle-class families—of "bride burning," whereby a husband or mother-in-law fakes the accidental death (kitchen fires are favored) or suicide of a bride whose parents had defaulted in their dowry payments. Several thousand such deaths

Figure 5.32 Indian women in the informal sector Many women in South Asia are self-employed as small vendors in the daily markets. Others do home-based work such as weaving and dyeing cloth or embroider or sew garments. Nearly all workers in the informal sector, whether male or female, lack any sort of social protection such as health or unemployment insurance.

are reported in India each year, and this is almost certainly only a fraction of the real incidence.

The picture for women in South Asia, as elsewhere, is not entirely negative, however, and one of the most significant developments has been the emergence of women's self-help movements. Perhaps the best-known of these is the Grameen Bank, a grassroots organization formed to provide small loans to the rural poor in Bangladesh. Another is the Self-Employed Women's Association (SEWA) in India, which has made a major contribution to building self-confidence and self-reliance among poor working women by mobilizing and organizing them.

GLOBALIZATION AND CULTURAL CHANGE

The discussion of cultural geography in this chapter has raised one central question: How has globalization changed culture? We have seen that it affects different cultural groups differently and that different groups respond in different ways to these changes. With so much change occurring for so long, however, we must still ask ourselves what overall impact globalization is having on the multiplicity of culture groups that inhabit the globe.

Anyone who has ever traveled between major world cities—or, for that matter, anyone who has been attentive to the backdrops of movies, television news stories, and magazine photojournalism—will have noticed the many familiar aspects of contemporary life in settings that until recently were thought of as being quite different from one another. Airports, offices, and international hotels have become notoriously alike, and their similarities of architecture and interior design have become reinforced by the near-universal dress codes of the people who frequent them. The business suit, especially for males, has become the norm for office workers throughout much of the world. Jeans, T-shirts, and sneakers, meanwhile, have become the norm for young people, as well as those in lower wage jobs. The same automobiles can be seen on the streets of cities throughout the world (though sometimes they are given different names by their manufacturers); the same popular music is played on local radio stations; and many of the movies shown in local theaters are the same (*Spiderman, Signs, Goldmember,* for example). Some of the TV programming is also the same—not just the music videos on MTV, but CNN's news, the Internet, major international sports events, drama series like *West Wing* and *CSI*. The same brand names also show up in stores and restaurants: Coca-Cola, Perrier, Carlsberg, Nestlé, Nike, Seiko, Sony, IBM, Nintendo, and Microsoft, to name a few. Everywhere there is Chinese food, pita bread, pizza, classical music, rock music, and jazz.

Americanization and Globalization

It is these commonalities that provide a sense of familiarity among the inhabitants of the "fast world" described in Chapter 2. From the point of view of cultural nationalism, the "lowest common denominator" of this familiarity is often seen as the culture of fast food and popular entertainment that emanates from the United States. Popular commentators have observed that cultures around the world are being Americanized, or "McDonaldized" (see Chapter 2), which represents the beginnings of a single global culture that will be based on material consumption, with the English language as its medium (**Figure 5.33**).

There is certainly some evidence to support this point of view, not least in the sheer numbers of people around the world who view *Sesame Street,* drink Coca-Cola, and eat in McDonald's franchises or similar fast-food chains. Meanwhile, U.S. culture is increasingly embraced by local entrepreneurs around the world. Travel writer Pico Iyer, for example, describes finding dishes called "Yes, Sir, Cheese My Baby" and "Ike and Tuna Turner" in a local

Figure 5.33 McDonald's in Poland The American franchise restaurant McDonald's is becoming a fixture on the landscape of formerly communist eastern European countries like Poland. Although menu prices are quite high by local standards, frequenting places like McDonald's is a sign of status and personal prosperity in Poland and other recently communist countries like Romania or the Czech Republic.

buffeteria in Guanzhou, China.[4] It seems clear that U.S. products are consumed as much for their symbolism of a particular way of life as for their intrinsic value. McDonald's burgers, along with Coca-Cola, Hollywood movies, rock music, and NFL and NBA insignia, have become associated with a lifestyle package that features luxury, youth, fitness, beauty, and freedom.

It is important to recognize, however, that U.S. products often undergo changes when they travel across the globe. For instance, Revlon, a U.S.-based international public corporation and a leading mass market producer of cosmetics, skin care products, and fragrances, is very much aware of the need to vary its products for their consumption in non-U.S. markets. Not only are Revlon products manufactured and distributed in Asia, Africa, South America and Europe, as well as the United States, the company tailors its products and its advertizing to the particularities of those markets. For example, Modi Revlon, a joint venture between Revlon and the Indian company Modi Mundipharma, provides a product line of color cosmetics that complement Indian women's skin tones and appeal to upper income markets. The success of Modi Revlon, which has captured 80 percent of the Indian cosmetic market, has been remarkable.

The economic success of the U.S. entertainment industry has also helped reinforce the idea of an emerging global culture based on Americanization. Today the entertainment industry is the leading source of foreign income in the United States, with a trade surplus of over $25 billion. Similarly, the United States transmits much more than it receives in sheer volume of cultural products. Today the originals of over half of all the books translated in the world (more than 25,000 titles) are written in English, the majority of which are produced by U.S. publishers. In terms of international flows of everything from mail and phone calls to press-agency reports, television programs, radio shows, and movies, a disproportionately large share originates in the United States.

Neither the widespread consumption of U.S. and U.S.-style products nor the increasing familiarity of people around the world with global media and international brand names, however, adds up to the emergence of a single global culture. Rather, what is happening is that processes of globalization are exposing the inhabitants of both the fast world and the slow world to a common set of products, symbols, myths, memories, events, cult figures, landscapes, and traditions. People living in Tokyo or Tucson, Turin or Timbuktu may be perfectly familiar with these commonalities without necessarily using or responding to them in uniform ways. Equally, it is important to recognize that cultural flows take place in all directions, not just outward from the United States. Think, for example, of European fashions in U.S. stores; of Chinese, Indian, Italian, Mexican, and Thai restaurants in U.S. towns and cities; and of U.S. and European stores selling exotic craft goods from the periphery.

A Global Culture?

The answer to the question whether there is a global culture must therefore be no, or at least there is no indisputable sign of it yet. While people around the world share an increasing familiarity with a common set of products, symbols, and events (many of which originate in the U.S. culture of fast food and popular entertainment), these commonalties are configured in different ways in different places, rather than constituting a single global culture. The local interacts with the global, often producing hybrid cultures. Sometimes traditional, local cultures become the subject of global consumption; sometimes it is the other way around. This is illustrated very well by the case of two suqs (linear bazaars) in the traditional medieval city of Tunis, in North Africa. Both radiate from the great Zaytuna Mosque, which has always been the geographic focal point of the old city. Both are located near gateways from the medieval core to the French-built new city.

The older suq now specializes in Tunisian handicrafts, "traditional" goods, etc. It has kept its exotic architecture and multicolored colonnades. The plaintive sound of the ancient nose flute and the whining of Arabic music provide background for the European tourists in their shorts and T-shirts, who amble in twos and threes, stopping to look and to buy. Few natives, except for sellers, are to be seen. The newer suq, formerly less important, is a bustling madhouse. It is packed with partially veiled women and younger Tunisian girls in blouses and skirts, with men in knee-length tunic/toga outfits or in a variety of pants and shirts, with children everywhere. Few foreigners can be seen. The background to the din is blaring rock-and-roll music, and piled high on the pushcarts that line the way are transistor radios, watches, blue jeans (some prewashed), rayon scarves, and Lux face and Omo laundry soaps.[5]

A second illustration of the absence of a homogenous global culture is through **world music**, the musical genre defined largely in response to the sudden increase of non–English language recordings released in the United Kingdom and the United States in the 1980s. The term is one employed primarily by the media and record stores, and it includes such diverse sources as Tuvan throat singers and Malian griots. (Throat singing is a performance style originating in south-central Russia where the vocal sounds are projected from the throat, allowing two

[4]P. Iyer, *Video Nights in Kathmandu: Reports from the Not-So-Far East.* London: Black Swan, 1989.

[5]J. Abu-Lughod, "Going Beyond Global Babble," in A. D. King (ed.), *Culture, Globalization, and the World-System.* Basingstoke, England: Macmillan, 1991, p. 132.

Figure 5.34 World music Bands like *Gogol Bordello* combine punk with traditional Gypsy music to produce a sound that allows the two to come together not harmoniously, but chaotically. *Gogol Bordello's* objective is to produce what they call an "unstable hybrid" to demonstrate that culture is a living entity that is constantly confronted and challenged. The band sees its music not as celebrating multiculturalism, but as challenging listeners to understand contemporary alternative world music as "multi kontra culti," indicating that globalization is not conflict free.

to four tones to be simultaneously produced by the singer. Griots were ancient praise singers of West Africa and also important advisors to royalty). There are at least two major views on the effect of globalization on indigenous musical productions. The first emphasizes how the Western music industry has enabled indigenous music to be more widely disseminated and therefore more widely known and appreciated. This position sees local roots mixing with Western popular musical styles, with a hybrid sound resulting. The second view worries that the influence of Western musical styles and the Western music industry have transformed indigenous musical productions to the point where their authenticity has been lost and global musical heterogeneity diminished. Despite their fundamental disagreement, holders of both positions recognize that world music has enabled cultural diversity to flourish and hope that indigenous performers will be able to resist the power of the Western music industry to homogenize their work at the same time that they rely on it to disseminate their music (**Figure 5.34**).

CONCLUSION

Culture is a complex and exceedingly important concept within the discipline of geography. A number of approaches exist to understand culture. It may be understood through a range of elements and features, from single traits to complex systems. Cultural geography recognizes the complexity of culture and emphasizes the roles of space, place, and landscape and the ecological relationships between cultures and their environment. It distinguishes itself from other disciplinary approaches, providing unique insights that reveal how culture shapes the worlds we live in at the same time that the worlds we inhabit shape culture.

Two of the most universal forms of cultural identity are religion and language. Despite the secularization of many people in core countries, religion is still a powerful form of identity, and it has been used to mediate the impacts of globalization. Globalization has caused dramatic changes in the distribution of the world's religions, as well as interaction among and between religions. Perhaps most remarkable, religious conversion to religions of the periphery is now under way in the core.

While the number of languages that exist worldwide are threatened by globalization, some cultures have responded to the threat by providing special protection for regional languages. The 500-year history of globalization has resulted in the steady erosion of many regional languages and heavy contact and change in the languages that persist. It is not only religion and language, however, that are at risk from globalization; other forms of cultural expression such as art and film are as well.

Different groups in different parts of the world have begun to use cultural identities such as gender, race, ethnicity, and sexuality as a way of buffering the impacts of globalization on their lives. When the impacts of globalization are examined at the local level, some groups suffer more harm or reap more benefits than others. The unevenness of the impacts of globalization and the variety of responses to it indicate that the possibility of a monolithic global culture wiping out all differences is limited.

Finally, while globalization is undoubtedly reshaping the world and bringing different cultural groups closer together than they have ever been previously, there is no conclusive evidence that globalization leads to cultural homogenization. Instead, globalization seems to be a differential process, which means that it is deployed differently in different places and experienced and responded to differently by the people who live in those places.

MAIN POINTS REVISITED

- **Though culture is a central, complex concept in geography, it may be thought of as a way of life involving a particular set of skills, values, and meanings.**

  Culture includes youth styles of dress, as well as operatic arias and slang and ecclesiastical languages.

- **Geographers are particularly concerned about how place and space shape culture and, conversely, how culture shapes place and space. They recognize that culture is dynamic and is contested and altered within larger social, political, and economic contexts.**

  The places in which cultural practices are produced shape the cultural production as much as the cultural production shapes the places in which it occurs.

- **Like other fields of contemporary life, culture has been profoundly affected by globalization. However, globalization has not produced a homogenized culture so much as distinctive impacts in different societies and geographical areas as global forces come to be modified by local cultures.**

  Although U.S. culture, especially commercialized culture, is widely exported around the globe, foreign cultural practices affect the United States and other parts of the world as well. The French influence on Algeria, for instance, is much more pronounced than is that of the United States.

- **Contemporary approaches in cultural geography seek to understand the role played by politics and the economy in establishing and perpetuating cultures, cultural landscapes, and global patterns of cultural traits and complexes.**

  For example, the state often facilitates the import or export of cultural practices, such as movies or music, so that economic growth can be enhanced.

- **Cultural geography has been broadened to include analysis of gender, class, sexuality, race, ethnicity, and life-cycle stage, in recognition that important differences can exist within as well as between cultures.**

  What geographers find important about these identities are the ways in which they are constructed in spaces and places, and the ways those spaces and places shape the identities.

- **Globalization does not necessarily mean that the world is becoming more homogeneous.**

  In some ways globalization has made the local even more important than before. In other ways it has meant that local cultures have suffered, as in the case of the disappearance of indigenous languages. In still other cases, globalization has helped to foster hybrid forms that blend distant cultures in unique ways.

KEY TERMS

cultural complex (p. 178)
cultural geography (p. 175)
cultural hearths (p. 188)
cultural landscape (p. 176)
cultural nationalism (p. 193)
cultural region (p. 182)
cultural system (p. 183)
cultural trait (p. 177)
culture (p. 174)
dialects (p. 186)
diaspora (p. 183)
ethnicity (p. 204)
folk culture (p. 175)
gender (p. 205)
genre de vie (p. 177)
hajj (p. 198)
historical geography (p. 177)
Islam (p. 195)
Islamism (p. 198)
jihad (p. 198)
kinship (p. 190)
language (p. 186)
language branch (p. 186)
language family (p. 186)
language group (p. 186)
Muslim (p. 195)
popular culture (p. 175)
race (p. 205)
racialization (p. 205)
religion (p. 183)
rites of passage (p. 178)
sexuality (p. 201)
tribe (p. 192)
world music (p. 208)

ADDITIONAL READING

Anderson, K., *Vancouver's Chinatown: Racial Discourse in Canada, 1875–1980.* Montreal: McGill-Queens University Press, 1991.

Bentahila, A. and E. E. Davies, "Language Mixing in Rai Music: Localisation or Globalisation?" *Language & Communication,* Apr. 2002, vol. 22, issue 2, pp. 187–205.

Bonnett, A., "Constructions of 'Race,' Place and Discipline: Geographies of 'Racial' Identity and Racism." *Ethnic and Racial Studies,* 1996, vol. 19, pp. 864–883.

"Culture Wars," *Economist.* 09/12/98, vol. 348, pp. 97–99.

Cronon, W., *Changes in the Land: Indians, Colonists, and the Ecology of New England.* New York: Hill and Wang, 1983.

Davis, S., *Parades and Power: Street Theatre in Nineteenth Century Philadelphia.* Philadelphia: Temple University Press, 1986.

Duncan, J. C., N. C. Johnson and R. H. Schein (eds.), *A Companion to Cultural Geography.* Oxford: Blackwell, 2004.

Dwyer, O. J. and J. P. Jones, III, "White Socio-Spatial Epistemology." *Social and Cultural Geography,* 2000 vol. 1, no. 2, pp. 209–221.

Forman, M. and M. A. Neal (eds.), *That's the Joint! The Hip Hop Studies Reader.* New York: Routledge, 2004.

Hebdige, D., *Subculture: The Meaning of Style.* London: Routledge, 1989.

Hiro, D., *Holy Wars: The Rise of Islamic Fundamentalism.* New York: Routledge, 1989.

International Labor Office, *Women and Men in the Informal Economy: A Statistical Picture.* Geneva: International Labor Office, 2002.

Jackson, P., *Maps of Meaning: An Introduction to Cultural Geography.* London: Unwin Hyman, 1989.

Katz, C., and J. Monk, *Full Circles: Geographies of Women over the Life Course.* London: Routledge, 1993.

Marston, S., "Making Difference: Conflict over Irish Identity in the New York City St. Patrick's Day Parade." *Political Geography,* 2001, 21(3): 373–392.

McAnany, E. G. and K. T. Wilkinson, *Mass Media and Free Trade: NAFTA and the Cultural Industries.* Austin: University of Texas Press, 1996.

Mitchell, D., *Cultural Geography: A Critical Introduction.* Malden, MA: Blackwell, 2000.

NetLingo.Com, http://www.netlingo.com/emailsh.cfm

Papademetriou, D. G. and D. W. Meyers, *Caught in the Middle: Border Communities in the Era of Globalization.* Washington, DC: Carnegie Endowment for International Peace, 2001.

Ritzer, G., *McDonaldization: The Reader.* Thousand Oaks: Pine Forge Press, 2002.

Robinson, J. M., "The Information Revolution—Culture and Sovereignty—A Canadian Perspective." *Canada—United States Law Journal,* 1998, vol. 24, pp. 147–153.

Roy, O., *The Failure of Political Islam.* Cambridge, MA: Harvard University Press, 1994.

Saunders, R., "Kickin' Some Knowledge: Rap and the Construction of Identity in the African-American Ghetto." *The Arizona Anthropologist 10* (1993): 21–40.

Seago, A., "Where Hamburgers Sizzle on an Open Grill Night and Day: Global Pop Music." *American Studies,* Summer/Fall 2000, vol. 41, issue 2/3, pp. 119–136.

EXERCISES

On the Internet

The Internet exercises for this chapter will help you not only "see" where various types of sexualities are located (place) but also answer the question of "Why there?" Our review exercise focuses on such topics as race and place, ethnic symbols, gender, and language. The thinking-spatially exercise looks at ethnicity and space, as well as cultural and political ecology, and more. Throughout these drills, you will encounter examples of such concepts as cultural nationalism, cultural imperialism, and cultural borders, to name a few.

Unplugged

1. Using *Billboard Magazine* (the newsmagazine of the record industry), construct a historical geography of the top 20 singles over the last half century in order to determine how different regions of the country (or the globe!) have risen and fallen in terms of significance. You should also determine an appropriate interval for sampling—every three to five years is a generally accepted one. You may use the hometowns of the recording artists or the headquarters of the recording studios as your geographic variables. Once you have organized your data, you should be able to answer the following questions: How has the geography you have documented changed? What might be the reasons for these changes? What do these changes mean for the regions of the country (or globe) that have increased or decreased in terms of musical prominence?
2. Ethnic identity is often expressed spatially through the existence of neighborhoods or business areas dominated by members of a particular group. One way to explore the spatial expression of ethnicity in a place is to look at newspapers over time. In this exercise you are to look at ethnic change in a particular neighborhood over time. You can do this by using your library's holdings of local or regional newspapers. Examine change over at least a four-decade period. To do this, you must identify an area of the city in which you live or some other city for which your library has an extensive newspaper collection. You should trace the history of an area you know is now occupied by a specific ethnic group. How long has the group occupied that area? What aspects of the group's occupation of that area have changed over time (school, church, or sports activities, or the age of the households)? If different groups have occupied the area, what might be the reasons for the changes?
3. College and university campuses generate their own cultural practices and ideas that shape behaviors and attitudes in ways that may not be so obvious at first glance. For this exercise you are to observe a particular practice that occurs routinely at your college or university. For example, fraternity and sorority initiations are important rituals of college life, as are sports events and class discussions. Observe a particular cultural practice that is an ordinary part of your life at college. Who are the participants in this practice? What are their levels of importance? Are there gender, age, or status differences in those who carry out this practice? What are the time and space aspects of the practice? Who controls its production? What are the intended outcomes? How does the practice contribute to or detract from the maintenance of order in the larger culture?
4. Using your local library as your source, find a description of a coming-of-age ceremony for any part of the world other than the United States. Summarize that description and then compare it to one you have either experienced directly or observed in the United States. What are the differences and the similarities between your experience and the one you read about? What might be some of the reasons for these?

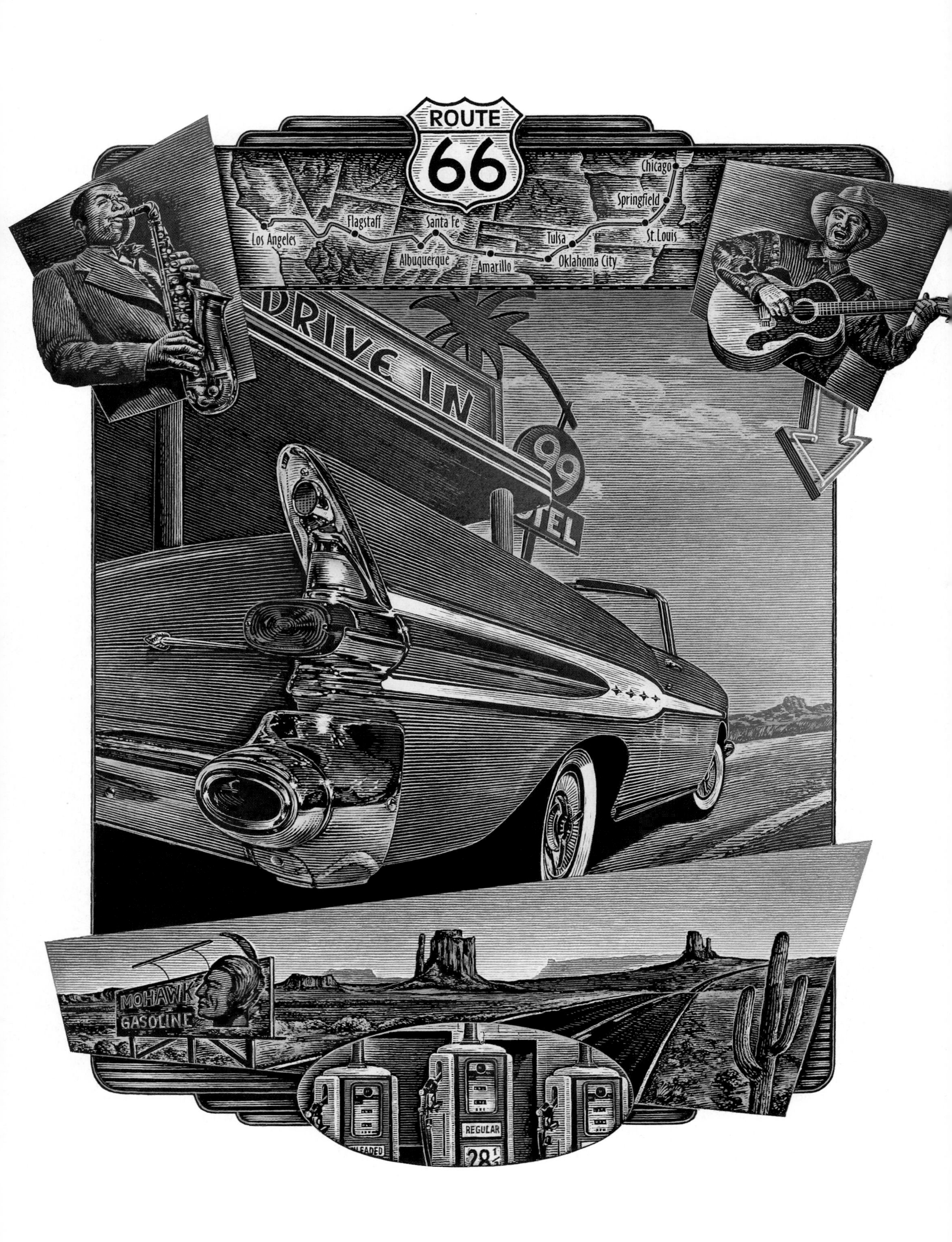
ROUTE
66
Chicago
Springfield
St. Louis
Los Angeles
Flagstaff
Santa Fe
Albuquerque
Amarillo
Tulsa
Oklahoma City
DRIVE IN
99
MOHAWK
GASOLINE
REGULAR

6 Interpreting Places and Landscapes

MAIN POINTS

- In addition to understanding how the environment shapes (and is shaped by) people, geographers seek to identify how it is perceived and understood by people.
- The language in which a landscape is written is a kind of code. The code includes signs that direct our attention toward certain features and away from others.
- The written code of landscape is also known as semiotics. Codes signify important information about landscapes, such as whether they are sacred or profane, accessible or off limits, or oriented toward work or play.
- Different cultural identities and status categories influence the ways in which people experience and understand their environments, as well as how they are shaped by—and able to shape—them.
- Landscape serves as a kind of archive of society. It is a reflection of our culture and our experiences. Like a book, landscape is a text that is written by individuals and groups, and read by them as well.
- The emergence of the most recent phase of globalization has occurred in parallel with a transition from modernity to postmodernity.

Between 1916 and 1925 the U.S. Congress created the National Highway System Program, which provided for the construction of the first public highways throughout the country. The official designation of the highway that would connect Chicago to Los Angeles with over 2,400 miles of roadway was Route 66. From the beginning, the diagonal course of the highway linked hundreds of predominantly rural communities in Illinois, Missouri, Kansas, Oklahoma, Texas, New Mexico, Arizona, and California to Chicago and Los Angeles, enabling farmers to transport grain and produce for redistribution. The diagonal configuration of Route 66 was also significant to the growing trucking industry, which by 1930 had come to rival the railroad for preeminence in U.S. shipping. Finally, the road also became important for motoring tourists as well as migrants seeking a better life in the West. John Steinbeck, in his famous novel *The Grapes of Wrath,* proclaimed U.S. Highway 66 the "Mother Road" because hundreds of thousands of migrants took it to California to escape the despair of the Oklahoma Dust Bowl in the 1930s.

The final stretches of the road were constructed during the height of the Depression by thousands of unemployed young men who were put to work in a federal program. By 1938 the road was completed and its surface continuously paved from east to west. The road was also an important part of the U.S. effort in World War II. The War Department needed improved highways to achieve rapid mobilization during the war and to promote national defense after the war. Because of its all-weather capability, Route 66 helped to facilitate the single greatest wartime mobilization in the history of the nation.

The end of the war found U.S. residents more mobile than ever before, and Route 66 helped to facilitate a new migration of easterners to the West, as well as to promote automobile travel as a tourist experience. Yet by 1960 road travel and commerce were so important that Route 66 had become inadequate to the task of moving people and goods eastward and westward. And by 1970 large parts of Route 66 had been bypassed or replaced by a new national interstate highway system.

Since the early 1990s, a loose coalition of federal agencies such as the National Park Service and the United States Forest Service, state entities, and private individuals has taken steps to interpret, preserve, and commemorate remaining portions of Route 66. Tourists from all over the United States and abroad flock to the road to celebrate and appreciate it. Songs and books have been written about it, and a highly popular television show used the road as its backdrop. *Route 66* magazine is celebrating its fourteenth year of publication. There are Route 66 clubs in the United Kingdom, Germany, Canada, Norway, Switzerland, France, and Italy. In 1999 President Clinton signed into law The Historic Route 66 Corridor Act, establishing a federal fund of $10 million for use over the next decade to support businesses and tourist attractions located along the way. And the Smithsonian Institution acquired old sections of the Oklahoma Route 66 for an "America on the Move."

Route 66, Amboy, California

So what does the story of Route 66 have to do with interpreting landscapes and places? Geographer Elaine Mariolle's research shows that since its decommissioning in the 1980s, Route 66 has taken on a new role that transcends simple transportation. It has become such an important part of the national and international popular imagination that it is now a national icon that represents ideals of optimism, movement, new opportunities, and indomitable spirit. Route 66, as a place that in many ways no longer exists, has become one of the most popular symbolic landscapes in the global imagination.

In this chapter we explore the relationships among people, places, landscapes, and spaces in order to assess how individuals and groups experience their environments, create places, and find meanings in the landscapes they create.

BEHAVIOR, KNOWLEDGE, AND HUMAN ENVIRONMENTS

In addition to attempting to understand how the environment shapes and is shaped by people, geographers also seek to identify how it is perceived and understood by people. Arguing that there is an interdependence between people and places, geographers explore how individuals and groups acquire knowledge of their environments and how this knowledge shapes their attitudes and behaviors. Some geographers have focused their research on natural hazards as a way of addressing environmental knowledge, while others have tried to understand how people ascribe meaning to landscape and places. In this chapter we take the key geographical concepts of place, landscape, and space and explore the ways in which people understand them, create them, and operate within them.

In their attempts to understand environmental perception and knowledge, geographers share a great deal with other social scientists, but especially with psychologists. Human cognition and behavior have always been at the center of psychology. What makes environmental knowledge and behavior uniquely geographical is their relation to both the environmental context and the humans who struggle to understand and operate within it. Much of what we as humans know about the environment we live in is learned through direct and indirect experience. Our environmental knowledge is also acquired through a filter of personal and group characteristics, such as race, gender, stage of the life cycle, religious beliefs, and where we live (**Figure 6.1**).

For instance, children have interesting and distinct relationships to the physical and cultural environment. How do children acquire knowledge about their environments? How do boys differ from girls in the ways they learn about and negotiate their environments? What kind of knowledge do they acquire, and how do they use it? What role do cultural influences play in the process? What happens when larger social, economic, and environmental changes take place? Geographer Cindi Katz conducted research in rural Sudan to find answers to these questions. Working with a group of 10-year-old children in a small village, she sought to discover how they acquired environmental knowledge. What she also learned through this work was how the transformation of agriculture in the region changed not only the children's relationships to their families and community but also their perceptions of nature.[1] In this Sudanese village, as in similar communities elsewhere in the periphery, children are important contributors to subsistence activities, especially planting, weeding, and harvesting. The villagers were strict Muslims and thus had strict rules about what female members of the community were allowed to do and where they were allowed to go. Many of the subsistence activities that required leaving the family compound were customarily delegated to male children. In fact, within the traditional subsistence culture, boys predominated in all agricultural tasks except planting and harvesting and were responsible for herding livestock as well. Many boys (and occasionally girls) were also responsible for fetching water and helping to gather firewood. Both boys and girls collected seasonal foods from lands surrounding the village. Work and play were often mixed together, and play, as well as work, provided a creative means for acquiring and using environmental knowledge and for developing a finely textured sense of the home area (**Figure 6.2**).

What happens when globalization reshapes the agricultural production system, as it did in the village when, through an international development scheme, irrigated cash-crop cultivation was introduced? With the financial assistance of outside donors, a Sudanese government development project transformed the agriculture of the region from subsistence-level livestock raising and cultivation of sorghum and sesame to cultivation of irrigated cash crops such as cotton. The new cash-crop regime, which required management of irrigation works, application of fertilizers, herbicides, and pesticides, and more frequent weeding, required children as well as adults to work longer and harder. Parents were often forced to keep their children out of school because many of the tasks had to be done during the school term.

[1]C. Katz, "Sow What You Know: The Struggle for Social Reproduction in Rural Sudan," *Annals of the Association of American Geographers*, 81, 1991, pp. 488–514, and C. Katz, *Growing Up Global: Economic Restructuring and Children's Everyday Lives*, Minneapolis: University of Minnesota Press.

Figure 6.1 Conflicting environmental perceptions Pictured here is a member of the Algonquin Indian Tribe attempting to prevent a logging truck from passing along a forest road. In conflict here are opposing perceptions of the forest, where some members of the tribe see it as a place of spiritual renewal while lumber companies see it as the source of harvestable commodities.

Loss of forests required children to range farther afield to gather fuelwood and to make gathering trips more frequently (**Figure 6.3**). Soon wealthier households began buying wood rather than increasing demands on their children or other household members. For the children of more marginalized households, the selling of fuelwood, foods, and other items provided a new means for earning cash to support their families but also placed increasing demands on their energies and resourcefulness and changed their whole experience of their world.

For the children of this village, globalization (in the form of the transition to cash-crop agriculture) changed their relationships with their environments and with their futures. The kinds of skills the children had learned for subsistence production were no longer useful for cash-crop production. As they played less and worked longer hours in more specialized roles, their experience of their environment became narrowed. As their roles within the family changed, they attended school less and learned less about their world through formal education. As a result, their perceptions of their environment changed, along with their values and attitudes toward the landscape and the place they knew as home.

Today many of the children face a considerable challenge, for there are simply not enough agricultural jobs to go around. Many of the boys—especially those whose parents are tenant farmers and lacking status—will be left with few options but to become agricultural wage laborers or to seek nonagricultural work outside the village. The girls, trained largely to assume their mothers' household roles, face another challenge—how to take over the agricultural tasks previously performed by the boys, but without the advantage of knowing how to do those tasks or use the proper tools. Traditional relationships within village society have been undermined by globalization, and traditional opportunities to acquire important environmental knowledge are being lost to new generations of children.

PLACE-MAKING

As we saw in Chapter 1, places are socially constructed—given different meanings by different groups for different purposes. Most people identify with places as part of their personal identity, drawing on particular images and particular histories of places in order to lend distinctiveness to both their individuality and their sense of community. But identifying with place may also imply the exclusion of other people and the stereotyping of other places. People often reinforce their sense of place and of who they are by contrasting themselves with places and people they feel are very different from them. Seen in this context, place-making stands at the center of issues of culture and power relations, a key part of the systems of meaning through which we make sense of the world.

Territoriality

Some social scientists believe that wanting to have a place where you feel you belong is a natural human attribute, part of a strong territorial instinct. Humans, it is argued, have an innate sense of territoriality, just like many other species. The concept of **territoriality** refers to the persistent attachment of individuals or peoples to a specific location or territory. The concept is important to geographers

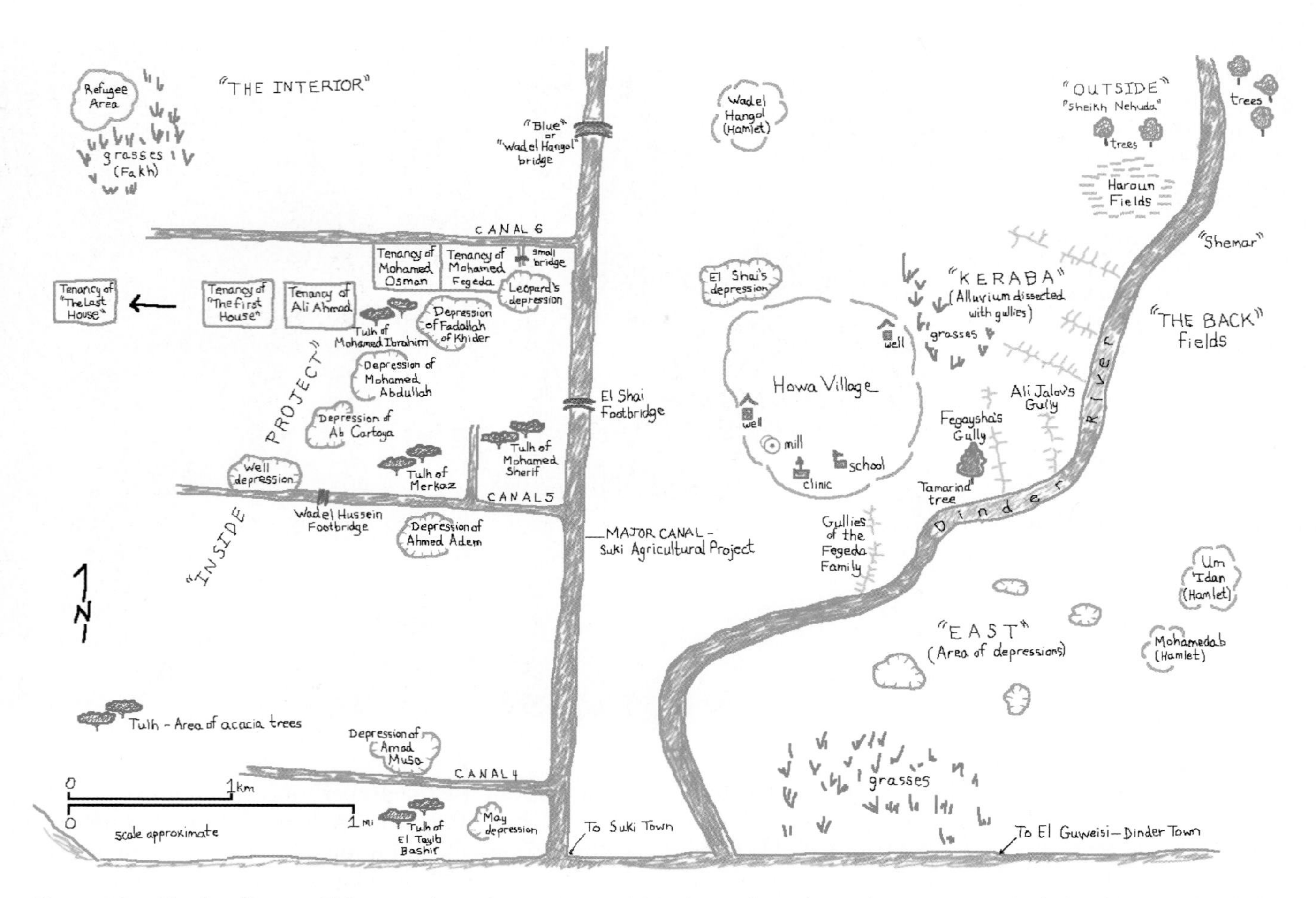

Figure 6.2 Shepherd's map This map, drawn by a 10-year-old Sudanese boy, shows the area over which the sheep are herded. It illustrates the detailed environmental knowledge Sudanese children possess of the landscape that surrounds their village. The village is an Islamic one, and norms determine the kinds of tasks in which boys and girls can participate. Only boys are allowed to tend sheep, which requires a particular environmental knowledge about grazing areas and water availability. (After C. Katz, "Sow What You Know: The Struggle for Social Reproduction in Rural Sudan," *Annals of the Association of American Geographers, 81,* 1991, pp. 488–514.)

Figure 6.3 Sudanese children working and playing The introduction of an agricultural development project brought the forces of globalization very directly to this small village in Sudan. The project dramatically changed both work and play for children and created a gap between the environmental knowledge they traditionally learned and the knowledge they would need to survive under different economic circumstance. In this photograph two 10-year-old girls are shown gathering firewood for domestic use. The wood is of very poor quality, and the girls had to range far to procure it. These branches were gathered in and around the irrigation canal, which for all members of the village is a risky environmental site. The waters in the canals contain the parasitic blood flukes that cause schistosomiasis, a chronic disease that can affect the liver, urinary bladder, lungs, or central nervous system.

because it can be related to fundamental place-making forces.

The specific study of people's sense of territoriality is part of the field of **ethology,** the scientific study of the formation and evolution of human customs and beliefs. The term is also used to refer to the study of the behavior of animals in their natural environments. According to ethologists, humans carry genetic traits produced by our species' need for territory. Territory provides a source of physical safety and security, a source of stimulation (through border disputes), and a physical expression of identity. These needs add up to a strong territorial urge that can be seen in the microgeography of people's behavior: claims to space in reading rooms or on beaches, for example, and claims made by gangs to neighborhood turf (**Figure 6.4**). Ethologists argue that the territorial urge also can be observed when people become frustrated because of overcrowding. They become stressed and, in some circumstances, begin to exhibit aggressive or deviant behavior. Ethologists and environmental psychologists link crowding to everything from vandalism and assault to promiscuity, listlessness, and clinical depression.

While such claims are difficult to substantiate, as is the whole notion that humans have an inborn sense of territoriality, the idea of territoriality as a product of *culturally* established meanings is supported by a large body of scientific evidence. Some of this evidence comes from the field of **proxemics,** the study of the social and cultural meanings that people give to personal space. These meanings make for unwritten territorial rules (rather than a biological urge) that can be seen in the microgeography of people's behavior. It has been established, for example, that people develop unwritten protocols about how to claim space. One common protocol is simply regular use (think of people's habits in classroom seating arrangements). Another is through the use of spatial markers such as a newspaper or a towel to fix a space in a reading room or on a beach. There are also bubbles, or areas, of personal space that we try not to invade (or allow to be invaded by others). Varying in size and shape according to location and circumstance, these bubbles tend to be smaller in public places and in busier and more crowded situations; they tend to be larger among strangers and in situations involving members of different social classes; and they tend to vary from one social class or cultural group to another.

At larger spatial scales, territoriality is mostly a product of forces that stem from political relations and cultural systems. This dimension of territoriality underpins a great deal of human geography. All social organizations and the individuals who belong to them are bound at some scale or another through formal or informal territorial limits. Many of them—nations, corporations, unions, clubs—actually claim a specific area of geographic space under their influence or control. In this context, territoriality can be defined as any attempt to assert control over a specific geographic

Figure 6.4 Graffiti as territorial markers Graffiti are used by neighborhood gangs to establish and proclaim their identity. Some graffiti, such as these from a Cambodian gang in Long Beach, California, also function as simple territorial markers that help to stake out turf in high-density environments where there exist few other clues about claims to territory.

area to achieve some control over other people, resources, or relationships. Territoriality also is defined as any attempt to fulfill socially produced needs for identity, defense, and stimulation. Territoriality covers many phenomena, including the property rights of individuals and private corporations; the neighborhood covenants of homeowners' associations; the market areas of commercial businesses; the heartlands of ethnic or cultural groups; the jurisdictions of local, state, and national governments; and the reach of transnational corporations and supranational organizations. Territoriality thus provides a means of meeting three social and cultural needs:

- The regulation of social interaction
- The regulation of access to people and resources
- The provision of a focus and symbol of group membership and identity

Territoriality meets these needs because, among other things, it facilitates classification, communication, and enforcement. We can classify people and/or resources in terms of their location in space much more easily than we can classify them in relation to personal or social criteria. All that is necessary to communicate territory is a simple marker or sign that constitutes a boundary. This in turn makes territory an efficient device for determining whether people are subject to the enforcement of a particular set of rules and conditions.

Territoriality also gives tangible form to power and control, but does so in a way that directs attention away from the personal relationships between the controlled and the controllers. In other words, rules and laws become associated with particular spaces and territories rather than with particular individuals or groups. Finally, territoriality allows people to create and maintain a framework through which to experience the world and give it meaning. Bounded territories, for example, make it easier to differentiate "us" from "them."

People and Places, Insiders and Outsiders

As we saw in Chapter 1, places are constantly under social construction by people responding to the opportunities and constraints of their particular locality. As people live and work in places, they gradually impose themselves on their environment, modifying and adjusting it to suit their needs and express their values. At the same time, they gradually accommodate both to their physical environment and to the values, attitudes, and comportment of people around them. People are constantly modifying and reshaping places, and places are constantly coping with change and influencing their inhabitants.

Places are both centers of meaning for people and the external frameworks for their actions and behavior. It is important to remember that places are constructed by their inhabitants from their own subjective point of view; and that they are simultaneously constructed and seen as an external "other" by outsiders. A neighborhood, for example, is both an area of special meanings to its residents as well as an area containing houses, streets and people that others may view from an outsider's ("decentered") perspective.

A fundamental element in individuals' construction of place is the **existential imperative** for people to define themselves in relation to their material world—their basic need and capacity to achieve a form of spiritual or psychic unity between themselves and their material worlds. People's subjective "creation" of space and place provides them with roots and a sense of identity—a phenomenon known to philosophers in terms of "dwelling." People's innate capacity for "dwelling" allows them to give meanings to places, through repeated experience, that are deepened and qualified over time with multiple nuances.

Yet the construction of place by individual insiders cannot take place independently of local societal norms that codify and frame the social construction of spaces and places. People's territoriality and sense of "dwelling," in other words, are shaped by locally shared notions of social distance, rules of comportment, forms of social organization, and so on. There is, therefore, an important two-way relationship between social structures and the everyday practices of the "insiders" of subjectively constructed spaces and places. We live both *in* and *through* places. As we saw in Chapter 1, a crucial concept here is that of the *lifeworld*, the taken-for-granted pattern and context for everyday living through which people conduct their day-to-day lives without having to make it an object of conscious attention. People's familiarity with one another's vocabulary, speech patterns, dress codes, gestures, and humor, and with shared experiences of their physical environment often carries over into people's attitudes and feelings about themselves and their locality and to the symbolism they attach to that place. When this happens, the result is a collective and self-conscious "structure of feeling": a socio-cultural frame of reference generated among people as a result of the experiences and memories that they associate with a particular place.

Over the past couple of decades, however, people and places have been confronted with change on an unprecedented scale and at an extraordinary rate. As we saw in Chapter 2, globalization has generated a "fast world" within which commonalities among places are intensifying. The fast world is a world of restless landscapes in which the more that places change, the more they seem to look alike, the less they are able to retain a distinctive sense of place, and the less they are able to sustain public social life. As a result, the experience of spectacular and distinctive places, physical settings, and landscapes has become an important element of consumer culture. Responding to this shift, developers have created theme parks, shopping malls, festival marketplaces, renovated historic districts, and neo-traditional villages and neighborhoods. But the more developers have competed to provide distinctive settings and the larger and more spectacular their projects, the more convergent the results. The inevitable result is that the "authenticity" of places is undermined. City spaces, in particular, become inauthentic and "placeless."

Experience and Meaning

The interactions between people and places raise some fundamental questions about the meanings that people attach to their experiences: How do people process information from external settings? What kind of information do they use? How do new experiences affect the way they understand their worlds? What meanings do particular environments have for individuals? How do these meanings influence behavior? How do people develop and modify their sense of a place, and what does it mean to them? The answers to these questions are by no means complete, but it is clear that people not only filter information from their environments through neurophysiological processes but also draw on personality and culture to produce cognitive images of their environment, pictures or representations of the world that can be called to mind through the imagination (**Figure 6.5**). Cognitive images are what people see in the mind's eye when they think of a particular place or setting.

Two of the most important attributes of cognitive images are that they both simplify and distort real-world environments. Research on the ways in which people simplify the world through such means has suggested, for example, that many people tend to organize their cognitive images of particular parts of their world in terms of several simple elements (**Figure 6.6**):

Paths: The channels along which they and others move; for example, streets, walkways, transit lines, canals.

Edges: Barriers that separate one area from another; for example, shorelines, walls, railroad tracks.

Districts: Areas with an identifiable character (physical and/or cultural) that people mentally "enter" and "leave"; for example, a business district or an ethnic neighborhood.

Nodes: Strategic points and foci for travel; for example, street corners, traffic junctions, city squares.

Landmarks: Physical reference points; for example, distinctive landforms, buildings, or monuments.

For many people individual landscape features may function as more than one kind of cognitive element. A freeway, for instance, may be perceived as both an edge and a path in a person's cognitive image of a city. Similarly, a railroad terminal may be seen as both a landmark and a node.

Distortions in people's cognitive images are partly the result of incomplete information. Once we get beyond our immediate living area, we know few spaces in complete detail. Yet our worlds—especially those of us in the fast world who are directly tied to global networks of communication and knowledge—are increasingly large in geographic scope. As a result, these worlds must be conceived, or understood, without many direct stimuli. We have to rely on fragmentary and often biased information from other people, from books, magazines, television, and the Internet. Distortions in cognitive images are also partly the result of our own biases. What we remember about places; what we like or dislike; what we think is significant; and what we impute to various aspects of our environments all are functions of our personalities, our experiences, and the cultural influences to which we have been exposed.

Images and Behavior

Cognitive images are compiled, in part, through behavioral patterns. Environments are "learned" through experience. Meanwhile, cognitive images, once generated,

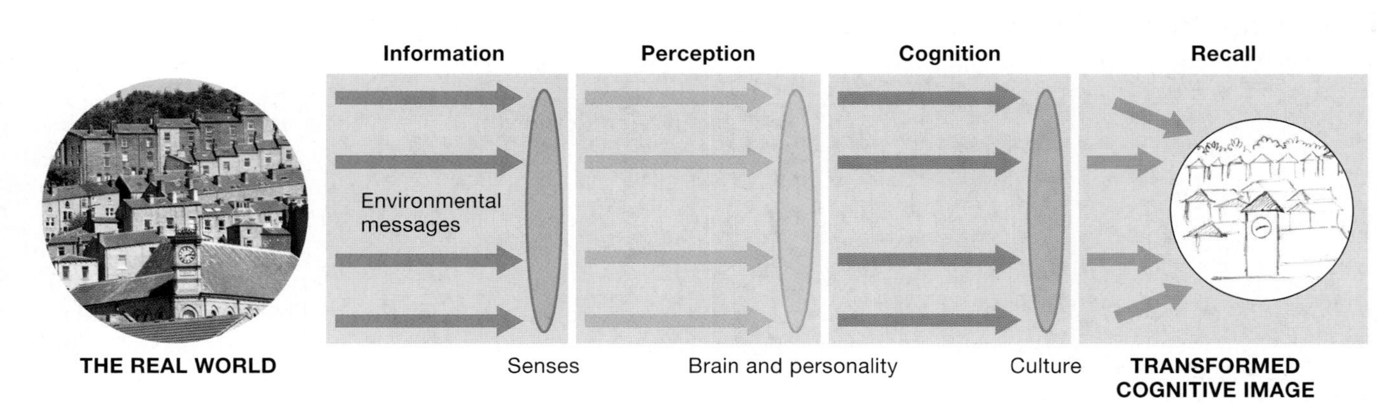

Figure 6.5 The formation of cognitive images People form cognitive images as a product of information about the real world, experienced directly and indirectly, and filtered through their senses, brain, personality, and the attitudes and values they have acquired from their cultural background. (After R. G. Golledge and R. J. Stimpson, *Analytical Behavior Geography.* Beckenham: Croom Helm, 1987, Fig. 3.2, p. 3.)

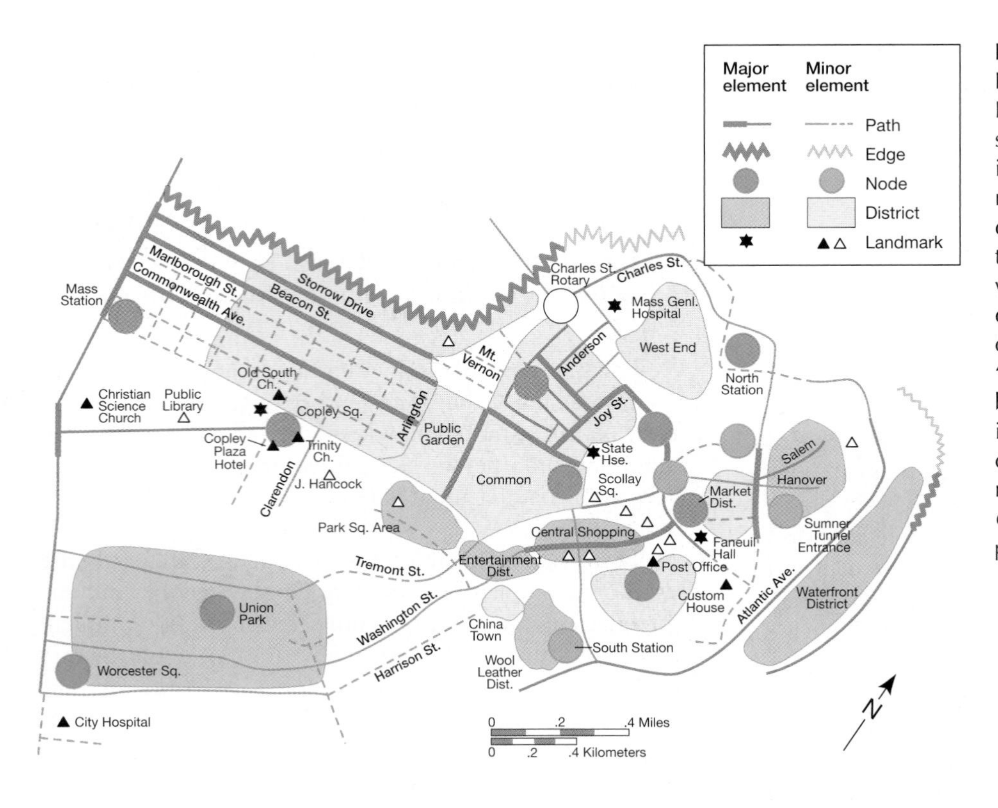

Figure 6.6 Cognitive image of Boston This map was compiled by Kevin Lynch, one of the pioneer researchers into cognitive images, from interviews with a sample of Boston residents. Lynch found that the residents of Boston tended to structure their cognitive images of the city with the same elements. He produced ingenious maps, such as this one, to demonstrate the collective "mental map" of the city, using symbols of different boldness or color to indicate the proportion of respondents who had mentioned each element. (After K. Lynch, *The Image of the City.* Cambridge, MA: M.I.T. Press, 1960, p. 146.)

influence behavior. In these two-way relationships, cognitive images are constantly changing. Each of us also generates—and draws on—different kinds of cognitive images in different circumstances.

Elements such as districts, nodes, and landmarks are important in the kinds of cognitive images that people use to orient themselves and to navigate within a place or region. The more of these elements an environment contains—and the more distinctive they are—the more legible that environment is to people and the easier it is to get oriented and navigate. In addition, the more firsthand information people have about their environment and the more they are able to draw on secondary sources of information, the more detailed and comprehensive their images will be.

This phenomenon is strikingly illustrated in **Figure 6.7**, which shows the collective image of Los Angeles as seen by the residents of three different neighborhoods: Westwood, an affluent neighborhood; Avalon, a poor, inner city neighborhood; and Boyle Heights, a poor, immigrant neighborhood. The residents of Westwood have a well-formed, detailed, and comprehensive image of the entire Los Angeles basin. At the other end of the socioeconomic spectrum, residents of the black ghetto neighborhood of Avalon, near Watts, have a vaguer image of the city, structured only by the major east–west boulevards and freeways and dominated by the gridiron layout of streets between Watts and the city center. The Spanish-speaking residents of Boyle Heights—even less affluent, less mobile, and somewhat isolated by language—have an extremely restricted image of the city. Their world consists of a small area around Brooklyn Avenue and First Street, bounded by the landmarks of city hall, the bus depot, and Union Station.

The importance of these images goes beyond people's ability simply to navigate around their environments. The narrower and more localized people's images are, for example, the less they will tend to venture beyond their home area. Their behavior becomes circumscribed by their cognitive imagery. People's images of places also shape particular aspects of their behavior. Research on shopping behavior in cities, for example, has shown that customers do not necessarily go to the nearest store or to the one with the lowest prices; they are influenced by the configuration of traffic, parking, and pedestrian circulation within their imagery of the retail environment. The significance of this clearly has not been lost on the developers of shopping malls, who always provide extensive space for free parking and multiple entrances and exits.

In addition, shopping behavior, like many other aspects of behavior, is influenced by people's values and feelings. A district in a city, for example, may be regarded as attractive or repellent, exciting or relaxing, fearsome or reassuring, or, more likely, a combination of such feelings. As with all other cognitive imagery, such images are produced through a combination of direct experience and indirect information, all filtered through personal and cultural perspectives. Images such as these often exert a strong influence on behavior. Returning to the example of consumer behavior, one of the strongest influences on shopping patterns relates to the imagery evoked by retail environments—something else that has not escaped the developers of malls, who spend large sums of money

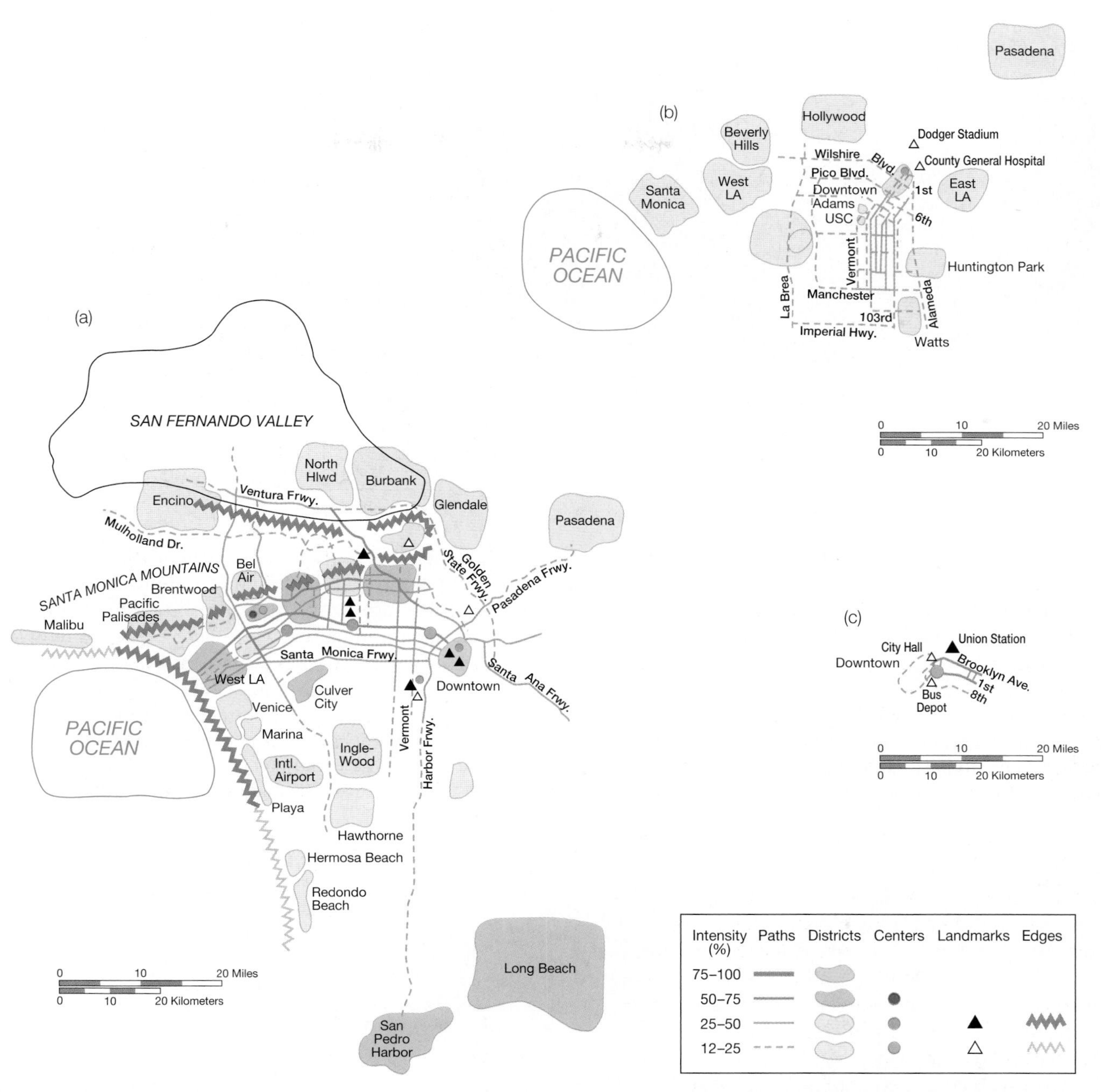

Figure 6.7 Images of Los Angeles These images, as seen by residents of (a) Westwood, (b) Avalon, and (c) Boyle Heights, are drawn to the same scale. The intensity of particular elements in the urban images of residents of different areas is measured by the percentage of residents who recall a particular element as being part of their own mental map of the city. (After P. Orleans, in R. Downs and D. Stea (eds.), *Image and Environment.* Chicago: Aldine, 1973, pp. 120–122.)

to establish the right atmosphere and image for their projects.

While shopping behavior is one narrow example of the influence of place imagery on behavior, other examples can be drawn from every aspect of human geography and at every spatial scale. The settlement of North America, for example, was strongly influenced by the changing images of the Plains and the West. In the early 1800s the Plains and the West were perceived as arid and unattractive, an image that was reinforced by early atlases. When the railroad companies wanted to encourage settlement in these regions, they changed people's image of the Plains and the West with advertising campaigns that portrayed them as fertile and hospitable regions. The images associated with different regions and localities continue to shape settlement patterns. People draw on their cognitive imagery, for example, in making decisions about migrating from one area to another. **Figure 6.8** shows the composite image of the United States held by a group of Virginia Tech students, based on the perceived attractiveness of cities and states as places in which to live.

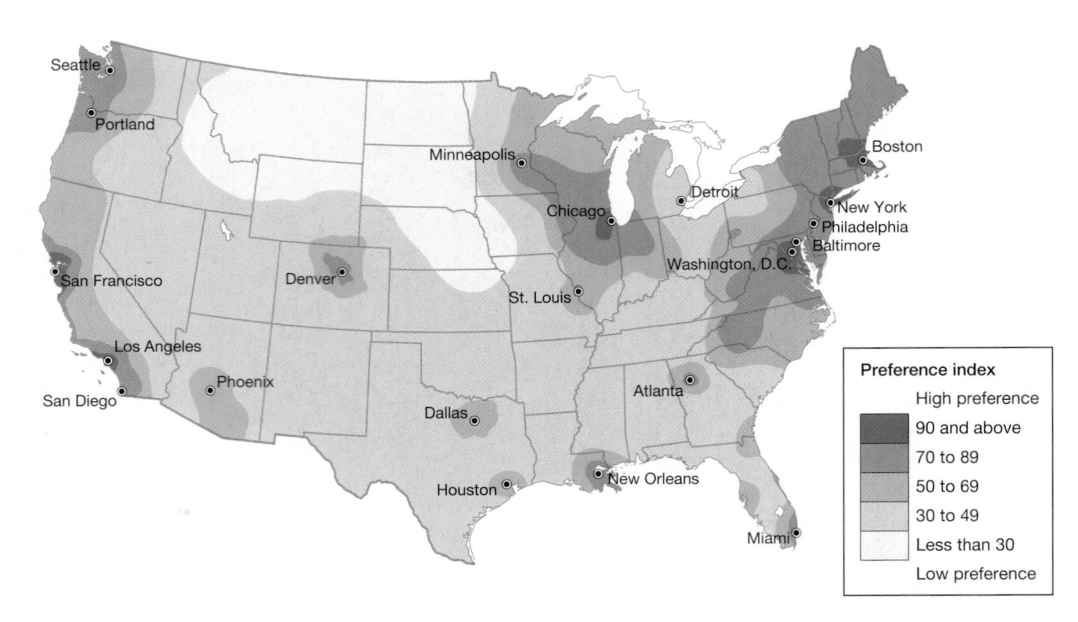

Figure 6.8 Preference map of the United States This isoline map illustrates collective preferences for cities in the coterminous United States as places in which to live and work, as expressed by architecture students at Virginia Tech in 1996. It is a generalization based on the scores the students gave to the 150 largest cities in the country. The higher the score, the stronger the preference for living and working there.

Another example of the influence of cognitive imagery on people's behavior is the way that people respond to environmental hazards, such as floods, droughts, earthquakes, storms, and landslides, and come to terms with the associated risks and uncertainties. Some people tend to change the unpredictable into the knowable by imposing order where none really exists (resorting to folk wisdom about weather, for example), while others deny all predictability and take a fatalistic view. Some tend to overestimate both the degree and the intensity of natural hazards, while others tend to underestimate them. These differences point to another important dimension of behavior: people's attitudes toward risk-taking. **Figure 6.9** illustrates diagrammatically the perceptions of individuals with different attitudes toward risk-taking. While the reckless person, for example, tends to exaggerate gains and discount losses, the economically disadvantaged person tends to exaggerate both. The cautious person tends to overestimate losses and undervalue gains. Most individuals adopt different risk-taking attitudes toward dif-

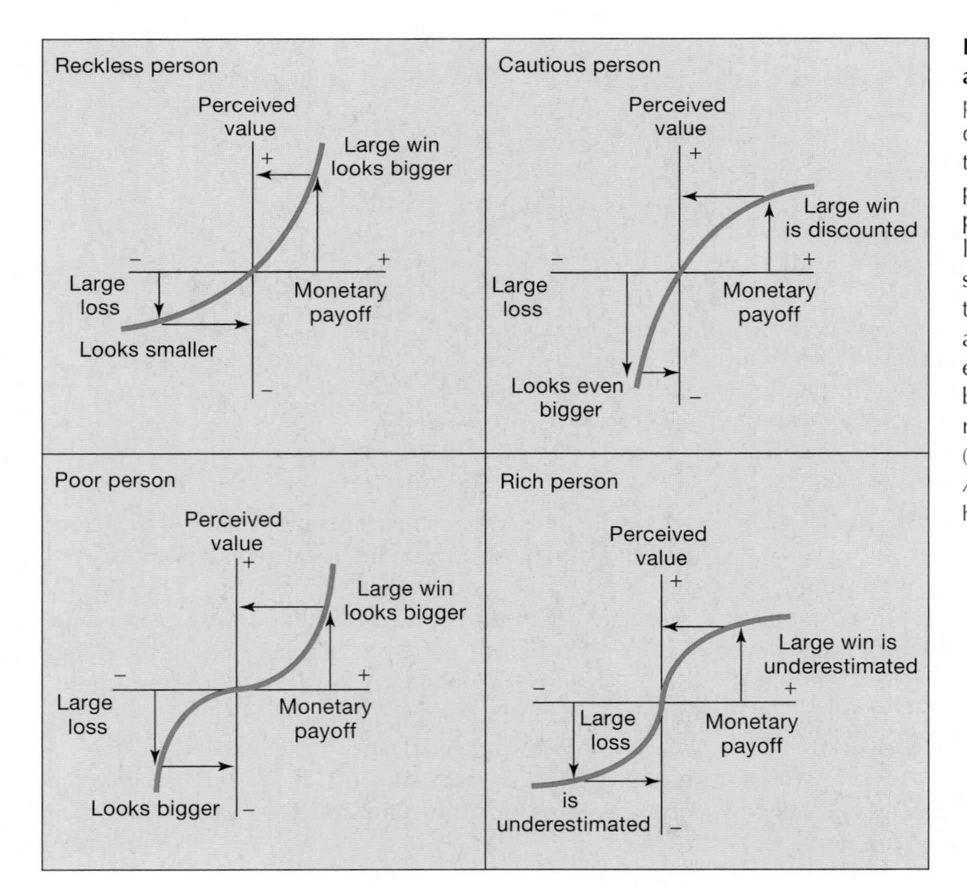

Figure 6.9 Individual utility functions and attitudes to risk taking Cautious persons tend to overestimate the consequences of a loss while underestimating the consequences of a gain. For reckless persons the opposite is true. Many people are not especially cautious or reckless, their utility functions being somewhere between these two extremes. Attitudes toward risk-taking can also depend on people's level of affluence. Poor persons tend to exaggerate both large gains and large losses, while rich persons tend to underestimate them. (After R. G. Golledge and R. J. Stimpson, *Analytical Behavioral Geography.* Beckenham: Croom Helm, 1987, Fig. 2.3, p. 48.)

ferent situations, making their decisions difficult to predict through models based on "rational" behavior.

Finally, one aspect of cognitive imagery is of special importance in modifying people's behavior: the sentimental and symbolic attributes ascribed to places. Through their daily lives and the cumulative effects of cultural influences and significant personal events, people build up affective bonds with places. They do this simultaneously at different geographic scales: from the home, through the neighborhood and locality, to the national state. The tendency for people to do this has been called topophilia. **Topophilia** literally means "love of place." Geographers use the term to describe the complex of emotions and meanings associated with particular places that, for one reason or another, have become significant to individuals. The result is that most people have a home area, hometown, or home region for which they have a special attachment or sense of identity and belonging.

LANDSCAPE AS A HUMAN SYSTEM

As we saw in Chapter 1, *landscape* is a term that means different things to different people. For some, the term brings to mind the design of formal gardens and parks, as in landscape architecture. For others, landscape signifies a bucolic countryside or even the organization of plantings around residences and public buildings. For still others, landscape calls to mind the artistic rendering of scenery, as in landscape painting.

Geographers study vernacular landscapes (or ordinary landscapes) because they reflect the distinctive attributes of particular places or regions. They study "landscapes of power," such as clusters of corporate tower blocks, and "landscapes of despair," such as homeless encampments and **derelict landscapes.** The latter are ones that have experienced abandonment, misuse, disinvestment, or vandalism. Geographers also study symbolic landscapes because they reflect certain values or ideals—either those intended by the builders or financiers of particular places or those perceived by other groups. Some individual buildings and structures are so powerfully symbolic that they come to stand for entire cities: the Eiffel Tower in Paris, Red Square in Moscow, and the Sugarloaf in Rio de Janeiro, for example (**Figure 6.10**). It is generic landscapes of different kinds that are most interesting to geographers, however.

Ordinary Landscapes

As we saw in Chapter 1, some landscapes come to symbolize entire nations or cultures. Some quite ordinary landscapes are also powerfully symbolic because they are understood as being a particular kind of place. The stereotypical New England townscape, for example (**Figure 6.11**), is widely taken to symbolize not just a certain type of regional architecture but the best that Americans have known "of an intimate, family-centered, Godfearing, morally conscious, industrious, thrifty, democratic *community*."[2] Another ordinary townscape with powerful symbolic connotations is the typical Main Street of Middle America (Figure 6.11). It is "middle" in several respects: between the frontier to the west and the cosmopolitan seaports to the east; between agricultural regions and industrial metropolises; between affluence and poverty. It has come to represent places with a balanced community, populated by property-minded, law-abiding citizens devoted to free enterprise and a certain kind of social morality.

A more recent example is provided by the landscapes of contemporary American suburbia. They are landscapes of bigness and ostentation, characterized by packaged developments, simulated settings, and conspicuous consumption—conservative utopias of themed and fortified subdivisions of private master-planned developments (**Figure 6.12**). They are landscapes of casual vulgarity, dominated by a presumed reciprocity between size and social supremacy: "Vulgaria."

> Vulgaria . . . is now a pathological condition of the American metropolis, characterized by inert and pretentious neighborhoods that are irradiated by bigness and spectacle. Vulgaria's homes are tract mansions and starter castles of 3,000 or 4,000 square feet and upwards, featuring two-story entrance halls, great rooms, and 3- or 4-car garages, huge kitchens, spa-sized bathrooms, his-and-hers room-sized master closets, media rooms, fitness centers, home offices, high-tech security systems; and perhaps even an au pair suite. SUVs, of course, are popular driveway accessories in Vulgaria—preferably "large" category SUVs like the Cadillac Escalade ESV (over 18 feet long; 3.6 tons gross weight), the Hummer H2 (nearly 16 feet long; 4.3 tons), and the Lincoln Navigator (17 feet long; 3.7 tons).
>
> Vulgaria's preferred exterior residential styling deploys any kind of neo-traditional motif as long as the street frontage is impressive, with high gabled roofs, unusual-shaped windows, and "architectural" features such as turrets, bays, and portes-cochère. . . . Place names are coy, rustic, unnaturally antiquated, and privileged-sounding, often with gruesome affectations of spelling.
>
> . . . The commercial components of Vulgaria provide some of its biggest and most spectacular elements: big-box retail stores of 250,000 square feet and more on a single level, luxurious mall complexes with themed restaurants and vast parking lots. Even the churches in Vulgaria are big and spectacular: enormous steel and glass complexes that accommodate facilities calculated to colonize every aspect of life, from aerobics classes, bowling alleys,

[2]D. W. Meinig (ed.) *Interpretations of Ordinary Landscapes: Geographic Essays*. New York: Oxford University Press, p. 27.

Figure 6.10 Famous landmarks Some cities are immediately recognizable because of certain buildings, landmarks, or cityscapes that have come to symbolize them. These examples are known worldwide.

Figure 6.11 Ordinary landscapes Some ordinary cityscapes are powerfully symbolic of particular kinds of places. The New England village and the Main Street of Middle America are in this category, so much so that they have been taken as symbolizing the United States itself, part of the "iconography of nationhood," the symbolic landscapes that give the country a sense of identity, both at home and abroad.

and aquatic centers with Christian themes, to multimedia bible classes and jumbotron screens that project the lyrics to happy-clappy pop-style religious songs. Congregations run into the thousands, and some megachurches have parking lots so big that shuttle buses circulate in order to collect visitors parked in their far corners.[3]

The point is this: ordinary landscapes, as geographers such as Don Mitchell have established, are instruments of social and cultural power that naturalize political-economic structures as if they were simply given and inevitable. As powerful complexes of signs, they perform vital functions of social regulation. The landscapes of Vulgaria have naturalized an ideology of competitive consumption, moral minimalism, and disengagement from notions of social justice and civil society—the peculiar mix of political conservatism and social libertarianism that is the hallmark of much of contemporary suburban America.

Geographers' interest in landscape originated in Carl Sauer's concept of the *cultural landscape,* described in Chapter 5. Since 1925, when Sauer advocated the study of the cultural landscape as a uniquely geographical pursuit, new generations of geographers have been expanding the concept. The fact that different people comprehend the landscape differently is central to the **humanistic approach** in geography, which places the individual—especially individual values, meaning systems, intentions, and conscious acts—at the center of analysis. As the Sudanese example given earlier suggests, children's perceptions of their worlds are different from those of their parents, and girls see their world differently from boys, even in the same family.

Environmental perception, and its close relative behavioral geography, are interdisciplinary, drawing together

[3]P. L. Knox, "Vulgaria: The Re-Enchantment of Suburbia." *Opolis, 1* (2), 2005, p. 44.

Figure 6.12 Vulgaria The dominant theme in upscale residential development in the United States is size and ostentation.

geographers, landscape architects, psychologists, architects, and others. Professionals in these disciplines investigate what preferences people have in landscapes, how they construct cognitive images of their worlds, and how they find (or fail to find) their way around in various settings. Recall from Chapter 1 that cognitive images are representations of the world that can be called to mind through the imagination. The humanistic approach's focus on the perceptions of individuals is an important counterweight to the tendency to talk about a social group or society more generally. Nevertheless, some critics argue that humanistic research has limited utility because individual attitudes and views do not necessarily add up to the views held by a group or a society.

One alternative to the humanistic approach explores both the role of larger forces, such as culture, gender, and the state, and the ways in which these forces enhance or constrain individuals' lives. Much recent cultural geographical work, therefore, conceptualizes the relationship of people and the environment as interactive, not one-way, and emphasizes the role that landscapes play in shaping and reinforcing human practices. This most recent conceptualization of landscape is more dynamic and complex than the one Carl Sauer advanced, and it encourages geographers to look outside their own discipline—to anthropology, psychology, sociology, and even history—to fully understand its complexity.

Landscape as Text

Such a dynamic and complex approach to understanding landscape is based on the conceptualization of **landscape as text,** by which we mean that, like a book, landscape can be read and written by groups and individuals. This approach departs from traditional attempts to systematize or categorize landscapes based on the different elements they contain. The landscape-as-text view holds that landscapes do not come ready-made with labels on them. Rather, there are "writers" who produce landscapes and give them meaning, and there are "readers" who consume the messages embedded in landscapes. Those messages can be read as signs about values, beliefs, and practices, though not every reader will take the same message from a particular landscape (just as people may differ in their interpretation of a passage from a book). In short, landscapes both produce and communicate meaning, and one of our tasks as geographers is to interpret those meanings.

CODED SPACES

Landscapes, then, are embedded with meaning, which can be interpreted differently by different people and groups. But in order to interpret or read our environment, we need to understand the language in which it is written. We must learn how to recognize the signs and symbols that go into the making of landscape. The practice of writing and reading signs is known as **semiotics.**

Semiotics in the Landscape

Semiotics assumes that innumerable signs are embedded or displayed in landscape, space, and place, sending messages about identity, values, beliefs, and practices. The signs that are constructed may have different meanings for those who produce them and those who read, or interpret, them. Some signs are so subtle as to be recognizable only when pointed out by a knowledgeable observer; others may be more readily available and more ubiquitous in their spatial range. For example, semiotics enables us to recognize that college students, by the way they dress, send messages to each other and the wider world about who they are and what they value. For some of us certain types, such as the "skater" or the "goth," are readily identifiable by their clothes, hairstyle, or footwear, by the books they carry, or even by the food they eat.

Semiotics, however, is not only about the signs that people convey by their mode of dress. Messages are also deployed through the landscape and embedded in places and spaces. Consider the very familiar landscape of the shopping mall. Although there is certainly a science to the size, scale, and marketing of a mall based on demographic research as well as environmental and architectural analysis, there is more to the mall than these concrete features. The placement and mix of stores and their interior design, the arrangement of products within stores, the amenities offered to shoppers, and the ambient music all combine to send signals to the consumer about style, taste, and self-image (**Figure 6.13**). Called by some "palaces of consumption," malls are complex semiotic sites, directing important signals not only about what to buy but also about who should shop there and who should not.

To fully appreciate the importance of shopping to life in the United States, consider the following statistics. In 1998, according to the National Research Bureau, there were 43,600 shopping centers in the United States with annual sales of $1,045 billion. In a typical month 189 million adults, or 94 percent of the population over age 18, shop at shopping centers. As much as Americans seem to enjoy shopping, a great many express disdain for shopping and the commercialism and materialism that accompany it. Thus, shopping is a complicated activity about which people feel ambivalence. It is not surprising, therefore, that developers have promoted shopping as a kind of tourism. The mall is a "pseudoplace" meant to encourage one sort of activity—shopping—by projecting the illusion that something else besides shopping (and spending money) is actually going on. Because of their important and complex function, malls are places with rich semiotic systems expressed through style, themes, and fantasy. The South Coast Plaza in Orange County, California, is the number one retail center in the United States with almost 3 million square feet of enclosed space covering 128 acres. It is one of the largest malls in the country and is probably the most profitable, boasting over $20 billion annually in sales. The mall contains luxury goods stores like Gucci, Versace, Chanel, Tiffany, Jimmy Choo, and Cartier—among a wide range of other upscale venues—as well as more middle-brow stores like Sears, Macy's, and Robinson's May. The latter stores anchor the mall at its outside corners, while the more luxurious shops and boutiques occupy interior locations. Thus the central stretches of the mall convey signs of affluence and luxury appealing to upper class patrons while the periphery is more oriented to necessity and practicality for middle and lower middle class patrons.

Figure 6.13 Shopping mall Developers of shopping malls know that consumer behavior is heavily conditioned by the spatial organization and physical appearance of retail settings. As a result, they find it worthwhile to spend large sums creating what they consider to be the appropriate atmosphere for their projects.

However complex the messages that malls send, one focus is consistent across class, race, gender, age, ethnicity, and other cultural boundaries: consumption, a predominant aspect of globalization. Indeed, malls are the early twenty-first century's spaces of consumption, where just about every aspect of our lives has become a commodity. Consumption—or shopping—defines who we are more than ever before, and what we consume sends signals about who we want to be. Advertising and the mass media tell us what to consume, equating ownership of products with happiness, a good sex life, and success in general. Within the space of the mall these signals are collected and resent. The architecture and design of the mall are an important part of the semiotic system shaping our choices and molding our preferences. As architectural historian Margaret Crawford writes:

> All the familiar tricks of mall design—limited entrances, escalators placed only at the end of corridors, fountains and benches carefully positioned to entice shoppers into stores—control the flow of

consumers through the numbingly repetitive corridors of shops. The orderly processions of goods along endless aisles continuously stimulate the desire to buy. At the same time, other architectural tricks seem to contradict commercial consideration. Dramatic atriums create huge floating spaces for contemplation, multiple levels provide infinite vistas from a variety of vantage points, and reflective surfaces bring near and far together. In the absence of sounds from the outside, these artful visual effects are complemented by the "white noise" of MUZAK and fountains echoing across enormous open courts.[4]

Malls, condominium developments, neighborhoods, university campuses, and any number of other possible geographic sites possess codes of meaning. By linking these sites with the forces behind globalization, it is possible to interpret them and understand the implicit messages they contain. And it is certainly not necessary to restrict our focus to sites in the core.

Consider Brasilia, the capital city of Brazil. As early as independence from Portugal in 1822, Brazilian politicians began suggesting that a new capital be established on the central plateau in the undeveloped interior of the country, but the government was not officially transferred until 1960. A symbol of the taming of the wild interior of the country and the conquest of nature through human ingenuity, Brasilia is also a many-layered system of signs conveying multiple and frequently contradictory messages. Interestingly, Brasilia, which was intended to symbolize a new age in Brazilian history, was also literally an attempt to construct one. That is why its plan and its architecture are so self-consciously rich with messages meant to transform Brazilian society through a new and radical form of architecture. To launch both the idea and the reality of a "city in the wilderness," the Brazilian government sponsored a design contest in the 1950s hoping to encourage the development of a new vision for the new capital (and by extension a new society). The winner of the contest was engineer Lucio Costa. His original plan, submitted to an international jury, was a simple sketch of a series of three crosses, each more elaborate than the previous one. In a semiotic reading of Costa's plan, as well as of the city, now more than 35 years old, the sign-of-the-cross ordering is an intentional use of a well-known mark. In its graphic form, the crossed axes represent the cross of Christianity. This aspect of the sign is important because it suggests that Brasilia was to be built on a sacred site, an important endorsement for the founding of a new capital.

Many observers of the plan and the completed city have said that it resembles an airplane. This observation seems especially apt when one looks at the master plan of the city (**Figure 6.14**). The plan shows that the residential districts were to be located along the north and south wings, and administrative government offices on the part corresponding with the fuselage. The commercial district was to be constructed at the intersection of the wings and the fuselage, with a cathedral and museum along the monumental axis. Like the sign of the cross, the significance of an airplane is obvious. Politicians and planners envisioned Brasilia as both the engine and the symbol of the rapid modernization of the country. The image of an airplane in flight was an exciting, soaring, and uplifting symbol of modernization.

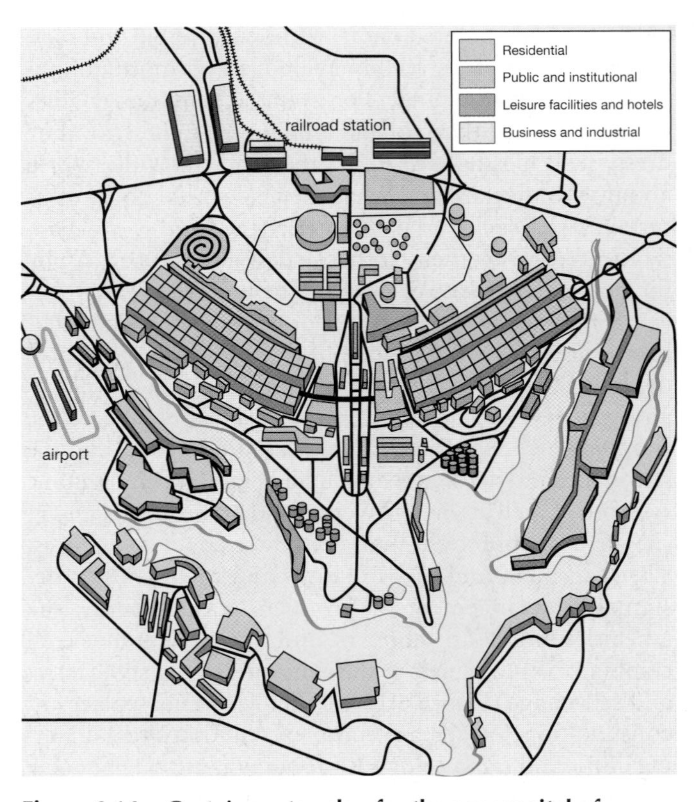

Figure 6.14 Costa's master plan for the new capital of Brasilia In 1957 Lucio Costa submitted his entry to the master plan competition for the new capital of Brazil on five medium-size cards, each containing a set of freehand sketches and some commentary. Three essential elements comprised his plan: a cross created by the intersection of two highway axes; an equilateral triangle superimposed on this cross that defined the geographical area of the city; and two terraced embankments and a platform. The more detailed rendition of the plan Costa drew for Brazil's new capital, illustrated here, reveals how he thought the various activities and material forms of city life should be arranged geographically within the city. (After B. Marshall (ed.), *The Real World.* London: Marshall Editions/Houghton Mifflin, 1991, p. 171.)

All of Brasilia's major public buildings were designed by the internationally famous Brazilian architect Oscar Niemeyer (**Figure 6.15**). The residential axes were designed with clusters of apartment buildings, each cluster surrounding a set of recreational facilities, school buildings, and shopping areas. The University of Brasilia was also part of the early vision of the city, as was the creation

[4]M. Crawford, "The World in a Shopping Mall," in M. Sorkin (ed.), *Variations on a Theme Park.* New York: Noonday Press, 1992, p. 14.

Figure 6.15 The landscape of Brasilia The landscape of Brasilia, as designed and constructed by architect Oscar Niemeyer, is that of a dynamic, modern world city. Aesthetic beauty, a sense of permanence, the expression of power, and the idiosyncrasies of the individual creative process are among the primary elements of this designed landscape.

of a lake and the official home of the president of the country, the Palacio de Alvorada, or the Palace of Dawn (**Figure 6.16**). The examples of the American shopping mall and the federal district of Brasilia illustrate the way that landscapes can be read, or decoded, by interpreting the signs and symbols they project. Not all the signs are consistent, even when planners and designers have complete control over their projects, because readers do not always interpret signs in ways the creators intended. In both cases social and political realities can disrupt the plan and send very different messages.

Figure 6.16 Performing Arts Center, Brasilia This structure combines the homogenized international style of "glass box" architecture with culturally distinctive artistic elements, such as the sweeping colonnades pictured here.

Sacred Spaces

Religious places can also be read and decoded. Indeed, most religions designate certain places as sacred, often because a special event occurred there. Sites are often designated as sacred in order to distinguish them from the rest of the landscape, which is considered ordinary or profane. **Sacred spaces** are areas of the globe recognized by individuals or groups as worthy of special attention because they are the sites of special religious experiences and events. They do not occur naturally; rather, they are assigned sanctity through the values and belief systems of particular groups or individuals. Geographer Yi-Fu Tuan insists that what defines the sacredness of a space goes beyond the obvious shrines and temples. Sacred spaces are those that rise above the commonplace and interrupt ordinary routine.

In almost all cases, sacred spaces are segregated, dedicated, and hallowed sites that are maintained as such generation after generation. Believers—including mystics, spiritualists, religious followers, and pilgrims—recognize sacred spaces as being endowed with divine meaning. The range of sacred spaces includes sites as different as a remote temple in central Bhutan, a holy spot for Bhutanese Buddhists, and the Black Hills of South Dakota, the sacred mountains of the Lakota Sioux (**Figures 6.17** and **6.18**).

Figure 6.17 Buddhist temple, central Bhutan An example of an elaborate, highly constructed and maintained sacred site. For Bhutanese Buddhists, temples are holy places that are sites of worship and important rituals.

Figure 6.18 Black Hills of South Dakota The Black Hills are a sacred site of the Sioux Indians. In 1877 the Lakota Sioux lost control of the Black Hills after a series of bloody battles with the United States Cavalry, including the Battle of Little Bighorn. The Sioux are still fighting to reclaim their sacred site—the Paha Sapa—and have turned down $105 million in compensation.

Often members of a specific religion are expected to journey to especially important sacred spaces to renew their faith or to demonstrate devotion. A pilgrimage is a journey to a sacred space, and a pilgrim is a person who undertakes such a journey. In India many of the sacred pilgrimage sites for Hindus are concentrated along the seven sacred rivers: the Ganges, the Yamuna, the Saraswati, the Naramada, the Indus, the Cauvery, and the Godavari. The Ganges is India's holiest river, and many sacred sites are located along its banks (**Figure 6.19**). Hindus visit sacred pilgrimage sites for a variety of reasons, including to seek a cure for sickness, wash away sins, and fulfill a promise to a deity.

Perhaps the most well-known pilgrimage is the hajj, the obligatory once-in-a-lifetime journey of Muslims to Mecca. For one month every year the city of Mecca in

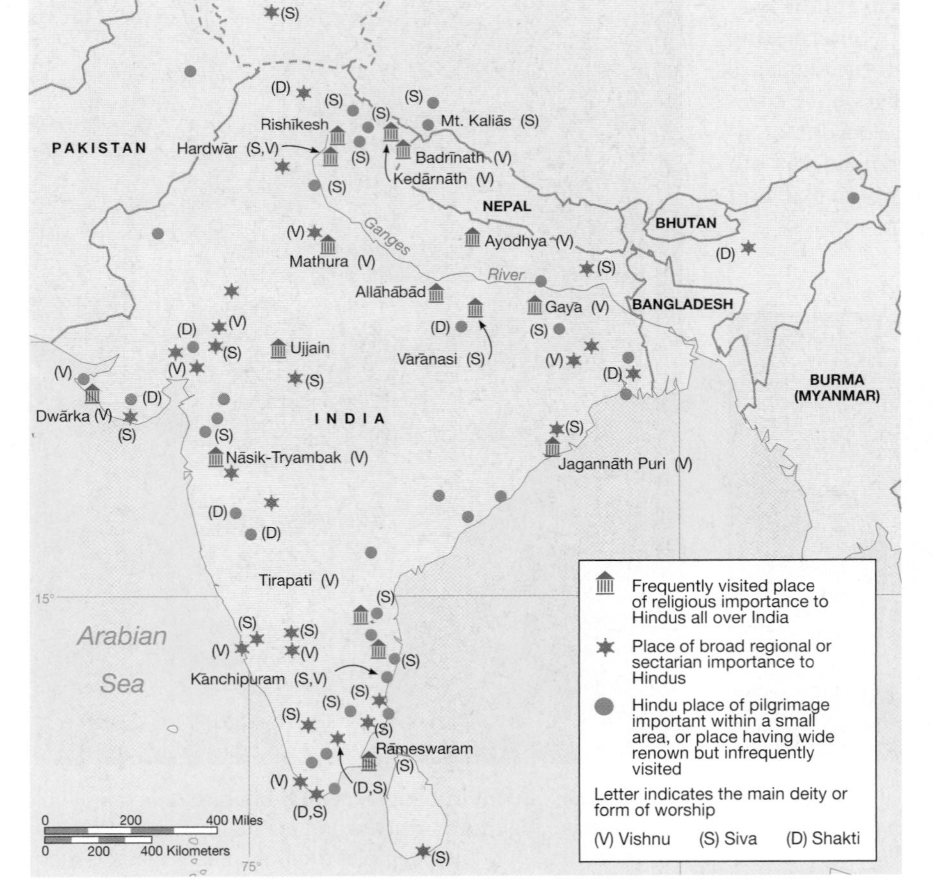

Figure 6.19 Sacred sites of Hindu India India's many rivers are holy places within Hindu religion, so sacred sites are located along the country's many riverbanks. Shrines closer to the rivers are regarded as holier than those farther away. (*Source:* Adapted from Ismail Ragi al Farugi and David E. Sopher, *Historical Atlas of the Religions of the World.* New York: Macmillan, 1974.)

Saudi Arabia swells from its base population of 150,000 to over 1,000,000 as pilgrims from all over the world journey to fulfill their obligation to pray in the city and receive the grace of Allah. **Figure 6.20** shows the principal countries that send pilgrims to Mecca.

Pilgrimages to sacred sites are made all over the world, and Christian Europe is no exception. The most visited sacred site in Europe is Lourdes, at the base of the Pyrenees in southwest France, not far from the Spanish border (**Figure 6.21**). Another sacred site that attracts pilgrims throughout the world is the city of Jerusalem, and the Holy Land more generally, which is visited by Jews, Orthodox, Catholics, Protestants, Christian Zionists, and followers of many other religions (see Box 6.1: "Geography Matters—Jerusalem, the Holy City"). As with most sacred spaces, the codes that are embedded in the landscape of the Holy Land may be read quite differently by different religious and even secular visitors. Two students of pilgrimage observed:

> Each group brings to Jerusalem their own entrenched understandings of the sacred; nothing unites them save their sequential—and sometimes simultaneous—presence at the same holy sites. For the Greek Orthodox pilgrims, indeed, the precise definition of the site itself is largely irrelevant; it is the icons on display which are the principal focus of attention. For the Roman Catholics, the site is important in that it is illustrative of a particular biblical text relating to the life of Jesus, but it is important only in a historical sense, as confirming the truth of past events. Only for the Christian Zionists does the Holy Land itself carry any present and future significance, and here they find a curious kinship with indigenous Jews.[5]

[5]J. Eade and M. Sallnow (eds.), *Contesting the Sacred*. London: Routledge, 1991, p. 14.

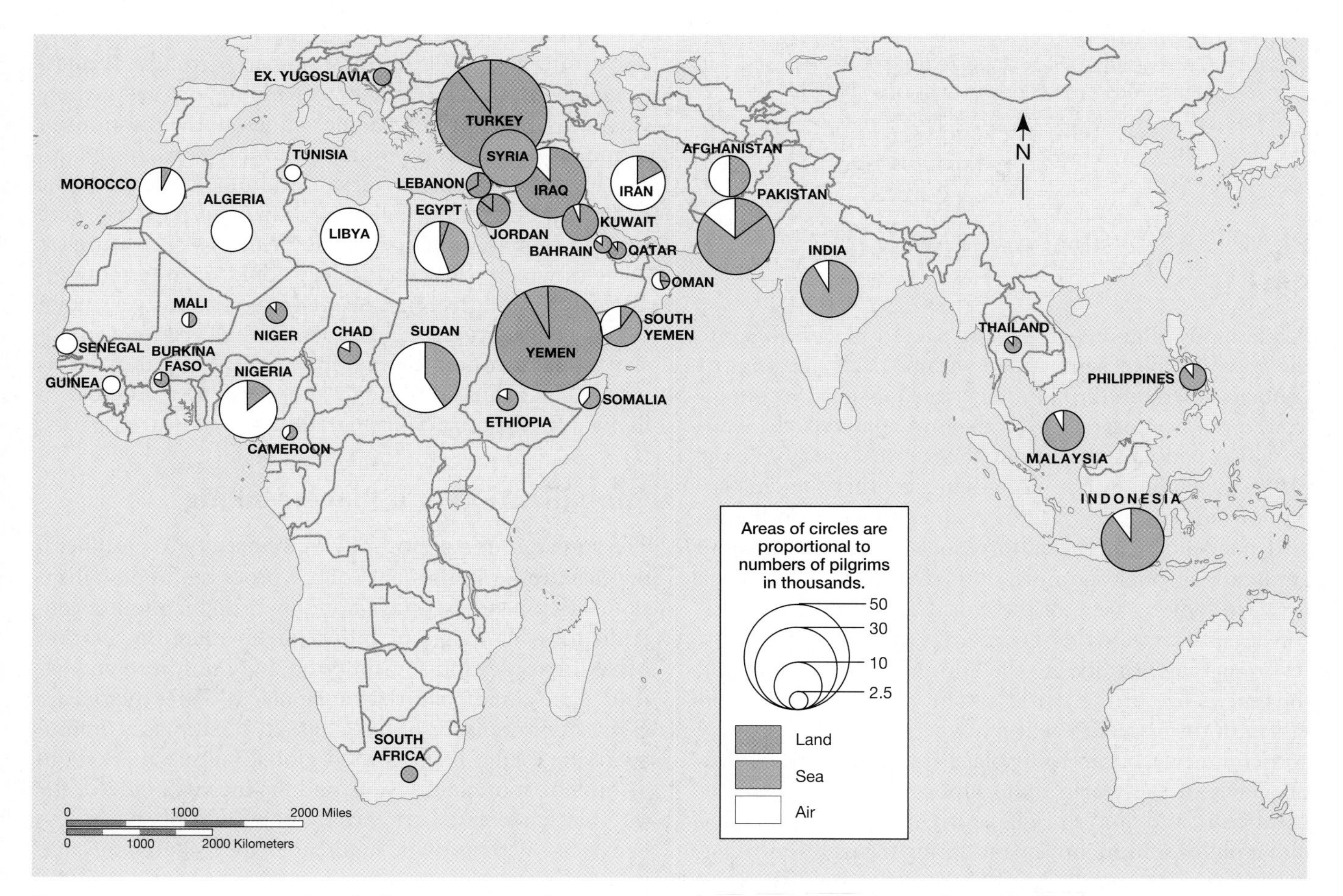

Figure 6.20 Source areas for pilgrims to Mecca Islam requires that every adult Muslim perform the pilgrimage to Mecca at least once in a lifetime. This obligation is deferred for four groups of people: those who cannot afford to make the pilgrimage; those who are constrained by physical disability, hazardous conditions, or political barriers; slaves and those of unsound mind; and women without a husband or male relative to accompany them. The pattern of actual pilgrimages to Mecca (which is located close to the Red Sea coast in Saudi Arabia) suggests a fairly strong distance-decay effect, with most traveling relatively short distances from Middle Eastern Arab countries. More distant source areas generally provide smaller numbers of pilgrims, though Indonesia and Malaysia are notable exceptions. (After C. C. Park, *Sacred Worlds*. London: Routledge, 1994, p. 268.)

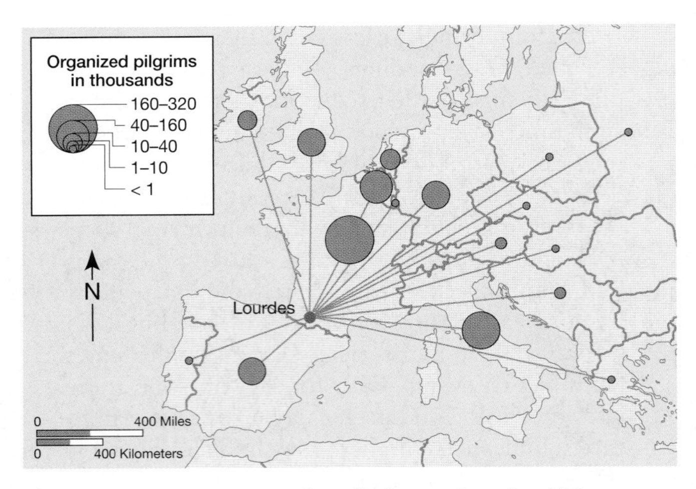

Figure 6.21 Source areas for pilgrims to Lourdes This map shows the points of origin of European group-organized pilgrims to Lourdes in 1978. These represent only about 30 percent of all pilgrims to Lourdes, most of whom travel to the shrine on their own. Improved transportation (mainly by train) and the availability of organized package trips have contributed to a marked increase in the number of pilgrims visiting. Many of the 5 million pilgrims who visit the town each year do so in the hope of a miraculous cure for medical ills at a grotto where the Virgin Mary is said to have appeared before 14-year-old Bernadette Soubirous in a series of 18 visions in 1858. (After C. C. Park, *Sacred Worlds*. London: Routledge, 1994, p. 284.)

PLACE AND SPACE IN MODERN SOCIETY

While malls, planned cities, and sacred places illustrate the way that places are coded within specific settings or contexts, there are broad dimensions of social, economic, and political organization that are related to the ways in which people think about themselves and about the places they inhabit. For more than a century the philosophy of modernity has been a major influence on the interdependencies among culture, society, space, place, and landscape throughout much of the world. **Modernity** is a forward-looking view of the world that emphasizes reason, scientific rationality, creativity, novelty, and progress. Its origins can be traced to the European Renaissance and the emergence of the world system of competitive capitalism in the sixteenth century, when scientific discovery and commerce began to displace traditional sociocultural views of the world that emphasized mysticism, romanticism, and fatalism. These origins were consolidated into a philosophical movement during the eighteenth century, when the so-called Enlightenment established the widespread belief in universal human progress and the sovereignty of scientific reasoning.

At the beginning of the twentieth century this philosophy developed into a more widespread intellectual movement. A series of sweeping technological and scientific developments not only triggered a new round of spatial reorganization (see Chapter 2) but also transformed the underpinnings of social and cultural life. These developments included the telegraph, the telephone, the X ray, the cinema, the radio, the bicycle, the internal combustion engine, the airplane, the skyscraper, relativity theory, and psychoanalysis. Universal human progress suddenly seemed to be a realistic prospect.

Nevertheless, the pace of economic, social, cultural, and geographic change was unnerving and the outcomes uncertain. The intellectual response, developed among a cultural avant-garde of painters, architects, novelists, and photographers, was a resolve to promote modernity through radical changes in culture. These ideas were first set out in the "Futurist Manifesto," published in 1909 in the Paris newspaper *Le Figaro* in the form of a letter from the Italian poet Filippo Marinetti. Gradually the combination of new technologies and radical design contributed to the proliferation of landscapes of modernity. Among the most striking of these were modernist urban landscapes, from Helsinki to Hong Kong and from New York to Nairobi (**Figure 6.22**). Indeed, in a general sense almost all of the place-making and landscapes of the early and mid-twentieth century are the products of Modernity.

Throughout this period a confident and forward-looking Modernist philosophy remained virtually unquestioned, with the result that places and regions everywhere were heavily shaped by people acting out their notions of rational behavior and progress. Rural regions, for example, bore the imprint of agricultural modernization. The hedgerows of traditional European field patterns were torn up to make way for landscapes of large, featureless fields in which heavy machinery could operate more effectively. At a different scale, peripheral countries sought to remake traditional landscapes through economic modernization. Economic development and social progress were to be achieved through a modern infrastructure of highways, airports, dams, harbors, and industrial parks.

Globalization and Place-Making

The spread of the philosophy of modernity to peripheral regions can be seen as part of the processes of globalization. These processes have not only brought about a generalization of forms of industrial production, market behavior, trade, and consumption, but reinforced and extended the commonalities among places. Three factors are especially important in this context. First, mass communications media have created global culture markets in print, film, music, television, and the Internet. Indeed, the Internet has created an entirely new *kind* of space—cyberspace—with its own "landscape" (or technoscape) and its own embryonic cultures (see Box 6.2: "Geography Matters—The Cultural Geography of Cyberspace"). The instantaneous character of contemporary communications has also made possible the creation of a shared, global consciousness from the staging of global events such as the Olympic Games and the World Cup. Second, mass communications media have diffused certain values and attitudes toward a wide spectrum of sociocultural issues,

Figure 6.22 Modern architecture This photograph of apartment buildings in Hong Kong reflects the pervasive influence of Modernist architecture in the skylines of the central areas of large cities throughout the world's core regions.

including citizenship, human rights, child rearing, social welfare, and self-expression. Third, international legal conventions have increased the degree of standardization and level of harmonization not only of trade and labor practices but of criminal justice, civil rights, and environmental regulations.

These commonalities have been accompanied by the growing importance of material consumption within many cultures. Increasingly, people around the world are eating the same foods, wearing the same clothes, and buying the same consumer products. Yet the more people's patterns of consumption converge, the more fertile the ground for counter-cultural movements. The more transnational corporations undercut the authority of national and local governments to regulate economic affairs, the greater the popular support for regionalism. The more universal the diffusion of material culture and lifestyles, the more local and ethnic identities are valued. The faster the information highway takes people into cyberspace, the more they feel the need for a subjective setting—a specific place or community—they can call their own. The faster the pace of life in search of profit and material consumption, the more people value leisure time. And the faster their neighborhoods and towns acquire the same generic supermarkets, gas stations, shopping malls, industrial estates, office parks, and suburban subdivisions, the more people feel the need for enclaves of familiarity, centeredness, and identity. The United Nations Center for Human Settlements (UNCHS) notes:

> In many localities, people are overwhelmed by changes in their traditional cultural, spiritual, and social values and norms and by the introduction of a cult of consumerism intrinsic to the process of globalization. In the rebound, many localities have rediscovered the "culture of place" by stressing their own identity, their own roots, their own culture and values and the importance of their own neighbourhood, area, vicinity, or town.[6]

New urbanism, a highly codified form of neotraditional design (**Figure 6.23**), represents the design professions' best-articulated response to this impulse. The use of traditional architectural styles along with urban elements such as alleys and public spaces surrounded by diverse housing types—all carefully controlled in a private regulatory framework—has had strong market appeal. But the proliferation of neotraditional developments (and watered-down developer look-alikes) has begun to make new urbanism an element of—rather than response to—globalization. More and more urban fabric in Europe and North America is manifest in an ersatz, sanitized and "Disneyfied" form, while most of the better-executed New urbanist developments remain artful fragments, exclusionary "privatopias" that are abstracted from the fabric—both physical and social—of their host city. Privately

[6]United Nations Center for Human Settlements, *Global Report on Human Settlements 2001*. London: Earthscan, 2001, p. 4.

6.1

GEOGRAPHY MATTERS

Jerusalem, the Holy City

Although many cities in the world have been the object of struggle and conflict over the centuries, none seems as endlessly beset by conflict as Jerusalem. Visitors, writers, and residents believe Jerusalem to be the most beautiful city in the world. If there are other serious competitors to that coveted beauty title, Jerusalem certainly has few rivals for the title of the most sacred city in the world, possessing as it does an unmatched Christian, Jewish, and Islamic history. Jerusalem began as a small settlement on the slopes of Mount Moriah. In 997 B.C. it was captured by David, king of the Israelites, who made Jerusalem the capital of Israel. Solomon, David's son and successor, built the Great (First) Temple on Mount Moriah to commemorate the place where Abraham offered to sacrifice his son. Though the temple was destroyed centuries ago, the site is central to the Jewish faith.

The history of the city reflects the history of the various empires that dominated and were succeeded by new and yet more powerful empires (**Figure 6.A**). Nebuchadnezzar, king of Babylon, destroyed the Great Temple in 586 B.C. and banished the Jews. But the Babylonian control of Jerusalem eventually gave way to the Persians, under whose rule the Jews were allowed to return and rebuild their temple, known as the Second Temple. The Persian occupation of Jerusalem was swept aside by Alexander the Great (356–323 B.C.), the king of Macedonia and one of the world's greatest military leaders, who helped spread the Greek Empire from the southern shores of the Caspian Sea into central Asia. The Romans entered the scene around 63 B.C., and Herod the Great was eventually installed to command the Roman Kingdom of Judea from Jerusalem.

During the early Christian period in Jerusalem, the Jews revolted openly against the Roman occupiers. In A.D. 132 the Romans responded by destroying the Second Temple and banishing all Jews from Jerusalem and Palestine. As a result, the Jews scattered north into Babylon and later into Europe and North Africa. The popular myth is that this ancient Jewish diaspora remained in exile until 1948, when the state of Israel was created.

The major Christian influence on Jerusalem, the "Holy City," began when Constantine I (A.D. 285?–337), the emperor of the Eastern Roman Empire, converted to Christianity in A.D. 313. This event led to the construction of churches and other build-

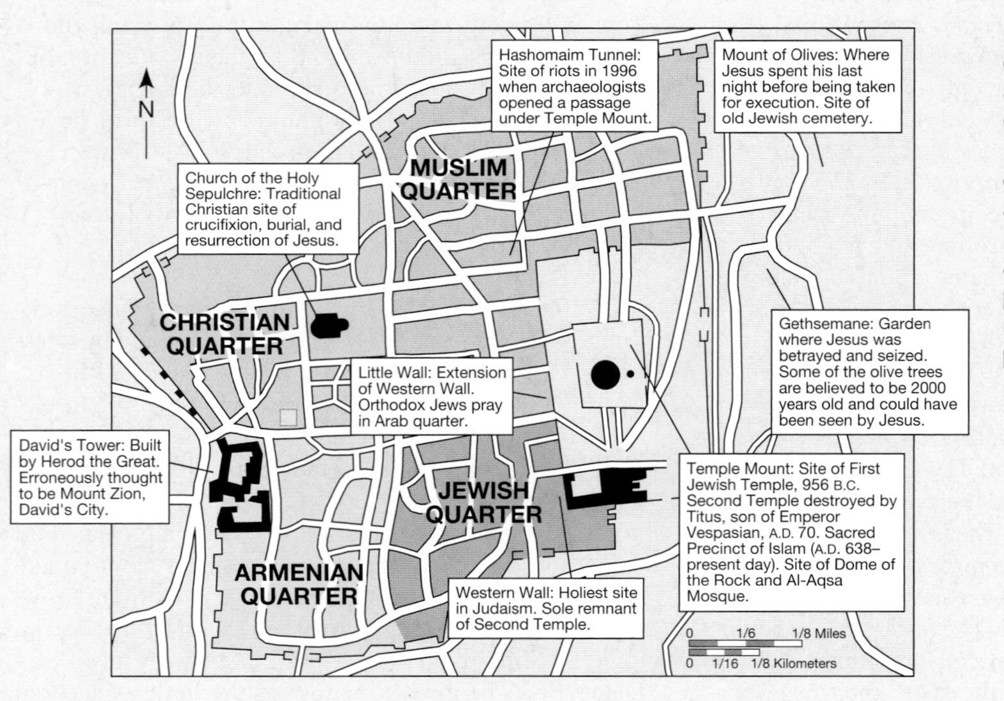

Figure 6.A Jerusalem, the Holy City This map of Jerusalem indicates the main sections of the city. Over the many years of the Israeli-Palestinian peace process, numerous proposals have been advanced about how to divide the city to satisfy the desire of both Palestinians and Israelis to possess it. The disposition of Jerusalem is one of the major issues in the ongoing peace process. (*Source:* Redrawn from *The Guardian,* October 14, 2000, p. 5.)

ings dedicated to celebrating the life of Jesus Christ. But Christian influence over the city ceased when Jerusalem eventually succumbed to Islam. In A.D. 638, Jerusalem was designated a holy city of Islam because it was believed that Muhammad had once visited heaven while in the city.

Although for several centuries Jews, Christians, and Muslims were all allowed access to the city of Jerusalem, by the 10th century, the persecution of non-Muslims became common. From the eleventh century until the thirteenth, European Christians undertook military expeditions—called the Crusades—to the Holy Land in an attempt to wrest control of Jerusalem from the Muslims. In 1099 Crusaders captured the city; Christians lost it again to the Muslim military leader Saladin in 1187. In 1517 Jerusalem was absorbed into the Ottoman empire, and the city was ruled from Istanbul for more than 400 years. The Ottomans, however, had little interest in Jerusalem, and Jewish immigrants began returning to the city and Palestine starting in the mid-nineteenth century.

The contemporary history of the city derives from political and geographical implications of the Balfour Declaration, which stipulated that Jerusalem should be an international city with no one state claiming it as entirely its own. Today, Jerusalem is a highly contested city as Palestinians, Christians, Muslims, and Israeli Jews fight for control of it. An example of this contest for control is the continuing dispute over the Dome of the Rock, which was constructed between A.D. 688 and 691. Muslims claim the Dome as their most sacred site (**Figure 6.B**). Yet the Dome sits on a site sacred to the Jews, the Temple Mount, the site where the Great Temple and Second Temple were built and later destroyed. Indeed, the Dome is believed to enclose the sacred rock upon which Abraham prepared to sacrifice his son, according to Jewish tradition, and, according to Islamic tradition, the same rock from which the prophet Muhammad launched himself to visit heaven. Also located at the Temple Mount is the Al-Aqsa Mosque, a central sacred site to Muslims.

While nationalist Israelis maintain that Jerusalem will be the "eternal and undivided capital" of Israel, Palestinians believe that Jerusalem is the future capital of the Palestinian state. In fact, the 1992 Oslo peace accords that lead to a Declaration of Principles between Israel and Palestine hint at the possibility of negotiating the future of Jerusalem. Further peace negotiations, including those at Wye River, Maryland, in 1998; at Sharm al-Shaykh in the Egyptian Sinai Desert in 1999; in Camp David, Maryland, in 2000; in Washington, D.C., in 2001; and in Geneva in 2003 are cause for hope that a resolution to the future of Jerusalem that includes some control by the Palestinians can be achieved. At present, Jerusalem is entirely controlled by Israel. Yet in accordance with the Oslo peace agreements, in May 2000 Israel conceded three villages to the control of the Palestinian Authority. All three villages are part of greater Jerusalem in the West Bank, and all but one, Abu Dis, are actually a few meters beyond the Jerusalem municipal border. From Abu Dis, the Old City of Jerusalem is clearly visible, particularly the Dome of the Rock. In late 2005 it is still unclear how the contested control of Jerusalem might eventually be resolved. What is clear is that both sides are equally passionate about their claims.

Figure 6.B Dome of the Rock Located in Haram Al-Shaif, or Temple Mount, the Dome of the Rock sits in part of Jerusalem that is technically neither Jewish nor Muslim, but both. The Muslims want this entire holy site. The Israelis also want it because the Wailing Wall, sacred to Jews, also occupies part of the Temple Mount. Palestinian anger was unleashed—and the delicate peace process derailed—in the fall of 2000 when Ariel Sharon, a nationalist Israeli politician who several months later became Israel's president, visited the Temple Mount during the first days of the Muslim holy season of Ramadan

Figure 6.23 New urbanism Well-appointed enclaves of private residential development in 'neo-traditional' style, intended to evoke the atmosphere and sociability of traditional small towns, are typical of New Urbanist design. This example is Kentland, near Gaithersburg, Maryland.

planned and developed communities in the new urbanist, neotraditional mold are popular with consumers because of their perceived social status, implied security, and real advantages in terms of the robustness of property values. Yet in spite of mandated mixed uses, diverse housing types, and careful detailing, they tend to be rather prim, somehow lacking the character that is essential to a sense of place, and with little evidence of social cohesion, identity, or vitality.

A rather different example of the impulse for people to recover a sense of place is provided by the slow city (CittaSlow) movement. The CittaSlow movement is more explicitly a grassroots response to globalization and is closely related to the longer established and better known slow food movement. The aims of the two movements are different but complementary: In broad terms, both organizations are in favor of local, traditional cultures, a relaxed pace of life, and conviviality. Both are a response to the increasing pace of everyday life associated with the acceleration of money around local, national, and global circuits of capital. Both are hostile to big business and globalization, though their driving motivation is not so much political as ecological and humanistic. Slow food is devoted to a less hurried pace of life and to the true tastes, aromas, and diversity of good food. The movement also serves as a rallying point against globalization, mass production, and the kind of generic fast food represented by U.S.-based franchises like McDonald's, Burger King, Pizza Hut, Taco Bell, and Kentucky Fried Chicken. By 2005 the slow food movement, based in Bra, near Turin in northern Italy, had established more than 450 slow food convivia (local branches) in more than 45 countries, with over 80,000 members worldwide. Its campaigns cover a range of specific causes, from protecting the integrity of chocolate to promoting the cultivation of traditional crop varieties and livestock breeds and opposing genetically engineered foods.

The slow city movement was formed in October 1999, when Paolo Saturnini, mayor of Greve-in-Chianti, a Tuscan hill town, organized a meeting with the mayors of three other municipalities (Orvieto, Bra and Positano) to define the attributes that might characterize a *città lente*—slow city. At their founding meeting in Orvieto, the four mayors committed themselves to a series of principles that included working toward calmer and less polluted physical environments, conserving local aesthetic traditions, and fostering local crafts, produce, and cuisine. They also pledged to use technology to create healthier environments, to make citizens aware of the value of more leisurely rhythms to life, and to share their experience in seeking administrative solutions for better living. The goal is to foster the development of places that enjoy a robust vitality based on good food, healthy environments, sustainable economies, and the seasonality and traditional rhythms of community life (see Box 6.3: "Window on the World—Slow Cities").

Many commentators on cultural change have noted this broad shift in cultural sensibilities within the world's more affluent countries. Most pronounced in the more affluent core countries, this shift is broadly characterized as a shift from modernity to postmodernity and has involved both avant-garde and popular culture. **Postmodernity** is a view of the world that emphasizes openness to a range of perspectives in social inquiry, artistic expression, and political empowerment. Postmodernity abandons modernity's emphasis on economic and scientific progress, focusing instead on living for the moment. Above all, postmodernity is consumption-oriented, with an emphasis on the possession of particular *combinations* of things and on the style of consumption.

Postmodern society has been called a "society of the spectacle" in which the symbolic properties of places and material possessions have assumed unprecedented importance. Since the mid-1970s, postmodernity has permeated every sphere of creative activity, including art, architecture, advertising, philosophy, clothing design, interior design, music, cinema, novels, television, and urban design. The shift to postmodern cultural sensibilities is of

6.2

GEOGRAPHY MATTERS

The Cultural Geography of Cyberspace

The rapid growth of the Internet is of great cultural significance, for it has created the basis for a massive shift in patterns of social interaction, a seedbed for new forms of human consciousness, and a new medium for cultural change. Culture is fundamentally based on communication, and in cyberspace we have an entirely new form of communication: uncensored, multidirectional, written, visual, and aural.

At face value the Internet represents the leading edge of the globalization of culture. In broad terms the culture propagated by the Internet is very much core-oriented. The Internet portends a global culture based on English as the universal world language, with a heavy emphasis on core-area cultural values, such as novelty, spectacle, fashionability, material consumption, and leisure. It is unlikely, however, that the Internet will simply be a new medium through which core-area values and culture are spread.

To begin with, the impact of the Internet is likely to be highly uneven because of the digital divide (see Figure 2.23). Moreover, there is resistance in some places and regions to the cultural globalization associated with cyberspace. The French and French-Canadian authorities, already sensitive about the influence of English-language popular culture on their own, have actively sought ways to give Francophone cybernauts access to the Internet without submitting to English, the dominant language of Web sites. The French government has also subsidized an all-French alternative to the Internet: Minitel, an online videotext terminal that plugs into French telecommunication networks. A free Minitel terminal is available to anyone who stops by a France Telecom office.

In much of Asia the Internet's basic function as an information-exchange medium clashes with local cultures in which information is a closely guarded commodity. Whereas many U.S. World Wide Web sites feature lengthy government reports and scientific studies, as well as lively debates about government policy, comparable Asian sites typically offer little beyond public relations materials from government agencies and corporations. In puritanical Singapore, political leaders, worried that the Internet will undermine morality, have taken to reading private e-mail as part of an all-out effort to beat back the menace of online pornography. Chinese authorities fear that the Internet will foment political rebellion, so officials have limited access to ensure that the Chinese portion of the Internet can easily be severed from the world in the event of political unrest. The reluctance of major Asian organizations to put important information on their Web sites—along with the need for Westerners to use special software to read any local language documents that do exist—has resulted in a largely one-way flow of information, from America to Asia.

Nevertheless, the greatest potential of the Internet in terms of cultural change resides in the liberating and empowering potential of its vast resource of knowledge and information. By its very nature—a decentralized and complex web of computer hosts—the Internet empowers individuals (rather than social groups or institutions), allowing millions to say whatever they want to each other, free (for the first time in history) from state control. As such, it is an important vehicle for the spread of participatory democracy to much of the world.

particular importance to cultural geography because of the ways in which changed attitudes and values have begun to influence place-making and the creation of landscapes. Some of the most striking postmodern landscapes are to be found in the redeveloped waterfronts, revitalized downtown shopping districts, and neotraditional suburbs of major cities (**Figure 6.24**).

Places as Objects of Consumption

In much of the world, people's enjoyment of material goods now depends not only on their physical consumption or use. It is also linked more than ever to the role of material culture as a social marker. A person's home, automobile, clothing, reading, viewing, eating and drinking preferences, and choice of vacations are all indicators of that person's social distinctiveness and sense of style. This pressures individuals to continuously search for new and distinctive styles. The wider the range of foods, products, and ideas from around the world—and from past worlds—the greater the possibilities for establishing such styles.

Given that material consumption is so central to the repertoire of symbols, beliefs, and practices of postmodern cultures, the "culture industries"—advertising, publishing, communications media, and popular entertainment—have become important shapers of spaces, places, and landscapes. Because the symbolic meanings of material culture must be advertised (in the broadest sense of the word) in order to be shared, advertising (in its narrower sense) has become a key component of contemporary culture and place-making. In addition to stimulating consumer demand, advertising has always had a role in

6.3

WINDOW ON THE WORLD

Slow Cities

In Western Europe a backlash is under way in reaction to the culture of speed imposed by the 24/7 global marketplace and the manic pace of life dictated by globalization. In France a 35-hour work week was introduced in 2002, an innovation that is also being considered by Belgium and Sweden. In Italy writer and activist Carlo Petrini was so offended by the temerity of McDonald's wanting to open a store in the Piazza di Spagna in 1989 that he started the "slow food" movement.

In 1999 the slow food movement spurred an offshoot called the CittaSlow (slow cities) movement, made up of municipalities whose leaders reject the North Americanization of European cities. In 2001, the first 28 slow cities were certified. All 28 charter members were Italian, the majority of them located in northern Italy, particularly in Tuscany and Umbria. By mid-2005, the list had grown to 45 (including Verteneglio in Croatia; Hersbruck (Figure 1.30), Überlingen and Waldkirch in Germany; Asolo, Chiavenna, Orvieto, Spilimbergo (**Figure 6.C**), Tevi, Todi, and Urbino in Italy; Sokndal and Levanger in Norway; and Ludlow and Aylsham in the United Kingdom) and a more than a dozen other towns were actively seeking certification through pilot programs. More than 100 other towns from around the world have inquired about joining. The movement's charter specifies that a slow city is one with a population of fewer than 50,000. Member towns must pledge to promote organically produced, traditional foods, a clean environment, quiet neighborhoods, urban charm, and the idea that the good life is an unhurried sensual experience. The slow city charter contains over 50 pledges, such as cutting noise pollution and traffic, increasing green spaces and pedestrian zones, supporting farmers who

Figure 6.C Even on a normal working day, Spilimbergo seems like the perfect place to get away from it all. Locals linger over coffee at sidewalk tables, gossiping with friends or watching the world drift by.

teaching people how to dress, how to furnish a home, and how to signify status through groupings of possessions.

In the 1970s and 1980s, however, the emphasis in advertising strategies shifted away from presenting products as newer, better, more efficient, and more economical (in keeping with modernist sensibilities) to identifying them as the means to self-awareness, self-actualization, and group stylishness (in keeping with postmodern sensibilities). Increasingly, products are advertised in terms of their association with a particular lifestyle rather than in terms of their intrinsic utility. Many of these advertisements deliberately draw on international or global themes, and some entire advertising campaigns (for Coca-Cola, Benetton, and American Express, for example) have been explicitly based on the theme of cultural globalization. Many others rely on stereotypes of particular places or kinds of places (especially exotic, spectacular, or "cool" places) in creating the appropriate context or setting for their product. Images of places therefore join with images of global food, architecture, pop culture, and consumer goods in the global media marketplace. Advertisements both instruct and influence consumers not only about products but also about spaces, places, and landscapes.

One result of these trends is that contemporary cultures rely much more than before not only on material consumption but also on *visual* and *experiential* consumption: the purchase of images and the experience of spectacular and distinctive places, physical settings, and landscapes. Visual consumption can take the form of magazines, television, movies, sites on the World Wide Web, tourism, window shopping, people-watching, or visits to galleries and museums. The images, signs, and experiences that are consumed may be originals, copies, or simulations.

The significance of the increased importance of visual consumption for place-making and the evolution of landscapes is that settings such as theme parks, shopping malls, festival marketplaces, renovated historic districts, produce local delicacies and the shops and restaurants that sell them, and preserving local aesthetic traditions. Promoting local distinctiveness and a sense of place is almost as important to the movement as the enjoyment of good local food, wine and beer. This means that the charter also covers many aspects of urban design and planning. Candidate cities must be committed not only to supporting traditional local arts and crafts, but to supporting modern industries whose products lend distinctiveness and identity to the region. They must also be committed to the conservation of the distinctive character of their built environment and must pledge to plant trees, create more green space, increase cycle paths and pedestrianized streets, keep public squares and piazzas free of advertising billboards and neon, ban car alarms, reduce noise pollution, light pollution, and air pollution, foster the use of alternative sources of energy, improve public transport, and promote eco-friendly architecture in any new developments.

Bra, one of the founding municipalities of the slow cities movement (along with Greve, Orvieto, and Positano) has banned cars, as well as supermarkets and lurid neon signs, from parts of its historic core. In order to promote a more leisurely pace of life, every small food shop in Bra is obliged to close on Thursdays and Sundays. Small family-run businesses selling locally produced crafts, fabrics, cheeses, roasted peppers, white truffles, fresh pasta, olive oil, and specialty meats are granted the best commercial real estate. The municipality subsidizes building renovations that use the honey-colored stucco typical of the Piedmont region. In schools, children are served organic fruit and vegetables grown by local producers.

One obvious critique of the CittaSlow movement is that it could all too easily produce enervated, backward-looking, isolationist communities: living mausoleums where the puritanical zealotry of slowness has displaced the fervent materialism of the fast world. Aware of the dangers of prescriptive slowness, the CittaSlow movement hopes to propagate vitality through farmers' markets, festivals, and the creation of inviting public spaces. It aims to deploy technology in air, noise and light pollution control systems, modern waste-cycling plants, and composting facilities. It seeks to encourage business through ecologically sensitive, regionally authentic, and gastronomically oriented tourism. Here, though, is another danger: that, paradoxically, slow city designation becomes a form of brand recognition within the heritage industry. Because they are small, the charming attractions of slow cities could all too easily be overwhelmed by tourism. So the more they flaunt their gentle-paced life, the faster they may end up changing. In this scenario, shop prices will rise, and cafés will lose their spilled-drink, smoky, messy, authenticity. The better known that slow cities become, the more affluent outsiders will choose to make their second homes in them. House prices will go up, and the poor and the young will be pushed out.

Nevertheless, whatever the eventual outcomes of the slow city movement *per se,* its principles speak directly to the concepts of "dwelling" and intersubjectivity that are key to the social construction of place. Fostering respect for seasonality and traditional rhythms of community life propagates recurring and interlocking patterns of events that make for cultural transactions and public sociability in the public realm.

museums, and galleries have all become prominent as centers of cultural practices and activities. The number of such settings has proliferated, making a discernible impact on metropolitan landscapes. The design of such settings, however, has had an even greater impact on metropolitan landscapes. Places of material and visual consumption have been in the vanguard of postmodern ideas and values, incorporating eclecticism, decoration, a heavy use of historical and vernacular motifs, and spectacular features in an attempt to create stylish settings that are appropriate to contemporary lifestyles.

One interesting aspect of the increasing trend toward the consumption of experiences is the emergence of restaurants as significant cultural sites. Restaurants often represent a synthesis of the global and the local, and they can be powerful cultural symbols in their own right (**Figure 6.25**). The dining experience in a particular restaurant can be an important symbolic good. By the same token, the social standing or celebrity status of customers contributes to the value of the dining experience offered by a restaurant. Restaurants themselves can be both theater and performance—particularly in big cities like New York and Los Angeles, where underpaid or out-of-work actors, dancers, and other artists form an important part of the restaurant labor force. Restaurants bring together a global and a local labor force (immigrant owners, chefs, and waiters, as well as locals) and clientele (tourists and business travelers, as well as locals). Finally, restaurant design also contributes to a city's visual style as architects and interior designers, restaurant consultants, and restaurant-industry magazines adapt global trends to local styles.

While the idea of the emergence of a single global culture is too simplistic, the postmodern emphasis on material, visual, and experiential consumption means that many aspects of contemporary culture transcend local and national boundaries. Furthermore, many residents of the fast world are world travelers—either directly or via the TV in their living room—so that they are knowledgeable

Figure 6.24 Landscape of postmodernity This photograph shows the Piazza d'Italia in New Orleans, one of the first public projects in the United States to consciously celebrate an eclectic, over-the-top decorative approach to design.

about many aspects of others' cultures. This contributes to **cosmopolitanism,** an intellectual and esthetic openness toward divergent experiences, images, and products from different cultures.

Cosmopolitanism is an important geographic phenomenon because it fosters a curiosity about all places, peoples, and cultures, together with at least a rudimentary ability to map, or situate, such places and cultures geographically, historically, and anthropologically. It also suggests an ability to reflect upon, and judge aesthetically between, different places and societies. Furthermore, cosmopolitanism allows people to locate their own society and its culture in terms of a wide-ranging historical and geographical framework. For travelers and tourists, cosmopolitanism encourages both the willingness and the ability to take the risk of exploring off the beaten track of tourist locales. It also develops in people the skills needed to be able to interpret other cultures and to understand their visual symbolism.

Place Marketing

Economic and cultural globalization has meant that places and regions throughout the world are increasingly seeking to influence the ways in which they are perceived by tourists, businesses, media firms, and consumers. As a result, places are increasingly being reinterpreted, re-imagined, designed, packaged, and marketed. Through place marketing, sense of place has become a valuable commodity and culture has become an important economic activity. Furthermore, culture has become a significant factor in the ability of places to attract and retain other kinds of economic activity. Seeking to be competitive within the globalizing economy, many places have sponsored extensive makeovers of themselves, including the creation of pedestrian plazas, cosmopolitan cultural facilities, festivals, and sports and media events—what geographer David Harvey has described as the "carnival masks" and "businessmen's utopias" of global capitalism. An increasing number of places have also set up home pages on the Internet containing maps, information, photographs, guides, and virtual spaces in order to promote themselves in the global marketplace for tourism and commerce. Meanwhile, the question of who does the re-imagining and cultural packaging, and on whose terms, can become an important issue for local politics.

Central to place marketing is the deliberate manipulation of material and visual culture in an effort to enhance the appeal of places to key groups. These groups include the upper level management of large corporations; the higher skilled and better educated personnel sought by expanding high-technology industries; wealthy tourists; and the organizers of business and professional conferences and other income-generating events. In part, this manipulation of culture depends on promoting traditions, lifestyles, and arts that are locally rooted (see Box 6.3: "Visualizing Geography—Place Marketing and Econom-

Figure 6.25 Consumption in style Restaurant design contributes to a city's visual style as architects and interior designers, restaurant consultants, and restaurant-industry magazines diffuse global trends adapted to local styles. Shown here is the interior of the Buddakan restaurant, in Philadelphia.

ic Development"); in part, it depends on being able to tap into globalizing culture through new cultural amenities and specially organized events and exhibitions. Some of the most widely adopted strategies for the manipulation and exploitation of culture include funding for facilities for the arts; investment in public spaces; the re-creation and refurbishment of distinctive settings like waterfronts and historic districts; the expansion and improvement of museums (especially with blockbuster exhibitions of spectacular cultural products that attract large crowds and can be marketed with commercial tie-ins); and the designation and conservation of historic landmarks.

The Dutch city of Amsterdam provides a good example of how investment in the arts can become a catalyst for economic development. The construction in the late 1980s of an arts and civic complex, the Stopera, in a declining neighborhood in the east-central part of the city led to recovery of the whole area in less than a decade. By the early 1990s nearly 2,000 people were working in the Stopera itself, and its presence had attracted bookstores, record and magazine stores, restaurants, cafés, and specialized food stores. Altogether, the neighborhood experienced a 40 percent increase in the number of shops in the first few years after the complex opened. The area attracts small businesses because of its new atmosphere, and increasing numbers of tourists, who previously avoided the area, now seek it out.

New York City also provides a good example of place marketing through investment in public space. Bryant Park, on 42nd Street in Manhattan (**Figure 6.26**), has become a small but celebrated component of the city's attempt to clean up its image and attract tourists and business investment. After a period of decline, disuse, and daily occupation by vagrants and drug dealers, the park was taken over by the Bryant Park Restoration Corp., a nonprofit business association of local property owners and their major corporate tenants. This group redesigned and renovated the park; installed new food services; hired private security guards; and organized a series of cultural events, including the showing, twice a year, of the fashion collections of New York designers (held in marquees set up inside the park). The result has been described as "pacification by cappuccino." The previous users of the space have been displaced, the image of the whole area around Bryant Park has been radically altered, and investments are once again flowing in.

Examples of the re-creation and refurbishment of distinctive settings like waterfronts and historic districts can be found in many cities. The popularity of waterfront redevelopments can be traced to the success of Harborplace in Baltimore (**Figure 6.27**). Harborplace is a waterfront complex that was redeveloped from a semiderelict wharf and docks area. Built in 1980, it has become one of the best-known examples of the kind of large, mixed-use developments that industrial cities have sponsored in an effort to bolster their image and attract tourists and shoppers. By 1995 Harborplace was attracting more than 30 million visitors a year. Not surprisingly, this kind of success has encouraged other cities to re-create themselves through major waterfront developments. Other examples include South Street Seaport in New York, the Marketplace and Harborwalk in Boston, and Darling Harbour in Sydney, Australia. The largest and most ambitious of all such developments is in London's Docklands (**Figure**

Figure 6.26 Bryant Park, New York City Bryant Park abuts the main building of the New York City Public Library in the heart of central Manhattan. Before renovation it was usually occupied by homeless people and drug dealers. Now, through the renovation efforts of a nonprofit volunteer group, the park has been transformed into a fashionable site for meeting and being seen. The transformation has been so successful that it is now the setting for New York's Fashion Week, when marquees are erected in the park and the fall collections of clothes designers are modeled. In its new incarnation it is not clear whether Bryant Park is a public space or a combination of public and private. Or perhaps the question is: Public for whom? The homeless are certainly no longer welcome to linger among the more affluent.

Figure 6.27 Harborplace, Baltimore The developer of this reclaimed waterfront, James Rouse, sought to provide a setting that would be attractive to daytime office workers from the city's nearby downtown; to visitors to the adjacent convention center and hotel; and to tourists visiting the nearby aquarium. Rouse provided a new waterfront promenade with decorative paving, benches, and streetlights; a food court offering a variety of ethnic, prepared, and fresh foods from restaurants, fast-food counters, stalls, and pushcarts; and a total of 142,000 square feet of shopping space.

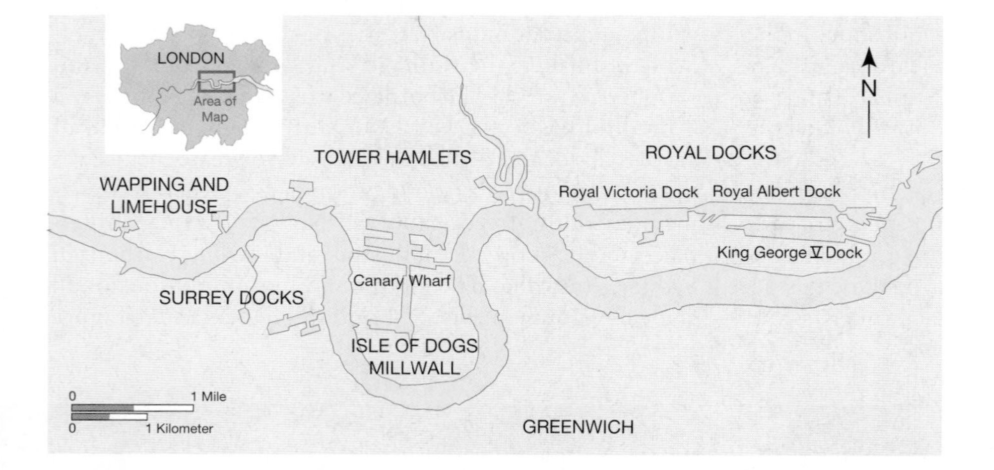

Figure 6.28 London's Docklands Once the commercial heart of Britain's empire, employing over 30,000 dockyard laborers, London's extensive Docklands fell into a sharp decline in the late 1960s because of competition from specialized ports using new container technologies. In 1981 the London Docklands Development Corp. was created by the central government and given extensive powers to redevelop the derelict dock areas. The Docklands are now recognized as the largest urban redevelopment scheme in the world, with millions of square feet of office and retail space and substantial amounts of new housing. The photo shows Canary Wharf.

6.28). The remaking of the Docklands in the 1980s was a deliberate attempt by Prime Minister Margaret Thatcher's government not simply to market this part of London to global tourists and investors but to sell the whole idea of the United Kingdom as a rejuvenated, postindustrial economy.

Examples of the re-creation and refurbishment of historic districts and settings are even more widespread—so widespread, in fact, that they have become a mainstay of the heritage industry. This industry, based on the commercial exploitation of the histories of peoples and places, is now worldwide, as evidenced by the involvement of the Economic, Social, and Cultural Organization of the United Nations (UNESCO) in identifying places for inclusion on World Heritage Lists. In countries such as the United Kingdom, with a high density of historic districts and settings, place marketing relies heavily on the heritage industry. In the United Kingdom more than 90 million people pay to view about 650 historic properties each year, and millions more visit free-of-charge heritage sites such as cathedrals, field monuments, remote ruins, and no longer useful industrial waterways. In 2001 more than 185 million tourists visited designated heritage sites in the United Kingdom, spending about $46 billion on entry fees, retail sales, travel, and hotel accommodations.

One important consequence of the heritage industry is that urban cultural environments are vulnerable to a debasing and trivializing "Disneyfication" process. The United Nations Center for Human Settlements (UNCHS) *Global Report on Human Settlements 2001* notes:

> The particular historic character of a city often gets submerged in the direct and overt quest for an international image and international business. Local identity becomes an ornament, a public relations artifact designed to aid marketing. Authenticity is paid for, encapsulated, mummified, located and displayed to attract tourists rather than to shelter continuities of tradition or the lives of its historic creators.[7]

As a result, contemporary landscapes contain increasing numbers of inauthentic settings—what David Harvey has called the "degenerative utopias" of global capitalism. These are as much the product of contemporary material and visual culture as they are of any cultural heritage. Particularly influenced by images and symbols derived from movies, advertising, and popular culture, examples include the re-creation of the Wild West in the fake cowboy town of Old Tucson in Arizona; the Bavarian village created in Torrance, California; and the simulation of European tourist hot spots in Las Vegas, Nevada (**Figure 6.29**). The town of Helen, Georgia, boasts a full-scale reconstruction of a Swiss mountain village, complete with costumed staff and stores selling Swiss merchandise. In Japan a reproduction village called British Hills, complete with church, pub, and school, has been constructed north of Tokyo.

[7] *Ibid.*, p. 38.

Place Marketing and Economic Development

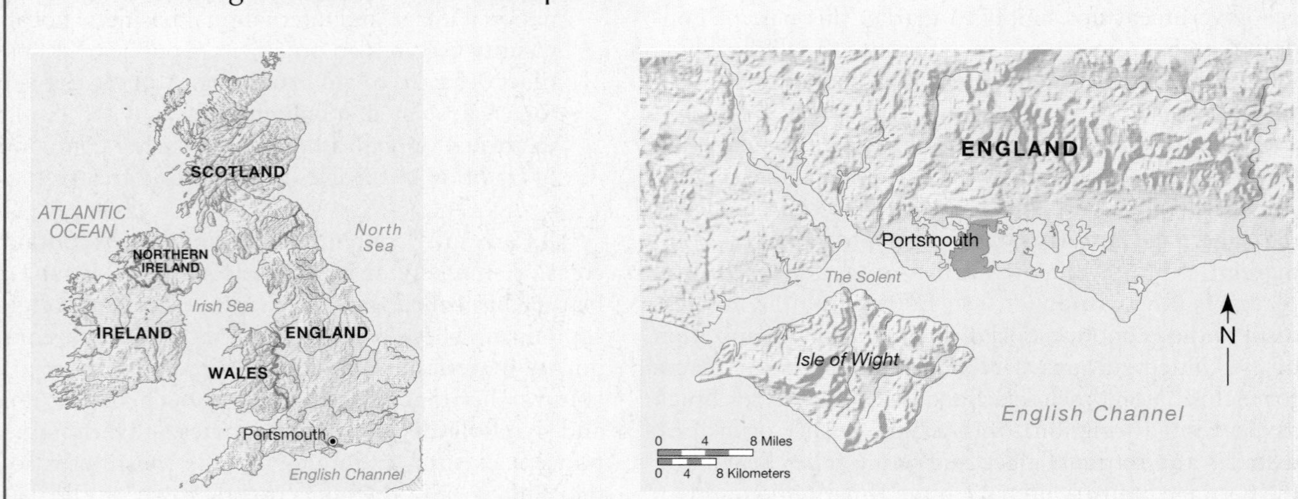

Portsmouth, England, with its superb natural harbor, has been England's chief naval port and dockyard since the reign of Henry VIII. At the height of the British Empire the growth of the dockyard fueled the development of the city, helping to create a specialized economy based on marine engineering. With the decline of the empire after the 1960s, the British navy was much reduced in size, and Portsmouth's dockyard functions shrank dramatically. After a period of economic adjustment, the city has begun to remake its image, exploiting its past to market itself as a distinctive conference center and tourist site.

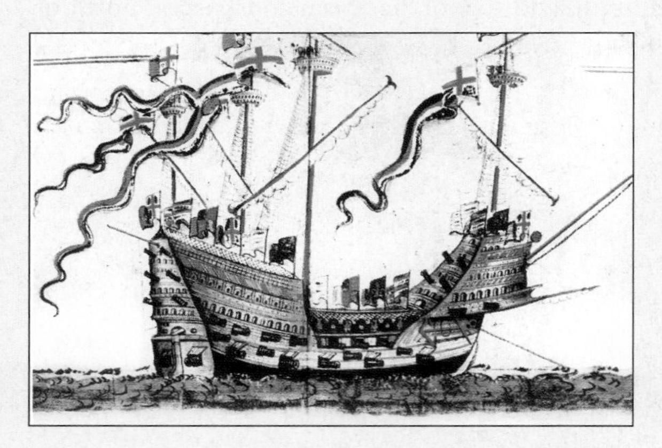

The Mary Rose. (*Source:* Pepys Library, Magdalene College, Cambridge.)

HMS Victory.

Entrance to the Mary Rose exhibit in the Heritage Area of Portsmouth naval base.

HMS Warrior.

The legacy of Portsmouth's naval history has allowed the city to develop a heritage industry that has helped to make up for the loss of jobs in the naval dockyard. HMS Victory, Nelson's flagship at the Battle of Trafalgar in 1805, had been a longstanding attraction in the Royal Dockyard. In 1979 a volunteer organization recovered the Mary Rose, a Tudor warship that had been sunk just off Portsmouth in

July 18, 1545: Henry VIII's fleet, in Portsmouth Harbour, fends off an attack by the French fleet.

1545 during an engagement with a French invasion fleet. It has since been restored and is a major attraction in the dockyard's heritage area. In 1988 HMS Warrior, built in 1860 as Britain's first ironclad battleship, was added to the dockyard's collection of major tourist attractions.

Throughout the second half of the nineteenth century and the first half of the twentieth, Portsmouth's economy included a lucrative resort and recreation function. When international travel became more affordable in the 1960s, many British vacationers chose the warmer beaches of the Mediterranean, and Portsmouth's tourist industry began to decline sharply. Since the 1980s, Portsmouth has repositioned itself as a different kind of resort, drawing on the city's naval history in order to attract overnight visitors traveling to and from France on the cross-Channel car ferries that now operate from an unused part of the naval dockyard. In 1996 the city formed a partnership with neighboring municipalities to redevelop Portsmouth Harbour. The £90 million ($135 million) redevelopment project is being substantially funded by proceeds from the National Lottery and will include a Baltimore-style wharf with a conference center, shops, cafés, hotels, and viewing terraces for maritime events.

Southsea Castle.

Sixteenth- and seventeenth-century coastal fortifications, formerly boarded up and fenced off as eyesores, have been landscaped in order to take advantage of their historic interest. In one formerly abandoned fort, Southsea Castle, a museum of Tudor life has been established exploiting the city's link with Henry VIII, who first developed the British Navy into a significant military force.

June 5, 1944: D-Day.

Restored naval cutter, in the Mast Pond, part of the Heritage Area within Portsmouth naval base.

Royal Marines Museum, Portsmouth.

D-Day Museum, Portsmouth.

During World War II, Portsmouth was the main port of departure for Allied forces invading German-held France on D-Day. In 1982 the city opened the D-Day museum, reinforcing the city's emerging identity as a specialized heritage site. A few years later the closure of a Royal Marine barracks provided the opportunity for the development of another specialized military museum.

(a)

(b)

Figure 6.29 Las Vegas The boundaries between the heritage industry and the leisure and entertainment industries have become increasingly blurred, with the result that a great deal of investment has been channeled toward the creation of inauthentic "historic" settings whose characteristics owe as much to movies and popular stereotypes as to historic realities. Shown here are the Venetian Hotel (a) and the Paris Hotel (b) in Las Vegas.

CONCLUSION

Geographers study the interdependence between people and places and are especially interested in how individuals and groups acquire knowledge of their environments and how this knowledge shapes their attitudes and behaviors. People ascribe meanings to landscapes and places in many ways, and they also derive meanings from the places and landscapes they experience. Different groups of people experience landscape, place, and space differently. For instance, the experience that rural Sudanese children have of their landscapes and the ways in which they acquire knowledge of their surroundings differ from how middle-class children in an American suburb learn about and function in their landscapes. Furthermore, both landscapes elicit a distinctive sense of place that is different for those who live there and those who simply visit.

As indicated in previous chapters, the concepts of landscape and place are central to geographic inquiry. They are the result of intentional and unintentional human action, and every landscape is a complex reflection of the operations of the larger society. Geographers have developed categories of landscape to help distinguish the different types that exist. Ordinary landscapes, such as neighborhoods and drive-in movie theaters, are ones that people create in the course of their everyday lives. By contrast, symbolic landscapes represent the particular values and aspirations that their developers and financiers want to impart to a larger public. An example is Mount Rushmore in the Black Hills of South Dakota, designed and executed by sculptor Gutzon Borglum. Chiseling the heads of George Washington, Thomas Jefferson, Theodore Roosevelt, and Abraham Lincoln into the granite face of the mountain, Borglum intended to construct an enduring landscape of nationalism in the wilderness.

More recently, geographers have come to regard landscape as a text, something that can be written and read, rewritten and reinterpreted. The concept of landscape as text suggests that a landscape can have more than one author, and different readers may derive different meanings from what is written there. The idea that landscape can be written and read is further supported by the understanding that the language in which the landscape is written is a code. To understand the significance of the code is to understand its semiotics, the language in which the code is written. The code may be meant to convey many things, including a language of power or of playfulness, a language that elevates one group above another, or a language that encourages imagination or religious devotion and spiritual awe.

The global transition from modernity to postmodernity has altered cultural landscapes, places, and spaces differently as individuals and groups have struggled to negotiate the local impacts of this widespread shift in cultural sensibilities. The shared meanings that insiders derive from their place or landscape have been disrupted by the intrusion of new sights, sounds, and smells as values, ideas, and practices from one part of the globe have been exported to another. The Internet and the emergence of cyberspace have meant that new spaces of interaction have emerged that have neither a distinct historical memory attached to them nor a well-established sense of place. Because of this, cyberspace carries with it some unique possibilities for cultural exchange. It remains to be seen, however, whether access to this new space will be truly open—or whether the Internet will become another landscape of power and exclusion.

MAIN POINTS REVISITED

- **In addition to understanding how the environment shapes (and is shaped by) people, geographers seek to identify how it is perceived and understood by people.**

  People not only filter information from their environments through neurophysiological processes, but draw on personality and culture to produce cognitive images of their environment, pictures or representations of the world that can be called to mind through the imagination.

- **The language in which a landscape is written is a kind of code. The code includes signs that direct our attention toward certain features and away from others.**

  To interpret our environment, we must learn how to read the codes that are written into the landscape.

- **The written code of landscape is also known as semiotics. Codes signify important information about landscapes, such as whether they are sacred or profane, accessible or off-limits, or oriented toward work or play.**

  Landscapes as different from each other as shopping malls and national capitals can be understood in terms of their semiotics.

- **Different cultural identities and status categories influence the ways in which people experience and understand their environments, as well as how they are shaped by—and able to shape—them.**

  Among the most important of these are the cultural identities of race, class, gender, ethnicity, and sexuality. Often these identities come together in a group, and their influence in combination becomes central to our understanding of how group identity shapes space and is shaped by it.

- **Landscape serves as a kind of archive of society. It is a reflection of our culture and our experiences. Like a book, landscape is a text that is written by individuals and groups and read by them as well.**

  It is therefore possible to have one landscape convey different meanings for different groups. Landscapes can be constructed to reflect the everyday words of social groups, as well as to represent power and the values of a particular society.

- **The emergence of the most recent phase of globalization has occurred in parallel with a transition from modernity to postmodernity.**

  This transition involves a shift in cultural sensibilities that affects every sphere of creative activity, from art and architecture to television and urban design. It seems to have originated in parts of the world's core countries and is currently spreading throughout the rest of the world.

KEY TERMS

cosmopolitanism (p. 239)
derelict landscapes (p. 223)
ethology (p. 216)
existential imperative (p. 218)
humanistic approach (p. 226)
landscape as text (p. 227)
modernity (p. 231)
postmodernity (p. 237)
proxemics (p. 216)
sacred space (p. 229)
semiotics (p. 227)
territoriality (p. 215)
topophilia (p. 223)

ADDITIONAL READING

Burnett, R. *The Global Jukebox: The International Music Industry.* New York: Routledge, 1996.

Carney, G. O. (ed.). *Fast Food, Stock Cars, and Rock'n' Roll: Place and Space in American Pop Culture.* Lanham, MD: Rowman & Littlefield, 1995.

Gold, J. R., and S. V. Ward (eds.). *Place Promotion: The Use of Publicity and Marketing to Sell Towns and Regions.* New York: Wiley and Sons, 1994.

Golledge, R. G., and R. J. Stimpson. *Spatial Behavior: A Geographic Perspective.* New York: Guilford Press, 1996.

Groth, P., and T. W. Bressi (eds.). *Understanding Ordinary Landscapes.* New Haven: Yale University Press, 1997.

Harvey, D. W. *Spaces of Hope.* Berkeley: University of California Press, 2000.

Honoré, C. *In Praise of Slowness. How a Worldwide Movement is Challenging the Cult of Speed.* New York: HarperCollins, 2004.

Jackson, J. B. *A Sense of Place, A Sense of Time.* New Haven: Yale University Press, 1994.

Jackson, J. B. *Discovering the Vernacular Landscape.* New Haven: Yale University Press, 1984.

Judd, D. R., and S. S. Fainstein (eds.). *The Tourist City.* New Haven: Yale University Press, 1999.

Katz, C. Growing up Global: Economic Restructuring and Children's Everyday Lives. Minneapolis & St. Paul: University of Minnesota Press, 2004.

Knox, P. L. "Creating Ordinary Places: Slow Cities in a Fast World," *Journal of Urban Design, 10* (1), 3–13, 2005.

Knox, P. L. "Vulgaria: The Re-Enchantment of Suburbia." *Opolis, 1* (2), 34–47, 2005.

Kearns, G., and C. Philo (eds.). *Selling Places.* Oxford: Pergamon Press, 1993.

Massey, D., and P. Jess (eds.). *A Place in the World? Places, Cultures, and Globalization.* New York: Oxford University Press, 1995.

Mitchell, D. *Cultural Geography: A Critical Introduction.* Malden, MA: Blackwell, 2000.

Relph, E. *Place and Placelessness.* London: Pion, 1976.

Ritzer, G. *The Globalization of Nothing.* Thousand Oaks, CA: Pine Forge Press, 2004.

Schama, S. *Landscape and Memory.* New York: Vintage Books, 1995.

Shurmer-Smith, P., and K. Hannam. *Worlds of Desire, Realms of Power.* London: Edward Arnold, 1994.

Urry, J. *Consuming Places.* London: Routledge, 1995.

Vale, L. J., and S. B. Warner, Jr. (eds.). *Imaging the City.* New Brunswick, NJ: CUPR Press, 2002.

Ward, S. V. *Selling Places: The Marketing and Promotion of Towns and Cities 1850–2000.* New York: Routledge, 1998.

Zukin, S. *The Cultures of Cities.* Cambridge, MA: Blackwell, 1995.

EXERCISES

On the Internet

The Internet exercises for this chapter explore in some depth how we recognize the coded spaces of our landscape. Via the Internet, we travel to museums to view Europe's eighteenth-century landscape in classic paintings and the nineteenth-century U.S. landscape as viewed by painters of the Hudson River School in an effort to decipher the semiotics of these landscapes—how they were written and read. We examine the use of ZIP codes as identity markers. We also use the Internet to examine health concepts, climate, e-commerce, language, and population.

Unplugged

1. Write a short essay (500 words or two double-spaced, typed pages) that describes, from your personal perspective, the sense of place that you associate with your hometown or county. Write about the places, buildings, sights, and sounds that are especially meaningful to you.
2. Draw up a list of the top 10 places in the United States in which you would like to live and work, then draw up a list of the bottom 10. How do these lists compare with the map of student preferences shown in Figure 6.8? Why might your preferences be different from theirs?
3. On a clean sheet of paper and without reference to maps or other materials, sketch a detailed map of the town or city in which you live. When you have finished, turn to page 220 and compare your sketch to Figure 6.6. Does your sketch contain nodes? Landmarks? Edges? Districts? Paths? How does "your" city compare to the "real" city?
4. Using telephone books that can be accessed at your university library and a map of your local area, map the sacred landscape of your locality by identifying the locations of different churches, temples, mosques, and other places of worship. What patterns do you see? How do they fit with your knowledge of particular areas? What other sorts of sacred sites might exist beyond the institutionalized religious sites?
5. Your campus provides an institutional landscape that has been "written" to convey certain important relationships. "Read" your campus landscape and compare the more "powerful" sites with the less powerful, and discuss which groups or academic disciplines occupy them. Map what you think are the most powerful and important places and sites and the least powerful and important. Why do different places and sites fit into each category? Are there significant differences in the architecture, in the location (central or on the edges of campus), in accessibility? How might your reading of the campus landscape differ from readings created by other members of the campus community (as well as outsiders)?

WAL★MART
SUPERCENTER
沃尔玛购物广场
NW
世纪新浪 手机城
Century NewWave
热线：3192222
最时尚的商场——万达
购物休闲
逛万达！
万达购物广场欢迎您！
ROMON

7 The Geography of Economic Development

MAIN POINTS

- Geographically, the single most important feature of economic development is that it is highly uneven.
- Geographical divisions of labor have evolved with the growth of the world-system of trade and politics, and with the changing locational logic of successive technology systems.
- Regional cores of economic development are created cumulatively, following some initial advantage, through the operation of several basic principles of spatial organization.
- Spirals of economic development can be arrested in various ways, including the onset of disinvestment and deindustrialization, which follow major shifts in technology systems and international geopolitics.
- The globalization of the economy has meant that patterns and processes of local and regional economic development are much more open to external influences than before.

In the past quarter-century many of China's regions have gone through an extraordinary transformation as the Chinese economy has become increasingly integrated into the global economy. In Guangdong province, near Hong Kong, for example, industrial growth, real estate development, and investment in highway construction has increased by an average of 15 to 25 percent per year throughout the past decade. Once known mainly for its abundant and high-quality long-grain rice, Guangdong itself now imports fragrant jasmine rice from Thailand. Along Guangdong's Pearl River Delta, it is difficult to photograph one of the classic rice paddies without capturing a construction site as well. Incomes and consumer spending in Guangdong suddenly skyrocketed in the 1990s. In Shandong Province, 1,500 kilometers (930 miles) north, conditions were also changing. Traditionally a region of peasant agriculture and a modest prosperity based on winter wheat and summer corn, Shandong is now beginning to join in China's industrial surge. In the tiny village of Xishan, on the plain of the Yellow River, a slaughterhouse, a pharmaceuticals plant, and a furniture factory now employ workers who until recently were peasant farmers. "Not even a single villager grows grain now," boasts Shan Chujie, the village leader. "We're not country bumpkins here."[1]

As individual regions develop new specializations, China itself is changing from a self-sufficient agricultural producer to a more diversified, affluent country that is beginning to export substantial amounts of manufactured items as well as import substantial amounts of food. Meanwhile, in the southwestern province of Sichuan, there remain more than 20,000 state-owned enterprises, most of which are not able to keep pace with China's economic transformation. Exports from the region's textile factories have declined, and the economy has become increasingly depressed. Although the official unemployment rate is just 3 percent, the actual rate is closer to 20 percent.

Economic development is always an uneven geographic phenomenon, and countries and regions can follow various pathways to development. Places and regions in China are beginning to experience many of the processes of capitalist economic development that have shaped the fortunes of places and regions in most of the rest of the world. Explaining how and why these processes occur is an important goal of human geography.

[1]"Collectives Make a Comeback," *Wall Street Journal*, March 10, 1995, p. A1.

Wal-Mart superstore, Jinan, Shandong Province, China.

PATTERNS OF ECONOMIC DEVELOPMENT

Economic development is often discussed in terms of levels and rates of change in prosperity, as reflected in bottom-line statistical measures of productivity, incomes, purchasing power, and consumption. Increased prosperity is only one aspect of economic development, however. For human geographers and other social scientists, the term *economic development* refers to processes of change involving the nature and composition of the economy of a particular region as well as to increases in the overall prosperity of a region. These processes can involve three types of changes:

- Changes in the structure of the region's economy (for example, a shift from agriculture to manufacturing);
- Changes in forms of economic organization within the region (for example, a shift from socialism to free-market capitalism);
- Changes in the availability and use of technology within the region.

Economic development is also expected to bring with it some broader changes in the economic well-being of a region. The most important of these are changes in the capacity of the region to improve the basic conditions of life (through better housing, health care, and social welfare systems) and to improve the physical framework, or infrastructure, on which the economy rests.

The Unevenness of Economic Development

Geographically, the single most important feature of economic development is that it is *uneven*. At the global scale, this unevenness takes the form of core-periphery contrasts within the evolving world-system (see Chapter 2). These global core-periphery contrasts are the result of a competitive economic system that is heavily influenced by cultural and political factors. The core regions within the world-system—currently, the tripolar core of North America, Europe, and Japan—have the most diversified economies, the most advanced technologies, the highest levels of productivity, and the highest levels of prosperity. They are commonly referred to as *developed regions* (though processes of economic development are, of course, continuous, and no region can ever be regarded as fully developed). Other countries and regions—the periphery and semiperiphery of the world-system—are often referred to as *developing* or *less developed*. Indeed, the nations of the periphery are often referred to as LDCs (less developed countries). Another popular term for the global periphery, developed as a political label but now synonymous with economic development, is the *Third World*. As we saw in Chapter 2, this term had its origins in the early Cold War era of the 1950s and 1960s, when the newly independent countries of the periphery positioned themselves as a distinctive political bloc, aligned with neither the First World of developed, capitalist countries nor the Second World of the Soviet Union and its satellite countries.

Global Core-Periphery Patterns

At this global scale, levels of economic development are usually measured by economic indicators such as gross domestic product and gross national income. **Gross domestic product (GDP)** is an estimate of the total value of all materials, foodstuffs, goods, and services that are produced by a country in a particular year. To standardize for countries' varying sizes, the statistic is normally divided by total population, which gives an indicator, *per capita* GDP, that is a good yardstick of relative levels of economic development. **Gross national income (GNI)** is a measure of the income to a country from production wherever in the world it occurs. For example, if a U.S.-owned company operating in another country sends some of its income (profits) back to the United States, this adds to the United States' GNI. In making international comparisons, GDP and GNI can be problematic because they are based on each nation's currency. As a result, it is now common to compare national currencies based on *purchasing power parity* (PPP). In effect, PPP measures how much of a common "market basket" of goods and services each currency can purchase locally, including goods and services that are not traded internationally. Using PPP-based currency values to compare levels of economic prosperity usually produces lower GNI figures in wealthy countries and higher GNI figures in poorer nations, compared with market-based exchange rates. Nevertheless, even with this compression between rich and poor, economic prosperity is very unevenly distributed across nations.

As **Figure 7.1** shows, most of the highest levels of economic development are to be found in northern latitudes (very roughly, north of 30° N), which has given rise to another popular shorthand for the world's economic geography: the division between the "North" (the core) and the "South" (the periphery). Viewed in more detail, the global pattern of per capita GNI (measured in the "international dollars" of PPP) in 2003 is a direct reflection of the core-semiperiphery-periphery structure of the world-system. In many of the core countries of North America, northwestern Europe, and Japan, annual per capita GNI (in PPP) exceeds $20,000. The only other countries that match these levels are Australia, Hong Kong, and Singapore, where annual per capita GNI in 2003 was $28,290, $28,810, and $24,180, respectively.

Semiperipheral countries such as Brazil, Russia, and Thailand have an annual per capita GNI ranging between $7,000 and $9,000. In the rest of the world—the periphery—annual per capita GNI (in PPP) is typically less than $5,000. The gap between the highest per capita GNIs ($37,500 in the United States and $37,300 in Norway,) and the lowest ($530 in Sierra Leone, $600 in Malawi) is huge. The gap between the world's rich and poor is also getting wider (**Figure 7.2**). In 1970 the average GNI per

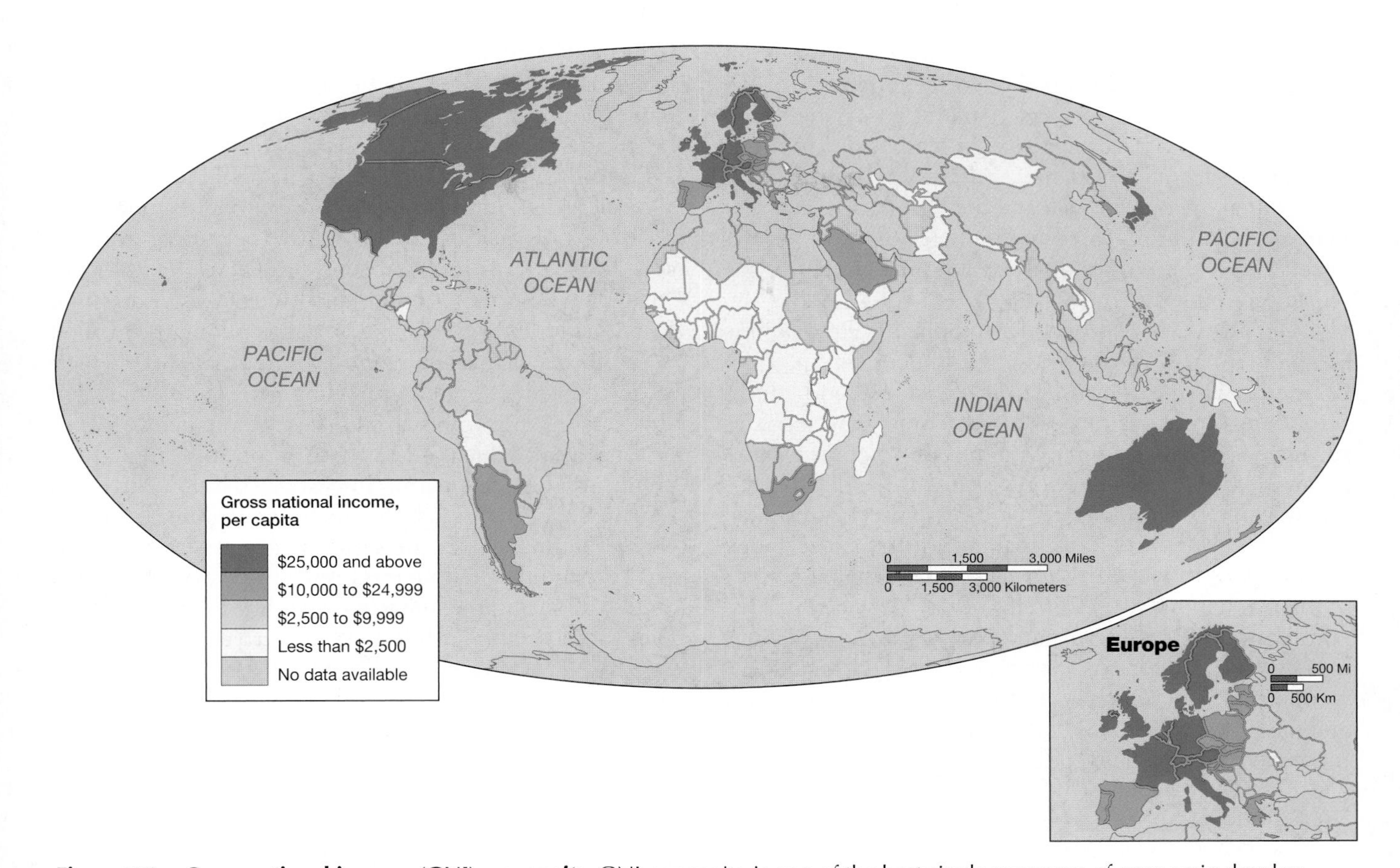

Figure 7.1 Gross national income (GNI) *per capita* GNI per capita is one of the best single measures of economic development. This map, based on 2003 data, shows the tremendous gulf in affluence between the core countries of the world economy—like the United States, Norway, and Switzerland, with annual per capita GNI (in PPP "international dollars") of more than $25,000—and peripheral countries like Angola, Haiti, and Mali, where annual per capita GNI was less than $2,500. In semiperipheral countries like South Korea, Brazil, and Mexico, per capita GNI ranged between $5,000 and $10,000. (After map projection, Buckminster Fuller Institute and Dymaxion Map Design, Santa Barbara, CA. The word *Dymaxion* and the Fuller Projection Dymaxion™ Map design are trademarks of the Buckminster Fuller Institute, Santa Barbara, CA, © 1938, 1967 & 1992. All rights reserved.)

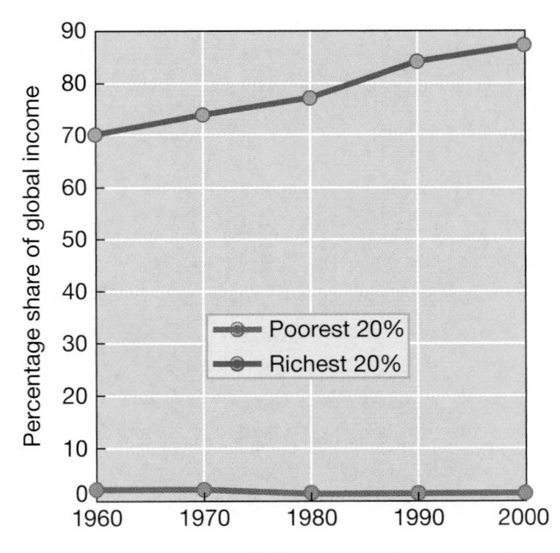

Figure 7.2 Long-term trends in per capita GNI This graph shows the steady divergence in international economic prosperity between the richest and poorest of the world's population. In 1960 the richest 20 percent of the world's population accounted for 70.2 percent of global income, while the poorest 20 percent accounted for 2.3 percent: a ratio of 30 to 1. By 1980 the ratio had increased to 45 to 1; by 1990 it was 64 to 1; and by 2000 it was 72 to 1.

capita of the 10 poorest countries in the world was just one-fiftieth of the average GNI per capita of the 10 most prosperous countries. By 1990 the relative gap had doubled, and by 2005 the average of the bottom 10 was approaching one two-hundredth of the average of the top 10.

Spatial Justice?

Inequality in income is reflected—and reinforced—by many aspects of human well-being. Patterns of infant mortality, a reliable indicator of social well-being, show the same steep core-periphery gradient (refer back to Figure 3.16). For adults in the industrial core countries, life expectancy is high and continues to increase. Life expectancy at birth in Australia and Switzerland in 2004 was 79 years, and in the United States it was 77. In contrast, life expectancy in the poorest countries is dramatically shorter. In Botswana and Malawi in 2004 life expectancy at birth was just 38 years, in Sierra Leone it was 34, and in Zambia it was 32. In most African countries only 60 to 75 percent of the population can expect to survive to age 40.

Patterns of food consumption, one of the most basic of all human needs, reflect another key dimension of inequality in human well-being. Every day around 800 million people—12 percent of the world's population—go hungry. In South Asia one person in four goes hungry, and in Sub-Saharan Africa the rate is as high as one in three. In terms of sheer numbers, however, there are more chronically hungry people in Asia and the Pacific. The fact that many of them are located in countries that are net exporters (by value) of foods is a telling indictment of the world economic system.

The United Nations Development Programme (U.N.D.P.) has devised an overall index of human development based on measures of life expectancy, educational attainment, and personal income. The index is calculated so that a country with the best combined scores among all countries in the world on all three indicators has a perfect index score of 1.0, while a country that ranked worst in the world on all three indicators has an index score of zero. **Figure 7.3** shows the international map of human development in 2002. Norway (0.956), Sweden and Australia (0.946), Canada (0.943) and the Netherlands (0.943) had the highest overall levels of human development, while Burkina Faso (0.302), Niger (0.292), and Sierra Leone (0.273) had the lowest levels. The same fundamental pattern is repeated across the entire array of indicators of human development: adult literacy, poverty, malnutrition, access to physicians, public expenditure on higher education, telephone lines, Internet users, and so on. Inequality on this scale poses the most pressing as well as the most intractable questions of spatial justice.

These questions are underscored by some simple comparisons between the needs of the periphery and the spending patterns in core countries. The UNDP has calculated that the annual cost of providing a basic education for all children in peripheral countries would be around $6 billion, which is less than the annual sales of cosmetics in the United States. Providing water and sanitation for everyone in peripheral countries is estimated at $9 billion per year, which is less than Europeans' annual expenditure on ice cream. Providing basic health and nutrition for everyone in the peripheral countries would cost an estimated $13 billion per year—less than the annual expenditure on pet foods in Europe and the United States. Reducing the military expenditures of core countries (in the region of $700 billion per year) by less than 10 percent each year would pay for the costs of basic education, water and sanitation, basic health and nutrition, and reproductive health programs for everyone in peripheral countries.

Figure 7.3 An index of human development, 2002 This index, calculated by the United Nations Development Program, is based on measures of life expectancy, educational attainment, and personal income. A country with the best scores among all countries in the world on all three measures would have a perfect index score of 1.0, while a country that ranked worst in the world on all three indicators would have an index score of zero. Most of the affluent core countries have index scores of 0.9 or more; the worst scores—those less than 0.4—are concentrated in Africa. (After map projection, Buckminster Fuller Institute and Dymaxion Map Design, Santa Barbara, CA. The word *Dymaxion* and the Fuller Projection Dymaxion™ Map design are trademarks of the Buckminster Fuller Institute, Santa Barbara, CA, © 1938, 1967 & 1992. All rights reserved.)

Development and Gender Equality

Core-periphery patterns are also reflected in indicators that measure economic development by *gender equality.* The U.N.D.P. has established a gender-sensitive development index that adjusts the overall human development index for gender inequality in life expectancy, educational attainment, and income. According to this index, in no country are women better off than men. Perhaps most revealing is the U.N.D.P.'s Gender Empowerment Index, which is based on measures of women's incomes, their participation in the labor force as administrators and managers, professional and technical workers, and the percentage of parliamentary seats held by women. As in the overall index of human development, a country with a perfect score would score 1.0, with zero representing the worst possible score. **Figure 7.4** shows the actual index values for 2002. The top four countries were Nordic: Norway (0.908), Sweden (0.854), and Finland (0.82). In many countries, data on women are insufficiently accurate to calculate the index, but the countries with the worst recorded index scores were Yemen (0.123), Saudi Arabia (0.207), and Bangladesh (0.218). Only 16 of the 78 countries in the study were able to record a Gender Empowerment Index of more than 0.70. Nevertheless, as Figure 7.4 demonstrates, high levels of economic development are not a prerequisite for creating economic opportunities for women. Costa Rica and Trinidad and Tobago, for example, both scored better than Italy and Japan, and the Bahamas scored better than Portugal.

Women are, in fact, playing a central and increasing role in processes of development and change in the global economy. In many peripheral countries, women constitute the majority of workers in the manufacturing sector created by the new international division of labor (**Figure 7.5**). In many poor countries it is women who keep households afloat in a world economy that has resulted in localized recession and intensified poverty. In these countries, on average, women earn 40 to 50 percent less than men for the same work. They also tend to work longer hours than men: 12 to 13 hours a week more (counting both paid and unpaid work) in Africa and Asia. Child labor also remains a serious problem around the world. According to International Labor Organization (ILO) estimates, more than 250 million children between the ages of 5 and 14 are sent out to work, many of them in dangerous conditions (**Figure 7.6**). The ILO also estimates that

Figure 7.4 An index of gender empowerment, 2002 The United Nations Development Programme's Gender Empowerment Index is based on measures of women's incomes; their participation in the labor force as administrators and managers, and as professional and technical workers; and the percentage of parliamentary seats held by women. As in the overall index of human development (Figure 7.3), a country with a perfect score would score 1.0, with zero representing the worst possible score (ranked worst on all measures). The map reflects a broad core-periphery pattern, though there is by no means a direct correlation between economic prosperity and gender empowerment: Creating economic opportunities for women does not necessarily require high levels of economic development. (After map projection, Buckminster Fuller Institute and Dymaxion Map Design, Santa Barbara, CA. The word *Dymaxion* and the Fuller Projection Dymaxion™ Map design are trademarks of the Buckminster Fuller Institute, Santa Barbara, CA, © 1938, 1967 & 1992. All rights reserved.)

Figure 7.5 Women in development The changing global economy has placed unprecedented demands on women. Much of the industrialization in peripheral countries depends on female labor because wage rates for women are significantly lower than those for men. In times of rapid economic change and adjustment, women are typically called upon to help sustain household incomes, in addition to their traditional household responsibilities. When employed, women are often more vulnerable than men, disproportionately concentrated in occupations with little or no job security and few opportunities for advancement.

some one million children work in small-scale mining and quarrying around the world. While most of these children live in peripheral countries, hundreds of thousands of underage workers are exploited in sweatshops, farm fields, and other workplaces in core countries, including up to 1.5 million children estimated by the United Nations Children's Fund (UNICEF) in the United States.

Regional Patterns

Inequality in economic development often has a regional dimension. Initial conditions are a crucial determinant of regional economic performance. Scarce resources, a history of neglect, lack of investment, and concentrations of low-skilled people all combine to explain the lagging performance of certain areas. In some regions, for example, initial disadvantages are so extreme as to constrain the opportunities of individuals born there. A child born in the Mexican state of Chiapas, for example, has much bleaker prospects than a child born in Mexico City. The child from Chiapas is twice as likely to die before age five, less than half as likely to complete primary school, and 10 times as likely to live in a house without access to running water. On reaching working age, he or she will earn 20 to 35 percent less than a comparable worker living in Mexico City and 40 to 45 percent less than one living in northern Mexico.

Figure 7.6 Child labor A 9-year-old Nepali girl harvests tea, Peshok, Nepal. One in twelve of the world's children (180 million young people aged five to 17) are involved in the worst forms of child labor—hazardous work, slavery, forced labor, in armed forces, commercial sexual exploitation and illicit activities. Of these children, 97 percent are in peripheral and semi-peripheral countries. Globally, an estimated 114 million children of primary school age are not enrolled in school, depriving one in five children of an education. They become exposed to exploitation and abuse and miss out on developing the knowledge and employable skills that could lift them and their own children out of the poverty cycle.

Other examples of regional inequality can be found throughout the world. Gansu, with an income per capita 40 percent below the national average, is one of the poorest and most remote regions in China. With poor soils highly susceptible to erosion, low and erratic rainfall, and few off-farm employment opportunities, a high proportion of its inhabitants live in poverty. Chaco Province, in Argentina, has a GDP per capita that is only 38 percent of the national average. Low educational attainment and lack of infrastructure, especially roads, explain much of this deviation.

Even when growth in other parts of a country's economy is strong, relative and absolute regional poverty can persist for long periods. In the seventeenth century, for example, the decline of the sugar economy pushed northeastern Brazil into a decline from which it has never fully recovered. In the United States the fortunes of coal-mining West Virginia waned with the collapse of the coal industry and the increased importance of oil and gas in energy production. It remains one of the poorest regions in the United States to this day. In Thailand rapid development has failed to reach the northern hill people. Fewer than 30 percent of their villages have schools, and only 15 percent of the hill people can read and write. Their average annual income is less than a quarter of the country's overall GNI per capita.

Globalization has been associated with increasing regional inequality within many countries since the 1980s. In China, for example, disparities are widening between the export-oriented regions of the coast and the interior. The transition economies of the countries of the former Soviet Union and its Eastern European satellites have registered some of the largest increases in regional inequality, while some core countries—especially Sweden, the United Kingdom, and the United States—have also registered significant increases in regional inequality since the 1980s. At this regional scale, as at the global scale, levels of economic development often exhibit a fundamental core-periphery structure. Indeed, within-country core-periphery contrasts are evident throughout the world: in core countries such as France and the United States, in semiperipheral countries such as South Korea, and in peripheral countries such as Nigeria and Indonesia.

Resources and Development

These patterns of economic development are the result of many different factors. One of the most important is the availability of key resources such as cultivable land, energy sources, and valuable minerals. Unevenly distributed across the world, however, are both key resources and—just as important—the *combinations* of energy and minerals crucial to economic development. A lack of natural resources can, of course, be remedied through international trade (Japan's success is a prime example of this). For most countries, however, the resource base remains an important determinant of development.

Energy

One particularly important resource in terms of the world's economic geography is energy. The major sources of commercial energy—oil, natural gas, and coal—are all very unevenly distributed across the globe. Most of the world's core economies are reasonably well off in terms of energy *production,* the major exceptions being Japan and parts of Europe. Most peripheral countries, on the other hand, are energy-poor. The major exceptions are Algeria, Ecuador, Gabon, Indonesia, Libya, Nigeria, Venezuela, and the Gulf states—all major oil producers. Because of this unevenness, energy has come to be an important component of world trade. Oil is, in fact, the most important single commodity in world trade, making up around 12 percent of the total by value in 2004.

For many peripheral countries the costs of importing energy represent a heavy burden. Consider, for example, the predicament of countries like India, Ghana, Paraguay, Egypt, and Armenia, where in 2004 the cost of energy imports amounted to more than one quarter of the total value of exported merchandise. Few peripheral countries can afford to consume energy on the scale of the developed economies, so patterns of commercial energy *consumption* tend to mirror the fundamental core-periphery cleavage of the world economy. In 2004 energy consumption *per capita* in North America was over 30 times higher than in India and nearly 60 times higher than in sub-Saharan Africa (**Figure 7.7**). The world's high-income countries, with 15 percent of the world's population, use half its commercial energy and 10 times as much *per capita* as low-income countries.

It should be noted that these figures do not reflect the use of firewood and other traditional fuels for cooking, lighting, heating, and sometimes industrial needs. In total, such forms of energy probably account for around 20 percent of total world energy consumption. In parts of Africa and Asia they account for up to 80 percent of energy consumption. This points to yet another core-periphery contrast. Whereas massive investments in exploration and exploitation are enabling more of the developed, energy-consuming countries to become self-sufficient through various combinations of coal, oil, natural gas, hydroelectric power, and nuclear power, 1.5 billion people in peripheral countries depend on collecting wood fuel as their principal source of energy. The collection of wood fuel causes considerable deforestation, and the problem is most serious in densely populated locations, arid and semi arid regions, and cooler mountainous areas, where the regeneration of shrubs, woodlands, and forests is particularly slow. Nearly 100 million people in 22 countries (16 of them in Africa) cannot meet their minimum energy needs even by overcutting remaining forests.

Cultivable Land

The distribution of cultivable land represents another important factor in international economic development. Much more than half of Earth's land surface is unsuitable

Figure 7.7 Rush-hour traffic, Alameda County, CA U.S. energy consumption is extremely high, partly as a result of government policies that keep gasoline prices low.

for any productive form of arable farming, as suggested by **Figure 7.8**. Poor soils, short growing seasons, arid climates, mountainous terrain, forests, and conservation limit the extent of agricultural land across much of the globe. As a result, the distribution of the world's cultivable land is highly uneven, being concentrated in Europe, west-central Russia, eastern North America, the Australian littoral, Latin America, India, eastern China, and parts sub-Saharan Africa. In detail, of course, some of these regions may be marginal for arable farming because of marshy soils or other adverse conditions, while irrigation, for example, sometimes extends the local frontier of productive agriculture.

We also have to bear in mind that not all cultivable land is of the same quality. This leads to the concept of the **carrying capacity** of agricultural land: the maximum population that can be maintained in a place with rates of resource use and waste production that are sustainable in the long term without damaging the overall productivity of that or other places. These issues have given some emphasis to the notion of sustainable development. **Sustainable development** is a vision of development that seeks a balance among considerations of economic growth, environmental impacts, and social equity. In practice, sustainable development means economic growth and change should occur only when the impacts on the environment are benign or manageable, and the impacts (both costs and benefits) on society are fairly distributed across classes and regions. Sustainable development is geared to meeting the needs of the present without compromising the ability of future generations to meet their needs. It envisages a future when improvements in the quality of human life are achieved within the carrying capacity of local and regional ecosystems (see Box 7.1: "Sustainable Development").

Industrial Resources

A high proportion of the world's key industrial resources—basic raw materials—are concentrated in Russia, the United States, Canada, South Africa, and Australia. The United States, for example, in addition to having 42 percent of the world's known resources of hydrocarbons (oil, natural gas, and oil shales) and 38 percent of the lignite ("brown coal," used mainly in power stations), has 38 percent of the molybdenum (used in metal alloys), 21 percent of the lead (used chiefly for batteries, gasoline, and construction), 19 percent of the copper (used mainly for electrical wiring and components and for coinage), 18 percent of the bituminous coal (used mainly for fuel in power stations, and in the chemical industry), and 15 percent of the zinc. Russia has 68 percent of the vanadium (used in metal alloys), 50 percent of the lignite, 38 percent of the bituminous coal, 35 percent of the manganese, 25 percent of the iron, and 19 percent of the hydrocarbons.

The concentration of known resources in just a few countries is largely a result of geology, but it is also partly a function of countries' political and economic development. Political instability in much of postcolonial Africa, Asia, and Latin America has seriously hindered the exploration and exploitation of resources. In contrast, the relative affluence and great political stability of the United States have led to a much more intensive exploration of its resources. We should also bear in mind that the significance of particular resources is often tied to particular technologies. As technologies change, so do resource requirements, and the geography of economic development is "rewritten." One important example of this was the switch in the manufacture of mass-produced textiles from natural fibers like wool and cotton to synthetic fibers in the 1950s and 1960s. When this happened, many farmers in the U.S. South, for example, had to switch from cotton to other crops.

Regions and countries that are heavily dependent on one particular resource are vulnerable to the consequences of technological change. They are also vulnerable to fluctuations in the price set for their product on the world market. These vulnerabilities are particularly important for countries whose economies are especially dependent on

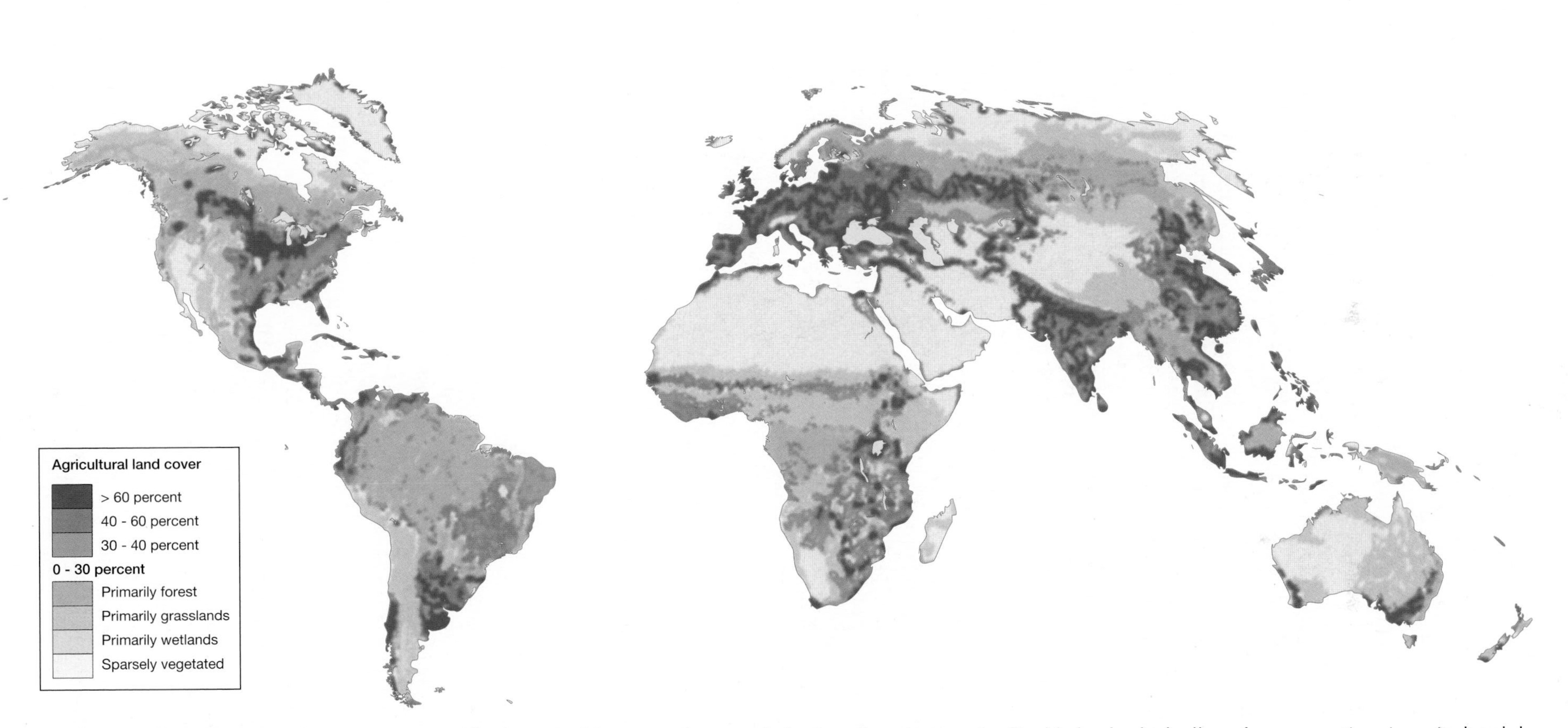

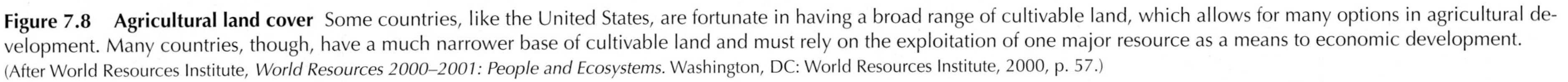

Figure 7.8 Agricultural land cover Some countries, like the United States, are fortunate in having a broad range of cultivable land, which allows for many options in agricultural development. Many countries, though, have a much narrower base of cultivable land and must rely on the exploitation of one major resource as a means to economic development. (After World Resources Institute, *World Resources 2000–2001: People and Ecosystems.* Washington, DC: World Resources Institute, 2000, p. 57.)

7.1 GEOGRAPHY MATTERS

Sustainable Development

In 1987 the World Commission on Environment and Development, chaired by former Norwegian Prime Minister Gro Harlem Brundtland, issued a report, *Our Common Future,* that stressed the intensification and interdependence of ecological and economic crises and made a strong plea for the principle of sustainable development—economic development that seeks to meet the needs and aspirations of the present without compromising the ability to meet those of the future.

The importance of sustainability is cogently illustrated by the concept of an **ecological footprint,** which is a measure of the human pressures on the natural environment from the consumption of renewable resources and the production of pollution. It represents a quantitative assessment of the biologically productive area required to produce the resources (food, energy, and materials) and to absorb the wastes of an individual, city, region, or country. The ecological footprint of a country or region changes in proportion to population size, average consumption per person, and the resource intensity of the technology being used. In the late 1970s, humanity's collective ecological footprint breached the sustainability mark for the first time, and has remained unsustainable ever since. By 2003 it was at around 30 percent greater than Earth's biologically productive capacity to sustain. At 9.57 hectares (23.6 acres) per person, the United States currently has the largest per capita ecological footprint on the planet. Other countries with extremely large ecological footprints include the United Arab Emirates, Canada, Norway, New Zealand, Kuwait, Sweden and Australia. Countries with the smallest ecological footprints, between 0.5 and 0.75 hectares (1.2 to 1.9 acres) per person, include Bangladesh, Burundi, Ethiopia, Haiti, Malawi, Nepal, and Pakistan.

Sustainable development means using renewable natural resources in a manner that does not eliminate or degrade them—by making greater use, for example, of solar and geothermal energy and recycled materials. It means managing economic systems so that all resources—physical and human—are used optimally. It means regulating economic systems so that the benefits of development are distributed more equitably (if only to prevent poverty from causing environmental degradation). It also means organizing societies so that improved education, health care, and social welfare can contribute to environmental awareness and sensitivity and an improved quality of life. A final and more radical aspect of sustainable development is to move away from wholesale globalization toward increased "localization": a desire to return to more locally based economies where production, consumption, and decision making can be oriented to local needs and conditions.

Put this way, sustainable development sounds eminently sensible yet impossibly utopian. A widespread discussion of sustainability took place in the early 1990s and was the focus of the "Earth Summit" (the United Nations Conference on Environment and Development) meeting in Rio de Janeiro in 1992. At-

nonfuel minerals, such as the Democratic Republic of the Congo (copper), Mauritania (iron ore), Namibia (diamonds), Niger (uranium—**Figure 7.9**), Sierra Leone (diamonds), Togo (phosphates), and Zambia (copper).

The Economic Structure of Countries and Regions

The relative share of primary, secondary, tertiary, and quaternary economic activities determines the *economic structure* of a country or region. **Primary activities** are those concerned directly with natural resources of any kind; they include agriculture, mining, fishing, and forestry. **Secondary activities** are those that process, transform, fabricate, or assemble the raw materials derived from primary activities or that reassemble, refinish, or package manufactured goods. Secondary activities include, for example, steelmaking, food processing, furniture making, textile manufacturing, automobile assembly, and garment manufacturing. **Tertiary activities** are those involving the sale and exchange of goods and services; they include warehousing, retail stores, personal services such as hairdressing, commercial services such as accounting and advertising, and entertainment. **Quaternary activities** are those dealing with the handling and processing of knowledge and information. Examples include data processing, information retrieval, education, and research and development (R&D).

Geographical Divisions of Labor

Variations in economic structure—according to primary, secondary, tertiary, or quaternary activities—reflect *geographical divisions of labor.* Geographical divisions of labor are national, regional, and locally based economic specializations that have evolved with the growth of the world-system of trade and politics (see Chapter 2) and tended by 128 heads of state, the meeting attracted intense media attention. At the conference, many examples were described of successful sustainable development programs at the local level. Some of these examples centered on the use of renewable sources of energy, as in the creation of small hydroelectric power stations to modernize Nepalese villages. Some examples centered on tourism in environmentally sensitive areas, as in the trips organized in Thailand's Phang Nga Bay, where tourists visit spectacular hidden lagoons by sea canoe under strict environmental regulation (no drinking, littering, eating, or smoking, and limits on the number of visitors allowed each day) and with a high level of social commitment (emphasizing respect for staff, local culture and family, proper training, and good pay). Most examples, however, centered on sustainable agricultural practices for peripheral countries, including the use of intensive agricultural features, such as raised fields and terraces in Peru's Titicaca Basin: techniques that had been successfully used in this difficult agricultural environment for centuries before European colonization. After the U.N. conference, however, many observers commented bitterly on the deep conflicts of interest between core countries and peripheral countries that were exposed by the summit.

One of the most serious obstacles to prospects for sustainable development is continued heavy reliance on fossil fuels as the fundamental source of energy for economic development. This not only perpetuates international inequalities but also leads to transnational problems such as acid rain, global warming, climatic changes, deforestation, health hazards and, many would argue, war. The sustainable alternative—renewable energy generated from the sun, tides, waves, winds, rivers, and geothermal features—has been pursued half-heartedly because of the commercial interests of the powerful corporations and governments that control fossil-fuel resources.

A second important challenge to the possibility of sustainable development is the rate of demographic growth in peripheral countries. Sustainable development is feasible only if population size and growth are in harmony with the changing productive capacity of the ecosystem. It is estimated that 1.2 billion of the world's 6.5 billion people are undernourished and underweight.

But the greatest single obstacle to sustainable development is the inadequacy of institutional frameworks. Sustainable development requires economic, financial, and fiscal decisions to be fully integrated with environmental and ecological decisions. In practice, national and local governments everywhere have evolved institutional structures that tend to separate decisions about what is economically rational and what is environmentally desirable. International organizations, while better placed to integrate policy across these sectors and better able to address economic and environmental "spillovers" from one country to another, have (with the notable exception of the European Union) not acquired sufficient power to promote integrated, harmonized policies. Without radical and widespread changes in value systems and unprecedented changes in political will, "sustainable development" is likely to remain an embarrassing contradiction in terms.

with the locational needs of successive technology systems. They represent one of the most important dimensions of economic development. For instance, countries whose economies are dominated by primary-sector activities tend to have a relatively low per capita GDP. The exceptions are oil-rich countries such as Saudi Arabia, Qatar, and Venezuela. Where the geographical division of labor has produced national economies with a large secondary sector, per capita GDP is much higher (as, for example, in Argentina and South Korea). The highest levels of per capita GDP, however, are associated with economies that are *postindustrial:* economies where the tertiary and quaternary sectors have grown to dominate the workforce, with smaller but highly productive secondary sectors.

As **Figure 7.10** shows, the economic structure of much of the world is dominated by the primary sector (that is, primary activities such as agriculture, mining, fishing, and forestry). In much of Africa and Asia, between 50 and 75 percent of the labor force is engaged in primary-sector activities. In contrast, the primary sector of the world's core regions is typically small, occupying only 5 to 10 percent of the labor force.

The secondary sector is much larger in the core countries and in semiperipheral countries, where the world's specialized manufacturing regions are located. In 2004, core countries accounted for almost three-quarters of world manufacturing value added (MVA). This share has been slowly decreasing, however. The core countries had an average annual growth rate for MVA of around 2 percent during 1990–2004, while the growth rate in the rest of the world was closer to 7 percent. This growth has been concentrated in semiperipheral, newly industrializing countries (NICs). **Newly industrializing countries** are countries, formerly peripheral within the world-system, that have acquired a significant industrial sector, usually through direct foreign investment. Of the 20 biggest manufacturing countries in 2004, seven were NICs: China, South Korea, Mexico, Brazil, India, Argentina, and

Figure 7.9 Commodity dependency Uranium mine, Arlit, Niger. Almost all of Niger's export earnings come from uranium mining.

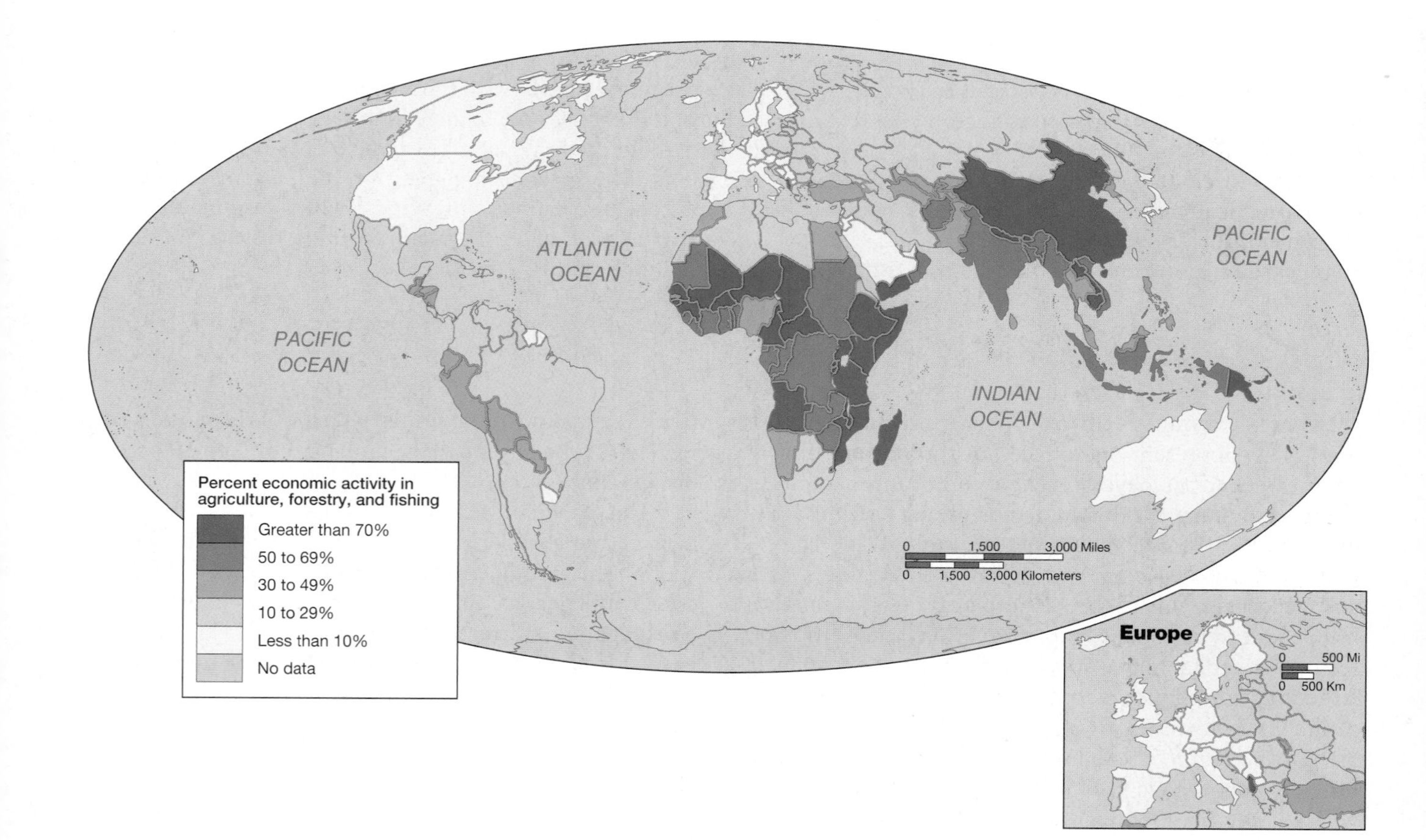

Figure 7.10 The geography of primary economic activities Primary economic activities are those that are concerned directly with natural resources of any kind. They include agriculture, mining, forestry, and fishing. The vast majority of the world's population, concentrated in China, India, Southeast Asia, and Africa, is engaged in primary economic activities. This map shows the percentage of the labor force in each country that was engaged in primary employment in 2002. In some countries, including China, primary activities account for more than 70 percent of the workforce. In contrast, primary activities always account for less than 10 percent of the labor force in the world's core countries, and often for less than 5 percent.

Thailand (in order of importance). The vast majority of peripheral countries have a very small manufacturing output. The share of world MVA for Africa has changed little over the last two decades, remaining at about 1 percent.

In terms of individual countries, the United States remains by far the most important source of manufactured goods, accounting for just over 25 per cent of global MVA in 2004. Just five countries—the United States, Japan, Germany, China, and the United Kingdom—together produced over 60 percent of the world total. Another important aspect of secondary activities concerns *productivity*. In general, the highly capitalized manufacturing industries of the developed countries have been able to maintain high levels of worker productivity, with the result that the contribution of manufacturing to their GDP has remained relatively high even as the size of their manufacturing labor forces has decreased.

Within the framework of this continuing dominance of the advanced industrial economies there are several important trends. Although the United States has retained its leadership as the world's major producer of manufactured goods, its dominance has been significantly reduced. In the 1960s its share of world manufacturing output was 40 percent, compared to its current share of around 25 percent. Meanwhile, Japan increased its share from less than 6 percent in the 1960s to about 18 percent by 2004. The emergence of a dozen or so NICs as settings for manufacturing is particularly striking. Several of these are in Latin America (Brazil, Mexico, and Argentina), but in terms of the rate of manufacturing growth, the Asian NICs are most impressive.

The growth of manufacturing in Pacific Asia has generated agglomerations of economic activity at a scale that sometimes crosses national boundaries, as with the Southern China—Hong Kong—Taiwan triangle and the Singapore–Batam–Johor triangle (**Figure 7.11**). The 1500 kilometer (932 mile) urban belt in North East Asia that runs from Beijing to Tokyo via Pyongyang and Seoul connects some 80 cities of over 200,000 inhabitants each, encompassing nearly 100 million urban dwellers altogether. In terms of individual countries, China has experienced a dramatic increase in manufacturing production, achieving annual average growth rates during the 1970s, 1980s, and 1990s of about 8 percent, 11 percent, and 14 percent, respectively (see Box 7.2: "China's Economic Development"). Of the four Asian "Tigers"—South Korea, Hong Kong, Taiwan, and Singapore—South Korea enjoyed the most spectacular increase in manufacturing production (**Figure 7.12**), achieving annual average growth rates of almost 18 percent in the 1960s, 17 percent in the 1970s, 12 percent in the 1980s, and over 7 percent in the 1990s (despite a financial crisis that occurred in Asia in the late 1990s). In fact, the four Asian "Tigers" have made remarkable progress up the world league table of exporters. In 2003 Hong Kong ranked 11th in the world (up from 26 in 1980), Taiwan ranked 15th (up from 22), South Korea ranked 12th (up from 27), and Singapore ranked 15th (up from 30). More recently, other Pacific Rim NICs, such as Malaysia and Thailand, have experienced rapid growth in manufacturing production.

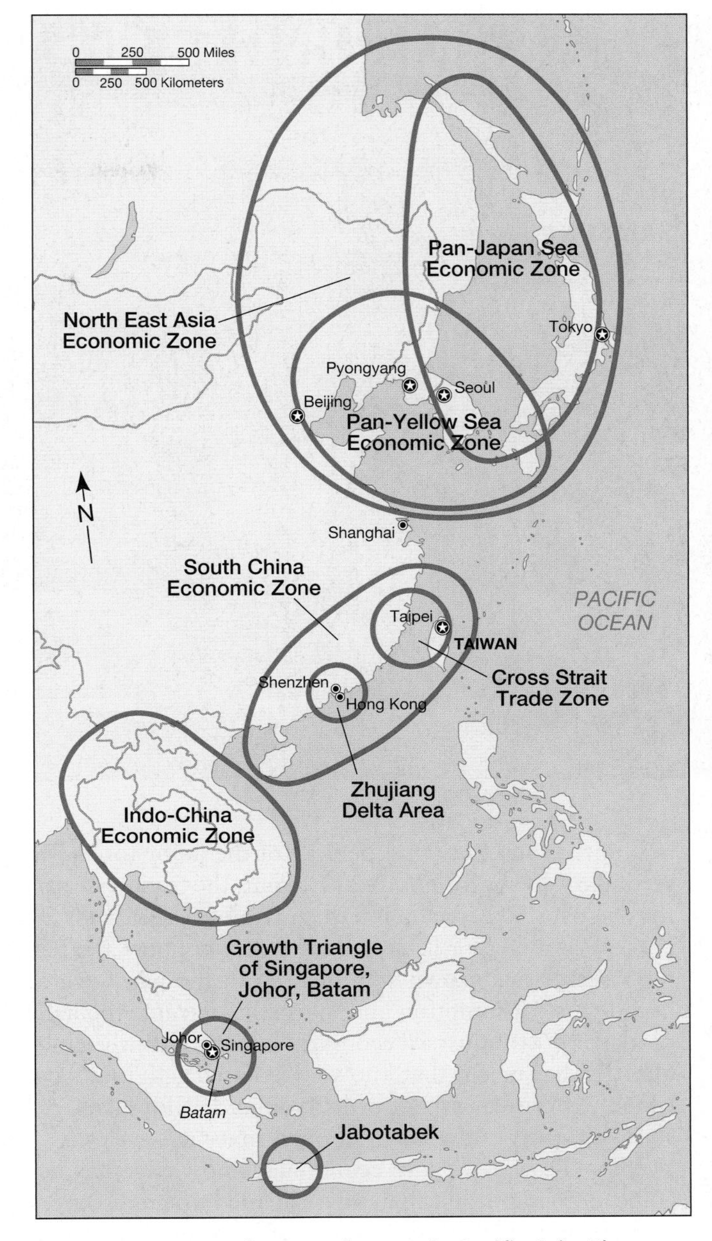

Figure 7.11 Emerging growth zones in Pacific Asia The globalization of manufacturing industries has spurred the growth of a series of extended metropolitan regions. (After P. Dicken, *Global Shift,* 4th ed.). New York: Guilford, 2003, Figure 3.28, p. 78.)

These shifts are part of a globalization of economic activity that has emerged as the overarching component of the world's economic geography. As we shall see, it has been corporate strategy, particularly the strategies of large transnational corporations (TNCs), that has created this globalization of economic activity.

The tertiary and quaternary sectors are significant only in the most affluent countries of the core. In the United States, for example, the primary sector in 2004

Figure 7.12 Manufacturing in South Korea A Korean businessman walks past a sign promoting Samsung electronic products in Seoul, South Korea.

accounted for less than 4 percent of the labor force, the secondary sector for about 22 percent, the tertiary sector for just over 50 percent, and the quaternary sector for 22 percent of the labor force. In every core country, the tertiary sector has grown significantly in the past several decades as consumption and marketing became the hallmarks of post-industrial economies. More recently, globalization has meant that knowledge-based activities have become a critical aspect of economic development, resulting in the rapid growth of quaternary industries.

For the world's core economies, knowledge has become more important than physical and human resources in determining levels of economic well-being. More than half of the GDP of major core countries is based on the production and distribution of knowledge. In the United States, more workers are engaged in producing and distributing knowledge than in making physical goods.

For the world's peripheral economies, lack of knowledge—along with a limited capacity to absorb and communicate knowledge—has become an increasingly important barrier to economic development. Poor countries have fewer resources to devote to research, development, and the acquisition of information technology. They also have fewer institutions that can provide high-quality education, fewer bodies that can enforce standards and performance, and only weakly developed organizations for gathering and disseminating the information needed for business transactions. As a result, economic productivity tends to fall relative to the performance of places and regions in core economies, where new knowledge is constantly generated and rapidly and effectively disseminated.

International Trade, Aid, and Debt

The geographical division of labor on a world scale means that the geography of international trade is very complex (**Figure 7.13**). One significant reflection of the increased economic integration of the world-system is that global trade has grown much more rapidly over the past few decades than global production. Between 1985 and 2005 the average annual growth rate of the value of world exports was twice that of the growth of world production and several times greater than that of world population growth.

The fundamental structure of international trade is based on a few **trading blocs**—groups of countries with formalized systems of trading agreements—with most of the world's trade taking place within four trading blocs:

- Western Europe, together with some former European colonies in Africa, South Asia, the Caribbean, and Australasia;
- North America, together with some Latin American states;
- the countries of the former Soviet world-empire; and
- Japan, together with other East Asian states and the oil-exporting states of Saudi Arabia and Bahrain.

Nevertheless, a significant number of countries exhibit a high degree of **autarky** from the world economy. That is, they do not contribute significantly to the flows of imports and exports that constitute the geography of trade. Typically, these are smaller peripheral countries,

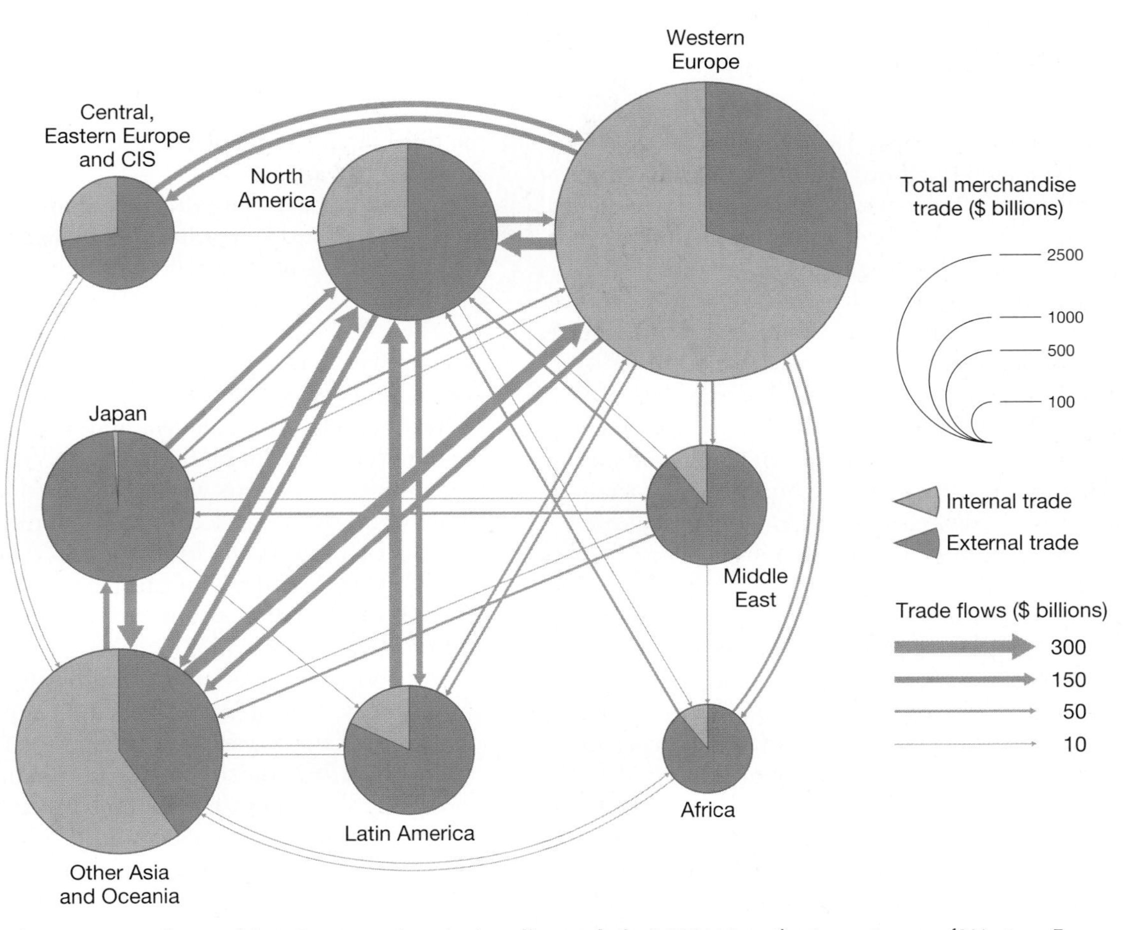

Figure 7.13 The world trade network: merchandise trade in 2003 Note the importance of Western Europe as a trading region. Two-thirds of Western Europe's merchandise trade is intra-regional—that is, between Western European countries themselves. Asia is the second most significant trade region, with North America third. (Updated from P. Dicken, *Global Shift,* 4th ed.). New York: Guilford, 2003, Figure 3.7, p. 41.

such as Bolivia, Burkina Faso, Ghana, Malawi, Samoa, and Tanzania.

Patterns of world trade have been shifting rapidly, however, in response to several factors. In general, the trend has been toward an intensification of the long-standing domination of trade within and between core regions at the expense of trade between core countries and peripheral countries—with the major exception of trade in oil. Second, innovations in transport, communications, and manufacturing technology have diminished the importance of the distance-based factors that have underpinned traditional trading blocs. Third, shifts in global politics have affected the geography of trade. One important shift was the breakup of the former Soviet Union. Other significant geopolitical changes include the trend toward the political as well as economic integration of Europe and the increasing participation of China in the world economy. But perhaps the most important shift in global politics in relation to world trade has been the shift toward open markets and free trade through neoliberal policies propagated by core countries. Fourth, the globalization of economic activity has created new flows of materials, components, information, and finished products. As a global system of manufacturing has emerged, significant quantities of manufactured goods are now imported *and* exported across much of the world through complex commodity chains; no longer do developed economies export manufactures and peripheral countries import them. African countries are an important exception, with many of them barely participating in world trade in manufactures.

The most striking aspect of contemporary patterns of trade is the persistence of the dependence of peripheral countries on trade with core countries that are geographically or geopolitically close. Thus, for example, the United States is the central focus for the exports and the origin of the bulk of the imports of most Central American countries, while France is the focus for commodity flows to and from French ex-colonies such as Algeria, Cambodia, Benin, and the Ivory Coast. These flows, however, represent only part of the action for the core economies, whose trading patterns are dominated by flows to and from other core countries.

One implication of this situation is that the smaller, peripheral partners in these trading relationships are highly dependent on levels of demand and the overall

China's Economic Development

Under the leadership of Deng Xiaoping (1978–1997), China embarked on a thorough reorientation of its economy, dismantling Communist-style central planning in favor of private entrepreneurship and market mechanisms, and began integrating itself into the world economy. Saying that he did not care whether the cat was black or white, as long as it caught mice, Deng Xiaoping established a program of "Four Modernizations" (industry, agriculture, science, and defense) and an "open-door policy" that allowed China to be plugged in to the interdependent circuits of the global economy. As a result, China has completely reorganized and revitalized its economy. Agriculture has been decollectivized, with Communist collective farms modified to allow a degree of private profit taking. State-owned industries have been closed or privatized, and centralized state planning has been dismantled in order to foster private entrepreneurship.

In the 1980s and early 1990s, when the world economy was sluggish, China's manufacturing sector grew by almost 15 percent each year. Almost all of the shoes once made in South Korea or Taiwan are now made in China. More than 60 percent of the toys in the world, accounting for more than $10 billion in trade, are made in China. Since 1992, China has extended its open-door policy, permitted foreign investment aimed at Chinese domestic markets, and normalized trading relationships with the United States and the European Union. In 2001, China was admitted to the World Trade Organization, allowing China to trade more freely than ever before with the rest of the world.

China's increased participation in world trade has created an entirely new situation within the world economy. Chinese manufacturers, operating with low wages, have imposed a deflationary trend in world prices for manufacturers. The Chinese economy's size makes it a major producer, and its labor costs stay flat year after year because there is an endless supply of people happy to work for 60 cents an hour. Meanwhile, the rapid expansion of consumer demand in China has begun to drive up commodity prices in the world market. Overall, China's economy is already the third largest in the world after those of the United States and Japan.

Nowhere have China's "open door" policies had more impact than in South China, where the Chinese government has deliberately built upon the prosperity of Hong Kong, the former British colony that was returned to China in 1997. The coastline of South China provides many protected bays suitable for harbors, and a series of ports has developed including Quanzhou, Shantou, Xiamen, and, on either side of the mouth of the Zhu Jiang (Pearl River), Macão and Hong Kong. These ports were a precondition for South China's emergence as a core manufacturing region, providing an interface with the world economy. The established trade and manufacturing of Macão—a Portuguese colony that was returned to China in 1999—and Hong Kong provided another precondition for success. When Deng Xiaoping established his "open door" policy, a third factor kicked in: capital investment from Hong Kong, Taiwan, and the Chinese diaspora. By 1993, more than 15,000 manufacturers from Hong Kong alone had set up businesses in neighboring Guangdong Province, and a similar number had established subcontracting relationships, contracting out processing work to Chinese companies. Today, the cities and special economic zones of South China's "Gold Coast" provide a thriving export-processing platform that has driven double-digit annual economic growth for much of the past two decades. The city of Shenzhen (**Figure 7.A**) has grown from a population of just 19,000 in 1975 to 1.04 million in 2000, with an additional 2 million in the surrounding municipalities.

Much of China's manufacturing growth has been based on a strategy of import substitution (see page 270). In spite of China's membership in the World Trade Organization (which has strict rules about intellectual property), a significant amount of China's industry is based on counterfeiting and reverse engineering (making products that are copied and then sold under different or altered brand names) and piracy (making look-alike products passed off as the real

Figure 7.A Shenzhen The population of the city of Shenzhen, just across the border from the Special Administrative Region of Hong Kong, grew from just 5000 in 1970 to more than 1 million in 2000.

thing). Copies of everything from DVDs, movies, designer clothes and footwear, drugs, motorcycles, and automobiles to high-speed magnetic levitation (maglev) cross-country trains save Chinese industry enormous sums in research and development and licensing fees, while saving the country even greater sums in imports.

Foreign investors, meanwhile, have been keen to develop a share of China's rapidly expanding and increasingly affluent market. The automobile market is particularly attractive to Western manufacturers. Volkswagen was the first to establish a presence in China, in 1985. By 2003, Volkswagen had claimed around 40 percent of China's annual production of almost 4 million cars and light trucks. General Motors, in partnership with Shanghai Automotive Industry Corporation, has a 10 percent market share. Other foreign manufacturers operating in China include Honda, Toyota, Nissan, and most recently, BMW and Mercedes.

Overall, most foreign investment in China comes from elsewhere within East Asia. Japan, Taiwan, and South Korea, having developed manufacturing industries that undercut those of the United States, now face deindustrialization themselves through the inexorable process of "creative destruction" (see p. 278). More than 50,000 Taiwanese firms have established operations in China, investing an estimated $80 billion. Pusan, the center of the South Korean footwear industry that in 1990 exported $4.3 billion worth of shoes, is full of deserted factories. South Korean footwear exports are down to less than $700 million, while China's footwear exports have increased from $2.1 billion in 1990 to $13 billion in 2002. Several Japanese electronics giants, including Toshiba Corp., Sony Corp., Matsushita Electric Industrial Co., and Canon Inc., have expanded operations in China even as they have shed tens of thousands of workers at home. Olympus manufactures its digital cameras in Shenzhen and Guangzhou. Pioneer has moved its manufacture of DVD recorders to Shanghai and Dongguan. Toshiba's factory in Dalian illustrates the logic. Toshiba is one of about 40 Japanese companies that have built large-scale production facilities in a special export-processing zone established by Dalian in the early 1990s with generous financial support from the Japanese government and major Japanese firms. By shifting production of digital televisions here from its plant in Saitama, Japan, in 2001, Toshiba cut labor costs per worker by 90 percent.

economic climate in developed economies. Another aspect of dependency, in this context, is the degree to which a country's export base lacks diversity. **Dependency** involves a high level of reliance by a country on foreign enterprises, investment, or technology. External dependence for a country can mean that it is highly reliant on levels of demand and the overall economic climate of other countries. Dependency for a peripheral country, for example, can result in a narrow economic base in which the balancing of national accounts and the generation of foreign exchange depend on the export of one or two agricultural or mineral resources. **Figure 7.14** shows one reflection of dependency: the index of commodity concentration of exports. Countries with low scores on this index have diversified export bases. They include Argentina, Brazil, China, India, and North and South Korea, as well as most of the core countries. At the other extreme are peripheral countries where the manufacturing sector is poorly developed and the balancing of national accounts and the generation of foreign exchange depend on the export of one or two agricultural or mineral resources, as in Angola, Chad, the Dominican Republic, Iran, Iraq, Libya, and Nigeria, for example.

Patterns of International Debt

In many peripheral countries, 20 percent or more of all export earnings are swallowed up by debt service—the annual interest on international debts (**Figure 7.15**). In 2002, for example, 9 peripheral countries, including Angola, Burundi, Sierra Leone, and Zambia, had total debts so large that they owed more than they produced.

At the root of the international debt problem is the structured inequality of the world economy. The role inherited by most peripheral countries within the **international division of labor** (the specialization, by countries, in particular products for export) has been one of producing primary goods and commodities for which both the elasticity of demand and price elasticity are low. The **elasticity of demand** is the degree to which levels of demand for a product or service change in response to changes in price. Where a relatively small change in price induces a significant change in demand, elasticity is high; where levels of demand remain fairly stable in spite of price changes, demand is said to be inelastic. Demand for the products of peripheral countries in their principal markets (the more developed countries) has a low elasticity: It tends to increase by relatively small amounts in response to significant increases in the incomes of their customers. Similarly, significant reductions in the price of their products tend to result in only a relatively small increase in demand. Think, for example, of the cocoa-producing regions of West Africa (**Figure 7.16**). No matter how they improve productivity in order to keep prices low, and no matter how much more affluent their customers in core

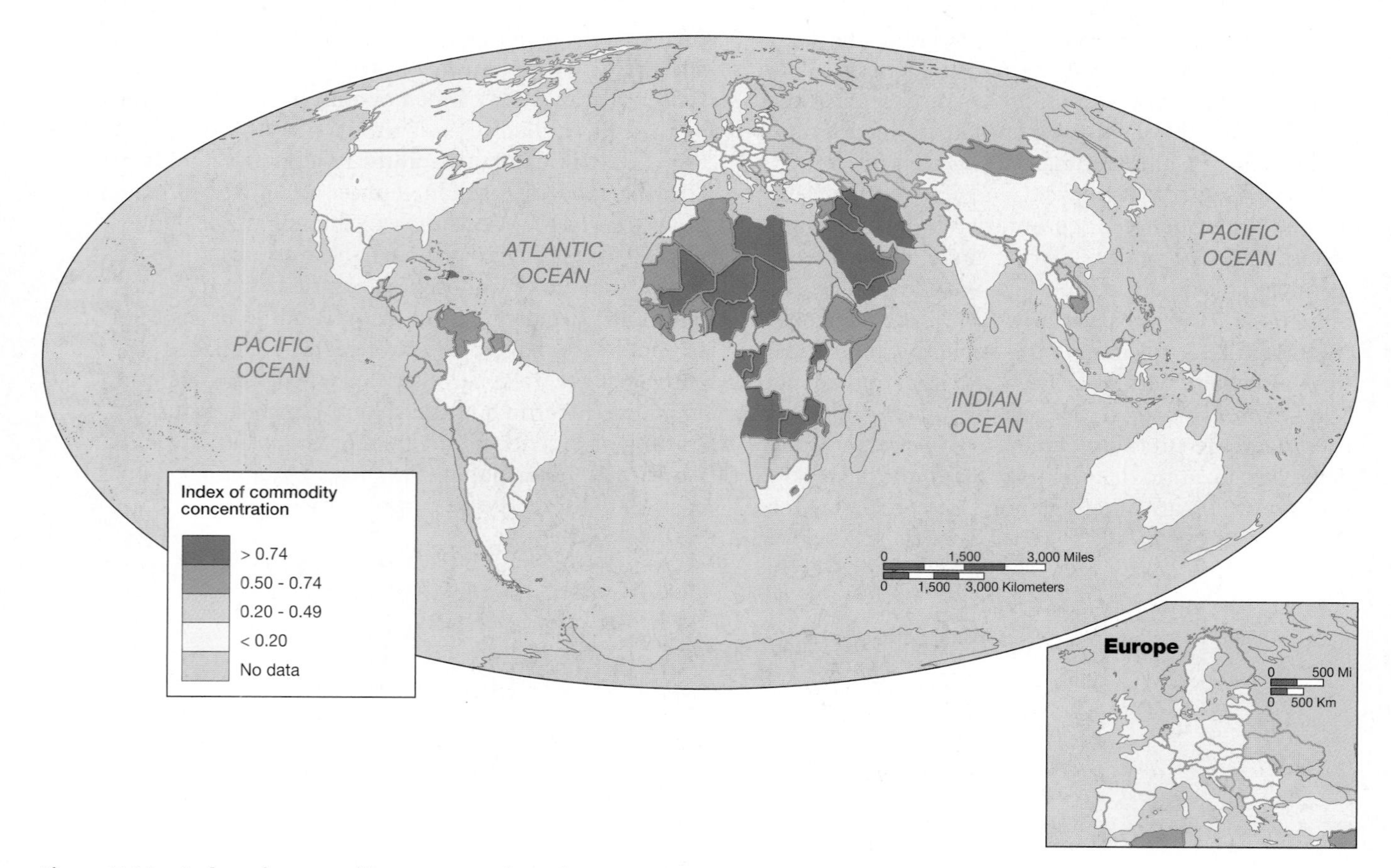

Figure 7.14 Index of commodity concentration of exports, 2002

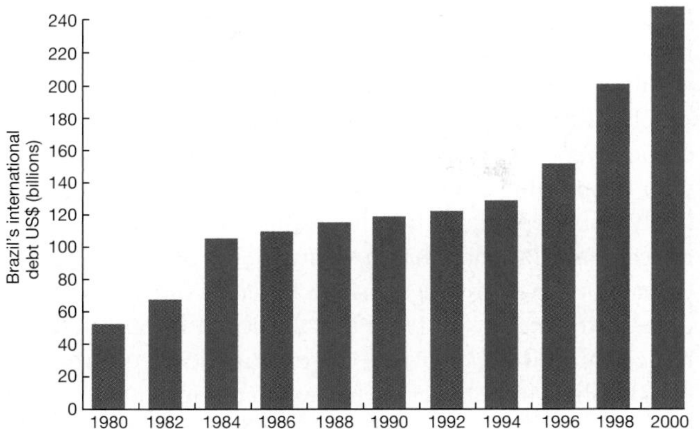

Brazil borrowed so much money in the 1970s and 1980s that it could no longer meet the interest payments. Between 1983 and 1989 the International Monetary Fund (IMF) bailed the country out but imposed austerity measures that were designed to curb imports. These included a 60 percent increase in gasoline prices and a reduction of the minimum wage to $50 a month, which gave workers half the purchasing power they had in 1940. Nevertheless, by 2002, the country's debt had reached nearly $228 billion, annual inflation rates in the 1990s having ranged between 500 percent and 2000 percent.

Mexico was forced to default on international debt repayments in 1982, triggering an international financial crisis that involved defaults the following year by Argentina, Brazil, Venezuela, the Philippines, Ghana, Nigeria, and many other peripheral countries. In 1996, Mexico was bailed out of another debt crisis by creditor countries, but was forced to devalue its currency, creating additional hardships for most of the population.

Since 1983, when a foreign exchange crisis left it unable to meet its debt repayments, the Philippine government has had to reschedule its debts several times with its creditor nations. By 2002 the Philippines had reduced its foreign debt to just over $59 billion, with a debt-service ratio of less than 12 percent of its GDP.

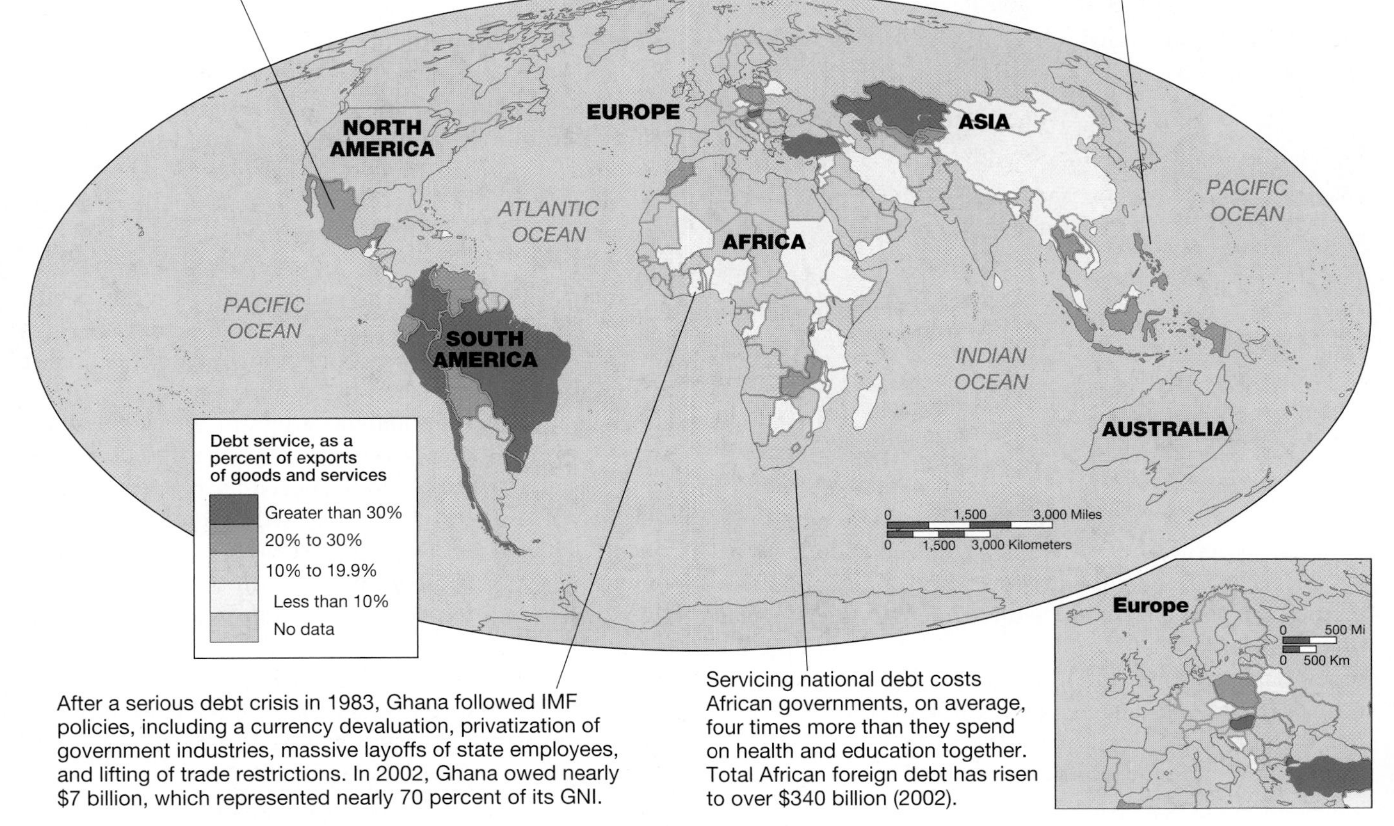

After a serious debt crisis in 1983, Ghana followed IMF policies, including a currency devaluation, privatization of government industries, massive layoffs of state employees, and lifting of trade restrictions. In 2002, Ghana owed nearly $7 billion, which represented nearly 70 percent of its GNI.

Servicing national debt costs African governments, on average, four times more than they spend on health and education together. Total African foreign debt has risen to over $340 billion (2002).

Figure 7.15 The debt crisis In some countries the annual interest on international debts (their "debt service") accounts for more than 20 percent of the annual value of their exports of goods and services. Many countries first got into debt trouble in the mid-1970s, when Western banks, faced with recession at home, offered low-interest loans to the governments of peripheral countries rather than being stuck with idle capital. When the world economy heated up again, interest rates rose and many countries found themselves facing a debt crisis. The World Bank and the International Monetary Fund (IMF), in tandem with Western governments, worked to prevent a global financial crisis by organizing and guaranteeing programs that eased poor countries' debt burdens. Western banks were encouraged to swap debt for equity stakes in nationalized industries, while debtor governments were persuaded to impose austere economic policies. These policies have helped ease the debt crisis, but often at the expense of severe hardship for ordinary people. In dark humor, the IMF became known among radical development theorists as "imposing misery and famine."

Figure 7.16 Cocoa production Workers opening cocoa beans on a plantation in Ivory Coast. Ivory Coast is the world's largest producer of cocoa, and heavily dependent on cocoa exports for its foreign earnings. But no matter how efficient they may become, the terms of trade will almost certainly be tilted against them.

countries become, there is a limit to the demand for cocoa products. In contrast, the elasticity of demand and price elasticity of high-tech manufactured goods and high-order services (the specialties of core economies within the international division of labor) are both high. As a result, the terms of trade are stacked against the producers of primary goods. The **terms of trade** are determined by the ratio of the prices at which exports and imports are exchanged. When the price of exports rises relative to the price of imports, the terms of trade reflect an improvement for the exporting country. No matter how efficient primary producers may become, or how affluent their customers, the balance of trade will be tilted against them. Quite simply, they must run in order to stand still.

An obvious counterstrategy for peripheral countries is to attempt to establish a new role in the international division of labor, moving away from a specialization in primary commodities toward a more diversified manufacturing base. This strategy is known as **import substitution.** It is a difficult strategy to pursue, however, because building up a diversified manufacturing base requires vast amounts of start-up capital. With the terms of trade running against them, it is extremely difficult for peripheral countries to accumulate this capital.

By 2002 the total debt owed by low- and middle-income countries to high-income core countries was just over \$2.3 trillion. Practically none of this money is ever likely to be repaid, but so long as the North American, European, and Japanese banks continue to receive interest on their loans, they will be satisfied. Indeed, core countries are doing extremely well from this aspect of international finance. In 2002 the world's core countries took in about \$300 billion in debt servicing, while paying out a total of less than \$50 billion in new loans. There is always a danger to the future well-being of the better off countries, however, in that debtor countries might act together in deliberately defaulting on their debts, which would cause a major disturbance to the global financial system.

The debt problem has led to calls for affluent lending countries to provide debt relief to some of the poorest countries. In 2005, the world's richest countries—the G8 group—agreed to write off \$40 billion in debts owed by 18 of the world's poorest countries, most of them in Africa. Addressing the full magnitude of LDC debt, however, will require continued substantial and sustained efforts if the poorest countries are to break free of the crushing financial obligations of their accumulated debts.

Patterns of International Aid

For many countries, the only alternative to crippling debt, short of opting for self-sufficiency (attempted, for example, by Tanzania and Burma) or opting out of the capitalist world economy altogether (e.g., Cuba), is to raise the capital as loans. International aid programs provide low interest rates for many of these loans. But if the borrowed capital is invested in economic development projects that do not yield sufficient returns, further loans have to be undertaken in order to service the original debt and/or to finance new development projects. In this way, international debt tends to pile up. The syndrome of having constantly to borrow in order to fund "development" has come to be known as the **debt trap.**

Large-scale movements of international development aid began shortly after World War II with the Marshall Plan, financed by the United States to bolster war-torn European allies whose economic weakness, it was believed, made them susceptible to communism. During the 1950s and 1960s, as more peripheral countries gained independence, aid became a useful weapon in Cold War offensives to establish and preserve political influence throughout the world. By the late 1960s the list of donor countries had expanded beyond the superpowers to include smaller countries such as Denmark and Sweden, whose motivation in aid-giving must be seen as more philanthropic than political. In addition, there was a greater geographic dispersal of aid, thanks largely to the activities of multilateral financial agencies such as the International Monetary Fund and the World Bank. Private aid organizations, meanwhile, have focused their efforts on emergency aid, on campaigning for debt-relief programs, and on fair trade practices (see Box 7.3: "Fair Trade").

Nevertheless, the geography of aid still has a strong political flavor, the United States in particular using foreign aid to support its international political agenda. In the past, this meant a disproportionately high level of support for a few strategically important countries, as the United States sought to limit the influence of the former Soviet bloc. Today the chief beneficiary of U.S. aid is Israel, which received $2.04 billion in military aid and $720 million in economic aid in 2002. Aid from several other countries reflects localized political aspirations and colonial ties. Much British and French aid, for example, is directed toward former African colonies, while Japanese aid is disbursed largely within Asia, and aid from the OPEC countries is directed mainly toward the "frontline" Arab countries.

The end of the Cold War resulted in diminished levels of international aid. Thus, whereas official development assistance from **OECD** countries (the Organization for Economic Cooperation and Development, an association of 30 industrialized countries) amounted to nearly 0.5 percent of their total GNI in 1965, it had fallen to 0.25 percent by 2003. The most striking decrease was in aid from the United States, whose overseas development assistance as a percentage of their GNI fell from 0.58 to 0.15. Japan, meanwhile, has never exceeded 0.33 percent of its GNI in aid. In contrast, Denmark, Luxembourg, the Netherlands, Sweden, and Norway have steadily increased their aid-giving to become the only ones to have met or surpassed the United Nations target of 0.7 percent of GNI (**Table 7.1.**). Perhaps more disturbing is the World Bank's assertion that a large proportion of developed countries' pledged aid is "phantom" aid because it ends up going to international consultants and donor countries' own firms. Altogether, around 60 percent of all international aid becomes phantom in this way; and in the case of aid from France and the United States, the figure is closer to 90 percent.

Nevertheless, the impact of aid on some countries is clearly significant. Mozambique, for example, received aid amounting to 57 percent of its GDP in 2002; and nine other countries—including Burundi, Mauritania, Rwanda, and Sierra Leone—received aid equivalent to more than 20 percent of their GDP. These figures, however, say as much about these countries' extremely small GDP as anything else. They are also exceptions. In general, the poorest countries are by no means the biggest recipients of aid; the amount of aid received per capita is generally very low, and a large share of it is deployed as emergency food aid (**Figure 7.17**). In 2002 low-income countries like Burundi, Chad, and Uganda received an average of around US$26 per capita in overseas development assistance, while middle-income developing economies like the Dominican Republic, Egypt, and Sri Lanka received around US$18 per capita. At these levels, aid cannot seriously be regarded as a catalyst for development or as an instrument for redressing core-periphery inequalities.

TABLE 7.1 International Economic Aid: Net Official Development Assistance Disbursed, 2003

	Total (US$ millions)	As % of GNI
Norway	2,042	0.92
Denmark	1,748	0.84
Netherlands	3,981	0.80
Sweden	2,400	0.79
France	7,253	0.41
Switzerland	1,299	0.39
Ireland	504	0.39
Belgium	1,853	0.39
Finland	558	0.35
United Kingdom	6,282	0.34
Germany	6,784	0.28
Australia	1,219	0.25
Canada	2,031	0.24
Spain	1,961	0.23
Japan	8,880	0.20
Austria	505	0.20
Italy	2,433	0.17
United States	16,254	0.15

Interpretations of Patterns of Development

The overall relationship between economic structure and levels of prosperity makes it tempting to interpret economic development in terms of distinctive stages. Each developed region or country, in other words, might be thought of as progressing from the early stages of development, with a heavy reliance on primary activities (and relatively low levels of prosperity), through a phase of industrialization and on to a "mature" stage of postindustrial development (with a diversified economic structure and relatively high levels of prosperity). This, in fact, has been a commonly held view of economic development, first conceptualized by a prominent economist, W. W. Rostow (**Figure 7.18**).

Rostow's model, however, is too simplistic to be of much help in understanding human geography. The reality is that places and regions are interdependent. The fortunes of any given place are increasingly tied up with those of many others. Furthermore, Rostow's model perpetuates the myth of "developmentalism," the idea that every country and region will eventually make economic progress toward "high mass consumption" provided that they compete to the best of their ability within the world economy. The main weakness of developmentalism is that it is simply not reasonable to compare the prospects of late starters to the experience of those places, regions, and countries that were among the early starters. For these early starters the horizons were clear: free of effective

Figure 7.17 Somali children eating at a relief center The United Nations estimated in 2004 that throughout Somalia 750,000 people were living in a state of chronic humanitarian need. Two decades of struggles between warlord factions ravaged the country with famine and violence until a peace and reconciliation conference resulted in a new federal government in 2004. Under such conditions, international aid was restricted to emergency food relief programs.

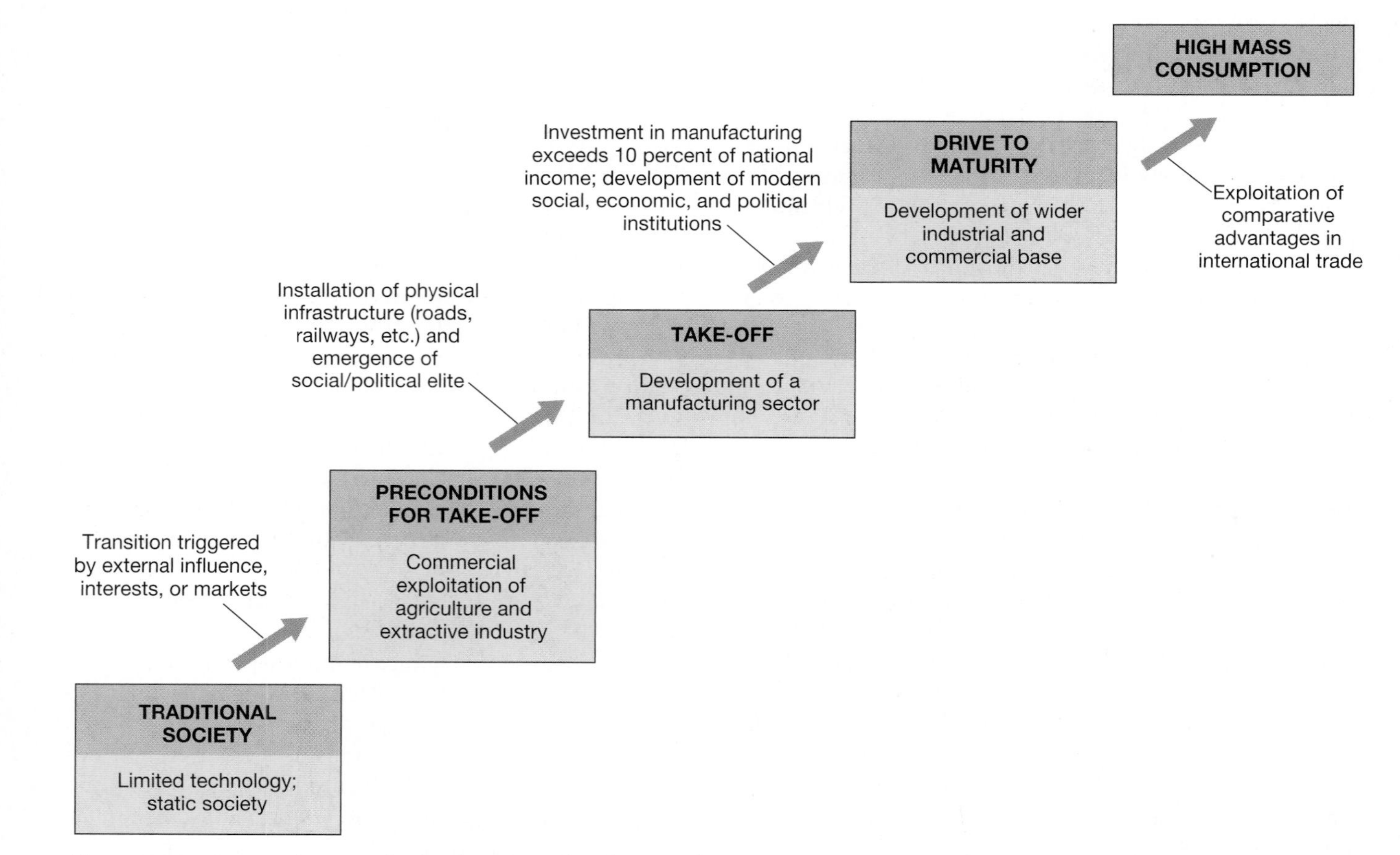

Figure 7.18 Stages of economic development This diagram illustrates a model of economic development based on the idea of successive stages of development. Each stage is seen as leading to the next, though some regions or countries may take longer than others to make the transition from one stage to the next. According to this view, now regarded as overly simplistic, places and regions can be seen as following parallel courses within a world that is steadily modernizing. Late starters will eventually make progress, but at speeds determined by their resource endowments, their productivity, and the wisdom of their people's policies and decisions.

competition, free of obstacles, and free of precedents. For the late starters the situation is entirely different. Today's less developed regions must compete in a crowded field while facing numerous barriers that are a direct consequence of the success of some of the early starters.

Indeed, many writers and theorists of international development claim that the prosperity of the core countries in the world economy has been based on *under*-development and squalor in peripheral countries. Peripheral countries, it is argued, could not "follow" the previous historical experience of developed countries in stages-of-development fashion because their underdevelopment (i.e., exploitation) was a structural requirement for development elsewhere. The development of Europe and North America, in other words, required the systematic underdevelopment of peripheral countries. By means of unequal trade, exploitation of labor, and profit extraction, the underdeveloped countries became increasingly rather than decreasingly impoverished.

The writing of André Gunder Frank exemplifies the explanations of international economic change that arose from this critique. Frank rejected the idea that underdevelopment is an original condition, equivalent to "traditionalism" or "backwardness." To the contrary, he argues, it is a condition created by integration into the worldwide system of exchange that is the world capitalist system. The world economy, Frank argues, has been unequally structured since Europeans first ventured out into the world in the sixteenth century. Although the form of the dominance of core over periphery has changed from colonialism and imperialism to neocolonialism, an overall transfer of wealth from periphery to core has continued to fuel growth in some places at the expense of others.

Frank's approach is an example of what can be called "dependency theory." This has been a very influential approach in explaining global patterns of development and underdevelopment. It states, essentially, that development and underdevelopment are reverse sides of the same process: *Development somewhere requires underdevelopment somewhere else.* Immanuel Wallerstein's world-system theory (see Chapter 2) takes this kind of dependency into account. According to this perspective, the entire world economy is to be seen as an evolving market system with an economic hierarchy of states—a core, a semiperiphery, and a periphery. The composition of this hierarchy is variable: Individual countries can move from periphery to semiperiphery, core to periphery, and so on.

PATHWAYS TO REGIONAL DEVELOPMENT

Patterns of *regional* economic development are also historical in origin and cumulative in nature. Recognizing this, geographers are interested in **geographical path dependence,** the relationship between present-day activities in a place and the past experiences of that place. When spatial relationships and regional patterns emerge through the logic of fundamental principles of spatial organization, they do so in ways guided and influenced by preexisting patterns and relationships.

These observations lead to an important principle of regional economic development, the principle of initial advantage. **Initial advantage** highlights the importance of an early start in economic development. It represents a special case of external economies. Other things being equal, new phases of economic development will take hold first in settings that offer **external economies:** existing labor markets, existing consumer markets, existing frameworks of fixed social capital, and so on. External economies are cost savings resulting from advantages that are derived from circumstances beyond a firm's organization and methods of production. These initial advantages will be consolidated by **localization economies**—cost savings that accrue to particular industries as a result of clustering together at a specific location—and so form the basis for continuing economic growth. Examples of localization economies include sharing a pool of labor with special skills or experience, supporting specialized technical schools, joining to create a marketing organization or a research institute, and drawing on specialized subcontractors, maintenance firms, suppliers, distribution agents, and lawyers. Where such advantages lead to a reputation for high-quality production, localization will be intensified, because more producers will want to cash in on the reputation. Among many examples are the electronics and software industries of Silicon Valley; recording companies in Los Angeles; Sheffield steel and steel products, especially cutlery; the U.S. auto industry in Detroit; Swiss watches made in the towns of Biel, Geneva, and Neuchâtel; Belgian lace from Bruges, Brussels, and Mechelen; English worsted in Bradford and Huddersfield; Stoke-on-Trent English china; Irish linen of Athlone; and French perfume made in Grasse.

For places and regions with a substantial initial advantage, therefore, the trajectory of geographical path dependence tends to be one of persistent growth—reinforcing the core-periphery patterns of economic development found in every part of the world and at every spatial scale. That said, geographers recognize there is no single pathway to development. The consequences of initial advantage for both core and peripheral regions can be—and often are—modified. Old core-periphery relationships can be blurred, and new ones can be initiated.

How Regional Economic Cores Are Created

Regional cores of economic development are created cumulatively, following some initial advantage, through the operation of several of the basic principles of economic geography. Commercial and industrial location decisions

7.3

GEOGRAPHY MATTERS

Fair Trade

The Fair Trade movement highlights the interdependencies involved in international trade. The movement is a result of increasing awareness within developed countries of the weak bargaining position of many small producers at the beginning of the commodity chains that undepin the global economy. Fair Trade has become part of the "mobilization against globalization" that we described in Chapter 2, an attempt to raise consumers' consciousness about the relationships embodied in their purchases.

The Fair Trade movement is a global network of producers, traders, marketers, advocates, and consumers focused on building equitable trading relationships between consumers and the world's most economically disadvantaged artisans and farmers. As such, it is fundamentally a strategy for poverty alleviation and sustainable development. The key principles of Fair Trade include (i) creating opportunities for economically disadvantaged producers, (ii) capacity building, (iii) ensuring that women's work is properly valued and rewarded, (iv) ensuring a safe and healthy working environment for producers, and (v) payment of a fair price—one that covers not only the costs of production but enables production that is socially just and environmentally sound. The result is often very modest in relation to the retail price of Fair Trade goods in developed countries—only about 9 cents of the $3.10 cost of an average fairly traded 100-gram bar of chocolate goes to producers in developing countries, the rest being accounted for by core-country food processors, designers, packagers, photographers, marketing staff, advertisers, shopkeepers, and tax authorities. Nevertheless, Fair Trade demonstrably helps producers in developing countries and typically involves democratic decision-making over how the extra money earned is to be distributed.

There are several types of Fair Trade organizations that perform various roles along the commodity chains linking producers to consumers. At the beginning of the commodity chain are producer organizations—village or community groups or cooperatives, for example, often joined together under export marketing organizations. In late 2004 there were 422 Fair Trade–certified producer groups (including many umbrella bodies) in 49 countries. These organizations typically sell their products to a second kind of Fair Trade organization: registered importers and wholesalers in more developed countries. In 2004 these existed in 19 countries. They in turn sell to Fair Trade retailers: "world shops" and catalog- or Internet-based retailers (**Figure 7.B**). In some countries, Fair Trade products are now mainstream, available in major supermarkets and independent shops and beginning to gain market share. In Switzerland, for example, Fair Trade bananas account for 20 percent of the retail market. In the United States, it is coffee that is the most important certified Fair Trade product, accounting for $131 million in sales in 2002.

The fourth category of Fair Trade organizations are labeling organizations that certify the chain of supply of certain commodities in order to guarantee adherence to Fair Trade practices. Fair Trade labeling was created in the Netherlands in the late 1980s when Max Havelaar launched the first Fair Trade consumer guarantee label in 1986 on coffee sourced from Mexico. Today, there are 19 organizations that together run the international standard setting and monitoring body Fairtrade Labeling Organizations

all take place within complex webs of *functional interdependence*. These webs include the relationships and linkages between different kinds of industries, different kinds of stores, and different kinds of offices. Particularly important here are the **agglomeration effects** that are associated with various kinds of economic linkages and interdependencies—the cost advantages that accrue to individual firms because of their location among functionally related activities. The trigger for these agglomeration effects can be any kind of economic development—the establishment of a trading port or the growth of a local industry or any large-scale enterprise. The external economies and economic linkages generated by such developments represent the initial advantage that tends to stimulate a self-propelling process of local economic development.

Given the location of a new economic activity in an area, a number of interrelated effects come into play. *Backward linkages* develop as new firms arrive to provide the growing industry with components, supplies, specialized services, or facilities. *Forward linkages* develop as new firms arrive to take the finished products of the growing industry and use them as inputs to their own processing, assembly, finishing, packaging, or distribution operations. Together with the initial growth, the growth in these linked industries helps to create a threshold of activity large enough to attract **ancillary industries** and activities (maintenance and repair, recycling, security, and business services, for example).

The existence of these interrelated activities establishes a pool of specialized labor with the kinds of skills

Figure 7.B Fair Trade retailing Transfair USA's founder and CEO, Paul Rice, inspects fair trade labels on foreign coffee products at company offices in Oakland, California. Transfair is a nonprofit group that audits the books of U.S. companies to ensure that the ingredients come from farmers who get fair prices for their goods.

International (FLO). Producers registered with FLO receive a minimum price that covers the cost of production and an extra premium that is invested in the local community.

Finally, there are organizations that are focused on labor practices, such as the Ethical Trading Initiative (ETI), which evolved from Fair Trade Campaigns run by British Aid organizations. ETI involves a multi-agency grouping of companies, non-governmental organizations (NGOs), and trades unions who have together devised a basic code of labor practices covering the right to collective bargaining, safe and hygienic working conditions, living wages, and a standard working week of no more than 48 hours.

Fair Trade sales are growing at double-digit rates and the movement is expanding rapidly. There are approximately 12,000 outlets in the United States where Fair Trade–labeled products can be purchased. In May 2000, the town of Garstang in Lancashire, England, declared itself the world's first Fair Trade Town, and by 2005 there were more than 200 towns and cities in Britain that had committed to Fair Trade principles and products. There were also several universities committed to Fair Trade, including the University of Birmingham, the University of Bristol, the University of Edinburgh, the London School of Economics and Political Science, and Sheffield University. Nevertheless, the total value of Fair Trade is miniscule in relation to the overall flows of international trade. Britain is the largest Fair Trade market in the world, with over $300 million in total sales in 2004. Total sales for the Fair Trade industry in North America in 2004 were just under $200 million, with worldwide sales amounting to less than $1 billion.

and experience that make the area attractive to still more firms. Meanwhile, the linkages between all these firms help to promote interaction between professional and technical personnel, and allow for the area to support R&D (research and development) facilities, research institutes, and so on, thus increasing the likelihood of local inventions and innovations that might further stimulate local economic development.

Another part of the spiral of local economic growth is a result of the increase in population represented by the families of employees. Their presence creates a demand for housing, utilities, physical infrastructure, retailing, personal services, and so on—all of which generate additional jobs. This expansion, in turn, helps to create populations large enough to attract an even wider variety and more sophisticated kinds of services and amenities. Last—but by no means least—the overall growth in local employment creates a larger local tax base. The local government can then provide improved public utilities, roads, schools, health services, recreational amenities, and so on, all of which serve to intensify agglomeration economies and so enhance the competitiveness of the area in being able to attract further rounds of investment.

Swedish economist Gunnar Myrdal was the 1974 Nobel prizewinner who first recognized that any significant initial local advantage tends to be reinforced through geographic principles of agglomeration and localization. He called the process **cumulative causation** (**Figure 7.19**), meaning the spiraling buildup of advantages that occurs in specific geographic settings as a result of the development

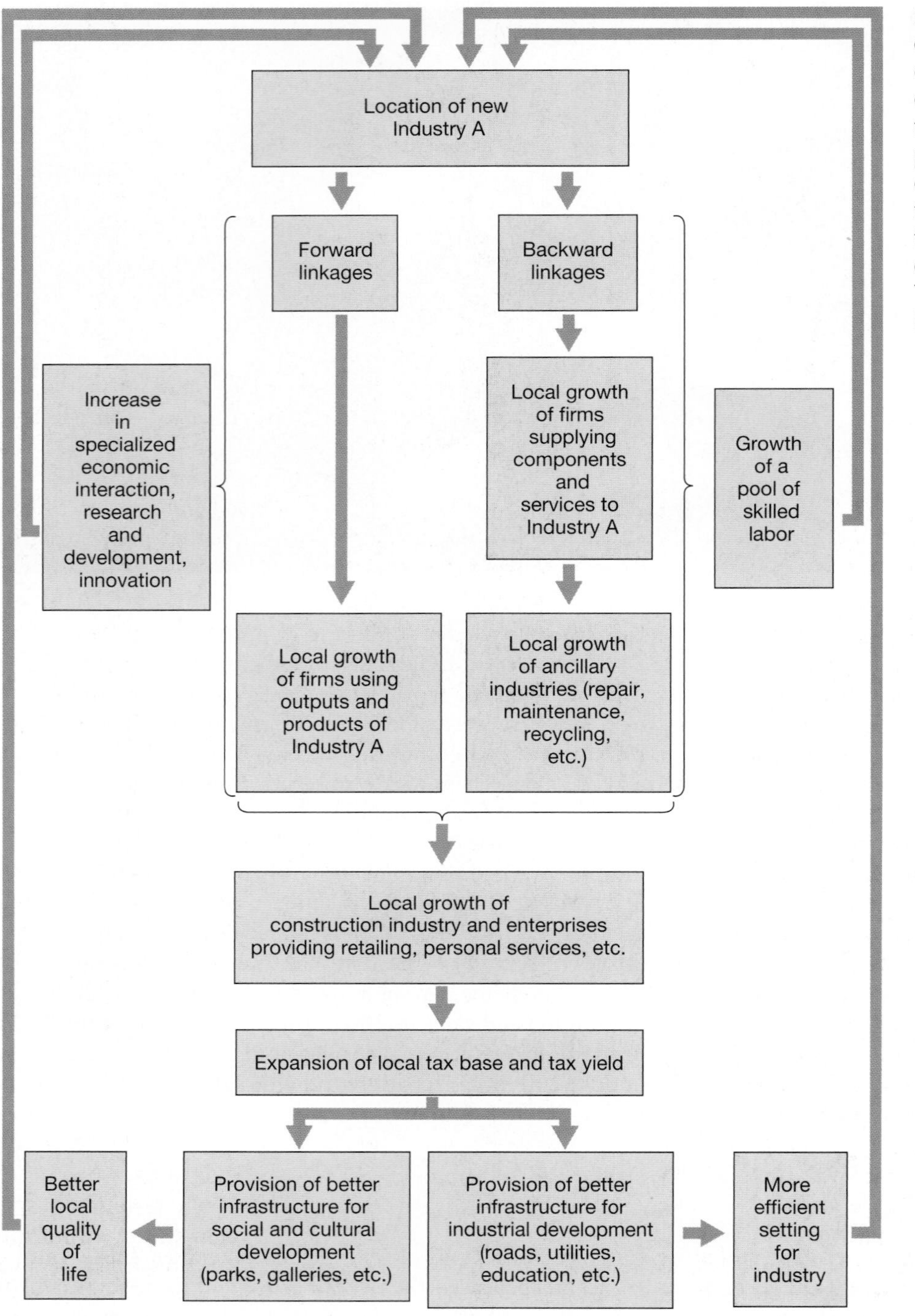

Figure 7.19 Processes of regional economic growth Once a significant amount of new industry becomes established in an area, it tends to create a self-propelling process of economic growth. As this diagram shows, the initial advantages of industrial growth are reinforced through geographic principles of agglomeration and localization. The overall process is sometimes known as *cumulative causation.*

of external economies, agglomeration effects, and localization economies. Myrdal also pointed out that this spiral of local growth tends to attract people—enterprising young people, usually—and investment funds from other areas. According to the basic principles of spatial interaction, these flows tend to be strongest from nearby regions, and those with the lowest wages, fewest job opportunities, or least attractive investment opportunities.

In some cases this loss of entrepreneurial talent, labor, and investment capital is sufficient to trigger a cumulative negative spiral of economic disadvantage. With less capital, less innovative energy, and depleted pools of labor, industrial growth in peripheral regions tends to be significantly slower and less innovative than in regions with an initial advantage and an established process of cumulative causation. This in turn tends to limit the size of the local tax base, making it difficult for local governments to furnish a competitive **infrastructure** of roads, schools, and recreational amenities. Myrdal called these disadvantages **backwash effects,** the negative impacts on a region (or regions) of the economic growth of some other region. These negative impacts take the form, for example, of out-migration, outflows of investment capital, and the shrinkage of local tax bases. Backwash effects are important because they help to explain why regional economic development is so uneven and why core-periphery contrasts in economic development are so common.

How Core-Periphery Patterns Are Modified

Although very important, cumulative causation and backwash effects are not the only processes affecting the geography of economic development. If they were, the world's economic geography would be even more starkly polarized than it is now. There would be little chance for the emergence of new growth regions, like Guangdong in Southeast China, and there would be little likelihood of stagnation or decline in once booming regions, like northern England.

Myrdal himself recognized that peripheral regions do sometimes emerge as new growth regions, and partially explained them in what he called *spread* (or trickle-down) effects. **Spread effects** are the positive impacts on a region (or regions) of the economic growth of some other region, usually a core region. This growth creates levels of demand for food, consumer products, and other manufactures that are so high that local producers cannot satisfy them. This demand gives investors in peripheral regions (or countries) the opportunity to establish a local capacity to meet the demand. Entrepreneurs who attempt this are also able to exploit the advantages of cheaper land and labor in peripheral regions. If strong enough, these spread effects can enable peripheral regions to develop their own spiral of cumulative causation, thus changing the interregional geography of economic patterns and flows. The economic growth of South Korea, for example, is partly attributable to the spread effects of Japanese economic prosperity.

Another way in which peripheral regions or countries can develop their own spiral of cumulative causation is through the process of *import substitution*. In this process, goods and services previously imported from core regions come to be replaced by locally made goods and locally provided services. Some things are hard to copy because of the limitations of climate or natural resources. However, many products and services *can* be copied by local entrepreneurs, thus capturing local capital, increasing local employment opportunities, intensifying the use of local resources, and generating profits for further local investment. The classic example is in Japan, where import substitution, especially for textiles and heavy engineering, played an important part in the transition from a peripheral economy to a major industrial power in the late nineteenth century. Import substitution also figured prominently in the Japanese "economic miracle" after World War II, this time featuring the automobile industry and consumer electronics. Today countries like Brazil, Peru, and Ghana are seeking to follow the same sort of strategy, subsidizing domestic industries and protecting them from outside competitors through tariffs and taxes.

Core-periphery patterns and relationships can also be modified by changes in the dynamics of core regions—internal changes that can slow or modify the spiral of cumulative causation. The main factor is the development of **agglomeration diseconomies**, the negative economic effects of urbanization and the local concentration of industry. Such effects include the higher prices that must be paid by firms competing for land and labor; the costs of delays resulting from traffic congestion and crowded port and railroad facilities; the increasing unit costs of solid waste disposal; and the burden of higher taxes that eventually have to be levied by local governments in order to support services and amenities previously considered unnecessary—traffic police, city planning, and transit systems, for example.

Diseconomies that are imposed through taxes can often be passed on by firms to consumers in other regions and other countries in the form of higher prices. Charging higher prices, however, decreases the competitiveness of a firm in relation to firms operating elsewhere. Agglomeration diseconomies that cannot be "exported"—noise, air pollution, increased commuting costs, and increased housing costs, for example—require local governments to tax even more of the region's wealth in attempts to compensate for a deteriorating quality of life. In California this issue has become so pressing that a report prepared jointly by the state, the Bank of America, the Greenbelt Alliance, and the Low Income Housing Fund concluded that "California businesses cannot compete globally when they are burdened with the costs of sprawl. . . . An attractive business climate cannot be sustained if the quality of life continues to decline and the cost of financing real estate development escalates."[2] California's response has been to enact some of the nation's toughest traffic and environmental regulations.

Deindustrialization and Creative Destruction

The most fundamental cause of change in the relationship between initial advantage and cumulative causation is to be found in the longer term shifts in technology systems and in the competition between states within the world-system. The innovations associated with successive technology systems generate new industries that are not yet tied down by enormous investments in factories or allied to existing industrial agglomerations. Combined with innovations in transport and communications, this creates *windows of locational opportunity* that can result in new industrial districts, with small towns or cities growing into dominant metropolitan areas through new rounds of cumulative causation.

Equally important as a factor in how core-periphery patterns are modified are the consequent shifts in the profitability of old, established industries in core regions compared to the profitability of new industries in fast-growing

[2]*Beyond Sprawl,* Sacramento, CA: State of California—Resources Agency, 1995.

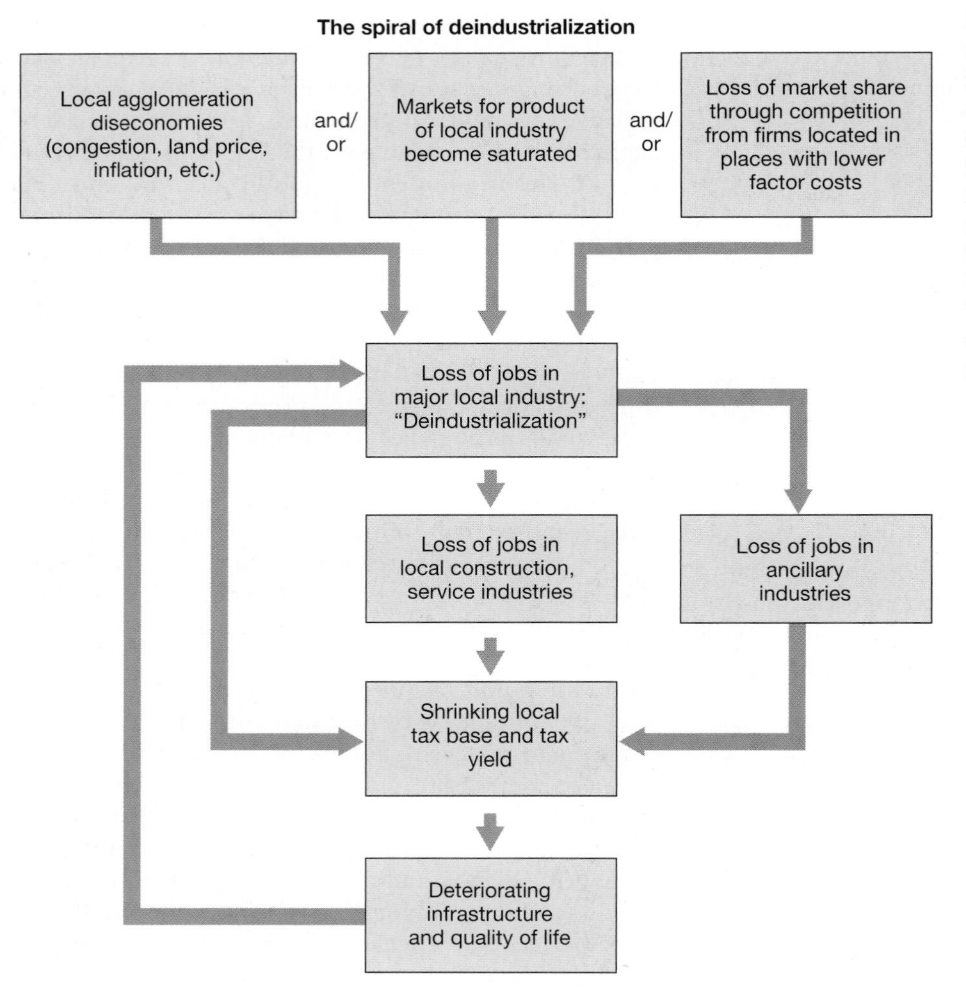

Figure 7.20 Regional economic decline When the locational advantages of manufacturing regions are undermined for one reason or another, profitability declines and manufacturing employment falls. This can lead to a downward spiral of economic decline, as experienced by many of the traditional manufacturing regions of Europe and North America during the 1960s, 1970s, and 1980s. (Reprinted with permission of Prentice Hall, from P. L. Knox, *Urbanization,* © 1994, p. 55.)

new industrial districts. As soon as the differential is large enough, some disinvestment will take place within core regions. This disinvestment can take place in several ways. Manufacturers can reduce their wage bill by cutting back on production; they can reduce their fixed costs by closing down and selling off some of their factory space and equipment; or they can reduce their spending on research and development for new products. This disinvestment, in turn, leads to *deindustrialization* in formerly prosperous industrial core regions.

Deindustrialization involves a relative decline (and in extreme cases an absolute decline) in industrial employment in core regions as firms scale back their activities in response to lower levels of profitability (**Figure 7.20**). This is what happened to the Manufacturing Belt (sometimes called the "Rustbelt") of the United States in the 1960s and 1970s (**Figure 7.21**). It also occurred in many of the traditional industrial regions of Europe during the 1960s, 1970s, and 1980s. In France, Belgium, the Netherlands, Norway, Sweden, and the United Kingdom, manufacturing employment decreased by between one-third and one-half between 1960 and 1990. The most pronounced example of this deindustrialization has been in the United Kingdom, where a sharp decline in manufacturing employment has been accompanied by an equally sharp rise in service employment.

Meanwhile, the capital made available from disinvestment in these core regions becomes available for investment by entrepreneurs in new ventures based on innovative products and production technologies. Old industries—and sometimes entire old industrial regions—have to be "dismantled" (or at least neglected) in order to help fund the creation of new centers of profitability and employment. This process is often referred to as creative destruction, something that is inherent to the dynamics of capitalism. **Creative destruction** involves the withdrawal of investments from activities (and regions) that yield low rates of profit in order to reinvest in new activities (and new regions). The concept of creative destruction provides a powerful image, helping us to understand the entrepreneur's need to withdraw investments from activities (and regions) yielding low rates of profit in order to reinvest in new activities (and, often, in new regions). In the United States, for example, the deindustrialization of the Manufacturing Belt provided the capital and the locational flexibility for firms to invest in the Sunbelt of the United States and in semiperipheral countries like Mexico and South Korea.

The process does not stop there, however. If the deindustrialization of the old core regions is severe enough, the relative cost of their land, labor, and infrastructure may decline to the point where they once again become at-

Figure 7.21
Deindustrialization This derelict steel mill in New Jersey is testament to the downward economic spiral in what was once one of the world's most important heavy manufacturing regions.

tractive to investors. As a result, a see-saw movement of investment capital occurs, which over the long term tends to move from developed to less developed regions—then back again, once the formerly developed region has experienced a sufficient relative decline. "Has-been" regions can become redeveloped and revitalized, given a new lease on life by the infusion of new capital for new industries. This is what happened, for example, to the Pittsburgh region in the 1980s, resulting in the creation of a postindustrial economy out of a depressed industrial setting. USX, the U.S. Steel group of companies, reduced its workforce in the Pittsburgh region from more than 20,000 to less than 5,000 between 1975 and 1995. These losses have been more than made up, however, by new jobs generated in high-tech electronics, specialized engineering, and finance and business services.

Government Intervention

In addition to the processes of deindustrialization and creative destruction, core-periphery patterns can also be modified by government intervention. National governments realize that regional planning and policy can be an important component of broad economic strategies designed to stabilize and reorganize their economies, as well as to maximize their overall competitiveness. Without regional planning and policy, the resources of peripheral regions can remain underutilized, while core regions can become vulnerable to agglomeration diseconomies. For political reasons, too, national governments are often willing to help particular regions adjust to changing economic circumstances. At the same time, most local governments take responsibility for stimulating economic development within their jurisdiction, if only in order to increase the local tax base.

The nature and extent of government intervention has varied over time and by country. In some countries, special government agencies have been established to promote regional economic development and reduce core-periphery contrasts. Among the best-known examples are the Japanese MITI (Ministry of International Trade and Industry), the Italian Cassa del Mezzogiorno (Southern Development Agency, replaced in 1987 by several smaller agencies), and the U.S. Economic Development Administration. Some governments have sought to help industries in declining regions by undertaking government investment in infrastructure and providing subsidies for private investment; others have sought to devise tax breaks that reduce the cost of labor in peripheral regions. Still others have sought to deal with agglomeration diseconomies in core regions through increased taxes and restrictions on land use.

While each approach has its followers, one of the most widespread governmental approaches to core-periphery patterns involves the exploitation of the principle of cumulative causation through the creation of growth poles. **Growth poles** are places of economic activity deliberately organized around one or more high-growth industries. Economists have noted, however, that not all industries are equal in the extent to which they stimulate economic growth and cumulative causation. The ones that generate the most pronounced effects are known as "propulsive industries," and they have received a great deal of attention from geographers and economists who are interested in helping to shape strategic policies that might promote regional economic development. In the 1920s shipbuilding was a propulsive industry. In the 1950s and 1960s automobile manufacturing was a propulsive industry, and today biotechnology and digital technologies are propulsive industries. The basic idea is for governments to promote regional economic growth by fostering propulsive industries in favorable locations. These locations are intended to become growth poles—places that, given an artificial start, develop a self-sustaining spiral of economic prosperity.

Many countries have used the growth-poles approach as a basis for regional development policies. For example, French governments have designated certain locations as *technopoles*—sites for the establishment of high-tech industries (such as computers and biotechnology)—under the assumption that these leading-edge activities will stimulate further development. In southern Italy various heavy industries were located in a number of remote areas after World War II in order to stimulate ancillary development. In the United States a whole series of growth centers were designated within Appalachia in an attempt to establish a manufacturing base in the region. The Appalachian Regional Commission, established by Congress in 1965 to promote the economic development of the region, initiated a massive program of roadbuilding in order to open up the region for modern economic development. The commission also designated over 200 growth centers that were intended to attract industrialists by furnishing a new physical infrastructure and offering economic incentives in the form of tax breaks. The new industries, it was hoped, would trigger the geographic process of cumulative causation.

The results of such policies have been mixed, however. The French technopoles have been fairly successful because the French government invested large sums of money in establishing propulsive industries in favorable locations. But the Italian and U.S. growth-pole efforts, like many others, have been disappointing. In practice, governments often fail to invest in the right industries, and they nearly always fail to invest heavily enough to kick-start the process of cumulative causation. In Appalachia, for example, the growth centers were too small and too numerous. As a consequence, none of them has been able to generate self-sustaining processes of local economic growth.

GLOBALIZATION AND ECONOMIC DEVELOPMENT

As we saw in Chapter 2, the globalization of the world economy involves a new international division of labor in association with the internationalization of finance; the deployment of a new technology system (using robotics, information technology, biotechnology, and other new technologies); and the homogenization of consumer markets. The first wave of corporate globalization, in the 1970s, was led by manufacturing giants like General Motors and General Electric, whose global reach had the triple objectives of reducing labor costs, outflanking national labor unions, and increasing overseas market penetration. In the 1980s, as the globalization of manufacturing spread and the information economy began to grow, the leading firms in advanced business services—accountancy, advertising, banking, and law—established global networks of their own. The globalization of advanced financial and business service firms was initially a reaction to the global practices of their most important clients, the global manufacturing corporations. If, say, a law firm or advertising agency wished to keep the business of a major corporation, it had to be able to provide its services in places where that corporation needed them. Thus advanced business service firms followed their clients along the globalization path in the late 1970s and especially in the 1980s. This meant creating an office network to match clients' needs. After a while some advanced business service firms used their global office network to win more clients in new markets. By the 1990s, the leading business service firms themselves became global corporations that could offer a seamless service with offices in key cities around the world.

The dynamics of economic globalization rest on the flows of capital, knowledge, goods, and services among countries. In 2004, companies around the world invested more than $651 billion in business ventures beyond their own shores. This level of **foreign direct investment** is 10 times that of the mid-1970s. Approximately 40 percent of foreign direct investment is targeted at peripheral and semiperipheral countries, the rest going to core countries. These investments are reflected in increased levels of world trade in goods and services. Between 1975 and 2004, world exports of goods and services increased threefold.

At a very general level, foreign direct investment can be expected to be good for the places and regions that are targeted for investment. Foreign direct investment increases competition among local producers, forcing them to improve their performance. At the same time, knowledge of new business practices and production technology spreads through the regional economy as regional manufacturers become suppliers to the enterprises funded through foreign investment and as personnel move from one firm to another. The overall effect is for higher levels of productivity all around.

Now that the world economy is much more globalized, the ability to acquire, absorb, and communicate knowledge is more important than ever before in determining the fortunes of places and regions. Patterns of local and regional economic development are much more open to external influences, and much more interdependent with economic development processes elsewhere. Shrinking space, shrinking time, and disappearing borders are linking people's lives more deeply, more intensely, and more immediately than ever before. This new framework for economic geography has already left its mark on the world's economic landscapes and has also meant that the lives of people in different parts of the world have become increasingly intertwined. You may recall Duong, Hoa, Françoise, and Jean-Paul, people mentioned in Chapter 1 (Box 1.1, p. 11). Here are sketches of the lives of three more people singled out by the World Bank in one of its *World Development Reports:*[3]

> Joe lives in a small town in southern Texas. His old job as an accounts clerk in a textile firm, where he had worked for many years, was not very secure. He earned $50 a day, but promises of promotion never came through, and the firm eventually went out of business as cheap imports from Mexico forced textile prices down. Joe went back to college to study business administration and was recently hired by one of the new banks in the area. He enjoys a comfortable living even after making the monthly payments on his government-subsidized student loan.
>
> Maria recently moved from her central Mexican village and now works in a U.S.-owned firm in Mexico's maquiladora sector. Her husband, Juan, runs a small car-upholstery business and sometimes crosses the border during the harvest season to work illegally on farms in California. Maria, Juan, and their son have improved their standard of living since moving out of subsistence agriculture, but Maria's wage has not increased in years: She still earns about $10 a day.
>
> Xiao Zhi is an industrial worker in Shenzhen, a Special Economic Zone in China. After three difficult years on the road as part of China's floating population, fleeing the poverty of nearby Sichuan province, he has finally settled with a new firm from Hong Kong that produces garments for the U.S. market. He can now afford more than a bowl of rice for his daily meal. He makes $2 a day and is hopeful for the future.

These examples begin to reveal a complex and fast-changing interdependence that would have been inconceivable just 15 or 20 years ago. Joe lost his job because of competition from poor Mexicans like Maria, and now her wage is held down by cheaper exports from China. Joe now has a better job, however, and the United States economy has gained from expanding exports to Mexico. Maria's standard of living has improved, and her son can hope for a better future. Joe's pension fund is earning higher returns through investments in growing enterprises around the world, and Xiao Zhi is looking forward to higher wages and the chance to buy consumer goods. Not everyone has benefited, however, and the new international division of labor has come under attack by some in industrial countries, where rising unemployment and wage inequality are making people feel less secure about the future. Some workers in core countries are fearful of losing their jobs because of cheap exports from lower cost producers. Others worry about companies relocating abroad in search of low wages and lax labor laws.

[3]World Bank, *World Development Report 1995: Workers in an Integrating World.* New York: Oxford University Press, 1995, p. 50.

Most of the world's population now lives in countries that are either integrated into world markets for goods and finance or rapidly becoming so. As recently as the late 1970s only a few peripheral countries had opened their borders to flows of trade and investment capital. About a third of the world's labor force lived in countries like the Soviet Union and China, with centrally planned economies, and at least another third lived in countries insulated from international markets by prohibitive trade barriers and currency controls. Today, with nearly half the world's labor force among them, three giant population blocs—China, the republics of the former Soviet Union, and India—are entering the global market. Many other countries, from Mexico to Thailand, have already become involved in deep linkages. According to World Bank estimates, fewer than 10 percent of the world's labor force remained isolated from the global economy in 2004. (The World Bank, properly called the International Bank for Reconstruction and Development, is a U.N. affiliate established in 1948 to finance productive projects that further the economic development of member nations.)

In this section we examine some specific impacts of three of the principal components of the global economy: the global assembly line, resulting from the operations of transnational manufacturing corporations; the global office, resulting from the internationalization of banking, finance, and business services; and the pleasure periphery, resulting from the proliferation of international tourism.

The Global Assembly Line

The globalization of the world economy represents the most recent stage in a long process of internationalization. At the heart of this process has been the emergence of private companies that participate not only in international trade but also in production, manufacturing, and/or sales operations in several countries. Almost 80 percent of Ford's workforce, for example, is employed overseas, and foreign sales account for 55 percent of its total revenues. Over 50 percent of IBM's workforce is employed overseas, and 61 percent of its revenues are derived from foreign sales; 82 percent of the workforce of Philips, the Dutch electronics firm, is employed overseas, and 95 percent of its revenues are derived from foreign sales. Many of these transnational corporations have grown so large through a series of mergers and acquisitions that their activities now span a diverse range of economic activities. **Transnational corporations** are companies that participate not only in international trade but also in production, manufacturing, and/or sales operations in several countries.

Corporations that consist of several divisions engaged in quite different activities are known as **conglomerate corporations,** those having diversified into various economic activities, usually through a process of mergers and acquisitions. Altria (formerly known as Philip Morris), for example, primarily known for its tobacco products (such as Marlboro cigarettes), also controls the single largest group of assets in the beverage industry (including Miller Brewing) and has extensive interests in real estate (the Mission Viejo company in California), import-export (Duracell do Brasil), publishing (E.Z. Editions, Zürich) and foods (including Kraft, General Foods, Tobler, Terry's, and Suchard chocolate). Kraft's brands alone include Cheese Whiz, Cool Whip, Country Time, Cream of Wheat, DiGiorno, Jell-O, Kool-Aid, Life Savers, Miracle Whip, Oscar Meyer, Nabisco, Oreo, Planters, Post cereals, Ritz, Sanka, Shake 'n Bake, Triscuit, and Yuban. Nestlé, the world's largest packaged-food manufacturer, is the largest company in Switzerland but derives less than 2 percent of its revenue from its home country. Its major U.S. product lines and brand names include beverages (Calistoga, Nescafé, Nestea, Perrier, Quik, Taster's Choice), chocolate and candy (After Eight, Butterfinger, Crunch, KitKat, Raisinets), culinary products (Buitoni, Carnation, Contadina, Libby, Toll House), frozen foods (Lean Cuisine), pet foods (Fancy Feast, Friskies, Mighty Dog), wine (Beringer Brothers), and drugs and cosmetics (Alcon optical, L'Oreal). In addition to its 438 factories in 63 countries around the world, Nestlé operates more than 40 Stouffer's hotels.

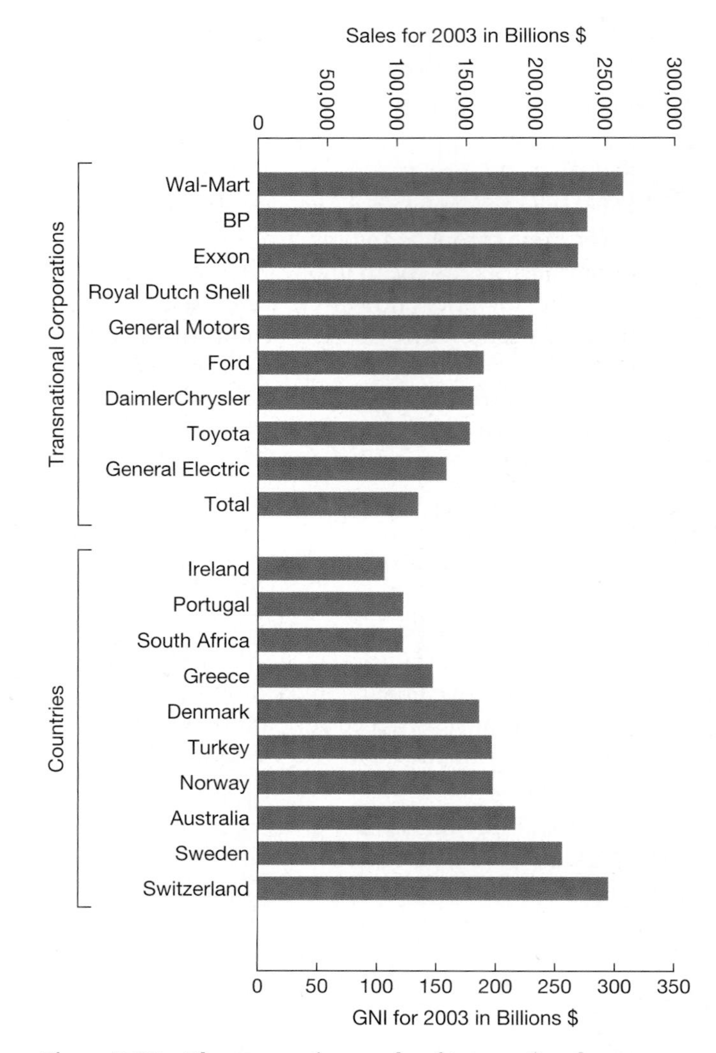

Figure 7.22 The economic muscle of transnational corporations The annual sales figures of many of the world's largest corporations exceed the Gross National Income of some countries.

Transnationals and Globalization

Transnational corporations first began to appear in the nineteenth century, but until the mid-twentieth century there were only a few, most of them U.S.- or European-based transnationals that were concerned with obtaining raw materials, such as oil or minerals for their domestic manufacturing operations. After World War II an increasing number of large corporations began to invest in overseas production and manufacturing operations as a means of establishing a foothold in foreign consumer markets. Between 1957 and 1967, 20 percent of all new U.S. machinery plants, 25 percent of all new chemical plants, and over 30 percent of all new transport equipment plants were located abroad. By 1970 almost 75 percent of U.S. imports were transactions between the domestic and foreign subsidiaries of transnational conglomerates. By the end of the 1970s overseas profits accounted for a third or more of the overall profits of the 100 largest transnational corporations. By the early 1980s, 40 percent of all world trade was in the form of intrafirm trade (that is, between different branches and companies of the same transnational conglomerate).

Beginning in the 1970s a sharp increase occurred in the growth of transnational conglomerates, not only in the United States but also in Europe, Japan, and even some semiperipheral countries. By 2003 there were almost 61,000 transnational corporations in the world. Of these, the top 300 controlled approximately one-quarter of the world's productive assets. Many of the largest transnational corporations are now more powerful, in economic terms, than most sovereign nations (**Figure 7.22**). General Motors' economy is larger than Portugal's; Toyota's is larger than Ireland's; and Wal-Mart's annual sales exceed Norway's gross domestic product.

The reason for such growth in the number and scale of transnational conglomerate corporations has been that international economic conditions have changed. A recession, triggered by a massive increase in the price of crude oil in 1973, meant that companies everywhere had to reexamine their strategies. At around the same time, technological developments in transport and communications provided larger companies with the flexibility and global reach to exploit the steep differentials in labor costs that exist between core countries and peripheral countries. Meanwhile, these same developments in transport and communications made for intensified international competition, which forced firms to search more intensely for more efficient and profitable global production and marketing strategies. Concurrently, a homogenization of consumer tastes (also facilitated by new developments in communications technologies) has made it possible for companies to cater to global markets.

It was the consequent burst of transnational corporate activity that formed the basis of the recent globalization of the world economy. In effect, the playing field for large-scale businesses of all kinds had been marked out anew. Companies have had to reorganize their operations in a variety of ways, restructuring their activities and redeploying their resources *between different countries, regions, and places* (see Box 7.4: "The Changing Geography of the Clothing Industry"). Local patterns of economic development have been recast again and again as these processes of restructuring, reorganization, and redeployment have been played out.

The advantages to manufacturers of a global assembly line are several. First, a standardized global product for a global market allows them to maximize economies of scale. Second, a global assembly line allows production and assembly to take greater advantage of the full range of geographical variations in costs. Basic wages in manufacturing industries, for example, are between 25 and 75 times higher in core countries than in some peripheral countries. With a global assembly line, labor-intensive work can be done where labor is cheap, raw materials can be processed near their source of supply, final assembly can be done close to major markets, and so on. Third, a global assembly line means that a company is no longer dependent on a single source of supply for a specific component, thus reducing its vulnerability to industrial troubles and other disturbances. Fourth, global sourcing allows transnational conglomerates better access to local markets. For example, Boeing has pursued a strategy of buying a significant number of aircraft components in China and has therefore succeeded in opening the Chinese market to its products.

The automobile industry was among the first to develop a global assembly line. In 1976 Ford introduced the Fiesta, a vehicle designed to sell in Europe, South America, and Asia as well as North America. The Fiesta was assembled in several locations from components manufactured in an even greater number of locations. It became the first of a series of Ford "world cars" that now includes the Focus, the Mondeo, and the Contour. The components of the Ford Escort, for example, are made and assembled in 15 countries across three continents. Ford's international subsidiaries, which used to operate independently of the parent company, are now being functionally integrated, using supercomputers and video teleconferences. The other automobile companies have organized their own global assembly systems (**Figure 7.23**). The global production networks of these companies

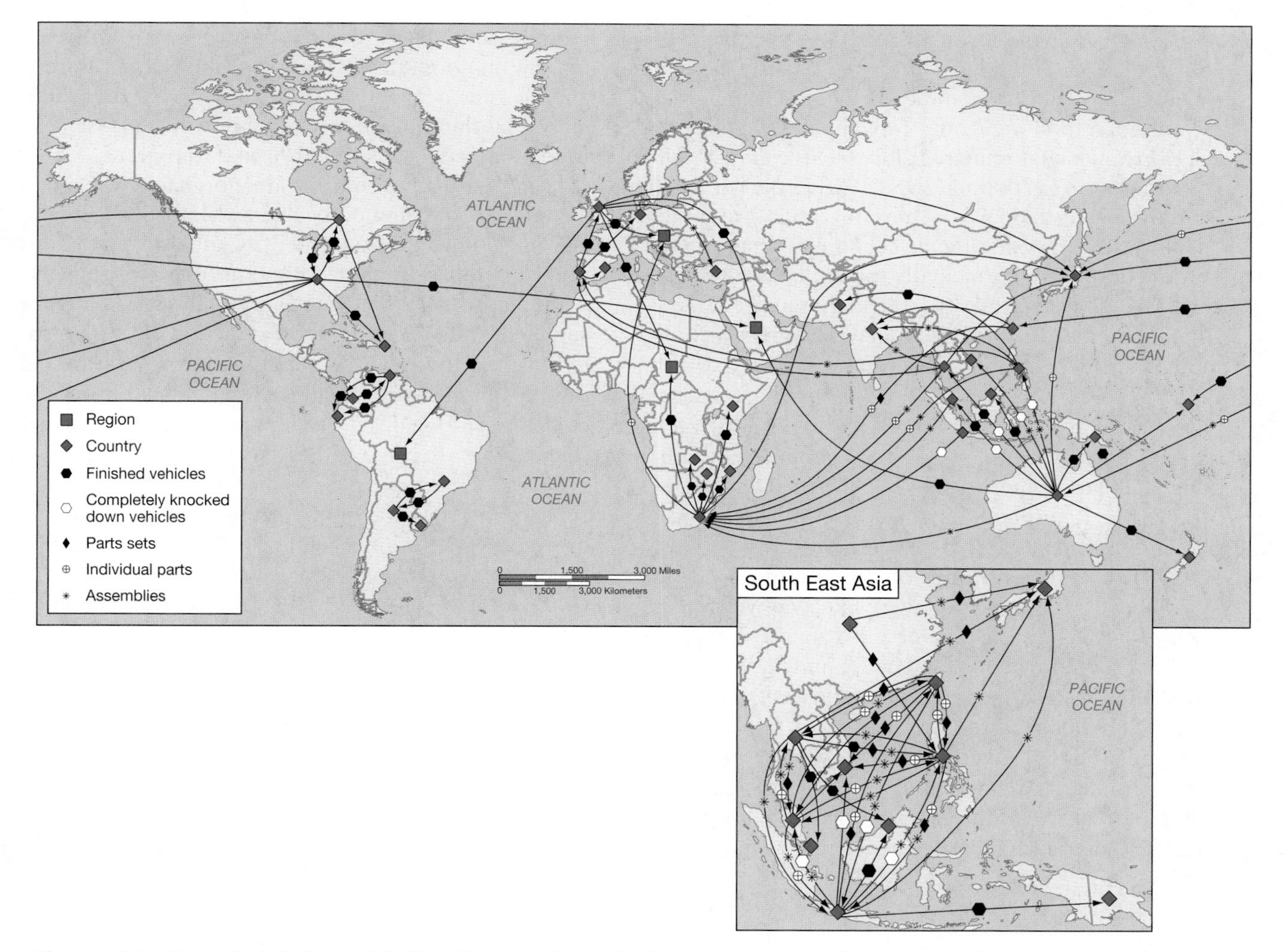

Figure 7.23 Toyota's global assembly line This map shows the flows of parts, sets of parts, assemblies, "completely knocked down" vehicles (that is, the unassembled parts of individual vehicles, shipped together) and finished vehicles between the countries and regions involved in Toyota's global production strategies. (Map after P. Knox, J. Agnew, and L. McCarthy, *The Geography of the World Economy*, 4th ed. London: Arnold, 2003.)

7.4

GEOGRAPHY MATTERS

The Changing Geography of the Clothing Industry

The clothing industry provides a good example of the way that local economic geographies are affected by an industry's response to globalization. In the nineteenth century the clothing industry developed in the metropolitan areas of core countries, with many small firms using cheap, migrant, or immigrant labor. In the first half of the twentieth century the industry, like many others, began to modernize. Larger firms emerged, their success based on the exploitation of mass-production techniques for mass markets and on the exploitation of principles of spatial organization within national markets. In the United States, for example, the clothing industry went through a major locational shift as a great deal of production moved out of the workshops of New York to big, new factories in smaller towns in the South, where labor was not only much cheaper but less unionized.

The global textile and apparel industry has undergone several shifts in production since the 1950s. The first was from North America and Western Europe to Japan in the 1950s and early 1960s, when western textile and clothing production was displaced by a sharp rise in imports from Japan. The second shift was from Japan to Hong Kong, Taiwan, and the Republic of Korea, which dominated global textile and clothing exports in the 1970s and early 1980s. In the late 1980s and the 1990s there was a third migration, from these countries to other developing economies. In the 1980s, production moved principally to mainland China, but also to Indonesia, Malaysia, the Philippines, Sri Lanka and Thailand. In the 1990s, new suppliers included India and Mexico but the largest newcomer in the 1990s was Turkey, whose total of $3.4 billion in clothing exports placed it fifth in world rankings. By 2000, important new clothing-producing countries (with a billion dollars or more of exports) included Bangladesh, the Czech Republic, Hungary, Mauritius, Morocco, Poland, Romania, and Viet Nam.

Much of this shifting globalization of production was "buyer-driven" by retailers like Wal-Mart, Sears and JC Penney, and fashion-oriented apparel companies like Liz Claiborne, Gap, and The Limited. Their cost-reduction strategies led them to retain design and marketing functions but to contract out the production of their apparel to firms in low-wage countries. In effect, they became "manufacturers without factories." In an attempt to protect domestic manufacturers, the United States, Canada, and 13 countries in Europe entered into a trade pact called the Multifiber Arrangement (MFA) in 1974. The MFA used quotas to regulate access to domestic clothing and textile markets, restricting the amount of imports from any one country. Designed to protect MFA signatories from competition from Japan, Hong Kong, Taiwan, and the Republic of Korea, the import quotas ended up working as a kind of affirmative action program for countries that had large workforces and low wage rates—hence the spread of production around the world.

The result was that by 1980 more than half of all apparel purchased in the United States was imported (compared to less than 7 percent in 1960). Leisure wear—jeans, shorts, T-shirts, polo shirts, and so on—was an important component of the homogenization of consumer tastes around the world at that time, and it could be produced most profitably by the cheap labor of young women in the peripheral metropolitan areas of the world. The same holds true today for fashion wear as well as leisure wear. The hourly compensation (including benefits) of apparel workers in the United States is $8–$10 for a 37-hour week, whereas for a 50- or 60-hour week, their counterparts in Hong Kong are paid around US$5 an hour, and in many parts of China the pay rate is US$0.20 an hour for 11–12 hour shift patterns for 6 days a week. Not surprisingly, the retail margin on domestically made clothing sold in Europe and the United States is 70 percent or so, the retail margin on clothing made in workshops in countries like Indonesia and Thailand is 100 to 250 percent.

This globalization of production has resulted in a complex set of commodity chains. Many of the largest clothing companies, such as Liz Claiborne, have most of their products manufactured through arrangements with many different independent suppliers, with no one supplier producing more than a fraction of the company's total output. These manufacturers are scattered throughout the world, making the clothing industry one of the most globalized of all manufacturing activities (**Figure 7.C**). In some countries, clothing manufacture is now the primary driver of their economy. Fifty percent of Sri Lanka's export earnings, for example, are derived from clothing, while for Bangladesh, El Salvador, and Mauritius the figure is 63 percent and for Cambodia it is 76 percent. As we have seen, the actual geography of commodity chains in the clothing industry is rather volatile, with frequent shifts in production and assembly sites as companies and their suppliers continuously seek out new locations with lower costs.

Although cheap leisure wear can be produced most effectively through arrangements with multiple suppliers in peripheral low-wage regions, higher end apparel for the global marketplace requires a different geography of production. These products—women's fashion, outerwear, and lingerie, infants' wear, and men's suits—are based on frequent style changes and high-quality finish. This requires short production runs and greater contact between producers and buyers. The most profitable set-

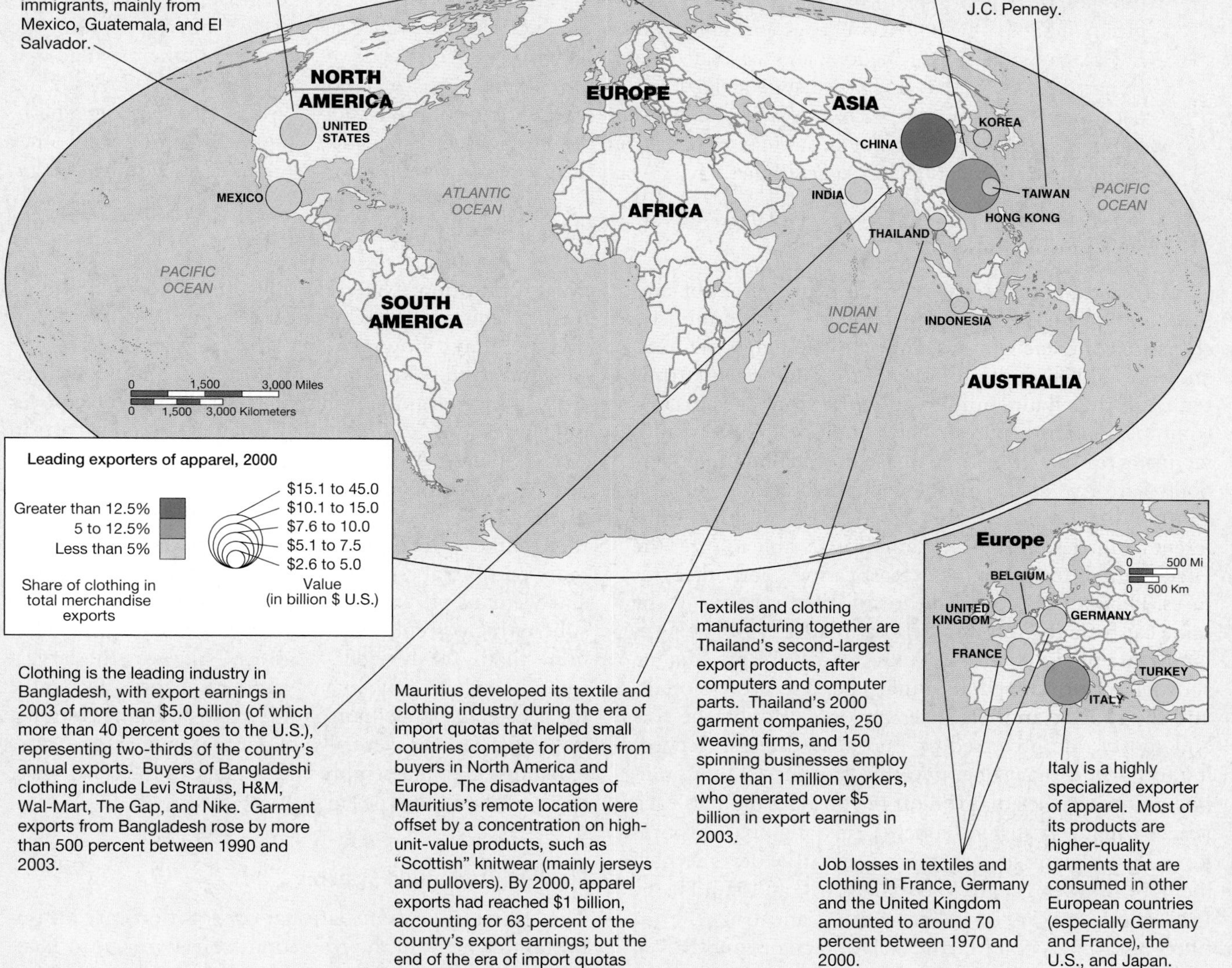

Figure 7.C The changing distribution of clothing manufacturing Most of the world's clothing exports come from just a few countries. In detail, however, the geography of clothing manufacturing changes rapidly in response to the changing patterns of costs and opportunities within the world economy.

tings for these products are in the core countries' metropolitan areas—London, Paris, Stuttgart, Milan, New York, and Los Angeles—where, once again, migrant and immigrant labor provides a workforce for "designer" clothing that can be shipped in small batches to upscale stores and shopping malls around the world.

The result is that commodity chains in the clothing industry are quite distinctive in the origins of products destined for different segments of the market. Fashion-oriented retailers in the United States who sell "designer" products to up-market customers obtain most of their goods from manufacturers in a small group of high-value-added countries, including France, Italy, Japan, the United Kingdom, and the United States. Department stores that emphasize "private label" products (that is, store brands, such as Nordstrom) and premium national brands obtain most of their goods from established manufacturers in semiperipheral East Asian countries. Mass merchandisers who sell lower priced brands buy primarily from a third tier of lower cost, mid-quality manufacturers, while large-volume discount stores like Wal-Mart import most of their goods from low-cost suppliers in peripheral countries like China, Bangladesh, and the Dominican Republic.

Currently, a further realignment of global production is taking place as a result of the phasing out of the MFA's network of quotas, which began in the mid-1990s. The last of the quotas expired on January 1, 2005, prompting what is likely to be one of the largest migrations of production in history as production moves increasingly to China and India, the two countries with the greatest comparative advantage in terms of labor costs—thereby jeopardizing hundreds of thousands of apparel jobs in Bangladesh, Cambodia, El Salvador, Lesotho, Mauritius and other countries that had prospered under the quota system.

allow them to process raw materials near sources of supply, to undertake labor-intensive work where labor is cheap, to complete final assembly close to major markets, and to establish multiple sources for components (thus reducing their vulnerability to work stoppages arising from local labor disputes). They employ modular manufacturing for their world cars based on a common underbody platform yet have the flexibility to adapt the interior, trim, body, and ride characteristics to conditions in different countries. Ford, for example, was able to offer the Focus with different engines, transmissions, and other features. Honda has produced three distinct versions of the same car from its Accord world car platform—the bigger, family-oriented Accord for American drivers, the smaller, sportier Accord aimed at young Japanese professionals, and the shorter and narrower Accord, offering the stiff and sporty ride preferred by European drivers. According to *Fortune* magazine, two-thirds of the 50 to 60 million motor vehicles that roll off the production lines each year are made by just six supergroups of global corporations. In order of size, they are General Motors (which includes Daewoo, Saab, Vauxhall, Opel, and Fuji Heavy Industries—makers of Subaru-Suzuki and Isuzu), Ford (incorporating Aston Martin, Volvo, Jaguar, Land-Rover, and Mazda), DaimlerChrysler (Mercedes-Benz, Chrysler, and Jeep, with Mitsubishi and Hyundai), Toyota (with Daihatsu and Lexus), Volkswagen (including SEAT, Skoda, Bentley, and Audi), and Renault and Nissan (with Samsung Motors).

The global assembly line is constantly being reorganized as transnational corporations seek to take advantage of geographical differences between places and regions and as workers and consumers in specific places and regions react to the consequences of globalization. Nike, the athletic footwear and clothing marketer, provides a good illustration. Nike once relied on its own manufacturing facilities in the United States and the United Kingdom. Today, however, most of its production is subcontracted to suppliers in East, South, and Southeast Asia. The geography of this subcontracting evolved over time in response to the changing pattern of labor costs in Asia. The first production of Nike shoes took place in Japan. The company then switched most of its subcontracting to South Korea and Taiwan. As labor costs rose there, Nike's subcontracting was spread across more and more peripheral countries—China, Indonesia, Malaysia, and Vietnam—in search of low labor costs. By 2004, Nike subcontractors employed more than 620,000 people in more than 700 different factories. Nike was the largest foreign employer in Vietnam, where its factories accounted for 5 percent of Vietnam's total exports. China, Indonesia, and Thailand were also major components in Nike's expanded global assembly line because of their low wage costs—around $60 per month.

Flexible Production Systems

The strategies of transnational corporations are an important element in the transition from Fordism to Neo-Fordism in much of the world. **Fordism** is named after Henry Ford, the automobile manufacturer who pioneered the principles involved: mass production, based on assembly-line techniques and "scientific" management, together with mass consumption, based on higher wages and sophisticated advertising techniques. In **neo-Fordism** the logic of mass production coupled with mass consumption has been modified by the addition of more flexible production, distribution, and marketing systems. This flexibility is rooted in forms of production that enable manufacturers to shift quickly and efficiently from one level of output to another and, more importantly, from one product configuration to another.

Flexible production systems involve flexibility both within firms and between them. *Within* firms, new technologies now allow a great deal of flexibility. Computerized machine tools, for example, are capable of producing a variety of new products simply by being reprogrammed, often with very little downtime between production runs for different products. Different stages of the production process (sometimes located in different places) can be integrated and coordinated through computer-aided design (CAD) and computer-aided manufacturing (CAM) systems. Computer-based information systems can be used to monitor retail sales and track wholesale orders, thus allowing producers to reduce the costs of raw materials stockpiles, parts inventories, and warehousing through sophisticated small-batch, just-in-time production and distribution systems. **Just-in-time production** employs vertical disintegration within large, formerly functionally integrated firms, such as automobile manufacturers, in which daily and even hourly deliveries of parts and other supplies from smaller (often non-union) subcontractors and suppliers now arrive "just in time" to maintain "last-minute" and "zero" inventories. The combination of computer-based information systems, CAD/CAM systems, and computerized machine tools has also given firms the flexibility to exploit specialized niches of consumer demand so that economies of scale in production can be applied to upscale but geographically scattered markets.

The Benetton clothing company provides an excellent case-study example of the exploitation of flexible production systems within a single firm. In 1965 the Benetton company began with a single factory near Venice. In 1968 it acquired a single retail store in the Alpine town of Belluno, marking the beginning of a remarkable sequence of corporate expansion. Benetton is now a global organization with over 5,000 retail outlets in more than 120 countries, and with its own investment bank and financial services organizations. It achieved this growth by exploiting computers, new communications and transportation systems, flexible outsourcing strategies, and new production-process technologies (such as robotics and CAD/CAM systems) to the fullest possible extent.

Only about 400 of Benetton's employees are located in the company's home base of Treviso, Italy (**Figure 7.24**). From Treviso, Benetton managers coordinate the activities of more than 250 outside suppliers in order to stock its worldwide network of retail outlet franchises. In Treviso the firm's designers create new shirts and sweaters on CAD terminals, but their designs are produced only for orders in hand, allowing for the coordination of production with the purchase of raw materials. In factories, rollers linked to a central computer spread and cut layers of cloth in small batches according to the numbers and colors ordered by Benetton stores around the world. Sweaters, gloves, and scarves, knitted in volume in white yarn, are dyed in small batches by machines similarly programmed to respond to sales orders. Completed garments are warehoused briefly (by robots) and shipped out directly (via private package delivery firms) to individual stores to arrive on their shelves within 10 days of manufacture.

Figure 7.24 Headquarters of the Benetton Group, Villa Minelli, Treviso, Italy.

Sensitivity to demand, however, is the foundation of Benetton's success. Niche marketing and product differentiation have been central to this sensitivity, which requires a high degree of flexibility in exploiting new product lines. Key stores patronized by trendsetting consumers (such as the store on the Rue Faubourg St. Honoré in Paris and the megastore on Omotesando Street in Tokyo) are monitored closely, and many Benetton stores' cash registers operate as point-of-sale terminals, so that immediate marketing data are available to company headquarters daily. Another notable feature of the company's operations is the way that different market niches are exploited with the same basic products. In Italy, Benetton products are sold through several different retail chains, each with an image and decor calculated to bring in a different sort of customer.

Between firms, the flexibility inherent to Neo-Fordism has been achieved through the externalization of certain functions. One way of doing this has been to reorganize administrative, managerial and technical functions into flatter, leaner, and more flexible forms of organization that can make increased use of outside consultants, specialists, and subcontractors. This has led to a degree of vertical disintegration among firms. **Vertical disintegration** involves the evolution from large, functionally integrated firms within a given industry toward networks of specialized firms, subcontractors, and suppliers. Another route to externalization has been for firms to participate in joint ventures, in the licensing or contracting of technology, and in strategic alliances involving design partnerships, collaborative R&D projects, and the like. **Strategic alliances** involve commercial agreements between transnational corporations, usually involving shared technologies, marketing networks, market research, or product development. They are an important contributor to the intensification of economic globalization.

For example, the Nestlé food company has a number of strategic alliances including a joint venture with General Mills called Cereal Partners Worldwide (CPW); an alliance since 1990 with the Walt Disney Company in Europe, the United States, Latin America and other markets that includes making Nestlé the main food company for Disneyland Paris; and a joint marketing venture with Pillsbury Häagen-Dazs since 1999 called Ice Cream Partners USA. Nestlé also has had a joint venture partnership with the Coca-Cola Company since 1991 in which the Swiss company cooperates with Coca-Cola in exchanging technologies and in marketing. Nestlé, for example, uses Coca-Cola's distribution network for products such as Nescafé instant coffee.

Maquiladoras and Export-Processing Zones

The type of subcontracting carried out by Nike is encouraged by the governments of many peripheral and semiperipheral countries, who see participation in global assembly lines as a pathway to export-led industrialization. They offer incentives such as tax "holidays" (not having to pay taxes for a specified period) to transnational corporations. In the 1960s Mexico enacted legislation permitting foreign companies to establish "sister factories"—*maquiladoras*—within 19 kilometers (12 miles) of the border with the United States for the duty-free assembly of products destined for re-export (see **Figures 7.25** and **7.26**). By 2005 more than 3500 such manufacturing and assembly plants had been established, employing around a million Mexican workers, most of them women, and accounting for more than 30 percent of Mexico's exports. But since 2000 over 350 *maquiladoras* have closed, resulting in the loss of more than 60,000 jobs in Ciudad Juarez alone. The tax breaks that favored the establishment of the *maquiladoras* are due to expire under the terms of the North American Free Trade Agreement (NAFTA), while lower wage rates and better incentives in other countries—particularly China—have proved more attractive to manufacturers.

Export-processing zones (EPZs) are small areas within which especially favorable investment and trading conditions are created by governments in order to attract export-oriented industries. These conditions include minimum levels of bureaucracy surrounding importing and exporting; the absence of foreign exchange controls; the availability of factory space and warehousing at subsidized rents; low tax rates; and exemption from tariffs and export duties. In 1985 it was estimated that there were a total of 173 EPZs around the world, which together employed 1.8 million workers. In 1998 the International Labor Organization (ILO) estimated that there were 850 EPZs, employing about 27 million people, 90 percent of whom were women. China alone had 124 EPZs, which together employed 18 million workers. The ILO's report on EPZs[4] criticized these "vehicles of globalization" because very few of them have any meaningful links with the domestic economies around them, and most trap large numbers of people in low-wage, low-skill jobs.

In addition to tax incentives and EPZs, many governments also establish policies that ensure cheap and controllable labor. Sometimes countries are pressured to

[4]International Labor Organization, *Labour and Social Issues Relating to Export Processing Zones*. Geneva: International Labor Office, 1998.

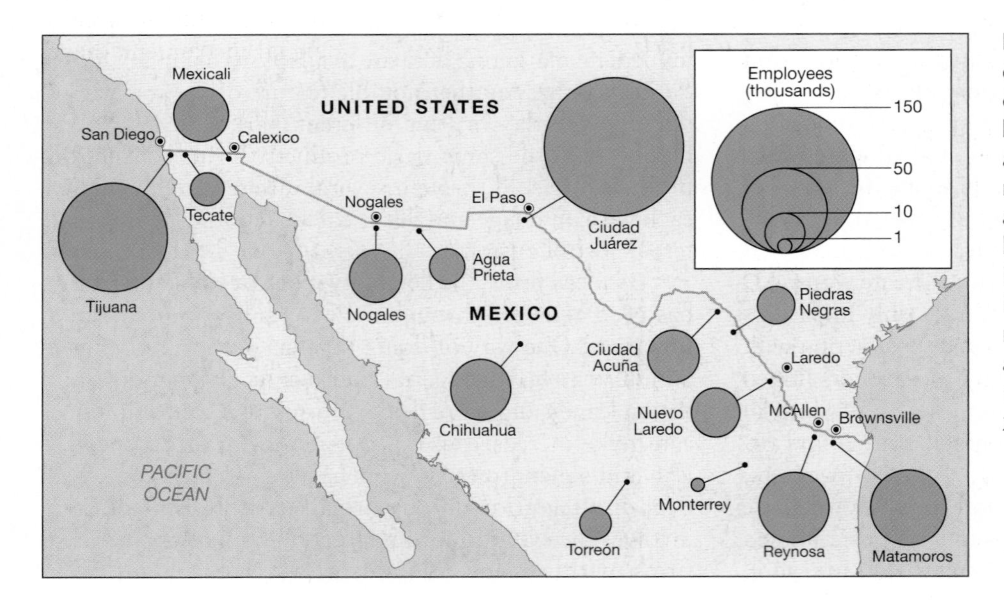

Figure 7.25 Principal maquiladora centers on the United States–Mexico border Cheap labor and tax breaks for firms manufacturing and assembling goods for re-export have made many Mexican border towns attractive to U.S. companies. Around half a million workers are employed in these maquiladora plants, producing electronic products, textiles, furniture, leather goods, toys, and automotive parts. (After P. Dicken, *Global Shift*, 4th ed. New York: Guilford, 2003.)

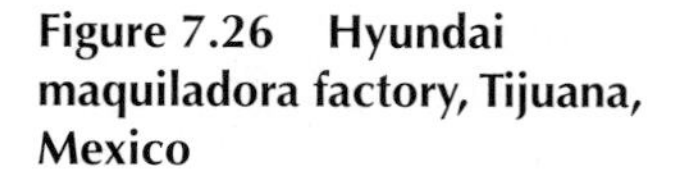

Figure 7.26 Hyundai maquiladora factory, Tijuana, Mexico

participate in global assembly lines by core countries and by the transnational institutions they support. The United States and the World Bank, for example, have backed regimes that support globalized production and have pushed for austerity programs that help to make labor cheap in peripheral countries. Countries pursuing export-led industrialization as an economic development strategy do not plan to remain the providers of cheap labor for foreign-based transnational corporations, however. They hope to shift from labor-intensive manufactures to capital-intensive, high-technology goods, following the path of semiperipheral Asian countries like Singapore and South Korea.

The Global Office

The globalization of production and the growth of transnational corporations have brought about another important change in patterns of local economic development. Banking, finance, and business services are no longer locally oriented ancillary activities, but important global industries in their own right. They have developed some specific spatial tendencies of their own—tendencies that have become important shapers of local economic development processes.

The new importance of banking, finance, and business services was initially a result of the globalization of manufacturing; an increase in the volume of world trade; and the emergence of transnational corporate empires. It was helped along by advances in telecommunications and data processing. Satellite communications systems and fiber-optic networks made it possible for firms to operate key financial and business services 24 hours a day around the globe, handling an enormous volume of transactions. Linked to these communications systems, computers permit the recording and coordination of the data. The world's fourth-largest stock market, the National Associated Automated Dealers Quotation System (NASDAQ), has no trading floor at all: Telephone and fiber-optic lines connect its half-million traders worldwide.

As banking, finance, and business services grew into important global activities, however, they were themselves transformed into something quite different from the old, locally oriented ancillary services. The global banking and financial network now handles trillions of dollars every day (estimates in 2004 ranged from five to eight trillion dollars)—no more than 10 percent of which has anything to do with the traditional world economy of trade in goods and services. International movements of money, bonds, securities, and other financial instruments have now become an end in themselves because they are a potential source of high profits from speculation and manipulation. Several factors have supported this development:

- In core countries the institutionalization of savings (through pension funds and so on) established a large pool of capital managed by professional investors with few local or regional allegiances or ties.
- The quadrupling of crude oil prices in 1973 (organized by OPEC—the Organization of Petroleum Exporting Countries) generated so much capital for oil-rich countries that their banks opened overseas branches in order to find enough borrowers. In many cases, the borrowers were companies and governments in underdeveloped, peripheral countries that had previously

been considered poor investment prospects. The internationalization of financial services soon paid off for the big banks. By the mid-1970s about 70 percent of Citibank's overall earnings came from its international operations, with Brazil alone accounting for 13 percent of the bank's earnings in 1976. For the borrowers, however, the result was a huge increase in international debt, and many peripheral countries became net exporters of capital as they started to make the interest payments on their loans.

- The initial response of many governments (including the U.S. government) to balance-of-payments problems in the 1960s and 1970s was to print more money—a short-term solution that eventually contributed to a significant surge in **inflation** in the world economy. This inflation, because it promoted rapid change and international differentials in financial markets, gave a further boost to speculative international financial transactions of all kinds.
- In many countries, governments have lifted restrictions and regulations relating to banking, finance, and business services in the hope of capturing more of their growth.
- The United States's persistent trade deficit vis-à-vis the rest of the world (a result of the postwar recovery of Europe and Japan) created a growing pool of dollars outside the United States, known as Eurodollars. This supply in turn created a pool of capital that was beyond the direct control of U.S. authorities.
- "Hot" money (undeclared business income, proceeds of securities fraud, trade in illegal drugs, and syndicated crime), easily laundered through international electronic transactions, also found its way into the growing pool of Eurodollars. It is estimated that 100 billion U.S. dollars are laundered each year through the global financial system.

Together these factors have amounted to a change so important that a deep-seated restructuring of the world economy has resulted. The effects of this restructuring have altered economic geography at every scale, affecting the lives of people everywhere. Banks and financial corporations with the size and international reach of Citigroup, Crédit Suisse, or Deutsche Bank are able to influence local patterns and processes of economic development throughout the world, just like the major transnational conglomerates involved in the global assembly line.

Electronic Offices, Decentralization, and Outsourcing

It is clear that an important shift has occurred in the economic structure of the world's core economies, with the rapid growth of banking, financial, and business services contributing to the expansion of the quaternary sector. What is particularly important from a *geographical* perspective is that this growth has been localized—that is, concentrated in relatively small and distinctive settings within major metropolitan centers. This phenomenon is surprising to some observers, who had expected that new communications technologies would allow for the dispersion of "electronic offices" and, with it, the decentralization of an important catalyst for local economic development. A good deal of geographic decentralization of offices has occurred, in fact, but it has mainly involved "back office" functions that have been relocated from metropolitan and business-district locations to small-town and suburban locations.

Back-office functions are record-keeping and analytical functions that do not require frequent personal contact with clients or business associates. The accountants and financial technicians of main street banks, for example, are back-office workers. Developments in computing technologies, database access, electronic data interchanges, and telephone call routing technologies are enabling a larger share of back-office work to be relocated to specialized office space in cheaper settings, freeing space in the high-rent locations occupied by the bank's front office. For example, the U.S. Postal Service uses optical character readers (OCRs) to read addresses on mail, which is then bar-coded and automatically sorted to its appropriate substation. Addresses that the OCRs cannot read are digitally photographed and transmitted to a computer screen, where a person manually types the address into a terminal. In Washington, D.C., OCR sorting takes place at the central mail facility, but the manual address entry is done in Greensboro, N.C., where wage rates are lower. Workers in Greensboro view images of letters as they are sorted in Washington and enter correct addresses, which are electronically transmitted back to be bar-coded on the piece of mail. Similarly, when residents of London, England, call to inquire about city parking-ticket fines, the calls are processed in a small city in northern England.

Some prominent examples of back-office decentralization from U.S. metropolitan areas have included the relocation of back-office jobs in American Express from New York to Salt Lake City, Fort Lauderdale, and Phoenix; the relocation of Metropolitan Life's back offices to Greenville, S.C., Scranton, Penn., and Wichita, Kan.; the relocation of Hertz's data-entry division to Oklahoma City, Dean Witter's to Dallas, and Avis's to Tulsa; and the relocation of Citibank's MasterCard and Visa divisions to Tampa and Sioux Falls. Some places have actually become specialized back-office locations as a result of such decentralization. Omaha and San Antonio, for example, are centers for a large number of telemarketing firms, while Roanoke, Va., has become something of a mail-order center.

Internationally this trend has taken the form of offshore back offices. By decentralizing back-office functions to offshore locations, companies save even more in labor costs. Several New York–based life-insurance companies, for example, have established back-office facilities in Ireland. Situated conveniently near Ireland's main international airport, Shannon Airport, they ship insurance claim documents from New York via Federal Express, process them, and beam the results back to New York via satellite or the transatlantic fiber-optics line.

The logical next step from decentralized back-office functions is outsourcing. The outsourcing of services is one of the most dynamic sectors of the world economy. Global outsourcing expenditures are expected to grow to US$827 billion by 2008 as small and medium-sized enterprises follow the example of large transnational companies in taking advantage of low wages in semi-peripheral and peripheral countries. Typically, international outsourcing in service industries involves the work of "routine producers" (who process data by following instructions, perform repetitive tasks and respond to explicit procedures) rather than "symbolic analysts" (who work with abstract images and are involved with independent problem-identifying, problem-solving, and strategic-brokering activities, and make decisions based on critical judgement sharpened by experience). Outsourced services range from simple business-process activities (e.g., data entry word processing, transcription) to more sophisticated, high-value-added activities (e.g., architectural drawing, product support, financial analysis, software programming, and human resource services). India has become one of the most successful exporters of outsourced service activities, ranging from call centers and business-process activities to advanced IT (information technology) services (**Figure 7.27**). More than 150 of the Fortune 500 companies, for example, now outsource software development to India. In the Philippines, special electronic "enterprise zones" have been set up with competitive international telephone rates for companies specializing in telemarketing and electronic commerce. Mexico, South Africa, and Malaysia have also become important locations for call centers and business-process activities.

Clusters of Specialized Offices

Decentralization is outweighed, however, by the tendency for a disproportionate share of the new jobs created in banking, finance, and business services to cluster in highly specialized financial districts within major metropolitan areas. The reasons for this localization are to be found in another special case of the geographical agglomeration effects that we discussed earlier in this chapter. Metropolitan areas such as New York City, London, Paris, Tokyo, and Frankfurt have acquired the kind of infrastructure—specialized office space, financial exchanges, teleports (office parks equipped with satellite Earth stations and linked to local fiber-optic lines), and communications networks—that is essential for delivering services to clients with a national or international scope of activity. These metropolitan areas have also established a comparative advantage both in the mix of specialized firms and expert professionals on hand and in the high-order cultural amenities available (both to high-paid workers and to their out-of-town business visitors). Above all, these metro areas have established themselves as centers of authority, with a critical mass of people in the know about market conditions, trends, and innovations—people who can gain one another's trust through frequent face-to-face contact, not just in business settings but also in the informal settings of clubs and office bars. They have become

Figure 7.27 Globalized office work A call center in Bangalore, India.

world cities, places that, in the globalized world economy, are able not only to generate powerful spirals of local economic development but also to act as pivotal points in the reorganization of global space: control centers for the flows of information, cultural products, and finance that collectively sustain the economic and cultural globalization of the world (see Chapter 10 for more details).

A good example of the clustering of business services in major world cities is provided by advertising services in Europe. In the early 1980s these services were distributed among major European cities, with Paris, London, Amsterdam, and Stockholm accounting for the headquarters of most firms, but with smaller concentrations in Brussels, Düsseldorf, Frankfurt, and Zürich. By 1990 the headquarters of most advertising agencies in Europe had moved to London (**Figure 7.28**), which, along with New York and Tokyo, has become one of the three most dominant world cities in the contemporary world-system.

Offshore Financial Centers

The combination of metropolitan concentration and back-office decentralization fulfills most of the locational needs of the global financial network. There are, however, some needs—secrecy and shelter from taxation and regulation, in particular—that call for a different locational strategy. The result has been the emergence of a series of **offshore financial centers:** islands and micro-states such as the Bahamas, Bahrain, the Cayman Islands, the Cook Islands, Luxembourg, Liechtenstein, and Vanuatu that have become specialized nodes in the geography of worldwide financial flows (**Figure 7.29**).

The chief attraction of offshore financial centers is simply that they are less regulated than financial centers elsewhere. They provide low-tax or no-tax settings for savings, havens for undeclared income and for hot money. They also provide discreet markets in which to deal currencies, bonds, loans, and other financial instruments without coming to the attention of regulating authorities or competitors. The U.S. Internal Revenue Service estimates that about $400 billion ends up in offshore financial centers each year as a result of tax-evasion schemes. Overall, about 60 percent of all the world's money now resides offshore.

The Cayman Islands provide the classic example of an offshore financial center. This small island state in the Caribbean has transformed itself from a poor, underdeveloped colony to a relatively affluent and modern setting for upscale tourism and offshore finance. With a population of just over 40,000, it has more than 30,000 registered companies, including 350 insurance companies and over 580 banks from all around the world. Yet fewer than 75 of the 580 or so registered banks in the Cayman Islands maintain a physical presence on the islands, and only half a dozen offer the kind of local services that allow local residents to maintain checking accounts or tourists to cash traveler's checks. Most exist as post office boxes, as nameplates in anonymous office buildings, as fax numbers, or as entries in a computer system. In 2004 over a trillion dollars passed through the Caymans, making it the most successful of all offshore financial centers.

The Pleasure Periphery: Tourism and Economic Development

Many areas of the world, including parts of the world's core regions, do not have much of a primary base (that is, in agriculture, fishing, or mineral extraction), are not

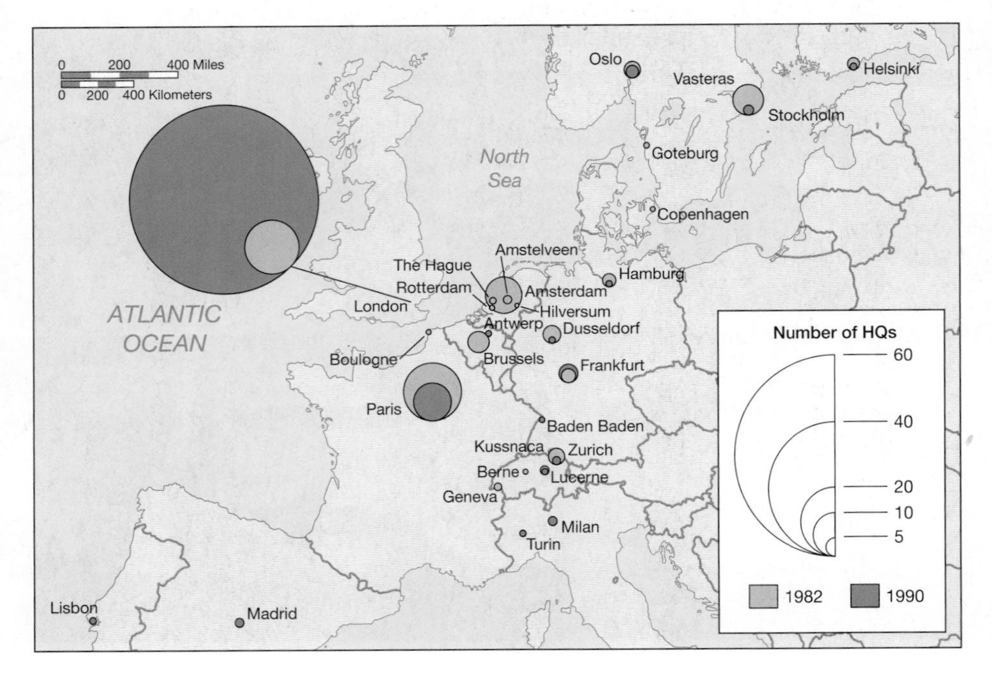

Figure 7.28 The clustering of advertising agencies in European cities This map shows the changing distribution of the headquarters offices of the top 100 advertising agencies in Europe between 1982 and 1990. Note the striking concentration of agency headquarters in London. (After P. Daniels, "Services in a Shrinking World," *Geography,* 80, 1995, Fig. 4, p. 102.)

THE CAYMAN ISLANDS, home to over 580 banks that together have assets of nearly US$1 trillion, are exempted from taxes by a royal decree granted by King George III in 1798, after islanders rescued a British prince.

The **BRITISH VIRGIN ISLANDS** set up as an offshore financial center with a 1984 law that authorized the operation of "International Business Companies" (IBCs) with a minimum of fuss and cost (there are no exchange controls or tax treaties, and registration of companies can be completed in one day. By 2004, over 300,000 IBCs had been established.

LABUAN, formerly a penal colony and pirates' lair, was set up as an International Offshore Financial Center by the Malaysian government in 1990 in order to attract business from Hong Kong and Japan. By 2000 more than 2,200 offshore and supporting companies had been established. Over 60 offshore banks, nearly 50 insurance and insurance related companies and 20 trust companies, as well as numerous legal and accounting firms have decided to take advantage of Labuan's zero tax on dividends, interest and royalties and its 3 per cent tax on dividends, interest, and royalties and its 3 per cent tax on net profits. The asset base of Labuan's offshore banking sector is estimated to be more than U.S.$50 billion.

NAURU, a Pacific island to the northwest of Papua New Guinea, is the smallest independent republic in the world. It operates about 400 offshore banks, all registered to one government mailbox. In 2002, it was the first country to face international banking sanctions (by the OECD countries) over its suspected role in global money laundering.

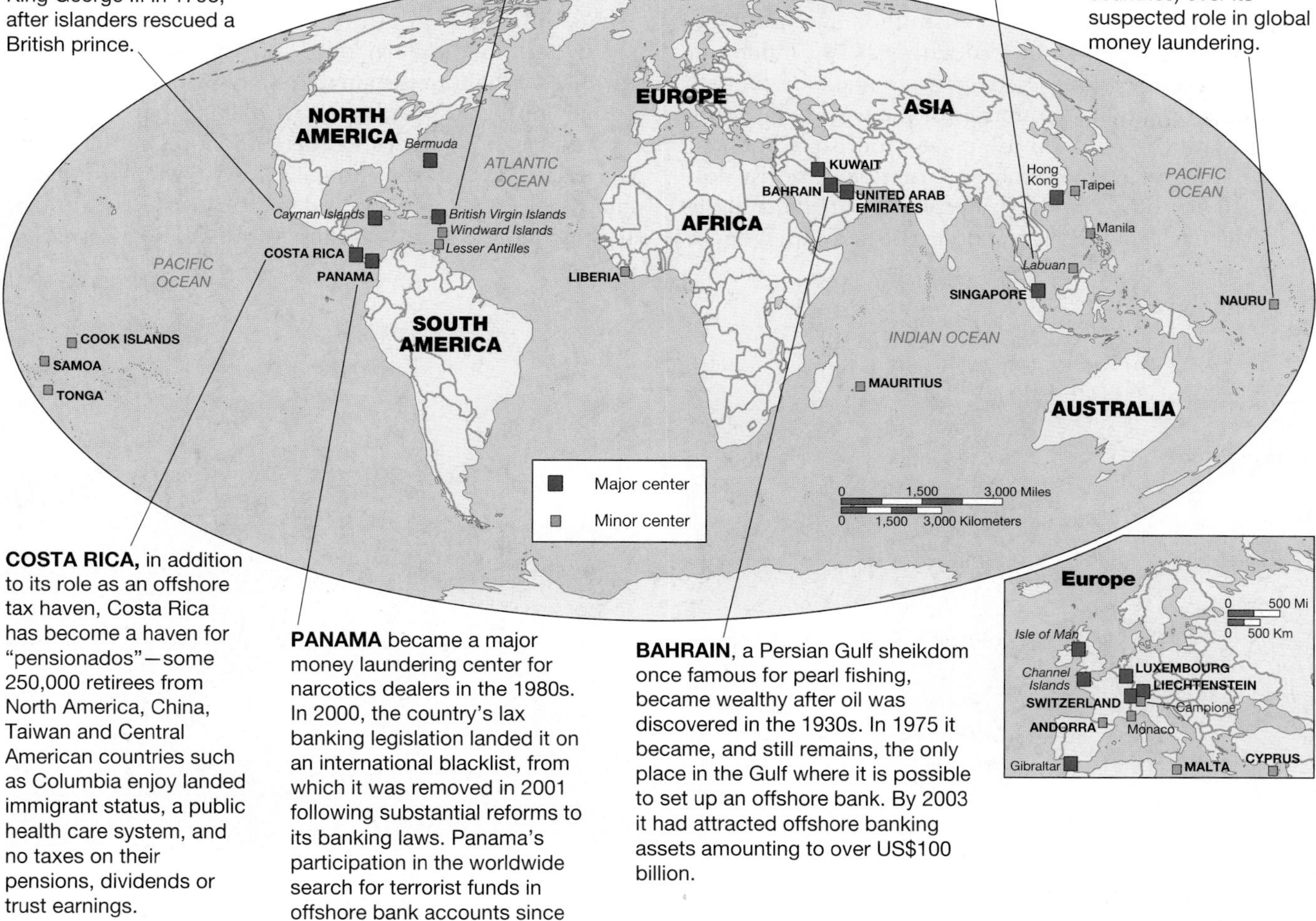

COSTA RICA, in addition to its role as an offshore tax haven, Costa Rica has become a haven for "pensionados"—some 250,000 retirees from North America, China, Taiwan and Central American countries such as Columbia enjoy landed immigrant status, a public health care system, and no taxes on their pensions, dividends or trust earnings.

PANAMA became a major money laundering center for narcotics dealers in the 1980s. In 2000, the country's lax banking legislation landed it on an international blacklist, from which it was removed in 2001 following substantial reforms to its banking laws. Panama's participation in the worldwide search for terrorist funds in offshore bank accounts since the September 11, 2001 attacks in the U.S. helped the country establish a stronger reputation.

BAHRAIN, a Persian Gulf sheikdom once famous for pearl fishing, became wealthy after oil was discovered in the 1930s. In 1975 it became, and still remains, the only place in the Gulf where it is possible to set up an offshore bank. By 2003 it had attracted offshore banking assets amounting to over US$100 billion.

Figure 7.29 Offshore financial centers Offshore financial centers offer confidentiality and shelter from the levels of taxation and regulation that prevail throughout most of the world. They have become important nodes in the global flows of finance that are required by the global economy. Some of the most important of these centers are not "offshore" at all in the literal sense (Bahrain, Luxembourg, and Switzerland, for example), but they are "islands" of financial secrecy and independence.

currently an important part of the global assembly line, and are not closely tied into the global financial network. For these areas tourism can offer the otherwise unlikely prospect of economic development. Tourism has, in fact, become enormously important. In 2003, international tourism earnings amounted to $514 billion. Tourism is already the world's largest nonagricultural employer, with one in every 12 workers worldwide involved in transporting, feeding, housing, guiding, or amusing tourists. The global stock of lodging, restaurant, and transportation facilities is estimated to be worth about $3 trillion.

The globalization of the world economy has been paralleled by a globalization of the tourist industry. In aggregate, there were almost 700 million international tourist trips in 2003, compared with just 147 million in 1970. The majority of these trips were made by tourists from the more affluent countries of the world. Spending by U.S., German, U.K., and Japanese tourists accounted for almost 40 percent of total international tourist dollars in 2003. Tourists from France, Italy, Canada, and the Netherlands accounted for another 12 percent. What is most striking, though, is not so much the growth in the number of international tourists as the increased range of international tourism. Thanks largely to cheaper long-distance flights, a significant proportion of tourism is now transcontinental and transoceanic. While Europe (58 percent) and the Americas (18 percent) continue to be the main tourist destinations, visits to countries in Africa, Asia, and the Pacific have grown to account for almost one-quarter of the industry. This, of course, has made tourism a central component of economic development in countries with sufficiently exotic wildlife (Kenya, for example), scenery (Nepal, Vietnam), beaches (the Seychelle Islands), shopping (Singapore and Hong Kong), culture (China, India, Japan, and Indonesia), or sex (Thailand).

In addition, "alternative" tourism has been widely advanced as a more sustainable strategy for economic development in peripheral regions (**Figure 7.30**). Alternative tourism emphasizes self-determination, authenticity, social harmony, preservation of the existing environment, small-scale development, and greater use of local techniques, materials, and architectural styles. Alternative

(a)

(b)

(c)

Figure 7.30 Ecotourism (a) Mala Mala game reserve, Eastern Transvaal, South Africa; (b) tourists in a Tundra Buggy, Churchill, Manitoba, Canada; (c) Acajatuba eco-lodge, on the Rio Negro, Brazil.

tourism includes ecotourism (e.g., birdwatching in Costa Rica; helping with programs for endangered sea turtles in Bali, "working" in elephant camps in Thailand), cultural tourism (e.g., visiting Machu Picchu, the lost city of the Incas), adventure tourism (e.g., 'exploring' the Amazon), and industrial tourism (e.g., touring the Potteries district of northern England by canal boat).

Costa Rica, despite being a poor country, has won high praise from environmentalists for protecting 30 percent of its territory in biosphere and wildlife preserves. Costa Rica has more bird species (850) than are found in the United States and Canada combined and more varieties of butterflies than in all of Africa. It has 12 distinct ecosystems that among them contain more than 6,000 kinds of flowering plants, over 200 species of mammals, 200 species of reptiles, and more than 35,000 species of insects. The payoff for Costa Rica is the escalating number of tourists who come to visit its active volcanoes, palm-lined beaches, cloud forests, and tropical parks. In 2004, Costa Rica received over 1.2 million tourists. Tourism is now the country's largest source of foreign exchange, followed closely by bananas, coffee, sugar, the textile industry, and more recently the production of microchips.

Ecuador is another country that has fostered alternative tourism. With six national parks, seven nature reserves, and 20 privately protected areas, Ecuador offers great variety for tourists. In a small area straddling the equator, it has some of the world's oldest rain forests, and the world's second-highest active volcano—Cotopaxi, at 5,897 meters (19,347 feet). In Ecuador, tourists can see Amazon tribes, spectacular wildlife in the Galapagos Islands, a thriving Andean culture, and a well-preserved legacy of Spanish colonialism. Two-thirds of all organized travel to Ecuador is handled by members of the Ecuadorean Ecotourism Association, an organization sponsored by the private sector as well as the government in an effort to ensure sustainable development through environmental awareness. Other examples of alternative tourism include guest-house developments in Papua New Guinea, bungalows in French Polynesia, packaged ecotourism in Belize, and "integrated rural tourism" in Senegal. Such developments must be aimed, however, at tourists who are both wealthy and environmentally conscious—not, perhaps, a large enough market on which to pin hopes of significant increases in levels of economic development.

While most tourists are relatively affluent people from the more developed parts of the world, tourism is by no means confined to the less developed ("unspoiled") peripheral regions. Only 10 percent of Americans, for instance, have passports. Most U.S. tourist dollars are spent in safe and predictable settings where English is spoken—in national parks, specialized resorts, theme parks, big cities, rural idylls, health spas, and renovated historic towns and districts. The growth of tourism and the economic success of places like Baltimore's renovated Harborplace (see Figure 6.27), which attracts about 30 million visitors a year, has meant that few localities exist in America—or anywhere else in the developed world, for that matter—that do not encourage tourism as one of the central planks of their economic development strategy. It has been estimated that because tourism requires only a basic infrastructure, no heavy plant, and little high-tech, equipment, the cost of creating one new job in tourism is less than one-fifth the cost of creating a job in manufacturing, and less than one-fiftieth the cost of creating a high-tech engineering job. Consequently, as we saw in Chapter 6, place marketing has become an extremely important aspect of local efforts to promote economic development.

Tourism can provide a basis for economic development, but it is often a mixed blessing. While it certainly creates jobs, they are often seasonal. Dependence on tourism also makes for a high degree of economic vulnerability. Tourism, like other high-end aspects of consumption, depends very much on matters of style and fashion. As a result, once-thriving tourist destinations can suddenly find themselves struggling for customers. Some places are sought out by tourists because of their remoteness and their "natural," undeveloped qualities, which are most vulnerable to shifts of style and fashion. Nepal and New Zealand are recent examples of this phenomenon; they are now too "obvious" as destinations and are consequently having to work hard to continue to attract sufficient numbers of tourists. Bhutan, Bolivia, Estonia, Patagonia, and Vietnam have been "discovered" and are coping with their first significant growth in tourism. China, meanwhile, is likely to become one of the most popular destinations of all by 2010.

It is not only exotic tourist locations that are vulnerable to changing tastes and fashions, however. The Mediterranean beach resort of Rimini, Italy, for example, has been all but deserted by its former northwest European, middle-class clientele, who now prefer more distant, exotic, and distinctive locations. Consequently, Rimini has had to look to the middle classes of Eastern Europe and the former Soviet Union in order to fill its hotel rooms. Unfortunately for Rimini, this is a much less affluent population that generates far fewer tourist dollars.

The vulnerability of tourism to political events and economic trends was demonstrated very emphatically in the wake of the 2001 terrorist attack on New York's World Trade Center. The immediate result was a significant drop in total worldwide air traffic. The immediate cause of the decline in travel was a loss of faith in airport security, especially within the United States. The second-order effect of the terrorist attack, however, was to intensify an economic downturn. As in any economic downturn, the tourist industry soon felt the adverse effects. The net result was not only localized economic recession in tourist-dependent places and regions but also a shift in the geography of tourism. Tourist destinations that were more distant from their principal markets, such as Australia and East Africa, tended to lose market share, while closer-to-home and less expensive destinations tended to gain.

Tourism in more exotic tourist destinations, meanwhile, is vulnerable in other ways: to political disturbances, natural disasters, outbreaks of disease or food poisoning, and atypical weather. For ski resorts, warm weather represents the equivalent of a harvest failure for an agricultural region. Moreover, although tourism is a multibillion-dollar industry, the financial returns for tourist areas are often not as high as might be expected. The greater part of the price of a package vacation, for example, stays with the organizing company and the airline. Typically, only 40 percent is captured by the tourist region itself. If the package involves a foreign-owned hotel, this number may fall below 25 percent.

The costs and benefits of tourism are not only economic, of course. On the positive side, tourism can help sustain indigenous lifestyles, regional cultures, arts, and crafts, and it can provide incentives for wildlife preservation, environmental protection, and the conservation of historic buildings and sites. On the negative side, tourism can adulterate and debase indigenous cultures and bring unsightly development, pollution, and environmental degradation. In the Caribbean, sewage has poisoned mangrove trees and polluted coastal waters, and boats and divers have damaged coral reefs. In Kenya the Maasai Mara National Reserve has been severely degraded by unregulated off-road driving by tour operators. In the United States, congestion in Yosemite National Park (with over 4 million visitors per year) has become so acute that Park Service rangers frequently have to turn away as many as 1,000 vehicles per day, and park administrators are considering the establishment of a strict advance reservation system. In the European Alps, where an incredible 40,000 ski runs attract a winter tourist population 10 times greater than the resident population, forests have been ripped up, pastures obliterated, rivers diverted, and scenic valleys and mountainsides covered with chalets, cabins, and hotels.

Tourism can also involve exploitative relations that debase traditional lifestyles and regional cultural heritages as they become packaged for outsider consumption. In the process, the behaviors and artifacts that are made available to an international market of outsiders can lose much of their original meaning. Traditional ceremonies that formerly had cultural significance for the performers are now enacted only to be watched and photographed. Artifacts like masks and weapons are manufactured not for their original use but as curios, souvenirs, and ornaments. In the process, indigenous cultures are edited, beautified, and altered to suit outsiders' tastes and expectations.

CONCLUSION

The growth of alternative tourism in Costa Rica, like the growth of the Cayman Islands as an offshore financial center, the emergence of Ireland as a center for back-office activities, and the decline of northern England as a manufacturing region, shows that economic development is not simply a sequential process of modernization and increasing affluence. Various pathways to development exist, each involving different ways of achieving increased economic productivity and higher incomes, together with an increased capacity to improve the basic conditions of life. Economic development means not just using the latest technology to generate higher incomes but also improving the quality of life through better housing, health care, and social welfare systems and enhancing the physical framework, or infrastructure, on which the economy rests.

Local, regional, and international patterns and processes of economic development are of particular importance to geographers. Levels of economic development and local processes of economic change affect many aspects of local well-being and so contribute to many aspects of human geography. Economic development is an important place-making process that underpins much of the diversity among regions and nations. At the same time, it is a reflection and a product of variations from place to place in natural resources, demographic characteristics, political systems, and social customs.

Economic development is always an uneven geographic phenomenon. Regional patterns of economic development are tied to the geographic distribution of resources and to the legacy of the past specializations of places and regions within national economies. A general tendency exists toward the creation of regional cores with dependent peripheries. Nevertheless, such patterns are not fixed or static. Changing economic conditions can lead to the modification or reversal of core-periphery patterns, as in the stagnation of once booming regions like northern England and the spectacular growth of Guangdong Province in Southeast China. Over the long term, core-periphery patterns have most often been modified as a result of the changing locational needs and opportunities of successive technology systems. Today economic globalization has exposed more places and regions than ever to the ups and downs of episodes of creative destruction—episodes played out ever faster, thanks to the way that telematics have shrunk time and space.

At the global scale the unevenness of economic development takes the form of core-periphery contrasts within the world-system framework. Most striking about these contrasts today are the dynamism and pace of change involved in economic development. The global assembly line, the global office, and global tourism are all making places much more interdependent and much faster changing. Parts of Brazil, China, India, Mexico, and South Korea, for example, have developed quickly from rural backwaters into significant industrial regions. The Cayman Islands have been transformed from an insignificant Caribbean colony to an upscale tourist resort and a major offshore financial center. Countries like Ecuador and Costa Rica, with few comparative advantages, suddenly find themselves able to earn significant amounts of foreign exchange through the development of ecotourism. This dynamism has, however, brought with it an expanding gap between rich and poor at every spatial scale: international, regional, and local.

MAIN POINTS REVISITED

- **Geographically, the single most important feature of economic development is that it is highly uneven.**

  At the global scale, this unevenness takes the form of core-periphery contrasts. These contrasts raise important issues of spatial justice that are closely bound up with gender inequality and social justice. Similar core-periphery contrasts—and equally important issues of spatial justice—exist at the regional scale.

- **Geographical divisions of labor have evolved with the growth of the world-system of trade and politics, and with the changing locational logic of successive technology systems.**

  Geographical divisions of labor are national, regional, and locally based economic specializations in primary, secondary, tertiary, or quaternary activities. The relationship between changing regional economic specialization and changing levels of prosperity has prompted the interpretation of economic development in distinctive stages. In reality, however, various pathways exist to development, as well as various processes and outcomes of development.

- **Regional cores of economic development are created cumulatively, following some initial advantage, through the operation of several basic principles of spatial organization.**

  Any significant initial local economic advantage—existing labor markets, consumer markets, frameworks of fixed social capital, and so on—tends to be reinforced through a process of cumulative causation, a spiral buildup of advantages that occurs in specific geographic settings as a result of the development of external economies, agglomeration effects, and localization economies.

- **Spirals of economic development can be arrested in various ways, including the onset of disinvestment and deindustrialization, which follow major shifts in technology systems and in international geopolitics.**

  The capital made available from disinvestment in core regions becomes available for investment by entrepreneurs in new ventures based on innovative products and innovative production technologies. Old industries—and sometimes entire old industrial regions—have to be "dismantled" (or at least neglected) in order to help fund the creation of new centers of profitability and employment. This process is often referred to as *creative destruction*.

- **The globalization of the economy has meant that patterns and processes of local and regional economic development are much more open to external influences than before.**

  The globalization of the world economy involves a new international division of labor in association with the internationalization of finance, the deployment of a new technology system, and the homogenization of consumer markets. This new framework for economic geography has meant that the lives of people in different parts of the world have become increasingly intertwined.

KEY TERMS

agglomeration diseconomies (p. 277)
agglomeration effects (p. 274)
ancillary activities (p. 274)
autarky (p. 264)
backwash effects (p. 276)
carrying capacity (p. 258)
conglomerate corporations (p. 281)
creative destruction (p. 278)
cumulative causation (p. 275)
debt trap (p. 270)
deindustrialization (p. 278)
dependency (p. 268)
ecological footprint (p. 260)
elasticity of demand (p. 268)
export-processing zones (EPZs) (p. 288)
external economies (p. 273)
flexible production systems (p. 287)
Fordism (p. 286)
foreign direct investment (p. 280)
geographical path dependence (p. 273)
gross domestic product (GDP) (p. 252)
gross national income (GNI) (p. 252)
growth poles (p. 279)
import substitution (p. 270)
inflation (p. 290)
infrastructure (fixed social capital) (p. 276)
initial advantage (p. 273)
international division of labor (p. 268)
just-in-time production (p. 287)
localization economies (p. 273)
neo-Fordism (p. 286)
newly industrializing countries (p. 261)
offshore financial centers (p. 292)
OECD (p. 271)
primary activities (p. 260)
quaternary activities (p. 260)
secondary activities (p. 260)
spread effects (p. 277)
strategic alliances (p. 288)
sustainable development (p. 258)
terms of trade (p. 270)
tertiary activities (p. 260)
trading blocs (p. 264)
transnational corporations (p. 281)
vertical disintegration (p. 287)
world cities (p. 292)

ADDITIONAL READING

Bryson, J., N. Henry, D. Keeble, and R. Martin (eds.), *The Economic Geography Reader.* New York: Wiley, 1999.

Castells, M., *The Information Age.* 3 vols. Oxford: Blackwell, 1996–1998.

Clark, G., M. Feldman, and M. S. Gertler (eds.), *The Oxford Handbook of Economic Geography.* New York: Oxford University Press, 2000.

Dicken, P., *Global Shift,* 4th ed. New York: Guilford Press, 2003.

Drainville, A., *Contesting Globalization: Space and Place in the World Economy.* New York: Routledge, 2004.

Gereffi, G., and O. Memedovic, *The Global Apparel Value Chain.* Vienna: United Nations Industrial Development Organization, 2003.

Hughes, A., and S. Reimer (eds.), *Geographies of Commodity Chains.* New York: Routledge, 2004.

Hugill, P., *Global Communications Since 1844: Geopolitics and Technology.* Baltimore: Johns Hopkins University Press, 1999.

Knox, P. L., J. Agnew, and L. McCarthy, *The Geography of the World Economy,* 4th ed. London: Edward Arnold, 2003.

Lechner, F. J., and J. Boli (eds.), *The Globalization Reader.* Malden, MA: Blackwell, 2000.

O'Loughlin, J., Staeheli, L., and Greenburg, E. (eds.), *Globalization and Its Outcomes.* New York: Guilford Press, 2004.

Potter, R. B., J. A. Binns, J. A. Elliott, and D. Smith, *Geographies of Development.* London: Longman, 1999.

Scott, A. J., *Regions and the World Economy.* New York: Oxford University Press, 1998.

Smith, N., *The Endgame of Globalization.* New York: Routledge, 2004.

Stiglitz, J. E., *Globalization and its Discontents.* New York: W. W. Norton, 2003.

U.N.D.P. (United Nations Development Programme), *Human Development Report 2004: Cultural Liberty in Today's Diverse World.* New York and Oxford: Oxford University Press, 2004.

United Nations, *Trade and Development Report 2004.* New York: UN Department of Economic and Social Affairs, 2004.

United Nations, *World Economic Situation and Prospects, 2005.* New York: UN Department of Economic and Social Affairs, 2005.

United Nations, *World Economic and Social Survey 2004: International Migration.* New York: UN Department of Economic and Social Affairs, 2005.

Wallach, B. *Understanding the Cultural Landscape.* New York: Guilford Press, 2005. See especially Chapters 10 ("Globalization") and 12 ("Manufacturing").

World Bank, *World Development Report 2005: A Better Investment Climate for Everyone.* Washington, DC: The World Bank, 2004.

World Bank, *Global Economic Prospects: Trade, Regionalism, and Development, 2005.* Washington, DC: The World Bank, 2005.

EXERCISES

On the Internet

The Internet exercises for this chapter compare and contrast the economic success of some capitalist countries that have interpreted capitalism differently. Through the Internet we examine the remarkable success of e-commerce, and we explore questions of infrastructure development that leapfrogs certain economic stages through technology. We examine successful and not-so-successful growth poles, and we compare recent eco-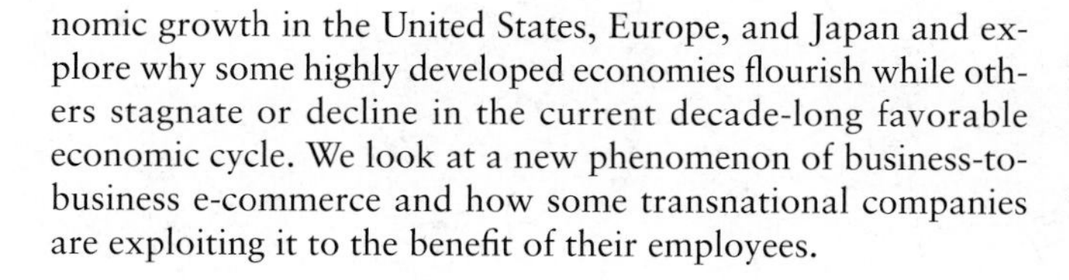nomic growth in the United States, Europe, and Japan and explore why some highly developed economies flourish while others stagnate or decline in the current decade-long favorable economic cycle. We look at a new phenomenon of business-to-business e-commerce and how some transnational companies are exploiting it to the benefit of their employees.

Unplugged

1. While India's per capita income is well below that of the United States (Figure 7.1), India has more people who earn the equivalent of $70,000 a year than the United States does. How can you explain this, and what might be some of the consequences from the point of view of economic geography?
2. Figure 7.4 suggests that creating economic opportunities for women does not necessarily require high levels of economic development. Which peripheral countries have high gender-empowerment index scores, and which core countries have relatively low gender-empowerment index scores? Can you think of explanations for these cases?
3. Write a short essay (500 words, or two double-spaced typed pages) on any specialized manufacturing region or office district with which you are familiar. Describe the different kinds of firms that are found there, and suggest the kinds of linkages between them that might be considered examples of agglomeration effects.

8 Agriculture and Food Production

MAIN POINTS

- Agriculture has been transformed into a globally integrated system; the changes producing this result have occurred at many scales and have had many sources.
- Agriculture has proceeded through three revolutionary phases, from the domestication of plants and animals to the latest developments in biotechnology and industrial innovation.
- The introduction of new technologies, political concerns about food security and self-sufficiency, and changing opportunities for investment and employment are among the many forces that have dramatically shaped agriculture as we know it today.
- The industrialized agricultural system of today's world has developed from—and largely displaced—older agricultural practices, including shifting cultivation, subsistence agriculture, and pastoralism.
- The contemporary agro-commodity system is organized around a chain of agribusiness components that begins at the farm and ends at the retail outlet. Different economic sectors, as well as different corporate forms, have been involved in the globalization process.
- Transformations in agriculture have had dramatic impacts on the environment, including soil erosion, desertification, deforestation, and soil and water pollution, as well as the elimination of some plant and animal species.

According to the most recent agricultural census, over 8.5 billion chickens were sold in the United States in 2002 (roughly 30 plus per person). In 1991 chicken consumption per capita exceeded beef for the first time, in a country that has had something of an obsession with red meat. The fact that each U.S. man, woman, and child currently consumes roughly 1.5 pounds of chicken each week reflects a complex vectoring of social forces in postwar United States. First, a change in taste was driven by a heightened sensitivity to health matters, especially the heart-related illnesses associated with red meat consumption. Second, the cost of chicken meat has in real terms *fallen* since the 1930s (a century ago U.S. residents would eat steak and lobster when they could not afford chicken). Finally, to a growing extent, chicken is consumed in a panoply of forms (Chicken McNuggets, say) which did not exist 20 years ago and which are now delivered to us by the massive fast-food industry—a fact pointing to the reality that people in the United States eat more and more food outside of the home (food consumption "away from home" is, by dollar value, 40 percent of the *average* household food budget). The United States is also the world's largest producer and exporter of poultry meat.

The vast majority of chickens sold and consumed are broilers (young chickens), which are rather extraordinary creatures. In the 1880s there were only 100 million chickens in the United States. The average live bird weight has almost *doubled* in the last 50 years, while the labor input in broiler production has fallen by 80 percent. Disease control and regulation of physiological development have fully industrialized the broiler to the point where it is really a cyborg: part nature, part machine (think *Terminator*). Our understanding of chicken nutrition now exceeds that of any other animal, *including* humans!

Broilers are overwhelmingly produced by family farmers in the United States, but they are farmers under contract to enormous transnational corporations—referred to as "integrators" in the chicken business—who provide the chicks and feed. The growers (who are not organized into unions and have almost no bargaining power) must borrow heavily in order to build the broiler houses and the infrastructure necessary to meet contractual requirements. Growers are not independent farmers at all. They are little more than underpaid workers—what we might call "propertied laborers"—of the corporate producers who also dominate the processing industry.

Jobs in the poultry processing industry, in which the broilers are slaughtered, dressed, and packaged into hundreds of products, are some of the most underpaid and dangerous in the country (a recent government report found almost two-thirds of all poultry processing plants violated overtime payment procedures). Immigrant labor—mostly Vietnamese, Laotian, and Hispanic—represents a substantial proportion of workers in the industry.

The largest 10 companies account for almost two-thirds of broiler production in the United States. Tyson Foods, Inc., the largest producer, accounts for 124 million

Broiler farm, Mississippi

■ The biggest issues food-policy experts, national governments, consumers, and agriculturalists face revolve around the availability and quality of food in a world where access to safe, healthy, and nutritious foodstuffs is unevenly distributed.

pounds of chicken meat per week and controls 21 percent of the U.S. market, with sales of over $5 billion (two-thirds of which go to the fast-food industry). Don Tyson, the CEO of Tyson Foods, says his aim is to "control the center of the plate for the American people."

The United States is the largest producer and exporter of broilers, with a sizable market share in Hong Kong, Russia, and Japan, but it faces intense competition from Brazil, China, and Thailand. The newly global chicken industry is driven by the lure of the massive Chinese market and by the newly emerging and unprotected markets of Eastern Europe and the post-Soviet states. Actually, the world chicken market is highly segmented: Americans prefer breast meat, while U.S. exporters take advantage of foreign preference for leg quarters, feet, and wings to fulfill the large demand from Asia. The chicken is a thoroughly global creature—in its own way not unlike the global car or global finance.[1]

In this chapter, we examine the history and geography of agriculture from the global to the household level. We begin by looking at traditional agricultural practices and proceed through the three major revolutions of agricultural change. Much of the chapter is devoted to exploring the ways geographers investigate the dramatic transformations in agriculture since the middle of the twentieth century and the effects of globalization on agricultural systems.

TRADITIONAL AGRICULTURAL GEOGRAPHY

The study of agriculture has a long tradition in geography. Because of geographers' interest in the relationships between people and land, it is hardly surprising that agriculture has been of primary concern. Research on agriculture is strongly influenced by geography's commitment to viewing the physical and human systems as interactively linked. Such an approach combines an understanding of spatial differentiation, the importance of place, and the fact that practices such as agriculture affect and are affected by processes occurring at different scales. It also provides geographers with a powerful perspective for understanding the dynamics of contemporary agriculture.

One of the most widely recognized and appreciated contributions that geographers have made to the study of agriculture is the mapping of the factors that shape agriculture. They have mapped soil, temperature, and terrain, as well as the areal distribution of different types of agriculture and the relationships among and between agriculture and other practices or variables.

Major changes in agriculture worldwide have occurred in the last four decades. Of these, the decline in the number of people employed in farming in both the core and the periphery is perhaps the most dramatic. Meanwhile, the use of chemical, mechanical, and biotechnological innovations and applications has significantly intensified farming practices (**Figure 8.1**). Agriculture has also become increasingly integrated into wider regional, national, and global economic systems at the same time that it has become more directly linked to other economic sectors, such as manufacturing and finance (**Figure 8.2**). The repercussions from these profound changes range from the structure of global finance to the social relations of individual households.

By examining agricultural practices, geographers have sought to understand the myriad ways humans have learned to modify the natural world around them to sustain themselves, their kin, and ultimately the global community. In addition to understanding agricultural systems, geographers are also interested in investigating the lifestyles and cultures of different agricultural communities. They and other social scientists often use the adjective **agrarian** to describe the way of life that is deeply embedded in the demands of agricultural production. *Agrarian* not only defines the culture of distinctive agricultural communities but also refers to the type of tenure (or landholding) system that determines who has access to land and what kind of cultivation practices will be employed there.

Agriculture is a science, an art, and a business directed at the cultivation of crops and the raising of livestock for sustenance and profit. The unique and ingenious methods by which humans have learned to transform the land through agriculture are an important reflection of the two-way relationship between people and their environments (**Figure 8.3**). Just as geography shapes our choices and behaviors, so we are able to shape the physical landscape. Most introductory textbooks give considerable attention to tracing the origins of agriculture and the distribution of different agricultural practices across the globe. Although agricultural origins are important, the impact of twentieth-century political and economic changes in agriculture are so transformative that in this textbook we focus on the state of global agriculture at the beginning of the new millennium.

[1]After M. Watts, "Commodities," *Introducing Human Geographies*, P. Clorke, P. Crang, M. Goodwin (eds.). Arnold: London, pp. 306–308.

Figure 8.1 Pesticide spraying, Nicaragua Pictured here is a plantation in Nicaragua where this group of workers is getting ready to apply pesticides to the crops.

While there is no definitive answer as to where agriculture originated, we know that before humans discovered the advantages of agriculture, they procured their food through hunting (including fishing) and gathering. **Hunting and gathering** characterizes activities whereby people feed themselves by killing wild animals and gathering fruits, roots, nuts, and other edible plants. Hunting and gathering are considered subsistence activities in that people who practice them procure only what they need to consume. Subsistence agriculture replaced hunting and gathering activities in many parts of the globe when people came to understand that the domestication of animals and plants could enable them to settle in one place over time rather than having to go off frequently in search of edible animals and plants (**Figure 8.4**). **Subsistence agriculture** is a system in which agriculturalists consume all they produce. While the practice of subsistence agriculture is declining, it is still practiced in many areas of the globe.

Figure 8.2 Agricultural Floor of the Chicago Board of Trade Farming has always been tied up with trading. Pictured here is the main site for the global trading of agricultural commodities in operation since 1885. While trade remains an important aspect of agricultural production, it is finance that has had one of the most significant impacts on farming over the last several decades in the core as well as in peripheral countries moving it from a household production form to a corporate one. Agricultural finance provides mortgages and credit for farm expansion, land acquisition, and refinancing of debt and is made available through everything from transnational commercial banks and local credit institutions to federal and international agencies and pension funds.

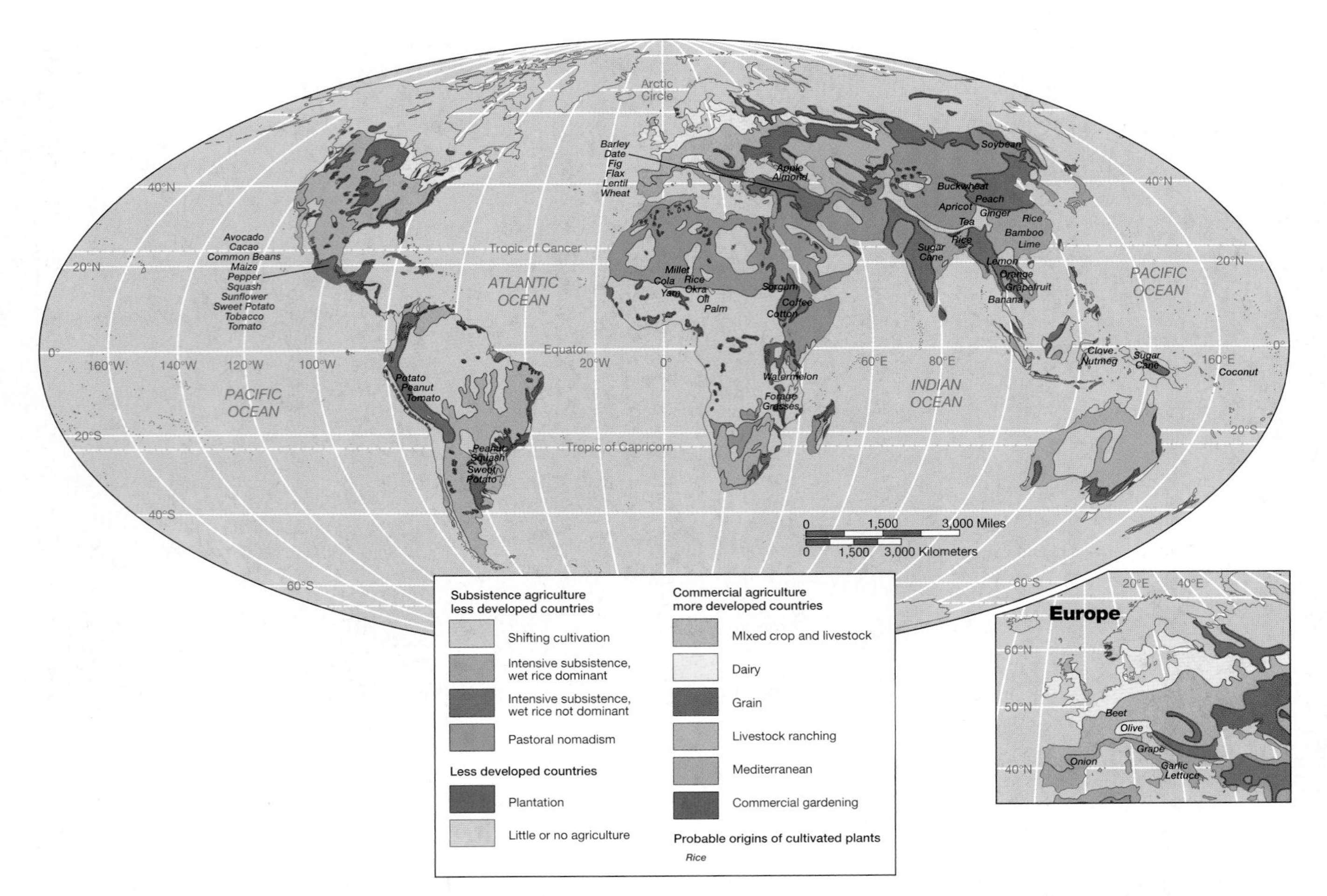

Figure 8.3 Global distribution of agriculture, 2005 The global distribution of agricultural practices is illustrated in this map. Notice the dramatic differences between core and periphery with respect to commercial versus subsistence agriculture. The periphery, though it does contain commercial agriculture, is largely dominated by forms of subsistence, while the core countries contain virtually none. The origins of cultivated plants can also be seen here as they are spread across both the Old World and the New. (After H. Veregin (ed.), *Goode's World Atlas*, 21st ed. Rand McNally, 2005, pp. 38–39.)

During the twentieth century, the dominant agricultural system in the core countries became **commercial agriculture,** a system in which farmers produce crops and animals primarily for sale rather than for direct consumption by themselves and their families. Worldwide, subsistence agriculture is diminishing as increasing numbers of places are irresistibly incorporated into a globalized economy with a substantial commercial agricultural sector. Still widely practiced in the periphery, however, subsistence activities usually follow one of three dominant forms: shifting cultivation, intensive subsistence agriculture, and pastoralism. Although many people in the periphery rely on these traditional practices to feed themselves, traditional practices are increasingly being abandoned or modified as peasant farmers convert from a subsistence and barter economy to a cash economy.

Shifting Cultivation

In **shifting cultivation,** a form of agriculture usually found in tropical forests, farmers aim to maintain soil fertility by rotating the fields they cultivate. Shifting cultivation contrasts with another method of maintaining soil fertility, **crop rotation,** in which the fields under cultivation remain the same but the crops planted are changed to balance the types of nutrients withdrawn from and delivered to the soil.

Shifting cultivation is globally distributed in the tropics—especially in the rain forests of Central and West Africa; the Amazon in South America; and much of Southeast Asia, including Thailand, Burma, Malaysia, and Indonesia—where climate, rainfall, and vegetation combine to produce soils lacking nutrients. The practices involved in shifting cultivation have changed very little over thousands of years (**Figure 8.5**). As a land rotation system, shifting cultivation requires less energy than modern forms of farming, though it can successfully support only low population densities.

The typical agrarian system that supports shifting cultivation is one in which small groups of villagers hold land in common tenure. Through collective agreement or a ruling council, sites are distributed among village families and then cleared for planting by family members. As villages grow, tillable sites must be located farther and far-

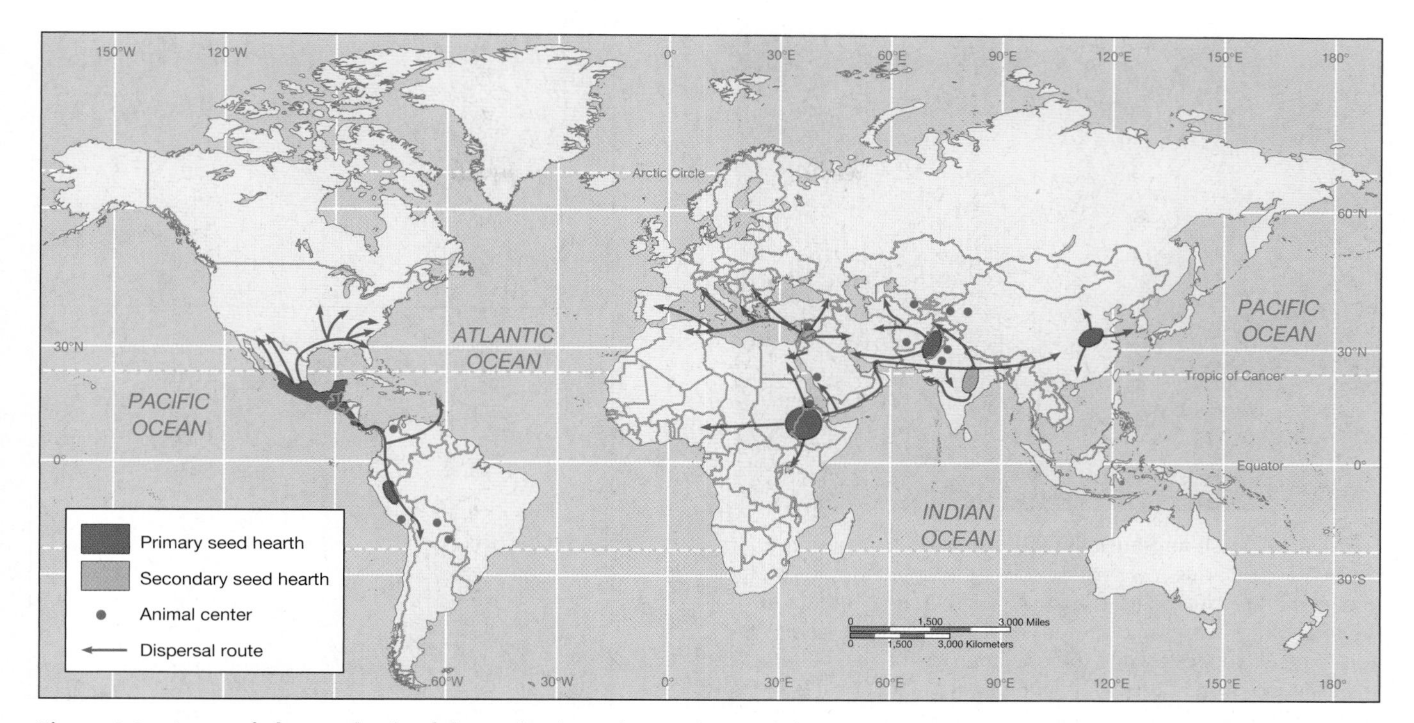

Figure 8.4 Areas of plant and animal domestication Plant and animal domestication did not predominate in any one continent but was spread out across the globe. The origins of plant and animal domestication, however, are not definitively known, and much of what is represented on this map is speculative. Archaeological evidence to date supports the distribution shown here and developed in the mid-twentieth century by Carl Sauer. (After J. M. Rubenstein, *The Cultural Landscape: An Introduction to Human Geography*, 7th ed., Prentice Hall © 2003, p. 319.)

ther away from their center. When population growth reaches a critical stage, several families within the village normally split off to establish another village in one of the more remote sites.

Because tropical soils are poor in nutrients, the problem of the rapid depletion of soil fertility through cultivation means that fields are actively planted for less than five years. The biggest culprits in soil depletion are the cultivated plants and the heavy tropical rains that draw off and wash out the few nutrients that are present in the soil. Once the soil nears exhaustion, a new site is identified and the process of clearing and planting, described in the next paragraph, begins again. It may take over two decades for a once cleared and cultivated site to become tillable again, after decomposition returns sufficient organic material to the soil. When this occurs, the site is reintegrated into cultivation.

Figure 8.5 Shifting cultivation Shifting cultivation is usually practiced in tropical forests. It is a system of agriculture that maintains soil fertility by rotating the fields within which cultivation occurs. This photograph shows a plot under cultivation as well as the burned stumps of the trees that used to occupy the site. Corn and bean shoots are scattered throughout the plot.

The typical method for preparing a new site is through slash-and-burn agriculture, in which existing plants are cropped close to the ground, left to dry for a period, and then ignited (**Figure 8.6**). The burning process adds valuable nutrients to the soil, such as potash, which is about the only readily available fertilizer for this form of agricultural practice. Once the land is cleared and ready for cultivation, it is known as **swidden** (see Chapter 2).

The practice of shifting cultivation usually occurs without the aid of livestock or plow to turn the soil. Thus, this type of agriculture relies largely on human labor, as well as extensive acreage for new plantings because old sites are abandoned frequently when soil fertility is diminished. Although a great deal of human labor is involved in cutting and clearing vegetation, once the site is planted, there is little tending of crops until harvest time.

From region to region the kinds of crops grown and their arrangement in the swidden varies depending upon local taste and plant domestication histories. In the warm,

Figure 8.6 Slash-and-burn agriculture Slash-and-burn is a process of preparing low-fertility soils for planting. In this practice, plants are cleared from a site through cutting, and then the remaining stumps are burned. The burning process helps to add minerals to the soil and thereby improve the overall fertility. Slash-and-burn is a form of agricultural practice that is most effective with low levels of population.

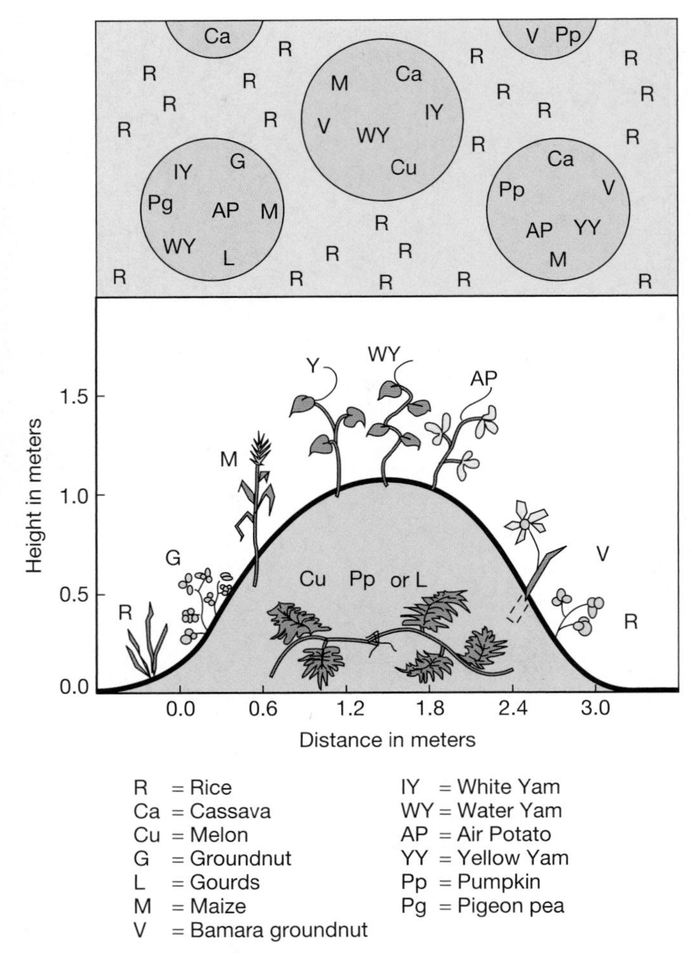

Figure 8.7 Intertillage Planting different crops together in the same field has many benefits, not the least of which are the spreading out of food production over the farming season, reduction of disease and pest loss, greater protection from loss of soil moisture, and control of soil erosion. This diagram provides an illustration of what an intertilled site might contain and how the planting is arranged. Hill-planted seeds have tall stalks and a deeper root system, while those planted on flat earth tend to be spreading plants that produce large leaves for shading.

humid tropics, tubers—sweet potatoes and yams—predominate, while grains such as corn or rice are more widely planted in the subtropics. The practice of mixing different seeds and seedlings in the same swidden is known as **intertillage** (**Figure 8.7**). Not only are different plants cultivated but their planting is usually staggered so that harvesting can continue throughout the year. Such staggered planting and harvesting reduce the risk of disasters from crop failure and increase the nutritional balance of the diet.

Shifting cultivation also involves a gender division of labor that may vary from region to region (**Figure 8.8**). For the most part, men are largely responsible for the initial tasks of clearing away vegetation, cutting down trees, and burning the stumps. Women are usually involved with sowing seeds and harvesting crops. Research on shifting cultivation indicates that the actual division of tasks between men and women (and sometimes children) results from traditional cultural practices, as well as the new demands placed upon households by globalization (recall the discussion of Sudanese children in Chapter 6). For instance, many women have found it necessary to complement their subsistence agricultural activities with craft production for local tourist markets. Their absence from routine agricultural activities means those tasks must be taken up by other household members.

Although heralded by many as an ingenious, well-balanced response to the environmental constraints of the tropics and subtropics, shifting cultivation is not without limitations. Its most obvious limitation is that it can be effective only with small populations. Increasing populations cause cultivation sites to be located farther from villages, with the result that cultivators expend as much energy traveling to sites as they garner energy from the crops they produce. Indeed, at any one time, it is not unusual for land closest to the village to be entirely fallow or unseeded because the soil is exhausted from previous plantings.

Increasingly, population pressures and ill-considered government policies are undermining the practicality of shifting cultivation, resulting in irreparable damage to the environment in many parts of the world. In Central and South America, for example, national governments have used rural resettlement programs to address urban population pressures. In some cases, individuals not familiar with shifting cultivation techniques have employed them improperly. In others, individuals have been relocated to areas unsuitable for such cultivation practices. In parts of the Brazilian Amazon, for example, shifting cultivators have acted in concert with cattle grazers, resulting in accelerated environmental degradation.

Figure 8.8 Gender division of labor Pictured here are women thinning rice plants by hand while a machine plow operated by a man is used to till the earth along the Mekong River, Vietnam. It is not unusual to find women having different access to technology—whether it be agricultural or information—than men in both the core and the periphery. Differential access to technology, training and jobs all help to constitute a gender division of labor in workplaces as well as at home.

Despite its negative impacts on the environment, shifting cultivation can be an elegant response to a fragile landscape. The fallow period, which is an essential part of the process, is a passive and perfectly effective way of restoring plant nutrients to the soil. The burning of stumps and other debris makes the soil more workable, and seeding can proceed with a minimum of effort. Intertillage mimics the natural pattern of differing plant heights and types characteristic of the rain forest. It also helps protect the soil from leaching and erosion. Shifting cultivation requires no expensive inputs (except possibly where native seeds are not available) because no manufactured fertilizers, pesticides, herbicides, or heavy equipment—mechanical or otherwise—are necessary. Finally, the characteristically staggered sowing allows for food production throughout the year. Although shifting cultivation was likely once practiced throughout the world, population growth and the greater need for increased outputs per acre have led to its replacement by more intensive forms of agriculture.

Intensive Subsistence Agriculture

The second dominant form of subsistence activity is **intensive subsistence agriculture,** a practice involving the effective and efficient use of a small parcel of land in order to maximize crop yield; a considerable expenditure of human labor and application of fertilizer is also usually involved. Unlike shifting cultivation, intensive subsistence cultivation often can support large rural populations. While shifting cultivation is more characteristic of low agricultural densities, intensive subsistence normally reflects high agricultural density. Consequently, intensive subsistence usually occurs in the region of the world where agricultural densities are especially high: Asia, and especially India, China, and Southeast Asia.

While shifting cultivation involves the application of a relatively limited amount of labor and other resources to cultivation, intensive subsistence agriculture involves fairly constant human labor in order to achieve high productivity from a small amount of land. With population pressures fierce and the amount of arable land limited, intensive subsistence agriculture also reflects the inventive ways in which humans confront environmental constraints and reshape the landscape in the process. In fact, the landscape of intensive subsistence agriculture is often a distinctive one, including raised fields and hillside farming through terracing (**Figure 8.9**).

Intensive subsistence agriculture is able to support large rural populations. Unlike shifting cultivation, fields are planted year after year as fertilizers and other soil enhancers are applied to maintain soil nutrients. For the most part, the limitations on the size of plots have more to do with demography than geography. In Bangladesh and southern China, for example, where a significant proportion of the population is engaged in intensive subsistence agriculture, land is passed down from generation to generation—usually from fathers to sons—so that each successive generation, where there are multiple male offspring, receives a smaller and smaller share of the family holdings. Yet each family must produce enough to sustain itself.

Under conditions of a growing population and a decreasing amount of arable land, it is critical to plant subsistence crops that produce a high yield per hectare. Different crops fulfill this need, depending on the regional climate. Generally speaking, the crops that dominate intensive subsistence agriculture are rice and other grains.

Rice production predominates in those areas of Asia—South China, Southeast Asia, Bangladesh, and parts of India—where summer rainfall is abundant. In drier climates and where the winters are too cold for rice production, other sorts of grains—among them wheat, barley, millet, sorghum, corn, and oats—are grown for subsistence. In both situations the land is intensively used. In fact, it is not uncommon in milder climates for intensive

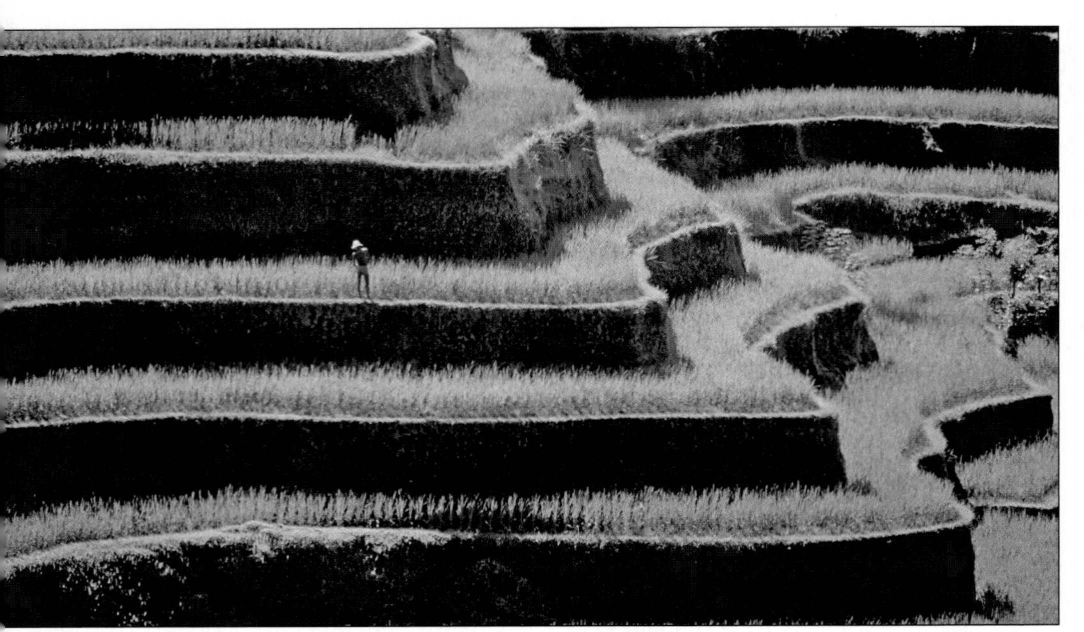

Figure 8.9 Intensive subsistence agriculture Where usable agricultural land is at a premium, agriculturalists have developed ingenious methods for taking advantage of every square inch of usable terrain. Landscapes like this one—of a terraced rice field in Bali, Indonesia—can be extremely productive when carefully tended, and can feed relatively large rural populations.

subsistence fields to be planted and harvested more than once a year, a practice known as **double cropping.**

Pastoralism

Although not obviously a form of agricultural production, pastoralism is a third, dominant form of subsistence activity associated with a traditional way of life and agricultural practice. **Pastoralism** involves the breeding and herding of animals to satisfy the human needs for food, shelter, and clothing. Usually practiced in the cold and/or dry climates of savannas (grasslands), deserts, and steppes (lightly wooded, grassy plains), where subsistence agriculture is impracticable, pastoralism can be either sedentary (pastoralists live in settlements and herd animals in nearby pastures) or nomadic (they travel with their herds over long distances, never settling in any one place for very long). Although forms of commercial pastoralism exist—the regularized herding of animals for profitable meat production, as among Basque Americans in the basin and range regions of Utah and Nevada and among the gauchos of the Argentinean grasslands—we are concerned here with pastoralism as a subsistence activity.

Pastoralism is largely confined to parts of North Africa and the savannas of central and southern Africa, the Middle East, and central Asia. Pastoralists generally graze cattle, sheep, goats, and camels, although reindeer are herded in parts of Eurasia. The type of animal herded is related to the culture of the pastoralists, as well as the animals' adaptability to the regional topography and foraging conditions (**Figure 8.10**).

As a subsistence activity, nomadism involves the systematic and continuous movement of groups of herders, their families, and the herds in search of forage. Most pas-

Figure 8.10 Pastoralism In this image, sheep forage near the summer settlement of yurts—circular tent of felt or skins on a collapsible framework—at the base of Tsaast Uul mountain in Mongolia, where pastoralism is the main livelihood. Note the dryness of the landscape. Pastoralism usually occurs where agriculture is not feasible.

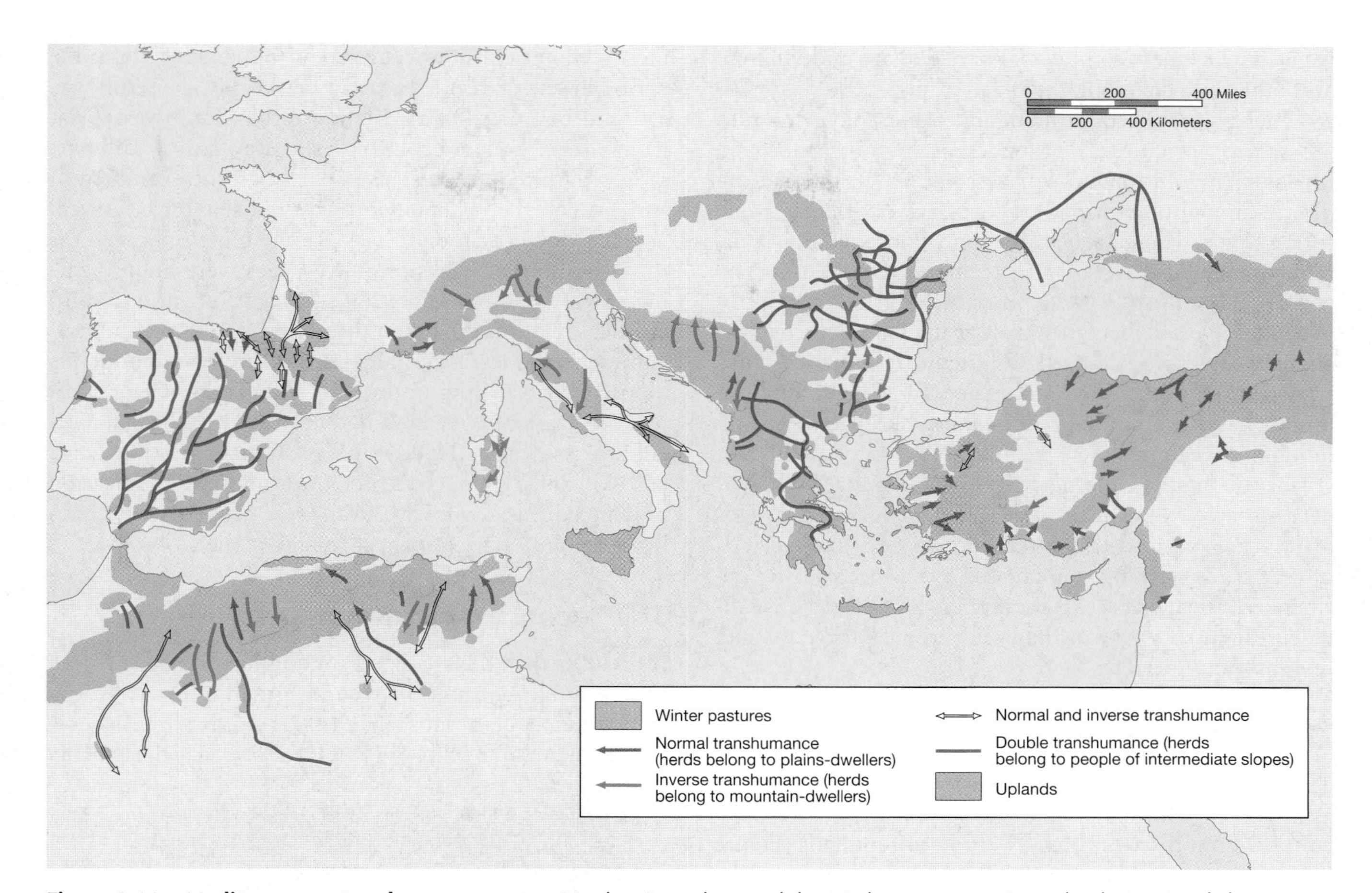

Figure 8.11 Mediterranean transhumance routes Herders in and around the Mediterranean region take their animals between summer and winter pastures by established seasonal routes. Transhumance is an effective adaptation to temporal rhythms. Just as environmental conditions shape herding practices, so do herding practices shape the landscape through the emplacement of identifiable trails. (After "Transhumance" (Figure 7) Map from *The Mediterranean and the Mediterranean World in the Age of Philip II,* Volume I, by Fernand Braudel. Copyright © Librairie Armand Colin, 1966. English translation copyright © by Wm. Collins Sons Ltd. and Harper & Row Publishers, Inc.)

toralists practice **transhumance,** the movement of herds according to seasonal rhythms: warmer, lowland areas in the winter, and cooler, highland areas in the summer (**Figure 8.11**). Although the herds are occasionally slaughtered and used directly for food, shelter, and clothing, often they are bartered with sedentary farmers for grain and other commodities. Women and children in pastoralist groups may also be involved with cultivation. They usually split off from the larger group and plant crops at fixed locations in the spring. The women and children may tend the crops throughout the growing season, or they may rejoin the group and return to the fields when the crops are ready for harvesting. The distinguishing characteristic of pastoralists is that they depend on animals, not crops, for their livelihood.

Like the two other traditional forms of agriculture previously mentioned, pastoralism is not simply a subsistence activity but part of a social system as well. Pastoralists consist of groups of families who are governed by a leader or chieftain. Groups are divided into units that follow different routes with the herds. The routes are well known, with members of the group intimately conversant with the landscape, watering places, and opportunities for contact with sedentary groups. Not surprisingly, pastoralism as a subsistence activity is on the decline as more and more pastoralists have become integrated into a global economy that requires more efficient and regularized forms of production. They have also been forced off the land by competition from other land uses and the state's need to track citizens for taxation and military reasons.

AGRICULTURAL REVOLUTION AND INDUSTRIALIZATION

For a long time human geography textbooks treated the differences in agricultural practices worldwide as systems to be described and cataloged, as we have just done. In the last 25 to 30 years, however, new conceptual approaches to the agricultural sector have transformed the ways we view it. Agriculture has become less a human activity to be described through classification and more a complex component of the global economic system to be explained. Indeed, while the importance and persistence of traditional agricultural forms are acknowledged, such description must be balanced with an understanding of the ways in which new commercial practices are undermining and otherwise changing the older forms.

Increasingly, geographers and others have come to see world agricultural practices as having proceeded through "revolutionary" phases, just as manufacturing did. As in manufacturing, practices have not been transformed everywhere at the same time; consequently, some parts of the world are still largely unaffected by certain aspects of agricultural change. By seeing agriculture in this new light, we can recognize that, as in manufacturing, the changes that have occurred in agricultural practices have transformed geography and society as the global community has moved from predominantly subsistence to predominantly capital-intensive, market-oriented practices.

To understand the new agricultural geography, it is necessary to review the history of world agriculture. This history has proceeded in alternating cycles: long periods of very gradual change punctuated by short, explosive periods of radical change. Geographers and others have divided the history of world agriculture into three distinct revolutionary periods.

The First Agricultural Revolution

The first agricultural revolution is commonly recognized as having been founded on the development of seed agriculture and the use of the plow and draft animals (**Figure 8.12**). Aspects of this transformation have been discussed in Chapters 2 and 4. The emergence of seed agriculture through the domestication of crops such as wheat and rice, and animals such as sheep and goats, replaced hunting and gathering as a way of living and sustaining life. In fact, seed agriculture occurred during roughly the same period in several regions around the world. The result was a broad belt of cultivated lands across Southwest Asia, from Greece in the west into present-day Turkey and part of Iran in the east, as well as in parts of Central and South America, northern China, northeast India, and East Africa.

The domestication of plants and animals allowed for the rise of settled ways of life. Villages were built, creating different types of social, cultural, economic, and political relationships than those that dominated hunter-gatherer societies. Especially important were floodplains along the Tigris, Euphrates, and Nile rivers, where complex civilizations were built upon the fruits of the first agricultural revolution (**Figure 8.13**). Over time the knowledge and skill underlying seed agriculture and the domestication of plants diffused outward from these original areas, having a revolutionary impact throughout the globe.

The Second Agricultural Revolution

A great deal of debate exists among historians as to the timing and location of the second agricultural revolution. Though most historians agree that it did not occur everywhere at the same time, they disagree over which elements

Figure 8.12 Yoked-oxen–drawn plow In many parts of the world, agriculturalists rely on draft animals to prepare land for cultivation. Using animals to assist in agricultural production was an important element in the first agricultural revolution. By expanding the amount of energy applied to production, draft animals enabled humans to increase food supplies. Many contemporary farmers view draft animals as their most valuable possessions. Pictured here are men and oxen tending to rice plants in Malaysia.

Figure 8.13 Present-day agriculture along the Nile Agriculture along the Nile River dates far back into prehistory. The Nile floodplain was one of the important cultural hearths for sedentary agriculture, providing the foundation for the growth of complex civilizations in Egypt. The Nile floodplain remains a remarkably productive area.

were essential to the fundamental transformation of subsistence agriculture. Important elements included:

- dramatic improvements in outputs, such as crop and livestock yields;
- such innovations as the improved yoke for oxen and the replacement of the ox with the horse; and
- new inputs to agricultural production, such as the application of fertilizers and field drainage systems.

The apex of the second agricultural revolution coincided historically and geographically with the Industrial Revolution in England and Western Europe. Although many important changes in agriculture preceded the Industrial Revolution, none had more of an impact on everyday life than the rise of an industrialized manufacturing sector, the effects of which spread rapidly to agriculture.

On the eve of the Industrial Revolution—in the middle of the eighteenth century—in Western Europe and England, subsistence peasant agriculture was predominant, though partial integration into a market economy was under way. Many peasants were utilizing a crop-rotation system that, in addition to the application of natural and semiprocessed fertilizers, improved soil productivity and led to increased crop and livestock yields. Additionally, the feudal landholding system was breaking down and yielding to a new agrarian system, based not on service to a lord but on an emerging system of private-property relations. Communal farming practices and common lands were being replaced by enclosed, individually owned land or land worked independently by tenants or renters.

Such a situation was logical in response to the demands for food production that emerged from the dramatic social and economic changes accompanying the Industrial Revolution. Perhaps most important of all these changes was the development—through the creation of an urban industrial workforce—of a commercial market for food. Many innovations of the Industrial Revolution, such as improvements in transportation technology, had substantial impacts on agriculture. Innovations applied directly to agricultural practices, such as the new types of horse-drawn farm machinery, improved control over—as well as the quantity of—yields.

By helping to usher in the second agricultural revolution, the Industrial Revolution changed rural life as profoundly as the sedentary requirements of seed agriculture had transformed a hunting and gathering society. As geographer Ian Bowler writes, this revolution moved rapidly from Europe to other parts of the world:

> From its origins in Western Europe, the new commercialized system of farming was diffused by European colonization during the nineteenth and twentieth centuries to other parts of the world. A dominant agrarian model of commercial capitalist farming was established, based on a structure of numerous, relatively small family farms. From this period can be traced both the dependence of agriculture on manufacturing industry for many farm inputs, and the increasing productivity of farm labor, which released large numbers of workers from the land to swell the ranks of factory workers and city dwellers. Moreover, the production of food surplus to domestic demand enabled international patterns of agricultural trade to be established.[2]

The Third Agricultural Revolution

The third agricultural revolution is a fairly recent one that, unlike the previous two, emanates mostly from the New World. Scholars identify the third agricultural revolution as beginning in the late nineteenth century and gaining momentum throughout the twentieth century. Each of its three important developmental phases originated in North America. Indeed, the globalization trends framing all of our discussions in this text are the very same ones that have shaped the third agricultural revolution. The difference between the second and third agricultural revolutions is mostly a matter of degree, so that by the late twentieth century, technological innovations have virtually industrialized agricultural practices (see Box 8.1: "The Blue Revolution and Global Shrimp").

The three phases of the third agricultural revolution are mechanization, chemical farming with synthetic fertilizers (**Figure 8.14**), and globally widespread food manufacturing. **Mechanization** is the replacement of human farm labor with machines. Tractors, combines, reapers, pickers, and other forms of motorized machines have, since the 1880s and 1890s, progressively replaced human and animal labor inputs to the agricultural production process in the United States (**Figure 8.15**). In Europe, mechanization did not become widespread until after World War II. **Figure 8.16** shows the global distribution of tractors as a measure of the mechanization of worldwide agriculture.

Chemical farming is the application of synthetic fertilizers to the soil—and herbicides, fungicides, and pesticides to crops—to enhance yields. Becoming widespread in the 1950s in the United States, chemical farming diffused to Europe in the 1960s and to peripheral regions of the world in the 1970s. The widespread application of synthetic fertilizers and their impact on the environment is what Rachel Carson wrote about in her pathbreaking book, *Silent Spring,* which we discuss in Chapter 4 (**Figure 8.17**).

Food manufacturing also had its origins in late-nineteenth-century North America. **Food manufacturing** involves adding economic value to agricultural products through a range of treatments—processing, canning, refining, packing, packaging, and so on—occurring off the farm and before the products reach the market (**Figure 8.18**). The first two phases of the third revolution affected inputs to the agricultural production process, whereas the final phase affects agricultural outputs. While the first

[2] I. Bowler (ed.), *The Geography of Agriculture in Developed Market Economies.* Harlow, England: Longman Scientific and Technical, 1992, pp. 10–11.

(text resumes on page 316)

8.1

GEOGRAPHY MATTERS

The Blue Revolution and Global Shrimp

Written by
Brian J. Marks

The first decade of the twenty-first century was a disastrous one for shrimp fishers across the U.S. Southeast, even before the devastating hurricanes of 2005. The price of shrimp has dropped precipitously: Between 2000 and 2003, average Gulf of Mexico dockside prices fell between 38 and 57 percent, depending on shrimp size. Much of this decline occurred in just a few months in 2001. As a result, almost half the Gulf shrimp fleet stopped fishing; ice houses, processing plants, and docks closed; and coastal communities dependent on shrimp for revenues were economically depressed. Fishers responded by intensifying their fishing effort, diversifying their labor into other means of earning income, and replacing hired deckhands on their boats with the unpaid labor of family members (**Figure 8.A**). Some people left the fishery altogether, and others held on, though at a much reduced standard of living.

The cause of this calamity was not because of declining shrimp populations in the Gulf or reduced consumer demand: shrimp landings have been stable and people in the United States are eating more shrimp than they used to. From 1999 to 2003 alone, per capita consumption increased from three to four pounds per person, and has almost tripled since 1980.[1] Why then the price collapse? To find the answer, one has to look at everything from the workings of the international economy to the microbiology of shrimp. The reasons for the price collapse are many, and speak to the global nature of the contemporary food system and how it is complexly affected by economics, politics, ecology, and technology.

Shrimp have been both fished from the ocean and farmed in ponds for hundreds of years. Commercial shrimp fisheries in the United States first developed in Louisiana as an export industry of dried shrimp to Asia and to Asian immigrants in the United States, then into canned products. Around World War II frozen shrimp began to predominate, tapping into the growing affluence of the U.S. domestic market as more people went out to eat more often and infrastructures for frozen foods (home freezers, refrigerated transport, cold storage warehouses, etc.) both for restaurant service and in homes became commonplace. This paralleled the overall shift in core countries' food systems at that time in terms of household labor (particularly of women), food processing, and production, a process researchers David Goodman and Michael Redclift refer to as "food into freezers: women into factories."[2]

Also after World War II, many countries in Latin America, Asia, and Africa received considerable investment from the United States, Western Europe, and Japan to modernize their economies. Much of this assistance took the form of loans and grants directed at modernizing agriculture (the so-called "Green Revolution" discussed in Box 8.2, pp. 318–320) which

[1]Statistics for this piece are drawn largely from the following: National Marine Fisheries Service *Draft Shrimp Business Options: Proposals to Develop a Sustainable Shrimp Fishery in the Gulf of Mexico and South Atlantic.* Silver Spring, MD: National Oceanic and Atmospheric Administration, 2004; and National Marine Fisheries Service *Current Fisheries Statistics No. 2004-2: Imports and Exports of Fishery Products—Annual Summary, 2004.* Silver Spring, MD: National Oceanic and Atmospheric Administration.

[2]Goodman, D. and M. Redclift, *Refashioning Nature: Food, Ecology, and Culture.* London: Routledge, 1991.

Figure 8.A Louisiana shrimpers Pictured here are fishers harvesting and sorting wild-caught shrimp near Bayou Lafourche, Louisiana.

Figure 8.B Shrimp farm Phang-Nqa Province, Thailand Shrimp aquaculture is based on the use of ponds where shrimp larvae can be introduced and tended. Many of the ponds are built on coastal farmlands, mangrove swamps, or other wetlands.

brought chemical fertilizers, pesticides, agricultural machinery, and new hybrid strains of crops to farmers. Alongside the Green Revolution, came a "**Blue Revolution**" which was the introduction of motorized and larger boats, processing technology and infrastructure, and new production techniques into peripheral country fisheries. A prominent component of the Blue Revolution is **aquaculture**, which is the growing of aquatic creatures in ponds on shore or in pens suspended in water (**Figure 8.B**). Both the Green and Blue Revolutions moved primary sector activities toward a greater dependence on capitalized inputs—like tractors, boats, diesel engines, and petroleum—instead of basic human labor. On the one hand, the Blue and Green Revolutions increased food production in many places; on the other they engendered conflict over how the new practices redistributed power and wealth among producers.

Aquaculture claimed to be an answer to feeding the periphery a cheap form of protein. Advances in farming fish like carp and tilapia in countries like China has increased the availability of fish for millions of people in producing countries. Yet, to date aquaculture has found its biggest economic successes in catering to the demand of affluent consumers in the core for products like shrimp and salmon. Modern shrimp aquaculture both created new shrimp farming industries where none had existed before and revolutionized wild shrimp fisheries and old forms of aquaculture already in place.

The growth of shrimp trade and aquaculture was rapid. From one-twentieth of global shrimp supply in 1980, aquaculture amounted to over one-third by 2001. The majority of world shrimp production of all types was exported in 2001, while approximately only one-quarter was exported in 1980. The major consumers of shrimp are the United States, Japan, and the European Union (**Figure 8.C**). In 2001 Thailand, India, Indonesia, Vietnam, and Mexico were the top five exporting countries by volume. Shrimp is big business, the most consumed seafood in the United States, and a major source of foreign exchange for many countries.

The global shrimp industry is also a major source of controversy. Fisherpeople and other coastal residents of exporting countries have long denounced shrimp exports, both caught at sea and raised in ponds, for damaging their catches of fish for local consumption, for destroying coastal wetlands and agricultural land to build shrimp ponds, and by using wild fish stocks to make fish meal to feed shrimp. There have been many incidents of violence, even killings, of people opposed to shrimp farming in their communities by shrimp farming interests. The so-called "Pink Gold Rush" of shrimp exports has come with a high social and ecological cost.

The industry also faces another problem: that of its own success. Aquaculture means that the limits on biological production from the ocean have been partially overcome so that production can increase much faster than demand, depressing prices. According to a National Marine Fisheries Service report, by volume world shrimp trade increased by 240 percent between 1980 and 2001, but only by 70 percent in terms of value. Global oversupply relative to demand, contributed to the fragility of the shrimp industry in the early 2000s that led to the price collapse felt on the Gulf Coast.

The immediate triggers of the decline in prices were many. In the mid- to late 1990s, the U.S. economy was strong, fueling high demand for shrimp. That demand drove global increases in production of shrimp at the same time that there were stagnant markets in Japan and the European Union. New exporting countries, like Vietnam and Brazil, dramatically accelerat-

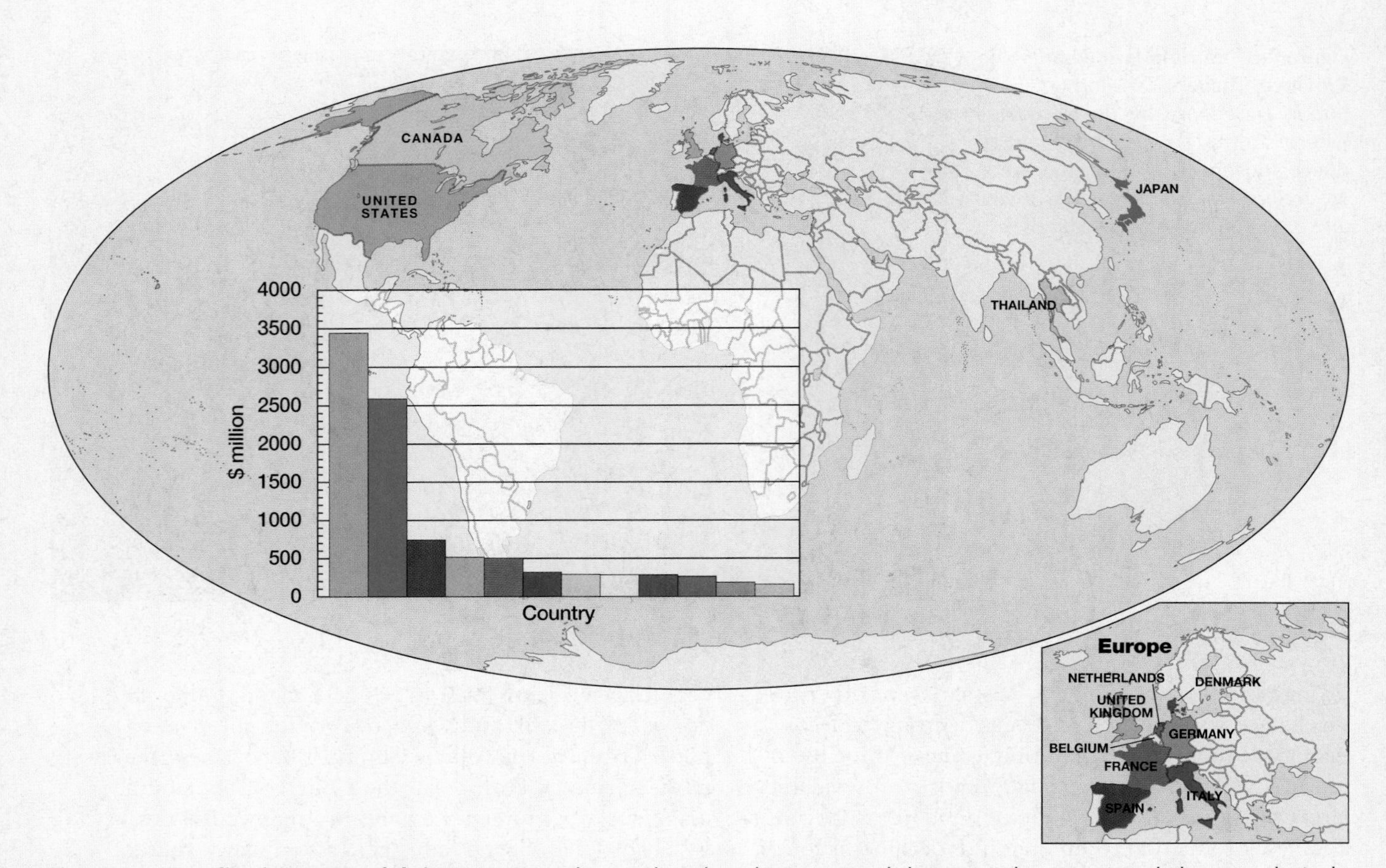

Figure 8.C Leading importers of shrimp, 2004 As this graph makes clear, imported shrimp is a luxury enjoyed almost exclusively by core countries.

ed their production to get their share of the profits. An international economic order that stressed earning foreign exchange in order to service debt and fund development projects was an added incentive to exporting countries to invest heavily in shrimp farming. Shrimp exports to the United States brought in dollars, which were very valuable at that time relative to other major world currencies, allowing these countries to buy more with the proceeds from each pound of shrimp they sold.

This growth in exports might have continued at stable prices if it were not for the faltering of the U.S. economy beginning in 2001 while demand remained flat in other consuming countries. As U.S. demand declined, shrimp exports had nowhere else to go, and as a result inventories started to pile up and prices began to fall. Other markets were not as desirable to shrimp exporters for a variety of reasons. For example, the U.S. market had no significant tariffs on imported shrimp, unlike Japan and Europe, which imposed between 6 and 12 percent tariffs on many countries.

Global economics does not explain everything that occurred in 2001 in the shrimp industry, however. Part of the answer can be found in changing attitudes about health and diet in the consuming countries. The "low-fat" and "low-carb" diet crazes fit well with increased shrimp consumption. Menus promoting healthier eating offered shrimp on salads, in fajitas, or on its own as an alternative to red meat or chicken. The promotion of seafood as a health food in the reconfiguration of the U.S. diet was a big part of the increase in demand for the product, and as prices fell, shrimp became even more attractive as a part of restaurant menus and frozen-food aisles.

The politics of food safety also played a part. The European Union (EU) had much more stringent food inspection requirements than the United States on imported seafood and detected and rejected numerous shipments of imported shrimp in 2001 and 2002 containing chloramphenicol, a powerful antibiotic used to control outbreaks of disease in shrimp ponds that is potentially dangerous to human health and illegal for use in food. In the wake of the chloramphenicol scare in Europe, much shrimp destined for the EU was rerouted to the United States, where standards were not as strict and the chances of detection were smaller.

The ecology of shrimp also contributed to the price collapse. Shrimp raised in ponds are susceptible to many forms of disease, which is why antibiotics are sometimes used on farmed shrimp, causing the food safety concerns mentioned above. This is especially true for farmers using "intensive" techniques, which means they stock high densities of shrimp, feed them processed feeds, and use other technological methods to increase production and profitability. Diseases enter ponds from wild caught or hatchery raised larval shrimp used to stock ponds and from the effluent from nearby already infected ponds. In an effort to reduce outbreaks caused by local wild caught larvae, farms increased their use of shrimp from hatcheries. This attempt backfired as hatcheries, which sell larvae internationally, helped rapidly spread diseases by unknowingly exporting infected shrimp to previously unaffected countries. The white spot syndrome virus outbreak in 1999–2001 is estimated to have reduced global farmed production for a time by about 25 percent and destroyed the majority of some countries' production. The unpredictable increases and decreases in supply caused by expansion of shrimp farming and disease led to considerable price volatility in the late 1990s. High prices caused by temporary disruptions in supply and strong demand led to waves of new investment in the industry, which flooded the market and caused price declines, leading to a boom–bust cycle that accelerated global overproduction. 2000 was one such boom year. The strong U.S. economy and declining production from some exporters due to disease drove prices to very high levels. Many U.S. fishers responded by making new investments in boats, sometimes taking on new debts in order to do so. Shrimp exporters also saw an opportunity to invest and increased production. In 2001, when the U.S. economy slowed down, disease outbreaks in shrimp ponds had been declined, and new exporters entered the market, the result was a crash in the price of shrimp that caught many fishers off guard, leading to the economic hardships that have been experienced since then.

The situation of shrimp is not very different from many other primary commodities, such as coffee, sugar, and cocoa, which have seen dramatic declines in price in recent decades. For example, nominal cocoa prices have declined from a $2,832/ton average in 1980 to $1,190/ton in 2002, a 58 percent decline. When 1980 prices are adjusted for inflation to 2002 dollars using the U.S. Consumer Price Index (reflecting the actual purchasing power of the good over time), the decline is more dramatic, $6,174 versus $1,190, or an 81.8 percent decrease in real value.[3] In comparison, average nominal Gulf of Mexico shrimp prices were $1.63 in 1980 and $1.43 in 2003. Real 1980 prices adjusted using the CPI (1982–2003 = 100) were $3.64 on average. Thus, the real buying power of the earnings from shrimp fell by over 60 percent; shrimp fishers were in effect working for less than half the money they were almost a quarter century earlier. Shrimp was one of several commodities (so-called nontraditional crops) people were encouraged to produce by governments and international development agencies to gain higher prices than those of established products whose values have collapsed. Now it appears that shrimp may be headed in the same direction as other commodities. While U.S. imports of shrimp increased by 49.8 percent from 345,000 tons to 517,000 between 2000 and 2004, the value of that shrimp actually *declined* by $100 million, from $3.8 billion to $3.7 billion. Per pound, the price of imports declined by 35 percent from $5.50 to $3.57 over the same period.

Prosperity from primary commodity production has been ephemeral, highly unequally distributed, and often lacking altogether for those that engage in it. This is true in both core and peripheral countries. Many proposals have been forwarded for addressing the widening gap between the price of primary commodities and cost of living and production for commodity producers, such as increasing international aid or market-based schemes, like fair trade coffee. Reestablishing supply management agreements to support prices, while it would run against deep-seated opposition from core country governments and international financial institutions strongly committed to the unfettered liberalization of trade and investment, may be one of the more promising options.

As domestic U.S. shrimp production accounts for barely 10 percent of U.S. shrimp supply (as of 2003, declining from around 20 percent in the mid-1990s), imports will continue to make up the majority of consumption. Stopping imports is not an option. Domestic shrimp fishers need better prices in order to maintain their livelihoods; fishers and farmers abroad likewise need the earnings from exports. Controlling supplies by means of quotas, floor prices, or other means could achieve higher prices for all producers, more stability in the international market, and an ability to address some of the negative consequences to people and the environment from shrimp production.

Brian Marks is a Ph.D. student in geography at the University of Arizona.

[3]P. Robbins, *Stolen Fruit: The Tropical Commodities Disaster.* London: Zed Books, 2003, pp. 8–9

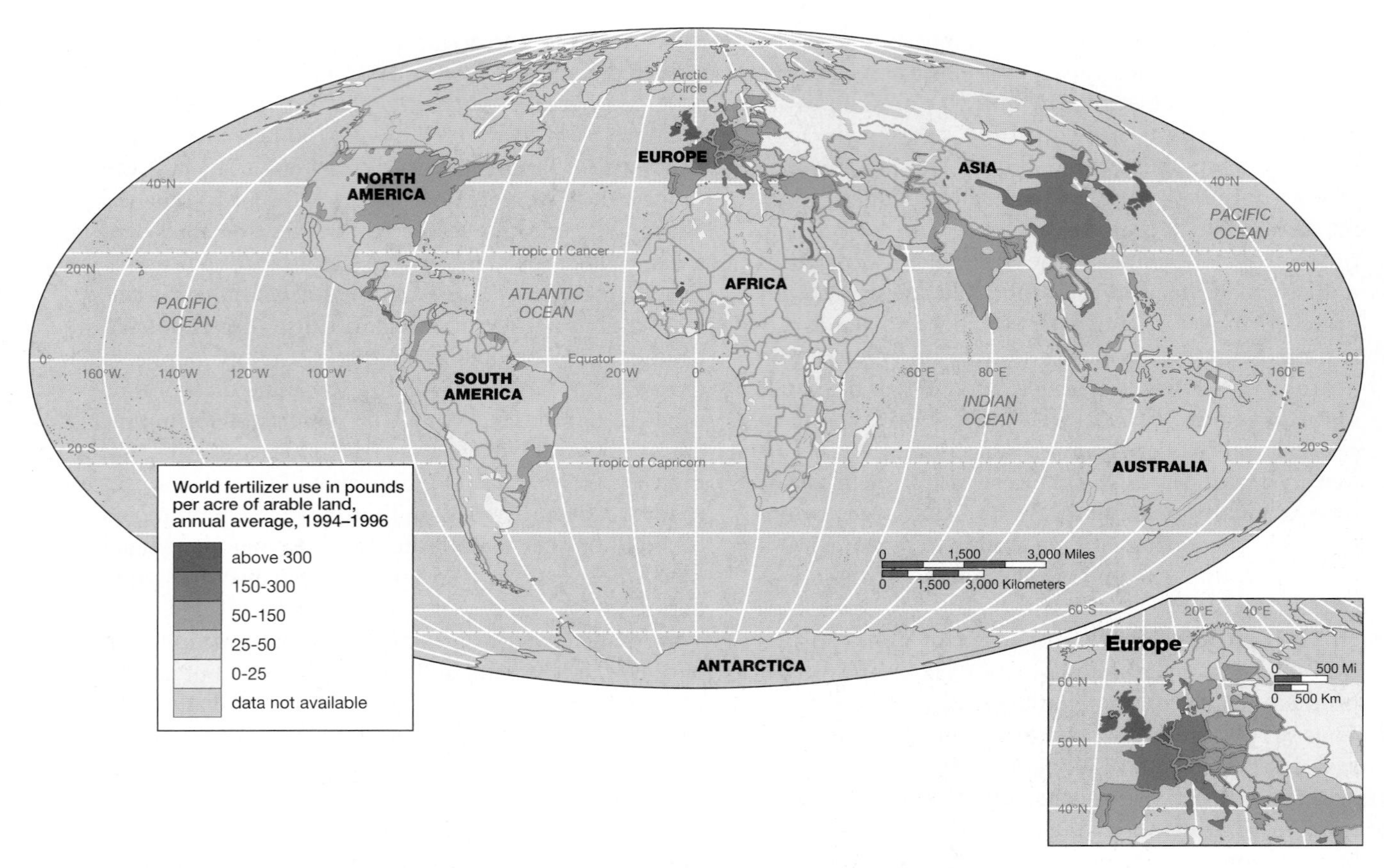

Figure 8.14 Global distribution of fertilizer use, 1992–1996 Western Europe, Egypt, Saudi Arabia, and Japan are among the largest users of fertilizers, with the United States and much of Asia close behind. While they have rapidly growing populations, many African countries cannot afford expensive agricultural inputs, such as fertilizers. The United States and other core European countries export a great deal of food products, and fertilizers enable them to do so. (After J. Hudson and E. Espanshade (eds.), *Goode's World Atlas,* 20th ed., Rand McNally, 2000, p. 49.)

two are related to the modernization of farming as an economic practice, the third involves a complication of the relationship of farms to firms in the manufacturing sector, which had increasingly expanded into the area of food early in the 1960s. Considered together, these three developmental phases of the third agricultural revolution constitute the industrialization of agriculture.

The Industrialization of Agriculture

Advances in science and technology—including mechanical as well as chemical and biological innovations—have determined the industrialization of agriculture over time. As with industrialization, more generally the industrialization of agriculture has unfolded as the capitalist economic system became more advanced and widespread. We regard **agricultural industrialization** as the process whereby the farm has moved from being the centerpiece of agricultural production to being one part of an integrated multilevel (or vertically organized) industrial process including production, storage, processing, distribution, marketing, and retailing. Experts in the study of agriculture have come to see it as clearly linked to industry and the service sector, thus constituting a complex agro-commodity production system.

Geographers have helped demonstrate the changes leading to the transformation of an agricultural product into an industrial food product. This transformation has been accomplished not only through the indirect and/or direct altering of agricultural outputs, such as tomatoes or wheat, but also through changes in rural economic activities. Agricultural industrialization involves three important developments:

- changes in rural labor activities as machines replace and/or enhance human labor;
- the introduction of innovative inputs—fertilizers and other agrochemicals, hybrid seeds, and biotechnologies—to supplement, alter, or replace biological outputs; and
- the development of industrial substitutes for agricultural products (Nutrasweet instead of sugar, and artificial thickeners instead of cornstarch or flour, for example).

Recall, however, that the industrialization of agriculture has not occurred simultaneously everywhere throughout the globe. Changes in the global economic

Figure 8.15 Old and new farm machines In the late nineteenth and early twentieth centuries, agriculture in the United States became increasingly linked with mechanization. (a) This photograph shows a young Vassar College student driving a tractor to plow a field in 1917. It helps to illustrate how transformations in personal transportation diffused to agriculture. (b) This photograph shows contemporary harvesting equipment on a farm in the midwestern United States. While both machines are operated by humans, the contemporary machinery relies on computer chips to send and receive information about the multiple operations of the vehicle.

system affect different places in different ways as different states and social groups respond to and shape these changes. For example, the use of fertilizers and high-yielding seeds occurred much earlier in core-region agriculture than in the periphery, where many people still farm without them. Beginning in the late 1960s, however, core countries exported a technological combination of fertilizers and high-yielding seeds to regions of the periphery (largely in Asia and Mexico) in an attempt to boost agricultural production. In a development known as the **green revolution,** this combination also included new machines and institutions, all designed to increase global agricultural productivity, as described in Box 8.2: "A Look at the Green Revolution."

GLOBAL RESTRUCTURING OF AGRICULTURAL SYSTEMS

When geographers talk about the globalization of agriculture, they are referring to the incorporation of agriculture into the world economic system of capitalism. A useful way to think about the term **globalized agriculture** is to recognize that, as both an economic sector and a geographically distributed activity, modern agriculture is increasingly dependent on an economy and set of regulatory practices that are global in scope and organization.

Forces of Globalization

Several forces, institutions, and organizational forms play a role in the globalization of agriculture. Recall that technology, the economy, and politics have played a central role in propelling national and regional agricultural systems to become global in scope. One important way these forces of change have been harnessed for global transactions is through new institutions, especially trade and financial organizations. The result has been a virtually unprecedented form of production that is built upon an integrated, globally organized agro-production system.

As a result of these forces, the globalization of agriculture has dramatically changed relationships among and within different agricultural production systems. Important outcomes of these changed relationships have been the elimination of some forms of agriculture and the erosion or alteration of some systems as they are integrated

A Look at the Green Revolution

The green revolution was an attempt by agricultural scientists to find ways to feed the world's burgeoning population. The effort began in 1943, when the Rockefeller Foundation funded a group of U.S. agricultural scientists to set up a research project in Mexico aimed at increasing that country's wheat production. Only seven years later scientists distributed the first green revolution wheat seeds. The project was eventually expanded to include research on maize as well. By 1967 green revolution scientists were exporting their work to other parts of the world and had added rice to their research agenda (**Figure 8.D**). Norman Borlaug, one of the founders of the green revolution, went on to win the Nobel Peace Prize in 1970 for an important component of the project: promoting world peace through the elimination of hunger.

The initial focus of the green revolution was on the development of seed varieties that would produce higher yields than those traditionally used in the target areas. However, in developing new, higher yielding varieties, agricultural scientists soon discovered that plants were limited in the amount of nitrogen they could absorb and use. The scientists' solution was to increase the nitrogen absorption capacity of plants by delivering nitrogen-based fertilizers in water (this led to the need to build major water and irrigation development projects). Then the scientists discovered that the increased nitrogen and water caused the plants to develop tall stalks. The tall stalks, with heavy heads of seed on top, fell over easily, thus reducing the amount of seed that could be harvested. The scientists went back to the drawing board and came up with dwarf varieties of grains that would support the heavy heads of seeds without falling over. Then another problem arose: The short plants were growing in very moist conditions, which encouraged the growth of diseases and pests. The scientists responded by developing a range of pesticides.

Thus, the green revolution came to constitute a package of inputs: new "miracle seeds," water, fertilizers, and pesticides. Farmers had to use all of the inputs—and use them properly—to achieve the yields the scientists produced in their experimental plots (**Figure 8.E**). Green revolution crops, if properly watered, fertilized, and treated for pests, can generate yields two to five times larger than those of traditional crops. In some countries, yields are high enough to engage in export trade, thus generating important sources of foreign exchange. Furthermore, the creation of varieties that produce faster maturing crops has allowed some farmers to plant two or more crops per year on the same land, thus increasing their individual production—and wealth—considerably.

Thanks to green revolution innovations, rice production in Asia grew 66 percent between 1965 and 1985. India, for example, became largely self-sufficient in rice and wheat by the 1980s. Worldwide, green revolution seeds and agricultural techniques accounted for almost 90 percent of the increase in world grain output in the 1960s and about 70 percent in the 1970s. In the late 1980s and 1990s at least 80 percent of the additional production of grains could be attributed to the use of green revolution techniques. **Figure 8.F**

Figure 8.D The CIYMMT headquarters The Centro International de Mejorimiento de Maiz y Trigo (CIMMYT) (International Center for the Improvement of Maize and Wheat) in Texcoco, Mexico, is involved in plant breeding and research. High-yield-variety seeds were developed here for the Green Revolution. The center holds the world's premiere collection of corn and wheat germplasm. Modern, refrigerated storage vaults store many thousands of varieties. Today, in addition to the static, cold storage, scientists are working with farmers to preserve seeds dynamically.

Figure 8.E Green revolution experimental plots The CIMMYT includes numerous plots for breeding and testing seed varieties.

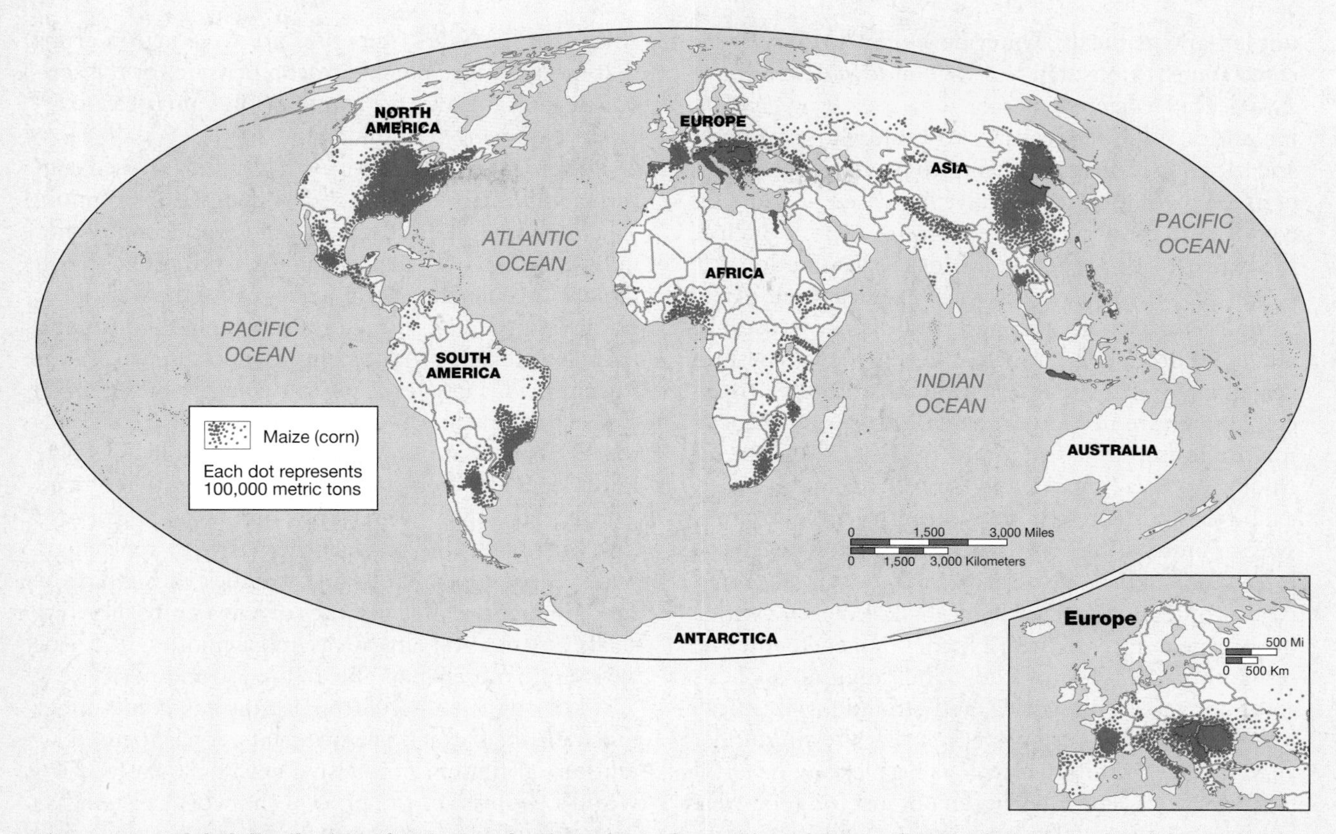

Figure 8.F Global distribution of maize production The widespread production of grains throughout the globe, particularly maize, has been one of the successes of the green revolution (After J. P. Goode, J. C. Hudson, and E. P. Espanshade, Jr., *Rand McNally's Goode's World Atlas,* 20th ed., Rand McNally, 2000, p. 41.)

shows the distribution of maize production worldwide. Thus, although hunger and famine persist, many argue that they would be much worse if the green revolution had never occurred.

The green revolution, however, has not been an unqualified success everywhere in the world. One important reason is that wheat, rice, and maize are unsuitable as crops in many areas, and research on more suitable crops, such as sorghum and millet, has lagged far behind. In Africa poor soils and lack of water make progress even more difficult to achieve. Another important factor is the vulnerability of the new seed strains to pest and disease infestation, often after only a couple of years of planting. Whereas traditional varieties often have a built-in resistance to the pests and diseases characteristic of an area, the genetically engineered varieties often lack such resistance.

Another problem is that green revolution technology has decreased the need for human labor. In southeastern Brazil machines replaced workers, creating significant unemployment. Green revolution technology and training have also tended to exclude women, who play important roles in food production. In addition, the new agricultural chemicals, especially pesticides, have contributed to ecosystem pollution and worker poisonings, and the more intensive use of irrigation has created salt buildup in soils (*salinization*) and water scarcity.

Yet another criticism is that the green revolution has magnified social inequities by allowing more wealth and power to accrue to a small number of agriculturalists while causing greater poverty and landlessness among poorer segments of the population. In Mexico a black market developed in green revolution seeds, fertilizers, and pesticides when poorer farmers, who were coerced into using them, accrued high debts that they could not begin to repay. Many ended up losing their lands and becoming migrant laborers or moved to the cities and joined the urban poor. Some critics who have monitored the effects of the green revolution suggest that political and economic conditions may, in fact, be more important than levels of production with regard to a country's food security.

Even regarding quality, the green revolution crops often fall short. The new seed varieties may produce grains that are less nutritious, less palatable, or less flavorful. The chemical fertilizers and pesticides that must be used are derived from fossil fuels—mainly oil—and are thus subject to the vagaries of world oil prices. Furthermore, the use of these chemicals, as well as monocropping practices, has produced worrisome levels of environmental contamination and soil erosion. In many countries these practices have posed substantial threats to public health, especially among farm workers who are frequently exposed to poisonous (if not lethal) chemicals. Water developments have benefited some regions, but less well-endowed areas have experienced a deterioration of already existing regional inequities. Worse, pressures to build water projects and to acquire foreign exchange to pay for importation of green revolution inputs have increased pressure on countries to grow even more crops for export, often at the expense of production for local consumption.

In recent years scientists have endeavored to develop seeds with greater pest and disease resistance and more drought tolerance. The new focus is best revealed in Africa. The International Institute of Tropical Agriculture in Ibadan, Nigeria, focuses on foods for the humid and subhumid tropics of Africa, including cassava (imported to Africa from South America by the Portuguese in the sixteenth century), yams, sweet potatoes, maize, soybeans, and cowpeas. The International Crops Research Institute for the Semi-Arid Tropics (located in Hyderabad, India, but with a major research center near Niamey, Niger) focuses on researching staples of the Sahel region, such as sorghum, millet, pigeonpea, and groundnut. Research in Africa on new varieties emphasizes testing under very adverse conditions (such as no plowing or fertilizing). New varieties are chosen not just for good yield but because they will provide stable yields over good and bad years. A focus also exists on developing plants that will increase production of fodder and fuel residues, as well as of food, and that give optimal yields when intertilled—a very common practice in Africa. In the Sahel, scientists are working on crops that mature more quickly to compensate for the serious drop in the average length of the rainy season recently experienced in the region.

There are two final criticisms that have raised concern about the overall benefits of the green revolution. The first is that it has decreased the production of biomass fuels—wood, crop residues, and dung—traditionally used in many peripheral areas of the world. For example, in India, as tractors have replaced draft animals, less dung is produced and thus less is available as fuel. Instead, a greater reliance is being placed upon oil to fuel both tractors and other energy needs; this means that if farmers are to be successful, they increasingly must depend upon the most costly of energy resources. The second is that the green revolution has contributed to a worldwide loss of genetic diversity by replacing a wide range of local crops and varieties with a narrow range of high-yielding varieties of a few crops. Planting single varieties over large areas (monocultures) has made agriculture more vulnerable to disease and pests.

Although the green revolution has come under much justified attack over the years, it has focused attention on finding innovative new ways to feed the world's peoples. In the process the world system has been expanded into hitherto very remote regions, and important knowledge has been gained about how to conduct science and how to understand the role that agriculture plays at all geographical scales of resolution, from the global to the local (**Figure 8.G**).

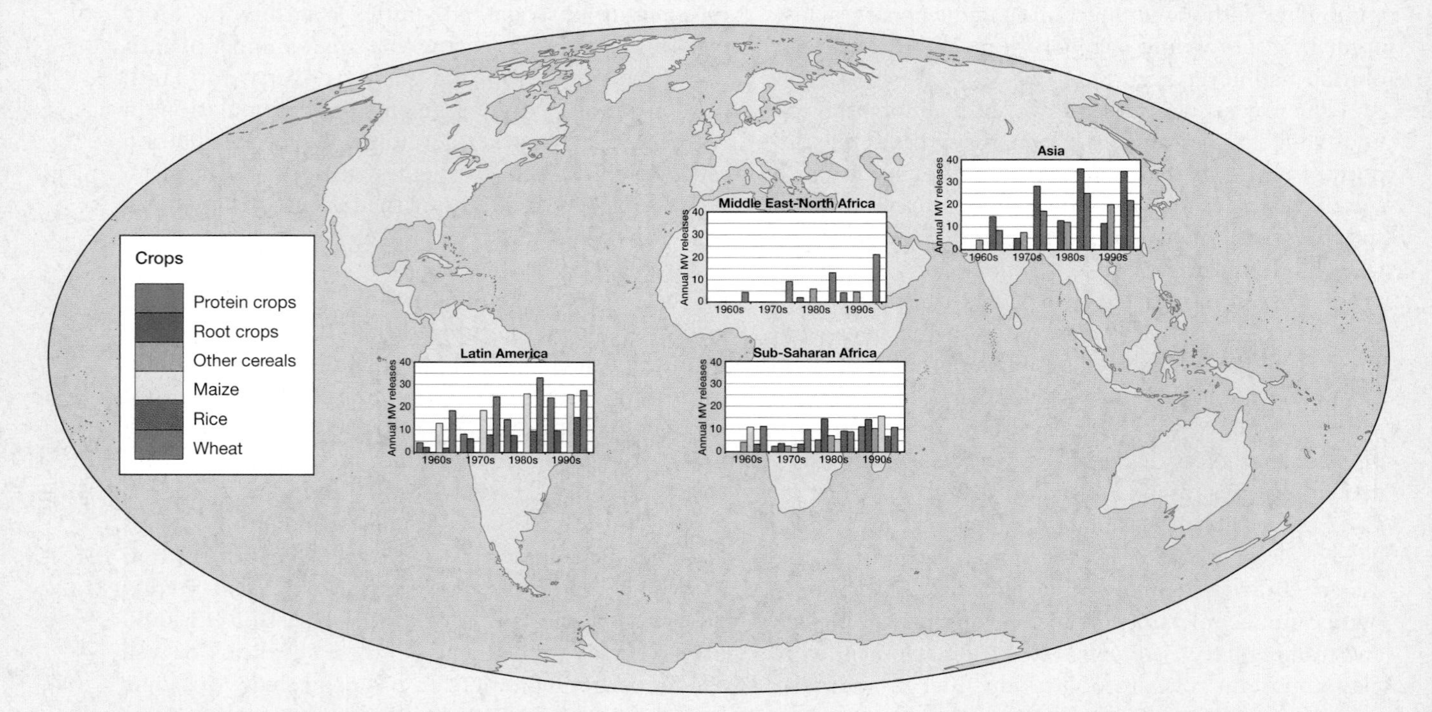

Figure 8.G Effects of the green revolution This map illustrates the increased yields of protein crops, root crops, other cereals, maize, rice and wheat brought about by the green revolution in selected countries in Latin America, Asia, Sub-Saharan Africa and the Middle East and North Africa. (Data from: R. E. Evenson and D. Gollin, "Assessing the Impact of the Green Revolution, 1960–2000," *Science, 300* (2 May 2005), p. 759.)

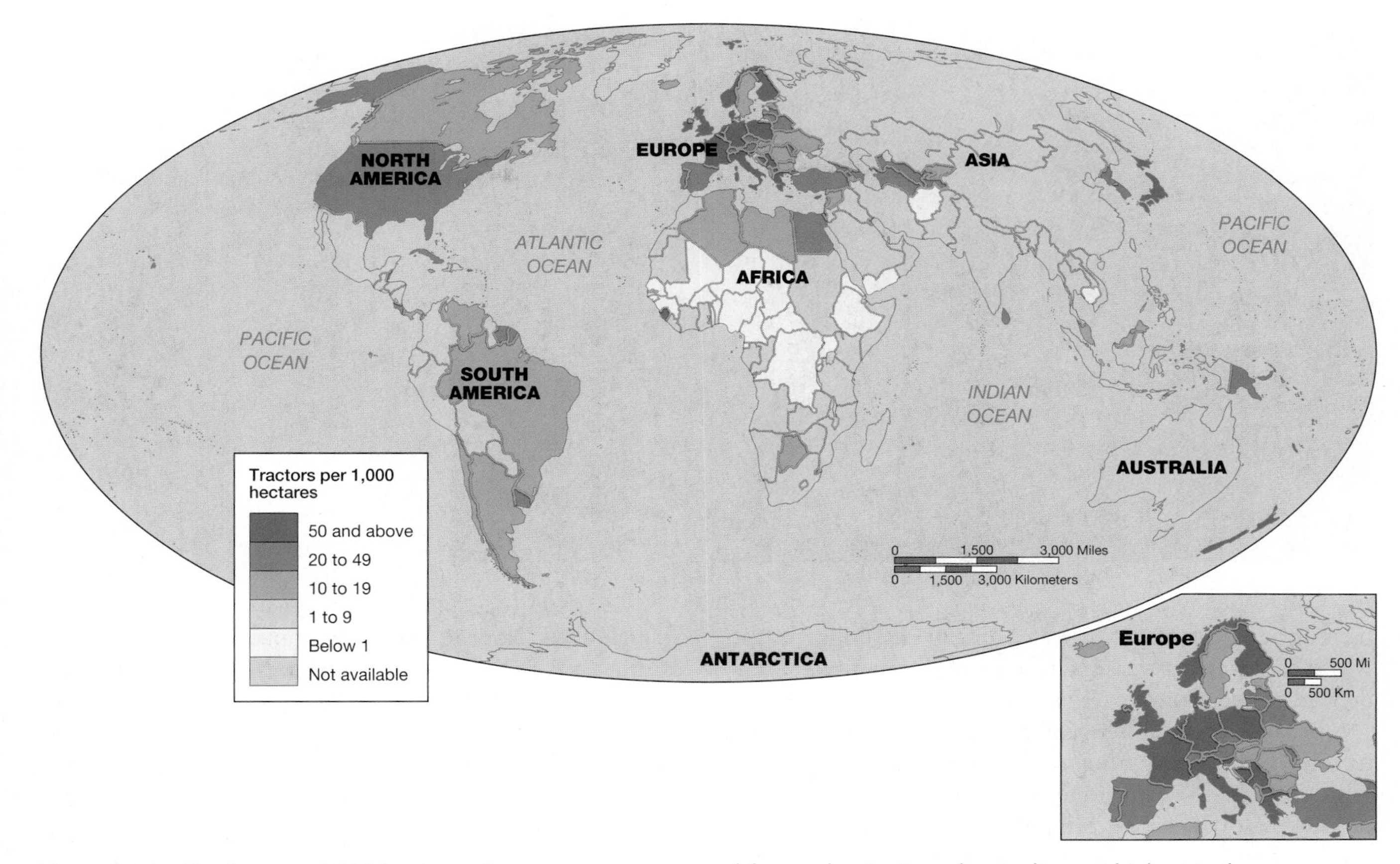

Figure 8.16 Tractors per 1,000 hectares Tractor use, a measure of the mechanization of agriculture, is highest in the core countries. Mechanized farming is an expensive undertaking requiring not only machines but the ability to afford fuels and repairs. Mechanization, however, also allows for more extensive areas of land to be brought into cultivation. (Reprinted with permission of Prentice Hall, from J. M. Rubenstein, *The Cultural Landscape: An Introduction to Human Geography,* 6th ed., © 1999, p. 341.)

into the global economy. Two examples include the current decline of traditional agricultural practices, such as shifting cultivation, and the erosion of a national agricultural system based on family farms (**Figure 8.19**).

Agriculture is one part of a complex and interrelated worldwide economic system. Consequently, important changes in the wider economy—whether technological, social, political, or otherwise—affect all sectors, including agriculture. National problems in agriculture, such as production surpluses, soil erosion, and food price stability, affect agriculture as well as other economic sectors globally, nationally, and locally in different ways. The same is true of more global factors—such as the price and availability of oil and other petroleum products critical to com-

Figure 8.17 Chemical farming An aircraft sprays fungicide on an orange grove near Ft. Pierce, Florida. The fungicide (trade name Kocide) is mixed with a solution to make it stick to the leaves. Planes spray the fungicide at windless times to minimize drift beyond the groves.

Figure 8.18 Food manufacturing Pictured here are tomatoes being processed through an assembly line operation. Food processing is one of the ways that economic value is added to agricultural products before they reach the market.

mercialized agriculture, the stability of the dollar in the world currency market, and recessions or inflationary tendencies in the economy.

Because of the systemic impact of many problems, integration and coordination of the global economy is needed to anticipate or respond to them. In the last several decades, global and international coordination efforts among states have occurred. These include policies advanced by the World Trade Organization (WTO), as well as the formation of supranational economic organizations such as the European Union (EU) and the Association of Southeast Asian Nations (ASEAN). Moreover, these new forms of cooperation are not without their opponents, as the protest against the WTO in Seattle in 1999 made very clear (see also Chapter 12).

At the same time that supranational organizations and coordination efforts have been addressing global problems, states continue to be essential in mediating crises at national levels. By changing public policy, the state attempts to regulate agro-industries in order to maintain production, consumption, and corporate profits. One way that states try to maintain the profitability of the agricultural sector while keeping food prices affordable is through direct and indirect subsidies to agricultural producers. For example, the U.S. government subsidizes agriculture in a number of ways. One is by paying farmers not to grow certain crops that are expected to be in excess supply. Another is by buying up surplus supplies and guaranteeing a fixed price for them.

Subsidies, however, while perhaps stabilizing agricultural production in the short term, can lead to problems within the larger national and international agricultural system. For instance, guaranteeing a fixed price for surplus food can act as a disincentive for producers to lower their production, so the problem of overproduction continues. Once in possession of the surplus, governments must find ways to redistribute it. The U.S. government often sells or donates its surplus to foreign governments, where the "dumping" of cheap foodstuffs may undermine the local price structure for food, as well as reduce economic incentives for farmers to farm. Billions of dollars are paid out each year in agricultural subsidies, the effects of which are complex and global in impact.

Many reasons exist for state intervention in agriculture. Economic interests can be both internally and externally driven. At the internal level, governments routinely intervene in one economic sector or another in order, for instance, to correct wider problems of inflation

Figure 8.19 Family farm Pictured is an idyllic scene of a prosperous U.S. family farm. Since the third agricultural revolution, the number of family farms in the United States and other core countries has declined dramatically as more corporate forms of farming have emerged. In 1920, about one in every three U.S. citizens lived on farms. By 1978 that number had dropped dramatically, to about one in 28. This drastic change in the U.S. farm population has caused some commentators to observe that the family farm in the United States is now just a myth, because many family farms have been bought out by larger corporations and family farms themselves have become incorporated.

or depression. In the 1930s the U.S. government attempted to address the problem of economic depression through policies intended to reduce overproduction—policies that were initiated over a century ago.

States can also intervene in the agricultural sector with respect to consumers' interests. Because of subsidies to farm income, the real cost of food may be quite high, but many states in both the developed and less developed world also subsidize the price of food in the marketplace. Such policies are meant to keep the workforce well fed and healthy, as well as to avoid problems of civil unrest should food prices exceed the general population's ability to pay. Nineteenth-century bread riots—a response to the high cost of flour and bread—were a common occurrence in Europe, and similar forms of civil unrest in response to food prices are not uncommon today in many peripheral countries, where states lack the capital to provide adequate subsidies or where subsidies have been eliminated.

In addition to the internal regulation and assistance of agricultural practices within a country, countries—especially in the core—are also involved directly and indirectly in the agricultural sectors of other countries, primarily those in the periphery. Food as well as agricultural development aid are widespread and popularly accepted ways in which core states intervene in the agricultural sector of peripheral states. Such intervention is one way that peripheral states are incorporated into the global economy. In addition to straightforward food aid, core states also attempt to improve the capacity of the agricultural sector of peripheral states (**Figure 8.20**). Unsuccessful agricultural development projects—whether because of poor design, implementation, or some other reason—as well as successful ones illustrate the many ways in which global forces produce different local consequences and reactions from local people. International development organizations and institutions like the World Bank and the Food and Agriculture Organization (FAO) have been involved in agricultural development projects in the periphery for nearly five decades.

A close look at Latin America helps explain how different international and national agricultural policies can shape productivity in a particular region. While Latin America provides a useful illustration of changing agricultural development policy, its experience is not necessarily the same as that of other regions of the world, whose histories, economies, politics, and culture have shaped the receipt and implementation of development programs in different ways.

Agricultural Change and Development Policies in Latin America

During the first part of the twentieth century, the yields of most agricultural crops in Latin America were very low (less than 1 ton per hectare), and farmers with small plots of land could not produce enough to feed themselves, let alone sell in the market. As population and urban consumption demands increased, countries such as Mexico and Brazil had to import basic food crops, such as wheat and corn. The legacy of large landholdings from the colonial period was compounded by the accumulation of land by the wealthy and by foreign companies in the late nineteenth century. This led to widespread rural poverty, landlessness, and frustration that aided

Figure 8.20 Ostrich rearing project, Kenya The men pictured here are Masai who once were nomadic but are now involved in an international development project focused on ostrich rearing and ecosystem management. The ostriches are a rare species that is under threat because of habitat destruction and bush meat trade.

uprisings such as the Mexican and Cuban revolutions and the election of socialist governments in Chile and Guatemala. In addition, many large landholdings were used for extensive ranching or for export or low-productivity crops and were not contributing to the food needs of the growing urban populations.

Land reform, which is the redistribution of land by the state with a goal of increasing productivity and reducing social unrest, was seen as a solution and was implemented by revolutionary governments and others seeking to reduce the risk of rural uprising. Mexico's post-revolutionary land reform redistributed expropriated and government lands to 52 percent of rural households between 1917 and 1980. In many cases, the land was distributed in the form of **ejidos,** communal lands given to groups of landless peasants who could farm collectively or as individuals but could not rent or sell the land outside the ejido. The government of Bolivia redistributed land to 79 percent of rural households between 1953 and 1975. The socialist governments of Guatemala (1952), Chile (1972), and Nicaragua (1979) distributed land to at least 20 percent of rural households, but some of those lands were subsequently returned to large landholders under military or more conservative governments.

Pressure for land reform continues throughout the region. For example, in Brazil the landless movement *Movimiento sim terra* has forced land redistribution by occupying more than 20 million hectares (nearly 50 million acres) and then demanding legal rights and political change with considerable public support. The question of whether land reform in Latin America has been successful or not is hotly debated. Some believe the reform sector is inefficient and that communal lands should be privatized, while others argue that land reform has increased rural stability and agricultural production. Most have recognized that land reform is ineffective unless it is part of an overall agrarian reform package that also provides technical advice, inputs, credit, and market access to the new landowners.

A second solution to low productivity and poverty in rural areas was the green revolution (see Box 8.2: "A Look at the Green Revolution"). Mexico was a global center for green revolution technology, hosting the International Center for Wheat and Maize Improvement near Mexico City. Scientists at the center, funded by the Rockefeller Foundation as well as the Mexican and U.S. governments, have used advanced plant-breeding techniques to produce new varieties of grains that resist disease and respond to fertilizer and irrigation with very high yields. Farmers, especially in irrigation districts in northern Mexico, were quick to adopt the new crop varieties, and national production of corn and wheat soared, turning Mexico into a major grain exporter by the 1970s. Other Latin American countries, such as Argentina and Brazil, also promoted green revolution agricultural modernization, including other key crops, such as rice and soybeans.

As described in Box 8.2, a second stage of the green revolution is now under way, involving crops engineered using biotechnology to resist pests and diseases and to produce even higher yields. This research is opposed by some who fear unanticipated consequences from such efforts, exemplified by concern that bioengineered corn pollen is harming monarch butterflies in Mexico.

Economic crises, a reduction in government programs, and reduced trade barriers have slowed the progress of the green revolution in many countries, not just in Latin America. Fertilizer use in countries such as Brazil and Mexico has declined with high prices, fewer subsidies, and increased competition from imported corn and wheat, especially from the United States. Many governments have shifted from giving top priority to self-sufficiency in basic grains to encouraging crops that are apparently more competitive in international trade, such as fruit, vegetables, and flowers. These **nontraditional agricultural exports (NTAEs)** (new export crops that contrast with traditional exports, such as sugar and coffee) have become increasingly important in areas of Mexico, Central America, Colombia, and Chile, replacing grain production and traditional exports, such as coffee and cotton. These new crops obtain high prices but also require heavy applications of pesticides and water to meet export-quality standards and fast refrigerated transport to market. They are vulnerable to climatic variation and to the vagaries of the international market, including changing tastes for foods and health scares about pesticide or biological contamination.

Fisheries are another important component of Latin American food and export systems, and activities range from subsistence fisheries in small coastal villages to large-scale commercial exploitation of offshore fisheries. The overall catch was more than 10 million metric tons in 1994, contributing on average about 10 percent of the overall food supply and making a significant contribution to exports in Chile, Ecuador, and Costa Rica. Aquaculture (the cultivation of fish and shellfish under controlled conditions, usually in coastal lagoons) has been growing rapidly and has resulted in the clearing of coastal mangroves and an increase in exports, especially of shrimp from countries such as Honduras. Food production sufficient to meet the needs of its population continues to be a significant political, economic, social, and environmental issue throughout Latin America and other parts of the periphery.

The Organization of the Agro-Food System

While the changes that have occurred in agriculture worldwide are complex, certain elements help reduce the complexity and serve as important indicators of change. Geographers and other scholars interested in contemporary agriculture have noted three prominent and nested forces that signal a dramatic departure from previous forms of agricultural practice: agribusiness, food chains, and integration of agriculture with the manufacturing, service, finance, and trade sectors. Box 8.3: "The New Geography of Food and Agriculture in New Zealand" illustrates the deployment of these practices in one country.

The concept of agribusiness has received a good deal of attention in the last two decades, and in the popular mind it has come to be associated with large corporations, such as ConAgra or DelMonte. Our definition of agribusiness departs from this popular conceptualization. Although multi- and transnational corporations (TNCs) are certainly involved in agribusiness, the concept conveys more than a corporate form. **Agribusiness** is a system rather than a kind of corporate entity. It is a set of economic and political relationships that organizes food production from the development of seeds to the retailing and consumption of the agricultural product. Defining agribusiness as a system, however, does not mean that corporations are not critically important to the food production process. In the core economies the transnational corporation is the dominant player, operating at numerous strategically important stages of the food production process. TNCs have become dominant for a number of reasons, but mostly because of their ability to negotiate the complexities of production and distribution in many geographical locations. That capability requires special knowledge of national, regional, and local regulations and pricing factors.

The concept of a food chain (a special type of commodity chain) is a way to understand the organizational structure of agribusiness as a complex political and economic system of inputs; processing and manufacturing; and outputs. A **food chain** is composed of five central and connected sectors (inputs, production, processing, distribution, and consumption) with four contextual elements acting as external mediating forces (the state, international trade, the physical environment, and credit and finance). **Figure 8.21** illustrates these linkages and relationships, including how state farm policies shape inputs, product prices, the structure of the farm, and even the physical environment.

The food chain concept illustrates the complex connections among producers and consumers and regions and places. For example, important linkages connect cattle production in the Amazon and Mexico, the processing of canned beef along the United States–Mexico border, the availability of frozen hamburger patties in core grocery stores, and the construction of McDonald's restaurants in Moscow. Because of complex food chains such as this, it is now common to find that traditional agricultural practices in peripheral regions have been displaced by expensive, capital-intensive practices (**Figure 8.22**).

That agriculture is not an independent or unique economic activity is not a particularly new realization. Beginning with the second agricultural revolution, agriculture began slowly but inexorably to be transformed by industrial practices. What is different about the current state of the food system is the way in which farming has become just one stage of a complex and multidimensional economic process. This process is as much about distribution and marketing—key elements of the service sector—as it is about growing and processing agricultural products.

Food Regimes

A **food regime** is a specific set of links that exists among food production and consumption and capital investment and accumulation opportunities. Like the agricultural revolutions already described, food regimes have developed out of different historical periods, during which different political and economic forces are in operation. While a food chain describes the complex ways in which specific food items are produced, manufactured, and marketed, the concept also indicates the ways in which a particular type of food item is dominant during a specific temporal period. Although hundreds of food chains may be in operation at any one time, agricultural researchers believe that only one food regime dominates a particular period.

During the decades surrounding the turn of the nineteenth century, an independent system of nation-states emerged and colonization expanded (see Chapters 2 and 9). At the same time, the industrialization of agriculture began. These two forces of political and economic change were critical to the fostering of the first food regime, in which colonies became important sources of exportable foodstuffs by supplying the industrializing European states with cheap food in the form of wheat and meat. The expansion of the colonial agriculture sectors, however, created a crisis in production. The crisis was the result of the higher cost-efficiency of colonial food production, which undercut the prices of domestically produced food, put domestic agricultural workers out of work, and forced members of the agricultural sector in Europe to look for new ways to increase cost-efficiency. The response was to industrialize agriculture, which helped to drive down operating costs and restabilize the sector (reducing even more the need for farm workers) while moving toward the integration of agriculture and industry (also known as agro-industrialization).

While a wheat and livestock food regime characterized global agriculture until the 1960s, researchers now believe that a fresh fruit and vegetable regime has emerged. This new pattern of food consumption and production has been called the "postmodern diet" because it represents an important shift away from grains and meats to the more perishable agro-commodities of fresh fruits and vegetables. Integrated networks of food chains, using integrated networks of refrigeration systems, deliver fresh fruits and vegetables from all over the world to the core regions of Western Europe, North America, and Japan. Echoing the former food networks that characterized nineteenth-century imperialism, peripheral production systems supply

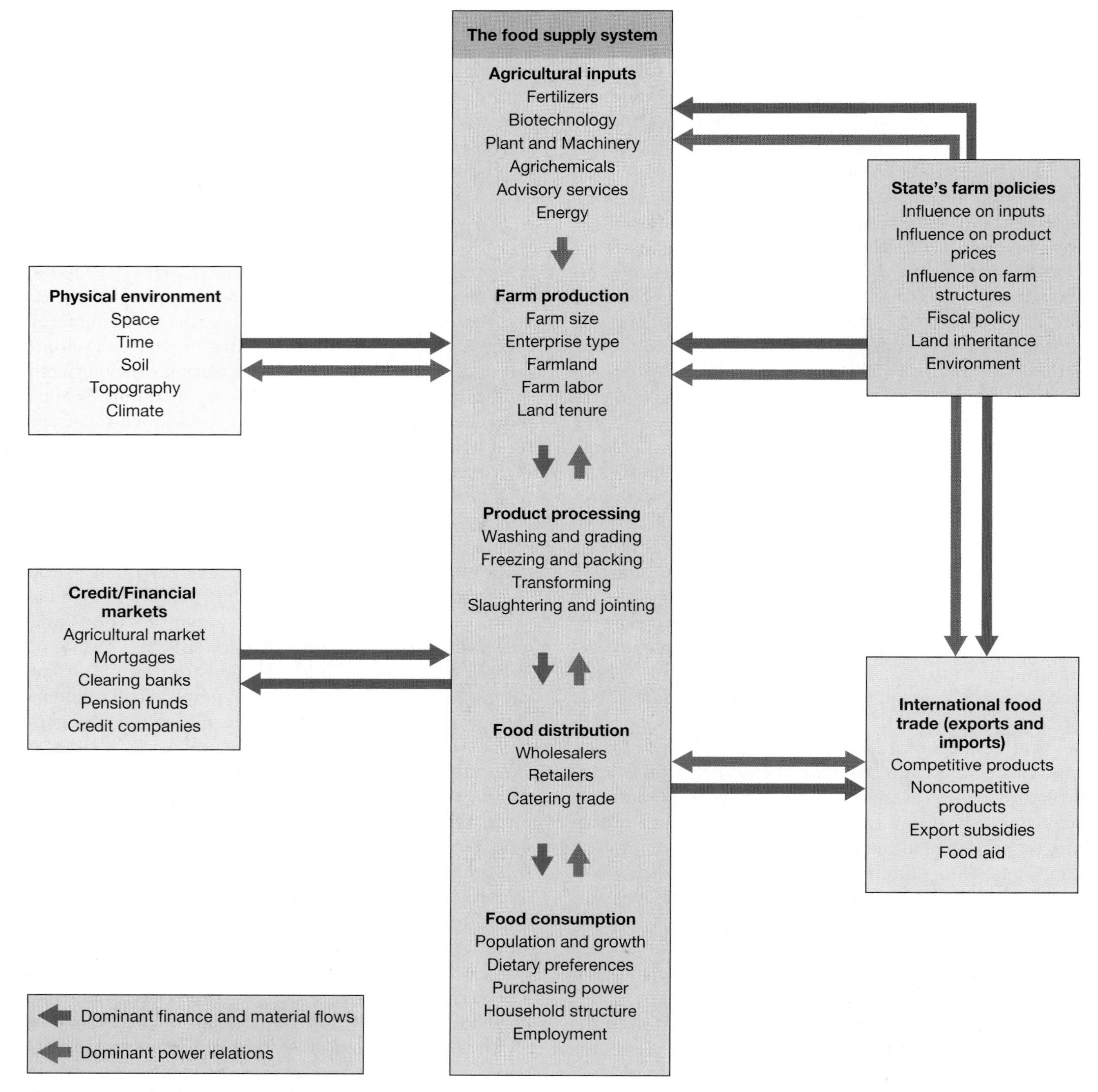

Figure 8.21 The food-supply chain The production of food has been transformed by industrialization into a complex system that comprises distinctly separate and hierarchically organized sectors. Mediating forces (the state, the structure and processes of international trade, credit and finance arrangements, and the physical environment) influence how the system operates at all scales of social and geographical resolution. (After I. Bowler (ed.), *The Geography of Agriculture in Developed Market Economies.* New York: J. Wiley & Sons, 1992, p. 12.)

core consumers with fresh, often exotic and off-season produce. Indeed, consumers in the core regions have come to expect the full range of fruits and vegetables to be available year-round in their produce sections, and unusual and exotic produce has become increasingly popular.

This emergence of a new food regime based on fresh fruit and vegetables has been helped by retailers, who provide symbolic cues and incentives to shoppers to consume the more exotic products. Store managers have introduced them by providing associations between the fruit or vegetable and prevailing ideas about health, class attachment, and epicurean eating. Thus, the transformation of agricultural practices at the global level has enabled the emergence of a new food regime, accompanied by new cultural

Figure 8.22 Kansas City Stockyards Grain-fed beef have become characteristic of beef production not only in the core but also in the periphery. Grain-fed beef is far more expensive to raise than cattle that graze in open fields on grasses. The grain itself must be produced through applications of seed, fertilizers, and water and then harvested and processed. Grasses grow naturally with little or no commercial inputs. The cattle pictured here come from all over the U.S. Midwest and are offered for auction at the Stockyards.

messages that promote and persuade at the local level. Furthermore, just as traditional agricultural practices worldwide have been affected by globalization trends, so too are the mainstream eating habits of consumers in core as well as peripheral regions.

SOCIAL AND TECHNOLOGICAL CHANGE IN GLOBAL FOOD PRODUCTION

In preceding sections of this chapter, we have tried to show how the globalization of agriculture has been accomplished through the same kinds of political and economic restructuring that have characterized the globalization of industry. Technological change was of particular importance to agriculture over the last half of the twentieth century as mechanical, chemical, and biological revolutions altered even the most fundamental agricultural practices. And just as restructuring in industry has not occurred without innumerable rounds of adjustment and resistance, the same is true of agriculture.

Besides generating economic competition, the newly restructured agro-commodity production system also fosters conflict and competition within sociocultural systems. For instance, in core and peripheral locations, men and women, landowners and peasants, different tribal groups, corporations, and family farmers struggle to establish or maintain control over production and over ways of life.

Two Examples of Social Change

The impact of a government development scheme to introduce irrigated rice production into the Gambia River Basin illustrates the many ways globalization of agricultural production has affected gender relations among and within households. The Gambian government, with the help of the West African Rice Development Association, launched a program in the 1980s to grow rice along the banks of the Gambia River.[3] The objective of the project was for Gambia to develop its own rice-producing sector and thereby decrease its dependence on imported rice. Through local agents employed by the project, the government distributed a package of high-yielding rice varieties, fertilizers, and pesticides in the hope that 2,000 peasant households distributed among 70 villages could attempt a double-cropping rice cultivation program (**Figure 8.23**). Husbands and wives were involved in the project.

The success of the project required a redistribution of labor, as well as the restructuring of land and crop rights. Incorrect assumptions that women's labor was constantly available and that there was no cost to that labor were made, however. These incorrect assumptions led to serious problems between spouses when it became apparent that women were not free to work during the season in which they were most needed to participate in the development scheme and that their participation required

[3]Adapted from J. Carney, "Converting the Wetlands, Engendering the Environment: The Intersection of Gender with Agrarian Change in the Gambia." *Economic Geography* 69(4), 1993, pp. 329–349.

Figure 8.23 Gambian women harvesting rice Development schemes have created problems within Gambian households because women wanted to control their own time and labor to produce rice for family consumption and local markets.

them to forgo other opportunities to increase household financial resources. Husbands and wives also disagreed about who controlled which pieces of land and the crops that were harvested. Thus, traditional ways of farming as well as gender relations were significantly challenged. The disagreements became so severe that the success of the project was compromised.

Lest it seem that only the economies and societies of the periphery are affected by the globalization of agriculture, a case closer to home can be instructive. In the 1980s the United States experienced what has become known as the **farm crisis,** the financial failure and eventual foreclosure of thousands of family farms across the Midwest. The agricultural sector of the state of Nebraska, highly specialized in grain production, was particularly hard hit when the international grain market collapsed in 1985. Soaring land prices, a decline in manufacturing employment, a rapid rise in farm bankruptcies, and decreasing revenues from crops sent entrepreneurs looking for ways to overcome the impact of this event on the state's economy. Meatpacking was identified as an alternative generator of economic growth. State-provided tax abatements and other corporate incentives caused IBP, Inc., a giant food conglomerate, to locate a meatpacking plant in Lexington, which at the time was a small rural town of 6,601 people.[4]

IBP, Inc., opened its plant at the end of 1992 with more than 2,000 workers employed. Most of these workers were immigrants, primarily Mexican and Central American men and women who were actively recruited by the company as a cost-reduction strategy. Such strategies are widespread among large American meatpacking companies. The arrival of so many new residents to the town created social and economic problems, however, which led to ethnic tension between the established residents—mostly of European descent—and the newcomers.

Also as a result of the new plant, the town experienced a severe housing crunch; had to build its first homeless shelter; had to raise new monies to expand the capacities of the school system; and had to build a new and larger jail. A rapid increase in the number of births, especially those not covered by health insurance, also occurred. In 1993 Lexington had the highest crime rate in Nebraska.

Examining the farm crisis is a useful way of demonstrating how core economies—especially farm households—experienced and responded to changes in U.S. agriculture.

Biotechnology Techniques in Agriculture

Ever since the nineteenth century, when Austrian botanist Gregor Mendel identified hereditary traits in plants and French chemist Louis Pasteur explained fermentation, the manipulation and management of biological organisms has been of central importance to the development of agriculture. The most recent manifestation of the influence of science over agriculture is exemplified by biotechnology. **Biotechnology** is any technique that uses living organisms (or parts of organisms) to make or modify products, to improve plants and animals, or to develop microorganisms for specific uses. Recombinant DNA techniques, tissue culture, cell fusion, enzyme and fermentation technology, and embryo transfer are some of the most-talked-about aspects of the use of biotechnology in agriculture (**Figure 8.24**).

The most common argument for applying biotechnology to agriculture is the belief that it helps reduce agricultural production costs, as well as acting as a kind of resource-management technique (where certain natural

[4]Adapted from L. Gouveia, "Global Strategies and Local Linkages: The Case of the U.S. Meatpacking Industry," in A. Bonanno et al. (eds.), *From Columbus to ConAgra: The Globalization of Agriculture and Food.* Lawrence: University of Kansas Press, 1994, pp. 125–148.

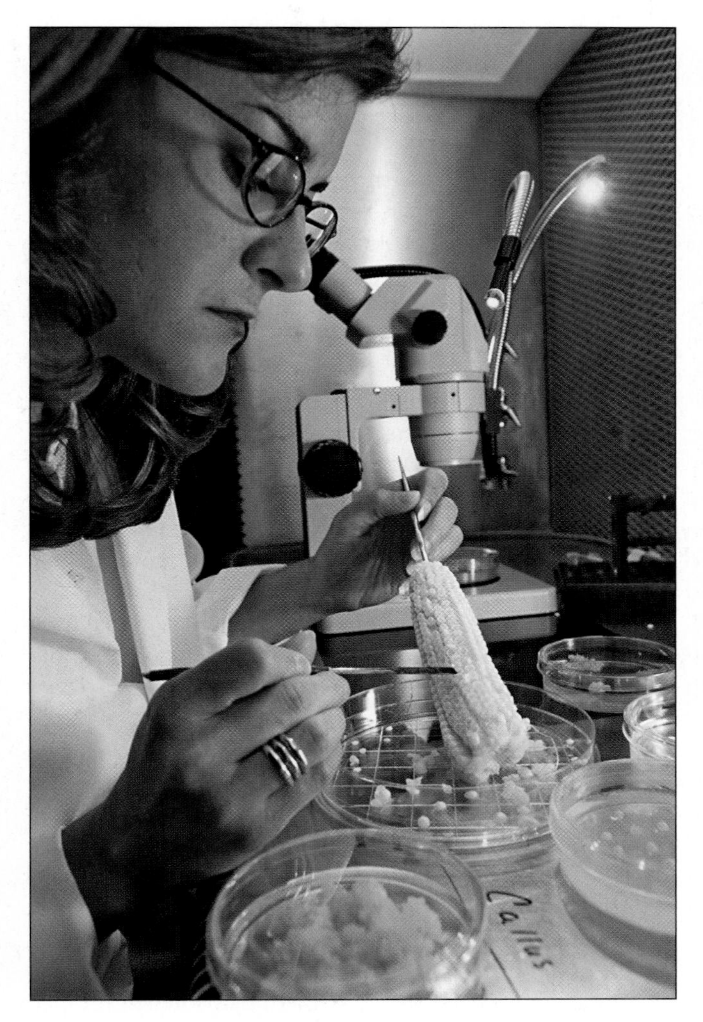

Figure 8.24 Biotechnology Biotechnology laboratories are typically high-technology greenhouses where plants are raised under carefully monitored conditions, from the implanting of special seeds to the applications of fertilizers and water and light. Biotechnology offers both benefits and costs. While the benefits include increased yields and more pest-resistant strains, too often the costs of such technology are too high for the world's neediest populations. In this photograph a Monsanto Corporation worker is extracting corn embryo for the development of a genetically modified crop.

resources are replaced by manufactured ones). Biotechnology has been hailed as a way to address growing concern for the rising costs of cash crop production; surpluses and spoilage; environmental degradation from chemical fertilizers and overuse; soil depletion; and related challenges now facing profitable agricultural production.

Indeed, biotechnology has provided impressive responses to these and other challenges. For example, biotechnological research is responsible for the development of "super plants" that produce their own fertilizers and pesticides, can be grown on nutrient-lacking soils, are high-yielding varieties, and are resistant to disease or the development of microorganisms. Additionally, biotechnologists have been able to clone plants—that is, take cells of tissues from one plant and use them to form new plants. The tissue culture may be no more than one cubic centimeter in size, but it has the potential to produce millions of identical plants. Such a procedure decreases the time needed to grow mature plants ready for reproduction.

While such technological innovations can seem miraculous, there is a downside to biotechnological solutions to agricultural problems. For example, cloned plants are more susceptible to disease than natural ones, probably because they have not developed tolerances. This susceptibility leads to an increasing need for chemical treatment. And while industry may reap economic benefits from the development and wide use of tissue cultures, farmers may suffer because they lack the capital or the knowledge to participate in biotechnological applications.

Biotechnology has truly revolutionized traditional agriculture. Its proponents argue that it provides a new pathway to the sustainable production of agricultural commodities. By streamlining the growth process with such innovations as tissue cultures, disease- and pest-resistant plants, and fertilizer-independent plants, optimists believe that the biorevolution can maximize global agricultural production to keep up with global requirements of population and demand.

Just as with the green revolution, however, biotechnology may have deleterious effects on peripheral countries (and on poor laborers and small farmers in core countries). For example, biotechnology has enabled the development of plants that can be grown outside of their natural or currently most suitable environment. Yet cash crops are critical to economic stability for many peripheral nations—such as bananas in Central America and the Caribbean, sugar in Cuba, and coffee in Colombia and Ethiopia (**Figure 8.25**). These and other export crops are

Figure 8.25 Workers in an Ethiopian coffee plantation harvesting beans For many peripheral countries, the production of cash crops is a way to boost exports and bring in needed income for the national economy. In Ethiopia, coffee has for decades been a cash crop grown for export. Luxury exports such as coffee generate some of the capital needed to import staple foods such as wheat.

8.3

WINDOW ON THE WORLD

The New Geography of Food and Agriculture in New Zealand

Geographers who study agriculture and food are using several new approaches to understand the way the restructuring of international trade, the activities of transnational corporations, and rapid shifts in government policies are affecting agriculture and the rural landscape in different countries. For example, agriculture has been transformed through horizontal integration in which smaller enterprises are merged to create larger units (for example, when adjacent farms are consolidated into one large landholding, resulting in the disappearance of small family-run farms) and through vertical integration, in which a single firm takes control of several stages in the production process (when a company owns the fertilizer and seed companies as well as the food-processing plant and supermarkets). The international corporation ConAgra, which owns grain companies, feedlots, meatprocessing, and wholesale distribution facilities is an agribusiness that organizes food production from the manufacturing of chemical inputs and the genetic manipulation of animal breeds or crop varieties to the processing, retailing, and consumption of the agricultural product.

Geographers such as Richard Le Heron and Guy Robinson have written extensively about how New Zealand agriculture has changed in response to the restructuring of the global food system. They document how New Zealand's agricultural system evolved during the nineteenth century with an orientation to exports of wool and lamb based on a pastoral landscape and a guaranteed market in the core economy of the United Kingdom. After World War II a second regime developed that included dairy cows on small farms and processing of products such as butter for export using refrigerated shipping. By the mid-twentieth century the New Zealand government was heavily involved in the agricultural system through "marketing boards" that mediated farmers' relationship with international markets through quality controls, price supports, and marketing.

In the 1970s the shock of the oil crisis (increasing the cost of agricultural inputs) and the loss of the imperial preference market when Britain joined the European Community resulted in further state support for producers, with price supports, incentives, and subsidies for inputs such as fertilizers providing more than a third of farm revenues. However, even these institutional supports could not fully buffer farmers against the increasing cost of inputs and loss of markets for the staples of wool, meat, and dairy, and some farmers began to diversify into nontraditional exports, such as venison, produced on deer farms, and fruit such as kiwi and Asian pears, responding to a new global food regime of specialty foods and the export of fruit and vegetables (**Figure 8.H** and **Figure 8.I**).

A dramatic change in agricultural policies in 1984 abruptly removed most price supports, trade protections, and farm subsidies, and required farms to pay for extension services, water, and quality inspections. Farm income fell by up to half, debt increased, 10 percent of farms were sold, herds were significantly reduced, and 10,000 farmers protested in front of Parliament. New Zealand agriculture was thrown into a global free market and the full impact of what has been called the "international farm crisis," while most other developed countries, including the United States, Canada, and those in Europe, maintained considerable state regulation and support for their agricultural systems. Although New Zealand farmers coped by adjusting herd sizes and changing crop mixes, some went out of business and their properties were horizontally integrated into larger farms. But New Zealand was also one of the first countries to adopt certification for organic agricultural products, and there is a thriving domestic market for

threatened by the development of alternate sites of production. Transformations in agriculture have ripple effects throughout the world system. As an illustration, **Table 8.1** compares the impacts of the biorevolution and the green revolution on various aspects of global agricultural production.

The availability of technology to peripheral nations is limited because most advances in biotechnology are the property of private companies. For example, patents protect both the process and the end products of biotechnological techniques. Utilizing biotechnological techniques requires paying fees for permission to use them, and the small farmers of both the core and the periphery are unlikely to be able to purchase or utilize the patented processes. The result of private ownership of biotechnological processes is that control over food production is removed from the farmer and put into the hands of biotechnology firms. Under such circumstances

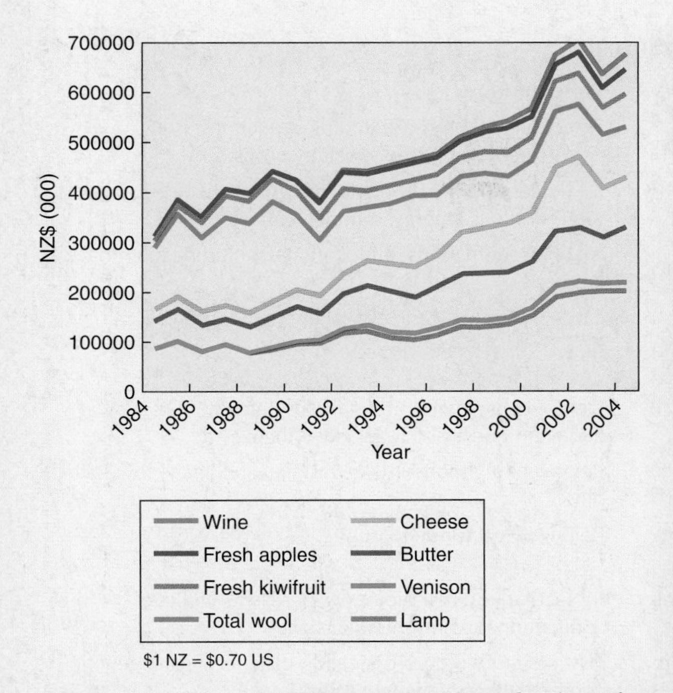

Figure 8.H New Zealand agricultural production The structure of New Zealand agricultural exports has changed in response to the restructuring of global agricultural trade and demand. Although the traditional exports of lamb meat and butter have remained high, wool exports have fallen. Exports of nontraditional crops such as kiwi, berries, avocados, wine, and venison have increased considerably.

Figure 8.I Kiwi production These kiwi orchards on the North Island of New Zealand are surrounded by lines of trees that protect the delicate fruit from strong winds.

sustainably grown foods. Transnational agribusiness firms, such as H.J. Heinz, purchased New Zealand agricultural processing enterprises with the goal of supplying growing Asian markets. Michael Moore, the New Zealander who had spearheaded his country's plunge into the free market as minister of trade, became the head of the World Trade Organization, charged with reducing barriers to trade worldwide.

Many geographers are reexamining agriculture in the context of global restructuring and changing government policies and contributing important insights into the ways in which the new international geography of food and agriculture is changing the economy and environments of countries such as New Zealand. The geographic perspective allows us to link international trade, regulation, and corporations to the decisions of national and local governments and to the impacts on and responses of agricultural regions, communities, and farm families.

it becomes possible for world food security to be controlled not by publicly accountable governments, but by privately held biotechnology firms. Finally, with the refinement and specialization of plant and animal species, laborers who are currently employed in ancillary activities could face the loss of their jobs. For example, if a grower chooses to plant a bioengineered type of wheat that does not require winnowing (the removal of the chaff, a normally labor-intensive process), laborers who once were involved in that activity are no longer needed.

The biorevolution in agriculture is so recent that we are just beginning to understand both its negative and positive impacts. At this point it seems quite clear that these impacts will be distributed unevenly across countries, regions, and places and certainly across class, race, and gender lines.

TABLE 8.1 Biorevolution Compared with Green Revolution

Characteristics	Green Revolution	Biorevolution
Crops affected	Wheat, rice, maize	Potentially all crops, including vegetables, fruits, agro-export crops, and specialty crops
Other sectors affected	None	Pesticides, animal products, pharmaceuticals, processed food products, energy, mining, warfare
Territories affected	Some developing countries	All areas, all nations, all locations, including marginal lands
Development of technology and dissemination	Largely public or quasi-public sector, international agricultural research centers (IARCs), R&D millions of dollars	Largely private sector, especially corporations, R&D billions of dollars
Proprietary considerations	Plant breeders' rights and patents generally not relevant	Genes, cells, plants, and animals patentable as well as the techniques used to produce them
Capital costs of research	Relatively low	Relatively high for some techniques, relatively low for others
Access to information	Restricted due to privatization and proprietary considerations	Relatively easy, due to public policy of IARCs
Research skills required	Conventional plant breeding and parallel agricultural sciences	Molecular and cell biology expertise as well as conventional plant-breeding skills
Crop vulnerability	High-yielding varieties relatively uniform; high vulnerability	Tissue culture crop propagation produces exact genetic copies; even more vulnerability
Side effects	Increased monoculture and use of farm chemicals, marginalization of small farmer, ecological degradation. Increased foreign debt due to decrease in biomass fuels and the increasing reliance on costly, usually imported, petroleum	Crop substitution replacing Third World exports; herbicide tolerance; increasing use of chemicals; engineered organisms might affect environment; further marginalization of small farmer

Source: Adapted from M. Kenney and F. Buttel, "Biotechnology: Prospects and Dilemmas for Third-World Development," *Development and Change* 16 (1995): 70; and H. Hobbelink, *Biotechnology and the Future of World Agriculture: The Fourth Resource* (London: Red Books, 1991).

THE ENVIRONMENT AND AGRICULTURAL INDUSTRIALIZATION

Agriculture always involves the interaction of biophysical as well as human systems. In fact, this relationship makes agriculture distinct from forms of economic activity that do not depend so directly on the environment. This relationship also requires determining how best to manage the environment in order to facilitate the continued production of food. Because the relationships between the human system of agriculture and the biophysical system of the environment are highly interactive, it is important to look at the ways each shapes the other.

The Impact of the Environment on Agriculture

Farmers have increasingly managed the environment over the course of the three agricultural revolutions. In fact, the widespread use of fertilizers, irrigation systems, pesticides, herbicides, and industrial greenhouses suggests that agriculture has become an economic practice that can ignore the limitations of the physical environment (**Figure 8.26**). Yet it is exactly because agriculture is an economic activity that management of the environment in which it occurs becomes critical. As geographer Martin Parry writes:

> Soil, terrain, water, weather and pests can be modified and many of the activities through the farming year, such as tillage and spraying, are directed toward this. But these activities must be cost-effective; the benefits of growing a particular crop, or increasing its yield by fertilizing, must exceed the costs of doing so. Often such practices are simply not economic, with the result that factors such as soil quality, terrain and climate continue to affect agriculture by limiting the range of crops and animals that can profitably be farmed. In this way the physical environment still effectively limits the range of agricultural activities open to the farmer at each location.[5]

Though the impact of the environment on industrialized agricultural practices may not at first seem obvious, the reverse is more readily observable. In fact, there are

[5]M. Parry, "Agriculture as a Resource System," in I. Bowler (ed.), *The Geography of Agriculture in Developed Market Economies*. Harlow, England: Longman Scientific and Technical, 1992, p. 208.

Figure 8.26 Modern irrigation system A self-propelling irrigation system can be electronically programmed to deliver different amounts of water at different times of the day or days of the week. Irrigation is just one way that humans can alter the environment to serve their agricultural needs. In many parts of the core, water prices are heavily subsidized for agricultural users in order to ensure food supplies. For many parts of the periphery, however, access to water is limited to the amount of rain that falls and can be stored behind small dams and in impoundments. (After Agricultural Research Service, USDA.)

Figure 8.27 Poisoned crane, Hungary In spring 2005, carcasses of more than hundred cranes were found in an area 150 kilometers southeast of Budapest. Experts say the birds of passage on their way to the northern regions of Europe were apparently poisoned by eating grain containing toxic pesticides.

many contemporary and historical examples of the ways that agriculture destroys, depletes, or degrades the environmental resources on which its existence and profitability depend.

The Impact of Agriculture on the Environment

As discussed in Chapter 4, one of the earliest treatises on the impact of chemical pesticides on the environment was Rachel Carson's *Silent Spring,* which identified the detrimental impacts of synthetic chemical pesticides—especially DDT—on the health of human and animal populations (**Figure 8.27**). Although the publication of the book and the environmental awareness that it generated led to a ban on the use of many pesticides in most industrialized nations, chemical companies continued to produce and market them in peripheral countries. While some of these pesticides were effective in combating malaria and other insect-borne diseases, many were applied to crops that were later sold in the markets of developed countries. Thus, a kind of "circle of poison" was set into motion, encompassing the entire global agricultural system.

One of the most pressing issues facing agricultural producers today is soil degradation and denudation, which are occurring at rates more than a thousand times natural rates. Although we in the United States tend to dismiss soil problems such as erosion as an artifact of the 1930s Dust Bowl, the effects of agriculture on worldwide soil resources are dramatic, as **Table 8.2** illustrates. Unfortunately, most forms of agriculture tend to increase soil degradation. Although severe problems of soil degradation persist in the United States—which has a federal agency devoted exclusively to managing soil conservation—more severe problems are occurring in peripheral countries.

The loss of topsoil worldwide is a critical problem because it is a fixed resource that cannot be readily replaced. It takes, on average, 100 to 500 years to generate 10 millimeters (one-half inch) of topsoil, and it is estimated that nearly 50,000 million metric tons (55,000 million tons) of topsoil are lost each year to erosion. The quantity and quality of soil worldwide are thus important determining factors in the quantity and quality of food that can be produced.

Soil erosion due to mismanagement in the semiarid regions of the world has led to desertification, in which topsoil and vegetation losses have been extensive and largely permanent. As explained in Chapter 4, desertification is the spread of desertlike conditions in arid or semiarid lands resulting from climatic change or human influences. Desertification means not only the loss of topsoil but possibly the deterioration of grazing lands and the decimation of forests (**Figure 8.28**). In addition to causing soil degradation and denudation problems, agriculture affects water quality and quantity through the over-withdrawal of groundwater and the pollution of the same water through agricultural runoff contaminated with herbicides, pesticides, and fertilizers. Deforestation can also result from poor agricultural practices.

Poor land-use practices and the destruction of complex ecosystems through over- or misuse led in the 1980s

TABLE 8.2 Global Degraded Soils (million hectares)

Region	Overgrazing	Deforestation	Agricultural Mismanagement	Other	Total	Degraded Areas as Share of Total Vegetated Land
Asia	197	298	204	47	746	20%
Africa	243	67	121	63	494	22%
South America	68	100	64	12	244	14%
Europe	50	84	64	22	220	23%
North and Central America	38	18	91	71	158	8%
Australia, New Zealand, and the South Pacific	83	12	8	0	103	13%
World	679	579	552	155	1,965	17%

Source: L. R. Brown et al., *State of the World.* New York: W. W. Norton and Company, 1994, p. 10.

to an innovation called "debt-for-nature swaps." In these swaps a core environmental organization, such as the World Wildlife Fund, retired some part of the foreign debt of a peripheral country, contingent upon the country's agreeing to implement a conservation program to save ecologically sensitive lands from abuse. This usually meant turning the land into a national park or extending the boundaries of an existing park. Funds generated by the swaps were to be used to administer the parks, train personnel, research habitats, and carry out environmental education.

For much of the 1980s, environmental organizations were extremely optimistic about the possible implications of debt-for-nature swaps for both the affected country and worldwide environmental degradation. It turns out, however, that the swaps were not able to address the fundamental causes of environmental degradation in peripheral regions—including extreme poverty, government subsidies for forest clearing, and insecure land tenure. As a result, the debt-for-nature swaps are now seen as mere Band-Aid solutions to extremely complex social, economic, political, and ecological problems.

The nature-society relationship discussed in Chapter 4 is very much at the heart of agricultural practices. Yet as agriculture has industrialized, its impacts on the environment have multiplied and in some parts of the globe have reached crisis stage. While in some regions the agricultural system leads to overproduction of foodstuffs, in others the quantity and quality of water and soil severely limit the ability of a region's people to feed themselves.

Figure 8.28 Desertification in Sahelian Africa Severe and largely permanent loss of vegetation and topsoil may result from human activities such as overgrazing or excessive deforestation. The ravaged landscapes of desertification are a compelling testimony to the need for humans to consider the implications of their actions more closely—not always an easy thing to do when ill-informed government policies and grinding hunger and poverty are daily facts of life.

PROBLEMS AND PROSPECTS IN THE GLOBAL FOOD SYSTEM

The future of the global food system is being shaped at this very moment in food science laboratories, in corporate boardrooms, on the street in organized protests, and in homely settlements throughout the world. The biggest issues food policy experts, national governments, consumers, and agriculturalists face revolve around the availability and quality of food in a world where access to safe, healthy, and nutritious foodstuffs is unevenly distributed. For the periphery, the most pressing concern is adequate food supplies to feed growing populations. For the core, there are concerns about food quality in a system that is increasingly industrialized and biologically engineered. Sometimes the solutions to one problem for one population become a new problem for another population.

In this final section we examine two problematic issues in the world food system, as well as an encouraging prospect. These cases certainly do not illuminate the myriad challenges and possibilities facing food producers and policymakers today, but they do provide a broad sense of the range.

Famine and Undernutrition

We have spent most of this chapter describing the ways food is cultivated, processed, engineered, marketed, financed, and consumed throughout the world. What we have yet to do is talk about access to this most essential of resources. While there is more than enough food to feed all the people who inhabit Earth, access to food is uneven, and many millions of individuals in both the core and the periphery have had their lives shortened or harmed because war, poverty, or natural disaster has prevented them from securing adequate nutrition. In fact, hunger is very likely the most pressing problem facing the world today.

Hunger occurs in two basic ways: chronic or acute. Chronic hunger is nutritional deprivation that occurs over a sustained period of time: months or even years. Acute hunger is short term and is often related to catastrophic events—personal or systemic. Chronic hunger, also known as **undernutrition,** is the inadequate intake of one or more nutrients and/or of calories. Undernutrition can occur in individuals of all ages, but its effects on children are dramatic, leading to stunted growth, inadequate brain development, and a host of other serious physical ailments.

Perhaps the most widely publicized examples of acute hunger are famines, especially those that have occurred in parts of the periphery over the last few decades. **Famine** is acute starvation associated with a sharp increase in mortality. The most widely publicized famines of the late twentieth century occurred in Bangladesh in 1974 and Ethiopia in 1984–1985. The causes of these two famines (and the other twentieth-century famines that preceded them) were complex. The crisis of starving people so often publicized by the news media is usually just the final stage of a process that has been unfolding for a far longer period, sometimes years or even decades. Experts who study famine argue that there are at least two critical factors behind longstanding vulnerability to famine. The first has to do with a population's command over food resources in terms of their livelihood (for instance, rural laborer or farmer). The second has to do with a trigger mechanism, which may be a natural phenomenon like drought or a human-made situation such as civil war.

The famine in Bangladesh resulted from a sharp rise in food prices combined with a drop in employment opportunities that resulted in part from a very serious flood that had wiped out crops in lowland areas (but left highland agricultural production intact). Adequate food supplies were still available, but individuals who labored in the countryside to produce, distribute, or process the lowland crops could not afford them. Exacerbating the problem was the fact that negotiations between the Bangladeshi government and foreign-aid donors had broken down to such an extent that the distribution of food aid to needy individuals was seriously impaired. The lesson to be drawn from the Bangladesh famine is that when access to livelihood collapses, famine can result, even when overall levels of food availability are adequate.

People who study famine and other forms of hunger have come to conceptualize vulnerability in terms of the notion of food security. **Food security** means that a person, a household, or even a country has assured access to enough food at all times to ensure active and healthy lives. And while famine is a dramatic reminder of the precarious nature of food security, it is important to appreciate that chronic hunger resulting from food insecurity is a far more widespread and devastating problem than famine, which tends to be shorter in duration and more contained geographically.

Globally, 24,000 people a day die from the complications brought about by undernutrition. Today in the United States, where food is abundant and overeating is a national problem, 10 percent of the population is undernourished or experiences food-security problems at one time or another each year. In the periphery, where food availability is more limited than in core countries like the United States, undernutrition is far more pervasive. Food-security experts agree that undernutrition is the result of an overarching set of factors ultimately caused by poverty. Wealthy people, whether they live in the core or the periphery, rarely if ever go hungry. But impoverished people everywhere are likely to have experienced short-term or long-term undernutrition, if not both. Thus, in a world where there is more than enough food to adequately and consistently nourish every man, woman, and child, many go hungry because they simply do not possess the livelihoods adequate to gain sufficient access to food resources. Furthermore, as the example of the famine in Hausaland in Nigeria in 1982 made clear, in some parts of the world, among some social classes, there are higher levels of undernutrition among women and girls than among men and boys. This is largely because different cultural and social norms favor men and boys, who eat first, leaving the leftovers for women, or who eat certain high-status foods, such as proteins like meat or fish, that women are not allowed to eat.

The most important point to take away from this discussion of hunger is that it is a problem that can be solved. Because neither short-term nor long-term hunger involves inadequate supply, the solution must lie in improving access to those supplies. This could occur, under a radical scenario, through a massive redistribution of wealth that would give all the world's people access to the same amount of food resources. Barring such a dramatic restructuring of the current economic system, the solution to the problem of hunger lies in improving access to livelihoods that pay well enough that adequate nutrition becomes a human right and not dependent upon the vagaries of economic or natural systems.

Genetically Modified Organisms and the Global Food System

In 1999 protesters in London, Chicago, and Seattle dressed up as monarch butterflies and informed passersby that genetically modified corn might be posing a threat to that insect. Shouting chants like, "Hey, hey, ho, ho, Frankenfoods have got to go!" these protesters represented a growing global movement that opposes the production and sale of genetically modified food products—or demands, at the very least, that these foods be labeled as such (**Figure 8.29**). In Brazil a federal judge banned the sale of Monsanto Corporation's Ready Roundup genetically engineered soybean seeds. Japan recently announced it would require that all genetically engineered food be labeled. In the United States, however, the Food and Drug Administration has made regulation of genetically engineered food voluntary: Companies can decide for themselves whether they want the agency to review their product. Since these early protests, popular opposition to GMOs has continued to grow.

A **genetically modified organism,** or **GMO** as it is more commonly known, is any organism that has had its DNA modified in a laboratory rather than through cross-pollination or other forms of evolution. Examples of GMOs are a bell pepper with DNA from a fish added to make it more drought-tolerant, a potato that releases its own pesticide, and a soybean that has been engineered to resist fungus.

Genetic modification has both critics and supporters. Proponents argue that it allows great advances in agriculture (for instance, making plants more resistant to certain diseases or of water shortages), as well as allowing other beneficial creations, such as the petroleum-eating bacteria that can help clean up oil spills. Opponents worry that genetically modified organisms may have unexpected and irreversible effects on human health and the environment, causing maturation problems in children or mutant plant and animal species.

In the United States genetic modification is permitted, on the principle that there is no evidence yet that it is dangerous. GMO foods are fairly common in the United States, and estimates of their market saturation vary widely. It is not, however, easy to recognize GMOs in the grocery store, as there are no labeling requirements. While the food-safety establishment in the United States maintains that GMOs are safe until proved otherwise, countries in Europe have taken the opposite position: that genetic modification has not been proved safe, so they will not accept genetically modified food from the United States or any other country. The World Trade Organization has determined that not allowing GMO food into a country creates an unnecessary obstacle to international trade.

GMOs represent perhaps the most highly technological way to date in which nature and society come together in the global food system. At present, so little is known about the wider impacts of GMOs on human health, the environment, or even the wider global economic system that it is difficult to sort the costs from the benefits of their increasing incorporation into global food production. What is clear is that genetic modification is no passing fad and the debate cannot be reduced to a simple "good" or "bad." While GMOs are neither entirely evil nor entirely good, certain applications may be widely beneficial, while others may not. Regulatory structures are crucial to protecting human health and the environment, as well as spreading whatever benefits may accrue from GMOs beyond the core and into the periphery, where food-security issues are at their most critical. But regulatory structures are not so easily accomplished.

Protests against GMO regulatory structures have been most effective in Europe and parts of Asia, Africa, and Latin America as well as in Canada and Mexico. In those regions, national governments have begun to devise regulations to control or publicize the entry of GMO commodities into the food system or are requiring more research in order to better understand the long-term effects of consuming GMOs on humans as well as the impact of GMOs on the food chain. While widespread protests against GMO foods have also been organized in the United States, the government has not responded with strong support for these movements. Although the U.S. govern-

Figure 8.29 Protest over GMOs in Seattle Protesters at the World Trade Organization meeting in Seattle, Washington in 1999, expressed their concern that genetically modified organisms may not be safe for humans or the environment.

ment has passed domestic legislation making the labeling of GMOs voluntary, at the international level it has threatened action against countries who are arguing for the mandatory labeling of GMO foods traded in the global marketplace. It is important to be aware that the position taken by the United States is a pro-trade one, rather than a strictly pro-GMO food one. In short, the U.S. government is mostly concerned with the economic impact GMO labeling is likely to have on trade and is reluctant to put the trade of U.S. agricultural products at risk. This strategy occurs because the United States is the largest exporter of agricultural goods in the world as well as the largest producer of GMO foods. If GMO labeling were mandated by the World Trade Organization, it is likely that many U.S. trading partners—both large and small—would be legally bound either to refuse GMO commodities or the commodities would be seen as less valuable to potential buyers. At present, there is no sign that the WTO is likely to take such a step. In fact, its current regulations specify that if a country imposes additional restrictions and safety concerns on products that are thought to be unsafe, the WTO can overrule that country. Issues such as this one threaten the independence of individual nations and create governance problems for the WTO. **Table 8.3** provides some insight into the national variation in response to GMO crops.

Global debate and activism—both popular and governmental—over GMOs is still in its early stages. Some of the most vocal opponents of GMOs are concerned that engineered food is destined for consumption by poor people in the periphery, while "real" food, produced by artisans and organic growers will only be available to rich people in the core with no attention paid to the importance of food to reducing world hunger or to ensuring a safe diet as a global human right.

Urban Agriculture

Although most people think of agriculture as a rural activity, urban agriculture made possible the emergence of the world's first cities. Until recently, however, urban agriculture was largely ignored in the development of urban economic policies, apparently because produce generated from it was seen to belong to the informal sector of the local economy and not significant in terms of income-generating potential. Most definitions of **urban agriculture** take it to mean the establishment or performance of agricultural practices in or near an urban or citylike setting. In countries like China, official policies have long recognized and even fostered urban agricultural practices. In many core countries, however, particularly since the Industrial Revolution, urban agriculture has been officially discouraged or made difficult as arable land has been taken for real estate development or seriously degraded through industrial processes.

Whether encouraged or discouraged by official policy, urban residents across the globe are increasing their participation in growing crops and raising livestock, for reasons ranging from food security to income production to taste and health concerns. Up to 30 percent of agricultural production occurs within metropolitan areas in the United States, for example, and up to 15 percent globally. Throughout the core, urban agriculture is largely a leisure activity that helps to supplement the routine purchase of commercial foodstuffs. In the periphery, however, it can often be the sole means of economic and personal survival. As wage cuts, inflation, job loss, civil strife, and natural disasters become more frequent, urban agriculture in the periphery has become a way to address greater food insecurity.

As we discuss in Chapter 10, urban populations throughout the world are growing more than twice as fast as rural populations. By the year 2015, according to the U.N. Center for Human Settlements, more than half of humanity will live in cities. And as development experts look to sustainable development as a way of maintaining economic growth without destroying the environment, urban agriculture has increasingly drawn their attention as a way of making cities sustainable. Proponents of urban agriculture contend that it should not be understood as an alternative to conventional agriculture but rather as a supplementary branch of modern agricultural systems. For most development experts, an ideal urban agricultural system would incorporate various elements of modern sustainable agriculture based on reusable, self-contained waste and nutrient cycles through resource conservation and management based on nonchemical fertilizers and pest-management techniques.

It is important to recognize that urban agriculture cannot solve the world's food-security problems. For example, small urban gardens will not replace agribusiness as the primary players in the global food system. Moreover, there are legitimate health concerns surrounding urban agriculture, particularly in terms of recycling urban wastes into agricultural inputs. In arid parts of the world, where water is scarce, the use of wastewater from domestic or commercial uses seems an obvious solution to the irrigation needs of urban agriculture. Yet while in some parts of the world wastewater is effectively treated and used for secondary applications, in others it is not treated and can easily carry disease, which can then be spread into the food system when applied to crops. Clearly, policies and practices need to be developed that very carefully address the health implications of urban agriculture in the very different settings in which it is being practiced.

Currently, urban agriculture is practiced in a variety of ways, including rooftop, hydroponic, and community gardens; roadside urban fringe agriculture; field-to-direct-sale farmers' markets; and livestock grazing in parks and feedlots (**Figure 8.30**). As a growing practice worldwide, urban agriculture may help establish sustainable food systems in predominately urban areas. While health concerns about urban agricultural practices should not be taken lightly, there is evidence that they are far outweighed by the current and potential benefits of urban agriculture. Particularly in developing countries and poorer inner city

TABLE 8.3 National regulatory responses to GMOs

	GMOs Banned with Few Exceptions	Strict Labeling Requirements	Cultivation of GMO Crops Allowed	Some Test Fields of GMO Cultivation Allowed	Some Varieties of GMO Banned	Significant Public Resistance to GMOs	Regions or Cities within Countries Where GMOs are Banned
Algeria	X			X			
Argentina			X	X			
Australia		X	X			X	
Austria					X		
Brazil	X	X	X			X	
Canada			X			X	
China		X	X		X		
Denmark		X					
Egypt					X		
E.U.		X	X	X	X	X	
France		X		X	X	X	
Germany		X			X	X	
Greece	X	X			X		
India			X		X	X	
Italy		X		X	X		X
Japan		X			X	X	
Korea		X					
Luxembourg		X			X		
Mexico		X					
Netherlands		X		X		X	
N. Zealand		X					
Norway					X		
Paraguay		X					
Philippines			X				X
Portugal				X		X	
Russia					X		
Saudi Arabia					X		
South Africa		X	X				
Spain				X		X	X
Sri Lanka	X						
Thailand	X	X				X	
U.K.				X		X	X
U.S.			X	X		X	
Vietnam		X					
Zimbabwe						X	

These bar graphs are from: http://pewagbiotech.org/resources/issuebriefs/feedtheworld.pdf

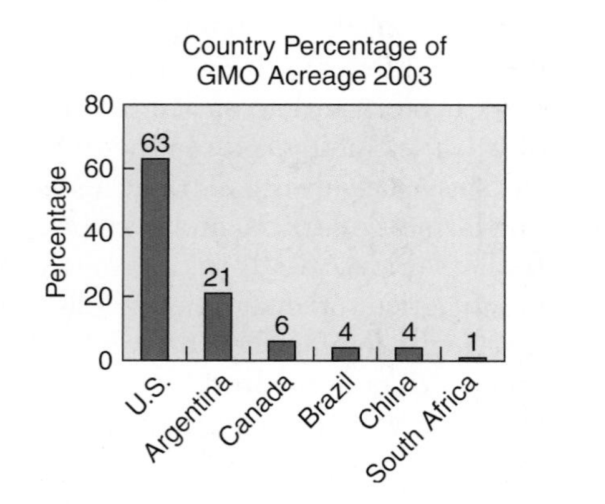

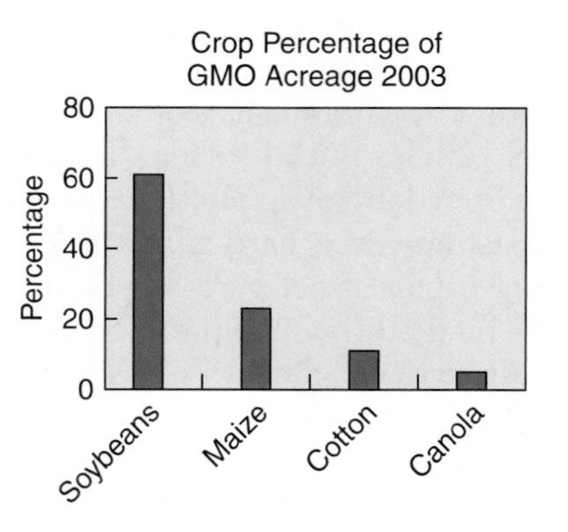

Figure 8.30 Urban garden, Hong Kong One of the most densely populated urban areas in the world, Hong Kong is also home to hundreds of urban gardens where residents of apartment blocks use the gardens to supplement market purchases. Throughout the world, in both the core and the periphery, private as well as communal gardens are becoming a more familiar feature of the urban landscape.

neighborhoods throughout the world, urban agriculture can be a crucial element in a family's survival. This ability of urban agriculture to enhance household food security seems to be currently drawing the most attention. Urban agriculture can also play a very important role in the absorption of labor—particularly women and youth—so that urban households are better able to take full advantage of their own human resources. It may even be a way of turning urban waste into a resource when incorporated safely into the food system.

CONCLUSION

Agriculture has become a highly complex, globally integrated system. While traditional forms of agricultural practices, such as subsistence farming, continue to exist, they have been overshadowed by the global industrialization of agriculture. This industrialization has included not only mechanization and chemical applications but also the linking of the agricultural sector to the manufacturing, service, and finance sectors of the economy. In addition, states have become important players in the regulation and support of agriculture at all levels, from the local to the global.

The dramatic changes that have occurred in agriculture have affected different places and different social groups. Households in both the core and the periphery have strained to adjust to these changes, often disrupting existing patterns of authority and access to resources. Just as people have been affected by the transformations in global agriculture, so have the land, air, and water.

The geography of agriculture at the turn of the twentieth century is a far cry from what it was 100 or even 50 years ago. As the globalization of the economy has accelerated in the last few decades, so has the globalization of agriculture. The changes in global agriculture do not necessarily mean increased prosperity in the core, nor are the implications of these changes simple. For example, the production of oranges in Florida is directly influenced by the newer Brazilian orange industry. Both industries, in turn, affect the prices of oranges in the marketplaces of Europe and Asia. Additionally, the European unease with genetically engineered foods will affect U.S. producers, agricultural research, trade, and a host of other factors that have repercussions throughout the world system.

MAIN POINTS REVISITED

- **Agriculture has been transformed into a globally integrated system; the changes producing this result have occurred at many scales and have had many sources.**

  In addition to the restructuring of entire national farming systems, farming households have been transformed as well in core, periphery, and semiperiphery regions.

- **Agriculture has proceeded through three revolutionary phases, from the domestication of plants and animals to the latest developments in biotechnology and industrial innovation.**

  These three revolutionary phases have not occurred simultaneously throughout the globe but have been adopted and adapted to differing degrees, based on levels of development, culture, and physical geography.

- **The introduction of new technologies, political concerns about food security and self-sufficiency, and changing opportunities for investment and employment are among the many forces that have dramatically shaped agriculture as we know it today.**

  Two of the most important forces behind these transformations in agriculture have been multinational and transnational corporations and states. The World Trade Organization is another important influence.

- **The industrialized agricultural system of today's world has developed from—and largely displaced—older agricultural practices, including shifting cultivation, subsistence agriculture, and pastoralism.**

  Although these systems no longer dominate agricultural practices on a global scale, they are still practiced in many areas of the world, in some cases alongside more mechanized forms.

- **The contemporary agro-commodity system is organized around a chain of agribusiness components that begins at the farm and ends at the retail outlet. Different economic sectors, as well as different corporate forms, have been involved in the globalization process.**

  The farm is no longer the central piece in this chain of agricultural organization, but one of several important components that includes seed and fertilizer manufacturers, food processors, food distributors, and consumers.

- **Transformations in agriculture have had dramatic impacts on the environment, including soil erosion, desertification, deforestation, and soil and water pollution, as well as the elimination of some plant and animal species.**

  While most of the core countries have instituted legislation to address some of the problems associated with environmental degradation, these problems exist throughout the global agricultural system to greater and lesser degrees. In peripheral countries, where governments are often too poor to monitor and enforce such legislation, they are being encouraged by international agencies and environmental organizations to limit their degradational practices through relief of part of their national debt.

- **The biggest issues food-policy experts, national governments, consumers, and agriculturalists face revolve around the availability and quality of food in a world where access to safe, healthy, and nutritious foodstuffs is unevenly distributed.**

  Genetic modification is one way of improving productivity, though it does not address issues of access to food. Increasing opportunities for the world's poor—who are increasingly residing in urban settings—to grow their own food is another way.

KEY TERMS

agrarian (p. 302)
agribusiness (p. 325)
agricultural industrialization (p. 316)
agriculture (p. 302)
aquaculture (p. 313)
biotechnology (p. 328)
blue revolution (p. 313)
chemical farming (p. 311)
commercial agriculture (p. 304)
crop rotation (p. 304)
double cropping (p. 308)
ejidos (p. 324)
famine (p. 335)
farm crisis (p. 328)
food chain (p. 325)
food manufacturing (p. 311)
food regime (p. 325)
food security (p. 335)
globalized agriculture (p. 317)
GMO (genetically modified organism) (p. 336)
green revolution (p. 317)
hunting and gathering (p. 303)
intensive subsistence agriculture (p. 307)
intertillage (p. 306)
land reform (p. 324)
mechanization (p. 311)
nontraditional agricultural exports (NTAEs) (p. 324)
pastoralism (p. 308)
shifting cultivation (p. 304)
subsistence agriculture (p. 303)
swidden (p. 305)
transhumance (p. 309)
undernutrition (p. 335)
urban agriculture (p. 337)

ADDITIONAL READINGS

Adeyemi, A., *Urban Agriculture: An Abbreviated List of References and Resource Guide 2000*. Beltsville, MD: USDA.

Brar, Karanjot K., *Green Revolution: Ecological Implications*. Delhi: Dominant Publishers, 1999.

Brown, A. D. *Feed or Feedback: Agriculture, Population Dynamics and the State of the Planet*. Utrecht: International Books, 2003.

Cheema, G. S., J. Smit, A. Ratta, and J. Nasr, *Urban Agriculture: Food, Jobs and Sustainable Cities.* NY: UNDP, 1996.

Conway, G., *The Doubly Green Revolution: Food for All in the Twenty-First Century.* Ithaca, NY: Comstock Publishers, 1998.

Food and Agriculture Organization of the United Nations, *The State of Food Insecurity in the World 2004: Monitoring Progress Towards the World Food Summit and Millennium Goals.* Rome, Italy, 2004.

Gruhn, P., F. Goletti, and M. Yudelman, *Integrated Nutrient Management, Soil Fertility, and Sustainable Agriculture: Current Issues and Future Challenges.* 2002 Vision Discussion Paper 32. Washington, DC: IFPRI.

Lang, T. and M. Heasman, *Food Wars: The Global Battle for Minds, Mouths, and Markets.* London: Earthscan, 2004.

Lebel, L., N. H. Tri, A. Saengnoree, S. Pasong, U. Buatama, and L. K. Thoa. "Industrial Transformation and Shrimp Aquaculture in Thailand and Vietnam: Pathways to Economic, Social and Ecological Sustainability?" *Ambio, 31*(4): 311–323, 2002.

Manning, R., *Food's Frontier: The Next Green Revolution.* NY: North Point Press, 2000.

Middleton, N. and D. Thomas (eds.), *World Atlas of Desertification,* 2nd edition. Oxford: Oxford University Press, 1997.

Public Citizen, *The Shrimp Stockpile: America's Favorite Imported Seafood.* Washington, DC: Public Citizen, 2005.

Raworth, K., *Trading Away Our Rights: Women Working in Global Supply Chains.* Oxford, England: Oxfam GB, 2004.

Scoones, I., *Agricultural Biotechnology and Food Security: Exploring the Debate.* Brighton, England: Institute of Development Studies, 2002.

Shiva, V., *Stolen Harvest: the Hijacking of the Global Food Supply.* Cambridge, MA: South End Press, 2000.

Shiva, V. and G. Bedi, *Sustainable Agriculture and Food Security: the Impact of Globalisation.* London: Sage Publications, 2002.

Waser, K., "The Newly Recognized Importance of Urban Agriculture." *Arid Lands Newsletter, 42:* Fall/Winter 1997.

EXERCISES

On the Internet

The Internet exercises for this chapter focus on the impact of agriculture on the environment. We look at water policies and their affect on agriculture, especially irrigation policies. Using a GIS on sustainable development, we survey chronic undernutrition, dietary patterns, food production growth, and the role of agricultural trade. Exploring the *World Agricultural Information Center* Web site, we appraise human-induced soil degradation, and we provide international time-series data sets for production, trade, chemicals, aid shipments, land use, and more. We also take a backward look (via the Internet) at the successes and failures of the Green Revolution. Finally, we examine biotechnology in agriculture and the public's changing attitude about the genetic engineering of our agricultural products.

Unplugged

1. Your neighborhood grocery store is a perfect location to begin to identify the "global" in the globalization of agriculture. Go to the produce section there, and document the source of at least 10 fruits and vegetables you find. You may need to ask the produce manager where they come from; once you have established that, illustrate those sources on a world map.
2. The Food and Agriculture Organization (FAO) has been publishing a range of yearbooks containing statistical data on many aspects of global food production since the mid-1950s. Using the *State of Food and Agricultural Production* yearbooks, compare the changes that have occurred in agricultural production between the core and the periphery since midcentury. You can use just two yearbooks for this exercise, or you may want to use several to get a better sense of when and where the most significant changes have occurred. Once you have identified where the changes have been most significant, try to explain why these changes may have occurred.
3. The U.S. Department of Agriculture (USDA) also provides statistics on food and agricultural production, though, of course, limited to the United States. Contained in volumes simply called Agricultural Statistics, a range of important variables are included, from what is being grown where, to who is working on farms, and what kinds of subsidies the government is providing. Using the USDA's annual publication *Agricultural Statistics,* examine the changing patterns of federal subsidies to agriculture over time. Using a map of the United States, show which states since the 1940s (just following the Great Depression) have received subsidies for the decades 1945, 1965, 1985, and the present. Have subsidies increased for some parts of the country and not others? If so or if not, why? Have subsidies increased or decreased overall for the entire country? Which farm sectors and, therefore, which regions have most heavily benefited from federal agricultural subsidies? Why?
4. Your breakfast is the result of the activities of a whole chain of producers, processors, distributors, and retailers whose interactions provide insights into both the globalization of food production and the industrialization of agriculture. Consider the various foods you consume in a typical breakfast and describe not only where (and by whom) they were produced—grown and processed—how they were transported (by whom) from the processing site—but also where and by whom they were retailed. Summarize how the various components of your breakfast illustrate the two concepts of globalization and the industrialization of agriculture.

9 The Politics of Territory and Space

Cambodia, Guatemala, East Timor, Bosnia, Rwanda, Sudan—all have experienced genocide, the deliberate and systematic killing of a people based on their ethnicity, nationality, race, religion, or politics. The deaths of over 40 million people in the past 100 years have been attributed to genocide and other acts of state violence. The term itself is a mid-twentieth century formulation coined by Jewish scholar Raphael Lemkin to describe the Nazi Holocaust. While genocide appears to be a widespread feature in the history of civilization, it wasn't until after the Holocaust, that steps were taken by the international community to create a set of laws forbidding it. In December, 1948 the U.N. General Assembly acted to prevent genocide through the Convention on the Prevention and Punishment of the Crime of Genocide. When the convention first became effective in 1951, only two of the five permanent members of the U.N. Security Council were parties to the treaty, France and the Republic of China. Eventually the Soviet Union ratified in 1954, the United Kingdom in 1970, the People's Republic of China in 1983 and the United States in 1988. But only in the 1990s did the international law on the crime of genocide begin to be enforced, this with respect to the atrocities in Rwanda.

According to Gregory Stanton, President of Genocide Watch in Washington D.C., genocide develops in eight stages:

1. Classification: Dividing people into "us and them" occurs.
2. Symbolization: Hateful symbols are imposed on "them," including hate speech.
3. Dehumanization: The pariah group is made to seem inhuman so that the normal human revulsion against murder can be overcome.
4. Organization: Special military units or militias are often trained and armed.
5. Polarization: Hate groups broadcast polarizing propaganda.
6. Identification: Victims are separated out because of their different identities.
7. Extermination: Murder, rape, torture and kidnapping are perpetrated.
8. Denial: The perpetrators deny they committed any crimes.

While eight stages characterize the unfolding of genocide, it is important to be aware that at any stage preventive measures are possible and can be undertaken by a state or combinations of states, though issues of sovereignty make some steps exceedingly difficult. With the arrival of the new century, it does not appear that either the international community or individual states have been able to use the Convention to prevent genocide nor has the process of enacting punishment for those guilty of genocide been swift. Today, in Darfur, Sudan, populations are still very much at risk from state-sponsored violence. And in Cambodia, no court has yet been convenved to try the members of the Khmer Rouge, who between 1975 and 1979 caused the death of nearly 2 million people (one-third of the total population) through starvation, forced labor, disease, and execution.

Temporary shelter at Sreef camps near Nyala in south Darfur

MAIN POINTS

- Political geography, a subfield of the discipline of geography, examines complex relationships between politics and geography (both human and physical).
- Political geographers recognize that the relationship between politics and geography is two-way. Political geography can be seen both as the geography of politics and the politics of geography.
- The relations between politics and geography are often driven by particular theories and practices of the world's states. Understanding imperialism, colonialism, heartland theory, domino theory, the end of the cold war, and the emergence of the new world order is key to comprehending how, within the context of the world system, geography has influenced politics and how politics has influenced geography.
- Political geography deals with the phenomena occurring at all scales of resolution, from the global to the body. Important East/West and North/South divisions dominate international politics, whereas regionalism, sectionalism, and similar divisions dominate intrastate politics.

A recent report by the International Rescue Committee has shown that for every death caused by violence in a genocidal war zone, another 62 nonviolent deaths occur due to malnutrition, respiratory disease and diarrhea, anemia, tuberculosis, accidents, and fever as well as other causes. And while institutions exist to prevent and punish, the world is still a very long way from making those practices a reality. Globalization has certainly created the conditions for economic cooperation across state territories, as international organizations and trade agreements like the World Trade Organization and North American Free Trade Agreement make clear. What is less clear is how cooperation among states will enable the end of genocidal political violence and the beginning of a global regime of fundamental, protected human rights.

This chapter explores how globalization continues to create new maps, politics, and geopolitical arrangements at the same time that established boundaries persist. Exploration, imperialism, colonization, decolonization, and the cold war between East and West are powerful forces that have created and re-created national boundaries. Much of the political strife that currently grips the globe involves local or regional responses to the impacts of globalization of the economy, aided by the practices of the state. The complex relationships between politics and geography—both human and physical—are two-way relationships. In addition, political geography is not just about global or international relationships. It is also about the many other geographic scales and political divisions, from the globe to the neighborhood, from large, far-reaching processes to the familiar sites of our everyday lives.

THE DEVELOPMENT OF POLITICAL GEOGRAPHY

Political geography is a long-established subfield in the wider discipline of geography. Aristotle is often considered the first political geographer because his model of the state is based upon factors such as climate, terrain, and the relationship between population and territory. Other important political geographers, from Strabo to Montesquieu, have promoted theories of the state that incorporated elements of the landscape and the physical environment, as well as the population characteristics of regions. From about the fourteenth through the nineteenth century, scholars interested in political geography theorized that the state operated cyclically and organically. They believed that states consolidated and fragmented based on complex relationships among and between factors such as population size and composition, agricultural productivity, land area, and the role of the city.

As these factors indicate, political geography at the turn of the century was influenced by two important traditions within the wider discipline of geography: the people-land tradition and environmental determinism. While different theorists placed more or less emphasis on each of these traditions in their own political geographic formulations, the traditions' effects are evident in the factors deemed important to state growth and change. Why these factors were identified as central undoubtedly had much to do with the widespread influence of Charles Darwin on intellectual and social life during this period. Darwin's theory of competition inspired political geographers to conceptualize the state as a kind of biological organism that grew and contracted in response to external factors and forces. It was also during the late nineteenth century that foreign policy as a focus of state activity began to be studied. This new field came to be called *geopolitics*.

The Geopolitical Model of the State

Geopolitics is the state's power to control space or territory and shape the foreign policy of individual states and international political relations. Within the discipline of geography, geopolitical theory originated with Friedrich Ratzel (1844–1904), a German trained in biology and chemistry. Ratzel was greatly influenced by the theories embodied in *social Darwinism* that emerged in the mid-nineteenth century. Ratzel employed biological metaphors adopted from Charles Darwin to describe the growth and development of the state, as well as seven laws of state growth:

1. The space of the state grows with the expansion of the population having the same culture.
2. Territorial growth follows other aspects of development.
3. A state grows by absorbing smaller units.
4. The frontier is the peripheral organ of the state that reflects the strength and growth of the state; hence, it is not permanent.
5. States in the course of their growth seek to absorb politically valuable territory.

6. The impetus for growth comes to a primitive state from a more highly developed civilization.
7. The trend toward territorial growth is contagious and increases in the process of transmission.[1]

Ratzel's model portrays the state as behaving like a biological organism; thus, its growth and change are seen as "natural" and inevitable. Although Ratzel advanced his model of the state at the turn of the nineteenth century, his views continue to influence state theorizing. What has been most enduring about Ratzel's conceptualization is the conviction that geopolitics stems from the interactions of power and territory.

Although it has evolved since Ratzel first introduced the concept, geopolitical theory has become one of the cornerstones of contemporary political geography and state foreign policy more generally. And, though the organic view of the state has been abandoned, the twin features of power and territory still lie at the heart of political geography. In fact, the changes that occurred in Europe and the former Soviet Union during the late twentieth century suggest that Ratzel's most important insights about geopolitics are still valid.

Figure 9.1 demonstrates Ratzel's conceptualization of the interaction of power and territory through the

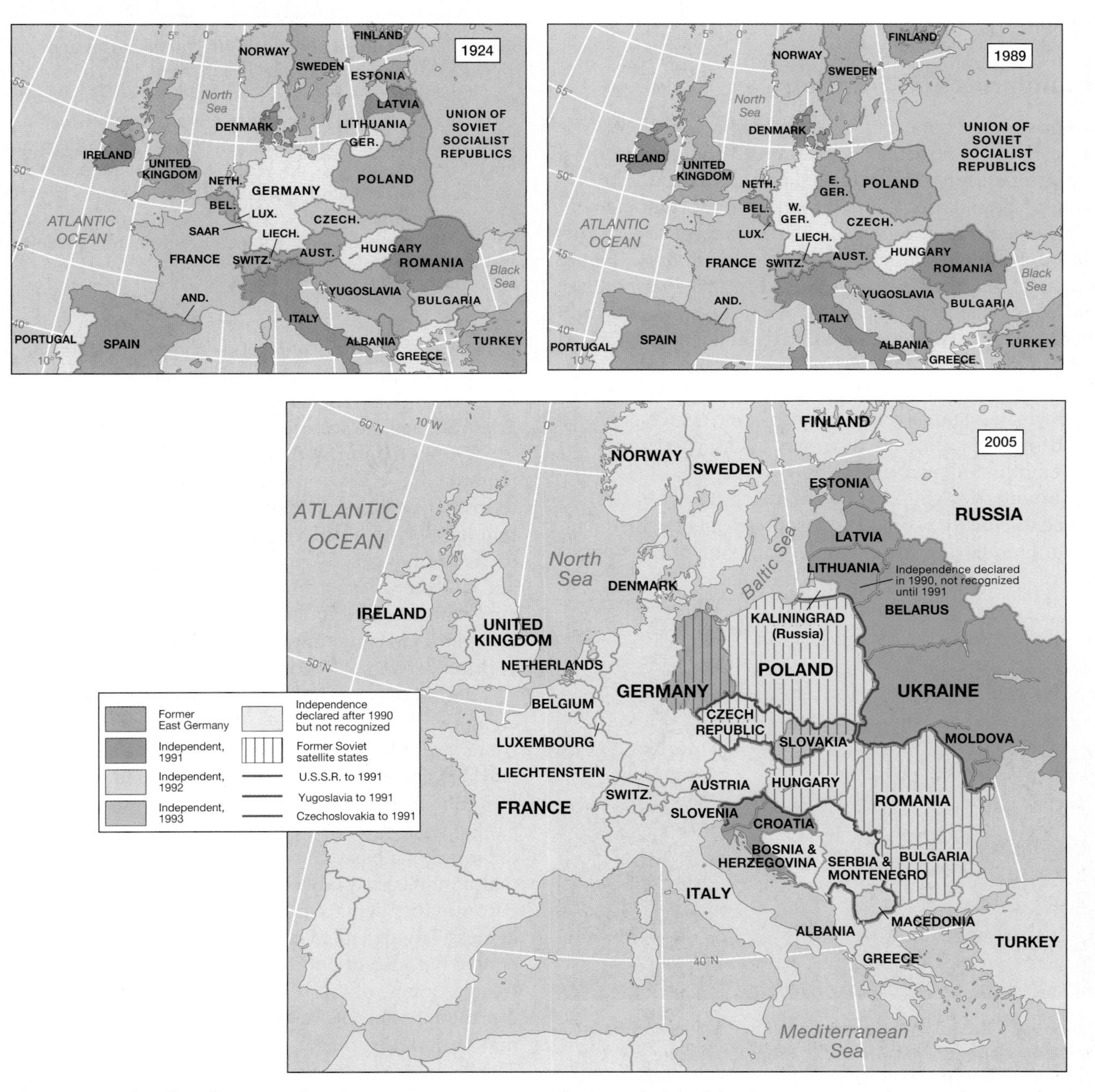

Figure 9.1 The changing map of Europe: 1924, 1989, 2005 The boundaries of the European states have undergone dramatic changes since World War I. The changing map of Europe illustrates the instability of international politics and the resultant dynamism in the geography of political boundaries.

[1]Adapted from Martin I. Glassner and Harm deBlij, *Systematic Political Geography,* 3rd ed. New York: J. Wiley & Sons, 1980, p. 164.

changing face of Europe from the end of World War I to the present. The fluidity of the maps reflects the unstable relationship between power and territory, especially some states' failure to achieve stability. The most recent map of Europe reflects the precarious nation-state boundaries in the post-cold war period. Estonia, Latvia, and Lithuania have regained their sovereign status. Czechoslovakia has dissolved into the Czech Republic and Slovakia. The former Soviet Union is now the Commonwealth of Independent States, with Russia the largest and most powerful. Yugoslavia has dissolved into four states, but not without much civil strife and loss of life. Indeed, the difference between Europe in 1989 and Europe now is far more dramatic than between any other two of the previous 50 years. The maps also illustrate the centrality of territorial boundaries to the operations of the state.

Figure 9.3 Boundary between the United States and Mexico The U.S.-Mexico border is exclusionary. It is heavily patrolled and lined with barbed-wire chain-link fences along the highly urbanized parts. Aerial surveillance is also extensive along the U.S.-Mexico border. In an effort to stem the flow of illegal immigration from Mexico, the U.S. government increased the number of border patrol agents from 5,176 in 1996 to 10,000 in 2000. This photo shows the U.S.-Mexico border along the Tijuana River estuary, with southern California on the left and Mexico on the right.

Boundaries and Frontiers

Boundaries are important phenomena because they allow territoriality to be defined and enforced and allow conflict and competition to be managed and channeled. The creation of boundaries is, therefore, an important element in place-making. It follows from the concept of territoriality that boundaries are normally inclusionary (**Figure 9.2**). That is, they are constructed in order to regulate and control specific sets of people and resources within them. Encompassed within a clearly defined territory, all sorts of activity can be controlled and regulated—everything, in fact, from birth to death. The delimited area over which a state exercises control and which is recognized by other states is **territory.** Such an area may include both land and water.

Boundaries can also be exclusionary. They are designed to control people and resources outside them. National boundaries, for example, can control the flow of immigrants or imported goods (**Figure 9.3**). Municipal boundaries and land-use zoning boundaries can regulate access to upscale residential neighborhoods, field boundaries can regulate access to pasture, and so on (**Figure 9.4**).

Figure 9.2 Boundary between the United States and Canada Most boundaries are established in order to regulate and control specific sets of people and resources within a given territory. Such boundaries need to be clearly identified but not necessarily fortified. The U.S.-Canadian border is a good example of an inclusionary boundary.

The key point is that once established, boundaries tend to reinforce spatial differentiation. This results partly from different sets of rules, both formal and informal, that apply within different territories. It is also partly because boundaries often restrict contact between people and so foster the development of stereotypes about "others." This restricted contact, in turn, reinforces the role of boundaries in regulating and controlling conflict and competition between territorial groups.

Boundaries can be established in many ways, however, and with differing degrees of permeability. At one extreme are informal, implied boundaries that are set by markers and symbols but never delineated on maps or in legal documents. Good examples are the "turf" of a city gang, the "territory" of an organized crime "family," and the range of a pastoral tribe. At the other extreme are formal boundaries established in international law, delimited on maps, demarcated on the ground, fortified, and aggressively defended against the movement not only of people but also of goods, money, and even ideas. An extreme example of this sort of boundary is the one between North and South Korea (**Figure 9.5**). In between are formal boundaries that have some degree of permeability. The boundaries between the states of the European Union, for example, have become quite permeable, and people and goods from member states can now move freely between them with no customs or passport controls.

Figure 9.4 Boundary between rural and urban places Some boundaries signal differences in settlement activities that may actually be governed by land-use regulations. The difference between agricultural activities and suburban living is clearly shown in this image.

Impermeability does not necessarily mean immutability, however. The boundary between East and West Germany, part of the Iron Curtain for more than 40 years, was as aggressively defended as the present boundary between North and South Korea, yet it was removed in 1989 when Germany was reunified (**Figure 9.6**). Similarly, the boundaries of the former Soviet Union have been dramatically redrawn since 1989, allowing states like Lithuania and Estonia to reappear. Boundaries are an important element of geopolitics and of the geography of domestic politics.

Frontier Regions

Frontier regions occur where boundaries are very weakly developed. They involve zones of underdeveloped territoriality, areas that are distinctive for their marginality rather than for their belonging. In the nineteenth century, for example, some vast frontier regions still existed—major geographic realms that had not yet been conquered, explored, and settled (such as Australia, the American West, the Canadian North, and sub-Saharan Africa). All of these are now subject to territoriality at various spatial

Figure 9.5 Border between North and South Korea Some boundaries are virtually impermeable. The border between North and South Korea shown here is highly fortified and heavily patrolled. It was established at the conclusion of the Korean War (1950–1953) between two states that still contest one another's territory.

Figure 9.6 Berlin Wall The boundary between East and West Germany was virtually impermeable for more than 40 years. The photograph here shows the scene on November 12, 1989, when Berliners tore the wall down in celebration of the reunification of Germany.

scales (that is, from individual land ownership to local and national governmental jurisdiction). Only Antarctica, virtually unsettled, exists today as a frontier region in this strict sense of the term.

There remain, nevertheless, many regions that are still somewhat marginal in that they have not been fully settled or do not have a recognized economic potential, even though their national political boundaries and sovereignty are clear-cut. Examples are the Rondonia region of the Amazon rain forest and the Sahelian region of Africa. Such regions often span national boundaries simply because they are inhospitable, inaccessible, and (at the moment) economically unimportant. Political boundaries are drawn through them because they represent the line of least territorial resistance.

At the local level many examples of frontier, or marginal, regions exist. Although the residents of most towns and cities recognize a series of distinctive districts and neighborhoods, these are often separated by zones or spaces that are marginal. Not fully integrated into the territorial realm of any one sociocultural group, these spaces are often transitional in nature, with a relatively rapidly changing pattern of land use and an equally rapidly changing profile of residents.

Boundary Formation

Generally speaking, formal boundaries tend first to follow natural barriers, such as rivers, mountain ranges, and oceans. Good examples of countries with important mountain-range boundaries include France with Spain (the Pyrenees); Italy with France, Switzerland, and Austria (the Alps); and India and Nepal with China (the Himalayas). Chile, though, provides the ultimate example: a cartographic freak, restricted by the Andes to a very long and relatively thin strip along the Pacific coast. Examples of countries with boundaries formed by rivers include China and North Korea (the Yalu Tumen), Laos and Thailand (the Mekong), and Zambia and Zimbabwe (the Zambezi). Similarly, major lakes divide Canada and the United States (along the Great Lakes), France and Switzerland (Lake Geneva), and Kenya and Uganda (Lake Victoria).

Where no natural features occur, formal boundaries tend to be fixed along the easiest and most practical cartographic device: a straight line. Examples include the boundaries between Egypt, Sudan, and Libya (**Figure 9.7**), between Syria and Iraq, and between the western United States and Canada. Straight-line boundaries are also characteristic of formal boundaries that are established through colonization, which is the outcome of a particular form of territoriality. The reason, once again, is practicality. Straight lines are easy to survey and even easier to delimit on maps of territory that remain to be fully charted, claimed, and settled. Straight-line boundaries were established, for example, in many parts of Africa during European colonization in the nineteenth century.

In detail, however, formal boundaries often detour from straight lines and natural barriers in order to accommodate special needs and claims. Colombia's border, for instance, was established to contain the source of the Orinoco River, the Democratic Republic of Congo's border was established to provide a corridor of access to the Atlantic Ocean, and Sudan's border detoured to include the settlement Wadi Halfa.

After primary divisions have been established, internal boundaries tend to evolve as smaller, secondary territories are demarcated. In general, the higher the population density, the smaller these secondary units tend to be. Their configuration tends to follow the same generalizations as for larger units, following physical features, accommodating special needs, and following straight lines where there are no appropriate natural features or where colonization has made straight lines expedient. It is this last reason, for example, that explains the rectilinear pattern of administrative boundaries in the United States to the west of the Mississippi (**Figure 9.8**).

Territories delimited by formal boundaries—national states, states, counties, municipalities, special districts, and so on—are known as *de jure* spaces or regions. *De jure* simply means "legally recognized." Historically, the

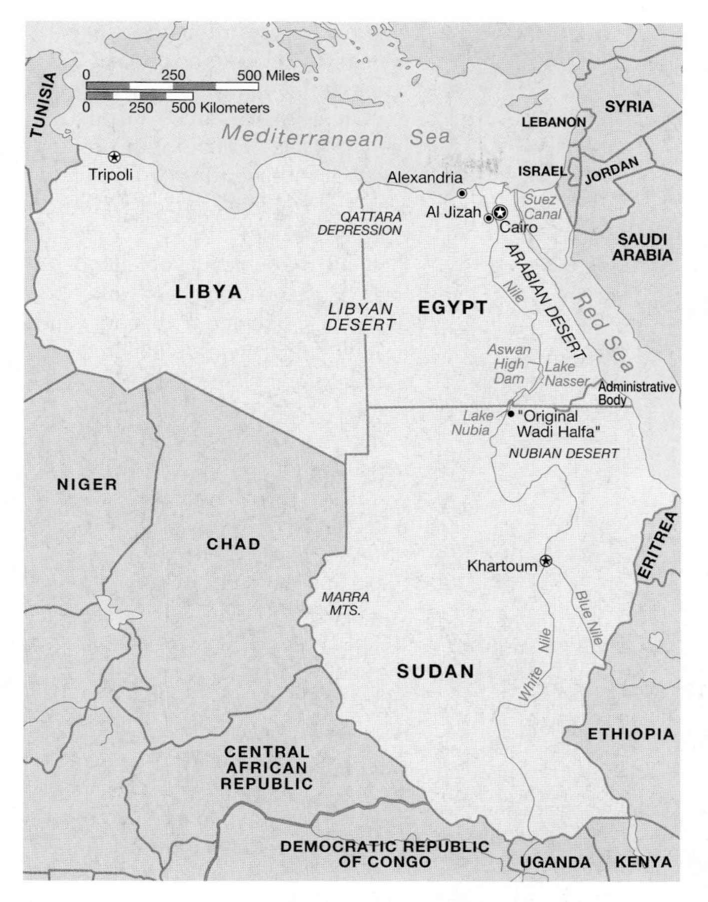

Figure 9.7 Borders between Egypt and Libya and Sudan Where no natural barriers to human activity exist, boundaries are often drawn as straight lines, easily described in a treaty or convention, as in the border between Egypt and Libya. Small natural barriers and isolated settlements sometimes require special detours, however. In the border between Sudan and Egypt, the small half-circle that straddles Lake Nubia in Sudan (Lake Nasser in Egypt) is the original location of the settlement of Wadi Halfa. When the Aswan Dam was under construction, the residents of the settlement were moved eastward as the rising waters of the Nile River flooded their land. That small area marks the location of the original settlement. The triangular area along the Red Sea on the eastern end of the border of Sudan and Egypt, called "Administrative Body," once belonged to Egypt but is now Sudan. That area was given to Sudan by Egypt in an attempt to induce it to join the now disbanded United Arab Republics (UAR). Sudan never joined the UAR but retained the land, which is a point of recurring controversy and sometimes outright conflict between the two countries.

world has evolved from a loose patchwork of territories (with few formally defined or delimited boundaries) to nested hierarchies (**Figure 9.9**) and overlapping systems of *de jure* territories.

These *de jure* territories are often used as the basic units of analysis in human geography, largely because they are both convenient and significant units of analysis. They are often, in fact, the only areal units for which reliable data are available. They are also important units of analysis in their own right because of their status as units of governance or administration. A lot of regional analysis and nearly all attempts at regionalization, therefore, are based on a framework of *de jure* spaces.

GEOPOLITICS AND THE WORLD ORDER

There is, arguably, no other concept to which political geographers devote more of their attention than the state. The state is one of the most powerful institutions—if not the most powerful—implicated in the process of globalization. The state effectively regulates, supports, and legitimates the globalization of the economy.

STATES AND NATIONS

As described in Chapter 2, the state is an independent political unit with recognized boundaries, even if some of these boundaries are in dispute. In contrast to a state, a **nation** is a group of people sharing certain elements of culture, such as religion, language, history, or political identity. Members of a nation recognize a common identity, but they need not reside within a common geographical area. For example, the Jewish nation refers to members of the Jewish culture and faith throughout the world, regardless of their places of origin. The term **nation-state** refers to an ideal form consisting of a homogeneous group of people governed by their own state. In a true nation-state, no significant group exists that is not part of the nation. **Sovereignty** is the exercise of state power over people and territory, recognized by other states and codified by international law. **Citizenship** is a category of belonging to a nation-state that includes civil, political and social rights.

Following the overthrow or decline of monarchies in Europe in the late eighteenth to mid-nineteenth centuries, a number of new republics were created. Republican government, as distinct from monarchy, requires the democratic participation and support of its population. Monarchical political power is derived from force and subjugation; republican political power derives from the support of the governed. By creating a sense of nationhood, the newly emerging states of Europe were attempting to homogenize their multiple and sometimes conflicting constituencies so that they could govern with their active cooperation according to a sense of common purpose. Modern citizenship as a political category was a product of the popular revolutions—from the English Civil War to the American War of Independence—that transformed monarchies into republics. In the process, these revolutions produced the need to reimagine the socially and culturally diverse population occupying the territory of the state. Where once they were considered subjects with no need for a unified identity, nationhood required of these same people a sense of an "imagined community," one that rose above divisions of class, culture,

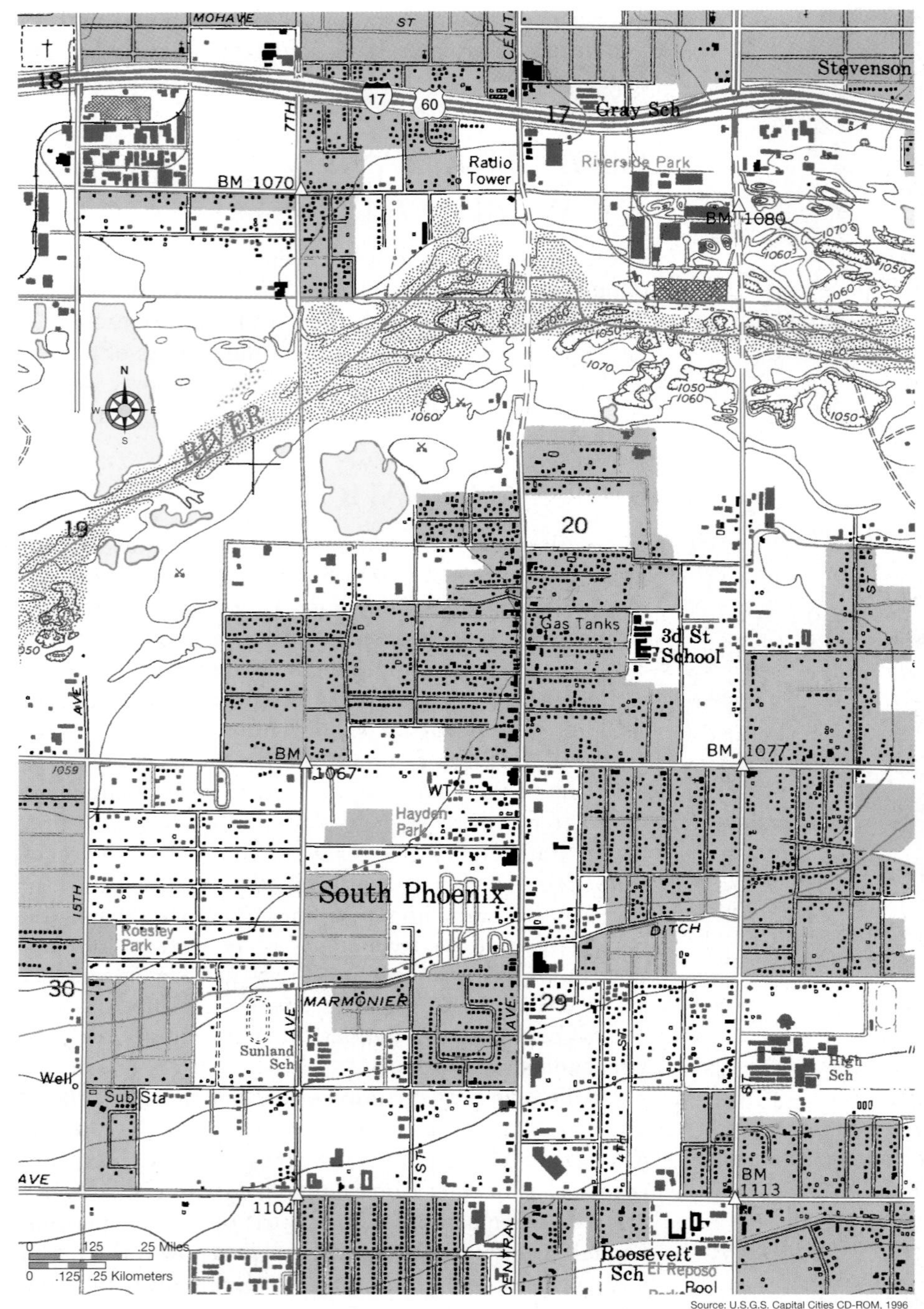

Source: U.S.G.S. Capital Cities CD-ROM, 1996
Original map was: *Phoenix Quadrangle* 1:24,000, U.S.G.S., 1982

Figure 9.8 Township-and-range system The checkerboard landscapes of the midwestern and western United States can be traced to the U.S. Land Ordinance of 1785 and the Northwest Territories Act of 1803, which divided much of the country into "townships" in order to facilitate the surveying of federal land. Most townships were laid out as squares, oriented north-south and east-west, with sides 8.9 kilometers (6 miles) long. Some were given irregular shapes that followed natural features such as rivers and mountains. The square-shaped townships were each divided into 36 "sections," each one being 1 square mile, or 640 acres (259 hectares). Sections, in turn, were divided into four "quarter-sections." The Homestead Act of 1863, designed to encourage the settlement of the Great Plains, stipulated that the quarter-section would be the basic unit of land ownership, giving each settler 160 acres of land. With this rectilinear pattern of land ownership, it also made sense for the boundaries of administrative areas—counties and states—to be rectilinear.

and ethnicity. This new identity was called citizen and citizenship came to be based on a framework of civil, political, and social rights and responsibilities.

Given that nations were created out of very diverse populations, it is not surprising that no entirely pure nation-states exist today. Rather, multinational states are the norm—states composed of more than one regional or ethnic group. Spain is such a multinational state (composed of Catalans, Basques, Gallegos, and Castilians), as are France, Kenya, the United States, and Bolivia. Since World War I, it has become increasingly common for groups of people sharing an identity different from the majority yet living within the same political unit with the majority to agitate to form their own state. This has been the case with the Québecois in Canada and the Basques in Spain. It is out of this desire for autonomy that the term "nationalism" emerges. **Nationalism** is the feeling of belonging to a nation, as well as the belief that a nation has a natural right to determine its own affairs. Nationalism can accommodate itself to very different social and cultural movements, from the white supremacy of the Aryan Nation movements in the United States and Europe to the movements for independence in Estonia, Latvia, and Lithuania. The impact of minority nationalism on the world map has been pronounced during the twentieth century.

Figure 9.9 Nested hierarchy of *de jure* territories *De jure* territories are constructed at various spatial scales depending on their origin and function. Administrative and governmental territories are often "nested," with one set of territories fitting within the larger framework of another, as in this example of states, districts, and municipalities in India.

Municipalities in Ahmednagar District

Districts in Maharashtra State

States in India

The history and the present status of the former Soviet Union also clearly illustrates the tensions among and between states, nations, and nationalism. Demonstrating how enduring are both nationalism and the desire for sovereignty is the history of the Russian Empire: the overthrow of the czar, the subsequent establishment of the Union of Soviet Socialist Republics, and the recent events that have resulted in the establishment of the Commonwealth of Independent States.

The Russian Empire, like other colonial empires in Europe such as Spain and Britain, had a long history. Although it had a medieval origin, most relevant to issues of nationalism was the territorial expansion of the Muscovite state beginning in the fifteenth century. In 1462 Muscovy was a principality of approximately 5,790 square kilometers (15,000 square miles) centered on the present-day city of Moscow. Over a 400-year period, the Muscovite state expanded at a rate of about 20 square kilometers (50 square miles) per day, so that by 1914, on the eve of the Russian Revolution, the empire occupied more than 3.25 million square kilometers (roughly 8.5 million square miles), or one-seventh of the land surface of Earth. The expansion of the Muscovite state was mostly eastward but also to the west and south. This remarkable expansion occurred in discernible stages (**Figure 9.10**).

By the end of the fifteenth century the Muscovite state had annexed most of the other principalities of present-day European Russia. During the sixteenth century, Moscow conquered the non-Russian Tartar states. Desirous of more forest resources—especially furs—Moscow expanded into Siberia. By the mid-seventeenth century, Russia had wrested the eastern and central parts of Ukraine from Poland. In the eighteenth century under Catherine the Great, Russia secured the territory of what would eventually become southern Latvia, Lithuania, Belarus, and western Ukraine. The majority of the remaining expansion of the Russian state occurred in the late eighteenth and nineteenth centuries with the acquisition of territory in the Transcaucasus, what is now Kazakhstan, and desert lands to the south bordering on China and Afghanistan.

Russia's imperial expansion was driven by the same forces that drove expansion of other European empires—the need for more territorial resources (especially a warm-water port) and additional subjects. Russia, however, annexed vast stretches of adjacent land on the Eurasian continent, whereas other empires established new territories overseas. What was strikingly similar about Russian (as well as other European) expansion was that different nations had to be incorporated into one state—and *not* as colonies. Numerous problems attendant on that fact came to challenge Russia's consolidation of its extensive empire.

To meet the challenge of incorporating different nationalities within one state, Russia needed to apply centripetal policies to counter centrifugal forces. **Centripetal**

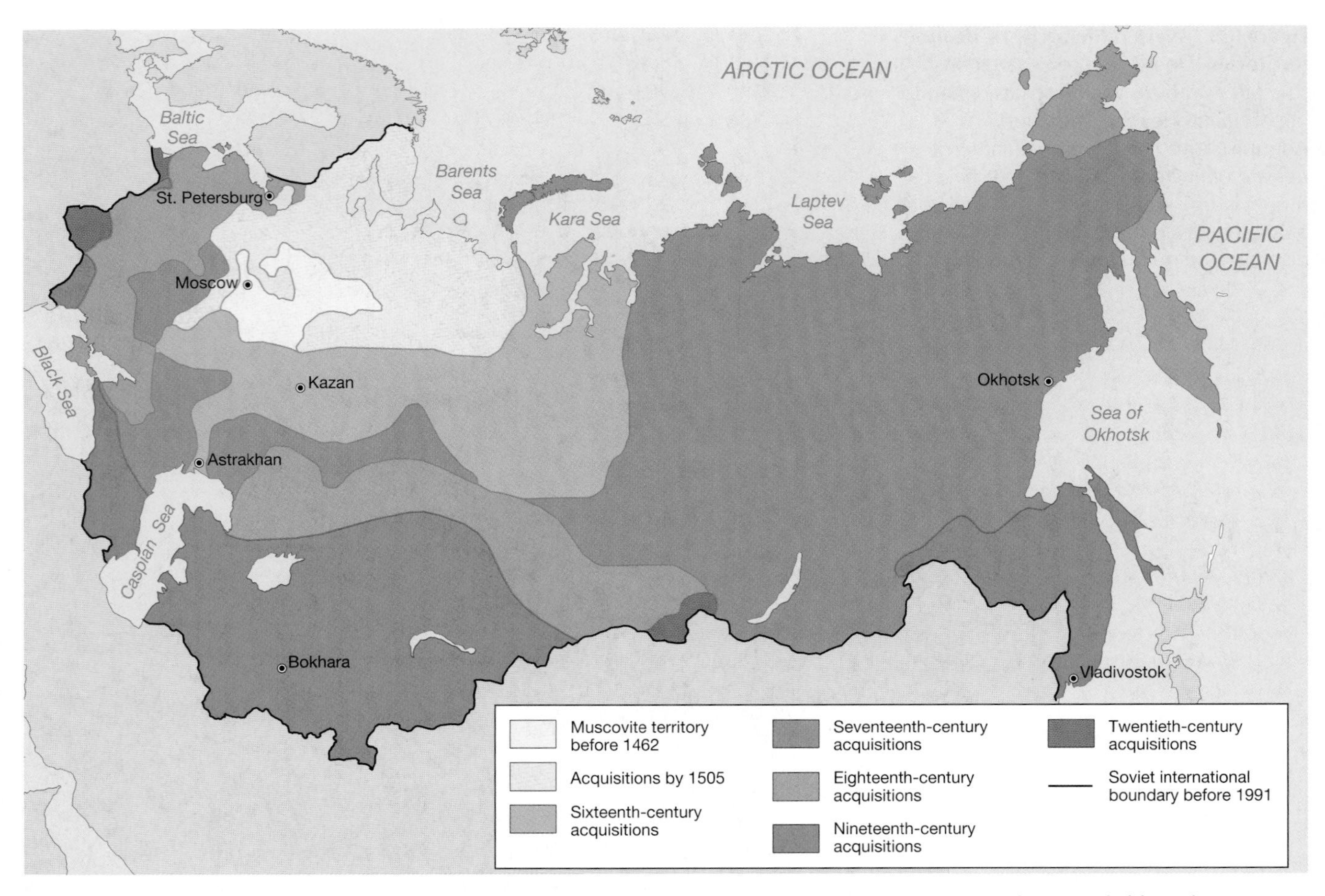

Figure 9.10 Territorial growth of the Muscovite/Russian state The vast Russian empire grew over the period (fifteenth century to the present) that corresponds with the globalization of the world economy. What makes the Russian empire different from many others is that the lands conquered were adjacent, not overseas. When the Bolsheviks came to power early in the twentieth century, some of the territory was lost. Eventually, however, the Bolsheviks were able to control most of the territories formerly ruled by the czars, and it was upon this that they built the Soviet state. (After D. J. B. Shaw, (ed.), *The Post-Soviet Republics: A Systematic Geography.* New York: John Wiley & Sons, 1995, p. 164.)

forces strengthen and unify something, while **centrifugal forces** divide or tend to pull something apart. Russia's centripetal strategies to bind the 100-plus nationalities (non-Russian ethnic peoples) into a unified Russian state were often punitive and not at all successful. Non-Russian nations were simply expected to conform to Russian cultural norms. Those that did not were more or less persecuted. The result was opposition and, among many if not most of the nationalities, sometimes rebellion and refusal to bow to Russian cultural dominance.

Such was the legacy that Lenin and the Bolsheviks inherited from the Russian Empire following the overthrow of the aristocracy in 1917. The solution to the "national problem" orchestrated by Lenin was *recognition* of the many nationalities through the newly formed Union of Soviet Socialist Republics (USSR). Lenin believed that a *federal system,* with *federal units* delimited according to the geographic extent of ethnonational communities, would ensure political equality among at least the major nations in the new state. This political arrangement recognized the different nationalities and provided them a measure of independence. Federation was also a way of bringing reluctant areas of the former Russian Empire into the Soviet fold. A **federal state** allocates power to units of local government within the country. The United States is a federal state with its system of state, county, and city/town government. A federal state can be contrasted with a **unitary state,** in which power is concentrated in the central government. The Russian state under the czar was a unitary state.

Lenin was optimistic that once international inequalities were diminished and the many nationalities became one Soviet people, the federated state would no longer be needed. Nationalism would be replaced by communism. Lenin's vision was short-lived, and following his death in 1924, a true federal system in the USSR also declined. Stalin came to power and enforced a new nationality policy. Although the federal system remained in place, nations increasingly lost their independence, and by the 1930s were punished for displays of nationalism. **Figure 9.11** shows the administrative units and nationalities that were part of the USSR during Stalin's tenure as premier. Figure 9.11 also shows how after World War II, Stalin expanded the power of

Figure 9.11 Soviet state expansionism, 1940s and 1950s World War II gave the Soviet state the opportunity to move westward for additional territories. Insisting that these countries would never again be used as a base for aggression against the USSR, Stalin retained control over Poland, East Germany, Czechoslovakia, Hungary, Romania, Bulgaria, Albania, and western Austria as well as Yugoslavia for a short time. (Mongolia gained its independence from China in 1921 with the help of the Soviets.) In 1945 Stalin promised democratic elections in these territories. After 1946, however, Soviet control over Eastern and Central Europe became complete as noncommunist parties were dissolved and Stalinist governments installed. (After *Atlas of Twentieth Century World History*. New York: HarperCollins Cartographic, 1991, pp. 86–87.)

the Soviet state westward to include domination over Albania, Austria, Bulgaria, Czechoslovakia, the German Democratic Republic, Hungary, Poland, Romania, and Yugoslavia.

While a federal system theoretically remained in place through the administration of Mikhail Gorbachev, the USSR actually operated as a unitary state with power concentrated in Moscow. When Gorbachev came to lead the USSR in 1985, he assumed that the diverse nationalities had been so "Sovietized" that an international worker orientation had replaced nationalism. This was not the case, however, and his economic and political policies created conditions that encouraged the reemergence of nationalism in the USSR (**Figure 9.12**).

Figure 9.12 Nationalism in Riga, Latvia The Baltic republics (Latvia, Estonia, and Lithuania) were the first to push for reestablishing their sovereignty in the wake of *perestroika* introduced by Gorbachev. All three had a history of independence movements that began with the early expansion of the Russian empire. One obvious reason for their resistance to Russian and later Soviet state incorporation was that all were largely populated by non-Russian people until their forced incorporation into the Soviet Union in 1940. These three republics were independent nation-states from 1918 to 1940. With a well-developed economic base and higher standards of living than most of the rest of the USSR, Latvia, Estonia, and Lithuania were quick to move from asserting their nationalism. Instead, they demanded separate nation-state status when the former Soviet Union crumbled in 1991, as illustrated by the Latvians pictured here.

Gorbachev's goal was a massive restructuring of the Soviet economy through radical economic and governmental reforms (*perestroika*) and the direct democratic participation of the republics in shaping those reforms through open discussions, freer dissemination of information, and independent elections (*glasnost*). Effectively, Gorbachev lifted the restrictions that had been placed on the legal formation of national identity. By 1988 grassroots national movements were emerging, first in the Baltic republics and later in Transcaucasia, Ukraine, and Central Asia. By 1991 the breakup of the Soviet Union was under way, and new nation-states had emerged to claim their independence. The federated structure that existed under the USSR made possible the relatively peaceful breakup of the Soviet Union. **Figure 9.13** is a recent map of the former USSR, including all the newly independent states now in the Commonwealth of Independent States (CIS). The CIS is a **confederation,** a group of states united for a common purpose. The newly independent states that chose confederation did so mostly for economic (and to a lesser extent for military) purposes. A similar case is that of the Confederate States of America, the 11 southern states that seceded from the United States between 1860 and 1861 for economic and political solidarity. This secession led ultimately to the Civil War, a bloody conflict that caused a massive loss of American lives for both the Union and the Confederacy.

With the fall of communism in Eastern Europe in 1991, regions such as the Balkans (the mountainous isthmus of land between the Danube River and the plains of northern Greece that includes Albania, Bulgaria, continental Greece, southeast Romania, European Turkey, and most of the territories formerly organized as Yugoslavia) have also experienced national movements resulting in the redrawing of political boundaries. And while the breakup of the former USSR has been mostly peaceful (with significant exceptions such as the Republic of Chechnya), the recent redrawing of national boundaries in the Balkans resulted in bitter ethnic conflict. The region is situated at a geopolitical crossroad where East meets West, Islam meets Christendom, the Ottoman Empire met the Austro-Hungarian Empire, and communism confronted capitalist democracy. In fact, conflict has been characteristic of the region for centuries. In the twentieth century, for instance, ethnic tensions spilled over the boundaries of the Balkan region, eventually pulling all of Europe (and later the United States as well as other more far-flung countries) into World War I in 1914–1915.

The recent conflicts have been concentrated in the region organized in 1918 as the Kingdom of Serbs, Croats, and Slovenes following the collapse of the Austro-Hungarian Empire. In 1929 this region was renamed Yugoslavia. After 1946 the country included six republics: Slovenia, Croatia, Bosnia-Herzegovina, Macedonia, Serbia, and Montenegro. The first four declared independence in 1991, while Serbia and Montenegro formed a new Yugoslavian state in 1992.

The most horrifying conflict of the late 1990s was the war in Kosovo, an autonomous region within Serbia that lost its autonomy in 1989 when Serbian nationalist leader Slobodan Milosevic placed it under military occupation. A year later Kosovo's Parliament was abolished and its political leaders fled. Milosevic orchestrated numerous attempts to rid the region of ethnic Albanians, and in 1995 Serb refugees from Croatia began flooding into Kosovo.

Figure 9.13 Independent states of the former Soviet Union For the most part, the administrative structure of the USSR has remained in place. The differences are that autonomous regions and republics now have more than nominal local control, and the former federal republics have become independent states (pictured here in color), as have the popular democracies of Eastern and Central Europe. Despite, or perhaps because of, recent democratic reforms, nationalist movements continue to plague the consolidation of the state in autonomous republics like Chechnya, located on the map in the yellow area north of Georgia. (*Source:* Reprinted with permission from Prentice Hall, from J. M. Rubenstein, *The Cultural Landscape: An Introduction to Human Geography,* © 1996, p. 318.)

Fueled by the racist rhetoric and military support of Milosevic, a civil war between Serbs and ethnic Albanians eventually erupted. By late 1999, 800,000 ethnic Albanians had fled Kosovo, and thousands more had been massacred. An 11-week air war by NATO helped to bring the war to a halt, though atrocities continued as ethnic Albanians retaliated against Serbs. An international war tribunal was set up in the Hague to try Milosevic and others on human rights violations, including the possible charge of genocide. Milosevic, acting in his own defense, began cross-examining witnesses in his case before the tribunal in the summer of 2002.

Theories and Practices of States

The definition of the state provided in the previous section is a static one. In fact, the state, through its institutions—such as the military or the educational system—can act to protect national territory and harmonize the interests of its people. Therefore, the state is also a *set of institutions* for the protection and maintenance of society. Thus, it is not only a place, a bounded territory, it is also an active entity that operates through the rules and regulations of its various institutions, from governing bodies to social-service agencies to the courts. Political geographers and related scholars have advanced numerous theories and models to explain state actions. For political geographers, theories and models of geopolitics have been their most prominent contributions to understanding the role and behavior of the state.

As mentioned previously, the most important influence on the field of geopolitics was the nineteenth-century German geographer Friedrich Ratzel. Ratzel's organic theory of the state used biological constructs to describe its actions. He believed that states, like living organisms, progress through stages of youth, maturity, old age, and, unlike living organisms, even a possible return to youth. He also believed that one could determine the general well-being of the state by regarding its size as measured according to its geographic expansion or contraction over time. Ratzel, though certainly an environmental determinist, did not believe that the state is an organism, only that it *acts* like one and grows—increases its territory—as its population grows.

Ratzel wrote during the late nineteenth century, a period of tremendous change in Europe, when the system of states that largely persists to this day was being solidified. It was also a period when European states, Japan, and the United States were maintaining or initiating imperialist practices. The states' efforts to expand were seen as a response to population pressures for more territory and resources. These pressures also led to the need for new markets for manufactured goods and increased colonization of less economically developed places. Eventually all-out war within Europe occurred as states fought each other over territories.

Imperialism, Colonialism, and the North/South Divide

Geopolitics may involve the extension of power by one group over another. Two ways in which it may occur are imperialism and colonialism, related processes. As was discussed in Chapter 1, imperialism is the extension of state authority over the political and economic life of other territories. As Chapter 2 describes, over the last 500 years imperialism has resulted in the political or economic domination of strong core states over the weaker states of the periphery. Imperialism does not necessarily imply formal governmental control over the dominated area. It can also involve a process by which some countries pressure the independent governments of other countries to behave in certain ways. This pressure may take many forms, such as military threat, economic sanctions, or cultural domination. Imperialism always involves some form of *authoritative control* of one state over another.

The process of imperialism begins with exploration (**Figure 9.14**), often prompted by the state's perception that there is a scarcity or lack of a critical natural resource. It culminates in development via colonization, the exploitation of indigenous people and resources, or both.

In the first phases of imperialism, the core exploits the periphery for raw materials. As the periphery becomes developed, colonization may occur and cash economies may be introduced where none previously existed. The periphery may also become a market for the manufactured goods of the core. Eventually, though not always, the periphery—because of the availability of cheap labor, land, and other inputs to production—can become a new arena for large-scale capital investment. Some peripheral countries improve their status to become semiperipheral or even core countries. **Figure 9.15** illustrates the colonies created by European imperialism in Africa.

Colonialism is a form of imperialism in that it involves the formal establishment and maintenance of rule by a sovereign power over a foreign population through the establishment of settlements. The colony does not have any independent standing within the world system, but it is considered an adjunct of the colonizing power. From the fifteenth to the early twentieth century, colonialism constituted an important component of core expansion. Between 1500 and 1900 the primary colonizing states were Britain, Portugal, Spain, the Netherlands, and France. **Figure 9.16** illustrates the colonization of South America, largely by Spain and Portugal.

Other important states more recently involved in both colonization and imperialist wars include the United States and Japan. Although colonial penetration often results in political dominance by the colonizer, such is not always the case. For example, Britain may have succeeded in setting up British colonial communities in China, but it never succeeded in imposing British administrative or legal structures in any widespread way. And at the end of the colonial era a few former colonies, such as the United States, Canada, and Australia, became core states themselves. Others, such as Rwanda, Bolivia, and Cambodia, remain

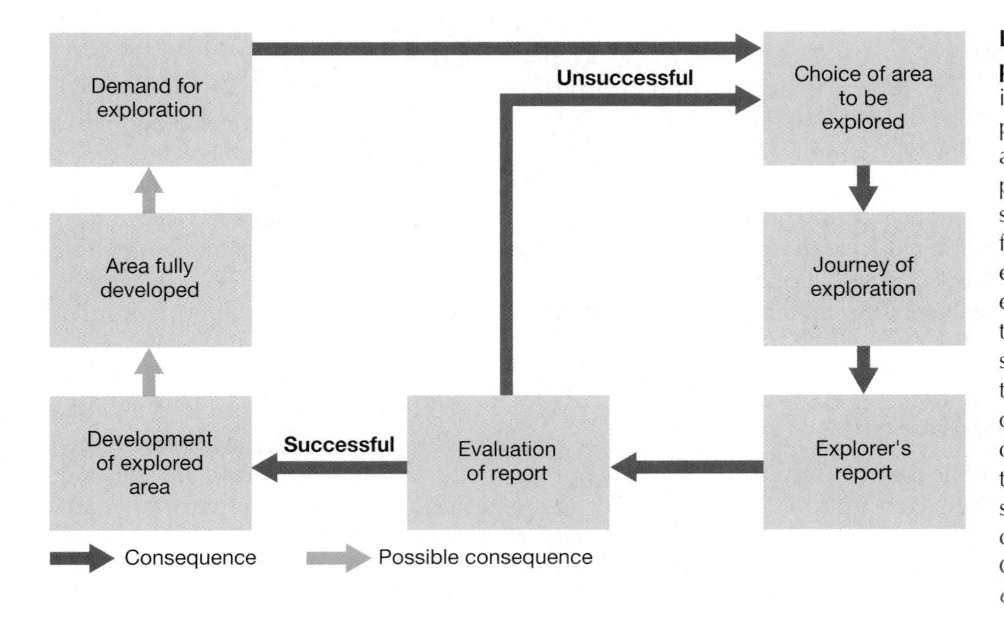

Figure 9.14 Principal steps in the process of exploration This diagram illustrates the main elements in the process of exploration, beginning with a need in the home country that prompts a desire to look outward to satisfy that need. Geographers have figured prominently in the process of exploration by identifying areas to be explored as well as actually traveling to those places and cataloging resources and people. Nineteenth-century geography textbooks are records of these explorations and the ways geographers conceptualized the worlds they encountered. Exploration is one step in the process of imperialism; colonization is another. (After J. D. Overton, "A theory of exploration," *Journal of Historical Geography* 7, 1981, p. 57.)

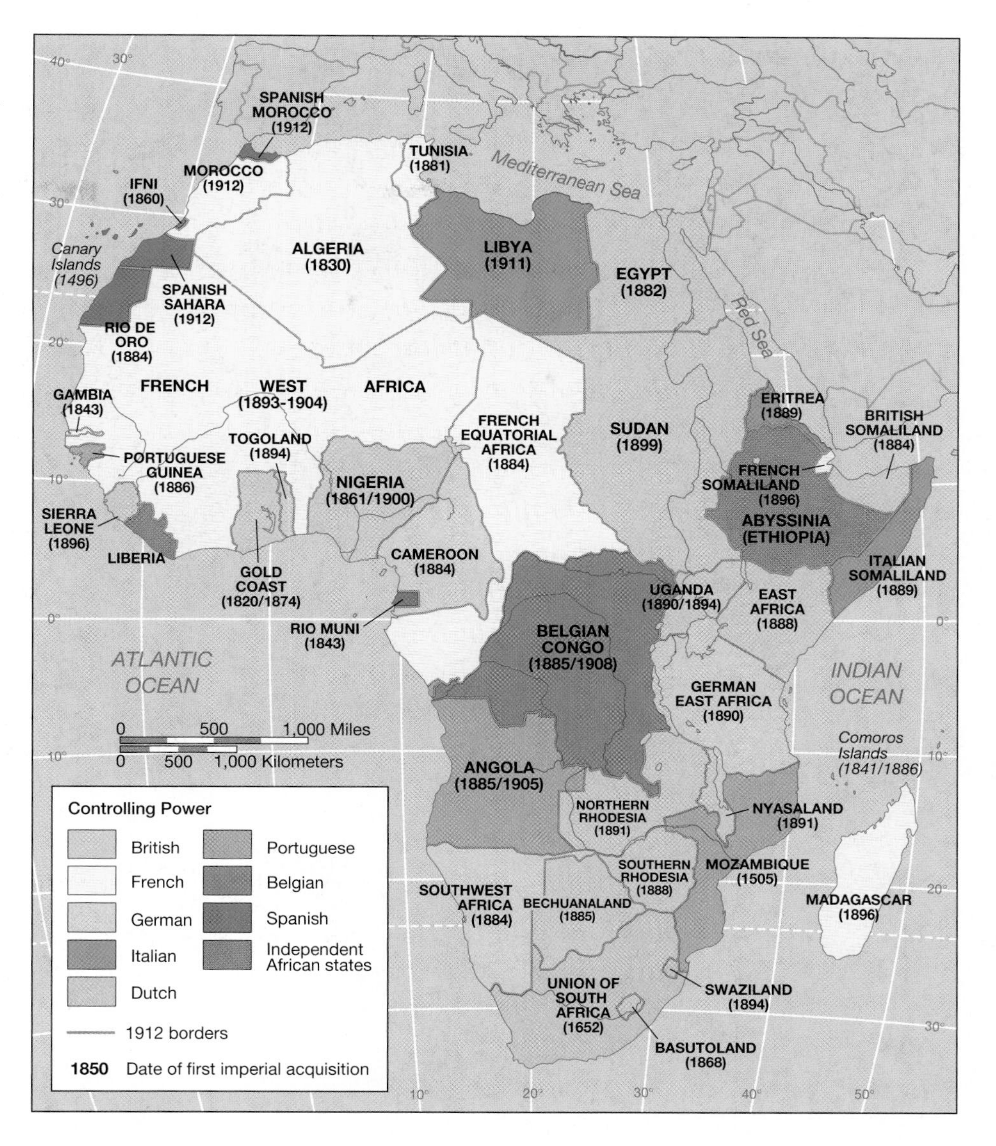

Figure 9.15 European colonies in Africa, 1496–1912 The partitioning of the African continent by the imperial powers created a crazy quilt that cross-cut preexisting affiliations and alliances among the African peoples. Lying directly within easy reach of Europe, Africa was the most likely continent for early European expansion. The Belgian, Italian, French, German, and Portuguese states all laid claim to various parts of Africa and in some cases went to war to protect those claims. (After *Harper Atlas of World History.* New York: HarperCollins, 1992, p. 139.)

firmly within the periphery. Some former colonies, such as Mexico and Brazil, have come close to the core but have not fully attained core status and therefore are categorized as being within the semiperiphery.

Two examples of colonialism are the extension of British rule to India and French rule to Algeria. The substantial British presence in India began with the establishment of the East India Trading Company in the mid-eighteenth century. The British government gave the company the power to establish forts and settlements, as well as to maintain an army. The company soon established settlements—including factories—in Mumbai (formerly Bombay), Chennai (formerly Madras), and Calcutta. What began as a small trading and manufacturing operation over time burgeoned into a major military, administrative, and economic presence by the British government, which did not end until Indian independence in 1947. During that 200-year period Indians were brutalized and killed and their society transformed by British influence. That influence permeated nearly every institution and practice of daily life—from language and judicial procedure to railroad construction and cultural identity (**Figure 9.17**).

The postcolonial history of the Indian subcontinent has included partition and repartition, as well as the eruption of regional and ethnic conflicts. In 1947 Pakistan split off from India and became a separate Muslim state. In 1971 Bangladesh, previously part of Pakistan, declared its independence. Regional conflicts include radical Sikh movements for independence in the states of Kashmir and Punjab. Ethnic conflicts include decades of physical violence between Muslims and Hindus over religious beliefs, and the privileging of Hindus over Muslims in the national culture and economy. It would be misleading, however, to attribute all of India's current strife to colonialism. The preexisiting caste system also plays a significant role in political conflict. The Hindu caste system, which distinguishes social classes based on heredity, preceded British colonization and persists to this day.

The French presence in Algeria is a story of 132 years of colonialism, with accompanying physical violence as well as cultural, social, political, and economic dislocation.

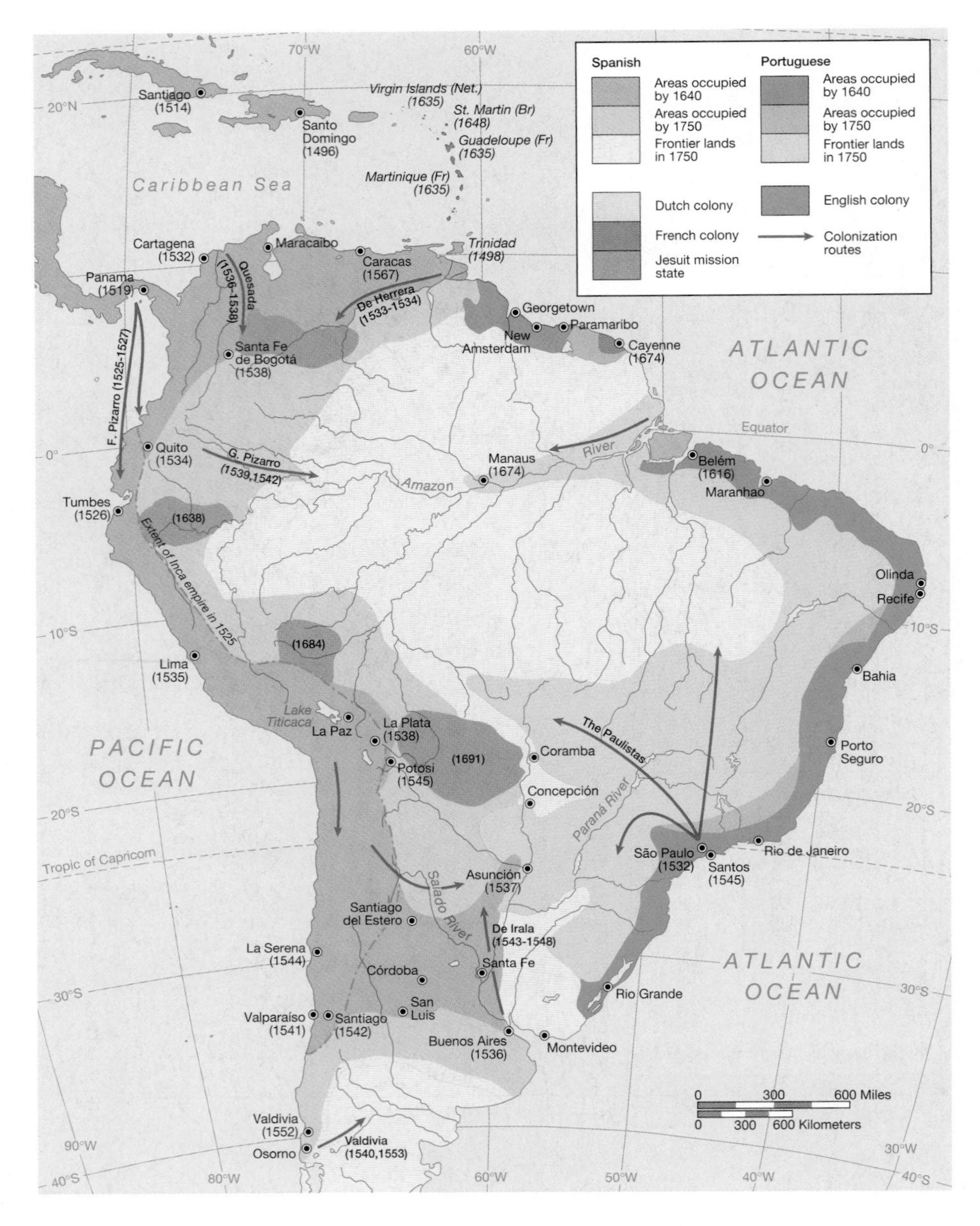

Figure 9.16 Colonization in South America and the Caribbean, 1496–1667 The Spanish and Portuguese dominated the colonization and settlement of South America, with the Dutch, French, and English maintaining only a minor and largely tentative presence. While African colonization focused upon new subjects and the simple acquisition of additional territories, South American colonization yielded rich commodity and mineral returns. (After *Rand McNally Atlas of World History.* Skokie, IL: Rand McNally, 1992, p. 85.)

Over the course of more than a century, the French appropriated many of the best agricultural lands and completely transformed Algiers, the capital, into a Westernized city. They also imposed a veneer of Western religious and secular practice over the native and deeply rooted Islamic culture.

In the cases of both India and Algeria, achieving independence has been a painful and sometimes bloody process for both colonizer and colonized. In Algeria, for example, in the aftermath of the protracted war of independence, there occurred a major exodus of French settlers, many of whose families had lived in Algeria for several generations. With few Westerners in positions of power, important aspects of Islamic society and culture were restored. More recently, Western-oriented politicians, bureaucrats, and citizens have been threatened and killed by radical Islamists who seek to return Algeria to a state governed by strict religious tenets (see Chapter 5 on cultural nationalism).

Since the turn of the nineteenth century, the effects of colonialism continue to be felt as peoples all over the globe struggle for political and economic independence. The 1994 civil war in Rwanda is a sobering example of the ill effects of colonialism. As occurred in India, where an estimated 1 million Hindus and Muslims died in a civil war when the British pulled out, the exit of Belgium from Rwanda left colonially created tribal rivalries unresolved and seething. Although the Germans were the first to colonize Rwanda, the Belgians, who arrived after World War I, established political dominance among the Tutsi by allowing them special access to education and the bureaucracy.

Previously a complementary relationship had existed between the Tutsi, who were cattle herders, and the more

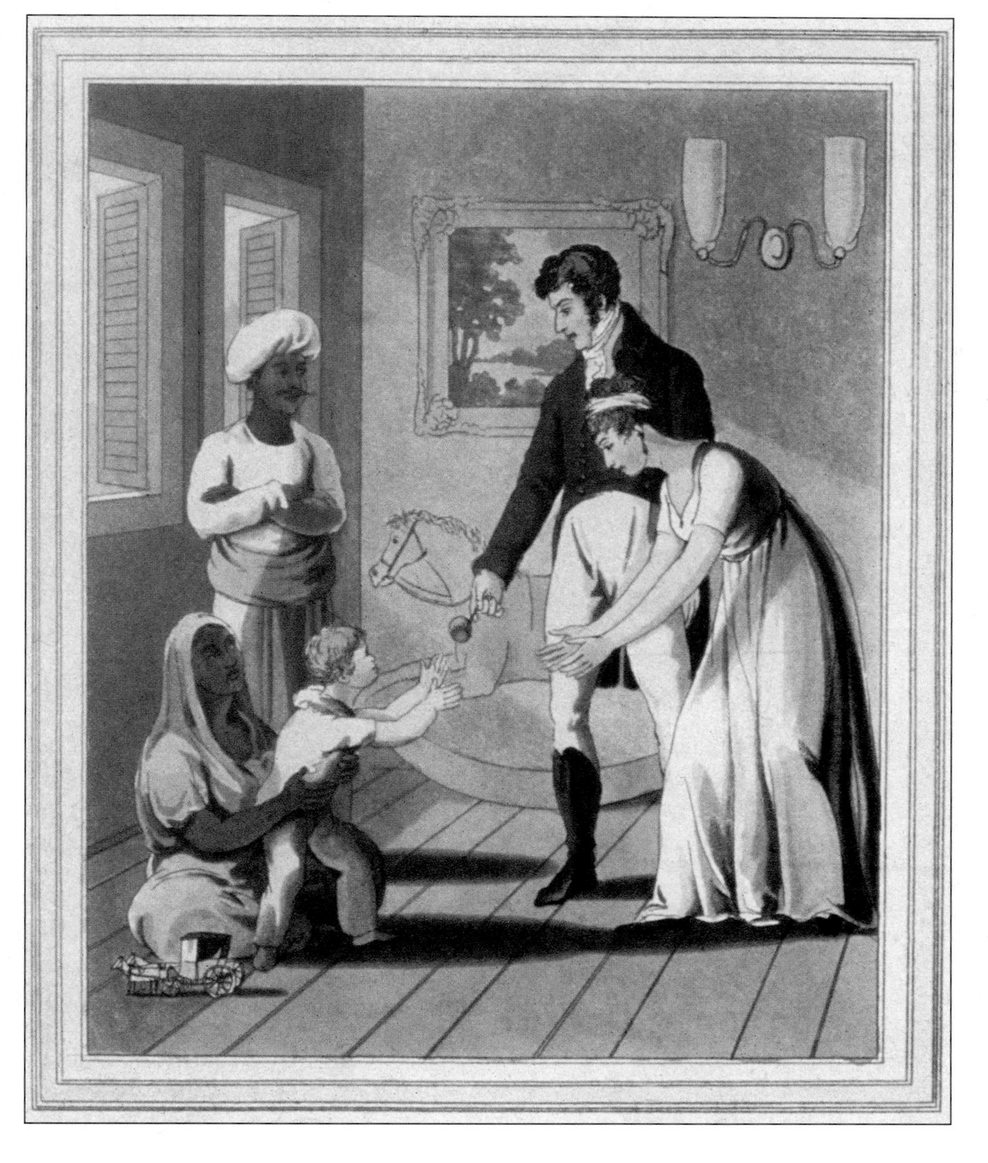

Figure 9.17 British colonialism in India The British presence in India affected culture, politics, the economy, and the layout of cities, as well as numerous other aspects of everyday life. This painting shows how Indian and British cultural practices intermingled, changing both in the process. (Note that the British child is being held by the Indian woman.) Indian society absorbed and remolded many British political and cultural practices so that contemporary Indian government, for example, embodies both British and Indian ideals and practices. British society continues to be shaped by its colonial history in India, most obviously through the large numbers of Indians who have migrated to Britain, affecting all aspects of society and culture.

numerous Hutu, who were agriculturists. In effect, colonialism introduced difference into an existing political and social structure that had operated more or less peacefully for centuries. In 1959 the Hutus rebelled and the Belgians abandoned their Tutsi favorites to side with the Hutus. In 1962 the Belgians ceded independence to Rwanda, leaving behind a volatile political situation that has erupted periodically ever since, most tragically in 1994's civil war. After a year of violence in which over half a million Tutsis were killed, the Hutus were driven across the border to the Democratic Republic of the Congo and a new Tutsi-led Rwandan government was formed. The Hutu refugees gathered in U.N. camps that gradually came to be controlled by armed extremists, who transformed them into virtual military bases and used them to attack the Tutsis in Rwanda (**Figure 9.18**).

When Rwanda's Tutsi-led military, with Uganda's support, invaded the Democratic Republic of Congo (DRC) to break up the camps, over a million refugees were released. Many have since fanned out across Central Africa. The most militant are instigating conflict and perpetrating atrocities in Uganda, DRC, and Burundi. Currently the political situation in Central Africa is one of extreme instability.

The colonization of Africa, South America, parts of the Pacific, Asia, and smaller territories scattered throughout the Southern Hemisphere resulted in a political geographic division of the world into North and South, known as the **North/South divide.** In the North are the imperialist states of Europe, the United States, Russia, and Japan. In the South are the colonized states. Though the equator has been used as a dividing line, some so-called southern territories, such as Australia and New Zealand, actually are part of the North in an economic sense.

The crucial point is that a relation of dependence was set up between countries in the South, or periphery, and those in the North, or the core, that began with colonialism and persists even today. Very few peripheral countries

Figure 9.18 Refugees returning to Rwanda Fleeing civil unrest in their own country, Rwandans from the Hutu tribe increasingly sought refuge in the Democratic Republic of Congo (formerly Zaire) when the Tutsi-led government assumed power in 1994. Two and a half years later, over half a million Rwandan refugees in the Democratic Republic of Congo occupied some of the largest refugee camps in the world. In late 1996 they began streaming back into Rwanda when the Tutsi-led government urged them to come help rebuild the country. Faced with two difficult alternatives—dire conditions in the camps or possible violence in Rwanda—many refugees chose to go home. Tens of thousands of Rwandans jammed the road between eastern Democratic Republic of Congo and Rwanda for over three days.

have become prosperous and economically competitive since achieving political autonomy. Political independence is markedly different from economic independence, and the South remains very much oriented to the economic demands of the North. An example of this one-way orientation from South to North is the transformation of agricultural practices in Mexico as increasing amounts of production have become directed toward consumption in U.S. markets rather than toward subsistence for the local peasant populations. Mexico has been described as the "salad bowl" of North America.

Twentieth-Century Decolonization

The reacquisition by colonized peoples of control over their own territory is known as **decolonization.** In many cases sovereign statehood has been achievable only through armed conflict. From the Revolutionary War in the United States to the twentieth-century decolonization of Africa, the world map created by the colonizing powers has repeatedly been redrawn. Today this map comprises an almost universal mosaic of sovereign states.

Many former colonies achieved independence after World War I. Deeply desirous of averting wars like the one that had just ended, the victors (excluding the United States, which entered a period of isolationism following the war) established the League of Nations. One of the first international organizations ever formed, the League of Nations had a goal of international peace and security. **Figure 9.19** shows the member countries of the League. An **international organization** is one that includes two or more states seeking political and/or economic cooperation with each other.

Within the League a system was designed to assess the possibilities for independence of colonies and to ensure that the process occurred in an orderly fashion. Known as the *colonial mandate system,* it had some success in overseeing the dismantling of numerous colonial administrations. **Figures 9.20, 9.21,** and **9.22** illustrate decolonization during the twentieth century of Africa, Asia, and the South Pacific and during the nineteenth century in South and Central America. (Although the League of Nations proved effective in settling minor international disputes, it was unable to prevent aggression by major powers and dissolved itself in 1946. It did, however, serve as the model for the more enduring United Nations.)

Decolonization does not necessarily mean an end to domination within the world system, however. Even though a former colony may exhibit all the manifestations of independence, including its own national flag, governmental structure, currency, educational system, and so on, its economy and social structures may continue to be dramatically shaped in a variety of ways by core states. Participation in foreign aid, trade, and investment arrangements originating from core countries subjects the periphery to relations that are little different from those they experienced as colonial subjects.

For example, core countries' provision of foreign aid monies, development expertise, and educational opportunities to selected individuals in Kenya has created a class of native civil servants that is in many ways more strongly connected to core processes and networks than those operating within Kenya. This relatively small group of men and women, often foreign educated, now comprises the first capitalist middle class in Kenyan history. Their Swahili name indicates their strong ties to core globaliza-

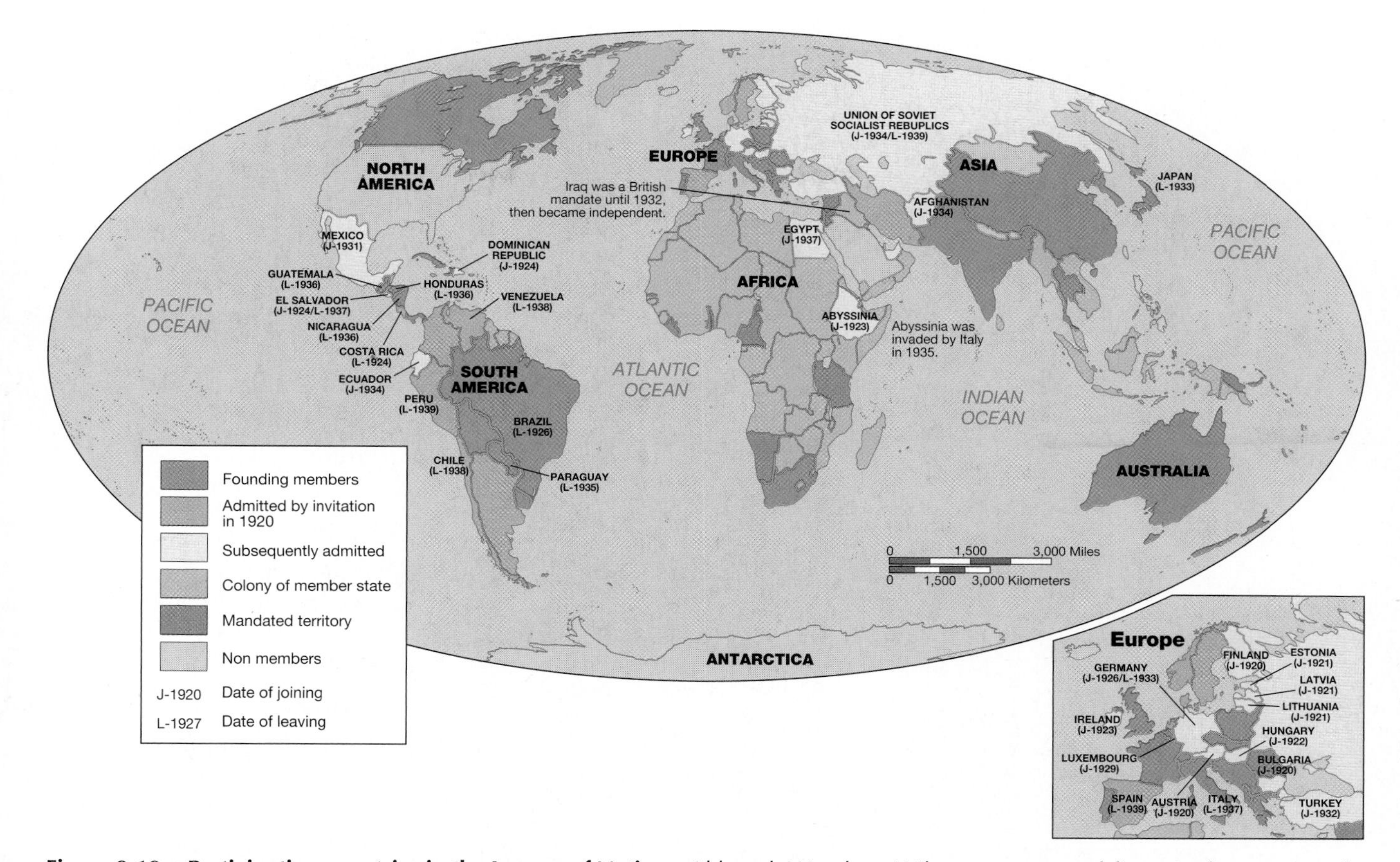

Figure 9.19 Participating countries in the League of Nations Although Woodrow Wilson was a central figure in the creation of the League of Nations, his inability to convince the U.S. Senate to join the League was a major blow to the effectiveness of this, the first international organization of the twentieth century. Britain and France played important roles in the League but were never able to secure arms limitations and security agreements among the membership. Perhaps its greatest success before it was dissolved in 1946 was pressing for the decolonization of Africa.

tion processes. They are called *wabenzi:* people (*wa*) who drive Mercedes-Benzes (*benzi*).

Commercial relations also enable core countries to exert important influence over peripheral, formerly colonized, countries. For example, *contract farming* has become the main vehicle around which agricultural production in the periphery is organized for core consumption. Dictated by core countries to growers in the periphery are the conditions of production of specified agricultural commodities. Whether the commodity is tea, processed vegetables, fresh flowers, rice, or something else, a contract sets the standards that are involved in producing the commodity.

For example, Japanese firms issue contracts that set the conditions of production for the Thai broiler (chicken) industry; the United Fruit Company, a U.S. firm, issues contracts for Honduran banana production. Thus, a core country can invoke a new form of colonialism in places it never formally colonized. As explained in Chapter 2, this new form, known as *neocolonialism,* is the domination of peripheral states by core states not by direct political intervention (as in colonialism) but by economic and cultural influence and control.

The spread of a capitalist world order has undeniably "modernized" traditional societies through education, health care, and other influences. This world order, however, based on imperialism and colonialism, has been financed with a great deal of bloodshed. An example is the imperial historical geography of South Africa, a country still laboring under the burdens of colonization. South Africa, as well as other formerly colonized locations around the globe, such as Egypt and Indonesia, show how the global process of imperialism and colonialism has unfolded locally.

Geographers have historically played very central roles in the imperialist efforts of European states. As Figure 9.14 illustrates, imperialism usually begins with exploration. Most, if not all, of the early geographical expeditions undertaken by Europeans were intended to evaluate the possibilities for resource extraction, colonization, and the expansion of empire. Organizations like the royal geographical societies in England and Scotland were explicitly formed to aid in the expansionary efforts of their home countries.

Exploration and colonization did not cease at the midpoint of the twentieth century, however. In fact,

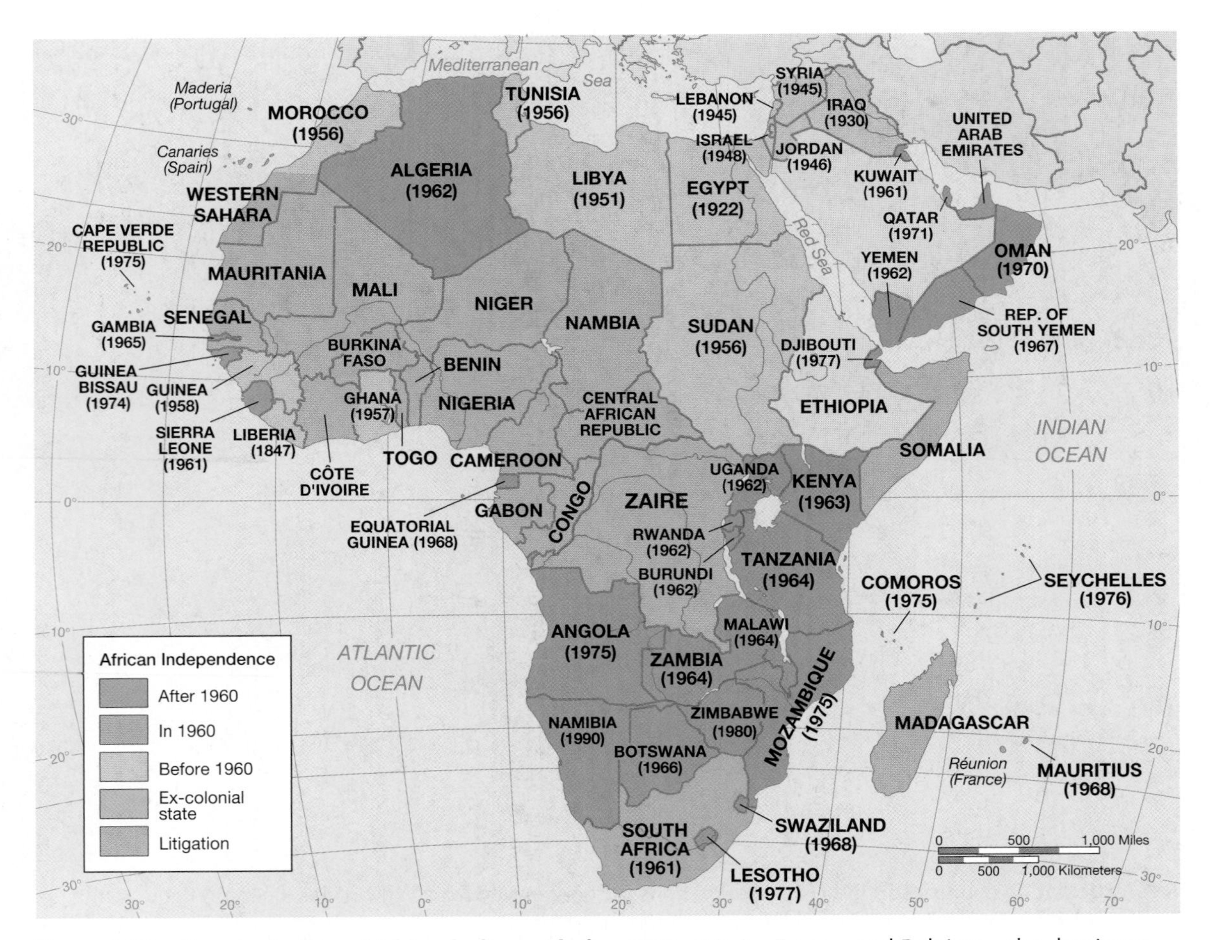

Figure 9.20 Decolonization of Africa, before and after 1960 Britain, France, and Belgium—the dominant European presences in African colonization—were also the first to divest themselves of their colonies. Britain was the first colonial power to grant independence. France granted independence to its African colonies soon after Britain made the first move. In the French-speaking former colonies, the transition to independence occurred largely without civil strife. Belgium's withdrawal as well as the withdrawal of Britain from the remainder of its colonial holdings did not go at all smoothly, with civil wars breaking out. Portugal did not relinquish its possession of Guinea Bissau, Mozambique, or Angola until 1974. (After *The Harper Atlas of World History, Revised Edition,* Librairie Hachette, p. 285. Copyright © 1992 by HarperCollins Publishers, Inc. Reprinted by permission of HarperCollins Publishers, Inc.)

exploration, and to a lesser extent, colonization are still occurring in Antarctica (**Figure 9.23**). This ice-covered land mass is an unusual example of twentieth-century imperialism in which strong states exerted power in an area where no people and, therefore, no indigenous state power existed. At present, while no one country exclusively "owns" the continent, 15 countries lay claim to territory and/or have established research stations there: Argentina, Australia, Belgium, Brazil, Chile, China, France, Germany, India, Japan, New Zealand, Norway, the United Kingdom, the United States, and Uruguay.

Heartland Theory

Because imperialism and colonialism shaped the existing world political map, it is helpful to understand one of the theories that drove them. By the end of the nineteenth century numerous formal empires were well established, and imperialist ideologies were dominant. To justify the strategic value of colonialism and explain the dynamic processes and possibilities behind the new world map created by imperialism, Halford Mackinder (1861–1947) developed a theory. Mackinder was a professor of geography at Oxford University and director of the London School of Economics. He later went on to serve as a member of Parliament from 1910 to 1922 and as chairman of the Imperial Shipping Committee from 1920 to 1945. With his background in geography, economics, and government, it is not surprising that his theory highlighted the importance of geography to world political and economic stability and conflict.

Mackinder believed that Eurasia was the most likely base from which a successful campaign for world conquest could be launched. He considered its closed heart-

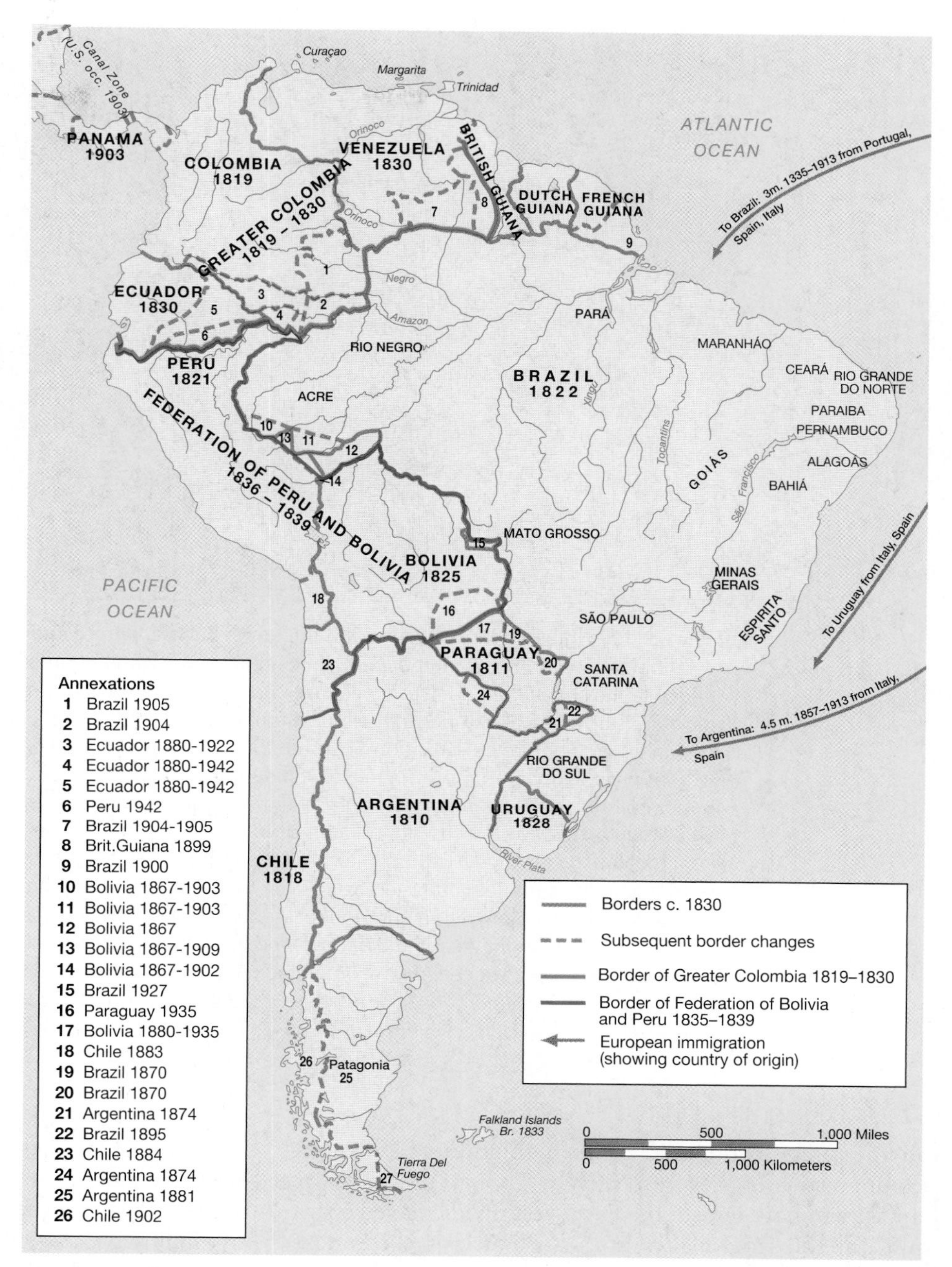

Figure 9.21 Independent South America, nineteenth century In comparison to Africa, independence came much earlier to South America (it had also been colonized much earlier than Africa). What was most influential in the independence movements there was the presence of local Spanish and Portuguese elites. This colonial ruling class became frustrated with edicts and tax demands from the home country and eventually waged wars—not unlike the U.S. Revolutionary War—against Spain and Portugal for independence. (After *Rand McNally Atlas of World History.* Skokie, IL: Rand McNally, 1992, p. 113.)

land to be the "geographical pivot," the location central to establishing global control. Mackinder premised his model on the conviction that the age of maritime exploration, beginning with Columbus, was drawing to a close. He theorized that land transportation technology, especially railways, would reinstate land-based power rather than sea prowess, as essential to political dominance. Eurasia, which had been politically powerful in earlier centuries, would rise again because it was adjacent to the borders of so many important countries, it was not accessible to sea power, and it was strategically buttressed by an inner and outer crescent of land masses (**Figure 9.24**).

When Mackinder presented his geostrategic theory in 1904, Russia controlled a large portion of the Eurasian land mass protected from British sea power. In an address to the British Royal Geographical Society, Mackinder suggested that the "empire of the world" would be in sight if one power, or combination of powers, came to control the heartland. He believed that Germany allied with Russia, and China organized by Japan, were alliances to be feared. Mackinder's theory was a product of the age of imperialism. To understand why Britain adopted this theory, remember that antagonism was increasing among the core European states leading to World War I a decade later.

The East/West Divide and Domino Theory

In addition to a North/South divide based on imperialism and colonialism, the world order of states could also be seen to divide along an East/West split. The **East/West**

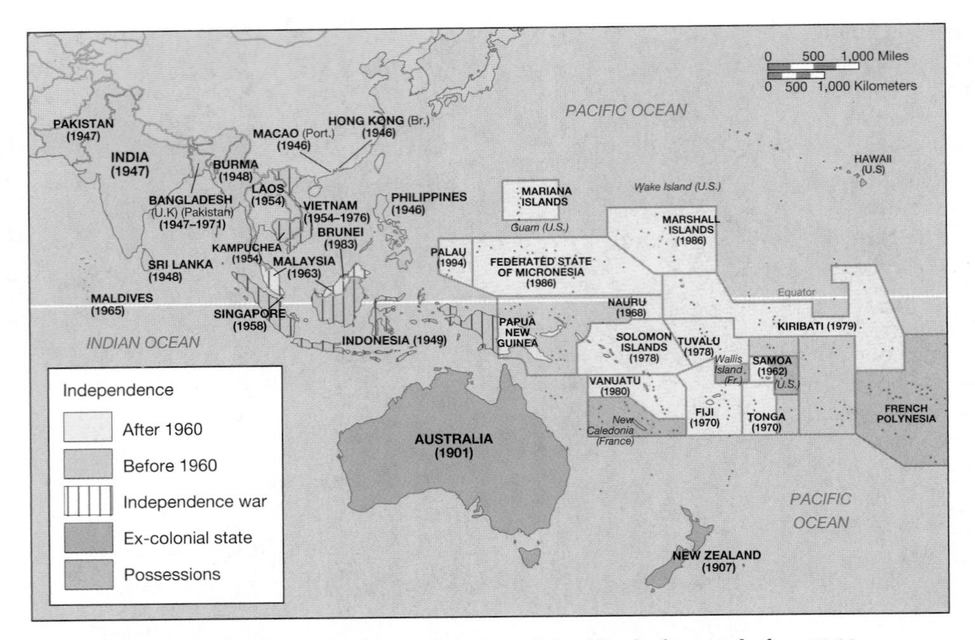

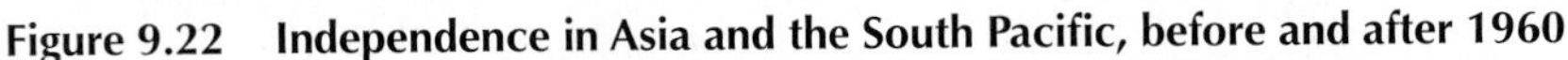
Figure 9.22 Independence in Asia and the South Pacific, before and after 1960
Decolonization and resultant independence are not uniform phenomena. Different factors influence the shape that independence takes. The form of colonial domination that was imposed is as much a factor as the composition and level of political organization that existed in an area before colonization occurred. Some former colonies gained independence without wars of liberation; in Asia these include India and Australia. In other places the former colonizers were prepared to surrender their colonies only after wars of liberation were waged. This was the case in Indochina, where the domino theory influenced the French colonizers—and later the United States—to go to war against the Vietcong. The war for independence in Vietnam lasted from 1954 until 1976, exacting huge costs for all involved. Mostly, decolonization and political independence forced societies into a nation-state mold for which they had little, if any, preparation. It is little wonder, then, that few former colonies have succeeded in competing effectively in a world economy. (After *The Harper Atlas of World History, Revised Edition,* Librairie Hachette, p. 283. Copyright © 1992 by HarperCollins Publishers, Inc. Reprinted by permission of HarperCollins Publishers, Inc.)

divide refers to the gulf between communist and noncommunist countries, respectively. Though the cold war appears to have ended, the East/West divide played a significant role in global politics since at least the end of World War II in 1945 and, perhaps more accurately, since the Russian Revolution in 1917. By the second decade of the twentieth century, the major world powers were backing away from colonization. Still, many were reluctant to accelerate decolonization for fear that independent countries in Africa and elsewhere would choose communist political and economic systems instead of some form of Western-style capitalism.

Cuba provides an interesting illustration of an East/West tension that persists despite the official end of the cold war. Although Cuba did not become independent from Spain until 1902, U.S. interest in the island dates back to the establishment of trade relations in the late eighteenth century. U.S. commercial expansion to Cuba in the first half of the twentieth century wrought major changes in the island's global linkages as U.S. economic imperialism replaced Spanish colonialism. Cuba, at that time, was experiencing a series of reform and revolutionary movements that culminated in the rise of Fidel Castro in the 1950s.

Following the end of World War II, anticommunist sentiment was at its peak in the United States, and U.S. fears about Cuba's "going communist" intensified. With Castro in power by 1959, the United States began training Cuban exiles in Central America for an attack on Cuba. This training proceeded despite the fact that in a 1959 visit to the United States, Castro had publicly declared his alliance with the West in the cold war. In 1961 an unsuccessful attempt by U.S.-trained Cuban exiles to invade Cuba and overthrow Castro further undermined Cuba's relations with the United States and paved the way for improved Cuban-Soviet relations.

Since that time, the United States's method of dealing with communism in Cuba has been economic, not military. Having failed to overthrow Castro militarily, the United States fell back upon a 1960 embargo designed to destabilize the Cuban economy. The intent of the embargo, which is still in place, was to create such popular dis-

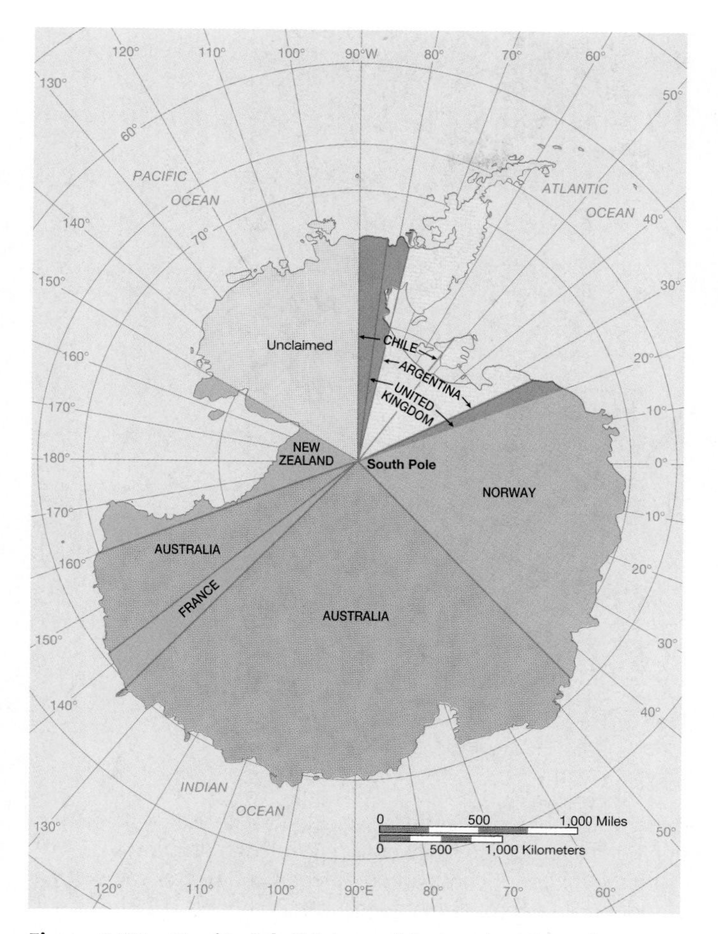

Figure 9.23 Territorial divisions of Antarctica Even the uninhabitable terrain of Antarctica has become a site for competition among states. The radial lines delineating the various claims bear no relationship to the physical geography of Antarctica; rather, they are cartographic devices designed to formalize and legitimate colonial designs on the region. (*Source:* Reprinted with permission from Prentice Hall, from J. M. Rubenstein, *The Cultural Landscape: An Introduction to Human Geography,* 5th ed., © 1996, p. 294.)

satisfaction with economic conditions that Cubans themselves would eventually overthrow Castro.

The embargo forbids economic trade between the United States and Cuba across a wide range of activities and products. While the embargo has not been successful—Castro remains in power—the country has become increasingly embattled because the collapse of the Soviet Union has left it without powerful and economically generous allies. Yet the standoff between Cuba and the United States persists, aggravated in the mid-1990s by Castro's tacit approval of the release of tens of thousands of refugees—many from Cuban prison populations or suffering from HIV infection—to the United States.

The end of World War II marked the rise of the United States to a dominant position among countries of the core. Following the war, the tension that arose between East and West translated into an American foreign policy that pitched it against the former Soviet Union. Domino theory underlay that foreign policy, which included economic, political, and military objectives directed at preventing Soviet world domination. **Domino theory** held that if one country in a region chose or was forced to accept a communist political and economic system, neighboring countries would fall to communism as well, just as one falling domino in a line of dominos causes all the others to fall. The means of preventing the domino-like spread of communism was often military aggression.

Domino theory first took hold in 1947, when the postwar United States feared communism would spread from Greece to Turkey to Western Europe. It culminated in U.S. wars in Korea, Vietnam, Nicaragua, and El Salvador. Yet preventing the domino effect was based not just on military aggression. Cooperation was also emphasized, as in the establishment of international organizations like NATO (North Atlantic Treaty Organization) in 1949, which had the stated purpose of *safeguarding* the West against Soviet aggression. After World War II, the core countries set up a variety of foreign aid, trade, and banking organizations. All were intended to open foreign markets and bring peripheral countries into the global capitalist economic system. The strategy not only improved productivity in the core countries but also was seen as a way of strengthening the position of the West in its confrontation with the East.

The Vietnam War and its aftermath were probably the most serious global manifestation of cold war competition and wrought terrible social and environmental effects on Southeast Asia as well as on U.S. domestic and international politics. More than a million Vietnamese died, together with 58,000 Americans (**Figure 9.25**). U.S. forces sprayed 2 million hectares (5 million acres) of Vietnam with defoliants, such as Agent Orange, that poisoned ecosystems and caused irreparable damage to human health. Cambodia and Laos were also bombed with napalm and defoliated to disrupt communist supply lines and camps.

The media images of destruction, the loss of U.S. lives, and the financial cost of the war created considerable opposition to the war in the United States, including protests on college campuses and marches on Washington. The United States began to withdraw its forces in the early 1970s and left South Vietnam altogether as the Vietcong approached Saigon in 1973; in 1975 Vietnam was unified under communist rule.

Two million people left South Vietnam fearing repression after unification, many (the so-called Vietnamese boat people) sailing away in small, fragile boats. The communist government confiscated farms and factories from owners to create state- and worker-owned enterprises, resettled hill tribes into intensive agricultural zones, and moved 1 million people into new economic development regions. But U.S.-led economic sanctions from 1973 to 1993 limited the potential for exports and restricted some critical imports, such as medicines. At present, sanctions

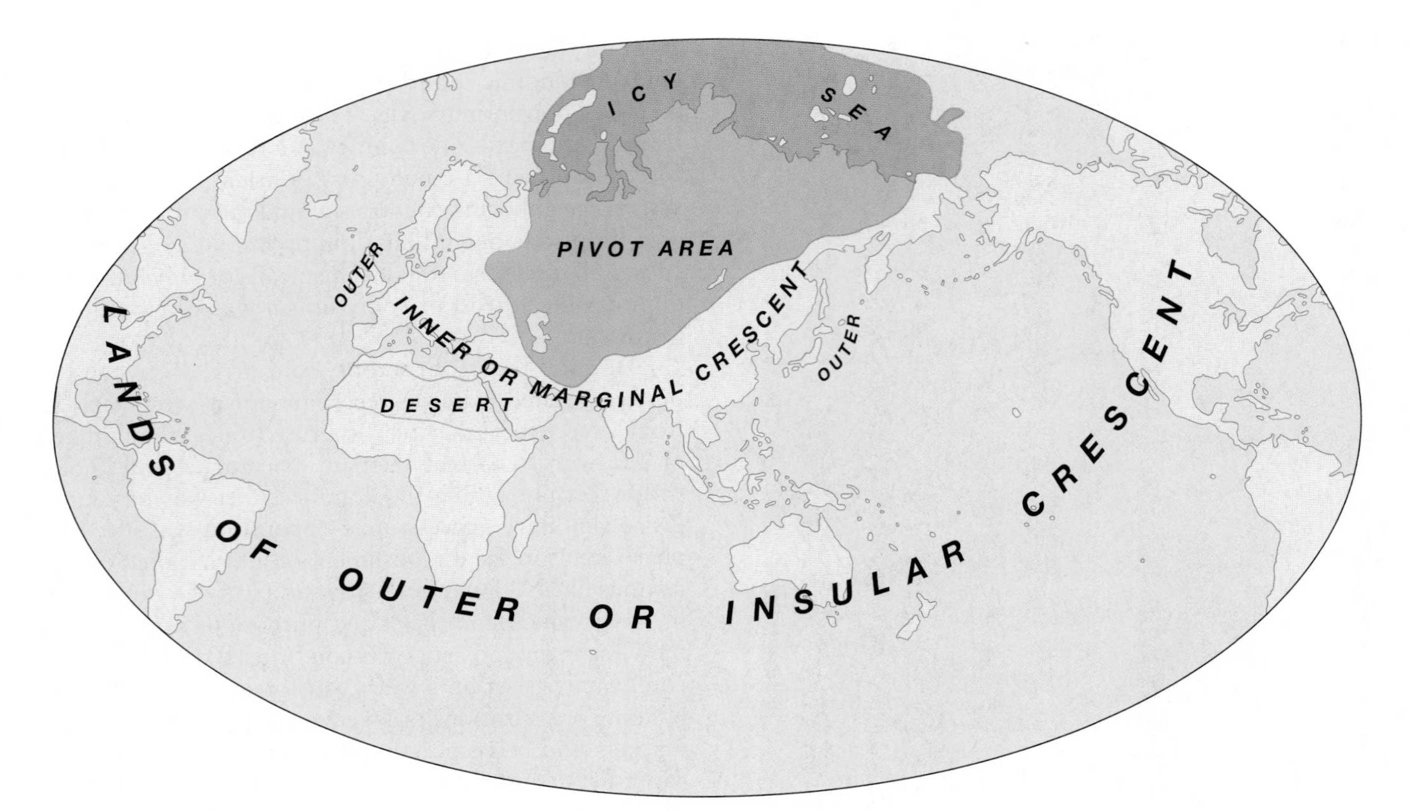

Figure 9.24 The heartland A quintessential geographical conceptualization of world politics, Mackinder's heartland theory has formed the basis for important geopolitical strategies throughout the decades since its inception. While the pivot area of Eurasia is wholly continental, the outer crescent is mostly oceanic and the inner crescent part continental and part oceanic. It is interesting to compare the Mercator map projection, which Mackinder used to promote his geostrategic theory, with the Dymaxion projection used in several maps in this text. This is a classic example of how maps can be used for ideological purposes. The Mercator projection decreases the size of the northern and southern oceans, which are vast and significant natural barriers. The spatial distortions inherent in the Mercator projection overemphasize the importance of Asia, and the splitting of North and South America so that they appear on both sides of the map exaggerates the centrality of Asia. The Dymaxion projection, as a northern polar representation, deemphasizes the centrality of any one landmass but exaggerates distances between continents (see Chapter 2). Mackinder's world-view map provides a good example of how cartographic representations can be employed to support ideological arguments. (After M. I. Glassner and H. DeBrij, *Systematic Political Geography,* 3rd ed. New York: J. Wiley & Sons, 1980, p. 291.)

continue to be lifted, and trade and diplomatic relations with the Socialist Republic of Vietnam has improved dramatically since the mid-1990s.

In Cambodia the Khmer Rouge communist revolutionaries overcame the U.S.-backed military government in 1975 and instituted a cruel regime under the leadership of Pol Pot. The Khmer Rouge suspended formal education, emptied the cities, and set out to eliminate the rich and educated and to isolate themselves from the world, renaming the country Kampuchea. It is estimated that nearly 2 million people—approximately 30 percent of Cambodia's population—died in the brutal death march out of the capital, Phnom Penh, in 1975–1976 and in the so-called killing fields of the countryside where civilians were tortured and executed. Conflict did not end even with the 1991 U.N. peace settlement and the return of the monarch, King Sihanouk, because the Khmer Rouge retook power in a coup in 1997. Today the Khmer Rouge have been defeated, and the Cambodian people, with the help of the international community, are building a democratic society as they recover from the emotional, physical, and economic brutality of the recent past.

The New World Order and the Increase in Terrorism

With the fall of the Berlin Wall in 1989 and the opening up of former socialist and communist countries, such as China and Russia, to Western-style capitalist economic development, the cold war is widely regarded as over. In March 1991 President George H.W. Bush made a speech referring to a "new world order" following the collapse of the Soviet Union. Francis Fukayama, author of *The End of History and the Last Man,* subjected the term to a more thorough analysis and put it into circulation in the popular imagination. The notion of the **new world order** assumes that with the triumph of capitalism over communism, the United States becomes the world's only superpower and therefore its policing force. With the political, economic, and cultural dominance of the Unit-

Figure 9.25 The Vietnam War U.S. troops waiting to be evacuated at Khe Sanh, Vietnam, in 1968—a year during which domestic opposition to the U.S. involvement in Southeast Asia grew dramatically as more and more people in the United States saw the war as immoral and unwinnable.

ed States comes the worldwide promotion of liberal democracy and the promotion of a global economy predicated on transnational corporate growth through organizations like the World Bank and the World Trade Organization.

However, the move toward liberal, Western-style democracies and the capitalist consumption practices necessary to the success of the new global economy have created instability in some parts of the world. This instability is especially problematic where the cold war struggle between the United States and the Soviet Union was once waged, in countries that appeared ripe to succumb to communism. The recent history of Afghanistan provides a telling illustration of this instability and its geopolitical implications (see Box 9.1: "Afghanistan: From the Cold War to the New World Order").

The emergence of a new world order has also meant that radical forms of warfare and political practices have replaced more conventional ones. The attacks on the World Trade Center and the Pentagon on September 11, 2001, and the resulting war on terrorism are prime examples of the political and economic restructuring impulses of the new world order. The terrorist attacks and the subsequent response by the United States and other governments make clear that terrorism is becoming an increasingly important factor in global geopolitics.

Terrorism is a complicated concept whose definition very much depends upon social and historical context. A very simple definition is that **terrorism** is the threat or use of force to bring about political change. It is most commonly understood as actions by individuals or groups of individuals against civilian populations to undermine state practices or institutional organizations. But the state can also be an agent of terrorism, as the original use of the term makes clear. Because terrorism involves violent acts directed against society—whether by antigovernment actors, governments themselves, angry mobs or militants, or even psychotic individuals—it will always mean different things to different people.

The term *terrorism* was first used during the French Revolution (1789–1795) to describe the new revolutionary government's repression of its people during the "reign of terror." Fifty years later the term began to be used to describe revolutionaries who violently opposed existing governments. As the nineteenth century came to a close, "terrorism" was often expanded to apply to militant labor and nationalist political organizations. By the mid-twentieth century the term was used to describe many left-wing groups, as well as subnationalist (minority groups within the nation-state) or radical ethnic groups. In the 1980s the violent activity of hate movements in the United States was defined as terrorist, while terrorism internationally

Afghanistan: From the Cold War to the New World Order

Afghanistan, known in ancient times as Gandhar, was once famous for its wealth, art, and culture (**Figure 9.A**). Its trading centers were important links on ancient trading routes between Central Asia and South Asia, and their wealth soon attracted invaders. Alexander the Great swept into Afghanistan—then a part of the Persian Empire—in 329 B.C. This invasion paved the way for a cultural awakening and the emergence of the Gandhar school of art, known for its amalgam of Indian and Greek styles.

But invading the mountain passes of Afghanistan is easier than maintaining control of them (**Figure 9.B**). The problem, put simply, is physical geography. The country is dominated by the rugged Hindu Kush mountains, which sweep from the east to the west, petering out near the northwestern city of Herat, where they sink into the desert. Tens of thousands of square kilometers of the Hindu Kush form an intricate and seemingly endless maze of valleys and ravines. Jagged scree-strewn mountains and rugged valleys and caves provide ideal territory in which to fight a guerrilla war against invaders or occupying forces. The problems of topography are compounded by the weather. By late October swirling snow descends on the mountains, sealing off many of the passes, valleys, and high plateaus and making the movement of troops almost impossible until late spring.

Figure 9.B The Khyber Pass Over the centuries, the Khyber Pass has been a gateway to Afghanistan's pivotal geopolitical position between Central Asia and South Asia. Persians, Greeks, Mongols, Afghans, and the British have all taken this route, and the surrounding mountains and gulches became the graveyard for many of them.

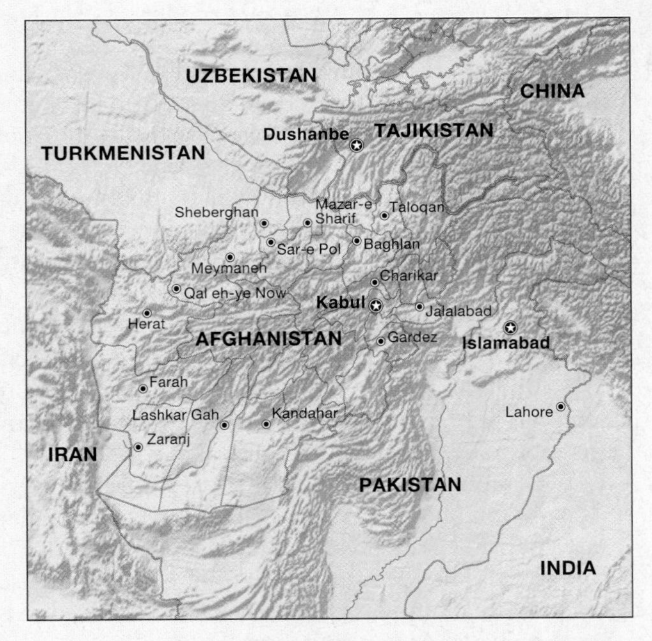

Figure 9.A Afghanistan As this map illustrates, Afghanistan is a land-locked, mountainous country sharing borders with six other countries. When war occurred here in the twentieth century, many of the people fled to these neighboring countries for safety.

Nevertheless, Afghanistan's geopolitical significance has attracted one invader after another. In A.D. 642, Arabs invaded the region and introduced Islam. Arabs quickly gave way to Persians, who controlled the region until they were conquered by Turkic Ghaznavids in A.D. 998. The Turks turned Ghazni into a great cultural center and a base for frequent forays into India. In A.D. 1219 a Mongol invasion, led by Genghis Khan, resulted in the destruction of many cities and the despoliation of fertile agricultural areas. Following Genghis Khan's death in 1227, a number of petty chieftains and princes struggled for supremacy until late in the fourteenth century, when one of his descendants, Timur, incorporated Afghanistan into his vast Asian empire. Babur, a descendant of Timur and founder of the Mughal dynasty, made Kabul the capital of his principality at the beginning of the sixteenth century. In 1747 Ahmad Shah Durrani consolidated the many chieftainships, petty principalities, and fragmented provinces of the region into one country. His rule extended from Mashhad in the west to Delhi in the east, and from the Amu Darya river in the north to the Arabian Sea in the south.

Late in the eighteenth century, Afghanistan's geopolitical significance increased still more. For the eastward-expanding Russian empire, Afghanistan represented the last barrier to a thrust toward the rich plains of India. For the British, who were establishing a hold on India, Afghanistan represented a bastion against Russian expansion. Both the Russians and the British desperately wanted to control Afghanistan, and so began the Great Game—the struggle between the two imperial powers for control of Afghanistan. The British were able to block the Russians in the Great Game but were not able to establish territorial control. After three wars with the stubborn Afghans, the British finally granted Afghanistan independence in 1921. The first Anglo-Afghan war (1839–1842) resulted in the destruction of the British Army and illustrated the ferocity of Afghan resistance to foreign rule and the difficulty of moving troops and supplies through the Hindu Kush. The second war (1878–1880) was sparked by the Afghans' refusal to accept a British mission in Kabul and resulted in the British and Russians together establishing the official boundaries of modern Afghanistan. The third war began in 1919 with an attack by Afghans on India and ended later that year when the war-weary British relinquished control over Afghan foreign affairs by signing the Treaty of Rawalpindi.

A brief period of Afghan independence followed, during which Mohammad Zahir Shah, who reigned from 1933 to 1973, established a relatively liberal constitution. Although Zahir's experiment in democracy produced few lasting reforms, it permitted the growth of extremist political movements, including the communist People's Democratic Party of Afghanistan (PDPA), which had close ideological ties to the Soviet Union, Afghanistan's northern neighbor.

In 1973, following charges of corruption against the royal family and poor economic conditions caused by a severe drought the previous year, a military coup, led by former prime minister Sardar Mohammad Daoud, abolished the monarchy and declared Afghanistan a republic with Daoud as its first president. His attempts to carry out badly needed economic and social reforms met with little success and were followed in April 1978 by a bloody coup led by the PDPA. During its first 18 months of rule, the PDPA brutally imposed a Marxist-style reform program that ran counter to deeply rooted Islamic traditions. As a result, opposition to the Marxist government emerged almost immediately.

The Soviet Union moved quickly to take advantage of the 1978 coup, signing a new bilateral treaty of friendship and cooperation with Afghanistan that included a military assistance program. Before long, as the opposition insurgency intensified, the PDPA regime's survival was wholly dependent upon Soviet military equipment and advisers. In December 1979, faced with a rapidly deteriorating security situation, the Soviet Union sent a large airborne force and thousands of ground troops to Kabul under the pretext of a field exercise. More than 120,000 Soviet troops were eventually sent to Afghanistan, but they were unable to establish authority outside Kabul. An overwhelming majority of Afghans opposed the Communist regime, and Islamic freedom fighters (*mujahideen*) made it almost impossible for the regime to maintain control outside major urban centers. Poorly armed at first, the *mujahideen* began receiving substantial assistance in the form of weapons and training from the United States, Pakistan, and Saudi Arabia in 1984. The *mujahideen* exploited Afghanistan's terrain expertly, firing from surrounding ridges to disable the first and last vehicles in the Soviet columns and then slowly picking off the soldiers trapped in the center. The guerrilla war ended in 1989 with the withdrawal of the Soviet Union, but as the victorious *mujahideen* entered Kabul to assume control over the city and the central government, a new round of fighting began between the various militia groups that had coexisted uneasily during the Soviet occupation.

With the demise of their common enemy, the militias' ethnic, clan, religious, and personality differences surfaced, and civil war ensued. Large-scale fighting in Kabul and in northern provinces caused thousands of civilian deaths and created new waves of displaced persons and refugees, hundreds of thousands of whom trekked across the mountains to Pakistan for sanctuary. Eventually, the hardline Islamist faction of the *mujahideen,* the Taliban (**Figure 9.C**), gained control of Kabul and most of Afghanistan. By 2001 the only remaining resistance to the Taliban regime was the Northern Alliance, a loose coalition of minority ethnic groups, including the Tajiks, who controlled between 5 and 10 percent of Afghan territory, including the beautiful Panjshir valley 65 miles (105 kilometers) to the northeast of Kabul and a small enclave around the far northern provinces of Badakhshan and Takhar. The Taliban regime not only imposed harsh religious

Figure 9.C Taliban militiamen The militia that dubbed itself the Taliban (which translates as "Islamic students") emerged in 1994 from the rural southern hinterlands of Afghanistan under the guidance of a reclusive former village preacher, Mullah Mohammed Omar. Fed by recruits from conservative religious schools across the border in Pakistan (most of whom were destitute refugees from the 1979–1989 war against the Soviet invasion), the Taliban won military and political support from Pakistan. It rose to power by promising peace and order to a country ravaged by corruption and civil war and the prospect of reestablishing the traditional dominance of the majority ethnic group, the Pashtun.

laws and barbaric social practices on the Afghan population but also harbored an entirely new geopolitical force with worldwide implications: Osama bin Laden and his Al Qaeda terrorist network, which was responsible for numerous attacks on the West, including the September 11, 2001, attacks on the Pentagon and the World Trade Center.

The United States, with support from the United Kingdom, Australia, Canada, and the Northern Alliance, invaded Afghanistan in October 2001 as part of its "War on Terrorism" campaign. The military campaign there is called Operation Enduring Freedom and the aim of the campaign has been to capture Osama bin Laden, who has been resident there since 1996, destroy the Al Qaeda network, and overthrow the Taliban regime. Bombing raids over Afghanistan ceased in 2002 but coalition forces are still operating in Afghanistan despite the recent democratic election of a new (transitional) president, Hamid Karzai, the first direct election in Afghanistan's history. While the Taliban regime has been displaced, the Al Qaeda network persists and Osama bin Laden has yet to be captured. Parliamentary elections, to fill seats in the Lower House and in provincial councils, took place in September 2005.

was seen as a brand of ethnic or subnational warfare sponsored by rogue regimes.

The Irish Republican Army (IRA) has been widely perceived as a rogue regime promoting terror in order to free Northern Ireland from British rule. It is important, however, to appreciate the complexity of the situation in Northern Ireland. For many Irish Catholics who yearn for a united Republic of Ireland, the IRA are freedom fighters who sacrifice their own lives, as well as the lives of others, often innocent civilians, to bring about this dream. Moreover, Protestant Orange Lodge organizations in Northern Ireland, which favor remaining part of Britain, are also considered by many to be a terrorist organization.

Ethnic and subnational terrorism affects many countries today, including Russia, Uzbekistan, India, China, Colombia, the Philippines, Israel, and Palestine. In Sri Lanka the Tamil minority rebelled in the early 1980s against the oppressive Sinhalese majority. While they started as a political movement, eventually the Liberation Tigers of Tamil Eelam launched a brutal campaign of terror through assassination and suicide bombings by dragooned children that has left 65,000 people dead. Another ethnic or subnational violent struggle involves the Russian region of Chechnya, where the population has demanded independence. For the Muslim Chechens, Russia is acting as a terrorist state by employing military force to keep them from gaining independence. Chechens believe they have little in common politically, historically, or culturally with Russia. The Russian state has identified the defiant Chechens as extremists and terrorists (see Box 9.2: "State Terrorism in Chechnya").

While subnational resistance organizations using terrorist tactics continue to operate throughout the world, the most widely recognized terrorism of the new century has religious roots. The September 11 attacks have helped to bring the realities of religious terrorism (or "faith-based" terrorism, to quote geographer Neil Smith) sharply into public focus. The connection between religion and terrorism is nothing new, as terrorism has been perpetrated by religious fanatics for more than 2000 years. Indeed, words like zealot, assassin, and thug all stem from fundamentalist religious movements of previous eras. And while the links between Muslims and terrorism in the world today are especially strong and geographically widespread, it is critical to understand that Muslim terrorism is not the only form of terrorism and that domestic terrorism sparked by fundamentalist Christian organizations has taken the lives of hundreds of innocent people and continues to be a real threat in the United States.

The most recent and widely known example of domestic, Christian, white supremacist–based terrorism in the United States is the bombing of the Alfred P. Murrah Federal Office Building in Oklahoma City in April 1995, when 168 people were killed. In June 1997 Timothy McVeigh, a U.S. Army veteran, was convicted in federal court of perpetrating the attack (**Figure 9.26**). McVeigh was executed by lethal injection in June 2001, while his accomplice, Terry L. Nichols, was sentenced to life in prison. Both McVeigh and Nichols were connected to the Christian Identity movement through the Michigan Militia, a 12,000-strong paramilitary survivalist organization. The Christian Identity movement is just one of several religious extremist groups in the United States, including other forms of white supremacy movements, apocalyptic cults, and Black Hebrew Israelism. The Christian Identity movement is based on a belief in the superiority of whiteness as ordained by God. Militant tax resistance and a form of regressive populism ("the people" are superior to government) are also characteristic of the U.S. militia groups. An estimated 800 militia groups are believed to exist in almost every state of the United States today. The orientations and objectives of these groups vary, but all are antigovernment and have a strong belief that God has ordained the superiority of white people. Conversely, members of the Black Hebrew Israelism movement believe that whites are evil incarnate and that blacks are the true Jews of old.

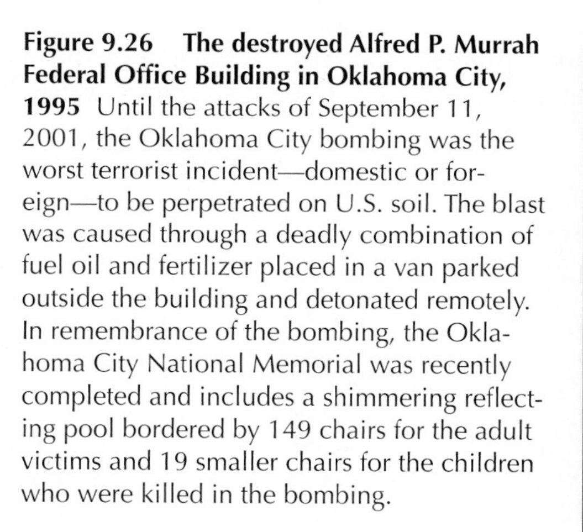

Figure 9.26 The destroyed Alfred P. Murrah Federal Office Building in Oklahoma City, 1995 Until the attacks of September 11, 2001, the Oklahoma City bombing was the worst terrorist incident—domestic or foreign—to be perpetrated on U.S. soil. The blast was caused through a deadly combination of fuel oil and fertilizer placed in a van parked outside the building and detonated remotely. In remembrance of the bombing, the Oklahoma City National Memorial was recently completed and includes a shimmering reflecting pool bordered by 149 chairs for the adult victims and 19 smaller chairs for the children who were killed in the bombing.

9.2 GEOGRAPHY MATTERS

State Terrorism in Chechnya

Of the many ethnic and subnational movements that surfaced with glasnost and the subsequent breakup of the Soviet Union, the Chechen independence movement has been the most bloody. In Chechnya, a region of the northern Caucasus mountain range (**Figure 9.D**), clans, not territory, is the traditional form of political organization. Ever since imperial Russia began expanding into the region in the late 1700s, the Sunni Muslim Chechens put up strong resistance, periodically waging holy wars against Christian Russia. When the Russian revolution occurred in 1917, they did not view the Bolsheviks as an improvement, not least because the newly created Soviet Union formally adopted scientific atheism as its state religion. Following a brief, failed attempt by the peoples of the North Caucasus and Transcaucasus (including the countries of Armenia, Azerbaijan, and Georgia) to resist Soviet domination, the Soviets decided to divide and conquer by creating administrative regions that encompassed a mixture of clans and ethnic groups. The anti-Soviet Chechens were put in the same region as the Ingush peoples to the south.

Figure 9.D The Northern Caucasus This reference map shows the principal features, political boundaries, and major cities of the northern Caucasus.

The Chechens remained defiant but paid a terrible price for doing so. In the late 1930s tens of thousands were liquidated by Stalin's purges aimed at suspected anti-Soviet elements. In 1944, after invading German forces were forced to retreat from the North Caucasus, Stalin accused the Chechens of having collaborated with the Nazis and ordered the entire Chechen population—then numbering about 700,000—exiled to Kazakhstan and Siberia. Brutal treatment during this mass deportation led to the deaths of more than 200,000 Chechens.

In 1957 Nikita Kruschev embarked on a program of de-Stalinization that included the rehabilitation of Chechens. But when the Chechens returned, they found newcomers had taken over many of their homes and possessions. Over the next 30 years many of these newcomers withdrew, while the Chechen population consolidated and grew to almost 1 million. When Mikhail Gorbachev initiated his policy of glasnost in 1985, Chechens finally saw a chance for self-determination, and with the breakup of the Soviet Union in 1989, they wasted no time unilaterally declaring their complete independence.

The most feared form of terrorism in the twenty-first century is bioterrorism. **Bioterrorism** is the deliberate use of microorganisms or toxins from living organisms to induce death or disease. Biological and chemical agents that could be used for bioterrorist purposes range from anthrax to West Nile virus. In 2002, the president of the United States signed into law the Bioterrorism Act, intended to protect the nation's food and drug supply from bioterrorist threats. Since 9/11, research about and public and commercial responses to possible bioterrorist threats have grown significantly. Legislation, such as the Bioterrorism Act of 2002, is the driving force behind this new growth industry.

The emergence and persistence of a wide range of extremist religious groups in the United States and throughout the world represents a potentially far more violent threat than terrorist groups of the early and mid-twentieth century. The rise in faith-based terrorism since the new millennium suggests that the potential for even greater acts of violence should not be ignored.

The United States responded to the terrorist attacks of September 11, 2001 by declaring a global war against terrorism and identifying first Afghanistan and then Iraq, as the greatest threats to U.S. security. Although the evidence of involvement in the 9/11 attacks by Iraq and its leader, Saddam Hussein, was highly questionable, on March 19, 2003 after amassing over 200,000 U.S. troops in the Persian Gulf region, President George W. Bush ordered the bombing of the city of Bagdad, Iraq. The declaration of war and invasion occurred without the explicit authorization of the U.N. Security Council, and some legal authorities take the view that the action violated the U.N.

Ingushetia decided to separate from Chechnya in 1992 and signed the Treaty of Federation with Russia. The Russian Federation chose at first to ignore Chechnya's declaration of independence but could not tolerate the possibility of losing the region, particularly since the area around Grozny is one of the Russian Federation's major oil-refining centers and has significant natural-gas reserves. In December 1994, Russian troops invaded Chechnya. The ensuing conflict brought terrible suffering to the Chechen population and resulted in mass migrations away from the fighting. Chechen resistance stiffened, with increased popular support because of the invasion. Russian forces were eventually worn down, and in 1996 the Russian Federation settled for peace, leaving Chechnya with *de facto* independence. For three years there were protracted negotiations over the nature of the peace settlement. In the summer of 1999, however, after Chechen rebels took the fight to the neighboring republic of Dagestan and to the Russian heartland with a series of terrorist bombings of apartment blocks, the Russian military effort was renewed. After bitter and intense fighting, during which the Russian army suffered over 400 deaths and nearly 1,500 wounded while hundreds of thousands of Chechens were made homeless and several thousand were dead or missing, Russian troops took the capital, Grozny, in February 2000. By that time Grozny was virtually uninhabitable (**Figure 9.E**). Since early 2000, Russian Federation troops have maintained control of Grozny, though they continue to be harassed by Chechen rebel guerrillas. In 2002, Chechen rebels seized a Moscow theater in an attack that left 129 hostages dead, almost all killed by a noxious gas

Figure 9.E Grozny refugees In December 1999, Chechens were advised by Russian troops to evacuate Grozny, the capital of Chechnya, before it was destroyed.

used by the Russian security forces to end the seige. Meanwhile, in Grozny, everything and everyone has been brutalized. Cars and buses line up behind the city's checkpoints, on which soldiers have scrawled the names of their hometowns and the warning "Stop 10 meters away or we shoot." Murders and kidnappings are commonplace, and organized crime flourishes through the Chechen mafiya. Chechyna provides just one example of state terrorism being practiced today. It also provides an illustration of the complexity of terrorism as a concept by showing that it can be practiced by both individuals as well as institutions; by rogue forces as well as by legitimate ones.

Charter. Some of the U.S.'s staunchest allies (Germany, France, and Canada) as well as Russia opposed the attack while hundreds of thousands of antiwar protestors repeatedly took to the streets throughout the world for the weeks and months preceding and following the onset of war launched by coalition forces of the United Kingdom and the United States. The motivation for the war, as expressed by British Prime Minister Tony Blair and George W. Bush, was that Iraq had stockpiled "weapons of mass destruction"—chemical and biological weapons capable of massive human destruction. In the days leading up to the war, the U.N. weapons inspector, Hans Blix, and his team were unable to locate any weapons despite an intensive search of the country. President Bush, however, proceeded to justify a dramatically stepped-up "war on terrorism" (following the war in Afghanistan) on the grounds that "neutralizing" Iraq's leader, Saddam Hussein, was necessary to global security.

On May 1, 2003, President Bush gave a speech after having landed a Lockheed S-3 Viking on the aircraft carrier USS *Abraham Lincoln,* where he announced the end of major combat operations in the Iraq war. The fact that "major combat" has ended, however, does not mean that peace has returned to Iraq. Iraq continues to experience violent conflict between U.S. and Iraqi soldiers and forces described by the occupiers as insurgents. The tactics in use include mortars, suicide bombers, roadside bombs, small arms fire, and rocket-propelled grenades, as well as sabotage against the oil infrastructure of the country. As of summer, 2005, the total number of deaths of U.S. soldiers as a direct result of the Iraq invasion reached 1,745, with 13,190 wounded, most of whom are young men

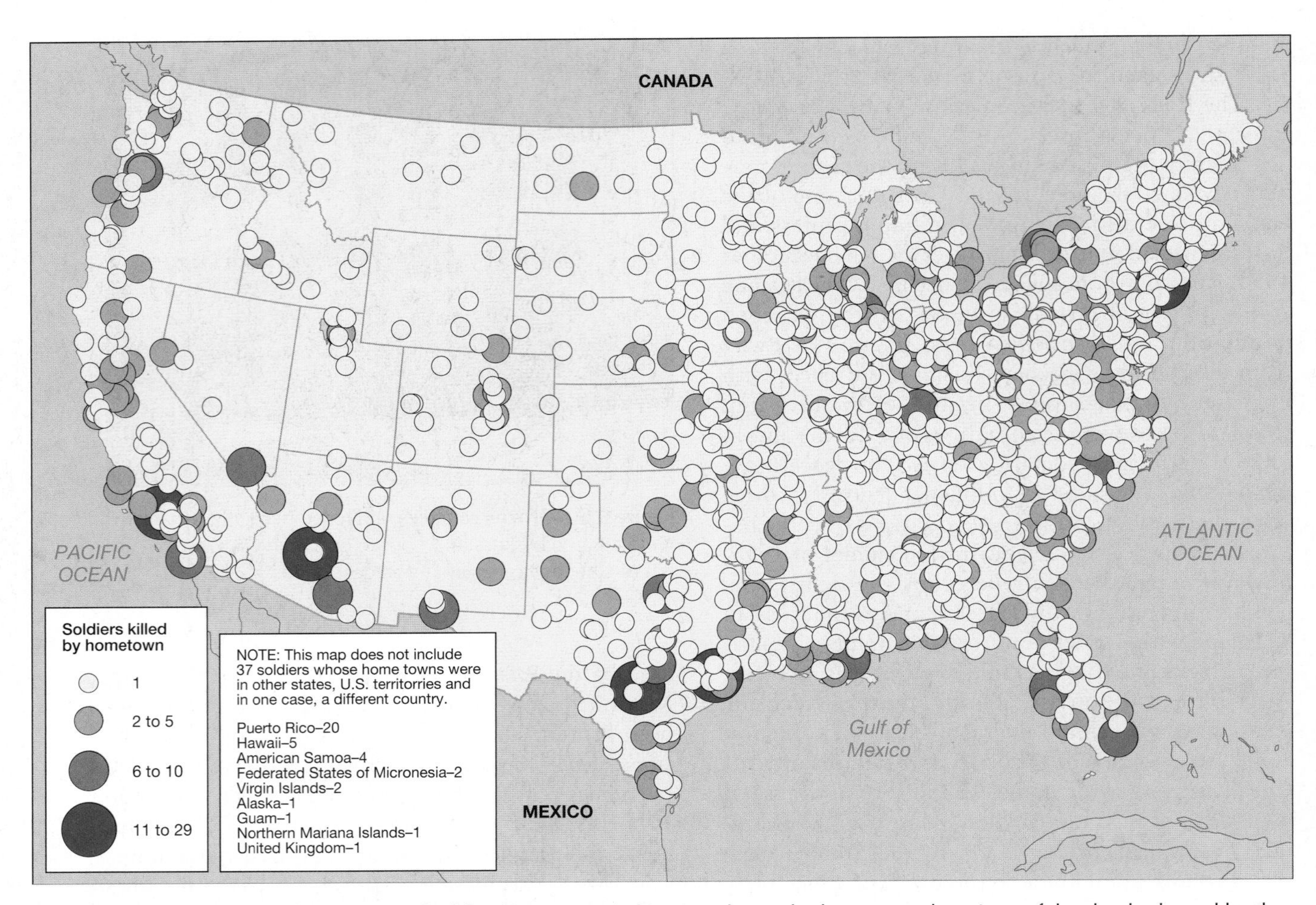

Figure 9.27 Military deaths in Iraq and Afghanistan, 2005 This map shows the hometown locations of the dead released by the Pentagon as of June 29, 2005. (*Source: Palm Beach Post,* Web site http://www.palmbeachpost.com/news/content/news/photos/war_casualties/map/m10000.html, accessed on July 5, 2005).

between the ages of 18 and 22 (**Figure 9.27**). Of these, over 1,318 were killed or died in accidents after the end of major combat was announced by President Bush. Official sources, including the United Nations, estimate over 125,000 Iraqi deaths from all causes as of summer 2005, with roughly three times as many injured. In addition, in early 2004, the 9/11 Commission (more formally known as the National Commission on Terrorist Attacks Upon the United States), concluded that there was no credible evidence that Saddam Hussein, the now captured former president of the country, had assisted the terrorist organization Al Qaeda in preparing for or carrying out the 9/11 attacks. There is also general agreement among U.S. intelligence and military personnel that Iraq had most likely completely destroyed its programs for biological and chemical weapons production before the U.N. team began its inspection.

From a geographic perspective, what is most striking about the war in Iraq is the way that it has fractured the global community both in terms of world leaders and citizens. The fact that the United States and United Kingdom joined forces while other high-profile core countries such as Germany and France opposed the war and struggling semiperipheral countries like Bulgaria, Mongolia, and Ethiopia supported it suggests that the global hegemony of the United States and its tendency to act unilaterally around issues of geopolitical significance is disturbing to other core countries. It also suggests that some peripheral and semiperipheral countries were looking to garner rewards (such as international aid from the United States) by lining up behind a controversial decision.

INTERNATIONAL AND SUPRANATIONAL ORGANIZATIONS AND NEW REGIMES OF GLOBAL GOVERNANCE

Just as states are key players in political geography, so too have international and supranational organizations become important participants in the world system in the last century. These organizations have become increasingly important ways of achieving goals that could otherwise be blocked by international boundaries. These goals include, among other things, the freer flow of goods and information and more cooperative management of shared resources, such as water.

Transnational Political Integration

Perhaps the best-known international organization operating today is the United Nations (**Figure 9.28**). The postwar period has seen the rise and growth not only of large international organizations but also of new regional arrangements. These arrangements vary from highly specific, such as the Swiss-French cooperative management of Basel-Mulhouse airport, to the more general, such as the North American Free Trade Agreement (NAFTA), which joins Canada, the United States, and Mexico into a single trade region. Regional organizations and arrangements now address a wide array of issues, including the management of international watersheds and river basins (such as the Great Lakes of North America and the Danube and Rhine Rivers in Europe). They also oversee the maintenance of health and sanitation standards, coordinated regional planning, and tourism management. Such regional arrangements seek to overcome the barriers to the rational solution of shared problems posed by international boundaries. They also provide larger arenas for the pursuit of political, economic, social, and cultural objectives.

Unlike international organizations, supranational organizations reduce the independence of individual states. A **supranational organization** is a collection of individual states with a common goal that may be economic and/or political in nature. By organizing and regulating designated operations of the individual member states, these organizations diminish, to some extent, individual state sovereignty in favor of the collective interests of the entire membership. The European Union (EU) is perhaps the best example of a supranational organization.

At the end of World War II, European leaders realized that Europe's fragmented state system was insufficient to

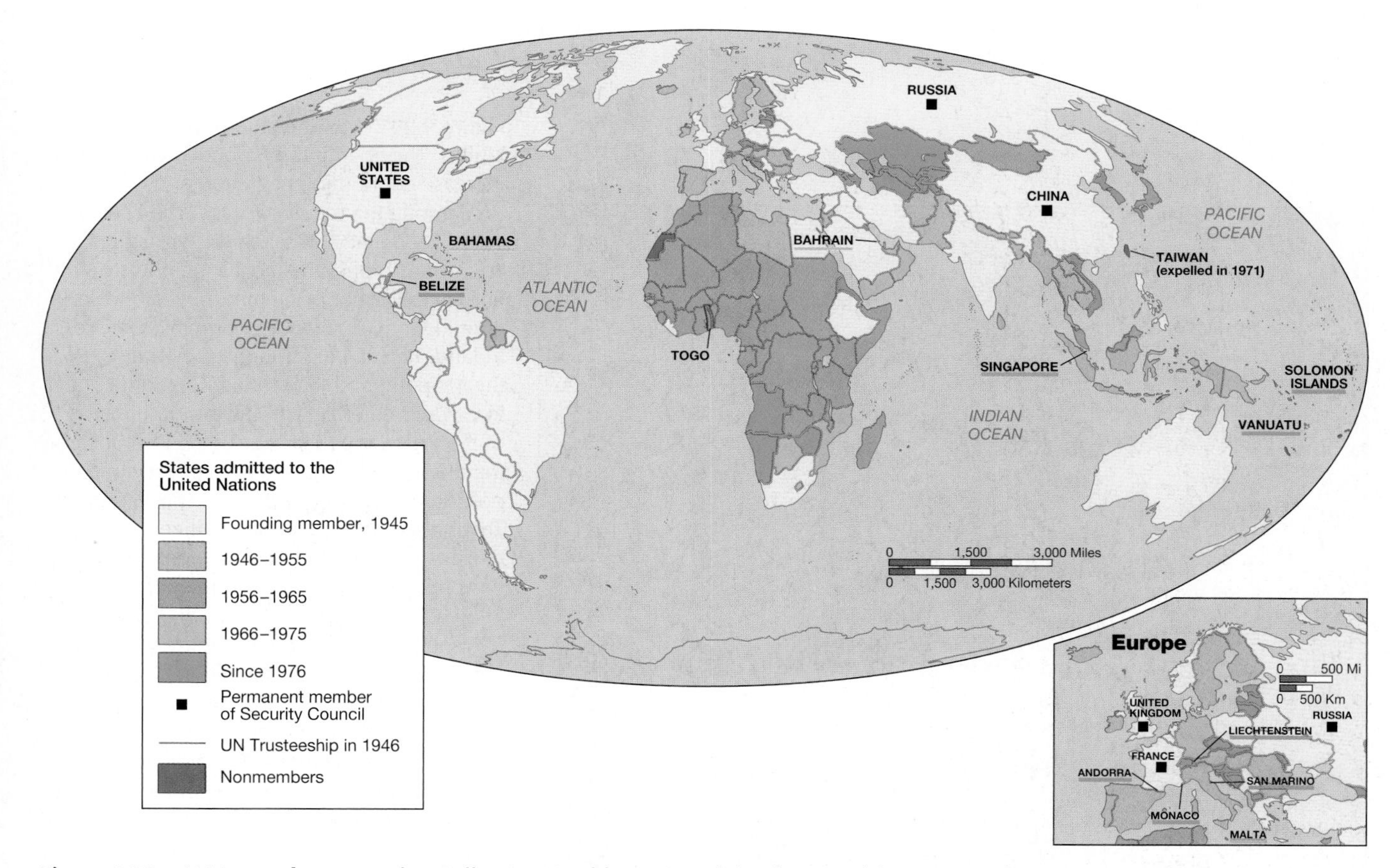

Figure 9.28 U.N. member countries Following World War II and the demise of the League of Nations, a renewed effort was made to establish an international organization aimed at instituting a system of international peace and security. The U.N. Charter was approved by the U.S. Senate in July 1945, raising hopes for a more long-lived organization than the ineffective League of Nations. Located in New York City, the United Nations is composed of a Security Council, which includes the permanent members of the United States, Britain, China, France, and Russia, and a General Assembly, which includes all those countries identified on the map. At the same time that the United Nations was set up, the United States lobbied for the creation of the International Monetary Fund (IMF) and the World Bank. The U.S. government had believed that World War II resulted from the collapse of world trade and financial dislocation caused by the Great Depression. The task of the IMF and the World Bank is to provide loans to stabilize currencies and enhance economic growth and trade.

the demands and levels of competition coalescing within the world political and economic system. They endeavored to create an entity that would preserve important features of state sovereignty and identity. They also intended to create a more efficient intra-European marketing system and a more competitive entity in global transactions. **Figure 9.29** shows the original member countries of the European Economic Community, which evolved into the EU in 1992, the existing members of the EU, and those countries that are currently applying for admission.

The EU holds elections, has its own parliament and court system, and decides whether and when to allow new members to join. Generally speaking, the EU aims to create a common geographical space within Europe in which goods, services, people, and information move freely and in which a single monetary currency prevails. Whether an EU foreign policy will ever be accomplished remains to be seen, but a common European currency, the euro, is now in circulation. Despite the members' concession to a common currency, indicators of nationalism within the individual member countries remain strong. For example, just as the European system of states is on the threshold of dissolving into the larger EU organizational form, national and regional movements (described in Chapter 5) have become potent forces resisting full integration.

Globalization, Transnational Governance, and the State

As we have already noted, globalization has been as much about restructuring geoeconomics as it has been about reshaping geopolitics. In fact, some globalization scholars believe that the impact of globalization on politics has been so profound that it is leading to the diminution of the powers of the modern state, if not its ultimate disappearance. These scholars, known as hyperglobalists, believe that because the modern state is organized around a bounded territory and because globalization is creating a new economic space that is transnational, the state is increasingly incapable of responding to the needs of the new transnational economy. Although we do not subscribe to this position, we do recognize that the state is undergoing

Figure 9.29 Membership in the European Union The goal of the European Union is to increase economic integration and cooperation among the member states. The EU was established in 1992, when the Maastricht Treaty was ratified by the 12 members of the European Economic Community, or EEC (Belgium, Denmark, France, Germany, Greece, Ireland, Italy, Luxembourg, Netherlands, Portugal, Spain, and the United Kingdom). The EEC was created in 1967. Upon ratification of the treaty, the countries of the EEC became members of the EU, and the EEC became the policymaking body of the EU. The Maastricht Treaty established European citizenship for citizens of each member state, enhanced EEC customs and immigration agreements, and allowed for the establishment of a common currency, the euro, which is currently in circulation among all of the original 12 members except for Denmark, the United Kingdom, and Greece. Newer members, including Finland, Austria, Slovakia, Latvia, Lithuania, and Malta also use the euro in addition to their own currencies. The EU is governed through both supranational European institutions (the European Commission and the European Parliament, both administered by the EU) and the governments of the member states, which send representatives to the Council of Ministers (the main lawmaking body of the EU). Membership in the EU is much sought after, and numerous European countries, such as Turkey, have applied and are on the waiting list for admission.

dramatic changes that are restructuring its role with respect to both local, domestic concerns as well as global, transnational ones. Moreover, these changes are very much part of recent history.

In the twentieth century, from the end of World War II until 1989, when the Berlin Wall was dismantled, world politics were organized around two superpowers. The capitalist West rallied around the United States, and the communist East around the Soviet Union. But with the fall of the Berlin Wall signaling the "end of communism" as a world force, the bipolar world order came to an end, and the new world order, which was organized around global capitalism, emerged and has increasingly solidified around a new set of political powers and institutions that have recast the role of the state.

We have referred throughout the text, and especially in Chapters 2 and 7, to the importance for the contemporary global economy of such regional and supranational organizations as the EU, NAFTA, the Association of Southeast Asian Nations (ASEAN), the Organization of Petroleum Exporting Countries (OPEC), and the World Trade Organization (WTO). We have also noted that these organizations are unique in modern history, as they aim to treat the world and different regional clusters as seamless trading areas unhindered by the rules that ordinarily regulate national economies. The increasing importance of these trade-facilitating organizations is the most telling indicator that the world, besides being transformed into one global economic space, is also experiencing global geopolitical transformations. But rather than disappearing altogether, the powers and roles of the modern state are changing as it is forced to interact with these sorts of organizations, as well as with a whole range of other political institutions, associations, and networks (**Figure 9.30**).

The point is not that the state is disappearing but rather that it must now contend with a whole new set of processes and other important political actors on the international stage as well as within its own territory. For instance, geographer Andrew Leyshon has shown how in the 1980s a transnational financial network was established that is far beyond the control of any one state, even a very powerful state like the United States, to regulate effectively. In fact, what the increasing importance of transnational flows and connections—from flows of capital to flows of migrants—indicates is that the state is less a container of political or economic power and more a site of flows and connections.

The increasing importance of flows and connections means that contemporary globalization has made possible a steadily shrinking world. In addition to allowing people and goods to travel farther faster and to receive and send information more quickly, time-space convergence—or the smaller world that globalization has created—means that politics and political action have also become global. In short, politics can move beyond the confines of the state into the global political arena, where rapid communications enable complex supporting networks to be developed and deployed, thereby facilitating interaction and decision making. For example, at the protests that occurred in Seattle, Washington, in 1999 over the WTO meeting, telecommunications, and especially the Internet, enabled the protest leaders to coordinate their actions with those of interested groups all over the world. The Seattle protests were an expression of a truly global politics that matched the global politics of the WTO itself. Thus, both institutionalized politics and popular political movements can be truly global in reach. One indication of the increasingly global nature of politics outside of formal political institutions is the increase in environmental organizations whose purview and membership are global, as discussed in Chapter 4.

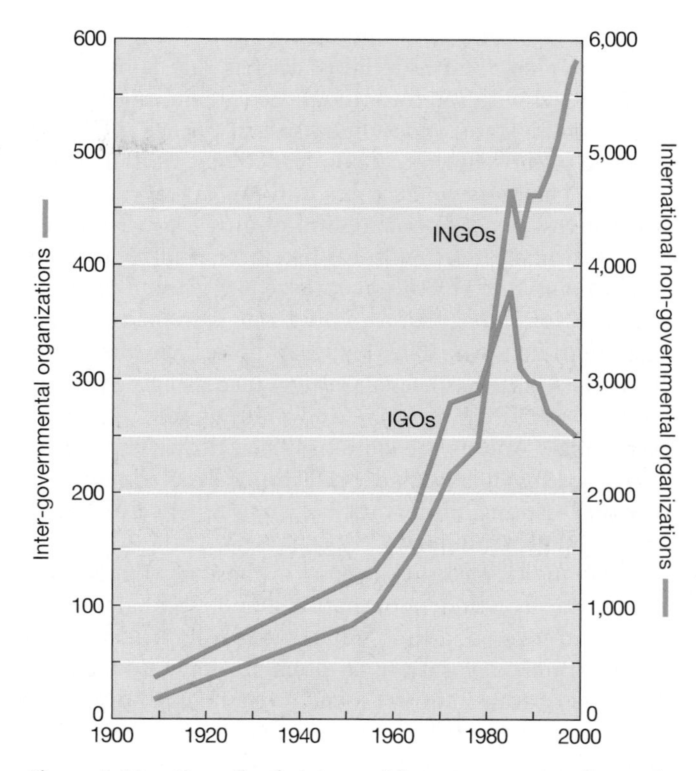

Figure 9.30 Growth of states and intergovernmental organizations and nongovernmental organizations in the twentieth century While the number of states has grown steadily over the twentieth century, intergovernmental organizations (IGOs) and international nongovernmental organizations (INGOs) have experienced dramatic growth, particularly since the 1960s. In 1909 there were 37 intergovernmental organizations and 176 international nongovernmental organizations. In 1996 there were 260 IGOs and 5,472 INGOs. Another important feature of the internationalization of governance is the number of international treaties in force between states. The number of international treaties increased from 6,351 to 14,061 between 1946 and 1975. While states remain the main forms of national government, they have turned over many of their governing responsibilities to international governing organizations and nongovernmental organizations. (After Union of International Associations, *Yearbook of International Organizations, 1996–1997*. Munich: K.G. Saur, 1996, p. 54.)

What has been most interesting about the increasing institutionalization of global politics is that it has been less involved with the traditional preoccupations of relations between states and military security issues and more involved with issues of economic, ecological, and social security. The massive increases in flows of trade, foreign direct investment, financial commodities, tourism, migration, crime, drugs, cultural products, and ideas have been accompanied by the emergence and growth of global and regional institutions whose role is to manage and regulate these flows. The impact of these twin forces on the modern state has been to draw it increasingly into this complex of global, regional, and multilateral systems of governance. And as the state has been drawn into these new activities, it has shed or deemphasized some of its previous responsibilities, such as maintenance of social welfare and the military.

The involvement of the state in these new global activities, the growth of supranational and regional institutions and organizations, the critical significance of transnational corporations to global capital, and the proliferation of transnational social movements and professional organizations are captured by the term **international regime.** The term reflects the fact that the arena of contemporary politics is now international, so much so that even city governments and local interests groups—from sister city organizations to car clubs—are making connections and conducting their activities both beyond and within the boundaries of their own states. An example of this is the human rights movement that has gained ascendancy over the last four to five decades or so. **Human rights** are considered by most societies to belong automatically to all people, including the rights to justice, freedom, and equality.

Until World War II, safeguarding human rights was the provenance of states whose rules and regulations legislated the proper treatment of its citizens, from prisoners to schoolchildren. Since the late 1940s and 1950s, nearly all states have come to accept the importance of a comprehensive political and legal framework that focuses on human rights and that allows an international organization to intervene in the operations of a sovereign state that is in violation of the International Bill of Human Rights adopted by the United Nations in 1948. In 1998 the United Nations realized another step in the protection of human rights by adopting a treaty to establish a permanent International Criminal Court (ICC). In establishing the court, Kofi Anan, Secretary-General of the United Nations, stated: "Our hope is that, by punishing the guilty, the ICC will bring some comfort to the surviving victims and to the communities that have been targeted. More important, we hope it will deter future war criminals, and bring nearer the day when no ruler, no State, no junta and no army anywhere will be able to abuse human rights with impunity." In forming the treaty, the United Nations aimed to create a permanent mechanism to bring to justice the perpetrators of such crimes as genocide, ethnic cleansing, sexual slavery, and maiming in order to put an end to the impunity so often enjoyed by those in positions of power. The court has the mandate to try individuals, not states, for crimes committed in the present (after July 2002). What makes the ICC unique is that its basic premise is the principle of complementarity, which means that the Court can only exercise its jurisdiction when a national court is unable or unwilling to genuinely do so itself. Thus, the importance of the international scale increased over the last 50 years of the twentieth century and appears likely to continue to do so into the twenty-first century. It is important to note, however, that not all states support the ICC. Seven U.N. members voted against the treaty to establish the ICC. These were the United States, China, Iraq, Israel, Libya, Qatar, and Yemen.

An additional and often overlooked aspect of human rights is **children's rights.** In 1989, the United Nations adopted the Convention on the Rights of the Child and despite the convention's nearly universal ratification (only the United States and Somalia have not ratified it) many of the most basic rights of children are still not being met. The Convention promises children around the world the fundamental right to life, liberty, education, and health care. Among the many other fundamental safeguards the Convention provides are protection of children in armed conflict; protection from discrimination; protection from torture or cruel, inhuman or degrading treatment or punishment; protection within the justice system; and protection from economic exploitation. And yet, while these promises are in place and every one but two of the world's nations have ratified the Convention, around the world street children are killed or tortured by police, they are recruited or kidnapped to serve as soldiers in military forces or labor under extremely difficult conditions, they are forced into prostitution, or they are imprisoned. Refugee children, often separated from their families, are vulnerable to exploitation, sexual abuse, or domestic violence. Keeping the promises made in the Convention on the Rights of the Child is one of the biggest challenges of the twenty-first century.

The emergence of human rights as a globally relevant issue has occurred as groups and organizations, both governmental and nongovernmental, have been able to debate and discuss issues that concern all people everywhere and can do so at the international level through conferences, email, listservs, and direct action at international events. The phenomenon of different people and groups across the world in common cause is known as global civil society. **Global civil society** is composed of the broad range of insitutions that operate between the private market and the state.

In the next section we move down from the scale of the international to more national, regional, and local scales. Our aim here is to show that political geography occurs at all levels of political organization and that each scale enables significant insights about the politics of geography and the geography of politics.

THE TWO-WAY STREET OF POLITICS AND GEOGRAPHY

Political geography can be viewed according to two contrasting orientations. The first orientation sees it as the *politics of geography*. This perspective emphasizes that *geography*—or the areal distribution/differentiation of people and objects in space—has a very real and measurable impact on politics. Regionalism and sectionalism, discussed later in this section, illustrate how geography shapes politics. The politics-of-geography orientation is also a reminder that politics occurs at all levels of the human experience, from the international order down to the neighborhood, household, and body.

The second orientation sees political geography as the *geography of politics*. This approach analyzes how *politics*—the tactics or operations of the state—shapes geography. Mackinder's heartland theory and the domino theory attempt to explain how the geography of politics works at the international level. In the heartland theory the state expands into new territory in order to relieve population pressures. In the domino theory, as communism seeks new members, it expands geographically to incorporate new territories. An examination of a series of maps of Palestine/Israel since 1923 reveals how the changing geography of this area is a response to changing international, national, regional, and local politics (see Box 9.3: "The Palestinian-Israeli Conflict").

The Politics of Geography

Territory is often regarded as a space to which a particular group attaches its identity. Related to this concept of territory is the notion of **self-determination,** which refers to the right of a group with a distinctive politico-territorial identity to determine its own destiny, at least in part, through the control of its own territory.

Regionalism and Sectionalism

Different groups with different identities—religious or ethnic—sometimes coexist within the same state boundaries. At times discordance between legal and political boundaries and the distribution of populations with distinct identities leads to movements to claim or reclaim particular territories. These movements, whether conflictual or peaceful, are known as *regional movements*. **Regionalism** is a feeling of collective identity based on a population's politico-territorial identification within a state or across state boundaries.

Regionalism often involves ethnic groups whose aims include autonomy from an interventionist state and the development of political power. For example, in spring 1993 several leading Basque guerrillas were arrested in France, raising hopes for an end to Basque terrorism in Europe. The Basque people are one of the oldest people of Europe, with a distinctive culture and language. Despite over a century of nationalism, they are still administered by both France and Spain, with only limited autonomy in selected parts.

For over 25 years the French, Spanish, and more recently the Basque regional police have attempted to undermine the Basque Homeland and Freedom movement through arrests and imprisonments. "Basquism" represents a regional movement that arose in response to industrialization and modernization in Spain beginning in the early twentieth century. The Basque people feared that *cultural forces* accompanying industrialization would undermine their preindustrial traditions. The Basque provinces of northern Spain and southern France have sought autonomy from those states for most of the twentieth century. Since the 1950s, agitation for political independence has included—especially for the Basques in Spain—terrorist acts. Not even the Spanish move to parliamentary democracy and the granting of autonomy to the Basque provinces could squelch the Basque thirst for self-determination (**Figure 9.31**). On the French side the Basque separatist movement is neither as violent nor as active as the movement in Spain.

We need only look at the long list of territorially based conflicts that have emerged in the post–cold war world to realize the extent to which territorially based ethnicity remains a potent force in the politics of geography. For example, the Kurds continue to fight for their own state separate from Turkey and Iraq. A significant proportion of Québec's French-speaking population, already accorded substantial autonomy, continues to advocate complete independence from Canada. In the 1996 plebiscite on the issue, the separatists were only very narrowly defeated. Consider also the former Yugoslavia, whose geography has fractured along ethnic and religious lines (**Figure 9.32**). Regionalism also underlies efforts to sever Scotland from the United Kingdom.

Regionalism may also be based on economics. For example, in a nonbinding referendum put before the California voters in 1993, drawing on sentiment over a century old, California's northern and mostly rural counties voted to separate from the southern, more urban counties. The desire for separation was based on the belief that political representation in the state legislature economically and politically advantaged the south over the north. Many citizens felt that the state had grown unmanageably large and that the government had become unresponsive to the people. Although many additional steps would have to be taken—such as state senate approval, approval by the governor and by the voters, and finally by the U.S. Congress—before separation could occur, the California separatist movement represents an interesting example of *economic* regionalism.

Although they sometimes coexist, **sectionalism,** an extreme devotion to local interests and customs, should not be confused with regionalism. Sectionalism has been identified as an overarching explanation for the U.S. Civil War.

Figure 9.31 Basque independence Posters and the raised fists signify the passionate opposition of the Basques toward the central government in Madrid. Acts of terrorism have occurred throughout Spain as the Basques maintain their desire for independence.

It was an attachment to the institution of slavery and to the political and economic way of life slavery made possible that prompted the southern states to secede from the Union. The Union went to war to ensure that sectional interests would not take priority over the unity of the whole; that is, that states' rights would not undermine the power of the federal government. Although the Civil War was waged around the real issue of permitting or prohibiting slavery, it also involved the power of the state. As **Figure 9.33** shows, the election of Abraham Lincoln to the presidency in 1860 reflected the sectionalism that dominated the country: He received no support from slave states.

Suburbs vs. Cities, and Rural vs. Urban

Sectionalism persists today, but in different forms. In the United States one of the most apparent manifestations of sectionalism may be found in the politics of differentiation

Figure 9.32 Map of the former Yugoslavia The former Yugoslavia now consists of five nations: Slovenia, Croatia, Bosnia and Herzegovina, the new Yugoslavia (made up of Serbia and Montenegro), and Macedonia. For the most part, the boundaries of the Yugoslav states were laid out only in the twentieth century, across segments of the Austro-Hungarian and Ottoman empires that had acquired a complex mixture of ethnic groups. The history of these boundaries has also been the history of ethnic conflict revolving around claims to territory, as well as intolerance for religious differences. As this map shows, with the exception of Slovenia, the new states are home to a mix of nationalities. (*Source:* Redrawn with permission from Prentice Hall, from J. M. Rubenstein, *The Cultural Landscape: An Introduction to Human Geography*, 6th ed., © 1999, p. 260.)

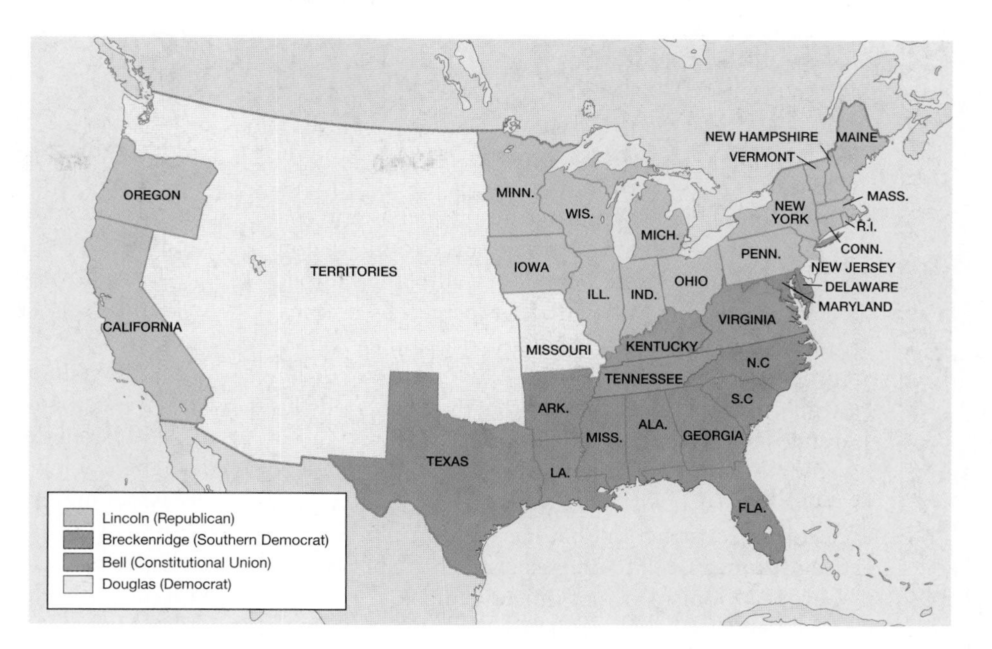

Figure 9.33 The 1860 presidential election The U.S. presidential election of 1860 graphically illustrates the role of sectionalism in determining who gets votes from which geographical regions. In a four-way race, Abraham Lincoln failed to win the support of any of the slave states. (After *Presidential Elections Since 1789,* 4th ed., Washington, DC, Congressional Quarterly Inc., 1987.)

between suburbs and cities. Two examples are the "taxpayer revolts," expressed in the passage of Proposition 13 in California and Proposition $2\frac{1}{2}$ in Massachusetts, and growth-control movements. The former were voter-approved mandates to cut property taxes and limit government spending, while the latter were government-enforced caps on population growth approved by residents to limit the density and extent of growth. Both types of actions have been dominated by suburban U.S. voters.

Proposition 13 in California was an initiative that had tremendous negative fiscal impacts on municipalities, especially on public education and social programs. What began as a localized movement of angry southern California taxpayers in South Bay and San Gabriel Valley grew into a statewide, suburban, antitax protest directed at lowering county government spending on social programs. Cutting across class lines and unified by the subculture of homeownership, Howard Jarvis's California Taxpayers League collected 1.5 million signatures to put Proposition 13 on the ballot in 1978, winning voter approval by a large majority. As a result, suburban homeowners became a powerful political force all over California and elsewhere, a force that continues to exert itself in local elections throughout the United States, as well as many European countries.

A child of the taxpayer revolts, the growth-control movements of the 1980s also drew their support from suburban homeowning constituencies. Arguing for the maintenance of open space, low density, large lots, and community control over land-use planning and development, the growth-control movements have attempted to keep the city out of the suburb by retaining a "country feel." Critics of growth control complain that it discriminates against newcomers to the community. Related to growth-control movements is *NIMBYism,* which stands for "Not In My Backyard" actions. NIMBYism is action by neighborhood residents against the introduction of unwanted land uses. These unwanted land uses can range from group homes for developmentally disabled adults or battered women's shelters to low-income housing and other residential forms. NIMBY supporters see these land uses as a threat to the composition and quality of the neighborhood and the market value of single-family residences.

The politics of geography, in terms of regionalism, also finds strong focus today in rural versus urban politics. In France, for example, attitudes about birth control (and birth rates themselves) are significantly different between the urbanized north of the country and the more rural south. Throughout the EU, farmers have fought the removal of farm subsidies and tariff arrangements advocated by urban-based policymakers because those arrangements have long protected agricultural productivity. Here the dispute pits the politics of local farmers against an international organization.

In Mexico the contest is between local and national levels. Chiapan peasants have forced the federal government in Mexico City to address profound rural poverty. This poverty is seen as aggravated by the government's largely urban (and industrial agriculture) orientation. Likewise, rural-to-urban migration throughout the periphery (grossly inflating the populations of cities such as Lima, Peru; Nairobi, Kenya; and Djakarta, Indonesia) has generated enormous social and political pressures and poses overwhelming challenges. Policymakers must ask themselves difficult questions: How much of the country's scarce resources should be devoted to slowing (or reversing) rural out-migration through development projects? What resources should be devoted to accommodating the throngs of new urban dwellers, most of whom have worse living conditions in the city than they did in the countryside?

Competition also exists among and between cities, as well as among and between states. The most ubiquitous form of this competition revolves around the desire by local and state authorities to attract corporate investment

The Palestinian-Israeli Conflict

The history of the Palestinian-Israeli conflict and the Palestinians's passionate desire for self-rule is a complex and highly volatile one despite persistent local and international efforts to bring peace to the region. The violence that erupted in fall of 2000 (and has persisted), just as the peace process seemed to be most promising, underscores the complexity of the problem and the difficulty of resolution. As with the Iran/Iraq/Kuwait case, the chief factors that have created this seemingly intractable political problem were exacerbated by British partitioning of the region.

The official Jewish state of Israel is a mid-twentieth-century construction that has its roots in the emergence of **zionism,** a late-nineteenth-century movement in Europe. Zionism's chief objective has been the establishment of a legally recognized home in Palestine for the Jewish people. Thousands of European Jews, inspired by the early zionist movement, began migrating to Palestine at the turn of the nineteenth century. When the Ottoman Empire was defeated in 1917, the British gained control over Palestine and the Transjordan area and issued the Balfour Declaration. The Balfour Declaration was highly problematic, however, because indigenous peoples, the Palestinians, already occupied the area and viewed the arrival of increasing numbers of Jews and European sympathy for the establishment of a Jewish homeland as an incursion into the sacred lands of Islam. In response to increasing Arab-Jewish tensions in the area, the British decided to limit Jewish immigration to Palestine in the late 1930s through the end of World War II. In 1947, with conflict continuing between the two groups, Britain announced that it had despaired of ever resolving the problems and would withdraw from Palestine in 1948, turning it over to the United Nations when that happened. The United Nations, under heavy pressure from the United States, responded by voting to partition Palestine into Arab and Jewish states and designated Jerusalem as an international city, preventing either group from having exclusive control. The Jewish state was to have 56 percent of the mandate of Palestine; an Arab state was to have 43 percent; and Jerusalem, a city sacred to Jews, Muslims, and Christians, was to be administered by the United Nations. The proposed U.N. plan was accepted by the Jews and angrily rejected by the Arabs, who argued that a mandated territory could not legally be taken from an indigenous population.

When Britain withdrew in 1948, war broke out. In an attempt to aid the militarily weaker Palestinians, combined forces from Egypt, Jordan, and Lebanon, as well as smaller units from Syria, Iraq, and Saudi Arabia confronted the Israelis. Their goal was not only to prevent the Jewish forces from gaining control over additional Palestinian territory but also to wipe out the newly formed Jewish state altogether. This war, however, which came to be known as the first Arab-Israeli war, resulted in the defeat of the Arab forces in 1949, and later armistice agreements enabled Israel to expand beyond the U.N. plan by gaining the western sector of Jerusalem, including the Old City. In 1950 Israel declared Jerusalem its national capital, though very few countries have recognized this.

Israel maintained the new borders gained during the first Arab-Israeli war for another 18 years until the Six-Day War in 1967, which resulted in further gains for Israel, including the Sinai Peninsula, the Golan Heights, and the southwestern corner of Syria. The eastern sector of Jerusalem, previously held by Jordan, was also annexed during the Six-Day War. As **Figure 9.F** shows, a long period of relatively little territorial change occurred until the 1970s and 1980s, when Israel moved toward reconciliation with Egypt through a series of withdrawals that eventually returned all of Sinai to Egyptian control by 1988.

The territorial expansion of Israel has meant that hundreds of thousands of Palestinians have been driven from their homelands, and the landscape of Palestine has been dramatically transformed. Today Palestinians live as refugees either in other Arab countries in the region, abroad, or under Israeli occupation in the West Bank, the Golan Heights, and the Gaza Strip (also known as the "Occupied Territories"). The Arabs of the Middle East and North Africa and many other international observers are convinced that Israel has no intention of allowing the diasporic Palestinian population to return to their homelands. By the late 1980s, in fact, Palestinians who had remained in their homeland had become so angered by Israeli territorial aggression that they rose up in rebellion. This rebellion, known as the **intifada** ("uprising"), has involved frequent clashes between fully armed Israeli soldiers and rock-throwing Palestinian young men. The intifada, more than anything, is a reaction against over three decades of Israeli occupation of the Palestinian homeland and increasing Israeli settlement, particularly in the West Bank and the Gaza Strip. In addition to the intifada, other Palestinian groups have coalesced in opposition to the Israeli occupation. The Palestinian Liberation Organization (PLO) was formed in 1964 as an organization devoted to returning Palestine to the Palestinians. Since its official recognition, the PLO has become the Palestinian Authority. The Palestinian Authority is seen as the only legitimate rep-

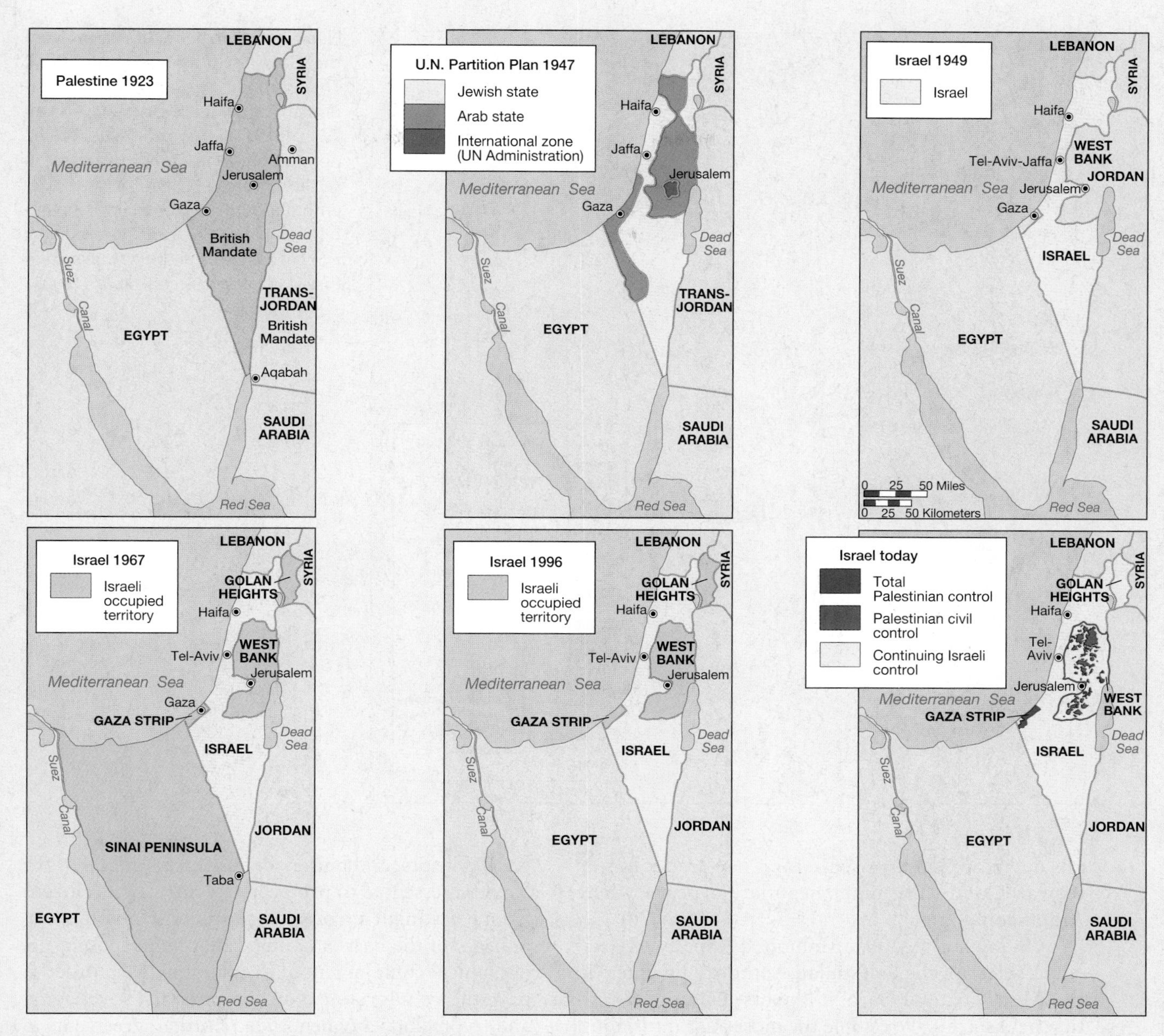

Figure 9.F Changing geography of Israel/Palestine, 1923–2005 Since the creation of Israel out of much of what had been Palestine in 1947, the geography of the region has undergone significant modifications. A series of wars between Israelis and Arabs and a number of political decisions regarding how to cope with both resident Palestinians and large volumes of Jewish people immigrating to Israel from around the world have produced the changing geographies we see here. (*Source:* Reprinted with permission from Prentice Hall, from J. M. Rubenstein, *The Cultural Landscape: An Introduction to Human Geography*, 5th ed., © 1996, p. 233.)

resentative of the Palestinian people. Besides the Palestinian Authority, however, other, more extreme groups exist claiming to represent the Palestinian cause. One of the most well known is Hamas (Harakat al-Muqawama al-Islmiyya, or the Islamic Resistance Movement), whose activities are largely centered in the West Bank and Gaza Strip.

Since the mid-1990s hopes for peace in the region have risen, fallen, risen, fallen, and most recently risen once again. In October 2000, after weeks of very difficult but promising U.S.-sponsored peace negotiations between Yasir Arafat, then chairman of the Palestinian Authority, and Ehud Barak, then Israeli prime minister, violence broke out again in the West Bank. This new violence left little hope in Israel, the Occupied Territories, or elsewhere that the Palestinian-Israeli conflict would be resolved anytime in the near future. Renewed hope emerged, however, following the death of Yasir Arafat in November 2004. Arafat was a controversial figure throughout his lengthy political career. While his supporters viewed him as a heroic freedom fighter who symbolized the national aspirations of the Palestinian people, his opponents often described him as a terrorist who promoted violence.

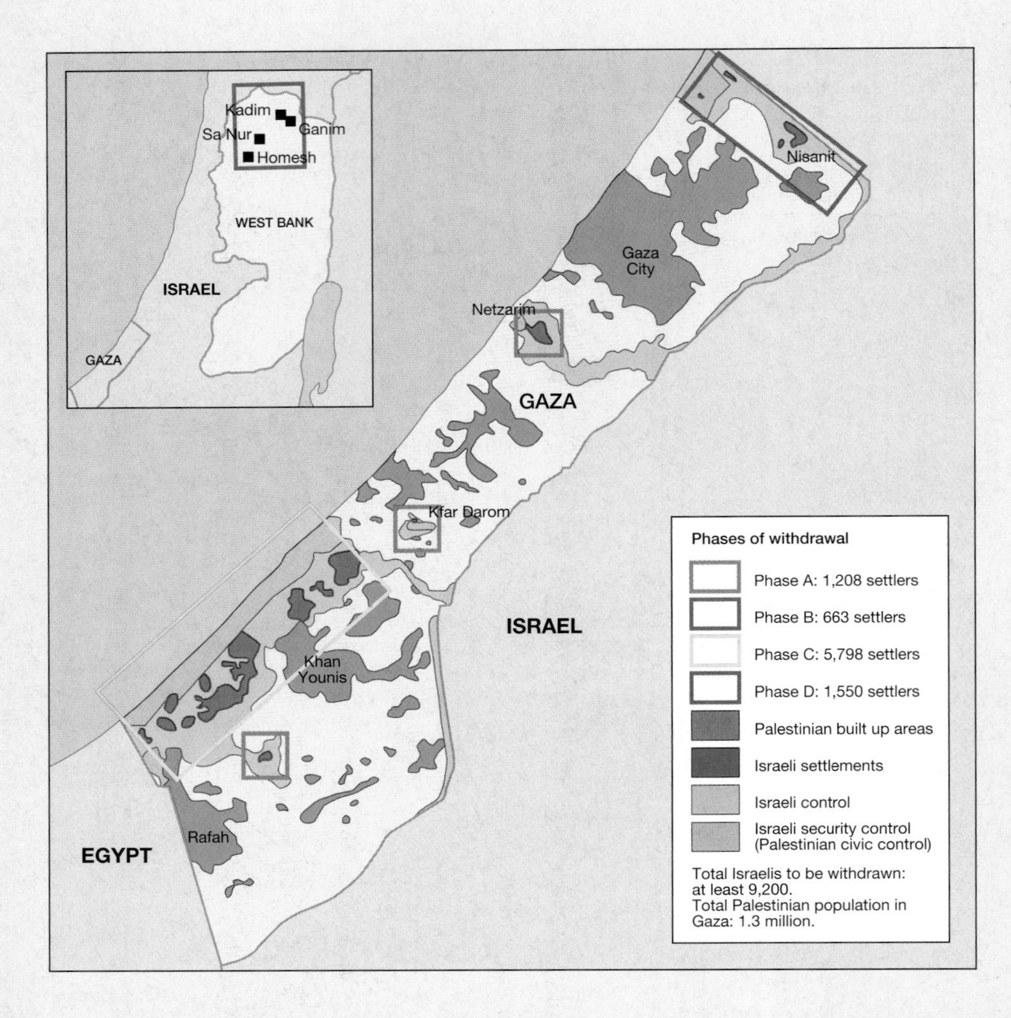

Figure 9.G Israeli withdrawal from the Gaza Strip and the West Bank This map shows the planned withdrawal of Israeli settlements from Palestinian territory. It should be pointed out that thousands of Israeli settlers protested against their government but were removed nonetheless. (*Source:* BBC News, Web edition; http://news.bbc.co.uk/1/hi/world/middle_east/3111159.stm; accessed July 4, 2005).

His death opened up a space for a new leader to step forward, one who might negotiate a peace where Arafat could not.

On January 9, 2005, Mahmoud Abbas was elected president of the Palestinian Authority by voters in the West Bank and Gaza, in the first Palestinian election held since 1996. While the most militant Islamist organizations, Hamas and the Islamic Jihad, boycotted the elections, it is estimated that about 66 percent of eligible voters went to the polls (compared to 42.45 percent in the 2005 U.S. presidential elections). Since the election, progress has occurred in moving toward Palestinian statehood at the same time that Israel has begun withdrawing Israeli settlers from Palestinian territory, many of whom are extremely resistant to leaving (**Figure 9.G**).

And yet, as Israel has begun to cede territory back to Palestine, it is continuing to construct a physical barrier between Israelis and Palestinians. The Gaza Strip barrier was constructed in 1994. It consists of 52 kilometers (30 miles) of mainly wire fence with posts, sensors, and buffer zones. Israel argues that the barrier is essential to protect the security of its citizens from Palestinian terrorism. Palestinians and other opponents of the barrier contend that its purpose is geographical containment of the Palestinians in order to pave the way for an expansion of Israeli sovereignty and to preclude any negotiated border agreements in the future. But Israel argues that the fence is purely a security obstacle, not a part of a future border. In 2002, the West Bank wall was begun and, when completed will seal off that portion of the Palestinian territories from Israel (**Figure 9.H**). The barrier continues to uproot and destroy Palestinian settlements and separate them from their livelihood. In October 2003, the U.N. General Assembly voted 144–4 that the wall was "in contradiction to international law" and therefore illegal. Israel called the resolution a "farce." As of May 2005, the barrier construction had already destroyed an estimated 102,320 Palestinian olive and citrus trees, demolished 75 acres (304,000 m^2) of greenhouses and 23 miles (37 km) of irrigation pipes.

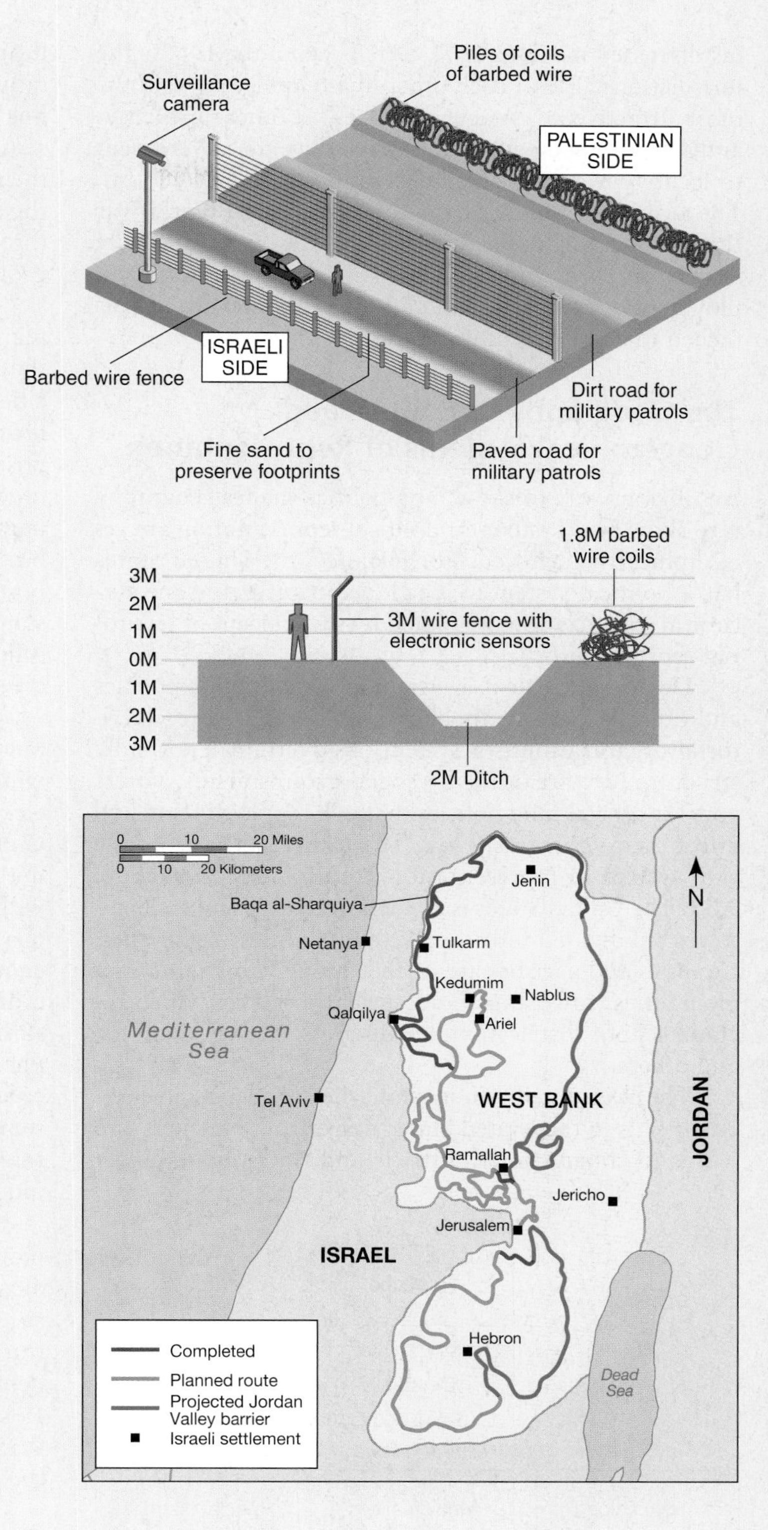

Figure 9.H Israeli security fence, 2005 Shown are the planned and completed portions of the security fence, called the "the wall" or the "apartheid wall" by Palestinians and other opponents. The physical barrier consists of a network of fences, walls, and trenches. Israel's stated purpose in constructing it is to create a zone of security between itself and the West Bank. (*Source:* BBC News, Web edition; http://news.bbc.co.uk/1/hi/world/middle_east/3111159.stm; accessed July 4, 2005).

(as discussed in Chapter 7). Often corporations play the jurisdictions against each other in attempts to obtain the most attractive investment packages. At other times cities and states compete in attempts to induce the government to locate government facilities within their jurisdictions. The location of military bases during and after World War II illustrates this point. Cities and states were especially keen to attract the military because it meant increased employment opportunities and all the economic growth attached to employment generation.

The Geography of Politics and Geographical Systems of Representation

An obvious way to show how politics shapes geography is to show how systems of political representation are geographically anchored. For instance, the United States has a political system in which democratic rule and territorial organization are linked by the concept of territorial representation.

Democratic rule is a system in which public policies and officials are directly chosen by popular vote. **Territorial organization** is a system of government formally structured by area, not by social groups. Thus, voters vote for officials and policies that will represent them and affect them *where they live*. The territorial bases of the U.S. system of representation are illustrated in **Figure 9.34**. The United States is a federation of 50 states, themselves subdivided into over 3,000 counties or parishes. Counties and parishes are further broken down into municipalities, townships, and special districts, which include school districts, water districts, library districts, and others.

The electoral divisions established for choosing elected officials in the United States range from precincts and wards to congressional districts and states. State power is applied within geographical units, and state representatives are chosen from geographical units. The bottom line is that in the United States—as in many other representative democracies—politics is geography. People and their interests gain representation in government through the location of their interests in particular places and through their relative ability to capture political control of *geographically based* political units.

For example, election of the president involves a popular vote carried out at the precinct level but totaled at the state level. Thus, even though particular precincts, cities, or counties may give a majority to one candidate, if the majority of votes at the state level supports the opposing candidate, then that person is declared the winner in that state. This arrangement is particularly important because the president is not elected by the popular vote, but by the electoral college. The electoral college is composed of a specified number of delegates allocated to each state based on that state's population as of the most recent official census. Thus, it is the state-level voting tally that drives the process.

A candidate may win the countrywide popular vote but lose in the electoral college if that candidate fails to win enough states to acquire the required majority of the electoral votes. This is what happened to candidate Al Gore in the 2000 presidential elections. Furthermore, as in the 1992 election of Bill Clinton, a candidate may win with considerably less than 50 percent of the popular vote if a third-party candidate (H. Ross Perot) siphons off enough of the popular vote to prevent the opposing candidate (in this case, George H. W. Bush) from winning a sufficient number of electoral votes. The geographical implications of the U.S. presidential voting arrangement are crucial to candidates' campaign strategies. To be a winner requires concentrating enormous time and energy to capture a majority of votes in some of the nation's most populous states.

The U.S. tradition of territorial representation promotes acceptance of the legitimacy of local interests. Such an arrangement also makes it difficult for third parties to succeed. Unless third parties are geographically concentrated, they can never win enough states to win a presidential election. The failed presidential campaign of H. Ross Perot in 1992 is an excellent example of this (**Figure 9.35**). Although Perot received 19 percent of the vote, he did not win a single state.

Other systems of representation exist throughout the world. For example, many electoral systems are based on representing special constituencies in the legislative branch of government. In Pakistan, for example, there are four seats for Christians, four seats for Hindus and people belonging to the scheduled castes, and one seat each for Sikh, Buddhist and Parsi communities. In Italy, 232 seats are attributed to single member constituencies on a regional basis. Seventy-seven seats must be distributed by proportional representation on a regional basis and six seats are distributed from abroad through a proportional quota.

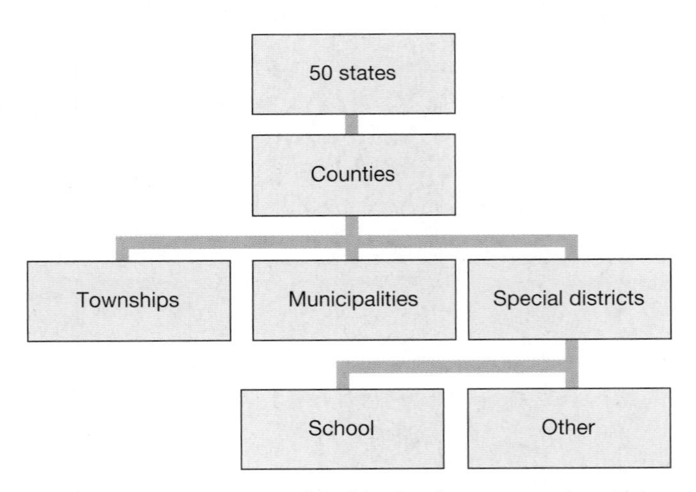

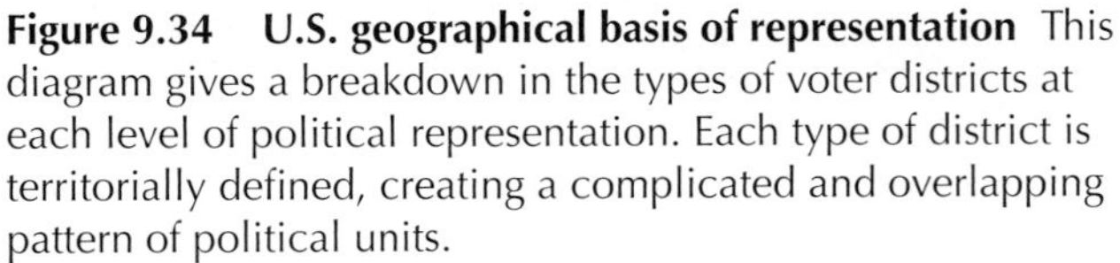

Figure 9.34 U.S. geographical basis of representation This diagram gives a breakdown in the types of voter districts at each level of political representation. Each type of district is territorially defined, creating a complicated and overlapping pattern of political units.

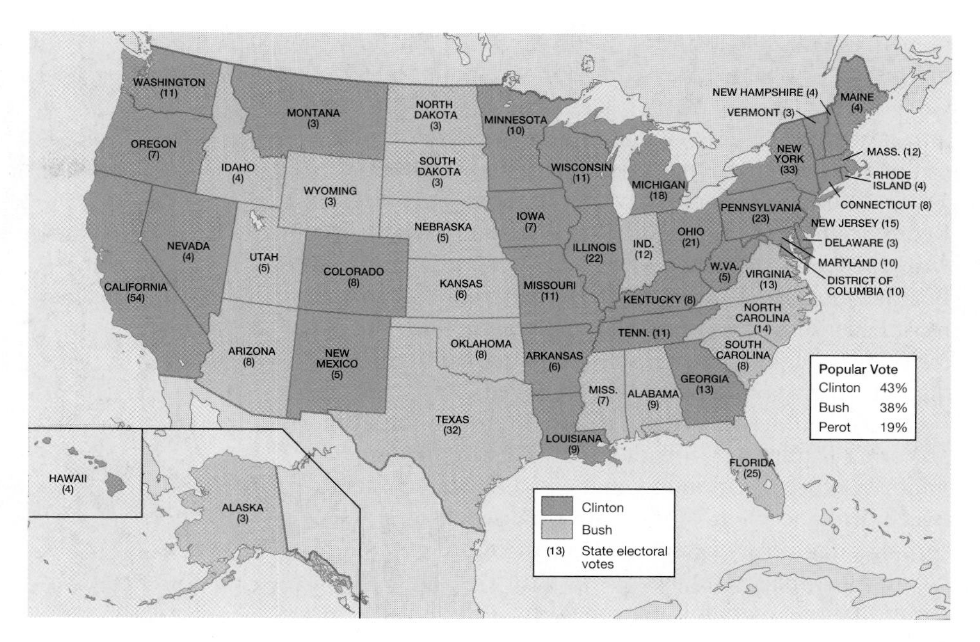

Figure 9.35 The vote for president, 1992 The total popular vote, aggregated at the state level, determines the number of electoral college votes for each candidate. Voter preferences show a clear geographical pattern, with George Bush running particularly strong in the West and South. H. Ross Perot won 19 percent of the popular vote but was unable to translate that into any electoral votes.

Systems of representation are very much tied to the history of a country with some very sensitive to the way that history and geography (who lives where) come together. These systems are both a product of and an important influence on the political culture of a country.

Reapportionment and Redistricting

The U.S. Constitution determines the allocation of legislators among states, guaranteeing that each state will have a representative system of government. For U.S. presidential and senatorial elections, candidates are elected at large within each state, not on the basis of electoral districts. U.S. representatives, however, are elected from congressional districts of roughly equal population size. This is also the case for state senators, representatives, and, often, other elected officials, from city council members to school board members. It is the responsibility of each state's legislature to create the districts that will elect most federal and state representatives. Other levels of government—from counties to special districts—also establish their own electoral districts. The result is that representatives are elected at any number of levels of government in a collection of districts that is complicated, extensive, and by no means systematic.

Problems of the proper "fit" between political representation and territory emerge when population changes. Because most forms of representation are based on population, it often becomes necessary to change electoral district boundaries to distribute the total population more evenly among districts. For example, the number of congressional representatives in the United States as a whole is fixed at 435. These 435 seats must be reapportioned in accordance with population change every 10 years. (Recall from Chapter 3 that the federal government is required to count the U.S. population every 10 years. One of the chief reasons for this is to maintain the proper match between population and representation.) **Reapportionment** is the process of allocating electoral seats to geographical areas. **Redistricting** is the defining and redefining of territorial district boundaries. Both are political, geographical, and statistical exercises. As geographer Richard Morrill writes:

> The process is *political* in that the design and approval of systems of districts is usually done by bodies of elected representatives; the balance of power between groups and areas is often involved; identification of citizens with a traditional electoral territory is altered; and the incumbency of individuals is usually at stake. . . . The process is *geographic* in that areas must be allocated to districts and boundaries drawn (or territories partitioned into districts); communities of interest which may have arisen in part from pre-existing systems of districts, may well be affected; restructuring of basic electoral geography is altered; and accessibility of voters to their representatives or centers of decision-making may be changed. . . . Redistricting is also *statistical* or mathematical in that there is a requirement for reasonably current and accurate data on population and its characteristics and, sometimes, of property and its valuation.[2]

Gerrymandering

The purpose of redistricting is to ensure the equal probability of representation among all groups. In the practice of redistricting for partisan purposes, known as **gerrymandering**, boundaries of districts are redrawn to advantage a particular political party or candidate or to

[2]R. Morrill, *Political Redistricting and Geographic Theory*. Washington, DC: Association of American Geographers, 1981, p. 1.

prevent or ensure a loss of power to a particular subpopulation (like African Americans). The term immortalized Governor Elbridge Gerry of Massachusetts, who signed into law a bill designed to maximize the election of Republican-Democrats over Federalists in the election of 1812. Although Federalists won the most votes overall, Republican-Democrats took 29 of 40 seats because they won the most districts. An electoral district north of Boston, illustrated in **Figure 9.36**, was redrawn so that most Federalist votes were contained there. With the majority of Federalist votes restricted to a small number of districts, their voting power was reduced.

Gerrymandering persists today. Although the Federal Voting Rights Act intended that redistricting enhance minority representation in Congress, a fine line exists between "enhancement" and creating a district solely to ensure that a minority person be elected. **Figure 9.37** shows North Carolina's reengineered Twelfth District, the constitutionality of which was upheld by a Supreme Court in 2001. Compare its tortured boundaries to the classic gerrymandering salamander in Figure 9.36.

Figure 9.36 Gerrymandering salamander, 1812 This strange beast was the result of political shenanigans that used the geography of electoral district boundaries to concentrate Federalist votes within a single "sacrifice" district in order to avert the possibility of Federalist supporters influencing elections in the other districts. Today laws attempt, with mixed success, to prevent such blatant manipulations of the geography of voting. (After *Boston Gazette,* 26 March 1812.)

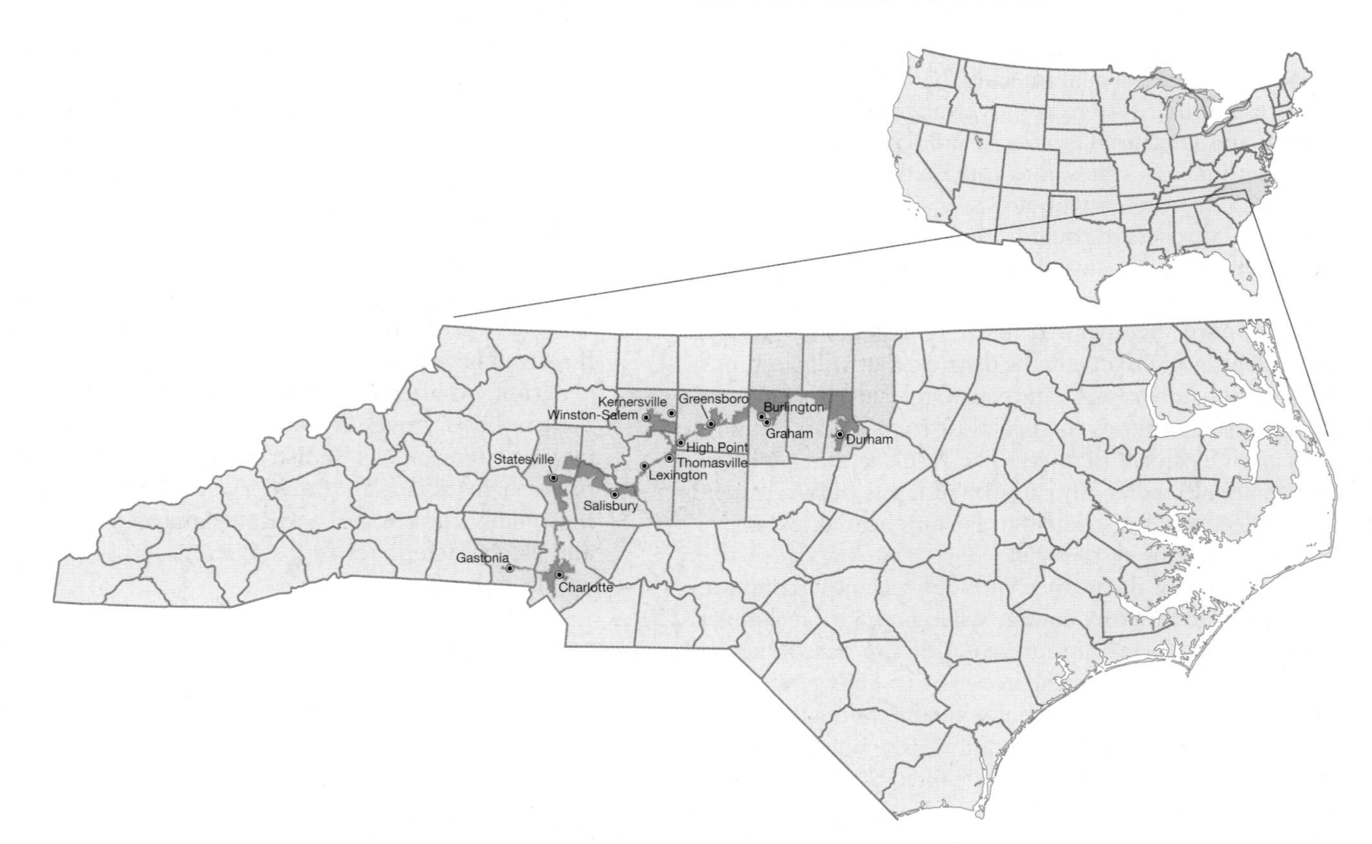

Figure 9.37 North Carolina's proposed Twelfth Congressional District The drawing of electoral district boundaries remains a politically volatile exercise. A 1996 case before the Supreme Court was the redrawn Twelfth Congressional District of North Carolina, whose shape (in red on the map) was as contorted as the original salamander. Accusations of gerrymandering circulated around the drawing of the district, which was intended to consolidate African American voting strength in the district. Although this redistricting was immediately legally challenged, the Supreme Court upheld its constitutionality in a 2001 decision.

CONCLUSION

The globalization of the economy has been largely facilitated by the actions of states extending their spheres of influence and paving the way for the smooth functioning of markets and industries. Political geography is as much about what happens at the global level as it is about what happens at other levels of spatial resolution, from the region to the neighborhood to the household and the individual.

Theories of the state have been one of geography's most important contributions to understanding politics. Ratzel's emphasis on the relationship between power and territory and Mackinder's model of the geographical pivot remind us that space and territory shape the actions of states in both dramatic and mundane ways. Time and space shape politics, and events distant in time and space—such as colonialism—continue to have impacts long after decolonization. The civil war in Northern Ireland, instigated by English colonial practices now centuries old, has only recently shown credible signs of ceasing. The impacts of English colonization have been felt in countries throughout the Northern Hemisphere, as well as by neighbors living unhappily for several generations side by side in cities like Belfast and Boston.

Continuing strife also characterizes the enduring North/South divide that pits core countries against peripheral, mostly formerly colonial, countries. Perhaps the most surprising political geographical transformation of this century has been the near dissolution of the East/West divide. Although it is too soon to tell whether communism has truly been superseded by capitalism, the distinctions between them are certainly more blurred than they once were.

Perhaps the most significant aspect of contemporary globalization for political geographical understandings is the emergence of a new world order and transnational institutions of governance. Both are reshaping not only governing structures and economic processes (such as creating new layers of rules, regulations, and policies, as well as new ways of political interaction among and between nation-states) but also the practices of everyday life (such as increased personal security measures and the transformation of human rights).

The pairing of the terms *politics* and *geography* serves to remind us that politics is clearly geographical at the same time that geography is unavoidably political. The simple divisions of area into states, counties, cities, towns, and special districts means that where we live shapes our politics, and vice versa. Geography is politics, just as politics is geography. And geographical systems of representation, as well as identity politics based on regional histories, confirm these interactive relationships.

MAIN POINTS REVISITED

- **Political geography, a subfield of the discipline of geography, examines complex relationships between politics and geography (both human and physical).**

  As societies are organized around territorial units, geography and access to it are often at the center of political conflicts and can also make possible the resolution of conflicts.

- **Political geographers recognize that the relationship between politics and geography is two-way. Political geography can be seen both as the geography of politics and the politics of geography.**

  The politics of geography is best illustrated through regional and sectional movements and conflicts, such as the India-Pakistan conflict or the American Civil War. The geography-of-politics approach recognizes that systems of political representation are geographically anchored and shape the opportunities of the people who live within them.

- **The relations between politics and geography are often driven by particular theories and practices of the world's states. Understanding imperialism, colonialism, heartland theory, domino theory, the end of the cold war, and the emergence of the new world order is key to comprehending how, within the context of the world system, geography has influenced politics and politics has influenced geography.**

  At present, theories about globalization and the interconnectedness of places are particularly important, whereas the domino theory and heartland theory have waned in their intellectual and popular appeal.

- **Political geography deals with the phenomena occurring at all scales of resolution, from the global to the body. Important East/West and North/South divisions dominate international politics, whereas regionalism, sectionalism, and similar divisions dominate intrastate politics.**

  No one scale necessarily dominates any other, and changes emanating from a locality may have international impacts and vice versa.

KEY TERMS

bioterrorism (p. 372)
centrifugal forces (p. 352)
centripetal forces (p. 351)
children's rights (p. 378)
citizenship (p. 349)
confederation (p. 354)
decolonization (p. 360)
democratic rule (p. 386)
domino theory (p. 365)
East/West divide (p. 363)
federal state (p. 352)
geopolitics (p. 344)
gerrymandering (p. 387)
global civil society (p. 378)
human rights (p. 378)
international organization (p. 360)
international regime (p. 378)
intifada (p. 382)
nation (p. 349)
nation-state (p. 349)
nationalism (p. 350)
new world order (p. 366)
North/South divide (p. 359)
reapportionment (p. 387)
redistricting (p. 387)
regionalism (p. 379)
sectionalism (p. 379)
self-determination (p. 379)
sovereignty (p. 349)
supranational organization (p. 375)
territorial organization (p. 386)
territory (p. 346)
terrorism (p. 367)
unitary state (p. 352)
zionism (p. 382)

ADDITIONAL READING

Agnew, J. A., *Hegemony: The New Shape of Global Power.* Philadelphia: Temple University Press, 2005.

Agnew, J., K. Mitchell and G. Toal, *A Companion to Political Geography.* Oxford: Blackwell, 2002.

Appelbaum, R. P. and W. I. Robinson, (eds.), *Critical Globalization Studies.* New York: Routledge, 2005.

Anderson, J., Brook, C., and A. Cochrane (eds.), *A Global World? Re-ordering Political Space.* Oxford: Oxford University Press, 1995.

Anderson, J., "Separation and Devolution: The Basques in Spain," *Environment and Planning D: Society and Space, 8*(4), 427–428, 1990.

Blomley, N., *Law, Space and the Geographies of Power.* New York: Guildford Press, 1994.

Bradshaw, York W., and M. Wallace, *Global Inequalities.* Thousand Oaks, CA: Pine Forge Press, 1996.

Cox, K. R., *Political Geography: Territory, State, Society.* Oxford: Blackwell, 2002.

Enloe, Cynthia H., *Bananas, Beaches and Bases: Making Feminist Sense of International Politics.* Berkeley: University of California Press, 2000.

Ferguson, Y. H. and R. J. B. Jones, *Political Space: Frontiers of Change and Governance in a Globalizing World.* Albany, NY: State University of New York Press, 2002.

Epic Project Web site, *Comparative and Country-by-Country Data on Election Systems, Laws, Management and Administration,* http://www.epicproject.org/en/ (accessed on September 24, 2005).

Harff, B., *Early Warning of Communal Conflict and Genocide: Linking Empirical Research to International Responses.* Boulder, CO: Westview Press, 2003.

Gettleman, M. E. and S. Schaar, (eds.), *The Middle East and Islamic World Reader.* New York: Grove Press, 2003.

Gregory, D., *The Colonial Present: Afghanistan, Palestine, and Iraq.* Malden, MA: Blackwell Publishers, 2004.

Held, D., A. McGrew, D. Goldblatt, and J. Perraton, *Global Transformations.* Malden: Blackwell Publishers, 1999.

Hirst, P. and G. Thompson, *Globalization in Question.* Cambridge: Polity Press, 1999.

Hoffman, B., *Inside Terrorism.* London: Victor Gollancz, 1998.

Juergensmeyer, M., *The New Cold War? Religious Nationalism Confronts the State.* Berkeley: University of California Press, 1993.

Riesebrodt, M., *Pious Passion: The Emergence of Modern Fundamentalism in the United States and Iran;* translated from the German *Fundamentalismus als patriarchalische Protestbewegung* by Don Reneau. Berkeley: University of California Press, 1993.

Salamon, L. M, S. Sokolowski, S. Wojciech, and R. List, *Global Civil Society: An Overview.* Baltimore, MD: Johns Hopkins University Institute for Policy Studies, 2003.

Staeheli, L. A., Kofman, E. and L. Peake, *Mapping Women, Making Politics: Feminist Perspectives on Political Geography.* New York: Routledge, 2004.

Taylor, P. and C. Flint, *Political Geography: World-Economy, Nation-State, and Locality,* 4th ed. Upper Saddle River, NJ: Prentice Hall, 2000.

Thurow, L., *The Future of Capitalism.* New York: Morrow, 1996.

Totten, S., W. S. Parsons, and I. W. Charny, (eds.), *Century of Genocide: Critical Essays and Eyewitness Accounts,* 2nd ed. New York: Routledge Press, 2004.

Tuathail, G., *Critical Geopolitics: The Politics of Writing Global Space.* Minnesota: University of Minnesota Press, 1996.

U.N. Office for Coordination of Humanitarian Affairs and U.N. Relief and Works Agency for Palestinian Refugees, *The Humanitarian Impact of the West Bank Barrier on Palestinian Communities,* 2005.

van den Anker, C., *The Political Economy of New Slavery.* New York: Palgrave Macmillan, 2004.

White, J., *Terrorism: An Introduction,* 3rd ed. Toronto: Wadsworth, 2002.

EXERCISES

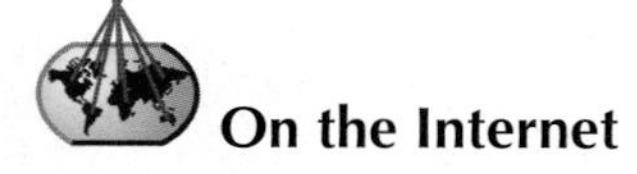

On the Internet

The Internet exercises for this chapter will help you to better understand political geography. We focus on such issues as the expansion of NATO membership, the reintegration of Hong Kong into China, the separation of East Timor from Indonesia, the increasing powers of the European Union, and the threatened breakup of states such as Canada, Spain, Russia, Indonesia, and the United Kingdom as a result of separatist movements. All of these issues combine questions of the organization of power with those of geographical definition. These and other issues will be considered using a number of geopolitical maps found in our thinking-spatially exercise, a concept-review exercise that revisits chapter ideas on the geopolitics of the world order, including nationalism, imperialism, colonization, and more.

Unplugged

1. International boundaries are a prominent feature of the political geography of the contemporary world. In this exercise you are asked to explore the impact of a boundary on nationalist attitudes and behaviors. You will need to use *The New York Times* Index to complete this assignment. Using the United States-Mexico border as your key word, describe the range of issues that derive from this juxtaposition of two very different nations. You should concentrate on a five-year period and show which issues grew in importance, which issues declined, and which issues continued to have a consistent news profile throughout the period.
2. National elections usually tell a story about the ways in which regional ideas and attitudes shape the national political agenda. In the 1996 U.S. presidential election, pollsters considered religion an important issue, with the Religious Right playing a crucial role in reelecting George Bush. Using national election result data available through the Government Documents Division of your college or university library, describe the political geography of fundamentalist or evangelical Christians. Did these Christians vote for Bush in all regions of the country? If not, which ones did not, and what might explain the regional distribution of this powerful voting block?
3. Using two maps of Europe (up to but not including Russia and the former Soviet Union), one from 1930 and one from the present, compare the differences. How do issues of ethnicity, religion, and political system help to explain these changes? Identify any areas on the map that you feel may be the sites of future border changes, and explain why.
4. U.S. presidential elections provide a snapshot of the changing political geography of the country. Using maps of the 1992 and 2004 presidential elections, compare the 1992 results with those for the year 2004. What are the most significant differences in the two maps? What are some reasons for these differences? If you are able to get maps that are disaggregated by race, ethnicity, or gender, what further explanations can you offer for the different maps based on these additional variables?

10 Urbanization

Urbanization is one of the most important geographic phenomena in today's world. The United Nations Center for Human Settlements (UNCHS)[1] notes that the growth of cities and the urbanization of rural areas are now irreversible because of the global shift to technological-, industrial-, and service-based economies. The proportion of the world's population living in urban settlements is growing at a rapid rate, and the world's economic, social, cultural, and political processes are increasingly being played out within and between the world's systems of towns and cities. The UNCHS has concluded that few countries are able to handle the urban population crush, which is causing problems on an unprecedented scale with everything from clean water to disease prevention. Already 10 million people are dying annually in densely populated urban areas from conditions produced by substandard housing and poor sanitation. About 600 million people worldwide are either homeless or living in unfit housing that is life-threatening. In this chapter we describe the extent and pattern of urbanization across the world, explaining its causes and the resultant changes wrought in people and places.

The United Nations International Children's Fund (UNICEF)[2] has blamed "uncontrollable urbanization" in less developed countries for the widespread creation of "danger zones" in which increasing numbers of children are forced to become beggars, prostitutes, and laborers before reaching their teens. Pointing out that urban populations are growing at twice the general population rate, UNICEF has concluded that too many people are being squeezed into cities that do not have enough jobs, shelter or schools to accommodate them. As a consequence, the family and community structures that support children are being destroyed, with the result that increasing numbers of children have to work. For hundreds of thousands of street kids in less developed countries, "work" means anything that contributes to survival: shining shoes; guiding cars into parking spaces; chasing other street kids away from patrons at an outdoor café; working as domestic help; making fireworks; selling drugs. In Abidjan, in the Ivory Coast, 15-year-old Jean-Pierre Godia, who cannot read or write, spends about six hours every day trying to sell 10-roll packets of toilet paper to motorists at a busy intersection. He buys the packets for about $1.20 and sells them for $2. Some days he doesn't sell any. In the same city, seven-year-old Giulio guides cars into parking spaces outside a chic pastry shop. He has been doing this since he was five, to help his mother and four siblings, who beg on a nearby corner.

MAIN POINTS

- The urban areas of the world are the linchpins of human geographies at the local, regional, and global scales.
- The earliest towns and cities developed independently in the various hearth areas of the first agricultural revolution.
- The expansion of trade around the world, associated with colonialism and imperialism, established numerous gateway cities.
- The Industrial Revolution generated new kinds of cities—and many more of them.
- Today the single most important aspect of world urbanization, from a geographical perspective, is the striking difference in trends and projections between the core regions and the peripheral regions.
- A small number of "world cities," most of them located within the core regions of the world-system, occupy key roles in the organization of global economics and culture.
- Many megacities of the periphery are primate and exhibit a high degree of centrality within their urban systems.
- There is a close relationship between globalization and urbanization, mediated by networked infrastructures of transportation, information, and communications technologies.

[1]United Nations Center for Human Settlements (HABITAT), *Cities in a Globalizing World: Global Report on Human Settlements, 2001.* London: Earthscan Publications, 2001.
[2]*The Progress of Nations.* New York: United Nations International Children's Fund (UNICEF), 1995.

Pudong financial district, Shanghai, China

URBAN GEOGRAPHY AND URBANIZATION

From small market towns and fishing ports to megacities of millions of people, the urban areas of the world are the linchpins of human geographies. They have always been a crucial element in spatial organization and the evolution of societies, but today they are more important than ever. Between 1980 and 2005 the number of city dwellers worldwide rose by 1.35 billion. Cities now account for almost half the world's population. Much of the developed world has become almost completely urbanized (**Figure 10.1**), while in many peripheral and semiperipheral regions the current rate of urbanization is without precedent (**Figure 10.2**). The United Nations Human Settlements Program (U.N.-Habitat) estimates that 60 percent of the world's population will live in cities by 2030, a trend that equals the addition of a city of 1 million residents every week. Around 80 percent of all city dwellers in 2030 will be in peripheral and semiperipheral countries. Urbanization on this scale is a remarkable geographical phenomenon—one of the most important processes shaping the world's landscapes.

Towns and cities are centers of cultural innovation, social transformation, and political change. They can also be engines of economic development. The Gross Product of large cities like London, Los Angeles, Mexico City, and Paris is roughly equivalent to that of entire countries like Australia and Sweden, while New York's Gross Product is almost as large as that of China. While they often pose social and environmental problems, towns and cities are essential elements in human economic and social organization. Experts on urbanization point to four fundamental aspects of the role of towns and cities in human economic and social organization:

- ***The* mobilizing function *of urban settlement.*** Urban settings, with their physical infrastructure and their large and diverse populations, are places where entrepreneurs can get things done. Cities, in other words, provide efficient and effective environments for organizing labor, capital, and raw materials, and for distributing finished products. In developing countries, urban areas produce as much as 60 percent of total gross domestic product with just one-third of the population.
- ***The* decision-making capacity *of urban settlement.*** Because urban settings bring together the decision-making machinery of public and private institutions and organizations, they come to be concentrations of political and economic power.

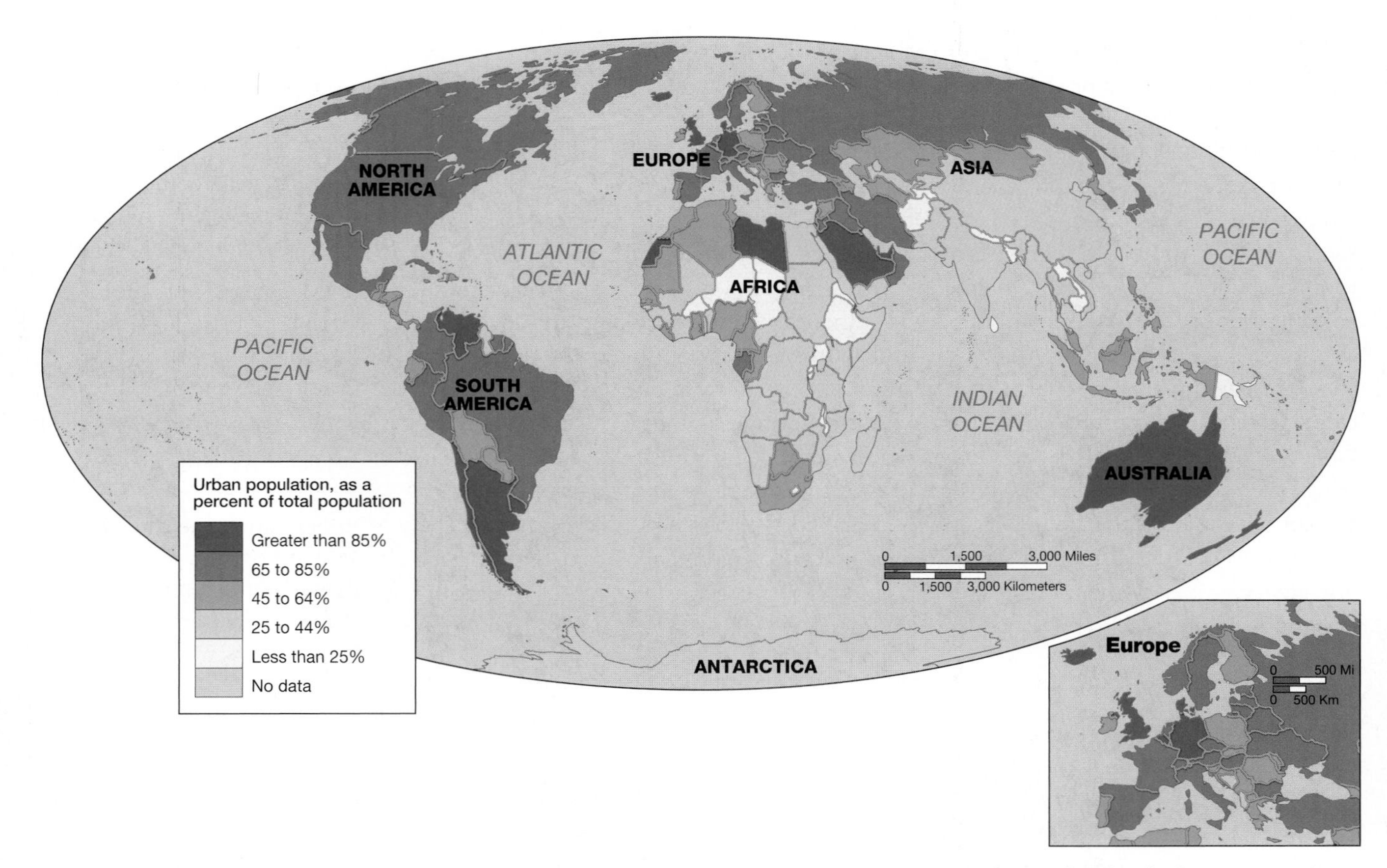

Figure 10.1 Percentage of each country's population living in urban settlements, 2003 The lowest levels of urbanization—less than 25 percent—are found in Central Africa and South and Southeast Asia. Most of the core countries are highly urbanized, with between 65 and 95 percent of their populations living in urban settlements. *Source:* Data from United Nations Department of Economic and Social Affairs, Population Division, *World Urbanization Prospects: The 2003 Revision,* 2004, pp. 25–29.

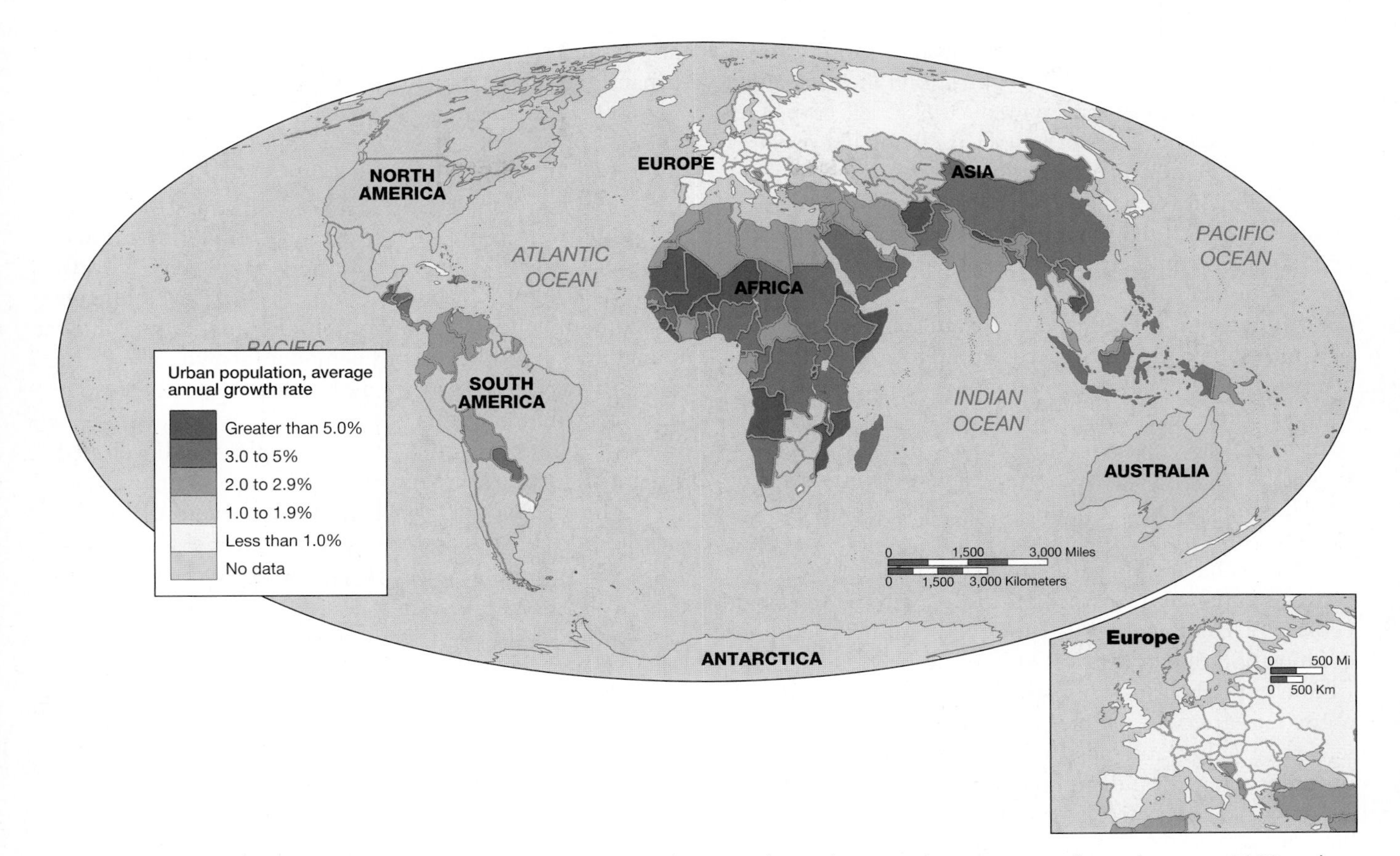

Figure 10.2 Rates of growth in urbanization, 2000–2005 This map shows the annual average growth rate between 2000 and 2005 in the proportion of people in each country living in urban settlements. Core countries, already highly urbanized, grew quite slowly. The urban populations of peripheral countries such as Angola, Afghanistan, Burundi, Liberia, Rwanda, and Somalia, on the other hand, grew by more than 5 percent each year, creating tremendous pressure on cities' capacity to provide jobs, housing, and public services. *Source:* Data from United Nations Department of Economic and Social Affairs, Population Division, *World Urbanization Prospects: The 2003 Revision,* 2004, pp. 70–78.

- ***The* generative functions *of urban settlement.*** The concentration of people in urban settings makes for much greater interaction and competition, which facilitates the generation of innovation, knowledge, and information.
- ***The* transformative capacity *of urban settlement.*** The size, density, and variety of urban populations tends to have a liberating effect on people, allowing them to escape the rigidities of traditional, rural society and to participate in a variety of lifestyles and behaviors.

The study of urban geography is concerned with the development of towns and cities around the world, with particular reference to the similarities and differences both *among* and *within* urban places. For urban geographers, some of the most important questions include: What attributes make towns and cities distinctive? How did these distinctive identities evolve? What are the relationships and interdependencies between particular sets of towns and cities? What are the relationships between cities and their surrounding territories? Do significant regularities exist in the spatial organization of land use within cities, in the patterning of neighborhood populations, or in the layout and landscapes of particular kinds of cities?

Urban geographers also want to know about the causes of the patterns and regularities they find. How, for example, do specialized urban subdistricts evolve? Why did urban growth occur in a particular region at a particular time? And why did urban growth exhibit a distinctive physical form during a certain period? In pursuing such questions, urban geographers have learned that the answers are ultimately to be found in the wider context of economic, social, cultural, and political life. In other words, towns and cities must be viewed as part of the economies and societies that maintain them.

Urbanization, therefore, is not simply the demographic growth of towns and cities. It also involves many other changes, both quantitative and qualitative. From the geographer's perspective, these changes can be conceptualized in several different ways. One of the most important of these is by examining the attributes and dynamics of urban systems. An **urban system,** or city system, is any interdependent set of urban settlements within a given region. Thus, for example, we can speak of the

French urban system, the African urban system, or even the global urban system. As urbanization takes place, the attributes of urban systems will, of course, reflect the fact that increasing numbers of people are living in ever larger towns and cities. They will also reflect other important changes, such as changes in the relative size of cities, changes in their functional relationships with one another, and changes in their employment base and population composition.

Another important aspect of change associated with urbanization processes concerns urban form. **Urban form** refers to the physical structure and organization of cities in their land use, layout, and built environment. As urbanization takes place, not only do towns and cities grow bigger physically, extending upward and outward, but they also become reorganized, redeveloped, and redesigned in response to changing circumstances.

These changes, in turn, are closely related to a third aspect of change: transformations in patterns of urban ecology. **Urban ecology** is the social and demographic composition of city districts and neighborhoods. Urbanization not only brings more people to cities, it also brings a greater variety of people. As different social, economic, demographic, and racial subgroups become sorted into different territories, distinctive urban ecologies emerge. As new subgroups arrive or old ones leave, these ecologies change.

A fourth aspect of change associated with urbanization concerns people's attitudes and behavior. New forms of social interaction and new ways of life are brought about by the liberating and transformative effects of urban environments. These changes have given rise to the concept of urbanism, which refers to the distinctive nature of social and cultural organization in particular urban settings. **Urbanism** describes the way of life fostered by urban settings, in which the number, physical density, and variety of people often result in distinctive attitudes, values, and patterns of behavior. Geographers are interested in urbanism because of the ways in which it varies both within and between cities.

URBAN ORIGINS

It is important to put the geographic study of towns and cities in historical context. After all, many of the world's cities are the product of a long period of development. We can only understand a city, old or young, if we know something about the reasons behind its growth, about the rate at which it has grown, and about the processes that have contributed to this growth.

In broad terms, the earliest urbanization developed independently in the various hearth areas of the first agricultural revolution. The very first region of independent urbanism was in the Middle East, in the valleys of the Tigris and Euphrates (in Mesopotamia) and in the Nile Valley from around 3500 B.C. (see Chapter 4). Together these intensively cultivated river valleys formed the so-called Fertile Crescent. In Mesopotamia, the growth in size of some of the agricultural villages located on the rich alluvial soils of the river floodplains formed the basis for the large rival city-states of the Sumerian empire. They included Ur (in present-day Iraq), the capital from about 2300 to 2180 B.C., as well as Eridu, Uruk, and Erbil (ancient Arbela—**Figure 10.3**). These fortified city-states contained tens of thousands of inhabitants; social stratification, with religious, political, and military classes; innovative technologies, including massive irrigation projects; and extensive trade connections. By 1885 B.C., the Sumerian city-states had been taken over by the Babylonians and then the Neo-Babylonians, who governed the region from their capital city, Babylon. In Egypt, which became a unified state as early as 3100 B.C., large irrigation projects controlled the Nile's waters for agricultural and other uses, sup-

Figure 10.3 Erbil Erbil (Ancient Arbela) in northeast Iraq is located atop a *tell,* a mound representing the remains of generations of sun-dried mud-brick buildings, visible as a hill rising high above the surrounding plain. The 100-foot high Erbil tell is believed to represent perhaps 6,000 years of continuous occupation.

porting a series of capital cities that included Thebes, Akhetaten (Tell el-Amarna), and Tanis. Internal peace in Egypt meant that there was no need for massive investments in these cities' defensive fortifications. Also, each Pharaoh was free to locate a new capital at any site he selected for his tomb, and after his death the city was usually abandoned to the priests.

By 2500 B.C. cities had appeared in the Indus Valley, and by 1800 B.C. they were established in northern China. Other areas of independent urbanism include Mesoamerica (from around 100 B.C.) and Andean America (from around A.D. 800). Meanwhile, the original Middle Eastern urban hearth continued to produce successive generations of urbanized world-empires, including those of Greece, Rome, and Byzantium.

Experts differ in their explanations of these first transitions from subsistence minisystems to city-based world-empires. The classical archaeological interpretation emphasizes the availability of an agricultural surplus large enough to allow the emergence of specialized, nonagricultural workers. Some urbanization, however, may have resulted from the pressure of population growth. This pressure, it is thought, disturbed the balance between population and resources, causing some people to move to marginal areas. Finding themselves in a region where agricultural conditions were unfavorable, these people either had to devise new techniques of food production and storage or establish a new form of economy based on services such as trade, religion, or defense. Any such economy would have required concentrations of people in urban settlements.

Most experts agree that changes in social organization were an important precondition for urbanization. Specifically, urbanization required the emergence of groups who were able to exact tributes, impose taxes, and control labor power, usually through some form of religious persuasion or military coercion. Once established, this elite group provided the stimulus for urban development by using its wealth to build palaces, arenas, and monuments in order to show off its power and status. This activity not only created the basis for the physical core of ancient cities but also required an increased degree of specialization in nonagricultural activities—construction, crafts, administration, the priesthood, soldiery, and so on—which could be organized effectively only in an urban setting. By A.D. 1000, city-based world empires had emerged in Europe, the Middle East, and China, including a dozen major cities with populations of 100,000 or more (**Figure 10.4**).

The urbanized economies of world-empires were a precarious phenomenon, however, and many of them lapsed into ruralism before being revived or recolonized. In a number of cases the decline of world-empires was a result of demographic setbacks associated with wars or epidemics. Such disasters left too few people to maintain the social and economic infrastructure necessary for urbanization. This lack of labor power seems to have been a major contributing factor to the eventual collapse of the Mesopotamian empire, and it may also have contributed to the abandonment of much of the Mayan empire more than 500 years before the arrival of the Spanish. Similarly, the population of the Roman empire began to decline in the second century A.D., giving rise to labor shortages, abandoned fields, and depopulated towns, and allowing the infiltration of "barbarian" settlers and tribes from the German lands of East-Central Europe.

The Roots of European Urban Expansion

In Europe the urban system introduced by the Greeks and reestablished by the Romans almost collapsed during the Dark Ages of early medieval Europe (A.D. 476–1000). During this period, feudalism gave rise to a fragmented landscape of inflexible and inward-looking world-empires. Feudalism was a rigid, rurally oriented form of economic and social organization based on the communal chiefdoms of Germanic tribes that had invaded the disintegrating Roman Empire. From this unlikely beginning, however, an elaborate urban system developed, its largest centers eventually growing into what would become the nodal centers of a global world-system.

Early medieval Europe, divided into a patchwork of feudal kingdoms and estates, was mostly rural. Each feudal estate was more or less self-sufficient regarding foodstuffs, and each kingdom or principality was more or less self-sufficient regarding the raw materials needed to craft simple products. Most regions, however, did support at least a few small towns. The existence of these towns depended mainly on their role:

- *Ecclesiastical or university centers*—Examples include St. Andrews in Scotland (**Figure 10.5**); Canterbury, Cambridge, and Coventry in England; Rheims and Chartres in France; Liège in Belgium; Bremen in Germany; Trondheim in Norway; and Lund in Sweden.
- *Defensive strongholds*—Examples include the hilltop towns of central Italy such as Foligno, Montecompatri, and Urbino; the bastide, or fortress, towns of southwestern France, such as Aigues-Mortes and Montauban; and gateway towns such as Bellinzona, Switzerland (**Figure 10.6**).
- *Administrative centers (for the upper tiers of the feudal hierarchy)*—Examples include Cologne (**Figure 10.7**), Mainz, and Magdeburg in Germany; Falkland in Scotland; Winchester in England; and Toulouse in France.

From the eleventh century onward, however, the feudal system faltered and disintegrated in the face of successive demographic, economic, and political crises that were caused by steady population growth in conjunction with only modest technological improvements and

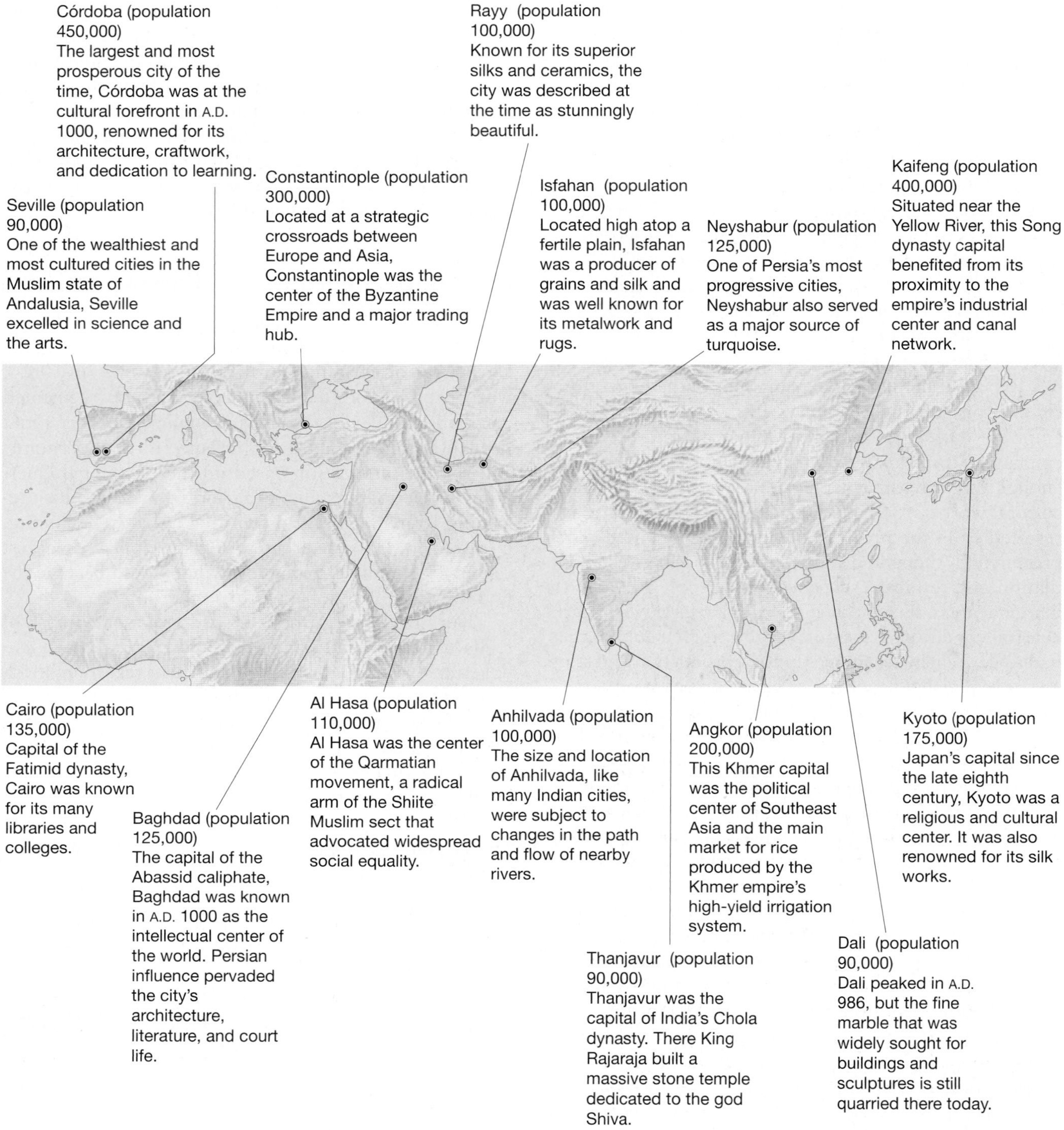

Córdoba (population 450,000)
The largest and most prosperous city of the time, Córdoba was at the cultural forefront in A.D. 1000, renowned for its architecture, craftwork, and dedication to learning.

Rayy (population 100,000)
Known for its superior silks and ceramics, the city was described at the time as stunningly beautiful.

Seville (population 90,000)
One of the wealthiest and most cultured cities in the Muslim state of Andalusia, Seville excelled in science and the arts.

Constantinople (population 300,000)
Located at a strategic crossroads between Europe and Asia, Constantinople was the center of the Byzantine Empire and a major trading hub.

Isfahan (population 100,000)
Located high atop a fertile plain, Isfahan was a producer of grains and silk and was well known for its metalwork and rugs.

Neyshabur (population 125,000)
One of Persia's most progressive cities, Neyshabur also served as a major source of turquoise.

Kaifeng (population 400,000)
Situated near the Yellow River, this Song dynasty capital benefited from its proximity to the empire's industrial center and canal network.

Cairo (population 135,000)
Capital of the Fatimid dynasty, Cairo was known for its many libraries and colleges.

Baghdad (population 125,000)
The capital of the Abassid caliphate, Baghdad was known in A.D. 1000 as the intellectual center of the world. Persian influence pervaded the city's architecture, literature, and court life.

Al Hasa (population 110,000)
Al Hasa was the center of the Qarmatian movement, a radical arm of the Shiite Muslim sect that advocated widespread social equality.

Anhilvada (population 100,000)
The size and location of Anhilvada, like many Indian cities, were subject to changes in the path and flow of nearby rivers.

Angkor (population 200,000)
This Khmer capital was the political center of Southeast Asia and the main market for rice produced by the Khmer empire's high-yield irrigation system.

Thanjavur (population 90,000)
Thanjavur was the capital of India's Chola dynasty. There King Rajaraja built a massive stone temple dedicated to the god Shiva.

Dali (population 90,000)
Dali peaked in A.D. 986, but the fine marble that was widely sought for buildings and sculptures is still quarried there today.

Kyoto (population 175,000)
Japan's capital since the late eighth century, Kyoto was a religious and cultural center. It was also renowned for its silk works.

Figure 10.4 Major cities in A.D. 1000 The most important cities in A.D. 1000 were the seats of world empires—the Islamic caliphates, the Byzantine Empire, the Chinese Empire, and Indian kingdoms—that had developed well-established civilizations with urban systems based on regional trade and protected by strong military rule. (*Source:* Data from T. Chandler, *Four Thousand Years of Urban Growth: A Historical Census.* Washington, DC: Worldwatch Institute, 1987; "The Year 1000," *U.S. News & World Report,* August 16, 1999, pp. 66–70.)

Figure 10.5 St. Andrews, Scotland St. Andrews was an important ecclesiastical center. The cathedral was built in the twelfth century, the castle (an episcopal residence) c. 1200. The university was founded in 1410.

limited amounts of cultivable land. To bolster their incomes and raise armies against one another, the feudal nobility began to levy increasingly higher taxes. Peasants were consequently obliged to sell more of their produce for cash on the market. As a result, a more extensive money economy developed, along with the beginnings of a pattern of trade in basic agricultural produce and craft manufactures. Some long-distance trade even began in luxury goods, such as spices, furs, silks, fruit, and wine. Towns began to increase in size and vitality on the basis of this trade.

The regional specializations and trading patterns that emerged provided the foundations for a new phase of urbanization based on merchant capitalism (**Figure 10.8**). Beginning with networks established by the merchants of Venice, Pisa, Genoa, and Florence (in northern Italy) and the trading partners of the Hanseatic League (a federation of city-states around the North Sea and Baltic coasts),

Figure 10.6 Bellinzona, Switzerland Bellinzona commands the narrow valley of the River Ticino that stands at the threshold of the great Alpine passes of the Novena (Nufenen), Gottardo (Gotthard), Lucomagno (Lukmanier) and San Bernadino. For nearly five centuries of Roman rule, the best of all possible military and trade routes between Rome and its colonies in the north led through Bellinzona. When the narrow, winding roads over the four passes met in Bellinzona, all it took was a single barrier to close the way. By the early Medieval period, the resumption of trans-Alpine trade re-established Bellinzona's strategic importance, and in the thirteen and fourteenth centuries the ruling families of the town built a series of castles and massive new ramparts right across the Ticino valley.

Figure 10.7 Cologne In the late 1400s, when this woodcut was made, Cologne had a population of less than 25,000 but was already an important commercial and manufacturing center, with an important cathedral and a university that was already more than 100 years old.

a trading system of immense complexity soon came to span Europe from Bergen to Athens and from Lisbon to Vienna. By 1400, long-distance trading was well established, based not on the luxury goods of the pioneer merchants but on bulky staples such as grains, wine, salt, wool, cloth, and metals. Milan, Genoa, Venice, and Bruges had all grown to 100,000 or more. Paris was the dominant European city, with a population of about 275,000. This was the Europe that stood poised to extend its grasp to a global scale.

Between the fifteenth and seventeenth centuries a series of changes occurred that transformed not only the cities and city systems of Europe but the entire world economy. Merchant capitalism increased in scale and sophistication; economic and social reorganization was stimulated by the Protestant Reformation and the scientific revolution. Meanwhile, aggressive overseas colonization made Europeans the leaders, persuaders, and shapers of the rest of the world's economies and societies. Spanish and Portuguese colonists were the first to extend the European urban system into the world's peripheral regions. They established the basis of a Latin American urban system in just 60 years, between 1520 and 1580. Spanish colonists founded their cities on the sites of Indian cities (in Oaxaca and Mexico City, Mexico; Cajamarca and Cuzco, Peru; and Quito, Ecuador) or in regions of dense indigenous populations (in Puebla and Guadalajara, Mexico; and Arequipa and Lima, Peru). These colonial towns were established mainly as administrative and military centers from which the Spanish Crown could occupy and exploit the New World. Portuguese colonists, in contrast, situated their cities—Recife, Salvador, São Paulo, and Rio de Janeiro—with commercial rather than administrative considerations in mind. They, too, were motivated by exploitation, but their strategy was to establish colonial towns in locations best suited to organizing the collection and export of the products of their mines and plantations.

In Europe, Renaissance reorganization saw the centralization of political power and the formation of national states, the beginnings of industrialization, and the funneling of plunder and produce from distant colonies. In this new context the port cities of the North Sea and Atlantic coasts enjoyed a decisive locational advantage. By 1700 London had grown to 500,000, while Lisbon and Amsterdam had each grown to about 175,000. The cities of continental and Mediterranean Europe expanded at a more modest rate. By 1700 Venice had added only 30,000 to its 1400 population of 110,000, and Milan's population had not grown at all between 1400 and 1700.

The most important aspect of urbanization during this period, however, was the establishment of gateway cities around the rest of the world (**Figure 10.9**). **Gateway cities** are those that serve as a link between one country or region and others because of their physical situation. They are control centers that command entrance to, and exit from, their particular country or region. European powers founded or developed literally thousands of towns in other parts of the world as they extended their trading networks and established their colonies. The great majority of them were ports. Protected by fortifications and European naval power, they began as trading posts and colonial administrative centers. Before long they developed manufacturing of their own to supply the pioneers' needs, along with more extensive commercial and financial services.

As colonies were developed and trading networks expanded, some of these ports grew rapidly, acting as gate-

Figure 10.8 The towns and cities of Europe, ca. 1350 Cities with more than 10,000 residents were uncommon in medieval Europe except in northern Italy and Flanders, such as Florence and Delft, where the spread of cloth production and the growth of trade permitted relatively intense urbanization. Elsewhere, large size was associated with a complex of administrative, religious, educational, and economic functions. By 1350, many of the bigger towns (for example, Barcelona, Cologne, Prague) supported universities as well as a variety of religious institutions. Most urban systems, reflecting the economic and political realities of the time, were relatively small. (After P. M. Hohenberg and L. H. Lees, *The Making of Urban Europe 1000–1950*.)

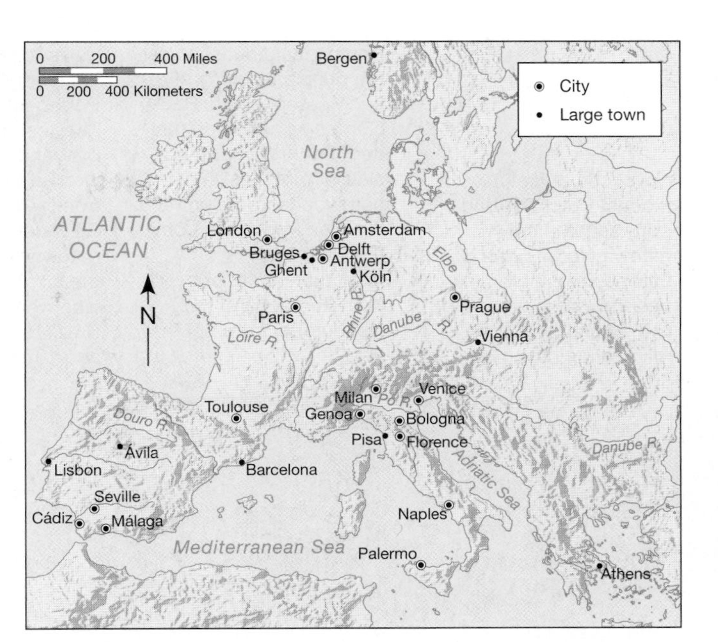

Ghent, Belgium

Venice, Italy

Florence, Italy

Prague, Czech Republic

New York, at first a modest Dutch fur-trading port, became the gateway for millions of European immigrants and for a large volume of U.S. agricultural and manufacturing exports.

Boston first flourished as the principal colony of the Massachusetts Bay Company, exporting furs and fish and importing slaves from West Africa, hardwoods from central America, molasses from the Caribbean, manufactured goods from Europe, and tea (via Europe) from South Asia.

Salvador, Brazil, was the landfall of the Portuguese in 1500. They established plantations that were worked by slave labor from West Africa. Salvador became the gateway for most of the 3.5 million slaves who were shipped to Brazil between 1526 and 1870.

Guangzhou was the first Chinese port to be in regular contact with European traders—first Portuguese in the sixteenth century and then British in the seventeenth century.

Nagasaki was the only port that feudal Japanese leaders allowed open to European traders, and for more than 200 years Dutch merchants held a monopoly of the import-export business through the city.

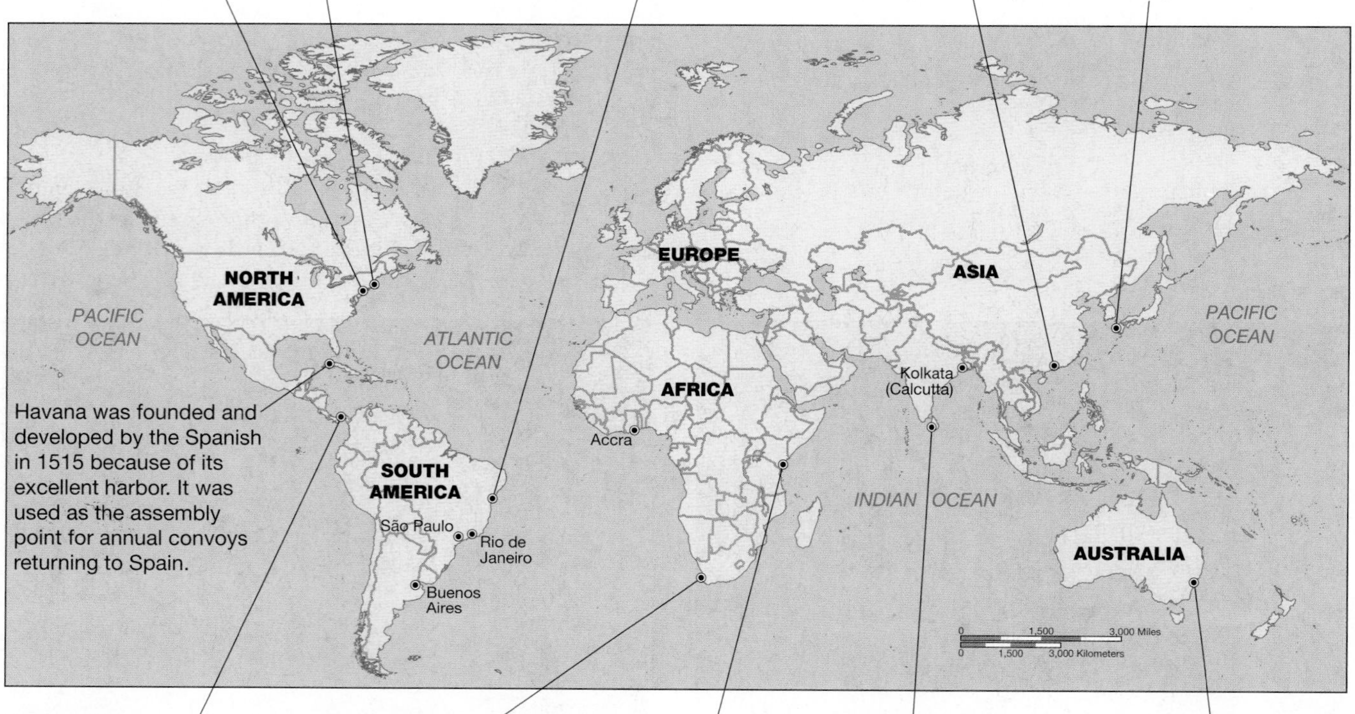

Panama City, founded by the Spanish in 1519, became the gateway for gold and silver on its way by galleon to Spain.

Cape Town was founded in 1652 as a provisioning station for ships of the Dutch East India Company. Later, under British rule, it developed into an import-export gateway for South Africa.

Mombasa (in present-day Kenya) was already a significant Arab trading port when Vasco da Gama visited it in 1498 on his first voyage to India. The Portuguese used it as a trading station until it was recaptured by the Arabs in 1698. It did not become an important gateway port until it fell under British Imperial rule in the nineteenth century, when railroad development opened up the interior of Kenya, along with Rwanda, Uganda, and northern Tanzania.

Colombo's strategic situation on trade routes saw it occupied successively by the Portuguese, the Dutch, and the British. It became an important gateway after the British constructed an artificial harbor to handle the exports from tea plantations in Ceylon (now Sri Lanka).

Sydney, Australia, was not settled until the late eighteenth century, and even then many of the settlers were convicts who had been forcibly transported from Britain. It soon became the gateway for agricultural and mineral exports (mostly to Britain) and for imports of manufactured goods and European immigrants.

Figure 10.9 Gateway cities in the world-system periphery Many of the world's most important cities grew to prominence as gateway cities because they commanded routeways into and out of developing colonies. Gateway cities are control centers that command entrance to and exit from their particular country or region.

ways for colonial expansion into continental interiors. Into their harbors came waves of European settlers; through their docks were funneled the produce of continental interiors. Rio de Janeiro (Brazil) grew on the basis of gold mining; Accra (Ghana) on cocoa; Buenos Aires (Argentina) on mutton, wool, and cereals; Kolkata (India, formerly Calcutta) on jute, cotton, and textiles; São Paulo (Brazil) on coffee; and so on. As they grew into major population centers, they became important markets for imported European manufactures, adding even more to their functions as gateways for international transport and trade.

Industrialization and Urbanization

It was not until the late eighteenth century, however, that urbanization began to become an important dimension of the world-system in its own right. In 1800 less than 5 percent of the world's 980 million people lived in towns and cities. By 1950, however, 16 percent of the world's population was urban, and more than 900 cities of 100,000 or more existed around the world. The Industrial Revolution and European imperialism had created unprecedented concentrations of humanity that were intimately linked in networks and hierarchies of interdependence.

Cities were synonymous with industrialization. Industrial economies could be organized only through the large pools of labor; the transportation networks; the physical infrastructure of factories, warehouses, stores, and offices; and the consumer markets provided by cities. As industrialization spread throughout Europe in the first half of the nineteenth century and then to other parts of the world, urbanization increased at a faster pace. The higher wages and greater variety of opportunities in urban labor markets attracted migrants from surrounding areas. The countryside began to empty. In Europe the *demographic transition* caused a rapid growth in population as death rates dropped dramatically (see Chapter 3). This growth in population provided a massive increase in the labor supply throughout the nineteenth century, further boosting the rate of urbanization not only within Europe itself but also in Australia, Canada, New Zealand, South Africa, and the United States as emigration spread industrialization and urbanization to the frontiers of the world-system.

The shock city of nineteenth-century European industrialization was Manchester, England, which grew from a small town of 15,000 in 1750 to a city of 70,000 in 1801; a metropolis of 500,000 in 1861; and a world city of 2.3 million by 1911 (see Box 10.1: "Shock City: Manchester"). A **shock city** is one that is seen at the time as the embodiment of surprising and disturbing changes in economic, social, and cultural life. As industrialization took hold in North America, the shock city was Chicago, which grew from under 30,000 in 1850 to 500,000 in 1880, 1.7 million in 1900, and 3.3 million in 1930 (**Figure 10.10**). When Chicago was first incorporated as a city in 1837, its population was only 4,200. Its growth followed the arrival of the railroads, which made the city a major transportation hub. By the 1860s, lake vessels were carrying iron ore from the Upper Michigan ranges to the city's blast furnaces, and railroads were hauling cattle, hogs, and sheep to the city for slaughtering and packing. The city's prime geographic situation also made it the nation's major lumber-distributing center by the 1880s.

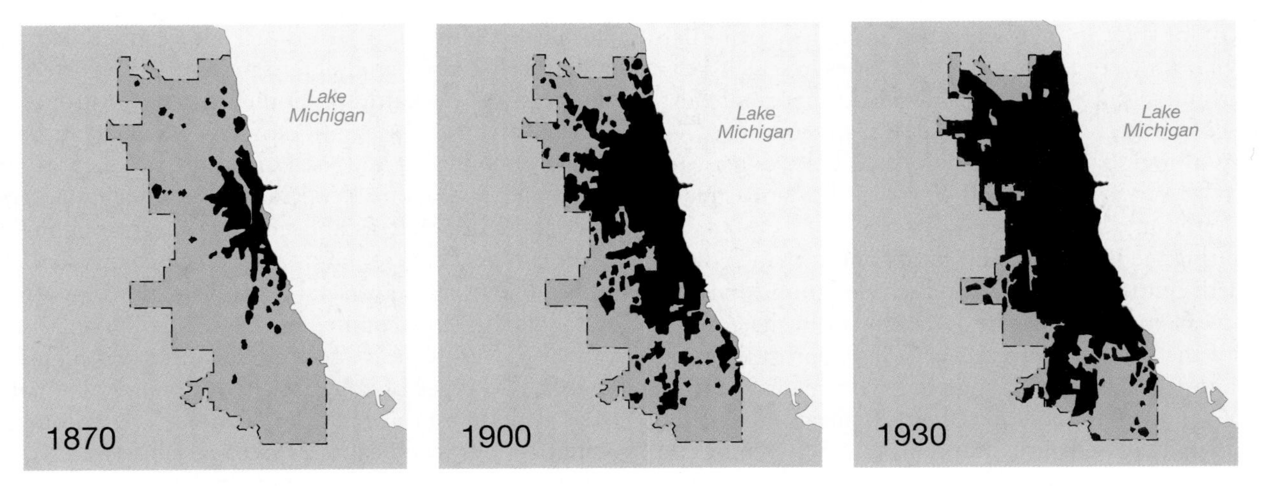

Figure 10.10 Growth of Chicago In 1870, when Manchester was already a thriving metropolis, Chicago was at the beginning of a period of explosive growth. A year later nine square kilometers (4 sq. mi.) of the city, including the business district, were destroyed by fire. It was rebuilt rapidly, with prosperous industrialists taking the opportunity to build impressive new structures in the downtown area. The city's economic and social elite colonized the Lake Michigan shore, while heavy industry, warehouses, and railyards crowded the banks of the Chicago River, stretching northwestward from the city center. To the south of the city center were the Union Stockyards and a pocket of heavy industry where the Calumet River met Lake Michigan. All around were the homes of working families in neighborhoods that spread rapidly outward as wave after wave of immigrants arrived in the city.

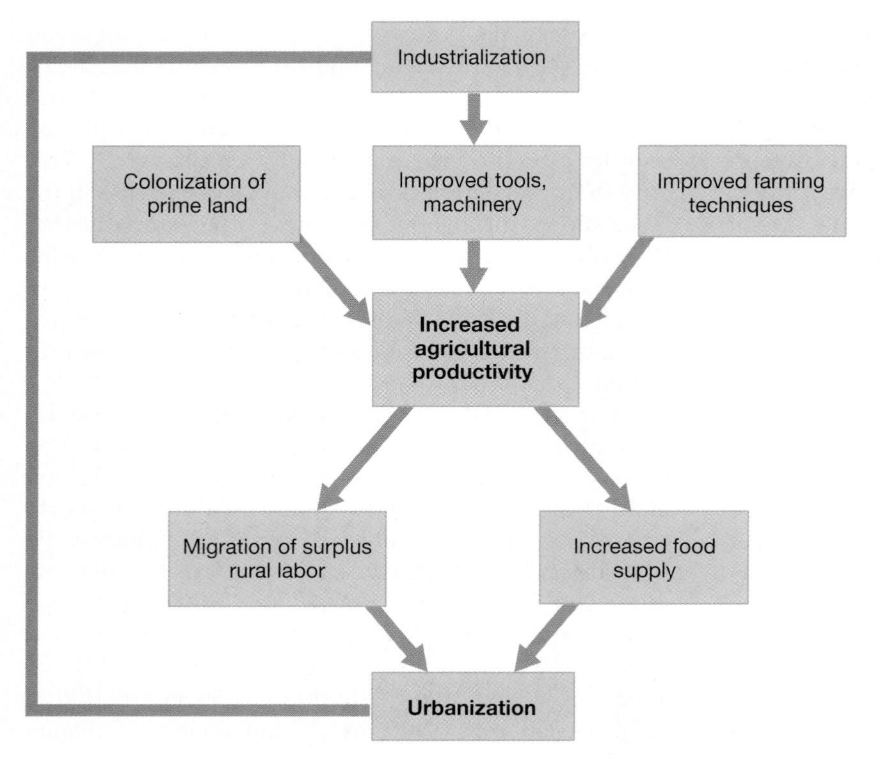

Figure 10.11 The urbanization process in the world's core regions Urbanization was stimulated by advances in farm productivity that (1) provided the extra food to support the increased numbers of townspeople, and (2) made many farmers and farm laborers redundant, prompting them to migrate to cities. Labor displaced in this way ended up consuming food rather than producing it, but this was more than compensated for by the increases in agricultural productivity and by the increased capacity of enlarged urban labor forces to produce agricultural tools, machinery, fertilizers, and so on that contributed further to agricultural productivity.

Both Manchester and Chicago were archetypal forms of an entirely new kind of city—the *industrial city*—whose fundamental reason for existence was not, as in earlier generations of cities, to fulfill military, political, ecclesiastical, or trading functions. Rather, it existed simply to assemble raw materials and to fabricate, assemble, and distribute manufactured goods. Both Manchester and Chicago had to cope, however, with unprecedented rates of growth and the unprecedented economic, social, and political problems that were a consequence of their growth. Both were also world cities, cities in which a disproportionate part of the world's most important business—economic, political, and cultural—is conducted. At the top of a global urban system, such cities experience growth largely as a result of their role as key nodes in the world economy.

During the Industrial Revolution and for much of the twentieth century, a close and positive relationship existed between rural and urban development in the core regions of the world (**Figure 10.11**). The appropriation of new land for agriculture, together with mechanization and the innovative techniques that urbanization allowed, resulted in increased agricultural productivity. This extra productivity released rural labor to work in the growing manufacturing sector in towns and cities. At the same time, it provided the additional produce needed to feed growing urban populations. The whole process was further reinforced by the capacity of urban labor forces to produce agricultural tools, machinery, fertilizer, and other products that made for still greater increases in agricultural productivity. This kind of urbanization is a special case of cumulative causation (Chapter 7), in which a spiral buildup of advantages is enjoyed by particular places as a result of the development of external economies, agglomeration effects, and localization economies.

Imperialism and Peripheral Urbanization

As we pointed out in Chapter 2, the industrialization of the core economies was highly dependent on the exploitation of peripheral regions. Inevitably, the new international division of labor that resulted from this relationship also had a significant impact on patterns and processes of urbanization in the periphery. European imperialism led to the creation of new gateway cities in peripheral countries and, as Europeans raced to establish economic and political control over continental interiors, colonial cities were established as centers of administration, political control, and commerce. **Colonial cities** are those that were deliberately established or developed as administrative or commercial centers by colonial or imperial powers. In fact, geographers often distinguish between two types of colonial city. The pure colonial city was usually established, or "planted," by colonial administrations in a location where no significant urban settlement had previously existed. Such cities were laid out expressly to fulfill colonial functions, with ceremonial spaces, offices, and depots for colonial traders, plantation representatives, and government officials; barracks for a garrison of soldiers; and housing for colonists. Subsequently, as these cities grew, they added housing and commercial land uses for local peoples drawn by the opportunity to obtain jobs such as servants, clerks, or porters. Examples of pure colonial cities were the original settlements of Mumbai (Bombay), Kolkata (Calcutta), Ho Chi Minh City (Saigon), Hong Kong, Jakarta, Manila, and Nairobi.

In the other type of colonial city, colonial functions were grafted onto an existing settlement, taking advantage of a good site and a ready supply of labor. Examples include Delhi, Mexico City, Shanghai, and Tunis. In these cities the colonial imprint is most visible in and around the city center in the formal squares and public spaces; the layout of avenues; and the presence of colonial architecture and monuments. This architecture includes churches, city halls, and railway stations (**Figure 10.12**); the palaces of governors and archbishops; and the houses of wealthy traders, colonial administrators, and landowners.

The colonial legacy can also be read in the building and planning regulations of many colonial cities. Often, colonial planning regulations were copied from those that had been established in the colonizing country. Because these regulations were based on Western concepts, many turned out to be inappropriate to colonial settings. Most colonial building codes, for example, are based on Western models of family and work, with a small family living in a residential area that is some distance from the adults' places of work. This is at odds with the needs of large, extended families whose members are involved with a busy domestic economy and with family businesses that are traditionally integrated with the residential setting. Colonial planning, with its gridiron street layouts; zoning regulations that do not allow for a mixture of land uses; and building codes designed for European climates ignored the specific needs of local communities and misunderstood their cultural preferences.

URBAN SYSTEMS

Every town and city is part of one of the interlocking urban systems that link regional-, national-, and international-scale human geographies in a complex web of interdependence. These urban systems organize space through hierarchies of cities of different sizes and functions. Many of these hierarchical urban systems exhibit common attributes and features, particularly in the relative size and spacing of individual towns and cities.

Geographers have long recognized the tendency for the functions of towns and cities as market centers to result in a hierarchical system of central places. A **central place** is a settlement in which certain types of products and services are available to consumers. **Central place theory** seeks to explain the tendency for central places to be organized in hierarchical systems, analyzing the relative size and geographic spacing of towns and cities as a function of consumer behavior. A fundamental tenet of central place theory is that the smallest settlements in an urban system will provide only those goods and services that meet everyday needs (bakery and dairy products, and groceries, for example) and that these small settlements will be situated relatively close to one another because consumers, assumed to be spread throughout the countryside, will not be prepared to travel far for such items. On the other hand, people will be willing to travel farther for more expensive, less frequently purchased items, so that the larger the settlement, with a broader variety of more specialized goods and services, the farther apart it will be from others of a similar size.

Figure 10.12 Colonial architecture and urban design Cities in the periphery of the world-system have grown very rapidly since the colonial era, but the legacy of the colonial period can still be seen in the architecture, monuments, and urban design of the period. This photograph shows the influence of Victorian British architecture in Mumbai (Bombay).

10.1

VISUALIZING GEOGRAPHY

Shock City: Manchester

"One day I walked with one of these middle-class gentlemen into Manchester. I spoke to him about the disgraceful unhealthy slums and drew his attention to the disgusting condition of that part of the town in which the factory worker lived. I declared that I had never seen so badly built a town in my life. He listened patiently and at the corner of the street he remarked: 'And yet there is a great deal of money to be made here. Good morning, Sir!'"

Friedrich Engels,
The Condition of the Working Class in England in 1844

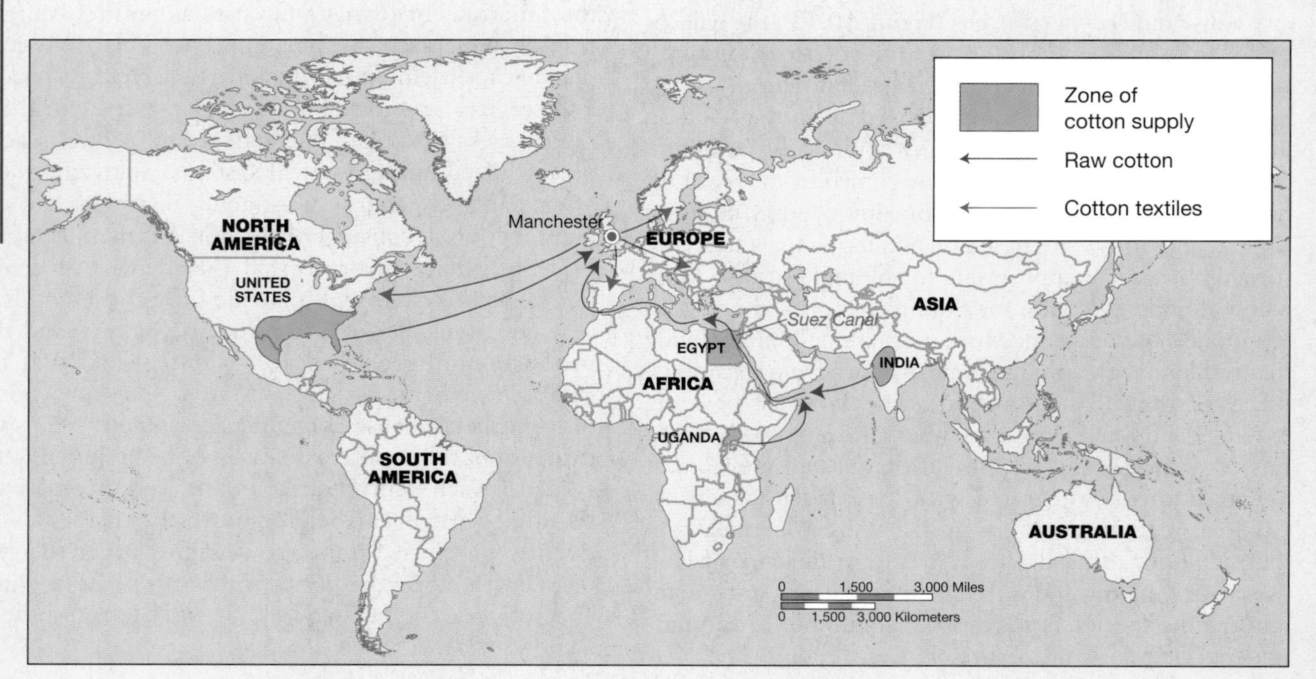

The opening of the Suez Canal in 1869 halved the traveling time between Britain and India. It ruined the Indian domestic cotton textile industry, but it allowed India to export its raw cotton to Manchester. Around the same time, British colonialists established cotton plantations in Egypt and Uganda, providing another source of supply.

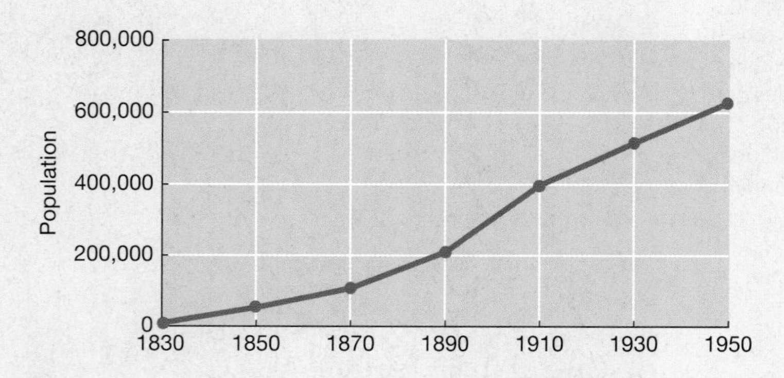

Migrants from Ireland and northern England contributed to Manchester's rapid growth from the mid-nineteenth century.

Manchester's first cotton mill was built in the early 1780s, and by 1830 there were 99 cotton-spinning mills. As the city grew, it spilled out into the surrounding countryside, bringing its characteristic landscape of red-brick terrace housing and "Dark Satanic Mills" with their tall brick chimneys.

Railway viaducts, like this one in Stockport, just outside Manchester, brought rail transportation to Manchester early in the nineteenth century and helped to make the city a major transportation hub.

Manchester City Hall, a classic example of Victorian Gothic architecture, was built to show the world that the city had arrived. Manchester in the nineteenth century was a city of enormous vitality not only in its economic life but also in its political, cultural, and intellectual life.

In the late nineteenth century, working-class housing was built to conform with local building codes—but only just. Much of it has now been replaced, but a good deal still remains.

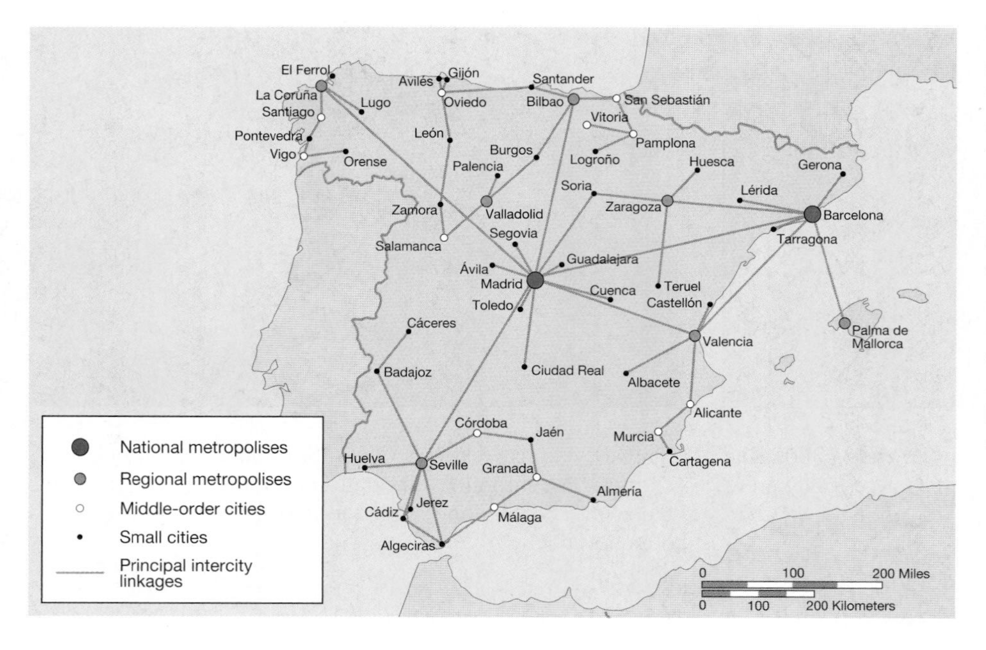

Figure 10.13 The Spanish urban system Note how the smaller cities tend to be linked to middle-order cities, while these in turn are linked to regional metropolises, which are linked to the national metropolises, Madrid and Barcelona. These linkages represent the major flows of capital, information, and goods within the Spanish urban system. (After L. Bourne, R. Sinclair, M. Ferrer and A. d'Entremont, (eds.), *The Changing Geography of Urban Systems.* Navarra, Spain: Department of Human Geography, Universidad de Navarra, 1989, fig. 2, p. 46.)

While consumer behavior certainly helps to explain some aspects of urban systems, there are relatively few regions today where the functions of towns and cities are dominated by local markets and shopping. Nevertheless, the urban systems of most regions do exhibit a clear hierarchical structure. This is partly a legacy of past eras, when towns and cities did function mainly as market centers for surrounding agricultural areas. **Figure 10.13** shows a typical example: the Spanish urban system, with smaller towns and cities functioning interdependently with successively larger ones, the whole system dominated by one or two metropolitan areas whose linkages are national in scope.

Urban systems also exhibit clear *functional* differences within such hierarchies, yet another reflection of the interdependence of places. The geographical division of labor resulting from such processes of economic development (Chapter 7) means that many medium- and larger size cities perform quite specialized economic functions and so acquire quite distinctive characters. Thus, there are steel towns (for example, Pittsburgh, Pennsylvania; Sheffield, England), textile towns (for example, Lowell, Massachusetts; Manchester, England), and automanufacturing towns (for example, Detroit; Turin, Italy; Toyota City, Japan). Some towns and cities, of course, do evolve as general-purpose urban centers, providing an evenly balanced range of functions for their own particular sphere of influence. **Figure 10.14** shows the urban system in the United States, where the top tier of cities consists of centers of global importance (including Chicago, New York, and Los Angeles) that provide high-order functions to an international marketplace. The second tier consists of general-purpose cities with diverse functions but only regional importance (including Atlanta, Miami, and Boston), and the third and fourth tiers consist of more specialized centers of subregional and local importance.

City-Size Distributions, Primacy, and Centrality

The functional interdependency between places within urban systems tends to result in a distinctive relationship between the population size of cities and their rank within the overall hierarchy. This relationship is known as the **rank-size rule,** which describes a certain statistical regularity in the city-size distributions of countries and regions. The relationship is such that the *n*th largest city in a country or region is $1/n$ the size of the largest city in that country or region. Thus, if the largest city in a particular system has a population of one million, the fifth-largest city should have a population one-fifth as big (that is, 200,000); the hundredth-ranked city should have a population one-hundredth as big (that is, 10,000), and so on. Plotting this relationship on a graph with a logarithmic scale for population sizes would produce a perfectly straight line. The actual rank-size relationship for the U.S. urban system has always come close to this (**Figure 10.15**). Over time the slope has moved to the right on the graph, reflecting the growth of towns and cities at every level in the urban hierarchy.

In some urban systems, the top of the rank-size distribution is distorted as a result of the disproportionate size of the largest (and sometimes also the second-largest) city. In Argentina, for example, Buenos Aires is more than 10 times the size of Rosario, the second-largest city. (According to the rank-size rule, the largest city should be just twice the size of the second-largest city.) In the United Kingdom, London is more than nine times the size of Birmingham, the second-largest city. In France, Paris is more than eight times the size of Marseilles, France's second-largest city. In Brazil, both Rio de Janeiro and São Paulo are five times the size of Belo Horizonte, the third-largest city. Geographers call this condition **primacy,** occurring when the population of the largest city in an urban

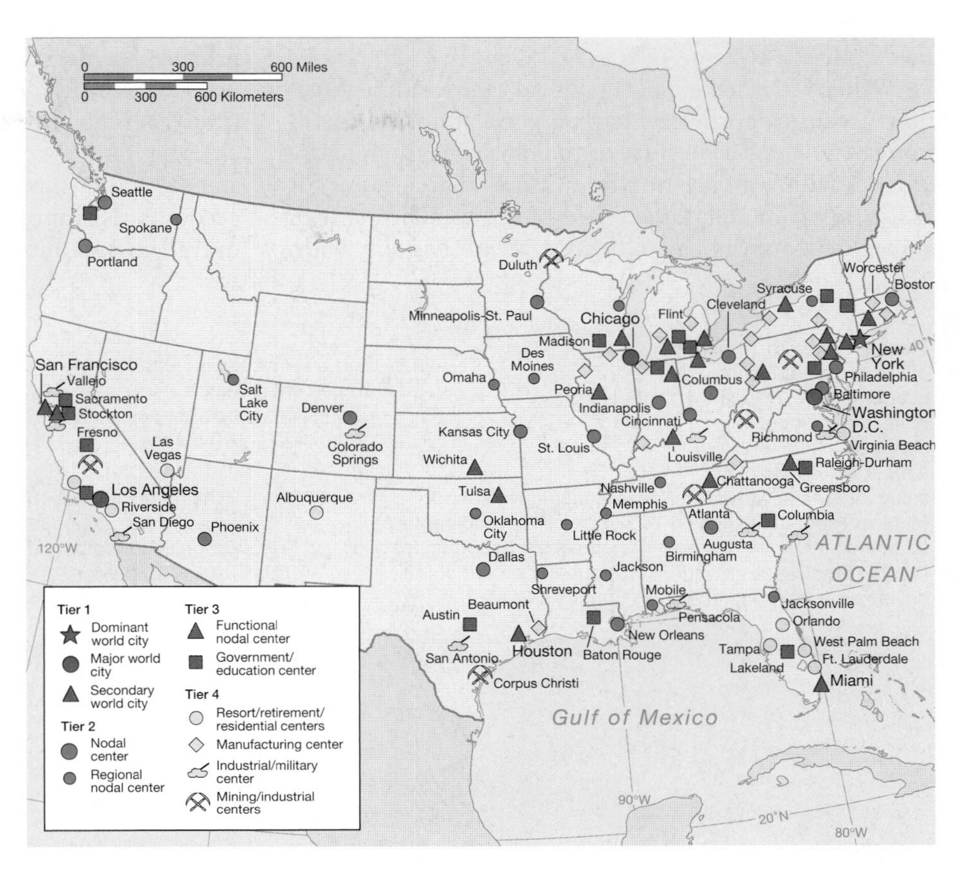

Figure 10.14 Functional specialization within an urban system Different cities tend to specialize in particular kinds of economic activities, and some provide a much greater range of functions than others. This map shows a functional classification of the U.S. urban system. The top tier of the system consists of "world cities" that provide high-order functions to a global marketplace. Among these, New York stands alone as the dominant metropolis. The next tier consists of cities with diverse functions but only regional importance (for example, Atlanta, Minneapolis). The third tier consists of more specialized centers of business, government, and producer services (for example Austin, Texas; Albany, New York; and Hartford, Connecticut). The fourth tier consists of still more specialized cities of various kinds: manufacturing centers (for example, Buffalo, New York; Chattanooga, Tennessee), mining/industrial centers (for example, Charleston, West Virginia; Duluth, Minnesota), industrial/military centers (e.g., Newport News, Virginia; San Diego, California), and resort/retirement centers (for example, Las Vegas, Nevada; Orlando, Florida). (*Source:* Reprinted with permission of Prentice Hall from P. L. Knox, *Urbanization,* © 1994, p. 64.)

system is disproportionately large in relation to the second- and third-largest cities in that system. Cities like London and Buenos Aires are termed *primate* cities.

Thus, primacy is not simply a matter of size. Some of the largest metropolitan areas in the world—Karachi, New York, and Mumbai (Bombay), for example—are not primate. Further, primacy is a condition that is found in both the core and the periphery of the world system. This suggests that primacy is a result of the roles played by particular cities within their own national urban systems. A relationship does exist to the world economy, however. Primacy in peripheral countries is usually a consequence of primate cities' early roles as gateway cities. In core countries it is usually a consequence of primate cities' roles

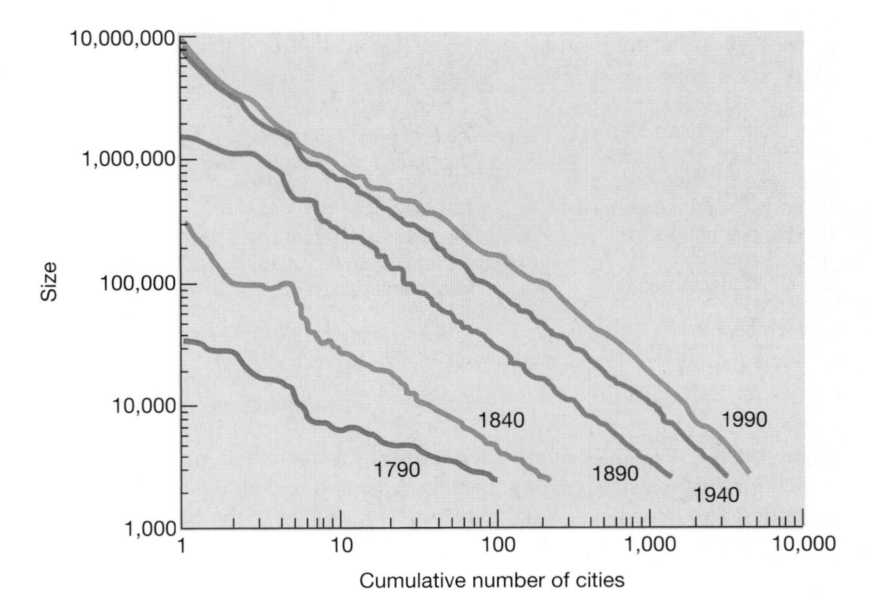

Figure 10.15 The rank-size distribution of cities in the U.S. urban system, 1790–1990 This graph shows that the U.S. urban system has conformed fairly consistently to the rank-size rule. As urbanization brought increased populations to cities at every level in the urban hierarchy, the rank-size graph has moved to the right. Meanwhile, the growth of some cities (for example, San Diego) has sent them from the lower end of the hierarchy to the very top; while other cities (for example, Savannah, Georgia) have declined, at least in relative terms, so that they have fallen down the hierarchy. (*Source:* Reprinted with permission of Prentice Hall from P. L. Knox, *Urbanization,* © 1994, p. 32.)

as imperial capitals and centers of administration, politics, and trade for a much wider urban system than their own domestic system.

When cities' economic, political, and cultural functions are disproportionate to their population, the condition is known as **centrality,** or the functional dominance of cities within an urban system. Cities that account for a disproportionately high share of economic, political, and cultural activity have a high degree of centrality within their urban system. Very often primate cities exhibit this characteristic, but cities do not necessarily have to be primate in order to be functionally dominant within their urban system. **Figure 10.16** shows some examples of centrality, revealing the overwhelming dominance of some cities within the world-system periphery. Bangkok, for instance, with around 12 percent of the Thai population, accounts for approximately 38 percent of the country's overall gross domestic product (GDP), over 85 percent of

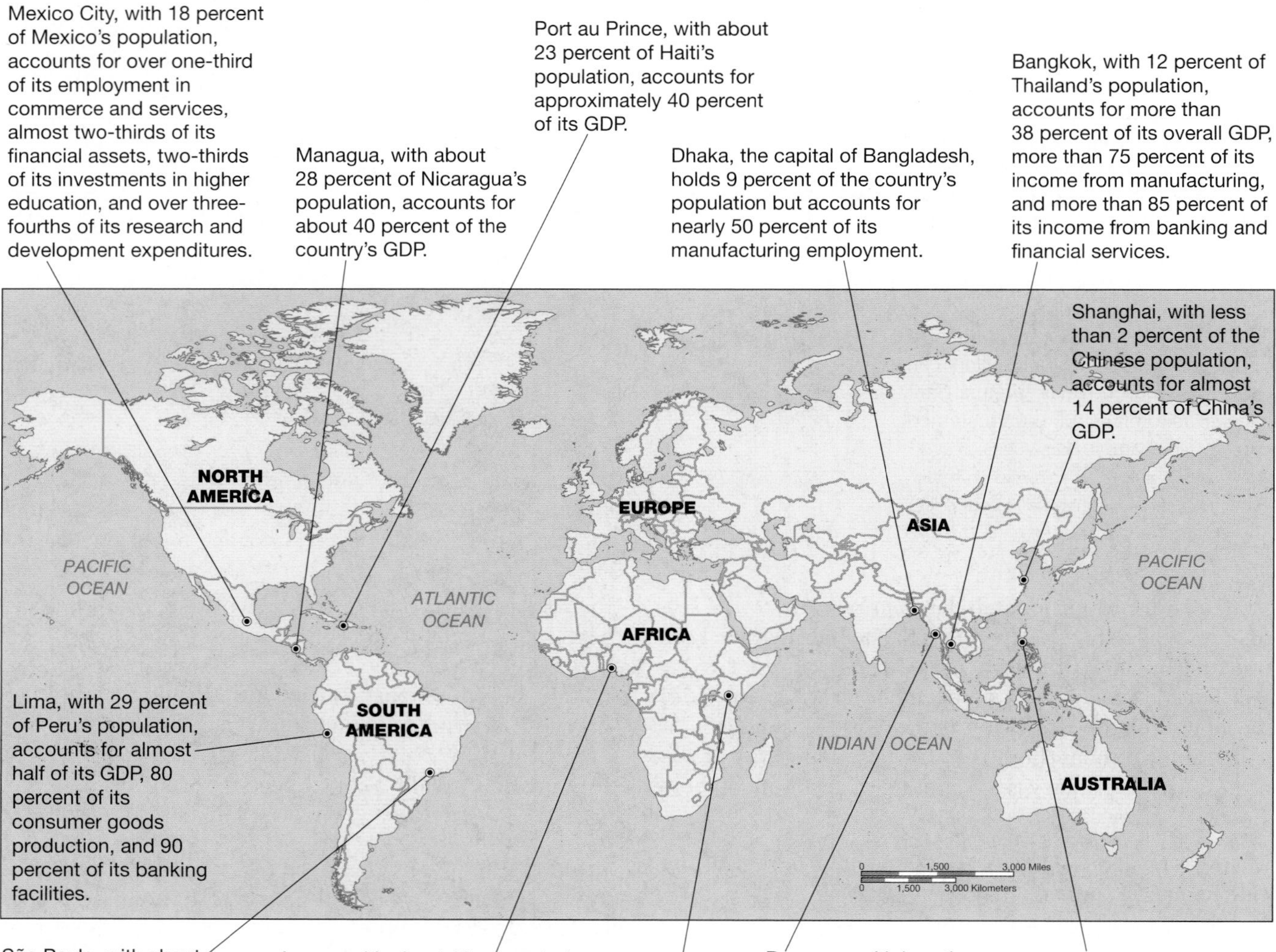

Figure 10.16 Examples of urban centrality The economic, political, and cultural importance of some cities is disproportionate to their population size, making them more central to their economies. This is a reflection of core-periphery differentials within countries, and often becomes a political issue because of the economic disparities. The centrality of these cities also leads to localized problems of congestion, land price inflation, and pollution.

the country's GDP in banking, insurance, and real estate, and 75 percent of its manufacturing.

World Cities and the Global Urban System

As explained in Chapter 7, ever since the evolution of a world-system in the sixteenth century, certain cities known as **world cities** (sometimes referred to as *global cities*) have played key roles in organizing space beyond their own national boundaries. In the first stages of world-system growth, these key roles involved the organization of trade and the execution of colonial, imperial, and geopolitical strategies. The world cities of the seventeenth century were London, Amsterdam, Antwerp, Genoa, Lisbon, and Venice. In the eighteenth century, Paris, Rome, and Vienna also became world cities, while Antwerp and Genoa became less influential. In the nineteenth century, Berlin, Chicago, Manchester, New York, and St. Petersburg became world cities, while Venice became less influential.

Today the globalization of the economy has resulted in the creation of a global urban system in which the key roles of world cities are concerned less with the deployment of imperial power and the orchestration of trade and more with transnational corporate organization, international banking and finance, supranational government, and the work of international agencies. World cities have become the control centers for the flows of information, cultural products, and finance that collectively sustain the economic and cultural globalization of the world.

World cities also provide an interface between the global and the local. They contain the economic, cultural, and institutional apparatus that channels national and provincial resources into the global economy, and that transmits the impulses of globalization back to national and provincial centers. As such, world cities possess several functional characteristics:

- They are the sites of most of the leading global markets for commodities, commodity futures, investment capital, foreign exchange, equities, and bonds.
- They are the sites of clusters of specialized, advanced business services, especially those that are international in scope and that are attached to finance, accounting, advertising, property development, and law.
- They are the sites of concentrations of corporate headquarters—not just of transnational corporations but also of major national firms and large foreign firms.
- They are the sites of concentrations of national and international headquarters of trade and professional associations.
- They are the sites of most of the leading nongovernmental organizations (NGOs) and intergovernmental organizations (IGOs) that are international in scope (for example, the World Health Organization; United Nations Educational, Scientific, and Cultural Organization (UNESCO); the International Labor Organization, and the International Federation of Agricultural Producers).
- They are the sites of the most powerful and internationally influential media organizations (including newspapers, magazines, book publishing, and satellite television); news and information services (including news wires and online information services); and culture industries (including art and design, fashion, film, and television).
- Because of their importance and visibility, world cities are also the setting for many terrorist acts—see Box 10.2: "Urban Terrorism."

A great deal of synergy exists among the various functional dimensions of world cities. A city like New York, for example, attracts transnational corporations because it is a center of culture and communications. It attracts specialized business services because it is a center of corporate headquarters and of global markets, and so on. These interdependencies represent a special case of the geographical *agglomeration effects* that we discussed in Chapter 7. Agglomeration is the clustering of functionally related activities. In the case of New York City, corporate headquarters and specialized legal, financial, and business services cluster together because of the mutual cost savings and advantages of being close to one another. At the same time, different world cities fulfill different roles within the world-system, making for different emphases and combinations (that is, differences in the nature of their world-city functions) as well as for differences in the absolute and relative localization of particular world-city functions (that is, differences in their degree of importance as world cities). For example, Brussels is relatively unimportant as a corporate headquarters location but qualifies as a world city because it is the administrative center of the European Union and has attracted a large number of nongovernmental organizations and advanced business services that are transnational in scope. To take another example, Milan is relatively dependent in terms of corporate control and advanced business services but has global status in terms of cultural influence (especially fashion and design) and is an important regional financial center.

Today the global urban system is dominated by London and New York, whose influence is "panregional"—that is, their world-city functions extend across all three main circuits of the global economy—the Americas, Europe/Africa/Middle East, and Asia/Oceania (**Figure 10.17**). A second tier of the global urban system consists of world cities with influence over large regions within the world-system. These are Miami (the regional center for Latin America), Hong Kong (the regional center for Northeast Asia), and Singapore (the regional center for Southeast

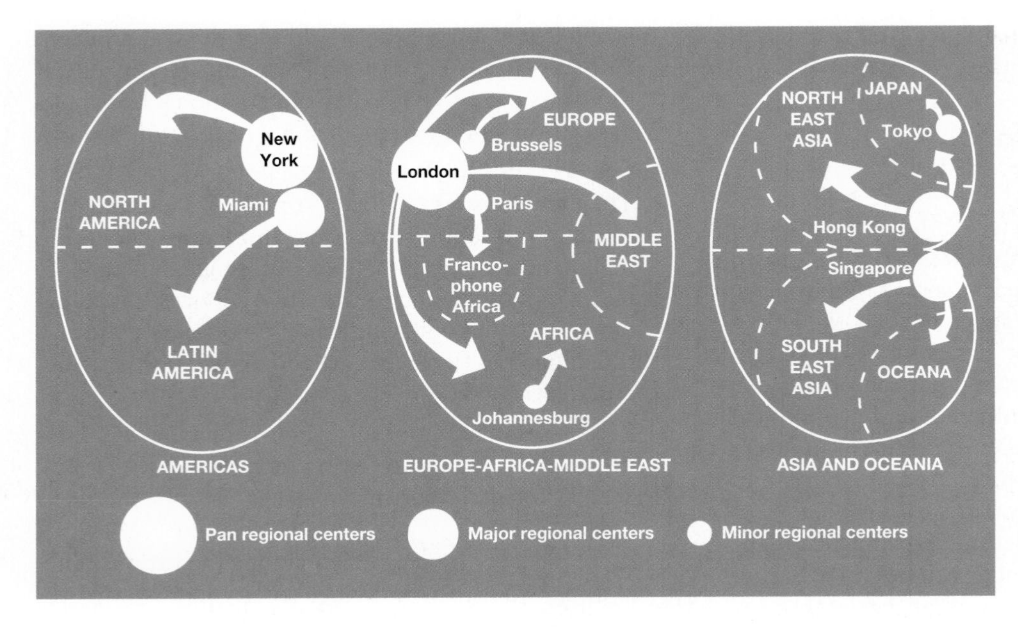

Figure 10.17 World cities in the global urban system This map shows the sphere of influence of world cities, based on an analysis of the regional headquarter functions of the world's largest advanced business services firms (accountancy, advertising, banking and finance, and law). (After P. J. Taylor, "World cities and territorial states under conditions of contemporary globalization," *Political Geography, 19,* 2000, pp. 5–32.)

Asia). A third tier consists of important international cities with more limited or more specialized international functions: Brussels, Paris, Johannesburg, and Tokyo.

World Urbanization Today

It is difficult to say just how urbanized the world has become. In many parts of the world, urban growth is taking place at such a pace and under such chaotic conditions that experts can provide only informed estimates. The most comprehensive source of statistics is the United Nations, whose data suggest that almost half of the world's population is now urban. These data incorporate the very different definitions of urban used by different countries. Some countries (Australia and Canada, for example) count any settlement of 1,000 people or more as urban, while others (including Italy and Jordan) use 10,000 as the minimum for an urban settlement, and Japan uses 50,000 as the cutoff. This, by the way, tells us something about the nature of urbanization itself: It is a relative phenomenon. In countries like Peru, where the population is thin and scattered, a settlement of 2,000 represents a significant center. In countries like Japan, however, with greater numbers, higher densities, and a tradition of centralized agricultural settlement, a much larger concentration of people is required to count as "urban."

Taking the definitions used in individual countries, almost one-half of the world's population is now urbanized. As **Table 10.1** shows, North America is the most urbanized continent in the world, with 80 percent of its population living in urban areas. In contrast, Africa and Asia are less than 40 percent urban.

To put these figures in perspective, only 29.7 percent of the world's population was urbanized in 1950, using the same definitions of urban settlements. In that year there were only 83 metropolitan areas of a million or more, and only eight of 5 million or more existed; in 2000 there were approximately 372 metropolitan areas of a million or more people, and 45 with over 5 million. Looking ahead, population projections for 2015 suggest that more than 53 percent of the world's population will be living in urban areas, and there will be around 541 cities

TABLE 10.1 Urbanization by Major World Regions, 2005

	Percent of Total Population in Urban Areas	Percent of Urban Population in Cities of Less Than 500,000	Percent of Urban Population in Cities of 5 Million or More
Africa	39.7	58.9	7.9
Asia	39.9	49.7	17.2
Latin America	77.6	48.0	20.6
North America	80.8	37.5	20.5
Europe	73.3	63.3	8.5
Oceania	73.3	41.7	0.0
World	**49.2**	**51.7**	**15.3**

Source: Data, United Nations, *World Urbanization Prospects: The 2003 Revision.* New York: U.N. Department of Economic and Social Affairs, 2004.

10.2

WINDOW ON THE WORLD

Urban Terrorism

Not only has terrorism taken on a global cast, with international linkages, it is also apparent that cities have become the stage on which this tragic drama is played. Between 1993 and 2000 there were more than 500 terrorist incidents in cities around the world. Even before the dramatic terrorist attacks on the World Trade Center and the Pentagon in 2001, on commuter trains in Madrid in 2004, and on London buses and underground trains in 2005 (**Figure 10.A**), cities had become the central venues of terror.

There are several reasons for this. First, cities—especially world cities—have considerable symbolic value. They are not only dense agglomerations of people and buildings but symbols of national prestige and military, political, and financial power. A blast in a mountain town or in the countryside may arouse local concern, but it is generally of little or no consequence for the rest of the world. But an attack on Wall Street (New York), a massacre in Piccadilly Circus (London), the bombing of the Eiffel Tower (Paris), or poison gas in a Tokyo metro arouses international alarm. Any such event will be instantly telegraphed to a larger world and alarm a much larger audience. Second, the assets of cities—densely packed and with a great mix of industrial and commercial infrastructure—make them rich targets. Third, cities have become nodes for a vast international network of communications. This is a reflection of their power, but it is also a vulnerability. A well-placed explosion can produce enormous reverberations, paralyze a city, and spread fear and economic dislocation. Finally, word gets around more quickly and socialization proceeds more rapidly in densely packed environments. This kind of environment provides an abundant source of recruitment for potential terrorists.

Figure 10.A Terrorist attack Railway workers remove debris from the wreckage of a public train near to Atocha train station in Madrid, Spain, March 11, 2004. Thirteen bombs on four packed commuter trains killed 191 people and wounded more than 1,500. The attack was attributed to the Islamic militant group al-Qaeda.

Heterogeneous urban settings, while providing rich synergies, can be a nesting ground for terrorist organizations under certain conditions. A sense of relative deprivation often sharpens as those struggling to get by come into closer proximity with other struggling groups and with the more affluent. Beirut provides a ready example of how different groups living under conditions of hopelessness and in proximity to one another turn on one another. Similar ecologies of terror pervade Belfast (Northern Ireland), Sarajevo (Bosnia-Herzegovina), Hyderabad (India), Karachi (Pakistan), and Baghdad (Iraq). Rather than directed from lower classes upward toward elites, conflict occurs between groups operating at the same level—Hindus fighting Muslims in Mumbai (India) or rival narco-gangs in Bogotá (Colombia).

Statistically, the cities with the greatest incidence of terrorist incidents in the eight-year period between 1993 and 2000 were Srinagar (India), Athens (Greece), Sanaa (Yemen), Paris (France), Istanbul (Turkey) and Lima (Peru), followed closely by Jerusalem (Israel), Algiers (Algeria) and Dushanbe (Tajikistan). The largest number of fatalities occurred in Nairobi, Kenya (291 killed in 1 incident); Colombo, Sri Lanka (108 killed in 3 incidents); and Jerusalem (77 killed in 9 incidents). Altogether, more than 250 cities around the world experienced one or more terrorist acts in that period.

Based on extracts from H. V. Savitch and Grigoriy Ardashev, "Does Terror Have an Urban Future?" *Urban Studies*, 2525–2533, 2001.

with a population of a million or more, including about 61 cities of 5 million or more.

Regional Trends and Projections

The single most important aspect of world urbanization, from a geographical perspective, is the striking difference in trends and projections between the core regions and the semiperipheral and peripheral regions. In 1950 two-thirds of the world's urban population was concentrated in the more developed countries of the core economies. Since then, the world's urban population has increased threefold, the bulk of the growth having taken place in the less developed countries of the periphery (**Figure 10.18**). In 1950, 21 of the world's largest 30 metropolitan areas were located in core countries—11 of them in Europe and 6 in North America. By 1980 the situation was completely reversed, with 19 of the largest 30 located in peripheral and semiperipheral regions. By 2010, all but 7 of the 30 largest metropolitan areas are expected to be located in peripheral and semiperipheral regions (**Table 10.2**).

Asia provides some of the most dramatic examples of this trend. From a region of villages, Asia is fast becoming a region of cities and towns. Between 1950 and 2003, for example, its urban population rose more than tenfold, to nearly 1.5 billion people. By 2020 about two-thirds of Asia's population will be living in urban areas. Nowhere is the trend toward rapid urbanization more pronounced than in China, where for decades the communist government imposed strict controls on where people were allowed to live, fearing the transformative and liberating effects of cities. By tying people's jobs, school admission, and even the right to buy food to the places where people were registered to live, the government made it almost impossible for rural residents to migrate to towns or cities. As a result, more than 70 percent of China's 1 billion people still lived in the countryside in 1985. Now, however, China is rapidly making up for lost time (see Box 10.3: "The Pearl River Delta: An Extended Metropolis"). The Chinese government, having decided that towns and cities can be engines of economic growth within a communist system, has not only relaxed residency laws but also drawn up plans to establish over 430 new cities. Between 1981 and 2003 the number of people living in cities in China more than tripled, from 162 million to 504 million, and the number of cities with a population of half a million or more increased from 16 to 97.

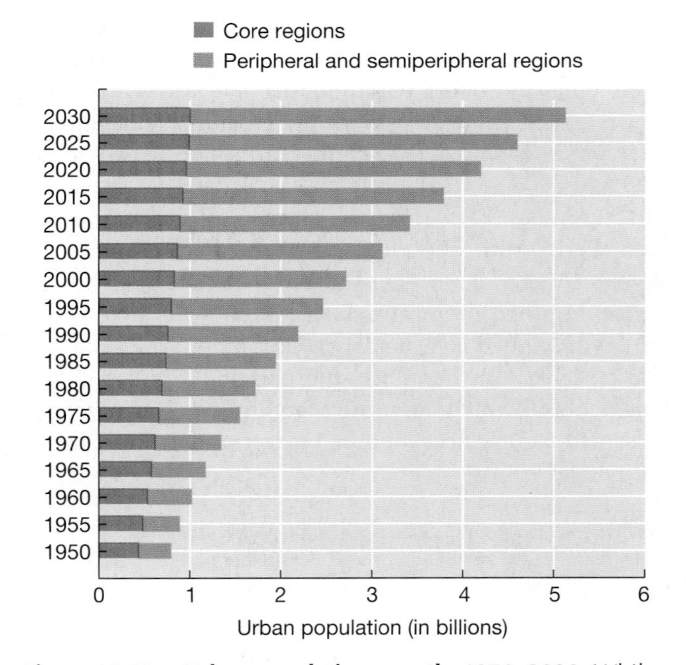

Figure 10.18 Urban population growth, 1950–2030 While the metropolitan areas of the world's core countries have continued to grow, most of them have been overtaken by the startling growth of the "unintended" metropolises of peripheral and semiperipheral countries. (*Source:* Data from United Nations, *World Urbanization Prospects.* New York: U.N. Department of Economic and Social Affairs, 1998.)

High levels of urbanization have existed in the world's core countries for some time. According to their own national definitions, the populations of Australia, Belgium, and Canada are more than 90 percent urbanized, while those of France, Germany, Norway, Spain, Sweden, and the United Kingdom are all more than 75 percent urbanized. In these core countries, however, rates of urbanization are relatively low, just as the overall rates of population growth are slow (see Chapter 3).

Levels of urbanization are also very high in many of the world's semiperipheral countries. Brazil, Hong Kong, Mexico, Taiwan, Singapore, and South Korea, for example, are all at least 75 percent urbanized. Unlike the core countries, however, rates of growth in semiperipheral countries are high. In peripheral countries, rates are even higher.

Whatever the current level of urbanization in peripheral countries, almost all are experiencing high rates of urbanization, with growth forecasts of unprecedented speed and unmatched size. Karachi, Pakistan, a metropolis of 1.03 million in 1950, had reached 8.5 million in 1995 and is expected to reach 16.2 million by 2015. Likewise, Cairo, Egypt, grew from 2.44 million to 9.7 million between 1950 and 1995 and is expected to reach 13.1 million by 2015. Mumbai (India; formerly Bombay), Delhi (India), Mexico City (Mexico), Dhaka (Bangladesh), Jakarta (Indonesia), Lagos (Nigeria), São Paulo (Brazil), and Shanghai (China) are all projected to have populations in excess of 17 million by 2015. The reasons for this urban growth vary. Wars in Liberia and Sierra Leone have pushed hundreds of thousands of people into their capitals, Monrovia and Freetown. In Mauritania, Niger, and other countries bordering the Sahara, deforestation and overgrazing have allowed the desert to expand and swallow up villages, forcing people toward cities. For the most part, though, urban growth in peripheral countries is a consequence of the onset of the demographic transition (see Chapter 3), which has produced fast-growing rural populations in regions that face increasing problems with

TABLE 10.2 The World's 30 Largest Metropolitan Areas, Ranked by Population Size, 1950, 1980, and 2010 (in millions)

1950	Population	1980	Population	2010	Population
New York	12.3	Tokyo	28.5	Tokyo	35.8
London	8.7	New York	15.6	Mumbai	20.4
Tokyo	6.9	Mexico City	13.0	Mexico City	19.8
Paris	5.4	São Paulo	12.1	São Paulo	19.3
Moscow	5.4	Shanghai	11.7	New York	19.1
Shanghai	5.3	Osaka	10.0	Delhi	18.2
Essen	5.3	Buenos Aires	9.9	Jakarta	15.5
Buenos Aires	5.0	Los Angeles	9.5	Kolkata	15.5
Chicago	4.9	Kolkata	9.0	Dhaka	15.2
Kolkata	4.4	Beijing	9.0	Lagos	14.0
Osaka	4.1	Paris	8.8	Buenos Aires	14.0
Los Angeles	4.0	Mumbai	8.7	Karachi	13.8
Beijing	3.9	Rio de Janeiro	8.6	Los Angeles	12.5
Milan	3.6	Seoul	8.3	Shanghai	12.4
Berlin	3.3	Moscow	8.1	Cairo	12.0
Mexico City	3.1	London	7.7	Rio de Janeiro	12.0
Philadelphia	2.9	Cairo	7.3	Manila	11.6
St. Petersburg	2.9	Tianjin	7.3	Osaka-Kobe	11.3
Mumbai	2.9	Chicago	7.2	Moscow	10.9
Rio de Janeiro	2.9	Rhein-Ruhr	6.3	Beijing	10.8
Detroit	2.8	Jakarta	6.0	Istanbul	10.6
Naples	2.8	Metro Manila	6.0	Paris	10.0
Manchester	2.5	Delhi	5.6	Tianjin	9.5
São Paulo	2.4	Milan	5.3	Seoul	9.4
Cairo	2.4	Tehran	5.0	Chicago	9.1
Tianjin	2.4	Karachi	5.0	Lima	8.8
Birmingham	2.3	Bangkok	4.7	Bogotá	8.3
Frankfurt	2.3	St. Petersburg	4.6	Tehran	7.8
Boston	2.2	Hong Kong	4.6	London	7.6
Hamburg	2.2	Philadelphia	4.5	Hong Kong	7.5

Source: Data, United Nations, *World Urbanization Prospects*. New York: U.N. Department of Economic and Social Affairs, 2003, pp. 120–123.

agricultural development (see Chapter 8). As a response, many people in these regions migrate to urban areas seeking a better life.

Many of the largest cities in the periphery are growing at annual rates of between 4 and 7 percent; at the higher rate their populations will double in 10 years, while at the lower rate they will double in 17 years. The *doubling time* of a city's population is the time needed for it to double in size, at current growth rates. To put the situation in numerical terms, metropolitan areas like Mexico City and São Paulo are adding half a million persons to their population each year: nearly 10,000 every week, even after making up for deaths and out-migrants. It took London 190 years to grow from half a million to 10 million. It took New York 140 years. By contrast, Mexico City, São Paulo, Buenos Aires, Kolkata (Calcutta), Rio de Janeiro, Seoul, and Mumbai (Bombay) all took less than 75 years to grow from half a million to 10 million inhabitants.

The Periphery and Semiperiphery: Overurbanization and Megacities

These statistics reflect the fact that urban growth processes in the world's peripheral regions differ widely from those in core regions. In contrast to the world's core regions, where urbanization has largely resulted from economic growth, the urbanization of peripheral regions has been a consequence of demographic growth that preceded economic development. Although the demographic transition is a fairly recent phenomenon in the peripheral regions of the world (see Chapter 3), it has generated large

The Pearl River Delta: An Extended Metropolis

The Pearl River Delta (**Figure 10.B**) is one of the fastest growing urban regions in the world. Anchored by the major metropolitan centers of Guangzhou, Hong Kong, Macao, Shenzhen, and Zhuhai, the Pearl River Delta is an extended metropolitan region of nearly 50 million people. It is one of three extended metropolitan regions—Beijing-Tianjin and Shanghai are the others—that have been fostered by the Chinese government as engines of capitalist growth since liberal economic reforms were introduced in the late 1970s.

Hong Kong (**Figure 10.C**), a British colony until 1997, is a metropolis of 6.8 million with a thriving industrial and commercial base that is recognized as a capitalist economic dynamo by the Chinese government, which has created a Special Administrative District for the metropolis. As a result, Hong Kong's citizens have retained their British-based legal system and its guaranteed rights of property ownership and democracy. Hong Kong is the world's largest container port, the third largest center for foreign-exchange trade, the seventh largest stock market, and the tenth largest trading economy.

Hong Kong's success encouraged the Chinese government to establish two of its first Special Economic Zones (SEZs) in nearby Shenzhen and Zhuhai. Designed to attract foreign capital, technology, and management practices, these SEZs were established as export-processing zones that offered cheap labor and land, along with tax breaks, to transnational corporations. Investors from Hong Kong and Taiwan responded quickly and enthusiastically. By 1993 more than 15,000 manufacturers from Hong Kong alone had set up businesses in Guangdong Province, and a similar number had established subcontracting relationships, contracting out assembly-line work to Chinese companies in the Pearl River Delta. Meanwhile, the entire delta region was subsequently designated an Open Economic Region, where local governments, individual enterprises and farm households could enjoy a high degree of autonomy in economic decision-making.

The relaxation of state control over the regional economy allowed the region's dense and growing rural population to migrate to urban areas in search of assembly-line jobs, or to stay in rural areas and diversify agricultural production from paddy-rice cultivation to more profitable activities such as market farming activities, livestock husbandry, and fishery. Economic freedom also facilitated rural industrialization—mostly low-tech, small-scale, labor-intensive, and widely scattered across the countryside—though the triangular area between Guangzhou, Hong Kong, and Macao has quickly emerged as an especially important zone because of its relatively cheap land and labor and because of significant levels of investment by regional and local governments in the transport and communications infrastructure. The result is a distinctive "ex-

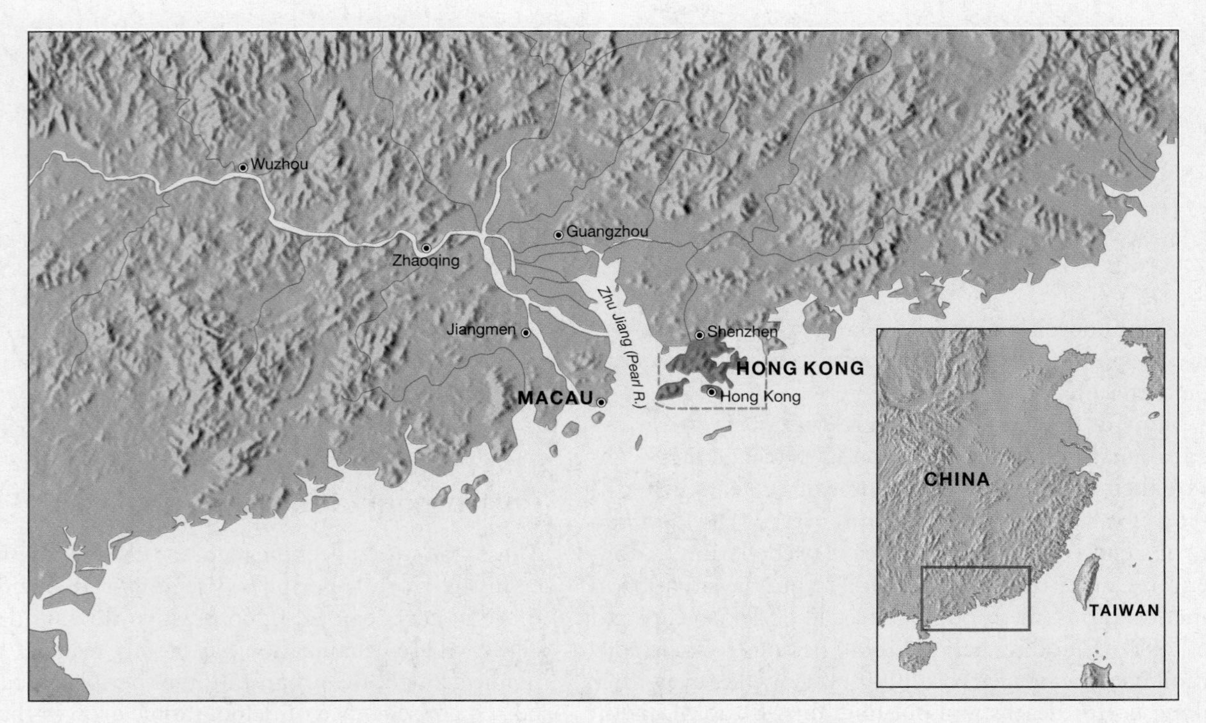

Figure 10.B Pearl River Delta One of the fastest-growing regions of the world, the Pearl River Delta is an extended metropolitan region of more than 50 million people.

Figure 10.C City of Hong Kong Although most of its manufacturing has been transferred to neighboring Guangdong Province, where wages are much lower, thousands of companies are located in Hong Kong simply for the purpose of doing business with China. As a result, Hong Kong remains a major world city: a major financial hub, a thriving commercial sector, with a population of 6.1 million.

tended metropolis" in which numerous small towns play an increasingly important role in fostering the process of urbanization, with an intense mixture of agricultural and nonagricultural activities and an intimate interaction between urban and rural areas.

The metropolitan cores of the region, aiming to increase their competitiveness and prominence in the globalizing world economy, have meanwhile invested heavily in infrastructure improvements. The Guangzhou municipal government, for example, invested more than 60 billion yuan ($7.23 billion) between 1998 and 2002 in infrastructure construction—including a metro system and an elevated railway network to link the city's new international airport, railway stations, and port. Throughout the region, enormous investments have been made in showpiece infrastructure projects geared to the needs of local and international capital. These include major airports, high-speed tolled highways, satellite ground stations, port installations, metro and light rail networks, and new water-management systems. These, in turn, have attracted business and technology parks, financial centers, and resort complexes in a loose-knit sprawl of urban development.

Today the Pearl River Delta provides a thriving export-processing platform that has driven double-digit annual economic growth for much of the past two decades. Guangzhou is a megacity with a population of 10.5 million in 2005 (**Figure 10.D**). Shenzhen has grown from a population of just 19,000 in 1975 to 4.6 million in 2005, with an additional 2 million in the surrounding municipalities. The southern border of the Shenzhen Special Economic Zone adjoins Hong Kong, but the northern border is walled off from the rest of China by an electrified fence to prevent smuggling and to keep back the mass of people trying to migrate illegally into Shenzhen and Hong Kong.

Figure 10.D Guangzhou, China

increases in population well in advance of any significant levels of industrialization or rural economic development.

The result, for the mainly rural populations of peripheral countries, has been more and more of worse and worse. Problems with agricultural development (see Chapter 8) have meant that fast-growing rural populations face an apparently hopeless future of drudgery and poverty. Emigration has provided one potential safety valve, but as the frontiers of the world-system closed, the more affluent core countries have put up barriers to immigration. The only option for the growing numbers of impoverished rural residents was—and still is—to move to the larger towns and cities, where at least there is the hope of employment and the prospect of access to schools, health clinics, piped water, and the kinds of public facilities and services that are often unavailable in rural regions. Cities also have the lure of modernization and the appeal of consumer goods—attractions that are now directly beamed into rural areas through satellite TV. Overall, the metropolises of the periphery have absorbed four out of five of the 1.2 billion city dwellers added to the world's population since 1970.

When natural disasters, environmental degradation, or civil war impacts highly populated rural regions in the periphery, rates of rural-urban migration increase dramatically:

> For several years now, relentless cycles of drought and flooding have wreaked havoc on the tiny country of Malawi, in the heart of southern Africa. In 2002 and 2003, torrential rains caused massive mudslides, washing away bridges and homes and devastating harvests of maize, the main food staple. Unable to eke a living from the ravaged countryside, rural residents have flocked in droves to the country's bourgeoning cities—giving Malawi the dubious distinction of being the world's fastest-urbanizing nation today.[3]

Rural migrants have poured into cities out of desperation and hope rather than being drawn by jobs and opportunities. Because these migration streams have been composed disproportionately of teenagers and young adults, an important additional component of urban growth has followed—exceptionally high rates of natural population increase. In most peripheral countries the rate of natural increase of the population in cities exceeds that of net in-migration. On average, about 60 percent of urban population growth in peripheral countries is attributable to natural increase.

The consequence of all this urban population growth has been described as **overurbanization,** which occurs when cities grow more rapidly than the jobs and housing they can sustain. In such circumstances, urban growth produces instant slums—shacks set on unpaved streets, often with open sewers and no basic utilities. The shacks are constructed out of any material that comes to hand, such as planks, cardboard, tarpaper, thatch, mud, and corrugated iron. Such is the pressure of in-migration that many of these instant slums are squatter settlements, built illegally by families who are desperate for shelter. **Squatter settlements** are residential developments on land that is neither owned nor rented by its occupants. Squatter settlements are often, but not always, slums. In Chile squatter settlements are called *callampas,* meaning "mushroom cities"; in Turkey they are called *gecekondu,* meaning that they were built after dusk and before dawn. In India they

[3]*State of the World 2005*. New York: W. W. Norton & Company, 2005, p. 29.

Figure 10.19 Slum housing in peripheral cities Throughout much of the world, the scale and speed of urbanization, combined with the scarcity of formal employment, have resulted in very high proportions of slum housing, much of it erected by squatters. (a) In Nairobi, Kenya, 40 percent of the city's population live in unauthorized settlements. The largest—the Mathere Valley squatter area—grew from 4,000 inhabitants in 1964 to 90,000 in 1979 and over 300,000 in 2004. (b) In Manila, less than 15 percent of the population can afford to buy or rent a legal house or apartment on the open market. Squatter settlements vary in size from the very large Tondo Foreshore area, with 30,000 families, to tiny groups of squatter homes on small sites.

Figure 10.20 Bangkok Bangkok's population of around 6 million is almost 50 times greater than that of the next largest city in Thailand, Chiang Mai. Many of Bangkok's population have moved to the city from the countryside, and most of the population are ethnic Thai, with immigrants being mainly Chinese or Indian.

are called *bustees;* in Tunisia, *gourbevilles;* in Brazil, *favelas;* and in Argentina, simply *villas miserias*. They typically account for well over one-third and sometimes up to three-quarters of the population of major cities (**Figure 10.19**). The world's worst city for housing, according to United Nations statistics, is Addis Ababa, the capital of Ethiopia—where, in 2000, over 80 percent of the population was homeless or living in unfit accommodations.

Megacities are very large cities characterized by both primacy and a high degree of centrality within their national economy. Their most important common denominator is their sheer size—most of them number 10 million or more in population. This, together with their functional centrality, means that in many ways they have more in common with one another than with the smaller metropolitan areas and cities within their own countries. Examples of such megacities include Bangkok (**Figure 10.20**), Beijing, Cairo (**Figure 10.21**), Kolkata (Calcutta), Dhaka, Jakarta, Lagos, Manila, Mexico City, New Delhi, São Paulo (**Figure 10.22**), Shanghai, and Teheran. Each has more inhabitants than 100 of the member countries of the United Nations. While most of them do not function as world cities, they do provide important intermediate roles between the upper tiers of the system of world cities and the provincial towns and villages of large regions of the world. They not only link local and provincial economies with the global economy but also provide a point of contact between the traditional and the modern, and between formal and informal economic sectors. The **informal sector** of an economy involves a wide variety of economic activities whose common feature is that they take place beyond official record and are not subject to formalized systems of regulation or remuneration. As we shall see in the next chapter, the slums and squatter settlements in

Figure 10.21 Cairo Cairo is Africa's most populous city, with more than 15 million people in the greater metropolitan area.

Figure 10.22 São Paulo About 20 million people live in the Greater São Paulo metropolitan area, with almost 11 million in the city itself.

megacities are often associated with severe problems of social disorganization and environmental degradation. Nevertheless, many neighborhoods are able to develop self-help networks and organizations that form the basis of community amid dauntingly poor and crowded cities.

The Core: Mature Metropolises

The high levels of urbanization and relatively slow rates of urban growth within the world's core regions are reflected in relatively stable urban systems. There is constant change, nevertheless, in patterns and processes of urbanization as the metropolises, cities, and towns adjust to the opportunities of new technologies and new industries, and to the constraints of obsolescent urban infrastructure and land-use conflicts. New rounds of urbanization are initiated in the places most suited to new technologies and new industries, while those least suited are likely to suffer a spiral of deindustrialization and urban decline.

Deindustrialization and Agglomeration Diseconomies

Deindustrialization involves a decline in industrial employment in core regions as firms scale back their activities in response to lower levels of profitability (see Chapter 7). Such adversity has particularly affected cities such as Pittsburgh and Cleveland, Sheffield and Liverpool (United Kingdom), Lille (France), and Liège (Belgium)—places where heavy manufacturing constituted a key economic sector. Cities like these have suffered substantial reductions in employment since the 1970s and 1980s. During that period, better and more flexible transport and communications networks allowed many industries to choose from a broader range of potential locations. In many instances deindustrialization has been intensified by the dampening effects of *agglomeration diseconomies* (Chapter 7) on the growth of larger metropolitan areas. Agglomeration diseconomies, the negative effects of urban size and density, include noise, air pollution, increased crime, commuting costs, the costs of inflated land and housing prices, traffic congestion, and crowded port and railroad facilities. They also include higher taxes levied to rebuild decaying infrastructure and to support services and amenities previously considered unnecessary—traffic police, city planners, and homeless shelters, for example.

The result has been a *decentralization* of jobs and people from larger to smaller cities within the urban systems of core countries, and from metropolitan cores to suburban and ex-urban fringes. In some cases routine production activities relocated to smaller metropolitan areas or to rural areas with low labor costs and more hospitable business climates. In other cases, these activities moved overseas—as part of the new international division of labor (see Chapter 2)—or were eliminated entirely.

Counterurbanization and Reurbanization

The combination of deindustrialization in core manufacturing regions, agglomeration diseconomies in major metropolitan areas, and the improved accessibility of smaller towns and rural areas can give rise to the phenomenon of counterurbanization. **Counterurbanization** occurs when cities experience a net loss of population to smaller towns and rural areas. This process results in the deconcentration of population within an urban system. This is, in fact, what happened in the United States, Britain, Japan, and many other developed countries as early as the 1970s and 1980s. Metropolitan growth slowed dramatically, while the growth rates of small and medium-size towns and of some rural areas increased. Counties that for decades had recorded stable populations grew by 15 or 20 percent. Some of the strongest gains were registered in counties that were within commuting range of metropolitan areas, but some remote counties also registered big population increases.

Counterurbanization was a major reversal of long-standing trends, but it seems to have been a temporary adjustment rather than a permanent change. The globalization of the economy and the growth of postindustrial activities in revamped and expanded metropolitan settings has restored the trend toward the concentration of population within urban systems. Most of the cities that were declining fast in the 1970s and 1980s are now either recovering (New York, London) or bottoming out (Paris, Chicago), while most of those that were growing only slowly (Tokyo, Barcelona) are now expanding more quickly. This trend is one of **reurbanization,** involving the growth of population in metropolitan central cores, following a period of absolute or relative decline in population. In the United States, two principal streams are driving reurbanization. One consists of new migrants, principally from Latin America and Asia, who have moved into the central districts of major metropolitan areas, especially New York and Los Angeles, and some medium-size metropolitan areas of the West, Southwest, and South. A second, very different migration stream to the central districts of metropolitan areas consists of "baby boomers" (see Chapter 3) electing to pursue urban rather than suburban lifestyles. This stream has been most pronounced in the central districts of metropolitan areas with expanding high-tech and defense-oriented economies, such as Boston and Seattle.

GLOBALIZATION AND SPLINTERING URBANISM

The major cities and metropolitan regions of the world are pivotal settings for the processes involved in economic and cultural globalization. As a result, they also reflect these processes in their own patterns of growth and change. Central to this close relationship between urbanization and globalization are networked infrastructures of transportation, information, and communications technologies such as telephone systems, satellite television, computer networks, electronic commerce, and business-to-business Internet services. According to the United Nations Center for Human Settlements (UNCHS),[4] information and communications technologies are intensifying global urbanization in three main ways. First, they allow specialist urban centers, with their high value-added services and manufacturing, to extend their powers, markets, and control to ever more distant regional, national, international, and global spheres of influence. Second, the growing speed, complexity, and riskiness of innovation in a global economy require a concentration of technological infrastructure and an associated, knowledgeable technology-oriented culture in order to sustain competitiveness. Third, demand for information and communications technologies is overwhelmingly driven by the growth of metropolitan markets. World cities, especially, are of disproportionate importance in driving innovation and investment in networked infrastructures of information and communications technologies. This is because of world cities' cultures of modernization, concentrations of capital, relatively high average disposable personal incomes and concentrations of internationally oriented firms and institutions.

In contrast to the infrastructure networks of earlier technology systems that underpinned previous phases of urbanization, these information and communications technologies are not locally owned, operated, and regulated. Rather, they are designed, financed, and operated by transnational corporations to global market standards. Detached from local processes of urban development, these critical networked infrastructures are highly uneven in their impact, selectively serving only certain neighborhoods, certain cities, and certain kinds of metropolitan settings. This important new tendency has been called splintering urbanism by geographers Stephen Graham and Simon Marvin.

Splintering urbanism is characterized by an intense geographical differentiation, with individual cities and parts of cities engaged in different—and rapidly changing—ways in ever broadening and increasingly complex circuits of economic and technological exchange. The uneven evolution of networks of information and communications technologies is forging new landscapes of innovation, economic development, and cultural transformation, while at the same time intensifying social and economic inequalities between the fast world and the slow world. The conclusion of the 2001 UNCHS report *Cities in a Globalizing World* is that traditional patterns of urbanization are rapidly giving way across the globe to a very new dynamic, one that is dominated by "enclaves of superconnected people, firms, and institutions, with their increasingly broadband connections to elsewhere via the Internet, mobile phones and satellite TVs and their easy access to information services, often exist[ing] cheek-by-jowl with much larger numbers of people with at most rudimentary access to modern communications technologies and electronic information."[5]

The chief beneficiaries of splintering urbanism are world cities and major regional metropolitan centers, particularly in core countries where large corporations are

[4]United Nations Center for Human Settlements (HABITAT). *Cities in a Globalizing World: Global Report on Human Settlements, 2001.* London: Earthscan Publications, 2001, p. 6.

[5]Ibid.

able to undertake the necessary capital investments in new networked infrastructures and where national governments and entrepreneurial urban agencies are able to subsidize and facilitate them. We can identify several specific kinds of urban settings that are most directly involved in this dimension of globalization:

- ***Enclaves of Internet and digital multimedia technology development, mostly in core-country world cities.*** Examples include "Multimedia Gulch" in the SOMA (South of Market Street) district of downtown San Francisco, and New York's "Silicon Alley" (just south of Forty-first Street in Manhattan).
- ***Technopoles and clusters of high-tech industrial innovation.*** These have emerged in campus-like suburban and ex-urban settings around core-country world cities (as in London, Paris, and Berlin); in new and renewed industrial regions within core countries (as in southern California, Baden-Württemburg in Germany, and Rhône-Alps in France); and in emerging high-tech production and innovation spaces in semiperipheral countries (as in Bangalore, India, and the Multimedia Super Corridor south of Kuala Lumpur, Malaysia).
- ***Places configured for foreign direct investment in manufacturing, with customized infrastructure, expedited development approval processes, tax concessions, and in some cases exceptions to labor and environmental regulations.*** Such places have emerged in economically depressed regions of core countries (including Northern England and parts of the U.S. Manufacturing Belt), but are mostly found in or near major cities in peripheral and semiperipheral countries (as in the Brazilian cities of Porto Alegre and Paranà, which have attracted foreign-owned automobile plants).
- ***Enclaves of international banking, finance, and business services in world cities and major regional centers.*** Examples include the business districts in Lower Manhattan, the City of London (**Figure 10.23**), Frankfurt, Hong Kong, and Kuala Lumpur.
- ***Enclaves of modernization in the megacities and major regional centers of peripheral countries.*** This modernization typically includes advanced telecommunications and satellite complexes, world trade centers, retail and commercial centers, and new university precincts. Examples include the Pudong development zone in Shanghai (**Figure 10.24**) and "growth corridors" and "new towns in town" in Bangkok, Thailand.
- ***Enclaves of back-office spaces, data-processing, e-commerce, and call centers.*** These have emerged in older industrial cities within core countries (e.g., Roanoke, USA, and Sunderland, England) and in many cities within semiperipheral countries, most notably in the Caribbean, the Philippines, and India.
- ***Spaces customized as "logistics zones:"*** airports, ports, export-processing zones—enclaves in major cities around the world within which the precise and rapid movement of goods, freight, and people are coordinated, managed, and synchronized between various transport modes.

Connected to one another through a complex dynamic of flows, these urban spaces and settings represent the spatial framework for the fast world. They are embedded within regions and metropolitan areas whose economic foundations derive from earlier technology systems and whose social and cultural fabric derives from more traditional bases. The result, as we shall see in Chapter 11, is that traditional patterns of land use and spatial organization in many parts of the world are being transformed by the local effects of splintering urbanism.

Figure 10.23 London's financial core London's status as a world city rests heavily on the "square mile" of the City of London, with its concentration of banking, insurance, and other advanced business services.

Figure 10.24 Pudong, Shanghai One of the most spectacular development zones in the world, Pudong was a 556 square kilometer (215 sq. mile) area of farmland until the 1990s. It has been developed around a new deep-water port, a new airport, and a new infoport, with massive investments in the rail network, highway network, and cross-river transportation network, and has become symbolic of China's economic reform and growth.

CONCLUSION

Urbanization is one of the most important geographic phenomena. Cities can be seedbeds of economic development and cultural innovation. Cities and groups of cities also organize space—not just the territory immediately around them, but in some cases national and even international space. The causes and consequences of urbanization, however, are very different in different parts of the world. The urban experience of the world's peripheral regions stands in sharp contrast to that of the developed core regions, for example. This contrast is a reflection of some of the demographic, economic, and political factors that we have explored in previous chapters.

Much of the developed world has become almost completely urbanized, with highly organized systems of cities. Today levels of urbanization are high throughout the world's core countries, while rates of urbanization are relatively low. At the top of the urban hierarchies of the world's core regions are world cities such as London, New York, Tokyo, Paris, and Zürich, which have become control centers for the flows of information, cultural products, and finance that collectively sustain the economic and cultural globalization of the world. In doing so, they help to consolidate the hegemony of the world's core regions.

Few of the metropolises of the periphery, on the other hand, are world cities that occupy key roles in the organization of global economics and culture. Rather, they operate as connecting links between provincial towns and villages and the world economy. They have innumerable economic, social, and cultural linkages to their provinces on one side and to major world cities on the other. Almost all peripheral countries, meanwhile, are experiencing high rates of urbanization, with forecast growth of unprecedented speed and unmatched size. In many peripheral and semiperipheral regions, current rates of urbanization have given rise to unintended metropolises and fears of "uncontrollable urbanization," with urban "danger zones" where "work" means anything that contributes to survival. The result, as we shall see in Chapter 11, is that these unintended metropolises are quite different from the cities of the core as places in which to live and work.

MAIN POINTS REVISITED

- **The urban areas of the world are the linchpins of human geographies at the local, regional, and global scales.**

  Towns and cities are engines of economic development and centers of cultural innovation, social transformation, and political change. They now account for almost half the world's population.

- **The earliest towns and cities developed independently in the various hearth areas of the first agricultural revolution.**

  The very first region of independent urbanism, in the Middle East, produced successive generations of urbanized world empires, including those of Greece, Rome, and Byzantium.

- **The expansion of trade around the world, associated with colonialism and imperialism, established numerous gateway cities.**

  European powers founded or developed literally thousands of towns as they extended their trading networks and established their colonies. The great majority of the towns were ports that served as control centers commanding entrance to, and exit from, their particular country or region.

- **The Industrial Revolution generated new kinds of cities—and many more of them.**

  Industrial economies could be organized only with the large pools of labor, the transportation networks, the physical infrastructure of factories, warehouses, stores, and offices, and the consumer markets provided by cities. As industrialization spread throughout Europe in the first half of the nineteenth century and then to other parts of the world, so urbanization increased at a faster pace.

- **Today the single most important aspect of world urbanization, from a geographical perspective, is the striking difference in trends and projections between the core regions and the peripheral regions.**

  In 1950 two-thirds of the world's urban population was concentrated in the more developed countries of the core economies. Since then, the world's urban population has increased threefold, the bulk of the growth having taken place in the less developed countries of the periphery.

- **A small number of "world cities," most of them located within the core regions of the world-system, occupy key roles in the organization of global economics and culture.**

  At the top of a global urban system, these cities experience growth largely as a result of their role as key nodes in the world economy. World cities have become the control centers for the flows of information, cultural products, and finance that collectively sustain the economic and cultural globalization of the world.

- **Many of the megacities of the periphery are primate and exhibit a high degree of centrality within their urban systems.**

  Primacy occurs when the population of the largest city in an urban system is disproportionately large in relation to the second- and third-largest cities in that system. Centrality refers to the functional dominance of cities within an urban system. Cities that account for a disproportionately high share of economic, political, and cultural activity have a high degree of centrality within their urban system.

- **There is a close relationship between globalization and urbanization, mediated by networked infrastructures of transportation, information, and communications technologies.**

  Because these critical networked infrastructures are highly uneven in their impact, selectively serving only certain kinds of metropolitan settings, globalization is exerting a "splintering" effect on patterns of urbanization.

KEY TERMS

central place (p. 405)
central place theory (p. 405)
centrality (p. 410)
colonial city (p. 404)
counterurbanization (p. 420)
gateway city (p. 400)
informal sector (p. 419)
megacity (p. 419)
overurbanization (p. 418)
primacy (p. 408)
rank-size rule (p. 408)
reurbanization (p. 421)
shock city (p. 403)
splintering urbanism (p. 421)
squatter settlements (p. 418)
urban ecology (p. 396)
urban form (p. 396)
urban system (p. 395)
urbanism (p. 396)
world city (p. 411)

ADDITIONAL READING

Abrahamson, M., *Global Cities.* New York: Oxford University Press, 2004.

Brunn, S., and J. F. Williams, (eds.), *Cities of the World,* 3rd ed. New York: Harper Collins, 2003.

Geyer, H. S., (ed.), *International Handbook of Urban Systems.* Northampton, MA: Edward Elgar, 2002.

Gugler, J., (ed.), *Cities in the Developing World: Issues, Theory, and Policy.* Oxford: Oxford University Press, 1997.

Hohenberg, P. M., and L. H. Lees, *The Making of Urban Europe 1000–1994.* Cambridge, MA: Harvard University Press.

Knox, P. L. and L. McCarthy, *Urbanization: An Introduction to Urban Geography.* Englewood Cliffs, NJ: Prentice Hall, 2005.

Kotkin, J., *The City: A Global History.* New York: Random House.

Marcuse, P. and R. van Kampen, eds., *Globalizing Cities.* Oxford: Blackwell, 2000.

Olds, K., *Globalization and Urban Change: Capital, Culture, and Pacific Rim Mega-Projects.* Oxford University Press, 2001.

Pacione, M., *Urban Geography: A Global Perspective.* New York: Routledge, 2001.

Potter, R. B., and S. Lloyd-Evans, *The City in the Developing World.* Harlow, Essex: Addison Wesley Longman, 1998.

Sassen, S., *Global Networks, Linked Cities.* New York: Routledge, 2002.

Scott, A. J., ed., *Global City-Regions: Trends, Theory, Policy.* Oxford: Oxford University Press, 2001.

Short, J. R., and Y-H. Kim, *Globalization and the City.* New York: Addison Wesley Longman, 1999.

Simmonds, R. and G. Hack, (eds.), *Global City Regions: Their Emerging Forms.* New York: Spon Press, 2000.

Taylor, P., *World City Network: A Global Urban Analysis.* New York: Routledge, 2004.

United Nations Center for Human Settlements (HABITAT), *Cities in a Globalizing World: Global Report on Human Settlements, 2001.* London: Earthscan Publications, 2001.

United Nations Human Settlements Program, *The State of the World's Cities 2004/2005.* Sterling, VA: Earthscan, 2005.

Verhulst, A. *The Rise of Cities in North-West Europe.* Cambridge: Cambridge University Press, 1999.

EXERCISES

On the Internet

The Internet exercises for this chapter will help you to better understand the forces of urbanization. Using the Internet, we examine some major world cities such as Paris, London, and Washington, D.C. City development was not carefully controlled, but instead was driven by the desire for land and by the hunger of the huge markets that arose in that quest. American cities are, perhaps to a greater extent than anywhere else in the world, the products of unrestrained capitalism. In this light we examine the growth of Los Angeles and Atlanta. We also compare and contrast regional growth during different time frames; we survey how our cities tend to distinguish themselves from their peers; we investigate counterurbanization; and we explore frontier urbanization in the United States.

Unplugged

1. Figures **10.1** and 10.2 show that the United States, like most core countries, is already highly urbanized and has a relatively low rate of urbanization. Nevertheless, some U.S. cities have been growing much faster than others. Which have been the fastest growing U.S cities in recent times, and what reasons can you suggest for their relatively rapid growth? *Hint:* The U.S. Bureau of the Census publishes data on population change by urban area, as does the Population Division of the U.N. Department of Economic and Social Affairs, in a volume entitled *World Urbanization Prospects*.
2. From census volumes in your library, find out the population of the town or city you know best. Do the same for every census year, going back from 2000 to 1990, 1980, and so on, all the way back to 1860. Then plot these populations on a simple graph. What explanations can you offer for the pattern that the graph reveals? Now draw a larger version of the same graph, annotating it to show the landmark events that might have influenced the city's growth (or decline).
3. The following cities all have populations in excess of 2 million. How many of them could you locate on a world map? Their size reflects a certain degree of importance, at least within their regional economy. What can you find out about each? Compile for each a 50-word description that explains its chief industries and a little of its history.

   Poona Ibadan Recife
   Bangalore Turin Ankara
4. Figure 10.9 features two colonial gateway cities on the Eastern Seaboard of the United States—Boston and New York. Two other colonial gateway cities were Charleston, South Carolina, and Savannah, Georgia. What can you find out about the commodities and manufactures that each imported and exported in pre-Revolutionary times, and about the origin and destination of these imports and exports? Which geographic concepts do you consider to be useful in explaining these facts?

仲見世
浅
13297

11 City Spaces: Urban Structure

MAIN POINTS

- The internal structure of cities is shaped heavily by competition for territory and location. In general, all categories of land users—commercial and industrial, as well as residential—compete for the most convenient and accessible locations within the city.
- Social patterns in cities are heavily influenced by territoriality, which provides a means of establishing and preserving group membership and identity.
- The typical North American city is structured around a central business district (CBD); a transitional zone; suburbs; secondary business districts and commercial strips; industrial districts; and, in larger metropolitan areas, edge cities.
- Traditionally, North American cities have experienced high rates of in-migration, which has resulted in their becoming structured into a series of concentric zones of neighborhoods of different ethnicity, demographic composition, and social status through processes of invasion and succession.
- The most acute problems of the postindustrial cities of the world's core regions are localized in the central city areas that have borne the brunt of restructuring from an industrial to a postindustrial economy.
- The problems of the cities of the periphery stem from the way in which their demographic growth has outstripped their economic growth.

"The neighborhood shopping street is at the end of the block. It is a narrow lane, barely wide enough for one car to pass, and is lined on both sides with small shops whose fronts open widely to the street . . . and invite customers in. There are more and more boutiques and other new arrivals on the street, including an extremely busy supermarket, but there are still quite a few of the older establishments left as well: fishmongers, rice sellers, a noodle maker, a cracker bakery, a cubby-hole that sells only buttons, a glazier's shop, and countless other, small places for the local market. Tucked away to the side is the neighborhood's Buddhist temple. It is a new building but designed in a traditional style, and has a welcome open space for community fairs and other gatherings in front, and a lovely Japanese garden at the back. The garden is such a contrast to the harsh lines and bustling activity of the surrounding city that at times it seems to me to be the most secluded and contemplative place in the world."[1]

We can recognize in this description of a Tokyo neighborhood several elements that are fairly common in central cities throughout the world's core regions: the mixture of old stores, new boutiques, and local supermarkets, for example. On the other hand, some elements are unique: the noodle maker and the Buddhist temple with its Japanese garden. In this chapter we turn our attention to the internal dynamics of cities, looking at the ways in which patterns and processes tend to vary according to the type of city and its history. Some of the most striking contrasts are to be found between the cities of the core regions and those of the periphery. The evolution of the unintended metropolis of the periphery has been very different from that of metropolitan areas in the world's core regions. Similarly, the problems they face are very different. Nevertheless, the splintering urbanism associated with processes of globalization is bringing similar new elements of urban structure to large cities in every world region.

[1]Roman Cybriwsky, *Tokyo*. London: Belhaven Press, 1991, p. 3.

Tokyo neighborhood

URBAN LAND USE AND SPATIAL ORGANIZATION

The internal organization of cities reflects the way they function, both to bring people and activities together and to sort them out into functional subareas and neighborhoods. While there are many complex processes at work in cities, there are some broad tendencies that go a long way toward shaping patterns of urban land use and spatial organization. We consider two of them here: people's need for accessibility and their sense of territoriality.

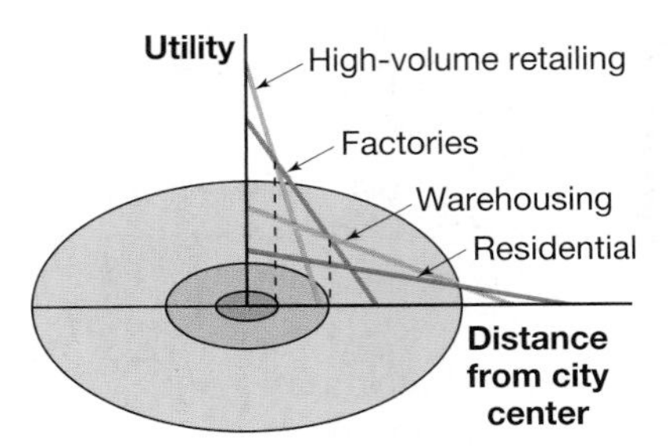

Figure 11.1 Accessibility, bid-rent, and urban structure Competition for accessible sites near the city center is an important determinant of land-use patterns. Different land users are prepared to pay different amounts—the bid-rents—for locations at various distances from the city center. The result is a tendency for a concentric pattern of land uses. (*Source:* Reprinted with permission of Prentice Hall, from P. L. Knox, *Urbanization* © 1994, p. 99.)

Accessibility and Land Use

Most urban land users want to maximize the *utility* they derive from a particular location. The utility of a specific place or location refers to its usefulness to particular persons or groups. The price they are prepared to pay for different locations—the bid-rent—is a reflection of this utility. In general, utility is a function of *accessibility.* Commercial land users want to be accessible to one another, to markets, and to workers; private residents want to be accessible to jobs, amenities, and friends; public institutions want to be accessible to clients. In an idealized city built on an isotropic surface, the point of maximum accessibility is the city center. An **isotropic surface** is a hypothetical, uniform plane: flat, and with no variations in its physical attributes. Under these conditions, accessibility decreases steadily with distance from the city center. Likewise, utility decreases, *but at different rates for different land users.* The result is a tendency toward concentric zones of different mixes of land use (**Figure 11.1**).

One counterintuitive implication of this model is that the poorest households will end up occupying the periphery of the city. While this is true in some parts of the world, we know that in the core countries the farthest suburbs are generally the territory of wealthier households, while the poor usually occupy more accessible locations nearer city centers. Some modification of the assumptions is clearly required. In this case, we must assume that wealthier households trade off the convenience of accessibility for the greater utility of being able to consume larger amounts of (relatively cheap) suburban space. Poorer households, unable to afford the recurrent costs of transportation, must trade off living space for accessibility to jobs so that they end up in high-density areas, at expensive locations near their low-wage jobs. Because of the presumed trade-off between accessibility and living space, this urban land-use model is often referred to as a *trade-off model.*

Territoriality, Congregation, and Segregation

In cities, as at other geographic scales, territoriality provides a means of establishing and preserving group membership and identity. The first step in forming group identity is to define "others" in an exclusionary and stereotypical way. **Congregation**—the territorial and residential clustering of specific groups or subgroups of people—enables group identity to be consolidated in relation to people and places outside the group. Congregation is thus a place-making activity and an important basis for urban structure and land use. It is particularly important in situations where there is one or more distinctive minority groups. Defined in relation to a general population or host community, **minority groups** are population subgroups that are seen—or that see themselves—as somehow different from the general population. Their defining characteristics can be based on race, language, religion, nationality, caste, sexual orientation, or lifestyle.

Several specific advantages of congregation exist for minority groups:

- *Congregation provides a means of cultural preservation. It allows religious and cultural practices to be maintained, and strengthens group identity through daily involvement in particular routines and ways of life.* Particularly important in this regard is the way that clustering fosters within-group marriage and kinship networks.
- *Congregation helps minimize conflict and provides defense against "outsiders."*
- *Congregation provides a place where mutual support can be established through minority institutions, businesses, social networks, and welfare organizations.*
- *Congregation helps establish a power base in relation to the host society.* This power base can be democratic, organized through local elections, or it can take the form of a territorial heartland for insurrectionary groups.

Congregation is not always voluntary, of course. Host populations are also impelled by territoriality, and they may respond to social and cultural differences by *discrimination* against minority groups. Discrimination can also have a strong territorial basis, the objective being to restrict the territory of minority groups and to resist their assimilation into the host society. This resistance can take a variety of forms. Social hostility and the voicing of "keep out" attitudes are probably the most widespread, although other forms of discrimination can have pronounced spatial effects. These forms include exclusion and prejudice in local labor markets; the manipulation of private land and housing markets; the steering of capital investment away from minority areas; and the institutionalization of discrimination through the practices and spatial policies of public agencies.

The combined result of congregation and discrimination is **segregation,** the spatial separation of specific subgroups within a wider population (**Figure 11.2**). Segregation varies a great deal in both intensity and form, depending on the relative degree and combination of congregation and discrimination. Geographers have identified three principal situations in terms of the spatial form of segregation:

- *Enclaves,* in which tendencies toward congregation and discrimination are long-standing but dominated by internal cohesion and identity. The Jewish districts of many of today's cities in Europe and the eastern United States are examples of enclaves.
- *Ghettos,* which are also long-standing, but are more the product of discrimination than congregation. Examples are the segregation of African Americans and Hispanics in American cities.
- *Colonies,* which may result from congregation, discrimination, or both, but in relatively weak and short-lasting ways. Their persistence over time depends on the continuing arrival of new minority-group members. For example, U.S. cities in the early twentieth century contained distinctive colonies of German, Scandinavian, Irish, and Italian immigrants that have now all but disappeared. Today there are colonies of Greek and Yugoslav neighborhoods in Australian cities, and colonies of Koreans in Japanese cities.

Figure 11.2 Segregation Congregation is often voluntary, but in many cases it is driven to some degree by discrimination. The combined result of congregation and discrimination is segregation, the spatial separation of specific subgroups within a wider population. This photograph is of a political mural in the Bogside district of Derry (Londonderry), Northern Ireland.

TRADITIONAL PATTERNS OF URBAN STRUCTURE

Economic competition for space and accessibility, along with the tendency toward social and ethnic discrimination, congregation, and segregation can be traced in many of the world's cities, particularly in affluent core regions where economic, social, and cultural forces are broadly similar. Nevertheless, urban structure varies considerably because of the influence of history, culture, and the different roles that cities have played within the world-system. In this section, we examine the typical characteristics of North American, European, and Islamic cities, and the unintended metropolises of the periphery.

North American Cities

Traditionally, the very center of North American cities has been the principal hub of shops and offices, together with some of the major institutional land uses such as the city hall, libraries, and museums. This center, known as the **central business district,** or **CBD,** is a city's nucleus of commercial land uses. It traditionally contains the densest concentration of shops, offices, and warehouses and the tallest nonresidential buildings. It usually developed at the nodal point of transportation routes, so that it also contains bus stations, railway terminals, and hotels. The CBD typically is surrounded by a zone of mixed land uses: warehouses, small factories and workshops, specialized stores, apartment buildings, public housing projects, and older residential neighborhoods (**Figure 11.3**). This zone is often referred to as the **zone in transition** because of its mixture of growth, change, and decline.

Beyond this zone are residential neighborhoods, suburbs of various ages and different social and ethnic composition. Just as different categories of land use attract and repel one another, so do different social and ethnic groups. In North America, where urban population growth has been fueled by streams of migrants and immigrants with very different backgrounds, sociologists have developed an ecological perspective to describe neighborhoods as being structured by the "invasion" of successive waves of migrants and immigrants.

When immigrants first arrive in the city looking for work and a place to live, they have little choice but to cluster in the cheapest accommodations, typically found in the zone in transition around the CBD. The classic example was provided by Chicago in the 1920s and 1930s. Immigrants from Scandinavia, Germany, Italy, Ireland, Poland, Bohemia (now part of the Czech Republic), and Lithuania established themselves in Chicago's low-rent areas, the only places they

Figure 11.3 The central city This photograph of Chicago shows the concentration of high-rise office buildings on the skyline that is typical of central business districts (CBDs) in U.S. metropolitan areas. In Chicago, as in other major cities, the modern CBD grew up around the point of maximum accessibility: near railway stations and at the intersection of the city's principal transit lines. It is surrounded by mixed land uses—older neighborhoods, abandoned lots, apartment complexes and housing projects—all pierced by freeway corridors that have taken the place of railway and transit lines.

could afford. By congregating in these areas, immigrants accomplished several things—they were able to establish a sense of security; continue speaking their native language; have familiar churches or synagogues, restaurants, bakeries, butcher shops, and taverns; and support their own community newspapers and clubs. These immigrants were joined in the city's zone in transition by African American migrants from the South, who established their own neighborhoods and communities. In Chicago, as in other U.S. cities of the period, the various ethnic groups formed a patchwork or mosaic of communities encircling the CBD.

These ethnic communities lasted from one to three generations, after which they started to break up. Many of the younger, city-born individuals did not feel the need for the security and familiarity of ethnic neighborhoods. Gradually, increasing numbers of them were able to establish themselves in better jobs and move out into newer, better housing. As the original immigrants and their families moved out, their place in the transitional zone was taken by a new wave of migrants and immigrants. In this way, Chicago became structured into a series of *concentric zones* of neighborhoods of different ethnicity and social status (**Figure 11.4**).

Throughout this process of invasion and succession, people of the same background tend to stick together—partly because of the advantages of residential clustering and partly because of discrimination. **Invasion and succession** is a process of neighborhood change whereby one social or ethnic group succeeds another in a residential area. The displaced group, in turn, invades other areas, creating over time a rippling process of change throughout the city. The result is that within each concentric zone there exists a mosaic of distinctive neighborhoods. Classic examples include the Chinatowns, Little Italys, Koreatowns, and African American ghettos of big North American cities (**Figure 11.5**). Such neighborhoods can be thought of as *ecological niches* within the overall metropolis—settings where a particular mix of people have come to dominate a particular territory and a particular physical environment, or habitat.

For decades the process of invasion and succession was a predominantly outward movement that gave North American cities their fundamental concentric structure. In the last quarter of the twentieth century, however, the flow of migrants and immigrants to many cities diminished significantly, prompting the gentrification of some central parts of many cities. **Gentrification** involves the invasion of older, centrally located, working-class neighborhoods by higher income households seeking the character and convenience of less expensive and centrally located residences. These invasions result in the physical renovation and upgrading of housing (**Figure 11.6**), but they also displace many existing occupants.

Problems of North American Cities

For all their relative prosperity, the cities of the world's core regions have their share of problems. The most acute are localized in central city areas and are interrelated: fiscal problems, infrastructure problems, and localized cycles of poverty and spirals of neighborhood decay. **Central cities** are the original, core jurisdictions of metropolitan areas.

Fiscal Problems

The term *fiscal* refers to taxes. Economic restructuring and metropolitan decentralization over the past several decades have left central cities with a chronic "fiscal squeeze." A **fiscal squeeze** occurs when increasing limitations on tax revenues combine with increasing demands for expenditures on urban infrastructure and city services. The rev-

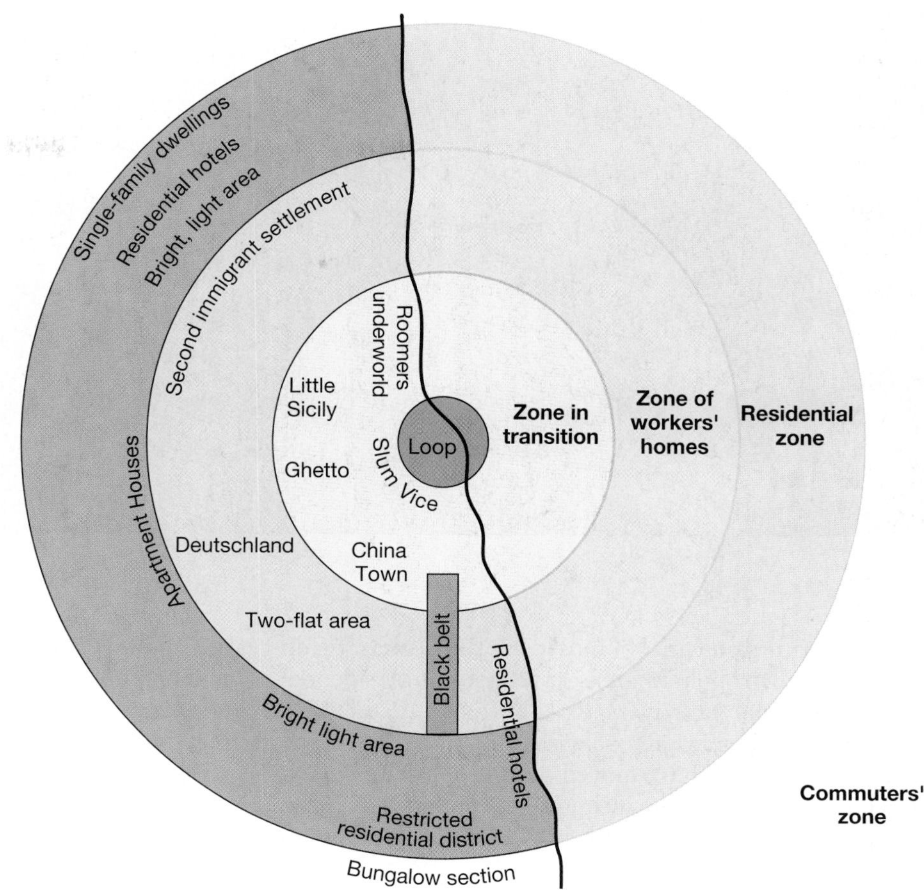

Figure 11.4 The ecological model of urban land use: Chicago in the 1920s Competition between members of different migrant and immigrant groups for residential space in the city often results in distinctive neighborhoods that have their own social "ecology." The classic example is Chicago of the 1920s, which had developed a series of concentric zones of distinctive neighborhoods as successive waves of immigrants established themselves. Over time most immigrant groups made their way from low-rent, inner-city districts surrounding the CBD (known in Chicago as the Loop) to more attractive and expensive districts farther out. (After R. E. Park, E. W. Burgess, and R. D. McKenzie, *The City.* Chicago: University of Chicago Press, 1925, p. 53.)

Figure 11.5 Chinatown In most larger cities there is a patchwork of distinctive neighborhoods that results from processes of congregation and segregation. Most distinctive of all are neighborhoods of ethnic minorities, such as the Chinatowns, Little Italys, and Little Koreas of major American cities. Shown here is part of San Francisco's Chinatown.

Figure 11.6 Gentrification In Old Town Alexandria, Virginia, a lengthy process of gentrification has established an elite area among neighborhoods of older town houses, most of which had previously been occupied by lower- and middle-income households.

enue-generating potential of most central cities has steadily fallen as metropolitan areas have lost both residential and commercial taxpayers to suburban jurisdictions. Growth industries, white-collar jobs, retailing, and more affluent households have moved out to suburban and exurban jurisdictions, taking their local tax dollars with them.

At the same time growth in property-tax revenues from older, decaying neighborhoods has slowed as the growth of property values has slowed. Yet these older, decaying neighborhoods cost more to maintain and service. Older streets, water and sewage lines, schools, and transit systems have high maintenance costs. The residual populations of these neighborhoods, with high proportions of elderly and indigent households, are increasingly in need of municipal welfare services. Large numbers of low-income migrants and immigrants also bring increased demands for municipal services. Added to all this, central city governments are still responsible for services and amenities used by the entire metropolitan population: municipal galleries and museums, sports facilities, parks, traffic police, and public transport, for example. The net result is that central city jurisdictions tend to be locked permanently into a precarious financial position. This in turn results in a constant drive to develop revenue-generating projects, with cities competing fiercely with one another to finance and attract tourist developments, museums, sports franchises, and business and conference centers.

Infrastructure Problems

As the rate of urban growth in North America has slowed and fiscal problems have intensified, public spending on the urban infrastructure of roads, bridges, parking spaces, transit systems, communications systems, power lines, gas supplies, street lighting, water mains, sewers, and drains has declined. Meanwhile, much of the original infrastructure, put in place 75 or 100 years ago with a design life of 50 or 75 years, is obsolete, worn out, and in some cases perilously near the point of collapse (**Figure 11.7**).

Infrastructure problems are easily overlooked because they build up slowly. Only when a bridge collapses or a water main bursts are infrastructure problems newsworthy. Yet cities' infrastructure is crucial not only to their economic efficiency but also to public health, safety, and the quality of life. One review of infrastructure problems in American cities estimated that between half and two-thirds of the country's cities are unable to support modernized investment until major new investments can be made in their basic infrastructure. This investment, of course, is unlikely, given the chronic fiscal problems they face. In Boston three-quarters of the sewer system was built a century or more ago and has now deteriorated to the point where about 20 percent of the system's overall flow is lost to leaks. In Cleveland, the District of Columbia, and Philadelphia the losses approach 25 percent. Hundreds of kilometers of water mains in New York City have been identified as being in need of replacement, at an estimated cost in 2004 of over $5 billion. Almost 50 percent of all wastewater treatment systems in America are operating at 80 percent or more of capacity, the level at which the federal government prohibits further industrial hookups in a community.

Freshwater supplies are also at risk: old water systems are unable to cope with the leaching of pollutants into city water. Common pollutants include chlorides, oil, phosphates, and nondegradable toxic chemicals from industrial wastewater; dissolved salts and chemicals from highway de-icing; nitrates and ammonia from fertilizers and sewage; and coliform bacteria from septic tanks and sewage. Many cities still use water-cleaning technology dating to World War I. About one-third of all towns and cities in the United States have contaminated water supplies, and about 8 million people are using water that is potentially dangerous.

Figure 11.7 Decaying Infrastructure A report issued in 2005 by the American Society of Civil Engineers (ASCE)[2] estimated that it would cost $1.6 trillion to bring the country's infrastructure up to acceptable standards. Most of the capital investment needed to repair and replace America's infrastructure of roads, bridges, transit systems, schools, water- and waste-treatment systems, and hazardous-waste disposal is needed in cities and metropolitan areas. In central cities, fiscal squeeze has continuously postponed attention to obsolescent infrastructure. In suburban areas, a political ideology dominated by low taxation and small government has consistently diverted attention from underdeveloped infrastructure.

Poverty and Neighborhood Decay

Inner city poverty and neighborhood decay have become increasingly pronounced in the past several decades as manufacturing, warehousing, and retailing jobs have moved out to suburban and edge-city locations, and as many of the more prosperous households have moved out to be near these jobs. The spiral of neighborhood decay typically begins with substandard housing occupied by low-income households who can afford to rent only a minimal amount of space. The consequent overcrowding not only causes greater wear and tear on the housing itself but also puts pressure on the neighborhood infrastructure of streets, parks, schools, and playgrounds. The need for maintenance and repair increases quickly but is rarely met. Individual households cannot afford it, and landlords have no incentive to do so, because they have a captive market. Public authorities face a fiscal squeeze and are in any case often indifferent to the needs of such neighborhoods because of their relative lack of political power.

Shops and privately run services such as restaurants and hair salons are afflicted with the same syndrome of decay. With a low-income clientele, profit margins must be kept low, leaving little to spare for upkeep or improvement. Many small businesses fail, or relocate to more favorable settings, leaving commercial property vacant for long periods. In extreme cases such property becomes abandoned, the owners unable to find either renters or buyers. Residential buildings may also be left derelict. Faced with escalating maintenance costs and rising property taxes and unable to increase revenues because of rent controls and the depressed state of inner-city housing markets, some landlords simply write off their property and abandon it.

[2]*Renewing America's Infrastructure.* American Society of Civil Engineers, 2005. http://www.asce.org/reportcard/

There is, meanwhile, a dismal cycle of poverty that intersects with these localized spirals of decay. The **cycle of poverty** involves the transmission of poverty and deprivation from one generation to another through a combination of domestic circumstances and local neighborhood conditions. This cycle begins with a localized absence of employment opportunities and, therefore, a concentration of low incomes, poor housing, and overcrowded conditions. Such conditions are unhealthy. Overcrowding makes people vulnerable to poor health, which is compounded by poor diets. This therefore contributes to absenteeism from work, which results in decreased income. Similarly, absenteeism from school through illness contributes to the cycle of poverty by constraining educational achievement, limiting occupational skills, and so leading to low wages. Crowding also produces psychological stress, which contributes to social disorganization and a variety of pathological behaviors, including crime and vandalism. Such conditions not only affect people's educational achievement and employment opportunities but also lead to the *labeling* of the neighborhood, whereby all residents may find their employment opportunities affected by the poor image of their neighborhood.

One of the most important elements in the cycle of neighborhood poverty is the educational setting. Schools, obsolete and physically deteriorated like their surroundings, are unattractive to teachers—partly because of the physical environment and partly because of the social and disciplinary environment. Because of fiscal squeeze, the schools are often resource-poor, with relatively small budgets for staff, equipment, and materials. Over the long term, poor educational resources translate into poor education, however positive the values of students and their parents. Poor education limits occupational choice and, ultimately, results in lower incomes. Students, faced with evidence all around them of unemployment or low-wage jobs at the end of school careers, find it difficult to be positive about school. The result becomes a self-fulfilling prophecy of failure, and people become trapped in areas of concentrated poverty (**Figure 11.8**).

Many poverty areas are also racial ghettos, although not all ghettos are poverty areas. As we have seen, ethnic and racial congregation can mitigate the effects of poverty. Nevertheless, discrimination is usually the main cause of ghettoization. In the United States, discrimination in housing markets is illegal, but it nevertheless takes place in a variety of ways. One example of housing-market discrimination by banks and other lending institutions is the practice of redlining. **Redlining** involves marking off bad-risk neighborhoods on a city map and then using the map to determine loans. This practice results in a bias against minorities, female-headed households, and other vulnerable groups, because they tend to be localized in low-income neighborhoods. Redlining tends to become another self-fulfilling prophecy, since neighborhoods starved of property loans become progressively run down and therefore increasingly unattractive to lenders. Discrimination affects education and labor markets as well as housing markets. In the case of ghetto poverty, all three types of discrimination come together, reinforcing the cycle of poverty and intensifying the disadvantages of the minority poor.

Social trends have compounded the problems of these areas in many instances. Increased divorce and teen pregnancy rates have led to more single-parent families and a feminization of poverty. These families have been portrayed as the core of a geographically, socially, and economically isolated underclass. The idea of an **underclass** refers to a class of individuals who experience a form of poverty from which it is very difficult to escape because of their isolation from mainstream values and the formal labor market. Isolated from the formal labor force and from the social values and behavioral patterns of the rest of society, the underclass has been seen as subject to increased social disorganization and deviant behavior. In cities in the United States, localized inner city poverty is now characterized by senseless and unprovoked violence; premeditated and predatory violence; domestic violence; the organized violence of street gangs; and epidemic levels of HIV/AIDS and other communicable diseases—all closely associated with drug taking and drug dealing.

One consequence of extreme poverty is *homelessness*. Chronic, long-term homelessness means not having customary and regular access to a conventional dwelling. This includes people who have to sleep in shelters, flophouse cubicles, and emergency dormitories and missions, as well as those sleeping in doorways, bus stations, cars, tents, temporary shacks, and cardboard boxes, and on park benches and steam grates.

The number of homeless persons in North American cities rose sharply in the mid-1980s. This was mainly a

Figure 11.8 Poverty areas Concentrations of poverty are found not only in decaying inner city areas but also in newer public housing projects and in first- and second-tier suburbs that have filtered down the housing scale, as in this example in the District of Columbia, a short distance from the Capitol.

consequence of the increased poverty and the economic and social dislocation caused by economic restructuring and the transition to a globalized, postindustrial economy. It was intensified by the fiscal squeeze confronting central cities and the tendency for governments to cut back on welfare programs of all kinds. It was also intensified by the adoption of revolving-door policies of mental health hospitals, which released large numbers of institutionalized patients. The homeless are now very visible throughout the major cities of North America (**Figure 11.9**): around the heating vents of shopping malls in Canadian cities, in the subterranean world of the New York subway system, under Miami's bridges and in Los Angeles's parks, in cardboard boxes by glass office towers in downtown Chicago, and near abandoned factories in Boston and St. Louis.

Estimates of the number of homeless people in U.S. cities vary a good deal. The estimate of the National Coalition for the Homeless is between 1.5 and 3 million. The National Law Center on Homelessness and Poverty estimates the number to be more like 700,000 at any one time, with about 2 million people experiencing homelessness for a time during any given year. The Urban Institute estimates that about 3.5 million people, 1.35 million of them children, are likely to experience homelessness in a given year. Estimates for Canadian cities suggest a total of more than 250,000 homeless persons. In Toronto alone more than 30,000 people—equivalent to 1.3 percent of the city's entire population—slept in shelters in 2000, a 40 percent increase from 1988.

What makes contemporary homelessness so striking a problem is not just the scale of the problem but also its nature. Whereas homelessness had previously involved white adult males, relatively few of whom actually had to sleep outdoors, the new homeless are of all races and include significant numbers of women, children, and the elderly. In Canada, for example, one quarter of the country's homeless are children. The new homeless are also much less likely to find shelter indoors, an have become a very visible feature of the public spaces of many cities.

European Cities

European cities, like North American cities, reflect the operation of competitive land markets and social congregation along ethnic lines. They also suffer from similar problems of urban management, infrastructure provision, and poverty. What makes most European cities distinctive in comparison with North American cities is that they are the product of several major epochs of urban development. As we saw in Chapter 2, because many of today's most important cities were founded in the Roman period, it is not uncommon for the outlines of Roman and medieval urban development to be preserved in their street plans. Many distinctive features of European cities derive from their long history. In the historic cores of some older cities, the layout of streets reflects ancient patterns of rural settlement and field boundaries. Beyond these historic cores, narrow, complex streets are the product of the long, slow growth of European cities in the pre-automobile era, when hand-pushed and horse-drawn carts were the principal means of transportation, and urban development was piecemeal and small-scale (**Figure 11.10**).

Plazas and squares are another important historical legacy in many European cities. Greek, Roman, and medieval cities were all characterized by plazas, central squares, and marketplaces, and those elements are still important nodes of urban activity (**Figure 11.11**). European history also means that its cities bear the accessories and scars of war. The legacy of defensive hilltop and clifftop sites and city walls has limited and shaped the growth of modern cities (**Figure 11.12**), while in more recent times the bombings and shellings of World War II destroyed many buildings in a large number of cities (**Figure 11.13**).

The legacy of a long and varied history includes a rich variety of symbolism in the built environment. Europeans are reminded of their past not only by large numbers of statues and memorials but also by cathedrals, churches, and monasteries; by guildhalls and city walls; by the

Figure 11.9 The new homeless In the 1980s and 1990s, changing conditions in U.S. cities created a huge increase in the number of homeless persons. The new homeless include significant numbers of women, children, and young adults, groups that had not previously been vulnerable to homelessness.

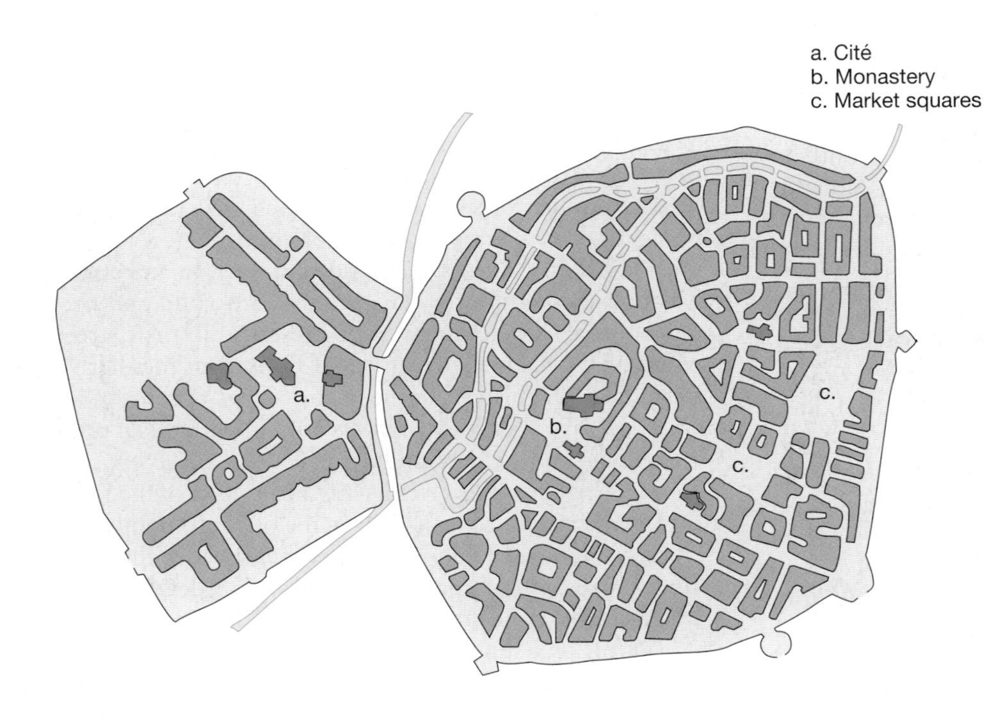

Figure 11.10 Medieval Arras The towns of medieval Europe were typically small, with compact outlines defined by defensive walls. In addition to their defensive function, these walls were signs of wealth and power. At their gates, tolls were collected and goods checked. They enclosed a separate administrative area in which residents were free and usually self-governing. Within the walls, towns were typically organized around a fortress, or cité, though the foci of activity and sources of growth were more likely to be religious institutions, market squares, and merchants' quarters. Arras, in northeastern France, shows the "organic" pattern of irregular-shaped blocks with narrow, winding streets that is typical of many of the medieval towns of Europe. Its growth was focused successively on three elements: (a) the cité, or fortress, (b) the monastery, and (c) two market squares. (After P. M. Hohlenberg and L. H. Lees, *The Making of Urban Europe 1000–1950.* Cambridge, MA: Harvard University Press, 1985, p. 32.)

palaces of royalty and the mansions of aristocracy; and by city halls and the libraries, museums, sports stadiums, and galleries that are monuments to civic achievement. European cities are also typically compact in form, resulting in high densities of population. A long history of pre-automobile urban development and the constraints of peripheral defensive walls all made urban land expensive and encouraged a tradition of high-density living in tenements and apartment houses.

Several other distinctive features of European cities should be noted:

- ***Low skylines*** —Although the larger European cities have a fair number of high-rise apartment buildings and a sprinkling of office skyscrapers, they all offer a predominantly low skyline, as in the example of Bonn (**Figure 11.14**). This is partly because much of their growth came before the invention of the elevator and

Figure 11.11 Tübingen, Germany This small piazza, one of several in the town, traditionally operated as the location of a market for wood and wood products.

Figure 11.12 Dubrovnik The ancient city of Dubrovnik, on the Mediterranean coast of present-day Croatia, was established on a promontory whose cliffs provided a natural defensive site. Fortifications were soon added, with continual improvements through the Middle Ages. The surviving walls date from the thirteenth century.

Figure 11.13 Cologne, Germany, 1945 About 90 percent of Cologne's central area was destroyed or severely damaged by Allied bombing during World War II. One of the few structures to survive intact was the solid, stone-built cathedral, built between the thirteenth and nineteenth centuries.

Figure 11.14 Bonn, Germany Although Bonn is a major center of administration and business services, it has a relatively low skyline, with only a handful of tall buildings.

the development of steel-reinforced, concrete building techniques, and partly because of master plans and building codes (some written as long ago as the sixteenth century) seeking to preserve the dominance of monumental buildings like palaces and cathedrals.

- ***Lively downtowns*** —The CBDs of European cities have retained their focal position in residents' shopping and social lives because of the relatively late arrival of the suburbanizing influence of the automobile and strong planning controls directed against urban sprawl (**Figure 11.15**).
- ***Neighborhood stability*** —Europeans change residence, on average, about half as often as Americans. In addition, the physical life cycle of city neighborhoods tends to be longer because of the past use of durable construction materials, such as brick and stone. As a result, European cities provide relatively stable socioeconomic environments.
- ***Municipal socialism*** —For decades European welfare states have provided a broad range of municipal services and amenities, from clinics to public transit systems. Perhaps the most important to urban structure is social housing (public housing), which accounts for 20 to 40 percent of all housing in most larger English, French, and German cities (**Figure 11.16**). In recent years neoliberal policies have resulted in a reduction in public service provision, especially in France and Great Britain.

We should note, however, that the richness of European history and the diversity of its geography mean that there are important regional variations: The industrial cities of northern England, northeastern France, and the Ruhr district of Germany, for example, are quite different in character from the cities of Mediterranean Europe. One of the most interesting regional variations is the Eastern European city, in which the legacy of an interlude of 44 years of socialism (1945–1989) was grafted onto cities that had already developed mature patterns of land use and social differentiation. Major examples include Belgrade, Budapest, Katowice, Kraków, Leipzig, Prague, and

Figure 11.15 Covent Garden, London The streets of central London still contain many of the city's best cafés and stores. As in many other European cities, people use these streets for evening and weekend strolls, window-shopping, and people-watching.

Figure 11.16 Municipal housing, Birmingham, England Public housing is a very important element in many European cities, often housing more than one in every five households, and in some cities three or four in every five households. [*photo credit:* S. Hall, P. Lee, and S. Sankey; *Source:* Kazepov, Y. (ed.), *Cities of Europe: Changing Contexts, Local Arrangements, and the Challenge to Urban Cohesion.* Malden, MA: Blackwell, 2005.

Figure 11.17 Social housing, Budapest, Hungary Large-scale developments of pre-cast, system-built apartment units—a legacy of the socialist era that spanned the 1950s through the 1980s—are a distinctive feature of most cities in East-Central Europe.

Warsaw. State control of land and housing meant that huge public housing estates and industrial zones were created in outlying districts (**Figure 11.17**). The structure of the older city was little altered, however, apart from the addition of socialist monuments and the renaming of streets.

Urban Design and Planning

European city planning and design have a long history. As we have seen, most Greek and Roman settlements were deliberately laid out on grid systems, within which the siting of key buildings and the relationship of neighborhoods to one another were carefully thought out.

The roots of modern Western urban planning and design can be traced to the Renaissance and baroque periods (between the fifteenth and seventeenth centuries) in Europe, when artists and intellectuals dreamed of ideal cities, and rich and powerful regimes used urban design to produce extravagant symbolizations of wealth, power, and destiny. Inspired by the classical art forms of ancient Greece and Rome, Renaissance urban design sought to recast cities in a deliberate attempt to show off the power and the glory of the state and the Church. Spreading slowly from its origins in Italy at the beginning of the fifteenth century, Renaissance design had diffused to most of the larger cities of Europe by the end of the eighteenth century. Dramatic advances in military ordnance (cannon and artillery) brought a surge of planned redevelopment that featured impressive fortifications, geometric-shaped redoubts, or strongholds (**Figure 11.18**), and an extensive *glacis militaire*—a sloping, clear zone of fire. Inside new walls, cities were recast according to a new aesthetic of grand design—fancy palaces, and geometrical plans, streetscapes, and gardens that emphasized views of dramatic perspectives. These developments were often of such a scale that they effectively fixed the layout of cities well into the eighteenth and even into the nineteenth century, when walls and/or glacis eventually made way for urban redevelopment in the form of parks, railway lines, or beltways.

As societies and economies became more complex with the transition to industrial capitalism, national rulers and city leaders looked to urban design to impose order, safety, and efficiency, as well as to symbolize the new seats of power and authority. One of the most important early precedents was set in Paris by Napoleon III, who presided over a comprehensive program of urban redevelopment and monumental urban design. The work was carried out by Baron Georges Haussmann between 1853 and 1870. Haussmann demolished large sections of old Paris to make way for broad, new, treelined avenues (**Figure 11.19**), with numerous public open spaces and monuments. In doing so, he made the city not only more efficient (wide boulevards meant better flows of traffic) and a better place to live (parks and gardens allowed more fresh air and sunlight in a crowded city and were held to be a "civilizing" influence) but also safer from revolutionary politics (wide

Figure 11.18 Sabbioneta, Italy Sabbioneta was built by Vespasiano Gonzaga in the mid-sixteenth century as an ideal town, with a central piazza, a ducal palace, churches, garden palace, theater, and residences all encompassed within a star-shaped plan, bounded by thick walls bearing the Gonzaga family crest.

Figure 11.19 Boulevard Montmartre, Paris Central Paris owes much of its character to the "grands boulevards" that were key to the urban renewal schemes of Baron Georges-Eugène Haussmann. (From *The Boulevard Montmartre, Paris,* by Camille Pisarro, 1830–1903.)

boulevards were hard to barricade; monuments and statues helped to instill a sense of pride and identity).

The preferred architectural style for these new designs was the **Beaux Arts** style, which takes its name from L'École des Beaux Arts in Paris. In this school, architects were trained to draw on Classical, Renaissance, and Baroque styles, synthesizing them in designs for new buildings for the Industrial Age. The idea was that the new buildings would blend artfully with the older palaces, cathedrals, and civic buildings that dominated European city centers. Haussmann's ideas were widely influential and extensively copied.

Early in the twentieth century there emerged a different intellectual and artistic reaction to the pressures of industrialization and urbanization. This was the **Modern movement,** which was based on the idea that buildings and cities should be designed and run like machines. Equally important to the Modernists was that urban design should not simply reflect dominant social and cultural values but, rather, help to create a new moral and social order. The movement's best-known advocate was Le Corbusier, a Paris-based Swiss who provided the inspiration for technocratic urban design. Modernist buildings sought to dramatize technology, exploit industrial production techniques, and use modern materials and unembellished, functional design. Le Corbusier's ideal city (*La Ville Radieuse*—**Figure 11.20**) featured linear clusters of high-density, medium-rise apartment blocks, elevated on stilts and segregated from industrial districts; high-rise tower office blocks; and transportation routes—all separated by broad expanses of public open space.

After World War II this concept of urban design became pervasive, part of what became known as the International Style: boxlike steel-frame buildings with concrete and glass façades. The International Style was avant-garde yet respectable and, above all, comparatively inexpensive to build. This tradition of urban design, more than anything else, has imposed a measure of uniformity on cities around the world. Globalization has brought the appearance of International Style buildings in big cities in every part of the world. Furthermore, the International Style has often been the preferred basis for large-scale urban design projects around the world. One of the best examples is Brasilia, the capital of Brazil, founded in 1956 in an attempt to shift the country's political, economic, and psychological focus away from the past; differentiate it from the former colonial cities on the coast; and orient the country toward the future and the interior (see Chapter 6).

Modern urban design has had many critics, mainly on the grounds that it tends to take away the natural life and vitality of cities, replacing varied and human-scale environments with monotonous and austere settings. In response to this, historic preservation has become an important element of urban planning in every city that can afford it.

Figure 11.20 La Ville Radieuse The modern era and the advent of new transportation and construction technologies encouraged the utopian idea that cities could be built as efficient and equitable "machines" for industrial production and progressive lifestyles. One of the most famous and influential examples was *La Ville Radieuse* (1933), a visionary design by Swiss architect Le Corbusier. His vision was for the creation of open spaces through collectivized, high-density residential areas, strictly segregated from industrial areas and highways through a geometric physical plan. (After Le Corbusier, *La Ville Radieuse*. Paris: Editions de L'Architecture D'Aujourd'hui.)

Islamic Cities

Islamic cities provide good examples of how social and cultural values and people's responses to their environment are translated into spatial terms through urban form and the design of the built environment. Indeed, it is because of similarities in cityscapes, layout, and design that geographers are able to talk about the Islamic city as a meaningful category. It is a category that includes thousands of towns and cities, not only in the Arabian Peninsula and Middle East—the heart of the Islamic Empire under the prophet Muhammad (A.D. 570–632)—but also in regions into which Islam spread later: North Africa, coastal East Africa, South-Central Asia, and Indonesia. Most cities in North Africa and South-Central Asia are Islamic, while many elements of the classic Islamic city can be found in towns and cities as far away as Seville, Granada, and Córdoba in southern Spain (the western extent of Islam), Kano in northern Nigeria and Dar-es-Salaam in Tanzania (the southern extent), and Davao in the Philippines (the eastern extent).

The fundamentals of the layout and design of the traditional Islamic city are so closely attached to Islamic cultural values that they are to be found in the Qur'an, the holy book of Islam. Although urban growth does not have to conform to any overall master plan or layout, certain basic regulations and principles are intended to ensure Islam's emphasis on personal privacy and virtue, on communal well-being, and on the inner essence of things rather than on their outward appearance.

The most dominant feature of the traditional Islamic city is the *Jami*—the city's principal mosque (**Figure 11.21**). Located centrally, the mosque complex is not only a center of worship but also a center of education and the hub of a broad range of welfare functions. As cities grow, new,

Figure 11.21 Al Hussein mosque, Baghdad, Iraq The dominant feature of traditional Islamic cities is the *Jami,* or main mosque.

smaller mosques are built toward the edge of the city, each out of earshot from the call to prayer from the Jami and from one another. The traditional Islamic city was walled for defense, with several lookout towers and a *Kasbah,* or citadel (fortress), containing palace buildings, baths, barracks, and its own small mosque and shops.

Traditionally, gates controlled access to the city, allowing careful scrutiny of strangers and permitting the imposition of taxes on merchants. The major streets led from these gates to the main covered bazaars or street markets *(suqs)* (**Figure 11.22**). The suqs nearest the Jami typically specialize in the cleanest and most prestigious goods, such as books, perfumes, prayer mats, and modern consumer goods. Those nearer the gates typically specialize in bulkier and less valuable goods such as basic foodstuffs, building materials, textiles, leather goods, and pots and pans. Within the suqs, every profession and line of business had its own alley, and the residential districts around the suqs were organized into distinctive quarters, or *ahya',* according to occupation (or sometimes ethnicity, tribal affiliation, or religious sect).

Privacy is central to the construction of the Islamic city. Above all, women must be protected, according to Islamic values, from the gaze of unrelated men. Traditionally, doors must not face each other across a minor street, and windows must be small, narrow, and above normal eye level (**Figure 11.23**). Cul-de-sacs (dead-end streets) are used where possible to restrict the number of persons needing to approach the home, and angled entrances are used to prevent intrusive glances. Larger homes are built around courtyards, which provide an interior and private focus for domestic life.

The rights of others are also given strong emphasis, the Qur'an specifying an obligation to neighborly cooperation and consideration—traditionally interpreted as applying to a minimum radius of 40 houses. Roofs, in traditional designs, are surrounded by parapets to preclude views of neighbors' homes, and drainage channels are steered away from neighbors' houses. Refuse and wastewater are carefully recycled. Public thoroughfares were originally designed to be wide enough to allow two fully laden camels to pass each other and high enough to allow a camel and rider to pass through. The overall effect is to produce a compact, cellular urban structure within which it is possible to maintain a high degree of privacy (**Figure 11.24**).

Because most Islamic cities are located in hot, dry climates, these basic principles of urban design have evolved in conjunction with certain practical solutions to intense heat and sunlight. Twisting streets, as narrow as permissible, help to maximize shade, as does latticework on windows and the cellular, courtyard design of residential

Figure 11.22 The suq The *suq,* a covered bazaar or open street market, is one of the most important distinguishing features of a traditional Islamic city. Typically, a suq consists of small stalls located in numerous passageways. Many important suqs are covered with vaults or domes. Within them is typically a marked spatial organization, with stalls that sell similar products clustered tightly together, as in these copper shops in Marrakech, Morocco.

Figure 11.23 Islamic architecture In Islamic societies, the privacy of individual residences is paramount, and elaborate precautions are taken through architecture and urban design to ensure the privacy of women. Entrances are L-shaped and staggered across the street from one another. Windows are screened and often placed high above pedestrian access, as in this example in Jeddah, Saudi Arabia. Architectural details also reflect climatic influences: window screens and narrow, twisting streets help to maximize shade, while air ducts and roof funnels help create dust-free drafts.

areas. In some regions, local architectural styles include air ducts and roof funnels with adjustable shutters that can be used to create dust-free drafts.

All these features are still characteristic of Islamic cities, though they are especially clear in their old cores, or *medinas*. Like cities everywhere, however, Islamic cities also bear the imprint of globalization. Although Islamic culture is self-consciously resistant to many aspects of globalization, it has been unable to resist altogether the penetration of the world economy and the infusion of the Western-based culture of global metropolitanism. The result can be seen in international hotels, skyscrapers and office blocks, modern factories, highways, airports, and stores (**Figure 11.25**). Moreover, Islamic culture and urban design principles have not always been able to cope with the pressures of contemporary rates of urbanization, so the

Figure 11.24 Village housing in the Ziz Valley, Morocco Seen from above, the traditional Islamic city is a compact mass of residences with walled courtyards—a cellular urban structure within which it is possible to maintain a high degree of privacy.

Figure 11.25 Dubai, United Emirates Like cities almost everywhere, Islamic cities reflect the imprint of globalization. Economic globalization is reflected in international hotels and the offices of transnational corporations; cultural globalization is reflected in the presence of Western consumer products and advertisements for Western popular culture.

larger Islamic cities—such as Algiers, Cairo, Karachi, and Teheran—now share with other peripheral cities the common denominators of unmanageable size, shanty and squatter development, and low-income mass housing. In the next section we examine the problems of large and rapidly growing metropolises throughout the periphery.

Cities of the Periphery: Unintended Metropolises

The cities of the world-system periphery, often still referred to as Third World cities, are numerous and varied. What they have in common is the experience of unprecedented rates of growth driven by rural "push"—overpopulation and the lack of employment opportunities in rural areas—rather than the "pull" of prospective jobs in towns and cities. Faced with poverty in overpopulated rural areas, many people regard moving to a city much like playing a lottery: You buy a ticket (in other words, go to the city) in the hope of hitting the jackpot (in other words, landing a good job). As with all lotteries, most people lose, and the net result is widespread underemployment. **Underemployment** occurs when people work less than full time even though they would prefer to work more hours. Underemployment is difficult to measure with any degree of accuracy, but estimates commonly range from 30 to 50 percent of the employed workforce in peripheral cities.

Because of their rapid growth and high underemployment, it is among the peripheral metropolises of the world—Mexico City (Mexico), São Paulo (Brazil), Lagos (Nigeria), Mumbai (India; formerly Bombay), Dhaka (Pakistan), Jakarta (Indonesia), Karachi (Pakistan), and Manila (the Philippines)—that we can find contenders for the title of shock city of the early twenty-first century—the city that embodies the most remarkable and disturbing changes in economic, social, and cultural life (see Box 11.1: "Shock City': Lagos, Nigeria").

As we have seen, the typical peripheral metropolis plays a key role in international economic flows, linking provincial regions with the hierarchy of world cities and, thus, with the global economy. Within peripheral metropolises, this role results in a pronounced **dualism**, or juxtaposition in geographic space of the formal and informal sectors of the economy. This dualism is very evident in the built environment, with high-rise modern office and apartment towers and luxurious homes contrasting acutely with slums and shantytowns (**Figure 11.26**).

The Informal Economy

In many peripheral cities more than one-third of the population is engaged in the informal sector, and in some—for example Chennai, India; Colombo, Sri Lanka; Delhi, India; Guyaquil, Ecuador; and Lahore, Pakistan—more than one half. People who cannot find regularly paid work must resort to various ways of gleaning a living. Some of these ways are imaginative, some desperate, some pathetic. Examples range from street vending, shoe shining, craft work, and street-corner repairs to scavenging in garbage dumps (**Figure 11.27**). The informal sector consists of a broad range of activities that represent an important coping mechanism. For too many, however, coping means resorting to begging, crime, or prostitution. Occupations such as selling souvenirs, driving pedicabs, making home-brewed beer, writing letters for others, and dressmaking may seem marginal from the point of view of the global economy, but more than a billion people around the world must feed, clothe, and house themselves entirely from such occupations. In many peripheral cities more than half the popu-

Figure 11.26 Rio de Janeiro This photograph, looking toward Rio de Janeiro's famous Ipanema Beach, shows very clearly the dualism of peripheral metropolises, with shanty housing (favelas) in the foreground and luxury apartments near the beach.

lation subsists in this way. Across Africa, the International Labor Office estimates, informal-sector employment is growing 10 times faster than formal-sector employment.

In most peripheral countries the informal labor force includes children. In environments of extreme poverty, every family member must contribute something, and so children are expected to do their share. Industries in the formal sector often take advantage of this situation. Many firms farm out their production under subcontracting schemes that are based not in factories but in home settings that use child workers. In these settings, labor standards are nearly impossible to enforce. In the Philippines, for example, batches of rural children are ferried by syndicates to work in garment-manufacturing sweatshops in urban areas.

Despite this side of the picture, the informal sector has a few positive aspects. Pedicabs, for example, provide an affordable, nonpolluting means of transportation in crowded metropolitan settings. Garbage picking, while it may seem desperate and degrading in Western eyes, provides an important means of recycling paper, steel, glass, and plastic products. One study of Mexico City estimated that as much as 25 percent of the municipal waste ends up being recycled by the 15,000 or so scavengers who work over the city's official dump sites. This positive contribution to the economy, though, scarcely balances the lives of poverty and degradation experienced by the scavengers.

Urban geographers also recognize that the informal sector represents an important resource to the formal sector of peripheral economies. The informal sector provides a vast range of cheap goods and services that reduce the cost of living for employees in the formal sector, thus enabling employers to keep wages low. Although this network does not contribute to urban economic growth or help alleviate poverty, it does keep companies competitive within the context of the global economic system. For export-oriented companies, in particular, the informal sector provides a considerable indirect subsidy to production. We must recognize, too, that this subsidy is often passed on to

11.1 VISUALIZING GEOGRAPHY

"Shock City": Lagos, Nigeria

Kate Adikiwe lives in the suburban district of Olaleye, Lagos, once a small village whose residents grew herbs, fruits, and vegetables, fished, trapped, made palm wine, and processed palm oil. The village grew rapidly when a railway line was constructed through it and as Lagos grew outwards after independence. In the mid-1960s Olaleye had about 2,500 residents; today there are about 25,000. Within its small site of some 35 hectares (86 acres) is an enormous range of economic activities—a large market, beer parlors, nightclubs, brothels, a makeshift cinema, tailors, shoemakers, blacksmiths, tinkers, watch repairers, knife sharpeners, mechanics, battery chargers, and itinerant barbers and beauticians. Many of the women produce and sell a great variety of cooked foodstuffs, while many of the men work outside the district in factories or offices.

Kate is one of six children. Her father is a clerical worker in one of the city's department stores. Her mother is a seamstress, working from the house. The house itself contains 12 families, each having a single room and sharing the one kitchen, toilet, and bathroom in the building. One of Kate's jobs is to draw water from the nearby well each morning before school. The water is stored in plastic buckets in the living room until needed. After school Kate has to complete her homework and help her mother prepare food for the family. Most of the cooking is done on kerosene stoves in the passageway. After the meal, Kate and her older sister help their mother with sewing. They do not expect to be able to get jobs after school, so they are learning to become seamstresses.

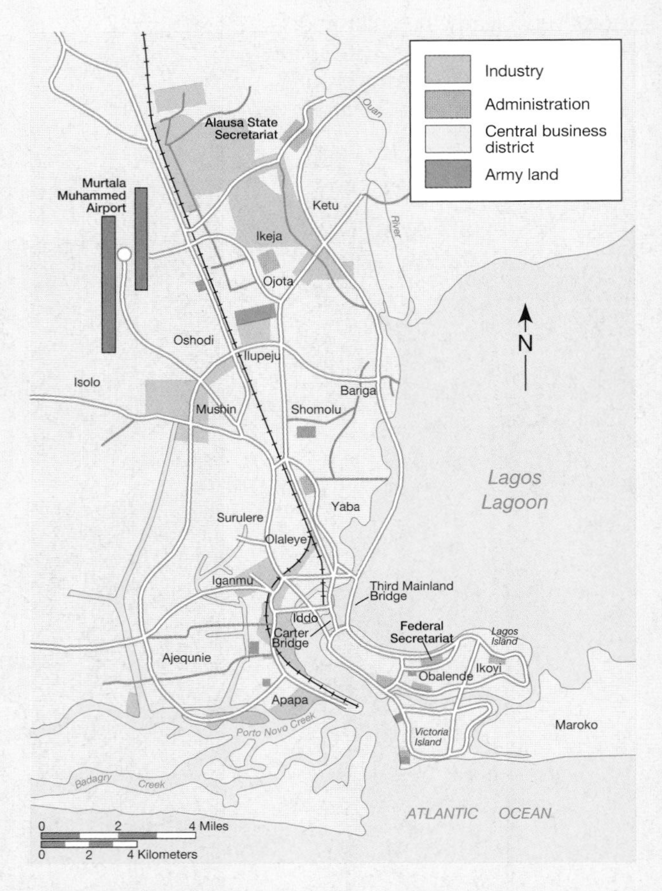

Lagos developed from an initial settlement at Iddo and on the northern shore of Lagos Island. Ikoyi, on Lagos Island, was laid out in 1918 as a government residential estate to house colonial officials. Most of the city's growth, however, has been unplanned and irregular, with swamps, coves, and canals impeding efficient development.

(After M. Peil, *Lagos*. London: Belhaven Press, 1991, p. 23.)

Lagos, like most metropolises in the world's periphery, grew relatively slowly until quite recently. The combination of the demographic transition, political independence, and an economic boom stimulated by the discovery of oil reserves in southeastern Nigeria triggered an explosive growth in population. Because of its difficult site on sand spits and lagoons, this growth has resulted in an irregular sprawl and, in the central area, a density of population higher than that of Manhattan Island in New York.

The cityscape on Lagos Island reflects both residential congestion and the postcolonial development of the city as a peripheral metropolis with important corporate functions.

For many people, life in the unintended metropolis is a matter of survival. This leads to a tremendous variety of informal economic activities, from street vending to home-brewed beer, and from prostitution to drug peddling. This photograph shows the most common form of informal activity: street trading, which takes place on almost every unoccupied sidewalk, street, or unclaimed space.

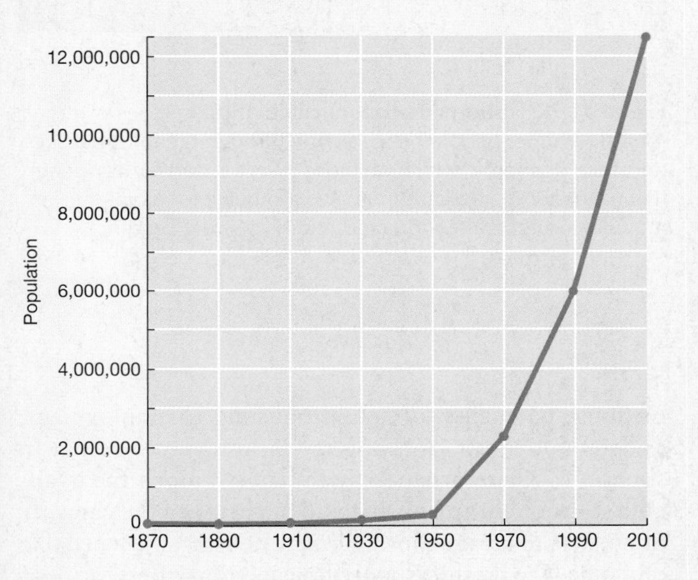

Population growth has far outstripped the city's capacity to deal with the daily movement of people, a problem that is worsened because the central city is trapped on an island site, with limited access by road bridges.

Overwhelmed by an unprecedented rate of urbanization, an economy that cannot provide regularly paid employment for a significant proportion of its residents, and a municipal government that has neither the financial resources nor the personnel to deal with the problems, Lagos has become emblematic of the problems of overurbanization. Shanty housing is a direct consequence of widespread poverty; open sewers are a consequence of limited or nonexistent municipal resources.

Figure 11.27 Informal economic activities In cities where jobs are scarce, people have to cope through the informal sector of the economy, which includes a very broad variety of activities, including agriculture (backyard hens, for example), manufacturing (craft work), and retailing (street vending).

consumers in the core regions in the form of lower prices for goods and consumer products made in the periphery.

Consider, for example, the paper industry in Cali, Colombia. This industry is dominated by one company, Cartón de Colombia, which was established in 1944 with North American capital and subsequently acquired by the Mobil Oil Company. Most of the company's lower quality paper products are made from recycled waste paper, and 60 percent of this waste paper is gathered by local garbage pickers. There are 1,200 to 1,500 garbage pickers in Cali. Some work the city's municipal waste dump, some work the alleys and yards of shopping and industrial areas, and some work the routes of municipal garbage trucks, intercepting trash cans before the truck arrives. They are part of Cali's informal economy, for they are not employed by Cartón de Colombia nor do they have any sort of contract with the company or its representatives. They simply show up each day to sell their pickings. This way the company can avoid paying both wages and benefits, while dictating the price it will pay for various grades of waste paper. The company can operate profitably while keeping the price of its products down—the arrangement is a microcosm of core-periphery relationships.

Slums of Hope, Slums of Despair

The informal labor market is directly paralleled in informal shantytowns and squatter housing: Because so few jobs with regular wages exist, few families can afford rent or house payments for sound housing. Unemployment, underemployment, and poverty mean overcrowding. In situations where urban growth has swamped the available stock of cheap housing and outstripped the capacity of builders to create affordable new housing, the inevitable outcome is makeshift shanty housing that offers, at best, precarious shelter. Such housing has to be constructed on the cheapest and least desirable sites. Often this means building on bare rock, over ravines, on derelict land, on swamps, or on steep slopes. Nearly always it means building without any basic infrastructure of streets or utilities. Sometimes it means adapting to the most extreme ecological niches, as in Lima, where garbage pickers actually live on the waste dumps, or in Cairo, where for generations the poor have adapted catacombs and cemeteries into living spaces (**Figure 11.28**). In many cities more than half of the housing is substandard. The United Nations estimated in 2004 that more than 1.1 billion people worldwide live in inadequate housing in urban areas.

Globalization and the accompanying trend toward neoliberal economic policies have intensified problems of poverty and slum housing in many cities. The United Nations Human Settlement Programme reported in 2004 that:

> The reduction of fiscal deficits has partly entailed reduction of public expenditure through downsizing of the civil service and privatization of state enterprises, resulting in the laying-off of large numbers of public-sector employees in many coun-

> tries. Trade liberalization has often resulted in the closure of some industries that have been unable to compete against cheap imports, again leading to massive retrenchment and higher unemployment levels. Rising urban unemployment and increasing poverty have forced large numbers of the urban poor into the informal sector. Underpaid formal-sector employees have also entered the informal sector as a survival strategy. This, in turn, leads to the erosion of the tax base and decreasing ability of national and local governments to assist the poor through social and basic services. The removal of price controls on subsistence goods, and increased utility charges through privatization . . . have resulted in rising inequalities and increasing poverty . . . As a result, cities end up with their prime resources largely appropriated by the affluent.[3]

Faced with the growth of slums, the first response of many governments has been to eradicate them. Encouraged by Western development economists and housing experts, many cities sought to stamp out unintended urbanization through large-scale eviction and clearance programs. In Caracas (Venezuela), Lagos (Nigeria), Bangkok (Thailand), Kolkata (India; formerly Calcutta), Manila (the Philippines), and scores of other cities in the periphery, hundreds of thousands of shanty dwellers were ordered out on short notice and their homes bulldozed to make way for public works, land speculation, luxury housing, urban renewal, and, on occasion, to improve the appearance of cities for special visitors. Seoul, South Korea, has probably had the most forced evictions of any city in the world. Since 1966, millions of people have been forced out of accommodations that they owned or rented as part of a sustained government clean-up campaign. Between 1983 and 1988, in preparation for the 1988 Olympic Games, nearly 750,000 people lost their homes in a beautification program.

Yet Seoul, more than most other cities, could afford to build new low-income housing to replace the demolished neighborhoods. Most peripheral cities cannot do so, which means that displaced slum dwellers have no option but to create new squatter and shanty settlements elsewhere in the city. Most cities, in fact, cannot evict and demolish fast enough to keep pace with the growth of slums caused by in-migration. The futility of slum clearance has led to a widespread reevaluation of the wisdom of such policies. The thinking now is that informal-sector housing should be seen as a rational response to poverty. Shanty and squatter neighborhoods can not only provide affordable shelter but also function as important reception areas for migrants to the city, with supportive communal organizations and informal employment opportunities that help them to adjust to city life. They can, in other words, be "slums of hope." City authorities, recognizing the positive functions of informal housing and self-help improvements, are now increasingly disposed to be tolerant and even helpful toward squatters rather than sending in police and municipal workers with bulldozers.

In fact, many informal settlements are the product of careful planning. In parts of Latin America, for example, it is common for community activists to draw up plans for invading unused land, then quickly build shanty housing before land owners can react. Part of the activists' strategy is usually to organize a critical mass of people large enough to be able to negotiate with the authorities to resist eviction. It is also common for activists to plan their invasions for public holidays so that the risk of early

[3]United Nations Human Settlements Programme, *The State of the World's Cities 2004/2005. Globalization and Urban Culture.* Sterling, VA: Earthscan, 2004, p. 102

Figure 11.28 Garbage picking Marisol and her mother, Eloisa, collect tins from the municipal dump in Matamoros, Mexico, where Marisol once found a dead woman. That event prompted Eloisa to finally move her children to "the other side" in order to join her husband, Vinicio, already a legal resident working in the United States.

detection is minimized. As the risk of eviction diminishes over time, some residents of informal housing are able to gradually improve their dwellings through self-help (**Figure 11.29**).

Nevertheless, there are many shanty and squatter neighborhoods where self-help and community organization do not emerge. Instead, grim and desperately miserable conditions prevail. These are "slums of despair," where overcrowding, lack of adequate sanitation, and lack of maintenance lead to shockingly high levels of ill health and infant mortality, and where social pathologies are at their worst. Consider, for example, the squatter settlement of Chheetpur in the city of Allahabad, India. The settlement's site is subject to flooding in the rainy season, and a lack of drainage means stagnant pools for much of the year. Two standpipes (outdoor taps) serve the entire population of 500, and there is no public provision for sanitation or the removal of household wastes. In this community, most people have food intakes of less than the recommended minimum of 1,500 calories a day; 90 percent of all infants and children under four have less than the minimum calories needed for a healthy diet. More than half of the children and almost half the adults have intestinal worm infections. Infant and child mortality is high—though nobody knows just how high—with malaria, tetanus, diarrhea, dysentery, and cholera as the principal causes of death among under-fives.

Transport and Infrastructure Problems

The informal labor market is also reflected in cities' inability to provide a basic infrastructure of highways, transportation, schools, utilities, and emergency services. Because the informal sector yields no tax revenues, municipal funds are insufficient to provide an adequate infrastructure or to maintain a safe and sanitary environment.

Even though the governments of peripheral cities typically spend nearly all of their budgets on transport and infrastructure in their race to keep up with population growth, conditions are bad and getting rapidly worse. Peripheral cities have always been congested, but in recent years the modernizing influence of formal-sector activities has turned the congestion into near gridlock. In many of the world's peripheral and semiperipheral metropolises, sharp increases have occurred in the availability and use of automobiles. One of the most dramatic examples is Taipei, Taiwan, where the number of automobiles increased from about 11,000 in 1960 to over 1.35 million in 2000. Not only are there more people and more traffic but the changing spatial organization of peripheral cities has increased the need for transportation. Traditional patterns of land use have been superseded by the agglomerating tendencies inherent in modern industry and the segregating tendencies inherent in modernizing societies. The greatest single change has been the separation of home from work, however, which has meant a significant increase in commuting.

In spite of many innovative responses to urban transportation needs, road transportation in many cities is breaking down, with poorly maintained roads, traffic jams, long delays at intersections, and frequent accidents. Many governments have invested in expensive new freeways and street-widening schemes, but because they tend to focus on city centers (which are still the settings for most jobs and most services and amenities), they ultimately fail, emptying vehicles into a congested and chaotic mixture of motorized traffic, bicycles, animal-drawn vehicles, and hand-drawn carts. Some of the worst traffic tales come from Mexico City—where traffic backups total more than 90 kilometers (60 miles) each day, on average—and Bangkok, where the 15-mile trip into town from Don Muang Airport can take three hours. In São Paulo,

Figure 11.29 Self-help as a solution to housing problems Self-help is often the only solution to housing problems because wages are so low and so scarce that builders cannot construct even the most inexpensive new housing and make a profit, and because municipalities cannot afford to build sufficient quantities of subsidized housing. One of the most successful ways of encouraging self-help housing is for municipal authorities to create the preconditions by clearing sites, putting in the footings for small dwellings, and installing a basic framework of water and sewage utilities. This "sites-and-services" approach has become the mainstay of urban housing policies in many peripheral countries. This photograph shows self-help housing in Ndola, Zambia.

Brazil, gridlock can span 160 kilometers (100 miles), rush-hour traffic jams average 85 kilometers (53 miles) in length, and 15-hour traffic jams are not unusual. The costs of these traffic backups are enormous. The annual costs of traffic delays in Singapore have been estimated at $305 million; in Bangkok, Thailand, they have been estimated at $272 million—the equivalent of around one percent of Thailand's gross national product.

Water supplies and sewerage also present acute problems for many cities (**Figure 11.30**). Definitions of what constitutes an adequate amount of safe drinking water and sanitation vary from country to country. Although many governments classify the existence of a water tap within 100 meters (328 feet) of a house as "adequate," such a tap does not guarantee that the household will be able to secure enough water for good health. Communal taps often function only a few hours each day, so residents must wait in long lines to fill even one bucket. In Rajkot, India, a city of 600,000 people, piped water routinely runs for only 20 minutes each day.

The World Bank estimates that around 65 percent of urban residents worldwide in less developed countries have access to a satisfactory water source, and only about 40 percent are connected to sewers (90 percent of which discharge their waste untreated into a river, a lake, or the sea). Hundreds of millions of urban dwellers have no alternative but to use contaminated water—or at least water whose quality is not guaranteed. A small minority, usually the residents of the most affluent neighborhoods, have water piped into their homes, while the majority have piped water nearby, which has to be collected. Those not served are obliged to carry water in small quantities over long distances or to use water from streams or other surface sources (**Figure 11.31**). In Colombo, Sri Lanka, about one-third of all houses have indoor piped water, and another one-fourth have piped water outside. In Dar es Salaam (Tanzania), Kinshasa (Democratic Republic of the Congo), and many other peripheral cities, almost half the population has no access to piped water, either indoors or outdoors.

Figure 11.30 Infrastructure problems Low incomes mean that city governments are unable to raise sufficient revenues through local taxes, which in turn means that infrastructure is neglected. Putting water and sewage lines into neighborhoods that were built without these basic utilities is arduous and expensive, yet without basic utilities, public health is seriously threatened. This photograph shows water lines being installed in a low-income neighborhood of Cartagena, Colombia.

In many cities, including Bangkok, Bogotá, Dar es Salaam, Jakarta, Karachi, and São Paulo, only one-fourth to one-third of all garbage and solid waste is collected and removed—the rest is partially recycled informally, tipped into gullies, canals, or rivers, or simply left to rot. Sewage services are just as bad. In Latin America, for example, only about 2 percent of collected sewage receives any treatment. In Mexico more than 90 percent of wastewater treatment plants are nonfunctional, and in cities like Bogotá, Buenos Aires, Mexico City, and Santiago some 50 to 60 million cubic meters of mostly untreated sewage is discharged every day into nearby bodies of water. São Paulo has over 1,600 kilometers (1,000 miles) of open sewers, and raw sewage from the city's slums drains into the Billings reservoir, a major source of the city's drinking water. In Bangkok less than 5 percent of the population is connected to a sewer system; human wastes are generally disposed of through septic tanks and cesspools with their effluents, as well as waste water from sinks, laundries, baths, and kitchens, discharged into stormwater drains or canals. Jakarta has no waterborne sewage system at all. Septic tanks serve about one-quarter of the city's population; others must use pit latrines, cesspools, and ditches along the roadside. A survey of over 3,000 towns and cities in India found that only eight had full sewage-treatment facilities, and another 209 had partial treatment facilities. Along one river, the Ganga (better known to English speakers as the Ganges), 114 towns and cities dump untreated sewage into the river every day, along with waste from DDT factories, tanneries, paper and pulp mills, petrochemical and fertilizer complexes, and other industrial pollutants. Each day the Yamuna River picks up 200 million liters of untreated sewage and 20 million liters of industrial effluents as it passes through Delhi. In China, inadequate sewerage and the meager 4.5 percent coverage of municipal wastewater plants have resulted in widespread water quality deterioration. Shanghai has had to move its water supply intake 40 kilometers (25 miles) upstream at a cost of US $300 million because of degradation of river water quality around the city.

Figure 11.31 Water-supply problems Many peripheral cities have grown so quickly and under such difficult conditions that large sections of the population do not have access to supplies of clean water. Where there is a public supply—a well or an outdoor standpipe—water consumption is limited by the time and energy required to collect water and carry it home. It is not rare for 500 or more people to have to share a single pump. Because low-income people work very long hours, the time spent waiting in line for water and then transporting buckets home is time that could have been used in earning an income. Limited quantities of water mean inadequate supplies for personal hygiene and for washing food, cooking utensils, and clothes. Where public agencies provide no water supply—as is common in squatter settlements—the poor often obtain water from private vendors and can pay 20 to 30 times the cost per liter paid by households with piped supplies. Water vendors probably supply about one-fourth of the population of peripheral metropolises.

These problems provide opportunities for the informal sector, however. Street vendors, who get their water from private tanker and borehole operators, sell water from two- or four-gallon cans. They typically charge five to 10 times the local rate set by public water utilities; in some cities they charge 60 to 100 times as much. Similarly, many cities have evolved informal-sector mechanisms for sewage disposal. In many Asian cities, for example, human waste is removed overnight by handcart operators. Unfortunately, it is rarely disposed of properly and often ends up polluting the rivers or lakes from which the urban poor draw their water.

Environmental Degradation

With pressing problems of poverty, slum housing, and inadequate infrastructure, it is not surprising that peripheral cities are unable to devote many resources to environmental problems. Because of the speed of population growth, these problems are escalating rapidly. Industrial and human wastes pile up in lakes and lagoons and pollute long stretches of rivers, estuaries, and coastal zones. Groundwater is polluted through the leaching of chemicals from uncontrolled dump sites, and the forests around many cities are being denuded by the demand of cities for timber and domestic fuels. This environmental degradation is, of course, directly linked to human health. People living in such environments have much higher rates of respiratory infections, tuberculosis, and diarrhea, and much shorter life expectancies than people living in surrounding rural communities. Children in squatter settlements may be 50 times as likely to die before the age of five as those born in affluent core countries.

In addition, air pollution has escalated to very harmful levels in many cities. With the development of a modern industrial sector and the growth of automobile ownership, but without enforceable regulations on pollution and vehicle emissions, tons of lead, sulfur oxides, fluorides, carbon monoxide, nitrogen oxides, petrochemical oxidants, and other toxic chemicals are pumped into the atmosphere every day in large cities. The burning of charcoal, wood, and kerosene for fuel and cooking in low-income neighborhoods also contributes significantly to dirty air. In cities where sewerage systems are deficient, the problem is compounded by the presence of airborne dried fecal matter. Worldwide, according to U.N. data, more than 1.1 billion people live in urban areas where air pollution exceeds healthful levels.

A U.N. study of 20 megacities found that every one of them had at least one major pollutant at levels exceeding World Health Organization (WHO) guidelines. Fourteen of the 20 had *two* major pollutants exceeding WHO guidelines, and seven had *three*. Such pollution is not only unpleasant but dangerous. In Manila, the Philippines, the Asian Development Bank found levels of suspended particulate matter in the air to be 200-400 percent above guideline levels. In Mexico City, where sulfur dioxide and lead concentrations are two to four times higher than the WHO guidelines, and where national ozone levels are exceeded on more than half of the days throughout the year, seven in 10 newborns have dangerously high levels of lead in their bloodstream. WHO studies demonstrate that it is unhealthy for human beings to breathe air with more than 100 to 120 parts per billion (ppb) of ozone contaminants for more than one day a year. Yet Mexico City residents breathe this level, or more, for over 300 days a year. In Bangkok, Thailand, where air pollution is almost as severe as in Mexico City, research has shown that lead-bearing air pollutants reduce children's IQ by an average of 3.5 points per year until they are seven years old. It has also been estimated that Bangkok's pall of dust and smoke causes more than 1,400 deaths each year and $3.1 billion each year in lost productivity resulting from traffic and pollution-linked illnesses.

Proximity to industrial facilities, often the result of the need and desire of the poor to live near places of employment, poses another set of risks. A notorious accident at the Union Carbide factory in Bhopal, India, in 1984 caused 2,988 deaths and more than 100,000 injuries, mostly among residents of the shantytowns near the chemical factory.

NEW PATTERNS: THE POLYCENTRIC METROPOLIS

Traditional patterns of land use and spatial organization in many parts of the world are being transformed by the local effects of splintering urbanism. Economic and cultural globalization, together with the uneven evolution of networked infrastructures of information and communications technologies, is forging new landscapes of innovation, economic development, and cultural transformation, while at the same time intensifying social and economic inequalities between the fast world and the slow world. As we noted in Chapter 10, these trends are most pronounced in world cities and major regional metropolitan centers, particularly in core countries. Nevertheless, fragments of this splintering urbanism are increasingly evident throughout the world as new technologies, new forms of economic organization, and new sociocultural norms spread through the global urban system.

North American cities were the first to break away from traditional patterns. Geographer Pierce Lewis coined the term "galactic metropolis" to capture the disjointed and decentralized urban landscapes of late-twentieth-century North America.[4] The galactic metropolis evolved from the traditional pattern of concentric zones as secondary business districts and commercial strips emerged in the suburbs to cater to neighborhood shopping and service needs and decentralized industrial districts developed around airports and freeway interchanges. Subsequently, edge cities grew into suburban hubs of shops and offices that sometimes overshadow the old central business districts. **Edge cities** are nodal concentrations of shopping and office space situated on the fringes of metropolitan areas, typically near major highway intersections. Tysons Corner, Virginia, just outside Washington's beltway, provides a good example (**Figure 11.32**).

The result is a polycentric metropolitan structure that now has variants around the world. Geographer Peter Hall has identified six common types of nodes within the polycentric metropolis:[5]

- ***The traditional downtown center,*** based on walking distances and served by a radial transportation center. The hub of the traditional metropolis, it has become the setting for the oldest informational services: banking, insurance, and government. Examples include the City of London, Châtelet–Les Halles (Paris), lower Manhattan, and Maronouchi/Otemachi (Tokyo).
- ***Newer business centers,*** often developing in an old prestigious residential quarter and serving as a setting for newer services such as corporate headquarters, the media, advertising, public relations, and design. Examples include London's West End, the 16th Arrondissment in Paris, midtown Manhattan, and Akasaki/Roppongi (Tokyo).
- ***Internal edge cities,*** resulting from pressure for space in traditional centers and speculative development in nearby obsolescent industrial or transportation sites. Examples include London's Docklands, La Défense (Paris), and Shinjuku (Tokyo).
- ***External edge cities,*** often located on an axis with a major airport, sometimes adjacent to a high-speed train station, always linked to an urban freeway system. Examples include Washington's Dulles corridor, London's Heathrow district, the O'Hare area in Chicago, Schipol (Amsterdam), and Arlanda (Stockholm).
- ***Outermost edge-city complexes*** for back offices and R&D operations, typically near major train stations 30 to 50 kilometers (18.6 to 31 miles) from the main core. Examples include Reading (outside London); St. Quentin-en-Yvelines (Paris); Greenwich, Connecticut (outside New York); and Shin-Yokohama (Tokyo).

[4]P. Lewis, "The Galactic Metropolis." In R. Platt and G. Macuriko, eds., *Beyond the Urban Fringe*. Minneapolis: University of Minnesota Press, 1983, pp. 23–49.

[5]P. Hall, "Global City-Regions in the Twenty-first Century." In A. J. Scott, ed., *Global City-Regions: Trends, Theory Policy.* New York: Oxford University Press, 2001, pp. 59–77.

Figure 11.32 Tysons Corner, Virginia An edge city located on the beltway outside Washington, D.C., Tysons Corner is an unincorporated area that contains 45,000 residents and over 100,000 jobs. Tysons Corner does not exist as a postal address: residents' mail must go either to Vienna or McLean, Virginia. But this anonymous city contains a huge concentration of commercial space (the eighth largest of all downtown CBDs in the United States in 2000), including more than 27 million square feet of office space, several million square feet of retail space, nine major department stores, more than 3,500 hotel rooms, and parking for more than 90,000 cars.

- *Specialized subcenters,* usually for education, entertainment, and sporting complexes, exhibition and convention centers. These take a great variety of forms and locations. Some are on reclaimed or recycled land close to the traditional core; some are older centers, formerly separate and independent, that have become progressively embedded in the wider metropolitan area.

The largest of the world's polycentric metropolises have become "100-mile cities"—metropolitan regions that are literally 100 miles or so across, consisting of a loose coalition of urban realms, or economic subregions bound together through urban freeways. In some regions, clusters of networked, polycentric metropolises have developed into cohesive "megapolitan" regions (see Box 11.2: "Megapolitan Regions."

Sprawl

Inherent in the polycentric metropolis and endemic to most contemporary urbanization is suburban sprawl. The United States is the exemplar (**Figure 11.33**). Between 1985 and 2000, when the population of U.S. metropolitan areas increased by 17 percent, about 25 million acres of farmland and open space (roughly the size of Indiana) was developed around these metropolitan areas: a 47 percent increase in developed land. Since 1985, the 100 largest urbanized areas have sprawled out over an additional 37,670 square kilometers (14,545 square miles). In the polycentric metropolis, suburbs are no longer just bedroom communities for workers commuting to traditional downtowns. Rather, many are now strong employment centers serving a variety of economic functions in their regions. As the polycentric metropolis has evolved, suburbs have become more economically and physically diverse.

At one end of the continuum lie suburbs built in the early or middle twentieth century that are experiencing central citylike challenges—aging infrastructure, deteriorating schools and commercial corridors, and inadequate housing. At the other end are the **boomburbs** of the western United States: new tracts of sprawling, low-density, and auto-dependent suburbs that are growing at a feverish pace at the fringe of metropolitan areas.

In many metropolitan areas the changing face of suburbia is fueling an intense debate about the quality, pace, and shape of growth. Central to the debate is the idea of smart growth. **Smart growth** is a package of suburban land-use planning principles designed to curb sprawl. Its advocates claim that growth restrictions raise the quality of life, increase the efficiency of urban infrastructure and protect the environment. Among the key principles of smart growth are the following:

- preserving large areas of open space and protecting the quality of the environment by setting aside large fringe areas where development is prohibited;
- redeveloping inner suburbs and infill sites with new and renovated structures to make them more attractive to middle- and upper income households;
- reducing dependency on private automotive vehicles—especially one-person cars—by requiring higher density development, clustering high density around transit stops, raising gas taxes, and increasing public investment in light-rail transit systems;
- encouraging innovative urban design and zoning regulations that create pedestrian-friendly communities, mixed land uses, and commercial centers located at transit stops;
- creating a greater sense of community within individual localities and a greater recognition of regional interdependence and solidarity.

Smart-growth advocates include anti- or slow-growth and environmental organizations, together with central city and inner suburban leaders and interest groups, such

Figure 11.33 Suburban sprawl Contemporary urban development in the United States is characterized above all by suburban sprawl. This photograph shows a new subdivision layout encroaching on farmland outside East Phoenix, Arizona.

as mayors, downtown business groups, and community-based organizations. Opposed to smart-growth principles are most developers, homebuilders, major landowners, and chambers of commerce. Their objections to smart growth are partly based on the contention that smart growth costs more, and partly on an ideological objection to planning controls that inhibit opportunities for profit through real estate development.

Resistance to smart growth and debate as to its effectiveness has meant that, in practice, sprawl continues to gather pace, consuming ever greater amounts of agricultural land. The scale of the development industry is now such that suburban sprawl occurs in large increments, cutting and within months filling swathes of rural land with residential subdivisions, condominium complexes, bleak access roads, strip malls, parking lots, and office parks. Most of the architecture and urban design is without merit, adding up to what architect Rem Koolhaas has called "Generica"[6] (**Figure 11.34**). Following the logic of a fast return on investment and flexibility in use, most commercial structures are simple boxes, while economies of scale dictate a cookie-cutter approach for all but the most upscale residential subdivisions—where "monster homes," "starter castles," and "McMansions" take over as the norm.

[6]R. Koolhaas, *Mutations*. Bordeaux: ACTAR, 2001, p. 524.

Packaged Landscapes

The generic landscapes of the polycentric metropolis are increasingly packaged in order to appeal to particular submarkets. As we have seen (Chapter 10), globalization has produced a splintering urbanism that requires the development of several kinds of functional enclaves within metropolitan areas that are active nodes within the global urban system. Globalization has also contributed to the emergence of a postmodern culture in which the symbolic properties of places and material possessions have assumed unprecedented importance, with places becoming important objects of consumption (Chapter 6).

In this new economic and cultural context, packaging has become an important marketing strategy. So, for example, the developers of business parks bundle office space with day-care centers, fitness centers, and integrated retail and entertainment spaces, surrounded by lush landscaping and perhaps a nine-hole golf course. Shopping malls are themed and packaged with movie theaters and dining, exhibit, and performance spaces. Condominium complexes are packaged with high-bandwidth Internet access, enhanced telephone services, movies on demand, party centers, pools, and fitness centers. Private, master-planned residential communities are packaged with security systems, concierge services, bike trails, "town" centers, and even elementary schools. Increasingly, master-planned communities are themed and packaged in different ways in order to appeal to different market segments and

11.2

WINDOW ON THE WORLD

Megapolitan Regions

The combination of rapid growth and massive decentralization has transformed once distant cities into galaxies and corridors of linked urban space. The scale of urbanization and economic integration is now such that some regions effectively consist of networks of polycentric metropolitan areas. Bound together by goods movement, business linkages, cultural ties, commuting patterns, and physical environment, these are the world's most important functional regions, the primary geographic units of a globalizing world economy.

Figure 11.A shows the ten megapolitan regions identified within the United States by researchers at the Metropolitan Institute. Between them, they account for almost 70 percent of the U.S. population in less than 20 percent of the land area. Comparable megapolitan regions include Japan's Tokaido megalopolis (**Figure 11.B**), the European Union's "global integration zone" that runs from London to Hamburg to Munich to Milan to Paris and back to London, the Piemonte-Lombardia region of northern Italy, and the Guangdong region of South China.

Megapolitan areas vary in spatial form. Some, such as the Tokaido megalopolis, the Northeast region of the United States, and the I-35 Corridor (**Figure 11.C**), show a clear corridor or linear form. The I-35 Corridor includes a string of polycentric metropolitan areas running from San Antonio, Texas, in the south to Kansas, City, Missouri, in the north. This linear region has a history of interconnectivity

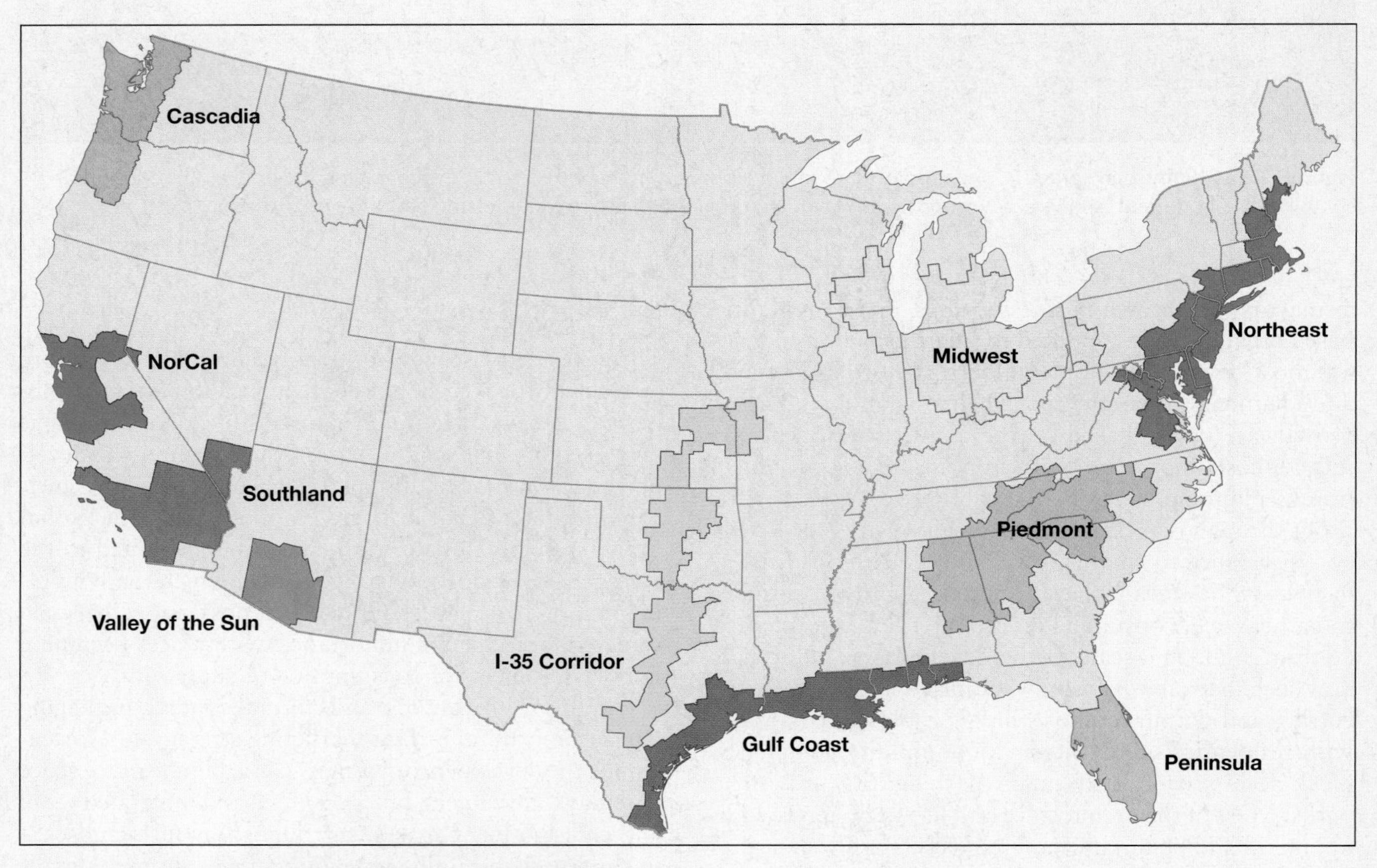

Figure 11.A The Megapolitan regions of the United States

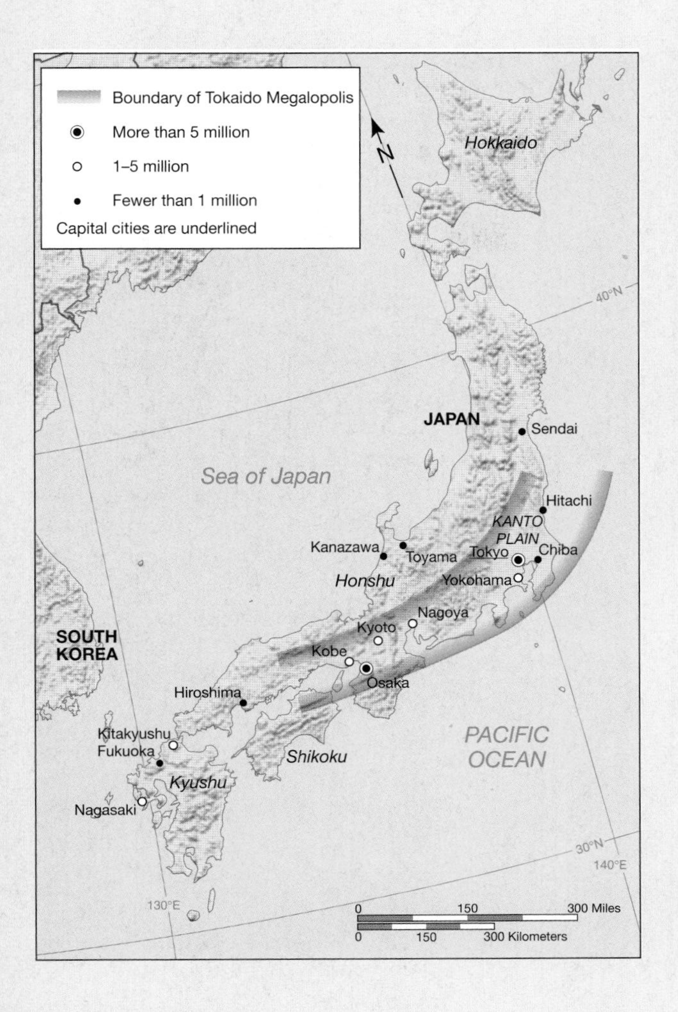

dating back to the Chisholm Trail. Because of increased trade and goods handling due to the North American Free Trade Agreement (NAFTA), this part of I-35 is now one of the busiest roads in the country. Other megapolitan areas, like the Piemonte-Lombardia region of northern Italy, the U.S. Midwest, and the Piedmont of the southeastern United States (**Figure 11.D**) spread out into vast urban galaxies, with major metropolitan areas connected by a web of limited-access highways and functionally linked to a net of smaller cities, micropolitan regions, and urbanized areas.

Based on R. E. Lang and D. Dhavale, "Beyond Megalopolis: Exploring America's new 'Megapolitan' Geography," *Metropolitan Institute Census Report Series*, 05:01, 2005.

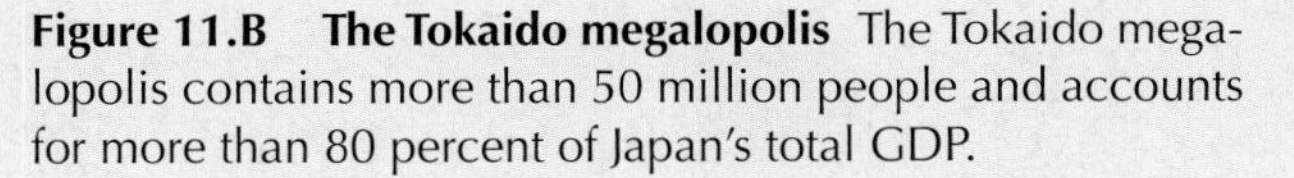

Figure 11.B The Tokaido megalopolis The Tokaido megalopolis contains more than 50 million people and accounts for more than 80 percent of Japan's total GDP.

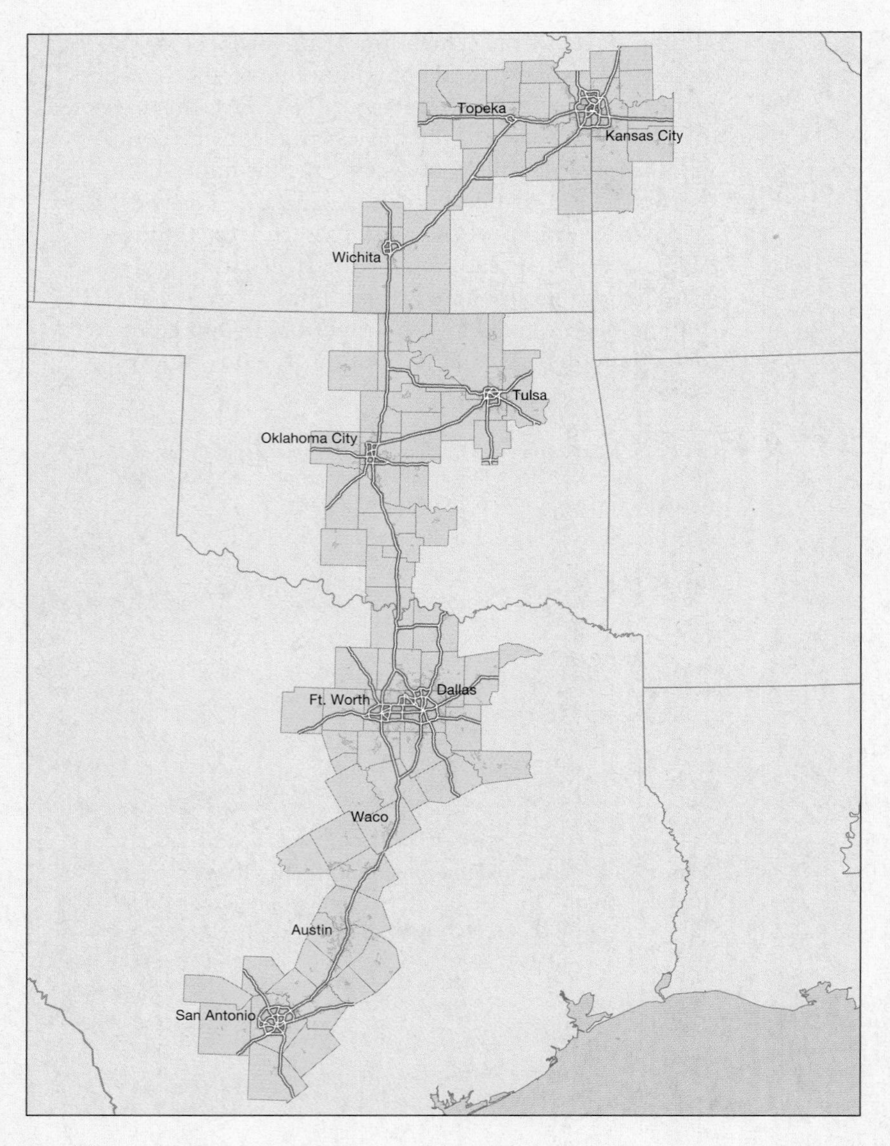

Figure 11.C The I-35 megapolitan corridor The light pink shaded area shows the region's megapolitan counties, while the darker red zones indicate the urbanized areas. The dark black lines are the interstate highways, and the light ones are the county boundaries.

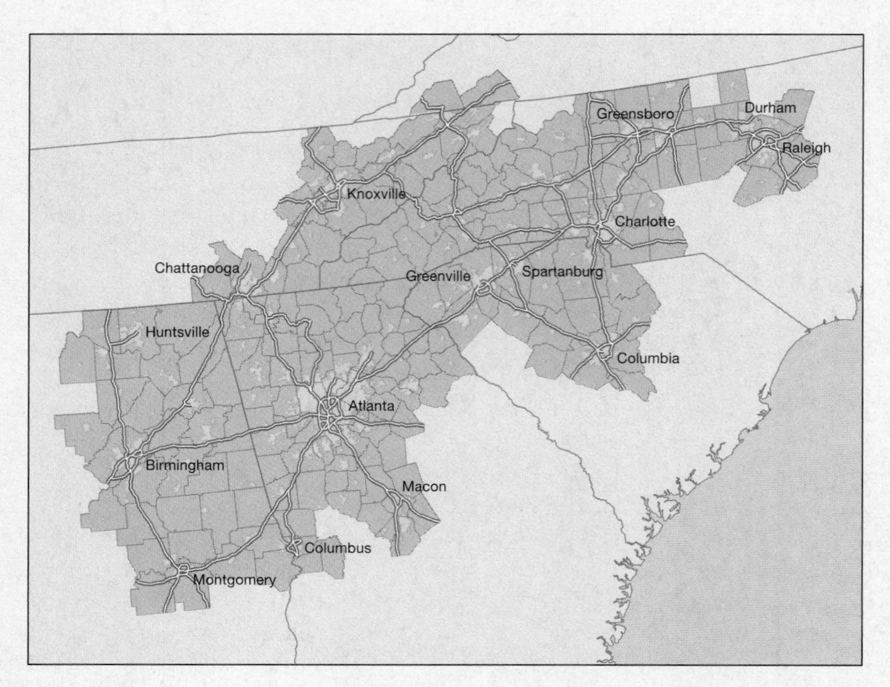

Figure 11.D The Piedmont megapolitan galaxy

A. Stamford, CT

B. Houston, TX

Figure 11.34 Generica American attitudes toward investment in the built environment, together with poor planning and weak regulations, have led to a generic landscape of office buildings, warehouses, shopping malls, and residential sprawl.

lifestyle communities. Some, like Anthem, north of Phoenix, Arizona (**Figure 11.35**), are packaged to appeal to young families. Anthem feels more like a luxury holiday resort than a town. It includes a water park with water slides, a children's railway, hiking trails, tennis courts, a rock-climbing wall, two golf courses, several spotless parks, a supermarket mall, two churches, a school, and a country club. Some master-planned residential communities are packaged to appeal to affluent retired people: The Del Webb development corporation's marketers have identified early-retiring baby-boomers—"Zoomers"—as the target market for their latest Sun City development, packaged accordingly with Starbucks cafés, Internet access, and multi-gyms, as well as the usual tennis courts, pools, and golf courses. Some are packaged as "green" communities or smart-growth developments. Others have a narrower focus: Front Sight, Nevada, is under construction with a guns-and-ammo theme, featuring streets with names like Second Amendment Drive and Sense of Duty Way, target-shooting ranges, a pro shop stocked with weapons, a martial arts gym, a defensive driving track, a kindergarten-through-12th-grade school where teachers will be allowed to carry concealed firearms, and sales inducements that include an Uzi machine gun and a game-hunting safari in Africa.

Marketing packaged landscapes like these has been very successful in metropolitan regions around the world. In Istanbul, Turkey, communities in the edge cities of Esenkent and Bogazköy are modeled on American suburbia. In Manila, in the Philippines, private master-planned communities are packaged with privatized infrastructure networks of roads, drainage, water supplies, power, and telecommunications and designed and marketed as fragments of urban Europe, with names like "Brittany" and "Little Italy." In the United States some 47 million people—one in six of the total population—live in 230,000 privately planned residential communities, and half the new home sales in big cities are in these communities.

The result is a mosaic of packaged developments and mega-projects: refurbished heritage and cultural zones, waterfront redevelopments, campuses and technopoles for high-tech industry, and airport complexes, as well as business centers, condominium complexes, master-planned communities, and shopping malls. The preferred architectural styling for much of this rebundled urban activity is postmodern. Postmodern urban design has brought a return to traditional and decorative motifs and introduced a variety of deliberately "playful" and "interesting" architectural styles in place of the functional designs of Modernism. **Postmodern urban design** is

The Globalization of Suburbia

Drive 20 miles out of the center of town and there is a series of cookie-cutter tract homes. Some are painted in pastels and are Italianate or Spanish in design, while others are more Georgian Revival. Manicured lawns and clubhouses dot the landscape. There is the occasional golf course. Neighborhoods are gated [see **Figure 11.E**] and carry names like Le Leman Lake, Capital Paradise, Yosemite, and River Garden. During the past several years, numerous gated subdivisions have sprung up outside mainland China's two largest cities, Beijing and Shanghai; and the trend is spreading to other cities, such as Tianjin and Shenzhen.

What is most remarkable is that most of them, visually and architecturally, look as if they belong in San Jose or Orange County. They are directly modeled on the tract homes that have defined American suburban growth in the past 30 years. Most of them carry few, if any, Asian influences in their design, layout, and ornamentation. The first ones were built to house mostly expatriates, such as diplomats or those heading the local offices of multinational companies. "In order to provide living facilities for foreigners (that match) the foreign standard, they copycat the Western style and Western standard," says Billie Chau, head of the Beijing office of FPD Savills, a British company that manages some of these compounds. "That's why they look and feel so Western."

Now, though, many more have just been built and more are in store, as an increasing number of local Chinese with money try to get out of the crowded and dirty milieu of the contemporary Chinese city. Many are also buying them for investment purposes. "As China's economy continues its breakneck upward trajectory, with high economic growth and suburbanization occurring in major cities such as Beijing and Shanghai, the wealthy inhabitants and white-collar workers of such cities are showing a preference for the purchase of villas or houses located on the suburban fringes of these cities," says Andrew Ness, Asia executive director for research for CB Richard Ellis. The global real estate company is now doing a lot of business in the selling and renting of these properties.

China seems to have caught onto sprawl. Yet, what is perhaps most distinctive about these developments is not the sprawl itself, but the way it looks [see **Figure 11.F**]. Vancouver Forest, for example, is

Figure 11.E Shanghai Suburb Thames Town, just outside Shanghai, has been built by developers to recreate an English town, complete with universities, hospitals, supermarkets, hotels, and even a pub. It is said to draw its inspiration for its Tudor, Georgian, and Victorian style homes from Bristol and Birmingham (neither of them on the River Thames).

Figure 11.F **Advertisements for suburban homes in Beijing**

a new subdivision of homes that mimics a typical neighborhood in British Columbia. It was built by Canadian architects, using Canadian materials to create a mini-Canada. "Vancouver Forest: A Showcase of Canadian Expertise in China," a banner headline from a recent Canadian Embassy newsletter proudly boasts.

Wish to live in Australia? Beijing residents can buy a home at Sydney Coast, a subdivision that offers its residents a "seven-day Australian-style villa life." "Designed by Australian experts, the project presents a kind of simple and fresh lifestyle," says a brochure for the new development. "Taking a walk along the streets in Sydney Coast, you will get a true sense of Australia." Sydney Coast is being developed by Beijing Capital Land, a venture partly owned by the municipality of Beijing. The company is also developing Upper East Side, a large series of apartment buildings in northeastern Beijing.

For those who would prefer to live in California, Beijingers have been able to set up shop in the Yosemite subdivision. Now there is the additional option of moving to Napa Valley, a new development under construction about 30 miles outside Beijing. Napa Valley attempts to capture a Californian Mediterranean lifestyle of laid-back, al fresco leisure. "Rustic stone is widely used, with rich stucco colors, along with wood shutters and wrought-iron accents, to create an intimate scale and village-like feel," according to Napa Valley's architects and planners, who are based in Palm Springs and Newport Beach (Orange County).

For those craving to recreate life in seventeenth century France, there is Chateau Regalia, located on Beijing's northern outskirts. Here, potential buyers can choose from several different models of homes: the Duke I, the Duke II, the Marquis, the Earl and the Viscount. In both form and decoration, Chateau Regalia's homes are an eccentric amalgam of French Baroque and neoclassical architecture.

The craze has also caught on in Shanghai, where there are many more tract homes built in foreign styles. Local officials recently announced plans to build a cluster of satellite towns built in different national styles outside Shanghai. There will be a French town, an Italian town, an English town and so on. Each will have its own commercial center.

From D. Elsea, "China's chichi suburbs: American-style sprawl all the rage in Beijing," *San Francisco Chronicle*, April 24, 2005.

Figure 11.35 Anthem, Arizona Anthem, which is planned to have 12,500 homes, opened in 1999. Its houses and roads look spotless. One reason for this is that everybody who buys a house in Anthem has to follow certain covenants, conditions, and restrictions (CC&Rs) governing everything from the color of your house to whether you can put your car up on blocks outside (you can't). (Photo courtesy of JEFF@JAGPPhotoInc.com © 2002.)

characterized by a diversity of architectural styles and elements, often combined in the same building or project. It makes heavy use of symbolism and of color and decoration. It is no coincidence that postmodern design has flourished in the most recent phase of globalization. Having emerged as a deliberate reaction to the perceived shortcomings of Modern design, its emphasis on decoration and self-conscious stylishness has made it a very convenient form of packaging for the new global consumer culture. It is geared to a cosmopolitan market, and it draws quite deliberately on a mixture of elements from different places and times. In many ways it has become the transnational style for the more affluent communities of the world's cities.

Globalization and the Quartering of Urban Space

Intensifying social polarization is another dimension of contemporary urbanization that is associated with globalization. The increased mobility of capital, together with the increasing span of control that is possible through advanced information and communications technologies, has resulted in an increasing concentration of wealth and an increasing gap between the fast world and the slow world (Chapter 2). In most of the world's metropolitan regions there has been an increase in the relative numbers of those who are rich and those who are poor, along with an increase in the financial distance between them. There has also been a greater differentiation among the intermediate groups, so that there is often a four- or five-part division of socioeconomic groups rather than a simple division into two. The United Nations Center for Human Settlements (UNCHS) has identified this social polarization as an indirect but crucial determinant of contemporary patterns of segregation of people and land uses around the world. The overall effect of globalization, according to the UNCHS *Global Report on Human Settlements 2001,*[7] is the "quartering" of cities into spatially partitioned, compartmentalized residential enclaves.

Although varying from city to city because of differences in national political and economic structures and in cities' roles in the international economy, their historical development, and their demographic composition, there are some basic features that characterize this quartering of urban space. Geographer Ronald van Kempen and planner Peter Marcuse describe six **socio-spatial formations—**

[7]United Nations Center for Human Settlements, *Cities in a Globalizing World: Global Report on Human Settlements 2001.* London: Earthscan, 2001, p. 33.

residential patterns that, they suggest, add up to a new spatial order that characterizes globalizing cities:[8]

- protected enclaves of the rich—"citadels" or exclusionary enclaves that generally consist of expensive apartments in favorable locations (see Box 11.3: "The Globalization of Suburbia");
- gentrified areas occupied by young professionals and managers, typically located in the inner districts of central cities;
- middle-class suburbs and medium- and high-rise apartment and condominium complexes;
- working-class neighborhoods, often of rented tenements, that are differentiated according to income, occupation, and ethnicity, sometimes with hardened spatial boundaries between them;
- ethnic enclaves—a specific form of tenement area where ethnic segregation is particularly pronounced;
- excluded ghettos—the slums inhabited by the very poor, the excluded, the never employed and permanently unemployed, and the homeless.

[8]P. Marcuse and R. van Kampen, "Introduction," in P. Marcuse and R. van Kampen, eds., *Globalizing Cities: A New Spatial Order?* Oxford: Blackwell, 2000, pp. 3–5.

These socio-spatial formations represent another aspect of splintering urbanism. The premium spaces within the quartered city are plugged in to the networked infrastructures of information and communications technologies that underpin the fast world. They are partitioned off from the slow world and its spaces of perceived danger, difference, and poverty through processes of segregation and congregation, assisted by the practices of designers and developers who deploy and market security as a key feature of the built environment. Urban designers have become adept at programming **defensible space** into the layout of new developments, while architects have adopted bunker- and fortress-style elements for their buildings.

Developers, meanwhile, have learned to include access control and surveillance technologies in the packaging of the built environment. Gated communities are the most striking example of the way in which the quartering of urban space is being splintered into hardened security zones. In the United States many of the private master-planned communities are gated: In all, about 8 million people live within the walls of gated communities. Gated communities are by no means confined to North American cities, however; they have become symptomatic of social polarization in globalizing cities everywhere.

CONCLUSION

Patterns of land use and the functional organization of economic and social subareas in cities are partly a product of economic, political, and technological conditions at the time of the city's growth, partly a product of regional cultural values, and partly a product of processes of globalization. Geographers can draw on several perspectives in looking at patterns of land use within cities, including an economic perspective that emphasizes competition for space and a sociocultural perspective that emphasizes ethnic congregation and segregation. Nevertheless, urban structure varies considerably because of the influence of history, culture, and the different roles that cities have played within the world-system.

The evolution of the unintended metropolis of the periphery has been very different from the evolution of metropolitan areas in the world's core regions. Similarly, the problems they have faced are very different. In the core regions, the consequences of an economic transformation to a postindustrial economy have dominated urban change. Traditional manufacturing and related activities have been moved out of central cities, leaving decaying neighborhoods and a residual population of elderly and marginalized people. New, postindustrial activities have begun to cluster in redeveloped CBDs and in edge cities around metropolitan fringes. In a few cases, metropolitan growth has become so complex and extensive that 100-mile cities have begun to emerge, with half a dozen or more major commercial and industrial centers forming the nuclei of a series of interdependent urban realms.

In other parts of the world, traditional patterns of land use and the functional organization of economic and social subareas have been quite different, reflecting different historical legacies and different environmental and cultural influences. A basic trend affecting the cities of the world's periphery is demographic—the phenomenal rates of natural increase and in-migration that have given rise to overurbanization. The example of Lagos provides some sobering insights into the human consequences of overurbanization. An ever growing informal sector of the economy, in which people seek economic survival, is reflected in extensive areas of shanty housing. High rates of unemployment, underemployment, and poverty generate acute social problems, which are overwhelming for city governments that are understaffed and underfunded. If present trends continue, such problems are likely to characterize increasing numbers of the world's largest settlements. Meanwhile, globalization processes are recasting metropolitan structure and intensifying social and economic inequalities between the fast world and the slow world. In the next chapter, we consider this question as part of a broader discussion of future geographies.

MAIN POINTS REVISITED

- **The internal structure of cities is shaped heavily by competition for territory and location. In general, all categories of land users—commercial and industrial, as well as residential—compete for the most convenient and accessible locations within the city.**

  An important exception is that wealthier households tend to trade off the convenience of accessibility for the greater utility of being able to consume larger amounts of (relatively cheap) suburban space. Poorer households, unable to afford the recurrent costs of transportation, must trade off living space for accessibility to jobs, thus ending up in high-density areas, at expensive locations near their low-wage jobs.

- **Social patterns in cities are heavily influenced by territoriality, which provides a means of establishing and preserving group membership and identity.**

  Territoriality involves processes of congregation and discrimination, often resulting in segregation, the spatial separation of specific subgroups within a wider population. Segregation varies a great deal in both intensity and form, depending on the relative degrees of congregation and discrimination.

- **The typical North American city is structured around a central business district (CBD); a transitional zone; suburbs; secondary business districts and commercial strips; industrial districts; and, in larger metropolitan areas, edge cities.**

  This internal organization of cities reflects the way that they function, both to bring certain people and activities together and to sort them out into neighborhoods and functional subareas.

- **Traditionally, North American cities have experienced high rates of in-migration, which has resulted in their becoming structured into a series of concentric zones of neighborhoods of different ethnicity, demographic composition, and social status through processes of invasion and succession.**

  Within each concentric zone a mosaic of distinctive neighborhoods tends to develop—ecological niches where particular mixes of people have come to dominate a particular territory or geographic setting.

- **Urban structure varies a good deal from one region of the world to another because of the influence of history, culture, and the different roles that cities have played within the world-system.**

  European cities have evolved under circumstances very different from those faced by American cities, and consequently European cities exhibit some distinctive characteristics in urban form. European urban ideals have influenced many colonial cities, while the new cities of the world's peripheral regions are distinctive because of their explosive growth.

- **The most acute problems of the postindustrial cities of the world's core regions are localized in the central city areas that have borne the brunt of restructuring from an industrial to a postindustrial economy.**

  In these areas there are several interrelated problems: fiscal problems, infrastructure problems, and localized spirals of neighborhood decay and cycles of poverty.

- **The problems of the cities of the periphery stem from the way in which their demographic growth has outstripped their economic growth.**

  The result is high rates of long-term unemployment and underemployment, low and unreliable wages of informal-sector jobs, chronic poverty, and slum housing. Their low rates of economic growth reflect their dependent position in the global economy.

- **Traditional patterns of land use and spatial organization in many parts of the world are being transformed by the local effects of splintering urbanism.**

  Economic and cultural globalization, together with the uneven evolution of networked infrastructures of information and communications technologies, is forging new landscapes of innovation, economic development, and cultural transformation, while at the same time intensifying social and economic inequalities between the fast world and the slow world.

KEY TERMS

Beaux Arts (p. 440)
boomburbs (p. 454)
central business district (CBD) (p. 429)
central cities (p. 430)
congregation (p. 428)
cycle of poverty (p. 433)
defensible space (p. 461)
dualism (p. 444)
edge cities (p. 453)
fiscal squeeze (p. 430)
gentrification (p. 430)
invasion and succession (p. 430)
isotropic surface (p. 428)
minority groups (p. 428)
Modern movement (p. 440)
Postmodern urban design (p. 456)
redlining (p. 434)
segregation (p. 429)
smart growth (p. 454)
socio-spatial formation (p. 457)
underclass (p. 434)
underemployment (p. 444)
zone in transition (p. 429)

ADDITIONAL READING

Biswas, R. K. (ed.), *Metropolis Now!* Vienna: Springer-Verlag, 2000.

Brunn, S., and J. F. Williams (eds.), *Cities of the World,* 3rd ed. New York: Harper Collins, 2003.

Goodman, D., and C. Chant (eds.), *European Cities and Technology: Industrial to Post-Industrial City.* London: Routledge, 1999.

Graham, S., and S. Marvin, *Splintering Urbanism.* New York: Routledge, 2001.

Gugler, J. (ed.), *Cities in the Developing World: Issues, Theory, and Policy.* New York: Oxford University Press, 1997.

Hall, P., *Cities in Civilization.* New York: Pantheon, 1998.

Kazepov, Y. (ed.), *Cities of Europe: Changing Contexts, Local Arrangements, and the Challenge to Urban Cohesion.* Malden, MA: Blackwell, 2005.

Knox, P. L. and L. McCarthy, *Urbanization: An Introduction to Urban Geography,* 2nd ed. Englewood Cliffs, NJ: Prentice Hall, 2005.

Knox, P. L., and S. Pinch, *Urban Social Geography,* 4th ed. London: Longman Scientific, 2000.

Marcuse, P. and R. van Kampen (eds.), *Globalizing Cities.* Oxford: Blackwell, 2000.

Morris, A. E. J., *History of Urban Form Before the Industrial Revolutions.* London: Longman, 1994.

Pacione, M., *Urban Geography: A Global Perspective.* New York: Routledge, 2001.

Potter, R. B., and S. Lloyd-Evans, *The City in the Developing World.* Harlow, Essex: Addison Wesley Longman, 1998.

Robbins, E. and R. El-Khoury (eds.), *Shaping the City: Studies in History, Theory, and Urban Design,* Spon Press, 2003.

Scott, A. J. (ed.), *Global City-Regions: Trends, Theory, Policy.* Oxford: Oxford University Press, 2001.

Simmonds, R. and G. Hack (eds.), *Global City Regions: Their Emerging Forms.* New York: Spon Press, 2000.

Smith, D. A., *Third World Cities in Global Perspective.* Boulder, CO: Westview Press, 1996.

United Nations Human Settlement Programme, *The State of the World's Cities 2004/2005: Globalization and Urban Culture.* Sterling, VA: Earthscan, 2004.

U.S. Department of Housing and Urban Development, *The State of the Cities 1999.* Washington, DC: U.S. Department of Housing and Urban Development, 1999.

Wallach, B., *Understanding the Cultural Landscape.* New York: Guilford, 2005. See Chapters 19 ("American Cities") and 21 ("Cities Aboad").

EXERCISES

On the Internet

The Internet exercises for this chapter focus on the diversity of urban areas and their importance in spatial organization and the evolution of societies. One observer described cities as "a place of our meeting with the other," while another critic observed that urban dwellers are always "people in the presence of otherness." Diversity is one thing certain in urban geographies, from the energy of everyday life in the cities to alienation and a loss of community. In describing the urban structure of contemporary cities, there are many realities that can be revealed through the Internet.

Unplugged

1. Collect a week's worth of local newspapers and review the coverage of urban problems. What kinds of problems are covered, and for what kinds of cities? Compile a list of such categories, and then carefully analyze the content of the week's coverage, calculating the number of column inches devoted to each category of problems.
2. On a tracing-paper overlay of a street map of your town or city, plot the distribution of houses and apartments for sale or rent in different cost brackets. (You can obtain the information from the real estate pages of your city's local newspaper; in smaller cities you may have to gather data from several issues of the paper a week or more apart in order for a pattern to emerge—your local library will likely have back issues.) What can you say about the spatial distribution that is revealed?
3. Most cities consist of "ordinary" cityscapes that are strongly evocative because they are widely understood as being a particular kind of place. Write a brief essay (500 words, or two double-spaced, typed pages) describing an "ordinary" cityscape with which you are familiar. What are its principal features, and how might it be considered typical of a particular kind of place?

12 Future Geographies

MAIN POINTS

- In some ways, the future is already here, embedded in the world's institutional structures and in the dynamics of its populations.
- New and emerging technologies that are likely to have the most impact in reshaping human geographies include advanced transportation technologies, biotechnology, materials technologies, and information technologies.
- The changes involved in shaping future geographies will inevitably bring some critical issues, conflicts, and threats, including important geographical issues that center on scale, boundaries, and territories; on cultural dissonance; and on sustainability.

Will the Internet bring about new patterns of human interaction? Will we be able to cope with the environmental stresses that increasing industrialization and rapid population growth will bring to many parts of the world? Will the United States retain its position as the world's most powerful and influential nation? Will more countries move up from peripheral status to join the semiperiphery and core of the future world-system? What kind of problems will the future bring for local, regional, and international development? What new technologies are likely to have the most impact in reshaping human geographies? Will globalization undermine regional cultures? Will technology and human determination be able to cope with the environmental stresses that industrialization in the periphery and rapid population increases will inevitably create? Will new regions emerge based on new types of connectivity such as trade, the Internet, or any number of political movements such as mobilizations against globalization or the human rights movement? These are just a few of the many questions that spring from the key themes in human geography that we have examined in Chapters 5 through 11. This chapter examines some scenarios for future geographies, drawing on the principles and concepts established in Chapters 1 through 4.

Tokyo waterfront

MAPPING OUR FUTURES

The most effective way to approach the questions listed above is to try to get a sense of how different aspects of globalization are changing the world and how they might continue to do so. As we discussed in Chapter 2, the globalization of the capitalist world-system involves processes that have been occurring for at least 500 years. But since World War II, world integration and transformation has been remarkably accelerated and dramatic. Among the forces driving integration and transformation are strengthening of regional alliances such as the European Union and the Organization of Petroleum Exporting Countries, the increasing connectivity of the most remote regions of the world due to telecommunications and transportation linkages, the emergence of the new economy in the core countries, and the rise of global institutions like the World Trade Organization. How will the forces of broadening global connectivity—and the popular reactions to them—change the fates and fortunes of world regions whose current coherence owes more to eighteenth- and nineteenth-century European colonialism than to forces of integration or disintegration in the twenty-first century? To answer this and related questions, we need to understand what the experts think about the processes behind globalization and how they might affect its future potential. But we also need to first understand the very risky issue of predicting the future, how predictions are made, and how useful predictive exercises can be.

There is no shortage of visionary scenarios (see, for example, **Figure 12.1** and Box 12.1: "Dark Age Ahead?"). Broadly speaking, futurists' projections can be divided into two kinds: optimistic and pessimistic. Optimistic futurists stress the potential for technological innovations to discover and harness new resources, to provide faster and more effective means of transportation and communication, and to make possible new ways of living. This sort of futurism is often characterized by science-fiction cities of mile-high skyscrapers and spaceship-style living pods, by bioecological harmony, and by unprecedented social and cultural progress through the information highways of cyberspace. It projects a world that will be stabilized and homogenized by supranational or even "world" governments. The sort of geography implied by such scenarios is rarely spelled out. Space and place, we are led to believe, will be transcended by technological fixes.

To pessimistic futurists, however, this is just "globaloney." They stress the finite nature of Earth's resources, the fragility of its environment, and population growth rates that exceed the capacity of peripheral regions to sustain them. Such doomsday forecasting is characterized by scenarios that include irretrievable environmental degradation, increasing social and economic polarization, and the breakdown of law and order. The sort of geography associated with these scenarios is rarely explicit, but it usually involves the probability of a sharp polarization between the haves and have-nots at every geographical scale.

The "Barbarism" scenarios (Figure 12.1), for example, are characterized by persistent poverty, population pressure, and increasingly disastrous environmental problems that lead to localized armed conflicts and violence in peripheral regions, as well as by rising unemployment, depressions, political instability, and outbreaks of civil disorder in core regions. As states lose relevance and power compared to multinational corporations, social welfare policies are increasingly abandoned in favor of corporate profitability. Throughout the world the rich and powerful become entrenched in militarized neighborhoods and "secure homelands," surrounded by extensive tracts of impoverished neighborhoods and regions. Such scenarios, of course, also invite the probability of more widespread state terrorism and terrorist organizations. One conceivable outcome of widespread terrorism is international political disorder and the collapse of economic globalization. This was evidently one goal of the perpetrators of the September 11, 2001, terrorist attacks on New York's World Trade Center and the Pentagon in Washington, D.C.

Fortunately, we don't have to choose between the two extreme scenarios of optimism and pessimism. Using what we have learned from the study of human geography, we can suggest a more grounded outline of future geographies. To do so, we must first glance back at the past. Then, looking at present trends and using what we know about processes of geographic change and principles of spatial organization, we can begin to map out the kinds of geographies that the future most probably holds.

Looking back at the way that the geography of the world-system has unfolded, we can see now that a fairly coherent period of economic and geopolitical development occurred between the outbreak of World War I (in 1914) and the collapse of the Soviet Union (in 1989). Some historians refer to this period as the "short twentieth century." It was a period when the modern world-system developed its triadic core of the United States, Western Europe, and Japan, when geopolitics was based on an East-West divide, and when geoeconomics was based on a North-South divide. This was a time when the geographies of specific places and regions within these larger frameworks were shaped by the needs and opportunities of technology systems that were based on the internal combustion engine, oil and plastics, electrical engineering, aerospace industries, and electronics. In this short century the modern world was established, along with its now familiar landscapes and spatial structures: from the industrial landscapes of the core to the unintended metropolises of the periphery; from the voting blocs of the West to the newly independent nation-states of the South.

Looking around now, much of the established familiarity of the modern world and its geographies seems to be disappearing. We have entered a period of transition, triggered by the end of the Cold War in 1989 and rendered more complex by the geopolitical and cultural repercussions of the terrorist attacks of September 11, 2001. The result is a series of unexpected developments and unsettling

Conventional Worlds

The two scenarios presented here have a common vision of a world in which development is governed by gradual and steady industrial growth. While the population grows, world economic output expands indefinitely as consumption and production practices in peripheral and semiperipheral regions converge toward those of the increasingly rich core. The regions of the world become progressively more interdependent as the competitive private market remains the main engine for economic growth. Transnational corporations increase their role as the dominant economic node of the global, borderless economy. The liberal state persists as the dominant unit of governance. The two variations on the Conventional World futures differ in terms of policy intervention and impacts.

(A) Reference Scenario

This scenario assumes that most of the world's regional economies will open and that largely unregulated markets will expand internationally. New and expanded markets foster rapid technological development. Population growth increases in the peripheral and semiperipheral regions of the world as most of the core regions grow slowly or not at all. As the core gets richer, the marginalized become increasingly poor and inequality increases (Figure 12.1a). Environmental quality improves in some of the core regions and gets worse in the periphery and overall global environmental conditions deteriorate. Social justice issues intensify.

(a) Homelessness in Leeds, England

Barbarism

The two scenarios that underpin the Barbarism future assume that the contemporary negative stresses present in Conventional Worlds scenarios overwhelm the coping capacity of markets and institutions. The world veers toward barbarism as regions with declining physical amenities experience breakdown. Growing populations, persistent poverty, increasingly disastrous environmental problems lead to the barbarization of the marginalized regions. In a future of Barbarism, social welfare policies are increasingly abandoned in favor of productivity and competitiveness policies as states lose relevance and power compared to multinational corporations. The marginalized experience growing environmental pollution and natural resource constraints. Rather than full-scale wars, the result is small-scale armed conflicts and violence.

(A) Social Breakdown

In this scenario, chaos ensues as random violence diverts resources from economic growth to security concerns. Civil order breaks down as states become too weak to set the global economy back on track (Figure 12.1c). As refugees fleeing from one disaster help to destabilize neighboring regions, states pour their resources into police powers, border fences and guards, monitoring of the movements and activities of citizens, and ultimately limiting or obstructing trade and travel. The result is the collapse of globalization, overburdened by rising unemployment, depressions, political instability, and outbreaks of civil disorder in elite, marginalized, and embattled regions as everyone gets poorer.

(c) Iragui Kurd refugees

Great Transitions

The most optimistic future is that of the Great Transition, where the world's regions evolve to a higher stage. Although these scenarios may seem naive and improbable, they are not impossible and may even be necessary to achieve the goals of sustainability and equity. While there is a range of possible Great Transitions scenarios, we consider two: Global Governance and the New Sustainability Paradigm, which differ in their means but not in their ends.

(A) Global Governance

This scenario is built upon a growing collective realization that individuals, institutions, and states must restrict certain activities and undertake others for the common global good. Cooperation is essential. The leadership for this effort comes from multinational and transnational corporations, intergovernmental global organizations, and nongovernmental organizations. Acting in concert, these three entities act as a counterforce to states and are given limited regulatory power to enforce voluntary guidelines and to tax (but not to restrict) international flows of currencies, goods, and telecommunications. In addition, international courts are strengthened, mediation bodies flourish, and the dispute resolution capability of international organizations is greatly enhanced (Figure 12.1e).

(e) U.N. General Assembly

Figure 12.1 Future scenarios In some ways the future is already here, embedded in the world's institutional structures and in the dynamics of its populations. Still, there are some aspects of the future that we can only guess at. Six scenarios are presented here, two each for three types of futures: Conventional Worlds, Barbarism, and Great Transitions. (After Global Scenario Group, Stockholm Environment Institute, Boston, MA).

(b) Grade-school children, Welkom, South Africa

(B) Balanced Growth Scenario
This scenario assumes the implementation of policy reforms that guide economic growth and sustain the environment. Balanced growth results from the introduction of better technology, reduction of subsidies for natural resource use, imposition of pollution taxes, land reform, development incentives, and increased foreign aid for education, health, and broader economic opportunities in the periphery and semiperiphery (Figure 12.1b). The gap between the elite and the marginalized is less than in the Reference Scenario; as a result, the poorer regions of the world are stabilized and widespread social conflict is avoided.

(d) U.S. air power

(B) Fortress World
In this scenario, the core regions recognize the crisis that is mounting and create alliances among themselves to protect their own interests. With the multinational corporations as their agents, the rich and powerful become entrenched, surrounded by oceans of misery (Figure 12.1d). The result is a society of elite and marginalized with entry into the elite by birth only. Strategic reserves of fossil fuels, minerals, freshwater, and genetic diversity are put under military control by the rich groups. While the impacts of pollution are limited for the elite, pollution and its noxious health effects increase for the marginalized.

(f) Wind farm

(B) New Sustainability
In this scenario, increases in technological and economic growth are concentrated in core regions and dominated by transnational corporations. The gap between the elite and the marginalized is extreme. Environmental problems in the periphery grow. Migration flows to the core increase. The result is a rise in global social movements opposing high-consumption lifestyles. Corporations follow the market and alter what they produce, how they produce it, even how they market it (Figure 12.1f). Sustainability becomes an economic and environmental goal. A global civil society is born based on social justice and open mechanisms for decision-making and consensus-seeking.

Figure 12.1 Continued

12.1

GEOGRAPHY MATTERS

Dark Age Ahead?

Jane Jacobs, recognized in 2004 by *Business Week* magazine as one of the most influential public intellectuals in North America in the last 75 years, argues that the United States is slipping toward the beginnings of a new "Dark Age" as a result of the deterioration of five pillars of modern society: community and family, higher education, the application of science and technology, the integrity of the professions, and the role of government in relation to society's needs and potential.

Jacobs has long been a passionate advocate of cities and urban life, and her views have been both influential and controversial. But Jacobs' fears for an incipient Dark Age go well beyond her past concerns for urban development, centering as they do on several important aspects of higher education, the application of knowledge, the role of the professions, and the role of government. "A culture is unsalvageable," writes Jacobs, "if stabilizing forces themselves become ruined."[1] The roots of her concerns are based on evidence of:

- corporate immorality in the marketplace instead of entrepreneurship bonded to social justice;
- universities that serve employers and act as credential factories, stripping the music, art, ethics, idealism, and notion of the public good out of education;
- scientific research increasingly and immorally being bought by corporations or suppressed and ignored by governments; and
- a neoliberal political economy that is intent on abandoning the stewardship of urban and regional development.

She does not pull her punches. Here is what she has to say about universities: They have become credential factories, stripping the art, poetry, ethics, idealism and notion of the public good out of education. They are complicit in allowing scientific research to be controlled by corporations and directed—and sometimes suppressed—by governments. Rather than serving as one of the cultural pillars of society, universities now serve employers and act as "colleges of heraldry," awarding graduates a "coat of arms" (i.e., university diploma) to distinguish them from those without marketable credentials.

She is equally direct about the professions. This time it is not so much planners and design professionals that are the focus of her concern but accountants, bankers, lawyers and other financial professions whose ethics and practices have serially been called into question in relation to the vast corporate scandals of the past two decades. Nevertheless, the design and public policy professions are implicated in her critique of urban trends. Cities—vital economic engines and crucibles of cultural change and innovation—are being starved of the money they need by national, state, and local governments. Neoliberal policies are responsible for sagging public transit and public education systems, increasing pollution, increasing social polarization, the erosion of community, and the burgeoning sullenness of citizens. Families, she argues, are rigged to fail by public policies that, unintentionally, force both parents to work to meet the financial needs for themselves and their children. Communities are rigged to fail by public policies that foster sprawling, placeless, and automobile-dependent suburbs—a reprise of her earlier critique in *The Death and Life of Great American Cities.*

It behooves us to reflect critically on Jacobs' extended essay. She confides that "I have written this cautionary book in hopeful expectations that time remains for corrective actions." If we accept that corrective actions are in fact warranted, then a foundation of geographic knowledge will be fundamental to the incisive analysis that will be necessary in order to contribute effectively to countervailing forces.

[1] J. Jacobs, *Dark Age Ahead: Caution.* New York: Random House, 2004.

juxtapositions. The United States, for example, has given economic aid to Russia; Eastern European countries have joined NATO and the European Union; Germany has unified, but Czechoslovakia and Yugoslavia have disintegrated; South Africa has been transformed, through an unexpectedly peaceful revolution, to black majority rule. Meanwhile, Islamist terrorists shoot up tourist buses, bomb office buildings, and sabotage aircraft; former communist Russian ultranationalists have become comradely with Austrian and German neo-Nazis; Hindus, Sikhs, and Muslims are in open warfare in South Asia; the United States has invaded Iraq; and Sudanese military factions steal food from aid organizations in order to sell it to the refugees for whom it was originally intended.

These examples show that we cannot simply project our future geographies from the landscapes and spatial structures of the past. Rather, we must map them out from a combination of existing structures and budding trends.

We have to anticipate, in other words, how the shreds of tradition and the strands of contemporary change will be rewoven into new landscapes and new spatial structures.

While this is certainly a speculative and tricky undertaking, we can draw with a good deal of confidence on what we know about processes of geographic change and principles of spatial organization. The study of human geography has taught us to understand spatial change as a composite of local place-making processes (see Chapter 6) that are subject to certain principles of spatial organization and that operate within the dynamic framework of the world-system (Chapter 2). It has also taught us that many important dimensions exist to spatial organization and spatial change, from the demographic dimension (Chapter 3) through to the urban (Chapter 11).

As we look to the future, we can appreciate that some dimensions of human geography are more certain than others (**Table 12.1**). In some ways the future is already here, embedded in the world's institutional structures and in the dynamics of its populations. We know, for example, a good deal about the demographic trends of the next quarter century, given present populations, birth and death rates, and so on. We also know a good deal about the distribution of environmental resources and constraints, about the characteristics of local and regional economies, and about the legal and political frameworks within which geographic change will probably take place.

On the other hand, we can only guess at some aspects of the future. Two of the most speculative realms are those of politics and technology. While we can foresee some of the possibilities (maybe a spread and intensification of ethno-nationalism; perhaps a new railway era based on high-speed trains), politics and technology are both likely to spring surprises at any time. The September 11, 2001, terror attack in the United States was a painful example of the sort of political surprises that can occur. In a matter of about 100 minutes, terrorists who crashed two commercial jetliners into the World Trade Center towers and one into the Pentagon were able to deliver a blow to the U.S. economy, draw the country into full military alert, and profoundly transform daily life in the United States. Other possible events—such as a political instability in Saudi Arabia, war between North Korea and its neighbors, or unanticipated breakthroughs in biotechnology—would cause geographies to be rewritten suddenly and dramatically.

RESOURCES, TECHNOLOGY, AND SPATIAL CHANGE

As we saw in Chapter 2, technological breakthroughs and the availability of resources have had a profound influence on past patterns of development, and the same factors will certainly be a strong influence on future geographies. The expansion of the world economy and the globalization of industry will undoubtedly boost the

TABLE 12.1 The 2020 Global landscape

Relative Certainties	Key Uncertainties
Globalization largely irreversible, likely to become less Westernized.	Whether globalization will pull in lagging economies; degree to which Asian countries set new "rules of the game."
World economy substantially larger.	Extent of gaps between "haves" and "have-nots"; backsliding by fragile democracies; managing or containing financial crises.
Increasing number of global firms facilitate spread of new technologies.	Extent to which connectivity challenges governments.
Rise of Asia and advent of possible new economic middle-weights.	Whether rise of China/India occurs smoothly.
Aging populations in established powers.	Ability of EU and Japan to adapt work forces, welfare systems, and integrate migrant populations; whether EU becomes a superpower.
Energy supplies "in the ground" sufficient to meet global demand.	Political instability in producer countries; supply disruptions.
Growing power of nonstate actors.	Willingness and ability of states and international institutions to accommodate these actors.
Political Islam remains a potent force.	Impact of religiosity on unity of states and potential for conflict; growth of jihadist ideology.
Improved WMD capabilities of some states.	More or fewer nuclear powers; ability of terrorists to acquire biological, chemical, radiological, or nuclear weapons.
Arc of instability spanning Middle East, Asia, Africa.	Precipitating events leading to overthrow of regimes.
Great power conflict escalating into total war unlikely.	Ability to manage flashpoints and competition for resources.
Environmental and ethical issues even more to the fore.	Extent to which new technologies create or resolve ethical dilemmas.
US will remain single most powerful actor economically, technologically, militarily.	Whether other countries will more openly challenge Washington; whether US loses S&T edge.

Source: National Intelligence Council, *Mapping the Global Future.* Washington, DC: US Government Printing Office, 2004, p. 8.

overall demand for raw materials of various kinds, and this will spur the development of some previously underexploited but resource-rich regions in Africa, Eurasia, and East Asia. Raw materials, however, will be only a fraction of future resource needs. The main issue, by far, will be energy resources. World energy consumption has been increasing steadily over the recent past, and as the periphery is industrialized and its population increases further the global demand for energy will expand rapidly. Basic industrial development tends to be highly energy-intensive. The International Energy Agency, assuming (fairly optimistically) that energy in peripheral countries will be generated in the future as efficiently as it is today in core countries, estimates that developing-country energy consumption will more than double by 2015, lifting total world energy demand by almost 50 percent. By 2020 peripheral and semiperipheral countries will account for more than half of world energy consumption. Much of this will be driven by industrialization geared to meet the growing worldwide market for consumer goods, such as private automobiles, air conditioners, refrigerators, televisions, and household appliances.

Without higher rates of investment in exploration and extraction than at present, production will be slow to meet the escalating demand. Many experts believe that current levels of production in fact represent "peak oil" and that by 2020 global oil production may be only 90 percent of its current peak. The result might well be a significant increase in energy prices. This would have important geographical ramifications: Companies would be forced to seriously reconsider their operations and force core-region households into a reevaluation of their residential preferences and commuting behavior; while peripheral-region households would be forced further into poverty. If the oil-price crisis of 1973 is anything to go by (after crude oil prices had been quadrupled by the OPEC cartel), the outcome could be a major revision of patterns of industrial location and a substantial reorganization of metropolitan form. Significantly higher energy costs may change the optimal location for many manufacturers, leading to deindustrialization in some regions and to new spirals of cumulative causation in others. Higher fuel costs will encourage some people to live nearer to their place of work, while others will be able to take advantage of telecommuting to reduce personal transportation costs. It is also relevant to note that almost all of the increase in oil production over the next 15 or 20 years is likely to come from outside the core economies. This means that the world economy will become increasingly dependent on OPEC governments, which control over 70 percent of all proven oil reserves, most of them in the Middle East.

In countries that can afford the costs of research and development, new materials will reduce the growth of demand both for energy and for traditional raw materials such as aluminum, copper, and tin. Japan, for instance, may be able to reduce motor-vehicle fuel consumption by 15 percent (and thereby reduce its total fuel oil consumption by 3 percent) by using ceramics for major parts of engines. It may also be possible to substitute ceramics for expensive rare metals in creating heat-resistant materials. Improved engineering and product design will also reduce the need for the input of some resources. American cars, for example, were 22 percent lighter in 2004 than they were in 1974. In addition, of course, the future may well bring technological breakthroughs that dramatically improve energy efficiency or make renewable energy sources (such as wind, tidal, and solar power) commercially more viable. As with earlier breakthroughs that produced steam energy, electricity, gasoline engines, and nuclear power, such advances would provide the catalyst for a major reorganization of the world's economic geographies.

Just what new technologies are likely to have the most impact in reshaping human geographies? Given what we know about past processes of geographic change and principles of spatial organization, it is clear that changes in transportation technology are of fundamental importance. Consider, for example, the impact of oceangoing steamers and railroads on the changing geographies of the nineteenth century, and the impact of automobiles and trucks on the changing geographies of the twentieth century (Chapter 2). Among the most important of the next generation of transportation technologies that will influence future geographies are high-speed rail systems, smart roads, and smart cars. Several emerging industrial technologies also exist whose economic impact is likely to be so great that they will influence patterns of international, regional, and local development. Studies by the U.S. government, the Japanese government, the European Union, the Organization for Economic Cooperation and Development (OECD), and the United Nations have all identified biotechnology, materials technology, and information technology as the most critical areas for future economic development.

Transportation Technologies

On land the most interesting developments seem likely to center on new high-speed rail systems. Improved locomotive technologies and specially engineered tracks and rolling stock will make it possible to offer passenger rail services at speeds of 275 to 370 kilometers per hour (180 to 250 mph). With shorter check-in times and in-town rail terminals, travel between some cities will be faster by rail than by air. Plans have already been proposed for such systems in Florida, Texas, and California. The most advanced plans, however, are in Europe, where the European Union has drawn up a master plan for the development of 30,000 kilometers (almost 20,000 miles) of high-speed track in a Trans-European Network (TEN) by 2010 (**Figure 12.2**). In addition, new tilt-technology rolling stock, designed to negotiate tight curves by tilting the train body into turns in order to counteract the effects of centrifugal force, is being introduced in many parts of Europe in order to raise maximum speeds on conventional rail tracks. German Railways (DB), for example,

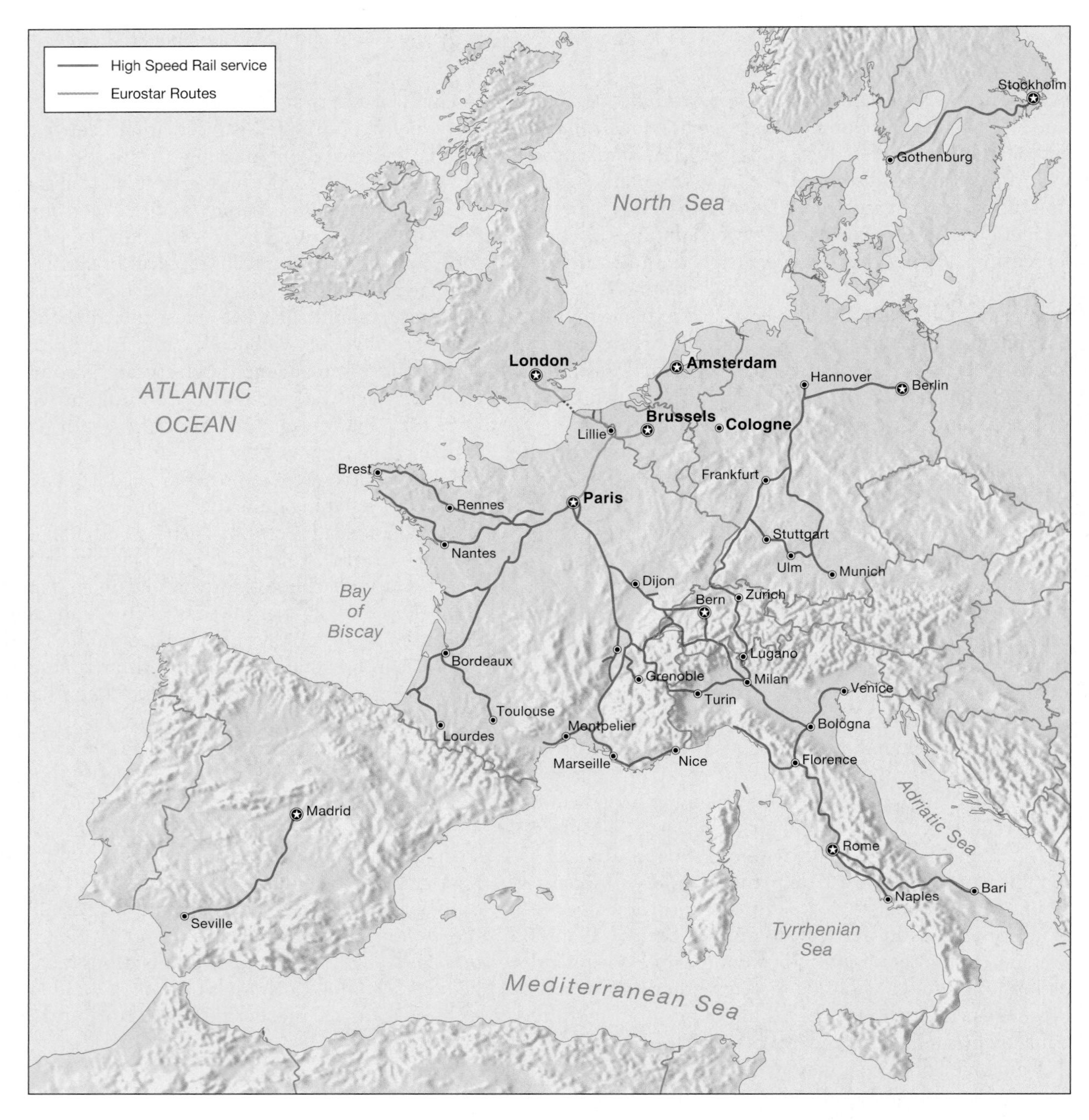

Figure 12.2 High-speed rail in Europe Europe, with its relatively short distances between major cities, is ideally suited to rail travel and less suited, because of population densities and traffic congestion around airports, to air traffic. Allowing for check-in times and accessibility to terminals, it is already quicker to travel between many major European cities by rail than by air. The European Union plans to coordinate and subsidize a $250 billion investment in 30,000 kilometers (almost 20,000 miles) of high-speed track, to be phased in through 2012. The heart of the system will be the "PBKAL web," which will connect Paris, Brussels, Cologne (Köln), Amsterdam, and London and will be completed by 2003. This web will bring about some restructuring of the geography of Europe. High-speed rail routes will have only a few scheduled stops because the time penalties resulting from deceleration and acceleration undermine the advantages of high-speed travel. Places with no scheduled stops will be less accessible, and so less attractive for economic development.

introduced third-generation ICE (intercity express) trains with a maximum speed of 330 kilometers per hour (205 mph) in 2000. The geographic implications of these systems are significant. Quite simply, once the systems become commercially viable, places that are linked to them will be well situated to grow in future rounds of economic development; places that are not will probably be left behind.

The same significance will attach to intelligent transportation systems (ITS), should they be developed from their current prototypes to become commercially viable. An ITS is a combination of so-called smart highways and smart cars. The basic ITS target concept is an interactive link of vehicle electronic systems with roadside sensors, satellites, and centralized traffic-management systems. This linkage allows for real-time monitoring of traffic conditions and enables drivers to receive alternate route information via two-way communications, on-board video screens, and mapping systems. The next step would be completely automated highway systems, on which groups of vehicles would be guided automatically, in closely packed platoons, with virtually no active driver control. With fewer gridlocked roads, driving would be safer, less polluting, and more efficient. Metropolitan areas and interstate corridors that have the infrastructure for ITS technology will be at a significant advantage in attracting new industries and their workers. The first generation of smart cars and trucks, with on-board, microproccesor-based electronics, is already on the road, using route-planning software, global positioning systems (GPS), and fleet management systems software and wireless communications. Some automobiles have adaptive cruise control that automatically maintains a minimum distance from the vehicle ahead; others have fully automated crash-notification systems that can sense, characterize, and survive a crash and deliver a message to a 911 center. Pilot projects are under way by Honda, Nissan, and Toyota to provide fleets of small electric vehicles at central points (such as train stations) for commuters to use via an automated, card-based rental system. Travelers will just plug in their preapproved card and go. The time they drive, as well as the time they have the vehicle, will be monitored, and they will be subsequently charged accordingly.

At sea no emerging technologies exist at present that might have an equivalent impact, though the European Union has been funding research on fuel cell systems for ocean-going ships. In the air, however, a large untapped market encourages big investments in research and development. The European Airbus 380 represents a new generation of wide-bodied jets that will carry up to 850 passengers (the present maximum is about 450) at a fuel consumption per passenger that is more efficient than most mid-size family automobiles. Quiet supersonic aircraft technology (QSAT) will feature an advanced new engine that will give aircraft supersonic performance without a sonic boom. Changes in transportation technology are potentially enormous and will have profound impacts on both the core and the periphery. But one question these innovations force us to ask is: How efficient do they allow us, as a global community, to be? For instance, does it make sense to consume large amounts of fossil fuels and generate more pollution in order to transport staple foodstuffs around the world when they could just as easily be grown locally? At present in France, which possesses a long-established apple-producing industry, it is possible to find in the markets as many New Zealand apples as French ones. In Mongolia, where milk-producing animals are ubiquitous, the butter in the markets is imported from Germany. Does such an arrangement make economic, environmental, and social sense?

Biotechnology

Although biotechnology is widely associated with both the genetic engineering of crops (through green revolution crops such as hybrid rice, corn, wheat, lettuce, tomatoes, sugarcane, and cotton [see Chapter 8]) and with its pharmaceutical potential (through products such as Interferon and growth hormones), biotechnology is also likely to have a profound effect on animal husbandry, industrial production, renewable energy, waste recycling, and pollution control. Genetic engineering, or genetic modification, is already being applied quite widely in animal husbandry. Examples include Japanese fish farms that use cell-fusion technology to produce algae in systems that are 350 times more efficient than other methods of raising brine shrimp; U.S. dairy farms that use recombinant bovine growth hormones, made with genetically engineered bacteria, in order to boost the production of milk at the same feed costs; a "frost-resistant" tomato that has been produced by splicing into its genetic code a gene that protects flounder from the cold; rapeseed that has had a bay-tree gene spliced into it to increase its oil content; and a potato that has been given a disease-resistant chicken gene.

Between 1996 and 2000, according to the United Nations Development Programme, the area planted with genetically engineered products increased from 2 to 44 million hectares worldwide, though 98 percent of that was in just three countries—Argentina, Canada, and the United States. These countries have developed more than 60 different genetically modified crops, most notably genetically engineered soybeans. The U.S. government has tested more than 4,500 genetically engineered plants, and more than 40—including 13 varieties of corn, 11 varieties of tomatoes, and four varieties of soybeans—have cleared government reviews. In 2004, more than 75 percent of the soy crop, more than 35 percent of the corn crop, and more than 70 percent of the cotton crop in the United States were grown with genetically engineered seeds, while 80 percent of all cheeses produced in the United States contained chymosin, which is produced with genetically engineered bacteria. Genetically modified products are now turning up in large quantities on produce shelves and also in processed foods from cookies to potato chips to baby food. In 2004 about 75 percent of the annual increase in world agricultural production resulted from applications of biotechnology.

The prospect, viewed optimistically, is for a scientific means to efficiently feed a hungry world. An additional advantage offered by genetically modified crops is that they can be engineered to be pest-resistant, thus saving both the dollar cost and the environmental cost of pesticides—conventional farming uses five or more broad-spectrum pesticide applications on crops each year. The implications for agricultural productivity are profound, as are the implications for the geography of agriculture. If crops can be genetically engineered to withstand pests, cold climates, and other adverse conditions, the whole geography of production may change, bringing the prospect of orange groves and avocado orchards in temperate climates, among other strange scenarios.

Nevertheless, the future growth in the production of genetically modified food is by no means certain. In Europe and Japan, consumers and their governments are already highly skeptical of genetically engineered foods, and environmental lobby groups have made genetically modified food a central issue in what promises to be one of the most critical cultural struggles of the early twenty-first century: local mobilization against transnational business (see Chapter 2, p. 80).

In geographical terms we can expect the long-term economic benefits of biotechnology to be greatest in the countries and regions that can afford the costs of research and development and the costs of installing and applying the new technologies. On the other hand, one of the principal advantages of many applications of biotechnology innovations is that they are economical to use on a small scale, without large infrastructure requirements. This should facilitate their use in peripheral regions. There is the prospect (but by no means the certainty) that these applications could not only bring commercial success to peripheral regions but also help solve the problem of food shortages. Similarly, biotechnology might reverse the environmental degradation of some parts of the world because it could provide economically viable ways of replacing chemical fertilizers and toxic sprays, recycling waste products, and cleaning up polluted water. It is also possible, unfortunately, that natural genetic diversity will be reduced as native seeds are replaced by clones and as locally adapted forms of agriculture are replaced by industrialized ones. The result could very easily be the disappearance of thousands of plant varieties and with them the sources of natural resistance to genetically adapted pests.

Materials Technologies

Materials technologies include new metal alloys, specialty polymers, plastic-coated metals, elastothermoplastics, laminated glass, fiber-reinforced ceramics, and nanotechnologies. They are important because they can replace scarce natural resources, reduce the quantity of raw materials used in many industrial processes, reduce the weight and size of many finished products, increase the performance of many products, produce less waste, and allow for the commercial development of entirely new products. Materials design has become so sophisticated that engineers will also be able to incorporate environmental criteria into their products rather than having to deal with environmental issues as an afterthought (though it may not be commercially realistic to do so). Like biotechnology, materials technology has been growing steadily for some time and is now set for a period of explosive growth in its applications, particularly in the automobile and aircraft industries. It has been estimated that specially engineered materials already account for around $1 trillion of the annual GDP of the United States.

Unlike biotechnology, applications of materials technologies will require a fairly close association with an expensive infrastructure of high-tech industry. As a result, their immediate geographical impact is likely to be much more localized within the core regions of the world-system. Peripheral regions and countries that are heavily dependent on the production and export of traditional raw materials—such as Guinea and Jamaica (bauxite), Zambia and the Democratic Republic of Congo (formerly Zaire; copper), Bolivia (tin), and Peru (zinc)—will probably be at the wrong end of the creative destruction prompted by these new technologies. In other words, as new materials technologies reduce the demand for traditional raw materials, production and employment in the latter industries will decline and investors will probably withdraw from producer regions in order to reinvest their capital in more profitable ventures elsewhere. In contrast, some peripheral regions and countries will benefit from the increasing demand for rare earth metals. Brazil, Nigeria, and the Democratic Republic of Congo, for example, together account for almost 90 percent of the world's production of niobium (used with titanium in making superconductive materials); Brazil, Malaysia, Thailand, Mozambique, and Nigeria together account for about 75 percent of the world's production of tantalum (used in making capacitors that store and regulate the flow of electricity in electronic components).

Information Technologies

Information technologies include all the components of information-based, computer-driven, and communications-related activities—a wide array of technologies that includes both hardware (silicon chips, microelectronics, computers, satellites, and so on) and the software that makes it operate. In addition to telematics—the automation of telecommunications and the linkage of computers by data transmission—information technologies include developments as diverse as real-time monitoring of traffic bottlenecks, computer-controlled manufacturing, chemical and biological sensors of effluent streams, 24-hour data-retrieval systems, bar-coded retail inventory control, telemetry systems for tracking parcels and packages, and geographic information systems. Information technologies have already found widespread applications in retailing, finance, banking, business management, and public administration, yet estimates indicate that even in the more developed countries, only about one-third of the benefits to be derived from information technology–based innovations have so far been realized.

As we have seen in earlier chapters, the dynamics of splintering urbanism have already transformed certain aspects of economic geography. In employment and production an overall concentration exists in core countries, where the detailed geography of information technologies takes the form of highly localized agglomerations of research activity; meanwhile, routine production, testing, and assembly functions have been decentralized to semiperipheral countries (see Chapter 11).

Future geographies of production and employment in information technologies will almost certainly follow the same pattern: Most R&D and high-end production will be localized within core countries, and most routine functions will be decentralized to peripheral and semiperipheral countries such as China, Indonesia, the Philippines, Sri Lanka, and Thailand. The geography of technological innovation and achievement has a well-established core-periphery pattern that is not likely to change a great deal in the foreseeable future (**Figure 12.3**). That is not to say that exceptions will not occur. India, for instance, has carved a successful niche in software development, mainly because of the Indian government's provision of generous tax breaks and liberal foreign-exchange regulations for the industry. A multimedia super corridor outside Kuala Lumpur, Malaysia, is emerging as a result of tax breaks that have attracted key investments from Microsoft, Sun Microsystems, Nippon Telegraph, and IBM. Silicon Plateau around Bangalore, India, has over 200 export-oriented software companies that are able to draw on a relatively cheap but highly educated labor pool. General Electric's biggest research center outside the United States is in Bangalore, with more than 2,000 Indian engineers and scientists. Altogether, India's software industries earned more than $700 billion in foreign exchange in 2004.

It is the spatial effects of information technologies' *applications* that are of most interest to geographers. As we have seen in previous chapters, information technologies have already greatly facilitated the globalization of

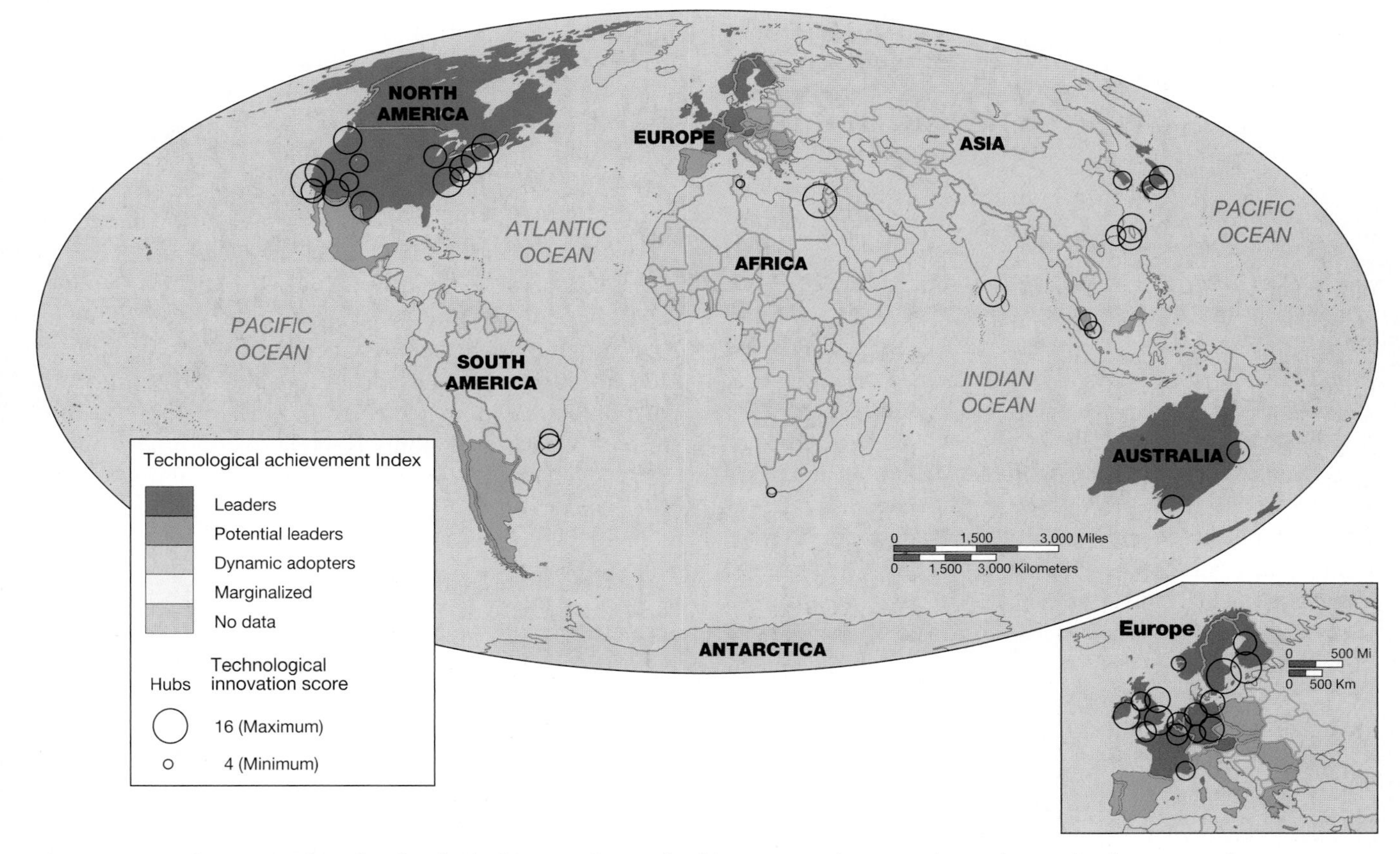

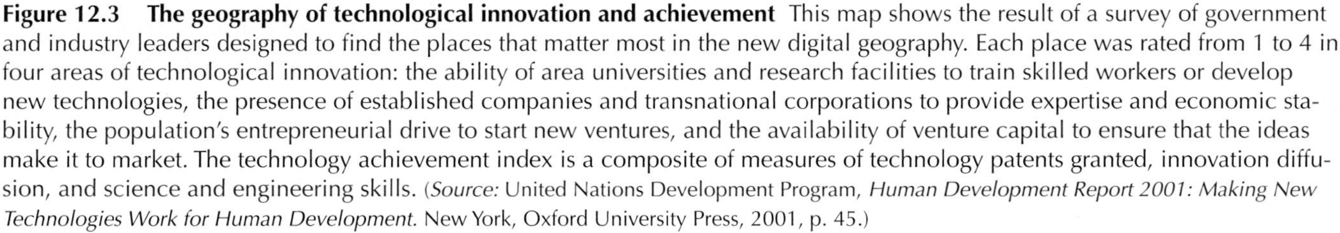

Figure 12.3 The geography of technological innovation and achievement This map shows the result of a survey of government and industry leaders designed to find the places that matter most in the new digital geography. Each place was rated from 1 to 4 in four areas of technological innovation: the ability of area universities and research facilities to train skilled workers or develop new technologies, the presence of established companies and transnational corporations to provide expertise and economic stability, the population's entrepreneurial drive to start new ventures, and the availability of venture capital to ensure that the ideas make it to market. The technology achievement index is a composite of measures of technology patents granted, innovation diffusion, and science and engineering skills. (*Source:* United Nations Development Program, *Human Development Report 2001: Making New Technologies Work for Human Development.* New York, Oxford University Press, 2001, p. 45.)

industry, finance, and culture. At local and regional scales, while they have been instrumental in decentralizing jobs and residences, their impact has been very *uneven*. Computers, for example, are mainly used in peripheral countries for standard functions of inventory control, accounting, and payroll, and even so they remain much too expensive for widespread use in homes, small businesses, and local governments. About 80 percent of the population of the world has never even used a telephone, let alone sent an e-mail message. We can expect future impacts of information technologies to exhibit the same unevenness. While the world will certainly shrink even further, a marked lag will occur in the diffusion of information technologies to many peripheral regions, thus perpetuating and even accentuating the digital divide between the fast world and the slow world. As the United Nations Development Programme observes:

> The Internet is creating parallel communications systems: one for those with income, education, and—literally—connections . . . the other for those without connections, blocked by barriers of time, cost, and uncertainty and dependent on out-of-date information. With people in these two systems living and competing side-by-side, the advantages of connection are overpowering. The voices and concerns of people already living in human poverty—lacking incomes, education, and access to public institutions—are being increasingly marginalized.[1]

It is clear that information technologies will bring about a great deal of change in the lifestyles and landscapes of the fast world. In 2005, about 8 percent of the workforce in the more affluent core countries was "distributed," working at home online for a significant part of the week. As telematics drop in cost, the distributed workforce may well increase to 20 percent by 2020. In addition to employees, there will be more people who operate small businesses out of their homes. Considering this growth of home-based work, there must be a point—perhaps when 30 or 40 percent of the workforce is no longer going to traditional workplaces—when there will be substantial changes in how people dress (much less need to "dress for success"), where they eat, how they play, who they socialize with, and how they use their time. In addition, how many will need a $20,000 or $30,000 automobile to travel only 4,500 or so miles per year?

Further advances in computational capacity and in telematics, GIS, and computer-driven surveillance will mean that many of the complex, hard-to-manage aspects of society, such as street crime and traffic, will increasingly be subject to automated surveillance and management systems. New buildings, from individual houses to mega-developments, will be designed and tested in virtual space, and more environments will be "smart." These same technologies will also likely increase the gap between the haves and the have-nots within the fast world. Robotics and computers have already displaced millions of workers in the fast world. As the Information Revolution unfolds and matures, further changes in labor markets can be expected, creating high-wage jobs for those equipped to participate in new, knowledge-based industries. On the other hand, there is no reason to expect any expansion in jobs for unskilled or poorly educated workers. The net result is likely to be an increase in socioeconomic polarization.

[1]United Nations Development Programme, *Cities in a Globalizing World: Global Report on Human Settlements 2001*. Sterling, VA: Earthscan Publications, 2001, p. 7.

REGIONAL PROSPECTS

For many years now, organizations such as the United Nations and the World Bank have prepared forecasts of the world economy. These forecasts are based on economic models that take data on macro-economic variables (for example, trends in countries' gross domestic product, imports and exports, economic structure, investment and saving performance, and demographic dynamism) and use known relationships between and among these variables to predict future outcomes. The problem is that economic projection is an inexact science. Economic models are not able to take into account the changes brought about by major technological innovations, significant geopolitical shifts, governments' willingness and ability to develop strong economic policies, or the rather mysterious longer term ups and downs that characterize the world economy. What is clear is that in *overall* terms the global economy is vastly richer, more productive, and more dynamic than it was just 15 or 20 years ago, and every prospect exists that, in the longer term, the world economy will continue to expand. The U.S. National Intelligence Council's 2004 report *Mapping the Global Future*—based on consultations with nongovernmental experts from around the world—concluded that the world economy is likely to continue growing impressively: By 2020, it is projected to be about 80 percent larger than it was in 2000, and average per capita income will be roughly 50 percent higher.

Uneven Development

This expansion, of course, will be uneven. The world in 2020 will change radically. Emerging powers—China, India, and perhaps others such as Brazil and Indonesia—have the potential to render obsolete the old categories of East and West, North and South, aligned and nonaligned, developed and developing. Traditional geographic groupings will increasingly lose salience in international relations. A state-bound world and a world of mega-cities, linked by flows of telecommunications, trade, and finance will co-exist. Competition for allegiances will be more open, less fixed than in the past.

In parallel with the world-system categorization of core, semiperipheral, and peripheral regions, we can think of the prospects of large sectors of the world's population as being in one of three categories: the elite, the embattled, and the marginalized. The elite are participants in—and beneficiaries of—the fast world of new transport and communications technologies, globalized production networks, and global consumer culture. The embattled are also participants, but in dependent roles, with fewer benefits and limited opportunities: assembly-line workers, for example, in offshore commodity chains. The marginalized have to survive within the slow world, largely disconnected from formal economies and the dynamics of globalization.

The most significant point to keep in mind about the stratification of the world economy through contemporary globalization is that the elite, the marginalized, and the embattled are likely to be less concentrated than they once were in particular regions. It is no longer accurate to see Nigeria, for instance, as a wholly embattled region or its population as exclusively embattled. Instead, Nigeria, like the United States and most other countries, contains a range of stratified regions and groups within its national boundaries such that elite, embattled, and marginalized social groups are all part of the larger whole. Moreover, the elite regions and social groups have more in common with elite regions and groups in other parts of the world than they do with the embattled and marginalized groups within their own national boundaries (see **Figure 12.4**).

Nevertheless, future geographies seem likely to be structured by an even greater gap between the haves and have-nots of the world. The gap between the world's core areas and the periphery has already begun to widen significantly. Even within the United States, the degree of income inequality ranks with that in China, Bolivia, Malaysia, Senegal, and Russia (**Figure 12.5**). There is also evidence of a growing disparity. U.S. Census Bureau data, for example, show that the gap between the rich and poor in the United States in 2001 was the widest it had been since World War II. The United Nations has calculated that the ratio of GDP per capita (measured at constant prices and exchange rates) between the developed and developing areas of the world increased from 10:1 in 1970 to 12:1 in 1985 and 13:1 in 2002. Little hope exists that any future boom in the overall world economy will reverse this trend. In spite of the globalization of the world-system (and in many ways *because* of it), much of the world has been all but written off by the bankers and corporate executives of the core.

A New World Order?

The same factors that will consolidate the advantages of the core of the world-system as a whole—the end of the Cold War, the availability of advanced telecommunications, the transnational reorganization of industry and finance, the liberalization of trade, and the emergence of a global culture—will also open the way for a new geopolitical and geoeconomic order. This is likely to involve some new relationships between places, regions, and countries. As we have suggested, the old order of the "short" twentieth century (1914–1989), dominated both economically and politically by the United States, is rapidly disappearing. In our present transitional phase, the new world order is up for grabs; we are coming to the end of a geopolitical leadership cycle. This does not necessarily mean that the United States will be unable to renew or extend its position as the world's dominant power. As we saw in Chapter 2, Britain had two consecutive stints as the dominant world power—the hegemony that was able to impose its political view on the world and set the terms for a wide variety of economic and cultural practices.

The United States

The United States is the reigning hegemon and its economy is the largest in the world, with a broad resource base; a large, well-trained, and very sophisticated work force; a domestic market that has greater purchasing power than any other single country; and a high level of technological sophistication. The United States also has the most powerful and technologically sophisticated military apparatus; and it has the dominant voice and last word in international economic and political affairs. It is at least as well placed as its rivals to exploit the new technologies and new industries of a globalizing economy. It also has a distinctive message as a global leader: free markets, personal liberty, private property, electoral democracy, and mass consumption.

For the moment, however, the United States is a declining hegemon, at least in relative terms. Its economic dominance is no longer unquestioned in the way it was in the 1950s, 1960s, and 1970s. A relatively sluggish rate of growth has brought the prospect of the U.S. economy being caught and overtaken by China. On some measures of economic development, the European Union has already overtaken the United States. More important, the globalization of the economy has severely constrained the ability of the United States to translate its economic might into the firm control of international financial markets that it used to enjoy.

The end of the Cold War, while a victory for the United States, robbed it of its image as Defender of the Free World and weakened the legitimacy of its role as global policeman. For a decade or so, the absence of a Cold War enemy and the globalization of economic affairs made it much more difficult for the United States to identify and define the national interest. The hesitancy in U.S. policies toward Kosovo, East Timor, Somalia, Sudan, Bosnia, and Haiti in the 1990s was symptomatic of such problems. The terrorist al-Qaida attacks of 2001 gave a new focus for U.S. geopolitical strategy. But the invasion of Iraq in the absence of evidence of weapons of mass destruction, together with U.S. refusal to participate in high-profile international agreements such as the Kyoto Protocol and its lack of cooperation with the International Criminal Court and the United Nations Organization, has resulted in a significant weakening of the political, cultural, and moral leadership of the United States in global affairs,

Figure 12.4 Global social hierarchy These photographs underscore the new social hierarchy that has emerged around globalization. Pictured here are (top) members of the marginalized stratum in Haiti; (middle) members of the middle stratum in Mexico; and (bottom) members of the elite stratum in the United Kingdom.

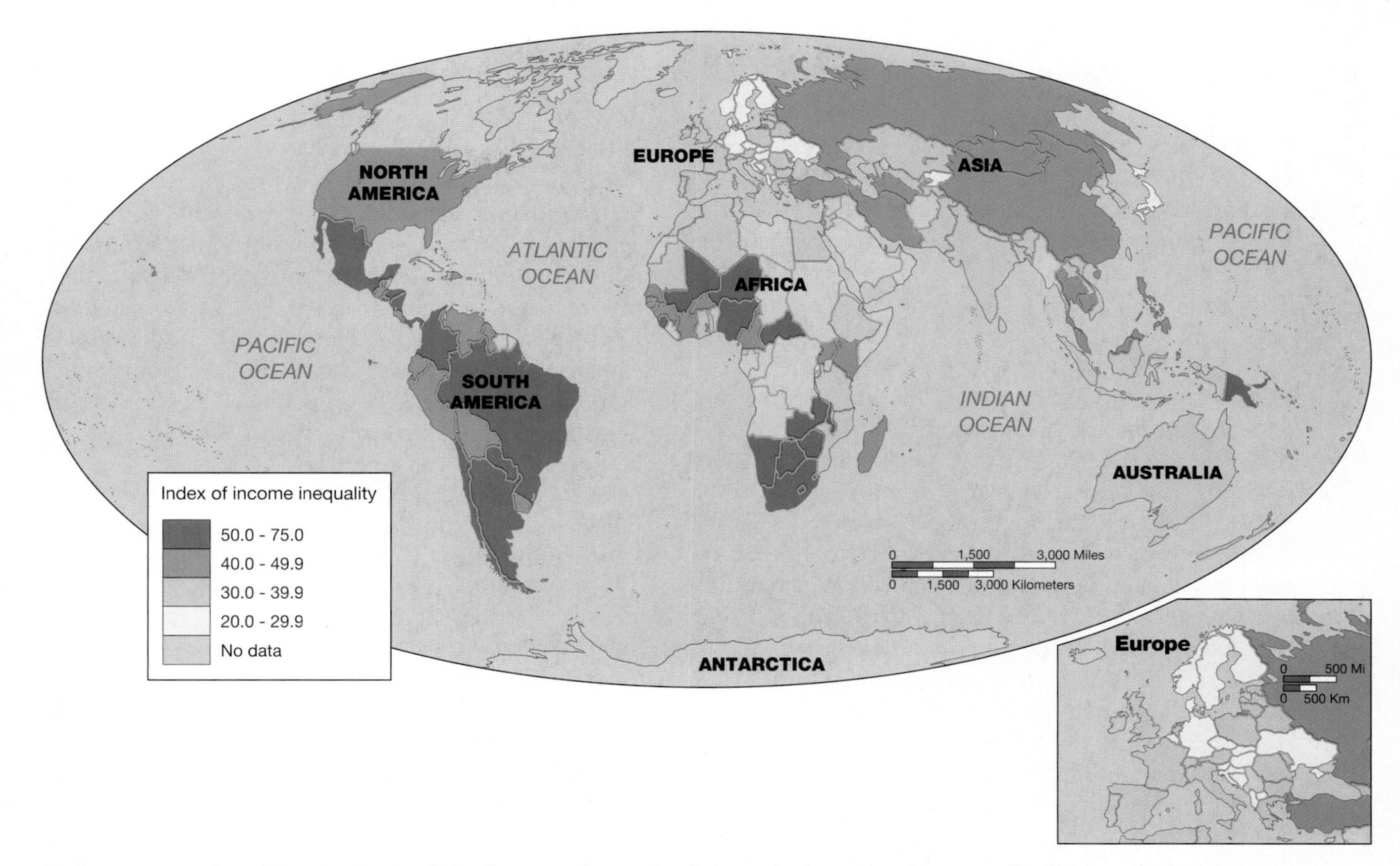

Figure 12.5 Index of income inequality This map shows the degree of inequality of income distributions within countries using the Gini coefficient, a commonly used indicator that compares actual income distributions to a hypothetical distribution of perfect equality. The higher the index score, the greater the degree of inequality within a country. (*Source:* United Nations Development Programme, *Cities in a Globalizing World.* Sterling, VA: Earthscan Publications, 2001, p. 18.)

even from the perspective of elite groups (see Box 12.2: "How the World Sees the United States"). Growing numbers of ordinary people around the world, especially in the Middle East and the broader Muslim world, believe the United States is bent on regional domination—or direct political and economic domination of other states and their resources. In the future, growing distrust could prompt governments to take a more hostile approach to the United States, including resistance to support for U.S. interests in multinational forums and the development of military capabilities as a hedge against the United States. Meanwhile, America's self-perception as a global leader and champion of benign universal values is contradicted by a deep well of national-greatness conservatism that is overwhelmingly self-referential. As geographer Neil Smith has argued, this nationalism has undercut the ability of the United States to impose its vision of globalization in the past, and may well do so again in the future.

In summary, while the United States must be considered the strongest contender, it is by no means a foregone conclusion that it will, in fact, become the leader for the next cycle of economic and political leadership. The United States has choices not dissimilar to those faced by the British at the end of the nineteenth century. It can either oppose its rivals or accommodate them. It can oppose by pressing for a seamless global system that remains under its own hegemony. Or it can try to accommodate by coaxing the others into a global sharing of power, with some mix of regional spheres of interest and collective world responsibilities.

The European Union

It is not stretching things too far to see a successful European Union as the main contender for world leadership. In geopolitical terms, the collapse of the Soviet Union advanced the prospects of the European Union significantly. The European Union, which began in 1952 as a trading bloc (the European Economic Community, or EEC), is now an economic union (with integrated economic policies among member States) and has moved a long way toward its goal of becoming a supranational political union (with a single set of institutions and policies). With the addition of Cyprus, the Czech Republic, Estonia, Hungary, Latvia, Lithuania, Malta, Slovakia, and Slovenia in 2004, the European Union had a total population of 458 million (the U.S. population is 293 million) and an overall economy of \$11.7 trillion (the GDP of the United States is just under \$11 trillion). The proposed addition of Bulgaria, Croatia, Romania, and Turkey would mean a total population of 527 million

12.2

GEOGRAPHY MATTERS

How the World Sees the United States

In 2004 the U.S. National Intelligence Council consulted with experts in six regional conferences, asking participants about their views of the role of the United States as a driver in shaping developments in their regions and globally.

Asia

Participants felt that U.S. preoccupation with the war on terrorism is largely irrelevant to the security concerns of most Asians. The key question that the United States needs to ask itself is whether it can offer Asian states an appealing vision of regional security and order that will rival and perhaps exceed that offered by China. U.S. disengagement from what matters to its Asian allies would increase the likelihood that they would climb on Beijing's bandwagon and allow China to create its own regional security order that excludes the United States. Participants felt that the rise of China need not be incompatible with a U.S.-led international order. The critical question is whether or not the order is flexible enough to adjust to a changing distribution of power on a global level. An inflexible order would increase the likelihood of political conflict between emerging powers and the United States. If the order is flexible, it may be possible to forge an accommodation with rising powers and strengthen the order in the process.

Sub-Saharan Africa

Sub-Saharan African leaders worry that the United States and other advantaged countries will "pull up the drawbridge" and abandon the region. Participants opined that the United States and other Western countries may not continue to accept Africa's most successful "export," its people. The new African diaspora is composed overwhelmingly of economic migrants rather than political migrants as in previous eras. Some participants felt that Africans worry that Western countries will see some African countries as "hopeless" over the next 15 years because of prevailing economic conditions, ecological problems, and political circumstances. Participants feared that the United States will focus only on those African countries that are successful.

Latin America

Conference participants acknowledged that the United States is the key economic, political, and military player in the hemisphere. At the same time, Washington was viewed as traditionally not paying sustained attention to the region and, instead of responding to systemic problems, as reacting primarily to crises. Participants saw a fundamentalist trend in Washington that would lead to isolation and unilateralism and undercut cooperation. Most shared the view that the U.S. "war on terrorism" had little to do with Latin America's security concerns. Latin American migrants are a

and a GDP of $12.1 trillion (Figure 9.29). If and when it achieves full political union, its size will enable it to outvote the United States in the International Monetary Fund and the World Bank.

The EU has evolved dramatically since its start with six member nations in 1952. Now there is a common EU passport and a single currency, the euro, is used by many of its members. The EU regulates trade as well as coordinating energy, communications, and transportation. It has a president, a parliament, foreign policy powers, and a court whose decisions are binding on member countries and individuals. The EU is the world's largest internal market and largest exporting power. Of the world's 20 largest commercial banks, 14 are European. The EU also has one of the world's most vibrant and sophisticated industrial core regions that stretches from Southeast England to Northern Italy; its economic policies and infrastructure investments have fostered emerging axes of economic growth that stretch east-west and north-south (**Figure 12.6**); and European industries are world leaders in chemicals, insurance, engineering, construction and aerospace industries. The EU also offers a distinctive message. In contrast to the American Dream,

> The European Dream emphasizes community relationships over individual autonomy, cultural diversity over assimilation, quality of life over the accumulation of wealth, sustainable development over unlimited material growth, deep play over unrelenting toil, universal human rights and the rights of nature over property rights, and global cooperation over the unilateral exercise of power.[2]

[2] J. Rifkin, *The European Dream*. Los Angeles: J. P. Tarcher, 2004.

stabilizing force in relations with the United States. An important part of the U.S. labor pool, migrants also remit home needed dollars along with new views on democratic governance and individual initiative that will have a positive impact on the region.

U.S. policies also can have a positive impact. Some participants said the region would benefit from U.S. application of regional mechanisms to resolve problems rather than punitive measures against regimes not to its liking, such as that of Fidel Castro.

Middle East

Participants felt that the role of U.S. foreign policy in the region will continue to be crucial. The perceived propping up of corrupt regimes by the United States in exchange for secure oil sources has in itself helped to promote continued stagnation. Disengagement is highly unlikely but would in itself have an incalculable effect. Regarding the prospects for democracy in the region, participants felt that the West placed too much emphasis on the holding of elections, which, while important, is only one element of the democratization process. There was general agreement that if the United States and Europe can engage with and encourage reformers rather than confront and hector, genuine democracy would be achieved sooner.

Some Middle East experts argued that Washington has reinforced zero-sum politics in the region by focusing on top Arab rulers and not cultivating ties with emerging leaders in and outside the government. Although the Middle East has a lot to gain economically from globalization, it was agreed that Arabs/Muslims are nervous that certain aspects of globalization, especially the pervasive influence of Western, particularly American, values and morality are a threat to traditional cultural and religious values.

Europe and Eurasia

Participants engaged in a lively debate over whether a rift between the United States and Europe is likely to occur over the next 15 years, with some contending that a collapse of the U.S.-EU partnership would occur as part of the collapse of the international system. Several participants contended that if the United States shifts its focus to Asia, the U.S.-EU relationship could be strained to breaking point. They were divided over whether China's rise would draw the United States and Europe closer or not. They also differed over the importance of common economic, environmental, and energy problems to the alliance. Participants agreed that the United States has only limited influence on the domestic policies of the Central Asian states, although U.S. success or failure in Iraq would have spillover effects in Central Asia. Countries in western Eurasia, they believed, will continue to seek a balance between Russia and the West. In their view, Ukraine almost certainly will continue to seek admission to NATO and the European Union, while Georgia and Moldova probably will maintain their orientation in the same direction.

Source: National Intelligence Council, *Mapping the Global Future.* Washington, DC: U.S. Government Printing Office, 2004, p. 114-5.

Successful enlargement, combined with already-successful monetary union and economic integration, would see Europe poised either to challenge U.S. hegemony or to become a senior partner among superpowers. EU citizens, however, have expressed reservations about establishing a Constitution to underpin a supranational political union. There are also concerns that the extension of the EU to 30 countries would risk destroying its own internal balance and cohesion. Europe's future international role also depends greatly on whether it undertakes major structural economic and social reforms to deal with the problem of its aging work-force. This will demand more immigration and better integration of workers, most of whom are likely to be coming from North Africa and the Middle East. Even if more migrant workers are not allowed in, Western Europe will have to integrate a growing Muslim population. Barring increased legal entry may only lead to more illegal migrants who will be harder to integrate, posing a long-term problem. It is possible to imagine European nations successfully adapting their workforces and social welfare systems to these new realities, but it is harder to see some countries—Germany, for example—successfully assimilating millions of new immigrant workers in a short period of time.

China and India

According to the U.S. National Intelligence Council (NIC), the greatest benefits of globalization will accrue to countries and groups that can access and adopt new technologies. The Council's report on *Mapping the Global Future* concludes that China and India are well positioned to become technology leaders, and that the expected next revolution in high technology—involving the convergence of nano-, bio-, information and materials technology—could further bolster China and India's prospects. Both

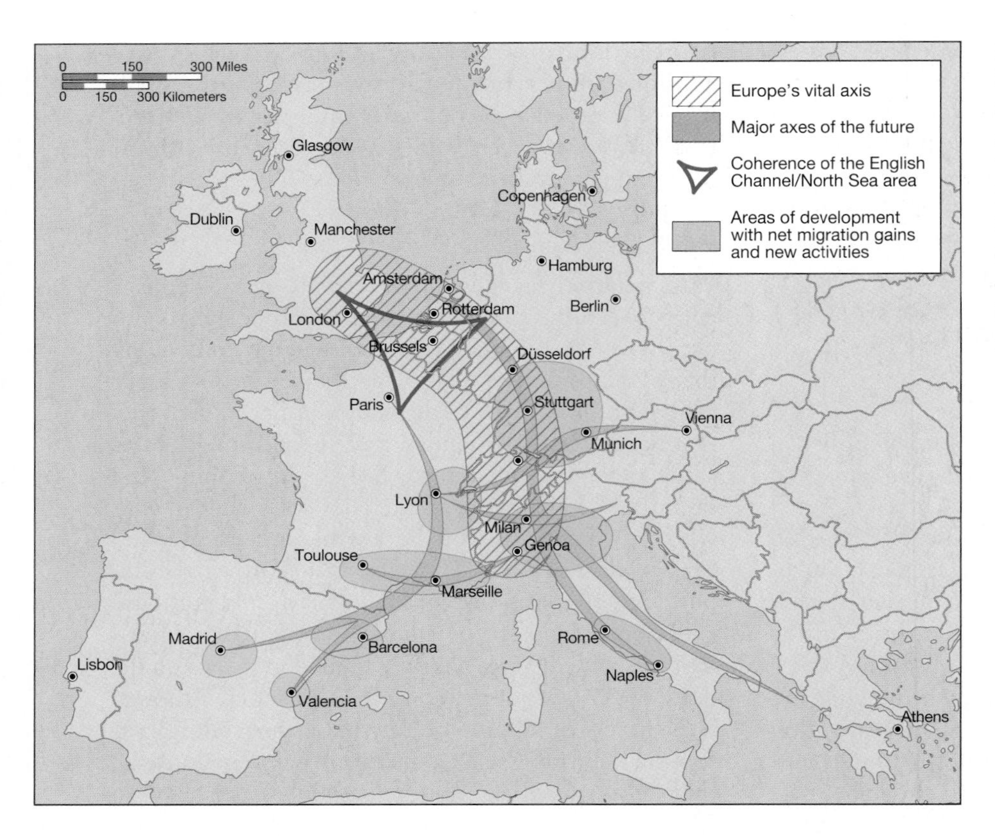

Figure 12.6 European growth axes Most of Europe's major cities and advanced manufacturing regions lie along a crescent-shaped axis that runs from Southeast England through Southwest Germany to Northern Italy, and along emerging axes that extend to Southern France, Spain, eastern Austria, and Southern Italy. Compare to the map of high-speed rail in Figure 12.2.

countries are investing in basic research in these fields and are well placed to be leaders in a number of key fields. As the NIC report puts it:

> The likely emergence of China and India, as well as others, as new major global players—similar to the advent of a united Germany in the 19th century and a powerful United States in the early 20th century—will transform the geopolitical landscape, with impacts potentially as dramatic as those in the previous two centuries. In the same way that commentators refer to the 1900s as the "American Century," the 21st century may be seen as the time when Asia, led by China and India, comes into its own. A combination of sustained high economic growth, expanding military capabilities, and large populations will be at the root of the expected rapid rise in economic and political power for both countries. . . . Barring an abrupt reversal of the process of globalization or any major upheavals in these countries, the rise of these new powers is a virtual certainty.[3]

Most forecasts indicate that by 2020 China's GNI will exceed that of individual Western economic powers except the United States. India's GNI will have overtaken or be on the threshold of overtaking the larger European economies. Because of the sheer size of China's and India's populations—projected by the U.S. Census Bureau to be 1.4 billion and almost 1.3 billion respectively by 2020—their standard of living need not approach Western levels for these countries to become important economic powers. How China and India exercise their growing power and whether they relate cooperatively or competitively to other powers in the international system, however, are key uncertainties.

So, although China is not currently part of the core of the world-system, many observers predict a "Pacific Destiny" for the twenty-first century. In this scenario, China will be the hub of a world economy whose center of gravity is around the rim of the Pacific rather than the North Atlantic. China certainly has the potential to be a contender. It has a vast territory with a comprehensive resource base and a long history of political, cultural, and economic integration. It has the largest population of any country in the world (1.3 billion in 2004), and an economy that has been growing very rapidly. Beginning in 1978, under the leadership of Deng Xiaoping, China completely reorganized and revitalized its economy. Agriculture was decollectivized, with communist collective farms modified to allow a degree of private profit taking. State-owned industries were closed or privatized, and centralized state planning was dismantled in order to foster private entrepreneurship. Since 1992, China has extended its "open door" policy (that is, allowing trade with the rest of the world) beyond its special economic zones and allowed foreign investment aimed at Chinese domestic markets. In the 1980s and early 1990s, when the world economy was sluggish, China's manufacturing sector grew by almost 15 percent each year, and has continued into the

[3]National Intelligence Council, *Mapping the Global Future: Report of the National Intelligence Council's 2020 Project.* Washington, DC: U.S. Government Printing Office, 2004, p. 11.

2000s with double-digit growth. China's increased participation in the world economy has created an entirely new situation, causing a deflationary trend in world prices for manufactures (**Figure 12.7**). Not only does the Chinese economy's size make it a major producer, but its huge labor force ensures that its wage levels will not approach Western levels for a long time, thus guaranteeing a key competitive advantage. Overall, China's economy is already the fourth largest in the world after the European Union, the United States, and Japan.

Nevertheless, China's economy will remain largely agrarian for some time yet. The task of feeding, clothing, and housing its enormous population will also constrain its ability to modernize its economy to the point where it can dominate the world economy. Meanwhile, a great deal of social and political reform remains to be accomplished before free enterprise can really flourish. Finally, before China can emerge as a hegemonic power it must resolve a major feature of its contemporary human geography: the dramatically uneven development that has been a consequence of its economic reforms. The positive spiral of cumulative causation has affected only the larger cities and coastal regions, while the vast interior regions of the country have become increasingly impoverished. This, ironically, is the very geographical pattern of spatial polarization that motivated Chinese communists to revolt in the 1940s. Today, it is a source of potential unrest and of political instability among the central government, the provincial governments of the interior regions, and the provincial governments of the coastal regions. In the short term, China mostly needs to make room inside itself: to develop its internal market and moderate its growth. In some respects, China has been following such a strategy with a big program of infrastructure investment. Recently, it has also been trying to boost domestic demand by easing credit. But it remains to be seen whether the Chinese government can continue to control the situation. Managing things is not easy when capitalist energy is turning life upside down, creating vast new wealth amid a poor society based on different principles.

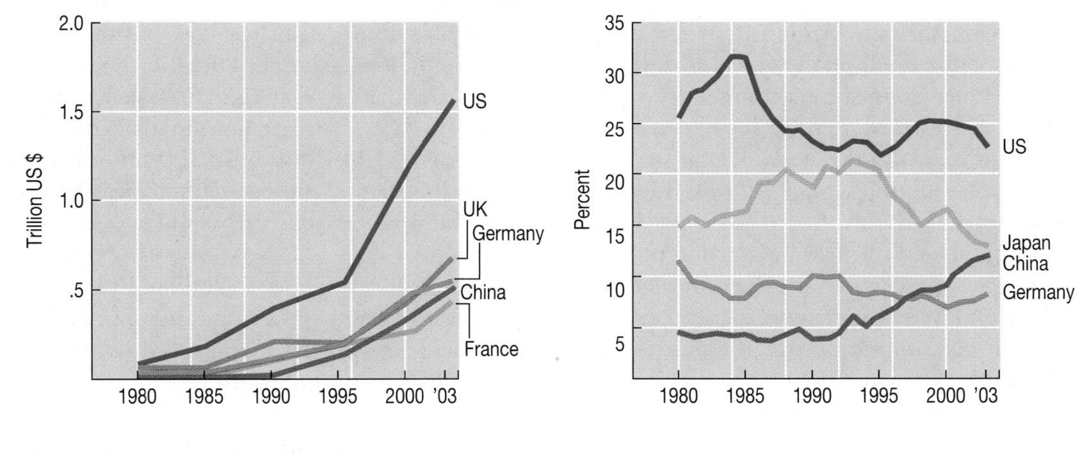

Stock of Inward Foreign Direct Investment, 1980 - 2003

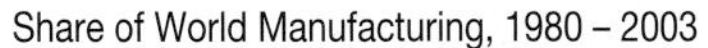

Share of World Manufacturing, 1980 - 2003

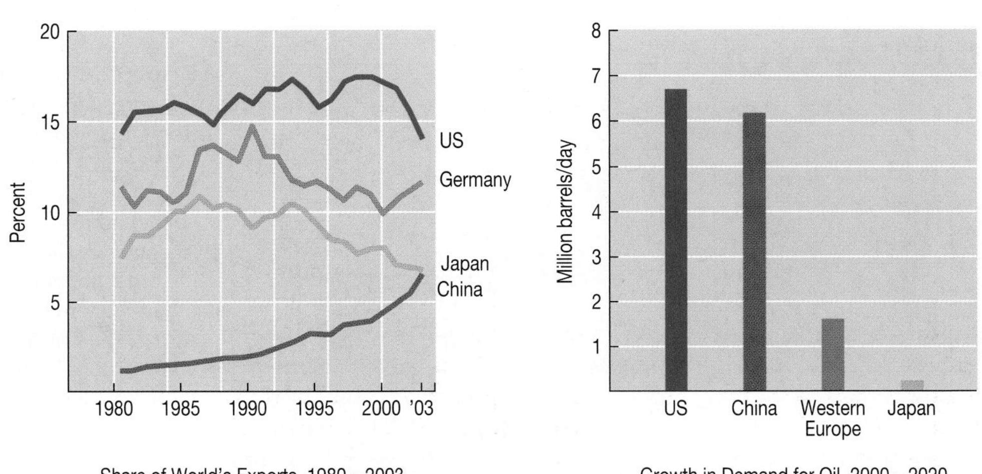

Share of World's Exports, 1980 - 2003

Growth in Demand for Oil, 2000 - 2020

Figure 12.7 China's changing role in the world economy China's emergence as a significant player in the world economy is reflected in trends of inward investment, exports, manufacturing output, and oil consumption.

Alternatively, we may not see the same kind of hegemonic power in the new world order of the twenty-first century—there may not be a new hegemony at all. Instead, the globalization of economics and culture may result in a polycentric network of nations, regions, and world cities bound together by flows of goods and capital. Order may come not from military strength rooted in national economic muscle but from a mutual dependence on *trans*national production and marketing, with stability and regulation provided by powerful international institutions (such as the World Bank, the IMF, the World Trade Organization, the European Union, NATO, and the United Nations).

At the Margins

Whatever the outcome of any new world order, the prospects for peripheral countries and marginalized populations are very bad indeed. Notwithstanding the debt-relief packages and proclamations of the G8 group of the world's most affluent countries and the grassroots popularity of the Live8 concerts in 2005 (**Figure 12.8**), peripheral countries and marginalized populations face near-insurmountable obstacles to significant progress. It is not just that they have already been dismissed by investors in the core, nor that their domestic economies are simply threadbare. They face unprecedented levels of demographic, environmental, economic, and societal stress. In the worst-off regions—including much of West and Central Africa, for example—the events of the next 15–20 years are going to be played out from a starting point of scarce basic resources, serious environmental degradation, overpopulation, disease, unprovoked crime, refugee migrations, and criminal anarchy. African countries will be further disadvantaged because the prices of commodities produced there and in other peripheral locations have been dropping, while imported goods from the core have become more expensive. What this means is that many peripheral countries, especially in Africa, have and will probably continue to have reduced purchasing power in the global marketplace because of the decline in the value of their exports. Continuing to disable the periphery's full participation in the global economy are the combined effects of external debt crisis, dwindling amounts of foreign aid, insufficient resources to purchase technology or develop indigenous technological innovations, and the high costs of marketing and transporting commodities.

Post-independence ideals of modernization and democracy now seem more remote than ever in these regions. Corrupt dictators, epitomized by The Democratic Republic of the Congo's former President Mobutu and Nigeria's former President General Abacha, have created "kleptocracies" (as in *kleptomania:* an irresistible desire to steal) in place of democracies. In early 2004, the anti-corruption watchdog Transparency International released a list of what it believes to be the 10 most self-enriching leaders in recent years. In order of amount allegedly stolen, they are: former Indonesian President Suharto (US$15 billion–$35 billion), former Philippine President Ferdinand Marcos ($5 billion–$10 billion), former Zairian President Mobutu Sésé Seko ($5 billion), former Nigerian President Sani Abacha ($2 billion–$5 billion), former Yugoslav President Slobodan Milosevic ($1 billion), former Haitian President Jean-Claude Duvalier ($300 million–$800 million), former Peruvian President Alberto Fujimori ($600 million), former Ukrainian Prime Minister Pavlo Lazarenko ($114 million–$200 million), former Nicaraguan President Arnoldo Alemán ($100 million), and former Philippine President Joseph Estrada ($78 million–$80 million)..

Some governments, including those of Liberia, Sudan, and Sierra Leone, have lost control of large parts of their territories: groups of unemployed youths plunder travelers; tribal groups war with one another; refugees trudge from war zones to camps and back again; environmental degradation proceeds unchecked (**Figure 12.9**). The U.S. State Department has estimated that during the 1990s, wars in Africa produced over 8 million refugees and claimed 7–8 million lives, including about 2 million chil-

Figure 12.8 Live8 concert Organized by former pop star Bob Geldorf, the Live8 concert in London's Hyde Park in June 2005 was one of a series of concerts held to support a campaign to convince G8 leaders to double aid, cancel debt, and improve the terms of trade for Africa. The concert featured, among others, Mariah Carey, Pink Floyd, Madonna, U2, and The Who.

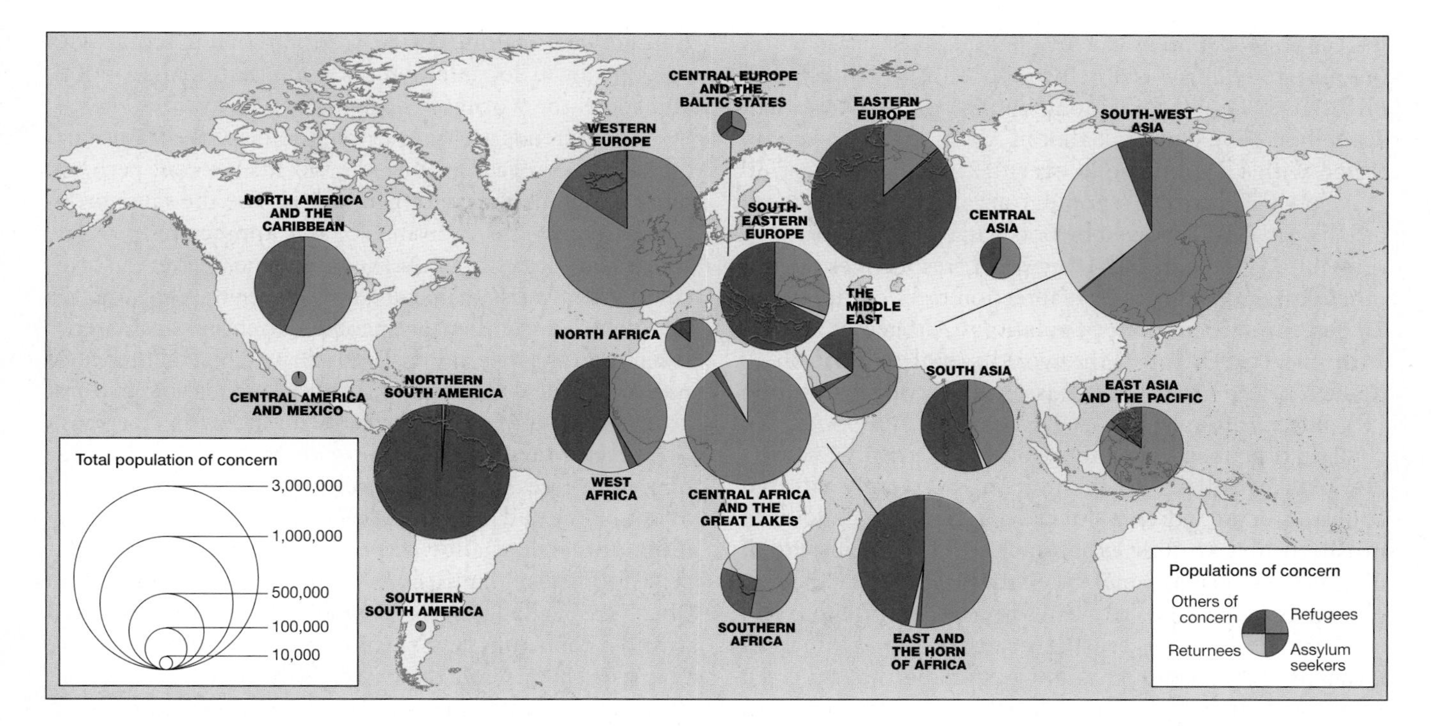

About 40 million people were at immediate risk of malnutrition or death in the mid 2000s because of the combination of war, famine, disease, and criminal anarchy. In this photograph, a vulture waits dispassionately as a young girl, weakened by hunger, collapses on the roadside. The photograph was taken during Sudan's civil war of the 1990s. After the photographer, Kevin Carter, had shooed away the vulture, the girl stood up and walked away toward a feeding camp.

In the Sudan, the brutal oppression, ethnic cleansing, and genocide sponsored by the central government allowed Arab rebels known as the Janjaweed to slaughter thousands of people in the Darfur region, causing the mass displacement of an estimated 1 million refugees. The people of the Darfur region resorted to popular political and military resistance for purposes of survival, forming a political movement known as the Sudan Liberation Movement. Five members of its military wing, the Sudan Liberation Army (SLA), stand in front of their transport in 2004.

Figure 12.9 Wild zones Parts of Africa, the Middle East, and South Asia have become "wild zones," places where national governments have lost control over economic development, ethnic conflict, and environmental degradation. In 2004, according to estimates supplied by the United Nations High Commissioner for Refugees, there were more than 17 million refugees, asylum seekers, and displaced persons in the world—more than 5 million of them in Africa. The map shows the areas of concern listed by the UNHCR in its Report for 2004.

dren. One of the bloodiest wars, between Eritrea and Ethiopia, was launched in May 1998 and centers on an ill-defined border region known as Badme. Most conflicts in Africa, however, are civil wars that have become complicated and intractable as guerrilla groups have proliferated and divided into warring factions. Among the worst-affected counties are Algeria, Angola, Burundi, Congo, Democratic Republic of Congo (formerly Zaire), Guinea-Bissau, Rwanda, Sierra Leone, Somalia, Sudan, and Uganda.

Amid this chaos, disease has prospered. Parts of Africa may be more dangerously unhealthy today than they were 100 years ago. Malaria and tuberculosis are out of control over much of sub-Saharan Africa, while AIDS is truly epidemic. In Uganda, where annual spending on health is less than $5 per person (compared with international debt repayments of almost $20 per person), one in five children die before their fifth birthday, and HIV/AIDS is epidemic among young adults. More than 23 million people in sub-Saharan Africa in 2002 were infected with HIV, or one in every 12 people. Almost 30 percent of Zimbabwe's population aged 15 to 49 is infected with HIV, followed by Botswana and Zambia, with rates of 25 percent and 19 percent, respectively. Eight other African countries have infection rates of more than 10 percent of their adult populations. Altogether, Africa, with just 10 percent of the world's total population, is home to more than two-thirds of the world's 33 million HIV-infected people. HIV/AIDS has left 6 million African children orphaned, accounting for 7 percent of the world's HIV/AIDS orphans. In the nine most severely affected countries in the world, all of them in Africa, a child born in 2003 will have a life expectancy of 43 years, instead of 60 years in the absence of HIV/AIDS. And while the rate of increase in HIV-positive cases in core countries has leveled off, it is still increasing dramatically in much of Africa. Given all this, it is not surprising that much of Africa is unattractive to the globalizing world economy, and it should not be surprising if it remains so.

CRITICAL ISSUES AND THREATS

We must add to our review of future prospects for spatial change the interdependent and less predictable dynamics of politics, culture, society, and environment. From what we have discussed already, it is clear that the immediate future will be characterized by a phase of geopolitical and geoeconomic transition, by the continued overall expansion of the world economy, and by the continued globalization of industry, finance, and culture. The processes of change involved in shaping this kind of future will inevitably bring forth critical issues, changes, conflicts, and threats. We can identify what several of these might involve: fault lines of cultural dissonance, complex security issues, and sustainability.

Globalizing Culture and Cultural Dissonance

At one level, globalization has brought a homogenization of culture through the language of consumer goods. This is the material culture of the fast world, enmeshed by Airbus jets, CNN, music video channels, cell phones, and the Internet; and swamped by Coca-Cola, Budweiser, McDonald's, GAP clothing, Nikes, Walkmans, Nintendos, Toyotas, Disney franchising, and formula-driven Hollywood movies. Furthermore, sociologists have recognized that a distinctive culture of "global metropolitanism" is emerging among the transnational elite. This is simply homogenized culture at a higher plane of consumption (French wines instead of Budweiser, Hugo Boss clothes instead of Levis, BMWs instead of Toyotas, and so on). The members of this new culture are people who hold international conference calls, who make decisions and transact investments that are transnational in scope, who edit the news, design and market international products, and travel the world for business and pleasure.

These trends are transcending some of the traditional cultural differences around the world. We can, perhaps, more easily identify with people who use the same products, listen to the same music, and appreciate the same sports stars that we do. At the same time, however, sociocultural cleavages are opening up between the elites and the marginalized. By focusing people's attention on material consumption, they are also obscuring the emergence of new fault lines—between previously compatible cultural groups and between ideologically divergent civilizations.

Several reasons account for the appearance of these new fault lines. One is the release of pressure brought about by the end of the Cold War. The evaporation of external threats has allowed people to focus on other perceived threats and intrusions. Another is the globalization of culture itself. The more people's lives are homogenized through their jobs and their material culture, the more many of them want to revive subjectivity, reconstruct we/us feelings, and reestablish a distinctive cultural identity. For the marginalized, a different set of processes is at work, however. The juxtaposition of poverty, environmental stress, and crowded living conditions alongside the materialism of the fast world creates a fertile climate for gangsterism. The same juxtaposition also provides the ideal circumstance for the spread and intensification of religious fundamentalism and for fundamentalist-inspired terrorism. This, perhaps more than anything else, represents a source of serious potential cultural dissonance.

The overall result is that cultural fault lines are opening up at every geographical scale. This poses the prospect of some very problematic dimensions of future regional geographies, from the scale of the city to the scale of the globe. The prospect at the metropolitan scale is one of fragmented and polarized communities, with outright cultural conflict suppressed only through electronic surveillance (**Figure 12.10**) and the "militarization" of urban space via security posts and "hardened" urban design using fences and gated streets. This, of course, presupposes a certain level of affluence in order to meet the costs of keeping the peace across economic and cultural fault lines. In the unintended metropolises of the periphery, where unprecedented numbers of migrants and refugees will be thrown together, the genuine prospect of anarchy and intercommunal violence exists—unless intergroup differences can be submerged in a common cause, such as religious fundamentalism.

At the subnational scale the prospect is one of increasing ethnic/racial rivalry, parochialism, and insularity. Examples of these phenomena can be found throughout the world. In North America we see increasing ethnic rivalry represented by the secessionism of the French-speaking Quebeçois and the insistence of some Hispanic groups in the United States on the installation of Spanish as an alternative official language. In Europe, examples are the secessionism (from Spain) of the Basques,

Figure 12.10 Surveillance camera Social, economic, and ethnic polarization in the cities of the world's core countries has led to an increase in the electronic surveillance of both public and private spaces. Another significant change has been the increased presence of private security personnel in upscale settings: Private security officers now outnumber police officers in the United States. This photo shows the interior of Chicago's Office of Emergency Management, the center for Chicago's criminal database, "Citizen Law Enforcement and Analysis" (CLEAR).

the separatist movement of the Catalans (also in Spain), the regional elitism of Northern Italy, and, recently, outright war among Serbs and Croats. In South Asia, examples are provided by the recurring hostility between Hindus and Muslims throughout the Indian subcontinent and between the Hindu Tamils and Buddhist Sinhalese of Sri Lanka. In Africa ethnic rivalry, parochialism, and insularity are reflected in the continuing conflict between the Muslim majority in northern Sudan and the Christian minority in the south, and widespread unrest in northeast Democratic Republic of Congo. Where the future brings prosperity, tensions and hostilities such as these will probably be muted; where it brings economic hardship or decline, they will undoubtedly intensify.

The prospect at the global scale is of a rising consciousness of people's identities in terms of their broader historical, geographical, and racial "civilizations:" Western, Latin American, Confucian, Japanese, Islamic, Hindu, and Slavic-Orthodox. According to some observers, deepening cleavages of this sort could replace the ideological differences of the Cold War era as the major source of tension and potential conflict in the world. And as many globalization experts have observed, they may also point the way toward new foundations for cooperation and unity as people see themselves less as citizens of a particular nation or state and more as world citizens sharing a common global cause (see Chapter 9).

Security

In addition to cultural fault lines, globalization will inevitably generate enormous economic, and, consequently, political upheavals that will translate into a complex set of security issues. With the gradual integration of China, India, and other emerging countries into the global economy, hundreds of millions of working-age adults will become available for employment in a more integrated world labor market. This enormous workforce—a growing portion of which will be well educated—will be an attractive, competitive source of low-cost labor at the same time that technological innovation is expanding the range of globally mobile occupations. The transition will hit the middle classes of the core regions in particular, bringing more rapid job turnover and requiring professional retooling, while outsourcing on an increasing scale will intensify the anti-globalization movement. Where these pressures lead will depend on how political leaders respond, how flexible labor markets become, and whether overall economic growth is sufficiently robust to absorb a growing number of displaced workers. In some regions, it is quite possible that weak governments, lagging economies, religious extremism, and youth bulges will align to create a perfect storm for internal conflict, with far-reaching repercussions for security elsewhere.

Such internal conflicts, particularly those that involve ethnic groups straddling national boundaries, risk escalating into regional conflicts and the failure of states, with expanses of territory and populations devoid of effective governmental control. These territories can become sanctuaries for transnational terrorists (such as al-Qaida in Afghanistan) or for criminals and drug cartels (such as in Colombia). According to the U.S. National Intelligence Council, al-Qaida itself will likely be superceded by similarly inspired Islamic extremist groups within the next 15 years, and there is a substantial risk that broad Islamic movements akin to al-Qaida will merge with local separatist movements.[4] Information technology, allowing for instant connectivity, communication, and learning, will enable the terrorist threat to become increasingly decentralized, evolving into an eclectic array of groups, cells, and individuals that do not need a stationary headquarters to plan and carry out operations. Training materials, targeting guidance, weapons know-how, and fund-raising will become virtual (i.e., online). As biotechnology information becomes more widely available, the number of people who can potentially misuse such information and wreak widespread loss of life will increase. Moreover, as biotechnology advances become more ubiquitous, stopping the progress of offensive biological warfare programs will become increasingly difficult. Over the next 10 to 20 years there is a risk that advances in biotechnology will allow the creation of advanced biological agents designed to target specific systems—human, animal, or crops.

Economic and political turbulence and instability will also be conducive to transnational crime. Transnational crime syndicates pose a considerable threat to global security. They distribute harmful materials, weapons, and drugs, exploit local communities, disrupt fragile ecosystems, and control significant economic resources. In 2003, transnational crime syndicates may have grossed up to $2 trillion—more than all national economies except the United States, Japan, and Germany.[5] The bulk of crime syndicates' revenue comes from drug trafficking, but other significant sources include environmental products—everything from protected plants and animals to hazardous waste and banned chemicals. Trafficking in humans—for labor, sex work, and even for the removal of kidneys or other organs for transplant purposes—is another aspect of transnational crime. The U.S. State Department estimates that at least 600,000 to 800,000 people are sold internationally each year. Weapons trafficking earns the syndicates comparatively little—estimated at less than $1 billion annually—but it contributes significantly to the complex security issues facing most countries. There are few international laws on arms trafficking, and U.N. arms embargoes often go unforced.

Another dimension of the security issues associated with globalization concerns the environmental and human health issues that have emerged as a result of increases in global commerce and tourism. As people and goods move around the world, so do other species, pathogens, and zoonotic diseases. New plant and animal species can invade ecosystems, choke pastures, disrupt water systems, drive other species to extinction, and result in expensive and unanticipated consequences. In the United States, the European zebra mussel has invaded freshwater systems across almost half of the country, displacing other marine species and costing an estimated $1 billion in eradication costs during the 1990s. The total costs of losses from all kinds of invasive species in the United States now exceeds $138 billion annually. The diffusion of contagious diseases such as influenza, and of zoonotic diseases such as anthrax, avian flu, Ebola, West Nile virus epidemics, also poses a much greater security risk as a result of the speed and intensity of global flows—another dimension of the "risk society" (see Chapter 1) associated with globalization.

[4] *Ibid.*, p. 9.

[5] E. Assadourian, "Transnational Crime," in Worldwatch Institute, *State of the World 2005*, New York: W.W. Norton, 2005, p. 20.

Sustainability

As we saw in Chapter 4, the world currently faces a daunting list of environmental threats: the destruction of tropical rain forests and the consequent loss of biodiversity; widespread, health-threatening pollution; the degradation of soil, water, and marine resources essential to food production; stratospheric ozone depletion; acid rain, and so on. Most of these threats are greatest in the world's periphery, where daily environmental pollution and degradation amounts to a catastrophe that will continue to unfold, slow motion, in the coming years (see Box 12.3: "The Asian Brown Cloud").

Future trends will only intensify the contrasts between rich and poor regions. We know that environmental problems will be inseparable from processes of demographic change, economic development, and human welfare. In addition, it is becoming clear that environmental problems are going to be increasingly enmeshed in matters of national security and regional conflict. The spatial *interdependence* of economic, environmental, and social problems means that some parts of the world are ecological time bombs. The prospect of civil unrest and mass migrations resulting from the pressures of rapidly growing populations, deforestation, soil erosion, water depletion, air pollution, disease epidemics, and intractable poverty is real. These specters are alarming not only for the peoples of the affected regions but also for the peoples of rich and faraway countries, whose continued prosperity will depend on processes of globalization that are not disrupted by large-scale environmental disasters, unmanageable mass migrations, or the breakdown of stability in the world-system as a whole. U.S. policy effectively denied such concerns when President George W. Bush announced in 2001 that the United States would no longer honor its commitment to the Kyoto agreement (see Box 4.3: "Global Climate Change and the Kyoto Protocol," page 162).

But despite the fact that the United States appears to be excluding itself from efforts to solve some of Earth's most pressing environmental problems, we cannot simply wait to see what the future will hold. If we are to have a better future (and if we are to *deserve* a better future), we must use our understanding of the world—and of geographical patterns and processes—to work toward more desirable outcomes. No discipline is more relevant to the ideal of sustainable development than geography. Where else, as British geographer W. M. Adams has observed, can the science of the environment (physical geography) be married with an understanding of economic, techno-

12.3

GEOGRAPHY MATTERS

The Asian Brown Cloud

At the heart of the World Summit on Sustainable Development in Johannesburg in September 2002 was the question: How can developing nations grow economically without overburdening Earth's environment and creating an uninhabitable planet for future generations? Emblematic of this issue is the Asian Brown Cloud, a blanket of air pollution 3 kilometers (nearly 2 miles) thick that hovers over most of the tropical Indian Ocean and South, Southeast and East Asia, stretching from the Arabian Peninsula across India, Southeast Asia, and China almost to Korea. The brown haze was first identified by U.S. Air Force pilots but is now clearly visible in satellite photographs and from the Himalayas (**Figure 12.A**).

The Asian Brown Cloud consists of sulfates, nitrates, organic substances, black carbon, and fly ash, along with several other pollutants. It is an accumulated cocktail of contamination resulting from a dramatic increase in the burning of fossil fuels in vehicles, industries, and power stations in Asia's megacities, from forest fires used to clear land, and from the emissions from millions of inefficient cookers burning wood or cow dung. A study of the Asian Brown Cloud sponsored by the U.N. Environment Program and involving more than 200 scientists suggests that the Asian Brown Cloud not only influences local weather but also may have worldwide consequences.

The smog of the Asian Brown Cloud reduces the amount of solar radiation reaching the Earth's surface by 10 to 15 percent, with a consequent decline in the productivity of crops. But it can also trap heat, leading to warming of the lower atmosphere. It suppresses rainfall in some areas and increases it in others, while damaging forests and crops because of acid rain. The haze is also believed to be responsible for hundreds of thousands of premature deaths from respiratory diseases.

Figure 12.A Asian Brown Cloud This view shows a dense haze over eastern China, looking eastward across the Yellow Sea toward Korea.

logical, social, political, and cultural change (human geography)? What other discipline offers insights into environmental change, and who but geographers can cope with the diversity of environments and the sheer range of scales at which it is necessary to manage global change?

Those of us in the richer countries of the world have a special responsibility for leadership in sustainable development, because our present affluence is based on a cumulative past (and present) exploitation of the world's resources that is disproportionate to our numbers. We also happen to have the financial, technical, and human resources to enable us to take the lead in developing cleaner, less resource-intensive technologies, in transforming our economies to protect and work with natural systems, in providing more equitable access to economic opportunities and social services, and in supporting the technological and political frameworks necessary for sustainable development in poor countries. We cannot do it all at once, but we will certainly deserve the scorn and resentment of future generations if we do not try.

CONCLUSION

The beginning of the twenty-first century is going to be a period of fluid and transitional relationships among places, regions, and nations. Nevertheless, we know enough about contemporary patterns and trends, as well as geographic processes, to map out some plausible scenarios for the future. Some aspects of the future, though, we can only guess at. Two of the most speculative realms are those of politics and technology.

The future of the worst-off peripheral regions could be very bad indeed. They face unprecedented levels of demographic, environmental, and societal stress, with the events of the next 50 years being played out from a starting point of scarce basic resources, serious environmental degradation, overpopulation, disease, unprovoked crime, refugee migrations, and criminal anarchy.

For the world-system core, however, the long-term question is one of relative power and dominance. The same factors that consolidate the advantages of the core as a whole—the end of the Cold War, the availability of advanced telecommunications, the transnational reorganization of industry and finance, the liberalization of trade, and the emergence of a global culture—will also open the possibility of a new geopolitical and geoeconomic order, within which the economic and political relationships among core countries might change substantially.

Many aspects of future geographies will depend on trends in the demand for resources and on the exploitation of new technologies. The expansion of the world economy and the globalization of industry will undoubtedly boost the overall demand for raw materials of various kinds, and this will spur the development of previously underexploited but resource-rich regions in Africa, Eurasia, and East Asia. Raw materials will be only a fraction of future resource needs, however; the main issue, by far, will be energy resources.

It also appears that the present phase of globalization has the potential to create such disparities between the haves and the have-nots (as well as between the core and the periphery) that social unrest will ensue. The evidence of increasing dissatisfaction with the contemporary distribution of wealth both within and between the core and the periphery is widespread. In Russia, where public demonstrations against the government had been outlawed for decades, people are taking to the streets to protest government policies that favor transnational economic development at the expense of workers' minimum wages. In India farmers damaged a Kentucky Fried Chicken restaurant to protest the company's role in dislocating the domestic poultry industry. In Mexico industrial workers in transnational plants are challenging the absence of health and safety regulations that leave some exposed to harmful chemicals.

At the same time that protests against globalization and new geographies are being waged, the products of a global economy and culture are being widely embraced. The market for blue contact lenses is growing in such unlikely places as Bangkok and Nairobi, and plastic surgery to make eyes more Western looking is increasing in many Asian countries. CNN as well as American TV soap operas are eagerly viewed in even the most remote corners of the globe. Highway systems, airports, and container facilities are springing up throughout the periphery.

In short, future geographies are being negotiated at this very moment—from the board rooms of transnational corporations to the huts of remote villagers. The outcomes of these negotiations are still in the making as we all, in our daily lives, make seemingly insignificant decisions about what to wear, what to eat, where to work, how to travel, and how to entertain ourselves. These decisions help to either support or undermine the larger forces at work in the global economy, such as where to build factories, what products to make, or how to package and deliver them to the consumer. In essence, future geographies can be very much shaped by us through our understanding of the relatedness of people, places, and regions in a globalized economy.

MAIN POINTS REVISITED

- **In some ways the future is already here, embedded in the world's institutional structures and in the dynamics of its populations.**

  We know, for example, a good deal about the demographic trends of the next quarter-century, given present populations and birth and death rates. We also know a good deal about the distribution of environmental resources and constraints, about the characteristics of local and regional economies, and the legal and political frameworks within which geographic change will probably take place.

- **New and emerging technologies that are likely to have the most impact in reshaping human geographies include advanced transportation technologies, biotechnology, materials technologies, and information technologies.**

  The evolution of the world's geographies has always been shaped by the opportunities and constraints presented to different places and regions by successive technology systems. Many aspects of future geographies will depend on trends in the demand for particular resources and on the exploitation of these new technologies.

- **The changes involved in shaping future geographies will inevitably bring some critical issues, conflicts, and threats, including important geographical issues that center on scale, boundaries, and territories; on cultural dissonance; and on the sustainability of development.**

  Many of these issues stem from the globalization of the economy, which is undermining the status of the territorial nation-state as the chief regulating mechanism of both global and local dimensions of the world-system. The implications for peripheral places and regions are dismal: No matter how strong governments may be in their apparatus of domestic power, they will be next to helpless in the face of acute environmental stress, increased cultural friction, escalating poverty and disease, and growing migrations of refugees.

ADDITIONAL READING

Atkinson, R. D., "Technological Change and Cities," *Cityscape: A Journal of Policy Development and Research 3*, 3 (1998): 129–170.

Castells, M., *The Information Age: Economy, Society and Culture*. Vol. 3: *End of Millenium*. Oxford: Blackwell, 1998.

Coates, J. F., J. B. Mahaffie, and A. Hines, *2025: Scenarios of U.S. and Global Society Reshaped by Science and Technology*. Greensboro, NC: Oakhill Press, 1997.

Hammond, A. *Which World? Scenarios for the 21st Century*. Washington, DC: Island Press, 1998.

Held, D., A. McGrew, D. Goldblatt, and J. Perraton, *Global Transformations*. Malden: Blackwell, 1999.

Huntington, S., "The Clash of Civilizations?" *Foreign Affairs* 72 (1993): 22–49.

Jacobs, J., *Dark Age Ahead: Caution*. New York: Random House, 2004.

Johnston, R. J., P. J. Taylor, and M. Watts (eds.), *Geographies of Global Change: Remapping the World*. Cambridge, MA: Blackwell, 2002.

Kaplan, R. D., "The Coming Anarchy," *Atlantic Monthly*, February 1994, 44–76.

National Intelligence Council, *Mapping the Global Future*. Washington, DC: U.S. Government Printing Office, 2004.

O'Meara, P., H. D. Mehlinger, and M. Krain (eds.), *Globalization and the Challenges of a New Century*. Bloomington: Indiana University Press, 2000.

Pearson, I. (ed.), *The Macmillan Atlas of the Future*. New York: Macmillan, 1998.

Smith, N. *The Endgame of Globalization*. New York: Routledge, 2005.

Worldwatch Institute, *State of the World 2005*. New York: W.W. Norton, 2005.

EXERCISES

On the Internet

The Internet exercises for this chapter focus on how future geographies may reshape our socioeconomic landscape. The Internet allows us to survey new and changing geographies, such as HIV/AIDS, biogenetics, citizenship, social relationships, geopolitical alliances, and more. Through our study exercises, we speculate on the changing geography of production and the power of the Internet to democratize.

Unplugged

1. Using census data, construct a population pyramid (see Chapter 3) for any county or city with which you are familiar. What does this tell you about the future population of the locality?
2. Drawing on what you know about the geography of this locality and its regional, national, and global contexts, construct two scenarios, about 200 words each (see Figure 12.1 for brief examples) for the future, each based on different assumptions about resources and technology.
3. Write a short essay (500 words, or two double-spaced, typed pages) in which you outline the possible effects of economic globalization on a particular place or region with which you are familiar.

ATLAS
OCEANUS
OCCIDEN
TALIS.
HIS PA NIA.
ITALIA
BARBARIA.
DEL
AFRI
LIBIA INF.
CA
NORT
GUINEA.
OCEANUS

Appendix A
MAPS AND GEOGRAPHIC INFORMATION SYSTEMS

Maps are representations of the world. They are usually two-dimensional graphic representations that use lines and symbols to convey information or ideas about spatial relationships. Maps express particular interpretations of the world, and they affect how we understand the world and how we see ourselves in relation to others. As such, all maps are "social products." In general, maps reflect the power of the people who draw them up. Just including things on a map—literally "putting something on the map"—can be empowering. The design of maps—what they include, what they omit, and how their content is portrayed—inevitably reflects the experiences, priorities, interpretations, and intentions of their authors. The most widely understood and accepted maps—"normal" maps—reflect the view of the world that is dominant in universities and government agencies.

Maps that are designed to represent the *form* of Earth's surface and to show permanent (or at least long-standing) features such as buildings, highways, field boundaries, and political boundaries are called *topographic maps* (see, for example, **Figure A1.1**). The usual device for representing the form of Earth's surface is the *contour,* a line that connects points of equal vertical distance above or below a zero data point, usually sea level.

Maps that are designed to represent the spatial dimensions of particular conditions, processes, or events are called *thematic maps.* These can be based on any one of a number of devices that allow cartographers or map makers to portray spatial variations or spatial relationships. One of these is the *isoline,* a line (similar to a contour) that connects places of equal data value (for example, air pollution, as in **Figure A1.2**). Maps based on isolines are known as *isopleth maps.* Another common device used in thematic maps is the *proportional symbol.* Thus, for example, circles, squares, spheres, cubes, or some other shape can be drawn in proportion to the frequency of occurrence of some particular phenomenon or event at a given location. **Figure A1.3** shows an example using proportional circles. Symbols such as arrows or lines can also be drawn proportionally in order to portray flows of things between particular places. Simple distributions can be effectively portrayed through *dot maps,* in which a single dot or other symbol represents a specified number of occurrences of some particular phenomenon or event. Yet another device is the *choropleth map,* in which tonal shadings are graduated to reflect area variations in numbers, frequencies, or densities (see, for example, Figures 7.1, p. 253 and 10.1, p. 394). Finally, thematic maps can be based on *located charts,* in which graphs or charts are located by place or region. In this way, a tremendous amount of information can be conveyed in one single map (**Figure A1.4**).

MAP SCALES

A *map scale* is simply the ratio between linear distance on a map and linear distance on Earth's surface. It is usually expressed in terms of corresponding lengths, as in "one centimeter equals one kilometer," or as a *representative fraction* (in this case, 1/100,000) or ratio (1:100,000). *Small-scale* maps are maps based on small representative fractions (for example, 1/1,000,000 or 1/10,000,000). Small-scale maps cover a large part of the Earth's surface on the printed page. A map drawn on this page to the scale of 1:10,000,000 would cover about half of the United States; a map drawn to the scale of 1:16,000,000 would easily cover the whole of Europe. *Large-scale* maps are maps based on larger representative fractions (e.g., 1/25,000 or 1/10,000). A map drawn on this page to the scale 1:10,000 would cover a typical suburban subdivision; a map drawn to the scale of 1:1,000 would cover just a block or two of it.

MAP PROJECTIONS

A map projection is a systematic rendering on a flat surface of the geographic coordinates of the features found on Earth's surface. Because Earth's surface is curved and it is not a perfect sphere, it is impossible to represent on a flat plane, sheet of paper, or monitor screen without

Antique print, c. 1790

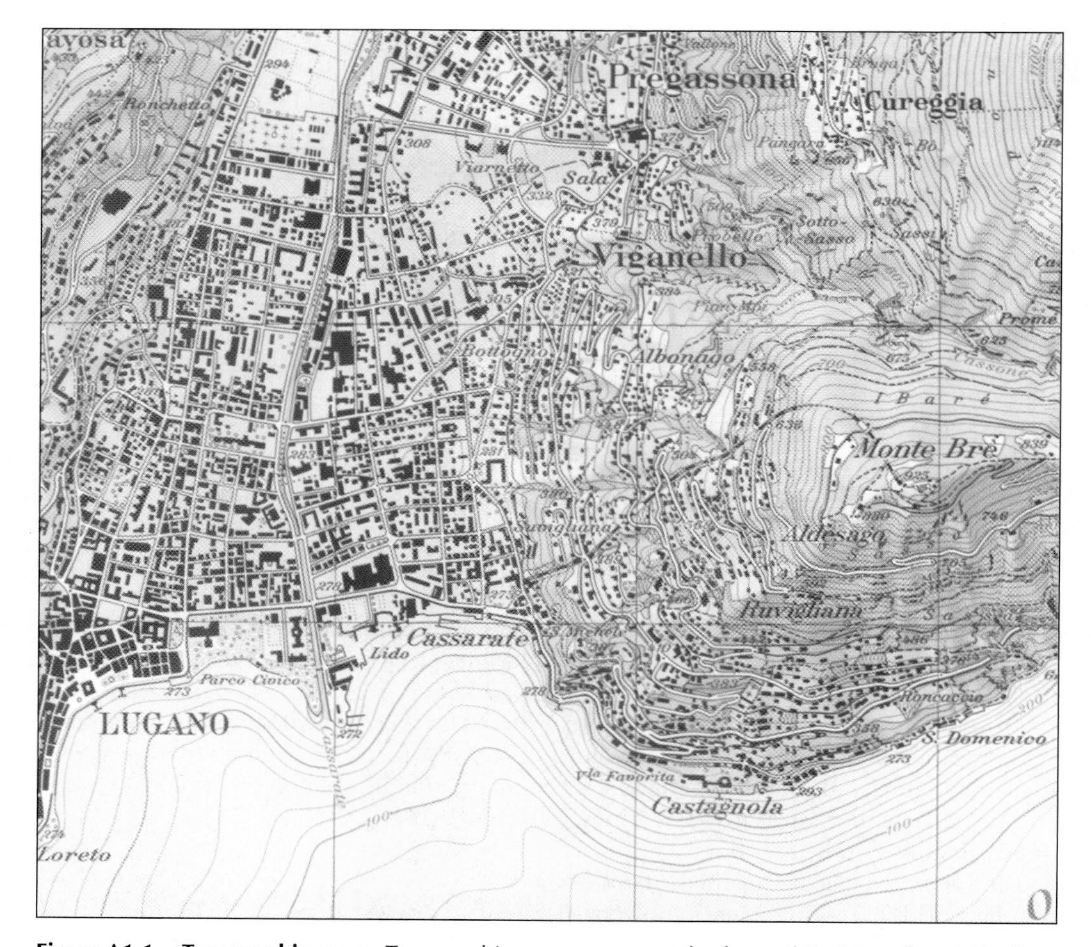

Figure A1.1 Topographic maps Topographic maps represent the form of Earth's surface in both horizontal and vertical dimensions. This extract is from a Swiss map of Lugano at the scale of 1:25,000. The height of landforms is represented by contours (lines that connect points of equal vertical distance above sea level), which on this map are drawn every 20 meters. Features such as roads, power lines, built-up areas, and so on are shown by stylized symbols. Note how the closely spaced contours of the hill slopes are able to represent the shape and form of the land. (*Source:* Extract from Carta Nazionale Della Svizzera, Sheet 1353, 1: 25,000 series, Ufficio Federale di Topografia, 3084 Wabern, Switzerland. Edizione 1998.)

some distortion. Cartographers have devised a number of different techniques of projecting latitude and longitude (see Figure 1.13) onto a flat surface, and the resulting representations of Earth each have advantages and disadvantages. None of them can represent distance correctly in all directions, though many can represent compass bearings or area without distortion. The choice of map projection depends largely on the purpose of the map.

Projections that allow distance to be represented as accurately as possible are called *equidistant projections*. These projections can represent distance accurately in only one direction (usually north-south), although they usually provide accurate scale in the perpendicular direction (which in most cases is the equator). Equidistant projections are often more aesthetically pleasing for representing Earth as a whole, or large portions of it. An example is the polyconic projection (**Figure A1.5**).

Projections on which compass directions are rendered accurately are known as *conformal projections*. On the Mercator projection (Figure A1.5) for example, a compass bearing between any two points is plotted as a straight line. As a result, the Mercator projection has been widely used in navigation for hundreds of years. The Mercator projection was also widely used as the standard classroom wall map of the world for many years, and its image of the world has entered deeply into general consciousness. Many Europeans and North Americans have an exaggerated sense of the size of the northern continents and are unaware of the true size of Africa. Some projections are designed such that compass directions are correct only from one central point. These are known as *azimuthal projections*. They can be equidistant, as in the

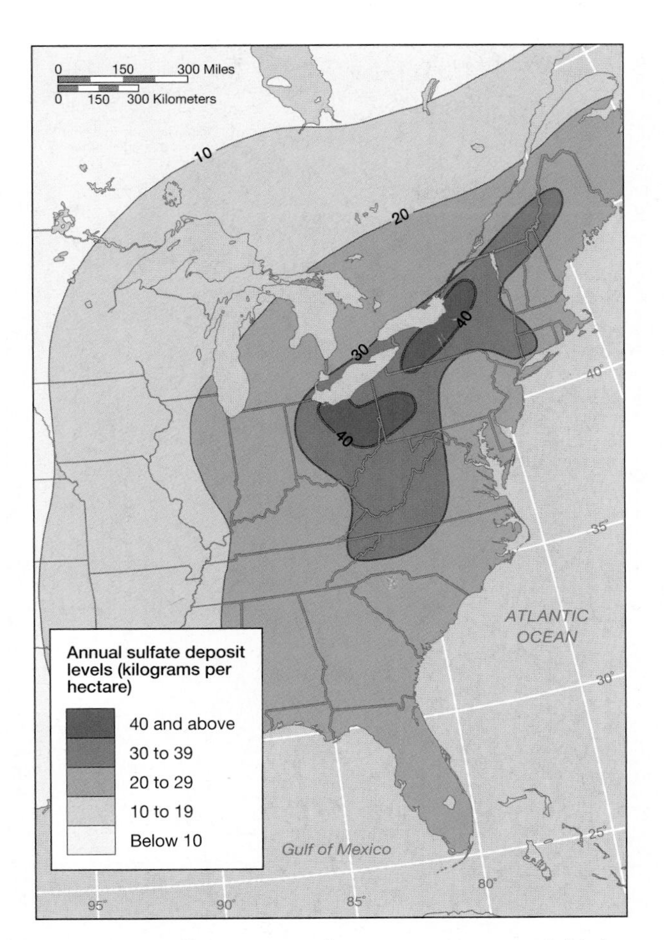

Figure A1.2 Isoline maps Isoline maps portray spatial information by connecting points of equal data value. Contours on topographic maps (see Figure A1.1) are isolines. This map shows air pollution in the eastern United States. (*Source:* Reprinted with permission of Prentice Hall, from J. M. Rubenstein, *The Cultural Landscape: An Introduction to Human Geography,* 1996, p. 584. Adapted from William K. Stevens, "Study of Acid Rain Uncovers Threat to Far Wider Area," *New York Times,* January 16, 1990, p. 21, map.)

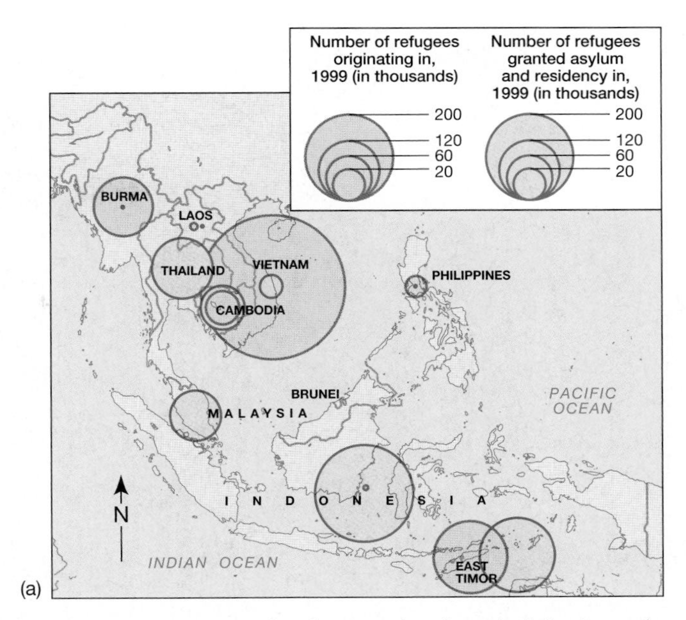

Figure A1.3 An example of proportional symbols in thematic mapping Proportional circles showing the number of refugees originating in Southeast Asian countries in 1999, and the number of refugees granted asylum in Southeast Asian countries in 1999. (After S. Marston, P. Knox, and D. Liverman, *World Regions in Global Context.* Upper Saddle River, NJ: Prentice Hall, 2002, p. 497.)

Azimuthal Equidistant projection (Figure A1.5), which is sometimes used to show air-route distances from a specific location, or equal-area, as in the Lambert Azimuthal Equal-Area projection.

Projections that portray areas on Earth's surface in their true proportions are known as **equal-area** or *equivalent projections.* Such projections are used where the cartographer wishes to compare and contrast distributions on the Earth's surface: the relative area of different types of land use, for example. Examples of equal-area projections include the Eckert IV projection, Bartholomew's Nordic projection (used in Figure 2.8) and the Mollweide projection (Figure A1.5). Equal-area projections such as the Mollweide projection are especially useful for thematic maps showing economic, demographic, or cultural data (see, for example, Figure 12.3). Unfortunately, preserving accuracy in terms of area tends to result in world maps on which many locations appear squashed and have unsatisfactory outlines.

For some applications, aesthetic appearance is more important than conformality, equivalence, or equidistance, so cartographers have devised a number of other projections. Examples include the Times projection, which is used in many world atlases, and the Robinson projection, which is used by the National Geographic Society in many of its publications. The Robinson projection (**Figure A1.6**) is a compromise projection that distorts both area and directional relationships but provides a general-purpose world map. There are also political considerations. Countries may appear larger and so more "important" on one projection rather than another. The Peters projection, for example (**Figure A1.7**), is a deliberate attempt to give prominence to the underdeveloped countries of the equatorial regions and the Southern Hemisphere. As such, it was officially adopted by the World Council of Churches and by numerous agencies of the United Nations and other international institutions. Its unusual shapes give it a shock value that gets people's attention. For some, however, its unusual shapes are ugly: It has been likened to laundry hung out to dry.

In this book we sometimes use another striking projection, the Dymaxion projection devised by Buckminster Fuller (**Figure A1.8**). Fuller was a prominent modernist architect and industrial designer who wanted to produce a map of the world with no significant distortion to any of the major landmasses. The Dymaxion

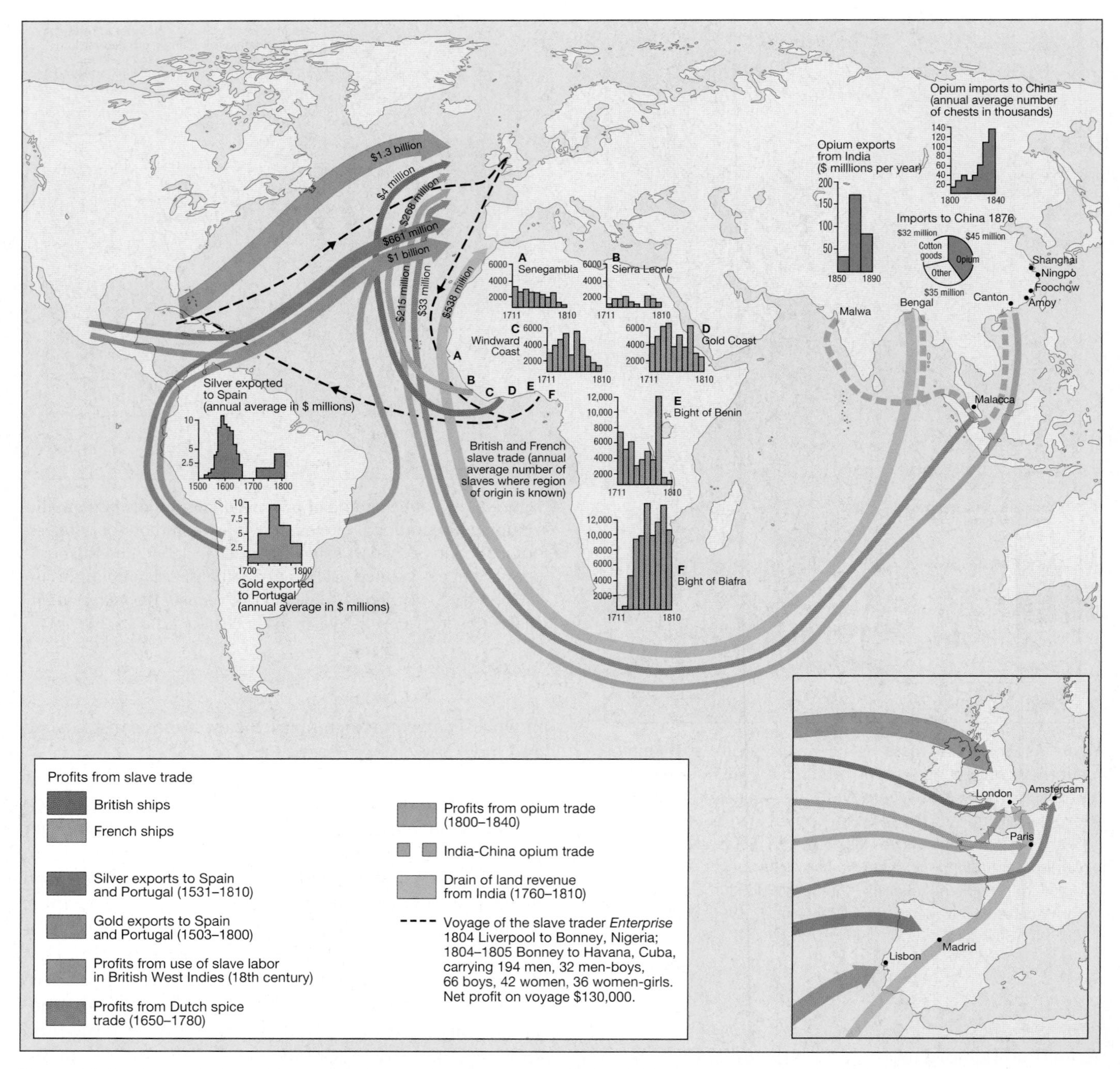

Figure A1.4 Located charts By combining graphs, charts, or symbols with base maps, a great deal of information can be conveyed in a single figure. This example illustrates the profits generated through European plunder of global minerals, spices, and human beings over a 300-year period. (After S. Marston, P. Knox, and D. Liverman, *World Regions in Global Context.* Upper Saddle River, NJ: Prentice Hall, 2002, p. 84.)

projection does this, though it produces a world that at first may seem disorienting. This is not necessarily such a bad thing, for it can force us to take a fresh look at the world and at the relationships between places. Because Europe, North America, and Japan are all located toward the center of this map projection, it is particularly useful for illustrating two of the central themes of this book: the relationships among these prosperous regions, and the relationships between this prosperous core group and the less prosperous, peripheral countries of the world. Fuller's projection shows the economically peripheral countries of the world as being cartographically peripheral, too.

One particular kind of map projection that is sometimes used in small-scale thematic maps is the *cartogram.* In this kind of projection, space is transformed according

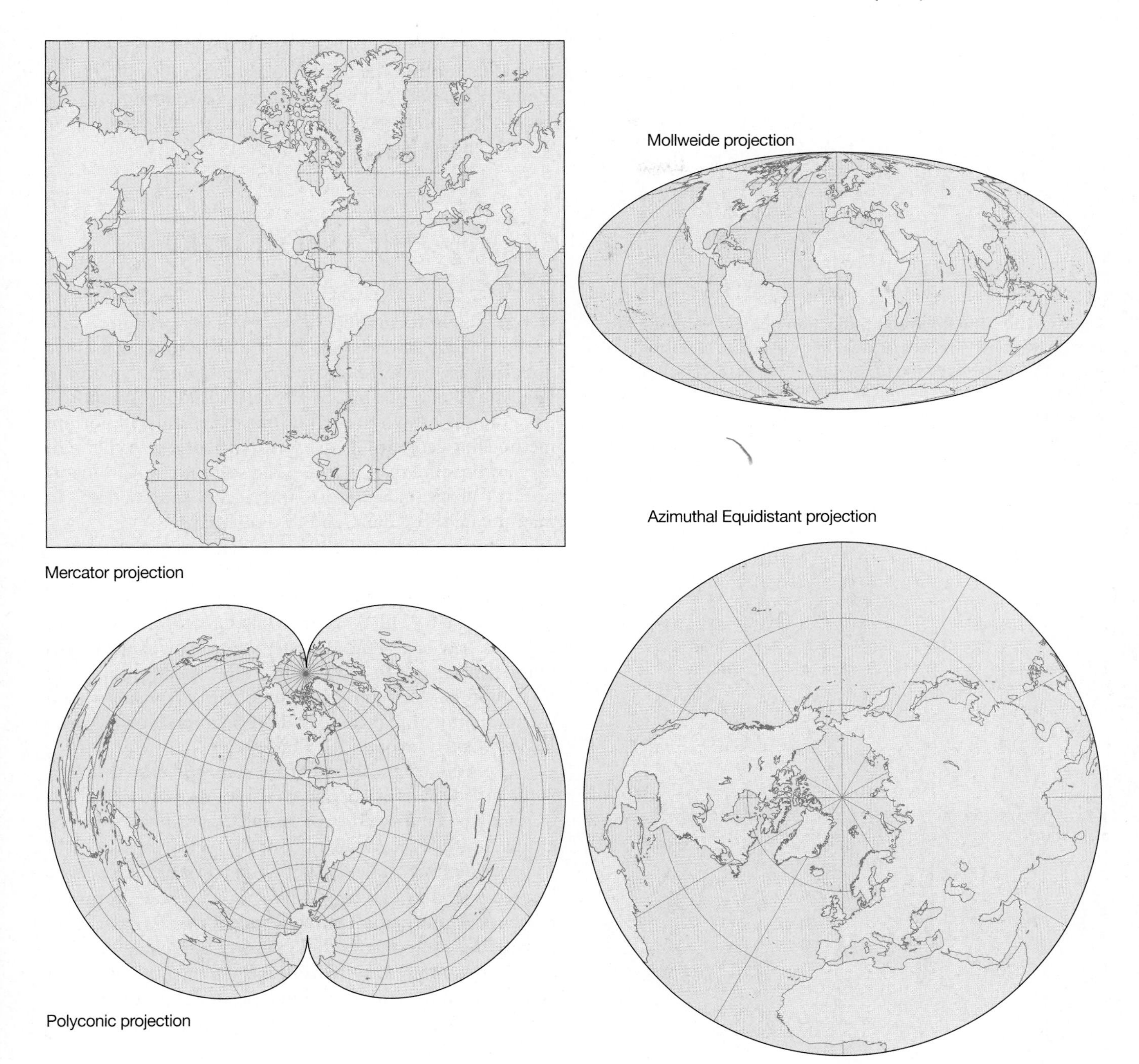

Figure A1.5 Comparison of map projections Different map projections have different properties. The polyconic projection is true to scale along each east-west parallel and along the central north-south meridian. It is neither conformal nor equal-area, and it is free of distortion only along the central meridian. On the Mercator projection, compass directions between any two points are true, and the shapes of landmasses are true, but their relative size is distorted. On the azimuthal equidistant projection, distances measured from the center of the map are true, but direction, area, and shape are increasingly distorted with distance from the center point. On the Mollweide projection, relative sizes are true, but shapes are distorted.

to statistical factors, with the largest mapping units representing the greatest statistical values. **Figure A1.9a** shows a cartogram of the world in which countries are represented as proportional to their population. This sort of projection is particularly effective in helping to visualize relative inequalities among the world's populations. **Figure A1.9b** shows a cartogram of the world in which the cost of telephone calls has been substituted for linear distance as the basis of the map. The deliberate distortion of the shapes of the continents in this sort of projection dramatically emphasizes spatial variations.

Finally, the advent of computer graphics has made it possible for cartographers to move beyond the use of maps as two-dimensional representations of Earth's surface.

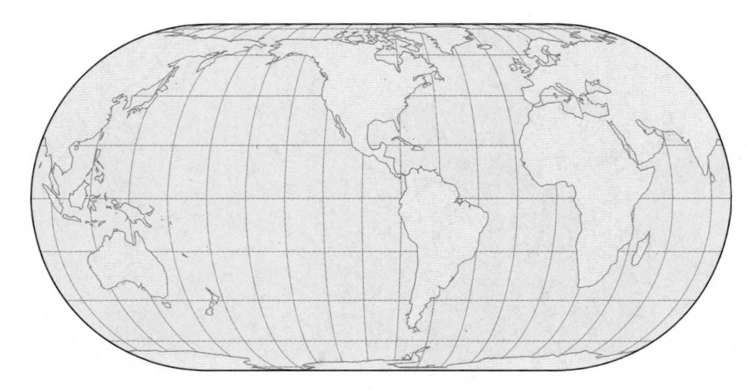

Figure A1.6 The Robinson projection On the Robinson projection, distance, direction, area, and shape are all distorted in an attempt to balance the properties of the map. It is designed purely for appearance and is best used for thematic and reference maps at the world scale. (*Source:* After E. F. Bergman, *Human Geography: Cultures, Connections, and Landscapes,* © 1995 by Prentice Hall, p. 12.)

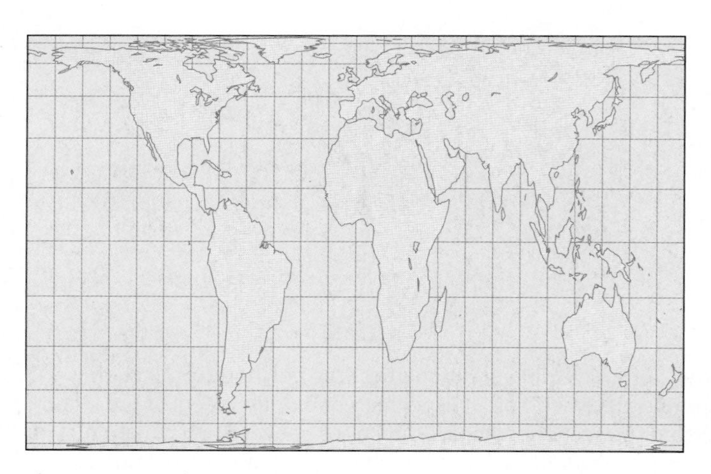

Figure A1.7 The Peters projection This equal-area projection was an attempt to offer an alternative to traditional projections which, Arno Peters argued, exaggerated the size and apparent importance of the higher latitudes—that is, the world's core regions—and so promoted the "Europeanization" of Earth. While it has been adopted by the World Council of Churches, the Lutheran Church of America, and various agencies of the United Nations and other international institutions, it has been criticized by cartographers in the United States on the grounds of aesthetics: One of the consequences of equal-area projections is that they distort the shape of landmasses. (*Source:* After E. F. Bergman, *Human Geography: Cultures, Connections, and Landscapes,* © 1995 by Prentice Hall, p. 13.)

Computer software that renders three-dimensional statistical data on to the flat surface of a monitor screen or a piece of paper facilitates the *visualization* of many aspects of human geography in innovative and provocative ways (**Figure A1.10**).

GEOGRAPHIC INFORMATION SYSTEMS

Geographic information systems (GIS)—organized collections of computer hardware, software, and geographic data that are designed to capture, store, update, manipulate, and display geographically referenced information—have rapidly grown to become one of the most important methods of geographic analysis, particularly in the military and commercial worlds. The software in GIS incorporates programs to store and access spatial data, to manipulate those data, and to draw maps.

Between 2000 and 2005, GIS services grew at a rate of around 10 percent per year. In 2002, estimates of global investment in GIS technologies ranged from $3.3 billion to more than $8 billion. In the United States, employment in GIS is now one of the 10 fastest-growing technical fields in the private sector. The United States spends more than $4 billion per year on geographic data acquisition, and the total annual market for GIS services in North America is valued at around $2.5 billion.

The primary requirement for data to be used in GIS is that the locations for the variables are known. Location may be annotated by *x, y,* and *z* coordinates of longitude, latitude, and elevation, or by such systems as ZIP codes or highway mile markers. Any variable that can be located spatially can be fed into a GIS. Data capture—putting the information into the system—is the most time-consuming component of GIS work. Different sources of data, using different systems of measurement, scales, and systems of representation, must be integrated with one another; changes must be tracked and updated. Many GIS operations in the United States, Europe, Japan, and Australia have begun to contract out such work to firms in countries where labor is cheaper. India has emerged as a major data-conversion center for GIS.

Applications of GIS

The most important aspect of GIS, from an analytical point of view, is that they allow data from several different sources, on different topics and at different scales, to be merged. This allows analysts to emphasize the spatial relationships among the objects being mapped. A geographic information system makes it possible to link, or integrate, information that is difficult to associate through any other means. For example, using GIS technology and water-company billing information, it is possible to sim-

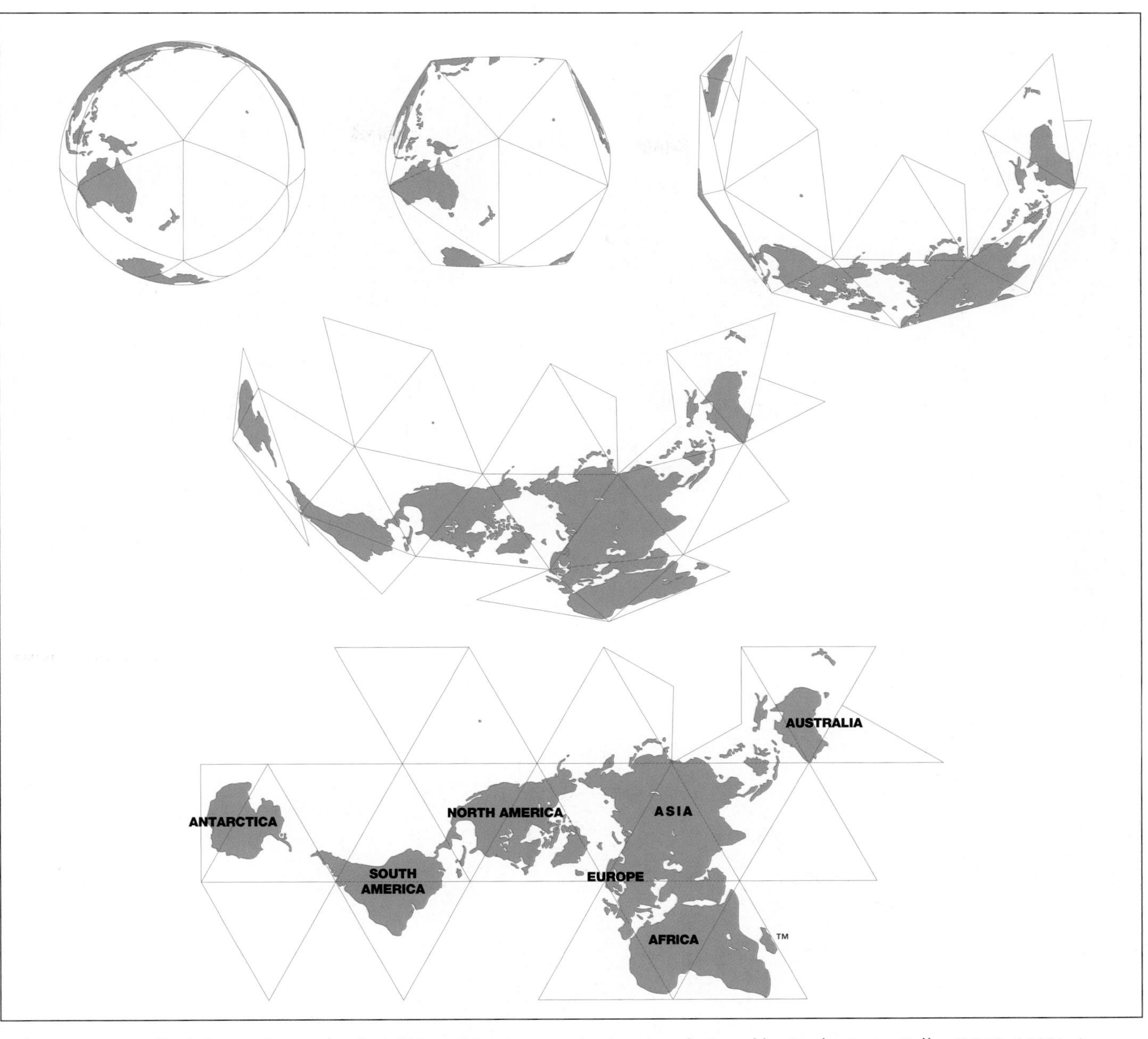

Figure A1.8 Fuller's Dymaxion projection This striking map projection was designed by Buckminster Fuller (1895–1983). As this figure shows, Fuller achieved his objective of creating a map with the minimum of distortion to the shape of the world's major landmasses by dividing the globe into triangular areas. Those areas not encompassing major landmasses were cut away, allowing the remainder of the globe to be "unfolded" into a flat projection. (*Source:* Buckminster Fuller Institute and Dymaxion Map Design, Santa Barbara, CA. The word *Dymaxion* and the Fuller Projection Dymaxion™ Map design are trademarks of the Buckminster Fuller Institute, Santa Barbara, CA, © 1938, 1967 & 1992. All rights reserved.)

ulate the discharge of materials in the septic systems in a neighborhood upstream from a wetland. The bills show how much water is used at each address. Because the amount of water a customer uses will roughly predict the amount of material that will be discharged into the septic systems, areas of heavy septic discharge can be located using a GIS.

GIS technology can render visible many aspects of geography that were previously unseen. GIS can, for example, produce incredibly detailed maps based on millions of pieces of information—maps that could never have been drawn by human hands. One example of such a map is the satellite image reconstruction of the vegetation cover of the United States shown in **Figure A1.11**. At the other extreme of spatial scale, GIS can put places under the microscope, creating detailed new insights using huge databases and effortlessly browsable media (**Figure A1.12**).

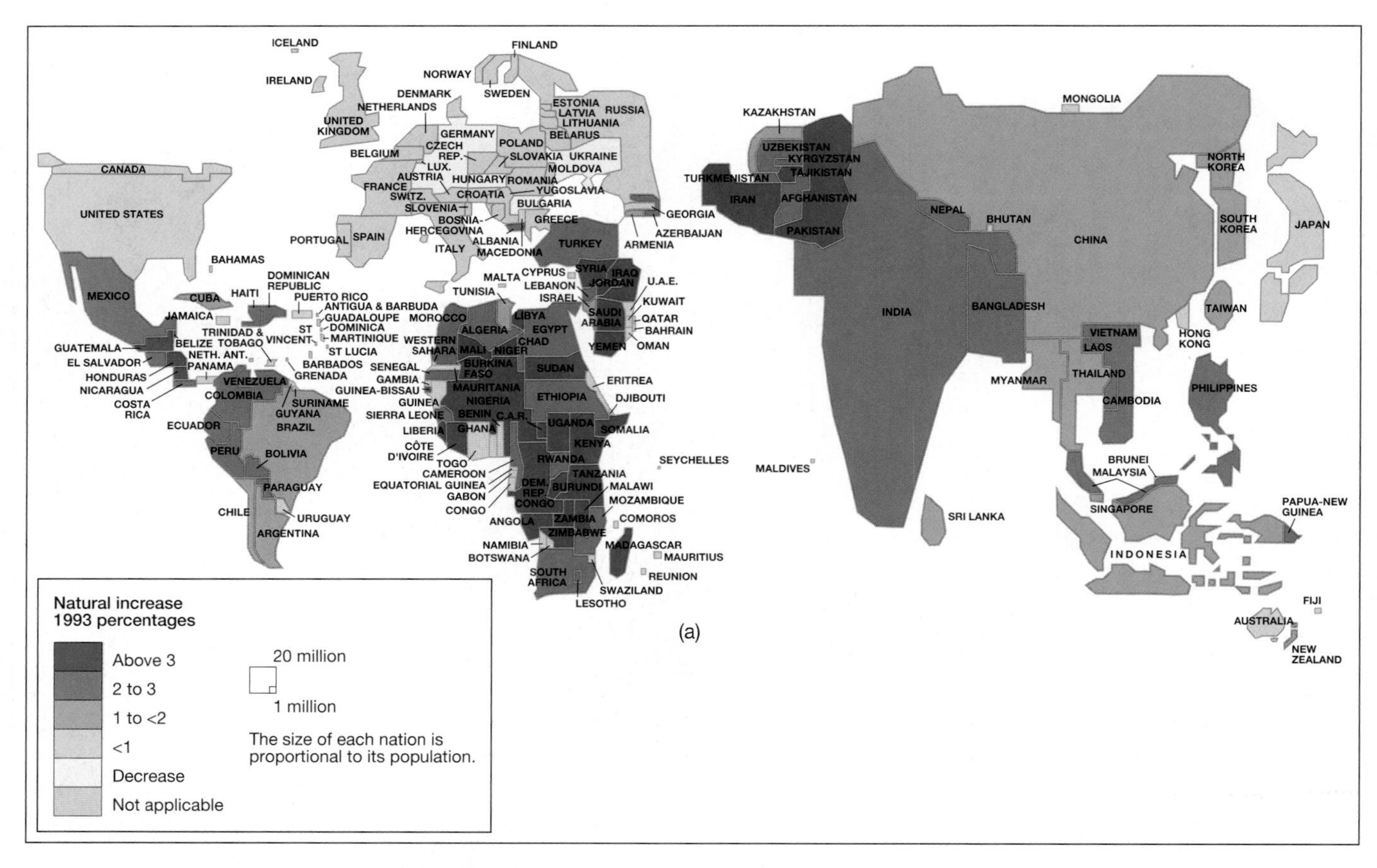

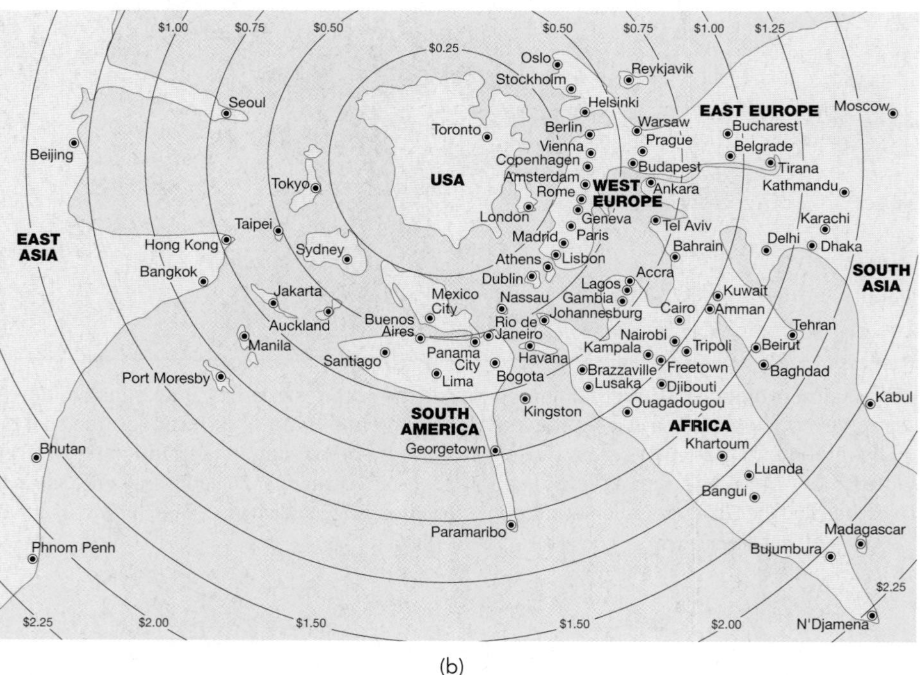

Figure A1.9 Example of a cartogram In a cartogram, space is distorted to emphasize a particular attribute of places or regions. (a) This example shows the relative size of countries based on their population rather than their area; the cartographers have maintained the shape of each country as closely as possible to make the map easier to read. As you can see, population-based cartograms are very effective in demonstrating spatial inequality. (b) In this example, the cost of telephone calls is substituted for linear distance as the basis of the map, thus deliberately distorting the shapes of the continents to dramatic effect. Countries are arranged around the United States according to the cost per minute of calls made from the United States in 1998. (After (a) M. Kidron and R. Segal (eds.), *The State of the World Atlas,* rev. 5th ed. London: Penguin Reference, 1995, pp. 28–29; (b) G. C. Staple (ed.), *TeleGeography 1999.* Washington, DC: TeleGeography, 1999, p. 82.)

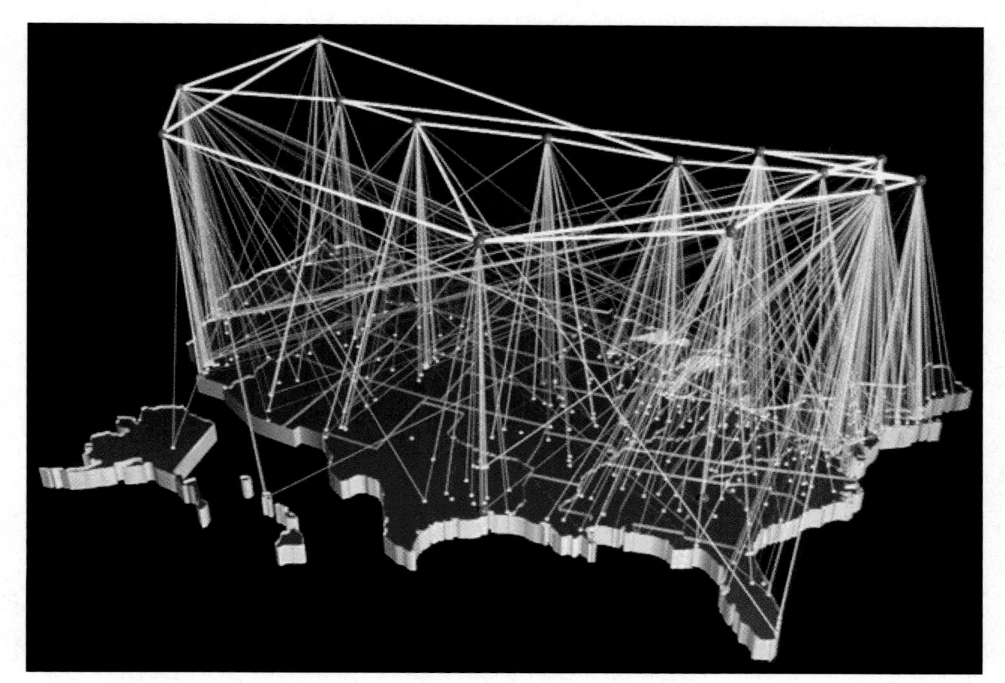

Figure A1.10 Visualization This example shows the spatial structure of the Internet backbone and associated traffic flows within the United States. (After Donna Cox and Robert Patterson, http://www.ncsa.uiuc.edu/SCMS/DigLib/text/technology/Visualization-Study-NSFNET-Cox.html)

Many advances in GIS have come from military applications. GIS allows infantry commanders to calculate line of sight from tanks and defensive emplacements, allows cruise missiles to fly below enemy radar, and provides a comprehensive basis for military intelligence. Beyond the military, GIS technology allows an enormous range of problems to be addressed. For instance, it can be used to decide how to manage farmland, to monitor the spread of infectious diseases, to monitor tree cover in metropolitan areas, to assess changes in ecosystems, to analyze the impact of proposed changes in the boundaries of legislative districts, to identify the location of potential business customers, to identify the location of potential criminals, and to provide a basis for urban and regional planning. Some of the most influential applications of GIS have resulted from geodemographic research. *Geodemographic research* uses census data and commercial data (such as sales data and property records) about the populations of small districts in creating profiles of those populations for market research. The digital media used by GIS make such applications very flexible. With GIS it is possible to zoom in and out, evaluating spatial relationships at different spatial scales. Similarly, it is possible to vary the appearance and presentation of maps, using different colors and rendering techniques.

Figure A1.11 Map of land cover This extract is from a map of land cover in the United States that was compiled from several data sets using GIS technology. These data sets included 1-kilometer resolution, Advanced Very High Resolution Radiometer (AVHRR) satellite imagery, and digital data sets on elevation, climate, water bodies, and political boundaries. Each of the 159 colors on the U.S. map represents a specific vegetation region. The purples and blues on this extract represent various subregions of western coniferous forests; the yellows are grasslands, and the reds are shrublands. The gray-brown region is the barren area of the Mojave Desert. (*Source:* United States Geological Survey, Map of Seasonal Land Cover Regions, 1993; see T. Loveland, J. W. Merchant, J. F. Brown, D. O. Ohlen, B. C. Reed, P. Olson, and J. Hutchinson, "Seasonal Land Cover Regions of the United States," *Annals, Association of American Geographers, 85* (1995), 339–355.)

Critiques of GIS

Within the past five years applications of GIS have resulted in the creation of more maps than were created in all previous human history. One result is that as maps have become more commonplace, more people and more businesses have become more spatially aware. Nevertheless, some critics have argued that GIS represents no real advances in geographers' understanding of places and regions. The results of GIS, they argue, may be useful but are essentially mundane. This misses the point that how-

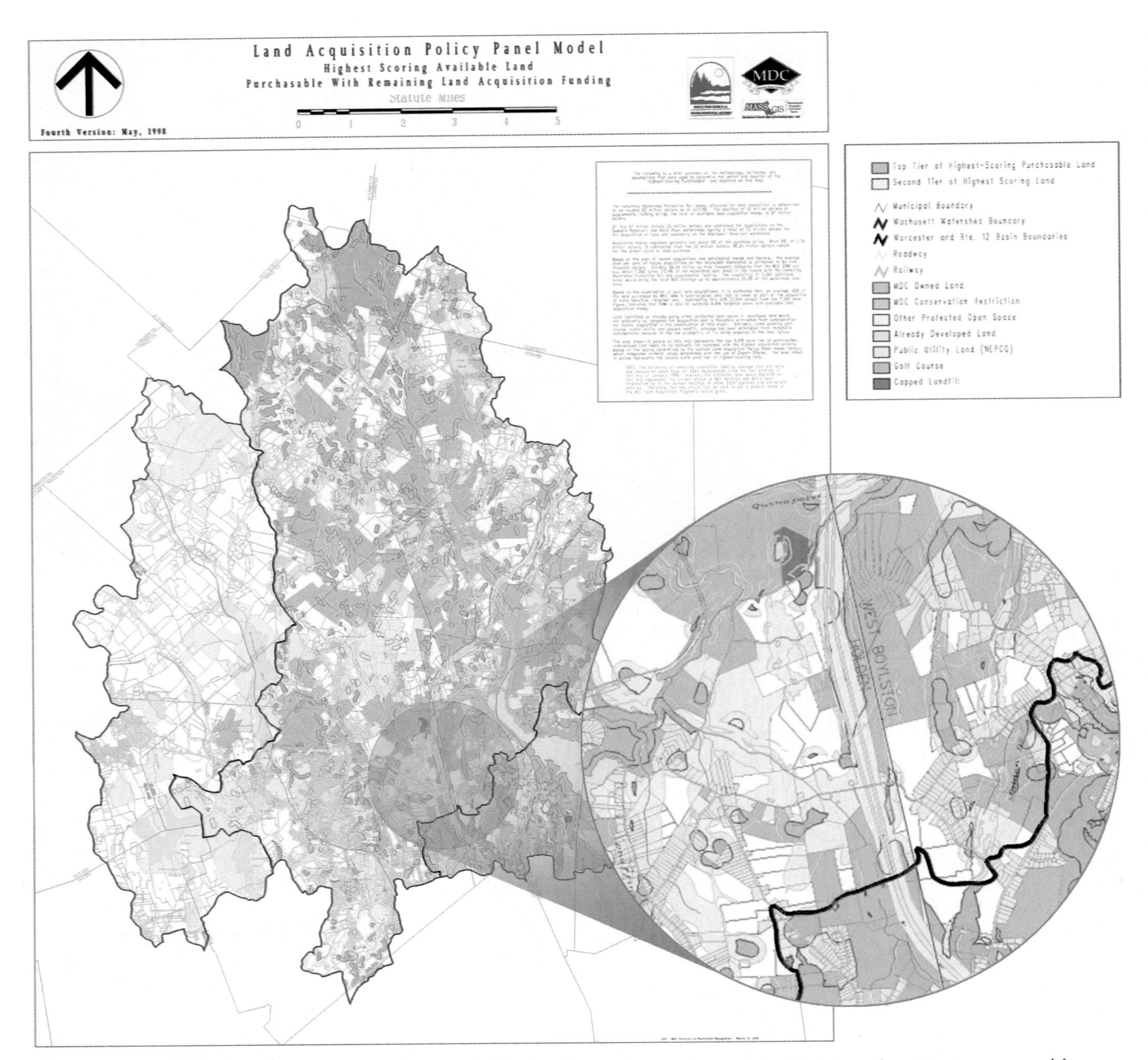

Figure A1.12 GIS-derived planning map This map of the Wachusett Reservoir watershed in Massachusetts was prepared from a composite of 12 layers of GIS maps, each depicting a key aspect of land use and topography. The map is used to show in detail the parcels of land that the Commonwealth of Massachusetts needs to buy in order to keep development from damaging the watershed and the quality of the reservoir water. (*Source:* R.W. Greene, *GIS in Public Policy.* Redlands, CA: ESRI Press, 2000, p. 67.)

ever routine their subject may be, all maps constitute powerful and influential ways of representing the world. A more telling critique, perhaps, is that the real impact of GIS has been to increase the level of surveillance of the population by those who already possess power and control. The fear is that GIS may be helping to create a world in which people are not treated and judged by who they are and what they do but more by where they live. People's credit ratings, ability to buy insurance, and ability to secure a mortgage, for example, are all routinely judged in part by GIS-based analyses that take into account the attributes and characteristics of their neighbors.

ADDITIONAL READING

Dorling, D. and D. Fairbairn, *Mapping: Ways of Seeing the World.* London: Addison Wesley Longman, 1977.

Thrower, N. J. W., Maps & Civilization. *Cartography in Culture and Society,* 2nd ed. Chicago: University of Chicago Press, 1999.

Wilton, J. N., *The Mapmakers.* New York: Vintage Books, 2000.

Glossary

A

accessibility: the opportunity for contact or interaction from a given point or location, in relation to other locations.

acid rain: the wet deposition of acids upon Earth created by the natural cleansing properties of the atmosphere.

age-sex pyramid: a representation of the population based on its composition according to age and sex.

agglomeration diseconomies: the negative economic effects of urbanization and the local concentration of industry.

agrarian: referring to the culture of agricultural communities and the type of tenure system that determines access to land and the kind of cultivation practices employed there.

agribusiness: a set of economic and political relationships that organizes agro-food production from the development of seeds to the retailing and consumption of the agricultural product.

agricultural density: ratio between the number of agriculturists per unit of arable land in a specific area.

agricultural industrialization: process whereby the farm has moved from being the centerpiece of agricultural production to become one part of an integrated string of vertically organized industrial processes including production, storage, processing, distribution, marketing, and retailing.

agriculture: a science, an art, and a business directed at the cultivation of crops and the raising of livestock for sustenance and for profit.

aquaculture: the cultivation of fish and shellfish under controlled conditions, usually in coastal lagoons.

autarky: insignificant contributions to the flows of imports and exports that constitute the geography of trade.

azimuthal projection: a map projection on which compass directions are correct only from one central point.

B

baby boom: population of individuals born between the years 1946 and 1964.

backwash effects: the negative impacts on a region (or regions) of the economic growth of some other region.

basic functions: economic activities that provide income from sales to customers beyond city limits.

Beaux Arts: a style of urban design that sought to combine the best elements of all of the classic architectural styles.

biotechnology: technique that uses living organisms (or parts of organisms) to make or modify products, to improve plants and animals, or to develop microorganisms for specific uses.

bioterrorism: deliberate use of microorganisms or toxins from living organisms to induce death or disease.

Blue Revolution: the introduction of new production techniques, processing technology, infrastructure, and larger, motorized boats into peripheral country fisheries. A prominent component of the Blue Revolution is the application of transgenics in *aquaculture*.

boomburbs: rapidly growing suburban jurisdictions at the fringe of metropolitan areas.

C

capitalism: a form of economic and social organization characterized by the profit motive and the control of the means of production, distribution, and the exchange of goods by private ownership.

carrying capacity: the maximum number of users that can be sustained, over the long term, by a given set of natural resources.

cartography: the body of practical and theoretical knowledge about making distinctive visual representations of Earth's surface in the form of maps.

census: count of the number of people in a country, region, or city.

central business district (CBD): the central nucleus of commercial land uses in a city.

central cities: the original, core jurisdictions of metropolitan areas.

central place: a settlement in which certain products and services are available to consumers.

central place theory: a theory that seeks to explain the relative size and spacing of towns and cities as a function of people's shopping behavior.

centrality: the functional dominance of cities within an urban system.

centrifugal forces: forces that divide or tend to pull the state apart.

centripetal forces: forces that strengthen and unify the state.

chemical farming: application of synthetic fertilizers to the soil—and herbicides, fungicides, and pesticides to crops—in order to enhance yields.

children's rights: the fundamental right of children to life, liberty, education, and health care. Codified by the United Nations Convention on the Rights of the Child in 1989 to include the protection of children in armed conflict; protection from discrimination; protection from torture or cruel, inhuman or degrading treatment or punishment; protection within the justice system; and protection from economic exploitation.

citizenship is a category of belonging to a nation-state that includes civil, political and social rights.

Clovis point: a flaked, bifaced projectile whose length is more than twice its width.

cognitive distance: the distance that people perceive to exist in a given situation.

cognitive images (mental maps): psychological representations of locations that are made up from people's individual ideas and impressions of these locations.

cognitive space: space defined and measured in terms of the nature and degree of people's values, feelings, beliefs, and perceptions about locations, districts, and regions.

cohort: a group of individuals who share a common temporal demographic experience.

colonial city: a city that was deliberately established or developed as an administrative or commercial center by colonial or imperial powers.

colonialism: the establishment and maintenance of political and legal domination by a state over a separate and alien society.

colonization: the physical settlement of a new territory of people from a colonizing state.

Columbian Exchange: interaction between the Old World—originating with the voyages of Columbus—and the New World.

commercial agriculture: farming primarily for sale, not direct consumption.

commodity chain: network of labor and production processes beginning with the extraction or production of raw materials and ending with the delivery of a finished commodity.

comparative advantage: principle whereby places and regions specialize in activities for which they have the greatest advantage in productivity relative to other regions—or for which they have the least disadvantage.

confederation: a group of states united for a common purpose.

conformal projection: a map projection on which compass bearings are rendered accurately.

conglomerate corporations: companies that have diversified into various economic activities, usually through a process of mergers and acquisitions.

congregation: the territorial and residential clustering of specific groups or subgroups of people.

conservation: the view that natural resources should be used wisely and that society's effects on the natural world should represent stewardship and not exploitation.

core regions: regions that dominate trade, control the most advanced technologies, and have high levels of productivity within diversified economies.

cosmopolitanism: an intellectual and aesthetic openness toward divergent experiences, images, and products from different cultures.

counterurbanization: the net loss of population from cities to smaller towns and rural areas.

creative destruction: the withdrawal of investments from activities (and regions) that yield low rates of profit in order to reinvest in new activities (and new places).

crop rotation: method of maintaining soil fertility in which the fields under cultivation remain the same, but the crop being planted is changed.

crude birthrate (CBR): ratio of the number of live births in a single year for every thousand people in the population.

crude death rate (CDR): the number of deaths in a single year for every thousand people in the population.

crude density (arithmetic density): total number of people divided by the total land area.

cultural adaptation: the complex strategies human groups employ to live successfully as part of a natural system.

cultural complex: combination of traits characteristic of a particular group.

cultural ecology: study of the relationship between a cultural group and its natural environment.

cultural geography: how space, place, and landscape shape culture at the same time that culture shapes space, place, and landscape.

cultural hearths: the geographic origins or sources of innovations, ideas, or ideologies.

cultural landscape: a characteristic and tangible outcome of the complex interactions between a human group and a natural environment.

cultural nationalism: an effort to protect regional and national cultures from the homogenizing impacts of globalization, especially from the penetrating influence of U.S. culture.

cultural region: the areas within which a particular cultural system prevails.

cultural system: a collection of interacting elements that taken together shape a group's collective identity.

cultural trait: a single aspect of the complex of routine practices that constitute a particular cultural group.

culture: a shared set of meanings that are lived through the material and symbolic practices of everyday life.

cumulative causation: a spiral buildup of advantages that occurs in specific geographic settings as a result of the development of external economies, agglomeration effects, and localization economies.

cycle of poverty: the transmission of poverty and deprivation from one generation to another through a combination of domestic circumstances and local, neighborhood conditions.

D

debt trap: syndrome of always having to borrow in order to fund development.

decolonization: the acquisition, by colonized peoples, of control over their own territory.

deep ecology: approach to nature revolving around two key components: self-realization and biospherical egalitarianism.

defensible space: a physical setting that allows residents to identify with, survey, and exert a degree of social control over public space.

deforestation: the removal of trees from a forested area without adequate replanting.

deindustrialization: a relative decline in industrial employment in core regions.

democratic rule: a system in which public policies and officials are directly chosen by popular vote.

demographic collapse: phenomenon of near genocide of native populations.

demographic transition: replacement of high birth and death rates by low birth and death rates.

demography: the study of the characteristics of human populations.

dependency ratio: measure of the economic impact of the young and old on the more economically productive members of the population.

dependency: high level of reliance by a country on foreign enterprises, investment, or technology.

derelict landscapes: landscapes that have experienced abandonment, misuse, disinvestment, or vandalism.

desertification: the degradation of land cover and damage to the soil and water in grasslands and arid and semiarid lands.

dialects: regional variations in standard languages.

diaspora: a spatial dispersion of a previously homogeneous group.

digital divide: inequality of access to telecommunications and information technology, particularly the Internet.

distance-decay function: the rate at which a particular activity or process diminishes with increasing distance.

division of labor: the specialization of different people, regions, or countries in particular kinds of economic activities.

domino theory: if one country in a region chose or was forced to accept a communist political and economic system, then neighboring countries would be irresistibly susceptible to falling to communism.

double cropping: practice used in the milder climates, where intensive subsistence fields are planted and harvested more than once a year.

doubling time: measure of how long it will take the population of an area to grow to twice its current size.

dualism: the juxtaposition in geographic space of the formal and informal sectors of the economy.

E

East/West divide: communist and noncommunist countries, respectively.

ecofeminism: the view that patriarchal ideology is at the center of our present environmental malaise.

ecological footprint: a measure of the human pressures on the natural environment from the consumption of renewable resources and the production of pollution indicating how much space a population needs compared to what is available.

ecological imperialism: introduction of exotic plants and animals into new ecosystems.

eco-migration: population movement caused by the degradation of land and essential natural resources.

economies of scale: cost advantages to manufacturers that accrue from high-volume production, since the average cost of production falls with increasing output.

ecosystem: a community of different species interacting with each other and with the larger physical environment that surrounds it.

edge cities: nodal concentrations of shopping and office space situated on the outer fringes of metropolitan areas, typically near major highway intersections.

ejidos: communal lands given to groups of landless peasants who could farm collectively or as individuals but could not rent or sell the land.

elasticity of demand: the degree to which levels of demand for a product or service change in response to changes in price.

emigration: a move from a particular location.

environmental determinism: a doctrine holding that human activities are controlled by the environment.

environmental ethics: a philosophical perspective on nature that prescribes moral principles as guidance for our treatment of it.

environmental justice: movement reflecting a growing political consciousness, largely among the world's poor, that their immediate environs are far more toxic than those in wealthier neighborhoods.

equal-area (equivalent) projection: a map projection that portrays areas on Earth's surface in their true proportions.

equidistant projection: a map projection that allows distance to be represented as accurately as possible.

ethnicity: a socially created system of rules about who belongs and who does not belong to a particular group based upon actual or perceived commonality.

ethnocentrism: the attitude that one's own race and culture are superior to others'.

ethology: the scientific study of the formation and evolution of human customs and beliefs.

export-processing zones (EPZs): small areas within which especially favorable investment and trading conditions are created by governments in order to attract export-oriented industries.

external arena: regions of the world not yet absorbed into the modern world system.

external economies: cost savings that result from circumstances beyond a firm's own organization and methods of production.

existential imperative: the tendency for people to define themselves in relation to their material world and their capacity to achieve a form of spiritual or psychic unity between themselves and their material worlds.

F

famine: acute starvation associated with a sharp increase in mortality.

farm crisis: the financial failure and eventual foreclosure of thousands of family farms across the U.S. Midwest.

fast world: people, places, and regions directly involved, as producers and consumers, in transnational industry, modern telecommunications, materialistic consumption, and international news and entertainment.

federal state: a form of government in which power is allocated to units of local government within the country.

fiscal squeeze: increasing limitations on city revenues, combined with increasing demands for expenditure.

flexible production systems: the ability of manufacturers to shift quickly and efficiently from one level of output to another, or from one product configuration to another.

folk culture: traditional practices of small groups, especially rural people with a simple lifestyle who are seen to be homogeneous in their belief systems and practices.

food chain: five central and connected sectors (inputs, production, product processing, distribution, and consumption) with four contextual elements acting as external mediating forces (the State, international trade, the physical environment, and credit and finance).

food manufacturing: adding value to agricultural products through a range of treatments—such as processing, canning, refining, packing, and packaging—that occur off the farm and before they reach the market.

food regime: a specific set of links that exists among food production and consumption and capital investment and accumulation opportunities.

food security: a person, a household, or even a country has assured access to enough food at all times to ensure active and healthy lives.

forced migration: movement by an individual against his or her will.

Fordism: principles for mass production based on assembly-line techniques, scientific management, mass consumption based on higher wages, and sophisticated advertising techniques.

foreign direct investment: the total of overseas business investments made by private companies.

formal region: groups of areal units that have a high degree of homogeneity in terms of particular distinguishing features.

friction of distance: the deterrent or inhibiting effect of distance on human activity.

functional regions: regions with some variability in certain attributes but with an overall coherence to the structure and dynamics of economic, political, and social organization.

G

gateway city: a city that serves as a link between one country or region and others because of its physical situation.

gender: the social differences between men and women rather than the anatomical differences that are related to sex.

genetically modified organism (GMO): any organism that has had its DNA modified in a laboratory rather than through cross-pollination or other forms of evolution.

genre de vie: a functionally organized way of life that is seen to be characteristic of a particular culture group.

gentrification: the invasion of older, centrally located working-class neighborhoods by higher-income households seeking the character and convenience of less expensive and well-located residences.

geodemographic analysis: practice of assessing the location and composition of particular populations.

geodemographic research: uses census data and commercial data (such as sales data and property records) about the populations of small districts in creating profiles of those populations for market research.

geographic information system (GIS): an organized collection of computer hardware, software, and geographic data that is designed to capture, store, update, manipulate, and display geographically referenced information.

geographical imagination: the capacity to understand changing patterns, changing processes, and changing relationships among people, places, and regions.

geographical path dependence: the historical relationship between the present activities associated with a place and the past experiences of that place.

geopolitics: the state's power to control space or territory and shape the foreign policy of individual states and international political relations.

gerrymandering: the practice of redistricting for partisan purposes.

global change: combination of political, economic, social, historical, and environmental problems at the world scale.

global civil society: the set of institutions, organizations and behaviour situated between the state, the business world, and the family. Specifically, this includes voluntary and non-profit organizations, philanthropic institutions, and social and political movements, as well as other forms of social participation.

Global Positioning System: a system of satellites that orbit the Earth on precisely predictable paths, broadcasting highly accurate time and locational information.

globalization: the increasing interconnectedness of different parts of the world through common processes of economic, environmental, political, and cultural change.

globalized agriculture: a system of food production increasingly dependent upon an economy and set of regulatory practices that are global in scope and organization.

green revolution: the export of a technological package of fertilizers and high-yielding seeds, from the core to the periphery, to increase global agricultural productivity.

gross domestic product (GDP): an estimate of the total value of all materials, foodstuffs, goods, and services produced by a country in a particular year.

gross migration: the total number of migrants moving into and out of a place, region, or country.

gross national income (GNi): similar to GDP, but also includes the value of income from abroad.

growth poles: economic activities that are deliberately organized around one or more high-growth industries.

guest workers: individuals who migrate temporarily to take up jobs in other countries.

H

hajj: a pilgrimage.

hearth areas: geographic settings where new practices have developed, and from which they have subsequently spread.

hegemony: domination over the world economy, exercised by one national state in a particular historical epoch through a combination of economic, military, financial, and cultural means.

hinterland: the sphere of economic influence of a town or city.

historical geography: the geography of the past.

human geography: the study of the spatial organization of human activity and of people's relationships with their environments.

humanistic approach: places the individual—especially individual values, meaning systems. intentions, and conscious acts—at the center of analysis.

human rights: people's individual rights to justice, freedom, and equality, considered by most societies to belong automatically to all people.

hunting and gathering: activities whereby people feed themselves through killing wild animals and fish and gathering fruits, roots, nuts, and other edible plants to sustain themselves.

I

identity: the sense that people make of themselves through their subjective feelings based on their everyday experiences and wider social relations.

immigration: a move to another location.

imperialism: the extension of the power of a nation through direct or indirect control of the economic and political life of other territories.

import substitution: the process by which domestic producers provide goods or services that formerly were bought from foreign producers.

infant mortality rate: annual number of deaths of infants under one year of age compared to the total number of live births for that same year.

inflation: the increase of printed currency that leads to higher prices and international financial differentials.

informal sector: economic activities that take place beyond official record, not subject to formalized systems of regulation or remuneration.

infrastructure (or fixed social capital): the underlying framework of services and amenities needed to facilitate productive activity.

initial advantage: the critical importance of an early start in economic development; a special case of external economies.

intensive subsistence agriculture: practice that involves the effective and efficient use—usually through a considerable expenditure of human labor and application of fertilizer—of a small parcel of land in order to maximize crop yield.

internal migration: a move within a particular country or region.

internally displaced person: individuals who are uprooted within the boundaries of their own country because of conflict or human rights abuse

international division of labor: the specialization, by countries, in particular products for export.

international migration: a move from one country to another.

international organization: group that includes two or more states seeking political and/or economic cooperation with each other.

international regime: the orientation of contemporary politics around the international arena, instead of the national.

intersubjectivity: shared meanings among people, derived from their lived experience of everyday practice.

intertillage: practice of mixing different seeds and seedlings in the same swidden.

intifada: uprising against Israel by the Palestinian people.

invasion and succession: a process of neighborhood change whereby one social or ethnic group succeeds another.

irredentism: the assertion by the government of a country that has a minority living outside its formal borders belongs to it historically and culturally.

Islam: an Arabic term that means submission to God's will.

Islamism: an anticolonial, anti-imperial, and generally anti-core political movement.

isotropic surface: a hypothetical, uniform plain: flat, and with no variations in its physical attributes.

J

jihad: a sacred struggle.

just-in-time production: manufacturing process where daily or hourly delivery schedules of materials allow for minimal or zero inventories.

K

kinship: a relationship based on blood, marriage, or adoption.

L

land reform: the redistribution of land by the state with a goal of increasing productivity and reducing social unrest.

landscape as text: the idea that landscapes can be read and written by groups and individuals.

language branch: a collection of languages that possess a definite common origin but have split into individual languages.

language family: a collection of individual languages believed to be related in their prehistorical origin.

language group: a collection of several individual languages that are part of a language branch, share a common origin, and have similar grammar and vocabulary.

language: a means of communicating ideas or feelings by means of a conventionalized system of signs, gestures, marks, or articulate vocal sounds.

latitude: the angular distance of a point on Earth's surface, measured north or south from the equator, which is 0°.

law of diminishing returns: the tendency for productivity to decline, after a certain point, with the continued application of capital and/or labor to a given resource base.

leadership cycles: periods of international power established by individual states through economic, political, and military competition.

life expectancy: average number of years an infant newborn can expect to live.

Lifeworld: the taken-for-granted pattern and context for everyday living through which people conduct their day-to-day lives.

localization economies: cost savings that accrue to particular industries as a result of clustering together at a specific location.

longitude: the angular distance of a point on Earth's surface, measured east or west from the prime meridian (the line that passes through both poles and through Greenwich, England, and that has the value of 0°).

M

map projection: a systematic rendering on a flat surface of the geographic coordinates of the features found on Earth's surface.

masculinism: the assumption that the world is, and should be, shaped mainly by men, for men.

mechanization: the replacement of human farm labor with machines.

medical geography: a subarea that specializes in understanding the spatial aspects of health and illness.

megacity: a very large city characterized by both primacy and high centrality within its national economy.

middle cohort: members of the population 15 to 64 years of age who are considered economically active and productive.

migration: a move beyond the same political jurisdiction, involving a change of residence—either as emigration, a move from a particular location, or as immigration, a move to another location.

minisystem: a society with a single cultural base and a reciprocal social economy.

minority groups: population subgroups that are seen—or that see themselves—as somehow different from the general population.

mobility: the ability to move, either permanently or temporarily.

Modern movement: an architectural movement based on the idea that buildings and cities should be designed and run like machines.

Modernity: a forward-looking view of the world that emphasizes reason, scientific rationality, creativity, novelty, and progress.

Muslim: a member of the Islamic community of believers whose duty is obedience and submission to the will of God.

N

nation: a group of people often sharing common elements of culture, such as religion or language or a history or political identity.

nationalism: the feeling of belonging to a nation as well as the belief that a nation has a natural right to determine its own affairs.

nation-state: an ideal form consisting of a homogeneous group of people governed by their own state.

natural decrease: the difference between CDR and CBR, which is the deficit of births relative to deaths.

natural increase: the difference between the CBR and CDR, which is the surplus of births relative to deaths.

nature: a social creation as well as the physical universe that includes human beings.

neocolonialism: economic and political strategies by which powerful states in core economies indirectly maintain or extend their influence over other areas or people.

neo-Fordism: economic principles in which the logic of mass production coupled with mass consumption is modified by the addition of more flexible production, distribution, and marketing systems.

neoliberal policies: economic policies that are predicated on a minimalist role for the state, assuming the desirability of free markets as the ideal condition not only for economic organization but also for political and social life.

neoliberalism: a reduction in the role and budget of government, including reduced subsidies and the privatization of formerly publicly owned and operated concerns, such as utilities.

net migration: the gain or loss in the total population of a particular area as a result of migration.

new world order: the triumph of capitalism over communism, wherein the United States becomes the world's only superpower and therefore its policing force.

newly industrializing countries (NICs): countries formerly peripheral within the world system that have acquired a significant industrial sector, usually through foreign direct investment.

nontraditional agricultural exports (NTAEs): new export crops that contrast with traditional exports.

North/South divide: the differentiation made between the colonizing states of the Northern Hemisphere and the formerly colonized states of the Southern Hemisphere.

nutritional density: ratio between the total population and the amount of land under cultivation in a given unit of area.

O

OECD: Organization for Economic Cooperation and Development.

offshore financial centers: islands or microstates that have become a specialized node in the geography of worldwide financial flows.

old-age cohort: members of the population 65 years of age and older who are considered beyond their economically active and productive years.

ordinary landscapes (vernacular landscapes): the everyday landscapes that people create in the course of their lives.

overurbanization: a condition in which cities grow more rapidly than the jobs and housing they can sustain.

P

Paleolithic period: the period when chipped-stone tools first began to be used.

pastoralism: subsistence activity that involves the breeding and herding of animals to satisfy the human needs of food, shelter, and clothing.

peripheral regions: regions with undeveloped or narrowly specialized economies with low levels of productivity.

physical geography: a subarea that studies the Earth's natural processes and their outcomes.

place: a specific geographic setting with distinctive physical, social, and cultural attributes.

plantation: a large landholding that usually specializes in the production of one particular crop for market.

political ecology: the approach to cultural geography that studies human-environment relations through the relationships of patterns of resource use to political and economic forces.

popular culture: the practices and meaning systems produced by large groups of people whose norms and tastes are often heterogeneous and change frequently, often in response to commercial products.

population policy: an official government policy designed to affect any or all of several objectives, including the size, composition, and distribution of population.

Postmodern urban design: a style characterized by a diversity of architectural styles and elements, often combined in the same building or project.

Postmodernity: a view of the world that emphasizes an openness to a range of perspectives in social inquiry, artistic expression, and political empowerment.

preservation: an approach to nature advocating that certain habitats, species, and resources should remain off-limits to human use, regardless of whether the use maintains or depletes the resource in question.

primacy: a condition in which the population of the largest city in an urban system is disproportionately large in relation to the second- and third-largest cities.

primary activities: economic activities that are concerned directly with natural resources of any kind.

producer services: services that enhance the productivity or efficiency of other firms' activities or that enable them to maintain specialized roles.

proxemics: the study of the social and cultural meanings that people give to personal space.

pull factors: forces of attraction that influence migrants to move to a particular location.

push factors: events and conditions that impel an individual to move from a location.

Q

quaternary activities: economic activities that deal with the handling and processing of knowledge and information.

R

race: a problematic classification of human beings based on skin color and other physical characteristics.

racialization: the practice of categorizing people according to race, or of imposing a racial character or context.

rank-size rule: a statistical regularity in city-size distributions of cities and regions.

reapportionment: the process of allocating electoral seats to geographical areas.

redistricting: the defining and redefining of territorial district boundaries.

redlining: the practice whereby lending institutions delimit "bad-risk" neighborhoods on a city map and then use the map as the basis for determining loans.

refugee: individual who crosses national boundaries to seek safety and asylum—are a significant global problem.

region: larger-sized territory that encompasses many places, all or most of which share similar attributes in comparison with the attributes of places elsewhere.

regional geography: the study of the ways in which unique combinations of environmental and human factors produce territories with distinctive landscapes and cultural attributes.

regionalism: a feeling of collective identity based on a population's politico-territorial identification within a state or across state boundaries.

regionalization: the classification of individual places or areal units.

religion: a belief system and set of practices that recognize the existence of a power higher than humans.

remote sensing: the collection of information about parts of the Earth's surface by means of aerial photography or satellite imagery designed to record data on visible, infrared, and microwave sensor systems.

reurbanization: the growth of population in metropolitan central cores, following a period of absolute or relative decline in population.

risk society: contemporary societies in which politics is increasingly about avoiding hazards.

rites of passage: the ceremonial acts, customs, practices, or procedures that recognize key transitions in human life such as birth, menstruation, and other markers of adulthood, such as marriage.

romanticism: the philosophy that emphasizes interdependence and relatedness between humans and nature.

S

sacred space: an area recognized by individuals or groups as worthy of special attention as a site of special religious experiences or events.

secondary activities: economic activities that process, transform, fabricate, or assemble the raw materials derived from primary activities, or that reassemble, refinish, or package manufactured goods.

sectionalism: extreme devotion to local interests and customs.

segregation: the spatial separation of specific population subgroups within a wider population.

self-determination: the right of a group with a distinctive politico-territorial identity to determine its own destiny, at least in part, through the control of its own territory.

semiotics: the practice of writing and reading signs.

semiperipheral regions: regions that are able to exploit peripheral regions but are themselves exploited and dominated by core regions.

sense of place: feelings evoked among people as a result of the experiences and memories that they associate with a place, and to the symbolism that they attach to it.

sexuality: the set of practices and identities that a given culture considers related to each other and to those things it considers sexual acts and desires.

shifting cultivation: a system in which farmers aim to maintain soil fertility by rotating the fields within which cultivation occurs.

shock city: a city that is seen as the embodiment of surprising and disturbing changes in economic, social, and cultural life.

siltation: the buildup of sand and clay in a natural or artificial waterway.

site: the physical attributes of a location—its terrain, its soil, vegetation, and water sources, for example.

situation: the location of a place relative to other places and human activities.

slash-and-burn: the system of cultivation in which plants are cropped close to the ground, left to dry for a period, and then ignited.

slow world: people, places, and regions whose participation in transnational industry, modern telecommunications, materialistic consumption, and international news and entertainment is limited.

society: the sum of the inventions, institutions, and relationships created and reproduced by human beings across particular places and times.

socio-spatial formation: a specific combination of demographic groups, social classes, cultural values, and local institutions at a particular time and place.

sovereignty: the exercise of state power over people and territory, recognized by other states and codified by international law.

spatial analysis: the study of geographic phenomena in terms of their arrangement as points, lines, areas, or surfaces on a map.

spatial diffusion: the way that things spread through space and over time.

spatial interaction: the movement and flows involving human activity.

spatial justice: the fairness of the distribution of society's burdens and benefits, taking into account spatial variations in people's needs and in their contribution to the production of wealth and social well-being.

splintering urbanism: the fragmentation of the economic, social, and material fabric of cities as a result of the selective impact of new technologies and networked information and communications infrastructures.

spread effects: the positive impacts on a region (or regions) of the economic growth of some other region.

squatter settlements: residential developments that take place on land that is neither owned nor rented by its occupants.

states: independent political units with territorial boundaries that are internationally recognized by other states.

strategic alliances: commercial agreements between transnational corporations, usually involving shared technologies, marketing networks, market research, or product development.

subsistence agriculture: farming for direct consumption by the producers; not for sale.

suburbanization: the growth of population along the fringes of large metropolitan areas.

supranational organizations: the collections of individual states with a common goal that may be economic and/or political in nature; such organizations diminish, to some extent, individual state sovereignty in favor of the group interests of the membership.

sustainable development: a vision of development that seeks a balance among economic growth, environmental impacts, and social equity.

swidden: land that is cleared using the slash-and-burn process and is ready for cultivation.

symbolic landscapes: representations of particular values or aspirations that the builders and financiers of those landscapes want to impart to a larger public.

T

technology systems: clusters of interrelated energy, transportation, and production technologies that dominate economic activity for several decades at a time.

technology: physical objects or artifacts, activities or processes, and knowledge or know-how.

terms of trade: the ratio of prices at which exports and imports are exchanged.

territorial organization: a system of government formally structured by area, not by social groups.

territoriality: the specific attachment of individuals or peoples to a specific location or territory.

territory: the delimited area over which a state exercises control and which is recognized by other states.

terrorism: the threat or use of force to bring about political change.

tertiary activities: economic activities involving the sale and exchange of goods and services.

threshold: the minimum market size required to make the sale of a particular product or service profitable.

time-space convergence: the rate at which places move closer together in travel or communication time or costs.

topological space: the connections between, or connectivity of, particular points in space.

topophilia: the emotions and meanings associated with particular places that have become significant to individuals.

total fertility rate (TFR): the average number of children a woman will have throughout the years that demographers have identified as her childbearing years, approximately ages 15 through 49.

trading blocs: groups of countries with formalized systems of trading agreements.

transcendentalism: a philosophy in which a person attempts to rise above nature and the limitations of the body to the point where the spirit dominates the flesh.

transhumance: the movement of herds according to seasonal rhythms: warmer, lowland areas in the winter; cooler, highland areas in the summer.

transnational corporations: companies with investments and activities that span international boundaries and with subsidiary companies, factories, offices, or facilities in several countries.

transnational migrant: migrants who set up homes and/or work in more than one nation-state.

tribe: a form of social identity created by groups who share a set of ideas about collective loyalty and political action.

U

underclass: a subset of the poor, isolated from mainstream values and the formal labor market.

underemployment: a situation in which people work less than full time even though they would prefer to work more hours.

undernutrition: inadequate intake of one or more nutrients and/or calories.

unitary state: a form of government in which power is concentrated in the central government.

urban agriculture: the establishment or performance of agricultural practices in or near an urban or citylike setting.

urban ecology: the social and demographic composition of city districts and neighborhoods.

urban form: the physical structure and organization of cities.

urban system: an interdependent set of urban settlements within a specified region.

urbanism: the way of life, attitudes, values, and patterns of behavior fostered by urban settings.

urbanization economies: external economies that accrue to producers because of the package of infrastructure, ancillary activities, labor, and markets typically associated with urban settings.

utility: the usefulness of a specific place or location to a particular person or group.

V

vertical disintegration: the evolution from large, functionally integrated firms within a given industry toward networks of specialized firms, subcontractors, and suppliers.

virgin soil epidemics: conditions in which the population at risk has no natural immunity or previous exposure to the disease within the lifetime of the oldest member of the group.

visualization: the computer-assisted representation of spatial data, often involving three-dimensional images and innovative perspectives, in order to reveal spatial patterns and relationships more effectively.

vital records: information about births, deaths, marriages, divorces, and the incidence of certain infectious diseases.

voluntary migration: movement by an individual based on choice.

W

world city: a city in which a disproportionate part of the world's most important business is conducted.

world music: the musical genre defined largely in response to the sudden increase of non-English language recordings released in the United Kingdom and the United States in the 1980s.

world region: large-scale geographic divisions based on continental and physiographic settings that contain major groupings of peoples with broadly similar cultural attributes.

world empire: minisystems that have been absorbed into a common political system while retaining their fundamental cultural differences.

world system: an interdependent system of countries linked by economic and political competition.

Y

youth cohort: members of the population who are less than 15 years of age and generally considered to be too young to be fully active in the labor force.

Z

zionism: the establishment of a legally recognized home in Palestine for the Jewish people.

zone in transition: an area of mixed commercial and residential land uses surrounding the CBD.

Photo Credits

T = top; M = middle; B = bottom; L = left; R = right

Frontmatter

Page ii: © Faith Saribas/Reuters/Corbis; Page v: Getty Images, Inc. Page vi: (c) Catherine Karnow / CORBIS All Rights Reserved. Page vii: AP Wide World Photos. Page viii: (c) Antar Dayal / Illustration Works / Getty Images, Inc. Page ix: Beatrice Mategwa / Reuters / (c) CORBIS All Rights Reserved. Page x: Getty Images, Inc. Page xi: (c) Michael S. Yamashita / CORBIS All Rights Reserved.

Chapter 1

Page 1: Paul L. Knox. Page 4: (c) Heinz Nissel, Vienna. Page 6: (T) AP Wide World Photos. Page 6: (B) Courtesy of Heike Mayer. Page 9: (M & B) Corbis/Bettmann. Page 14: Photo by Sean Smith, "Guardian Weekly", August 2002, p. 11. Page 15: Photo by Sean Smith, "Guardian Weekly", August 2002, p. 10. Page 16: (c) Dean Conger / CORBIS All Rights Reserved. Page 18: (c) Christopher J. Morris / CORBIS All Rights Reserved. Page AP Wide World Photos. Page 21: (T/L) Courtesy of Spaceimaging.com; (T/R) Joel Sartore/www.joelsartore.com; (M) NASA goddard Space Flight Center. Image by Reto Stockli (land surface, shallow water, clouds). Enhancements by Robert Simmon (ocean color, compositing, 3D globes, animation). Page 22: (M) Paul L. Knox. Page 29: (c) Brooks Kraft / CORBIS All Rights Reserved. Page 33: (T) (c) Bob Krist / CORBIS All Rights Reserved; (B) `(c) Michael Brennan / CORBIS All Rights Reserved. Page 34: (T) Corbis Royalty Free; (B) Paul L. Knox. Page 35: (T) Paul L. Knox; (B/R) (c) Danny Lehman / CORBIS All Rights Reserved. Page Paul L. Knox.

Chapter 2

Page 42: (c) Catherine Karnow / CORBIS All Rights Reserved. Page 45: DRK Photo. Page 46: (B/L) (c) Sandro Vannini / CORBIS All Rights Reserved; (B/R) (c) Ruggero Vanni / CORBIS All Rights Reserved. Page 49: Courtesy of the Library of Congress. Page 51: (c)Philip deBay / Historical Picture Archive / CORBIS All Rights Reserved. Page 52: Corbis/Bettmann. Page 53: The Granger Collection. Page 54: #27 (Chartres) from the Carte de France atlas, published in Paris in 1767. From the David Rumsey Historical Map Collection, available online in the Luna Insight Digital Image Database. Page 56: Paul L. Knox. Page Page 61: Paul L. Knox. Page 62: (c) Bettmann/CORBIS All Rights Reserved. Page 64: (M) Library of Congress; (B) Library of Congress. Page 65: (T) Library of Congress; (M) Library of Congress; (B) Corbis/Bettmann. Page 66: (T/R) (c) Swim Ink 2, LLC / CORBIS All Rights Reserved. Page 73: (c) Andreas Meier / CORBIS All Rights Reserved. Page 74: (c) Richard Hamilton Smith / CORBIS All Rights Reserved. Page 79: (c) Jim Hollander / Reuters / CORBIS All Rights Reserved. Page 80: (T/L): (c) Antoine Serra / In Visu / CORBIS All Rights Reserved; (T/R) Getty Images, Inc. - Agence France Presse. Page 80: (B) © Bazuki Muhammad / Reuters / CORBIS All Rights Reserved. Page 81: AP Wide World Photos.

Chapter 3

Page 84: AP Wide World Photos. Page 90: Getty Images, Inc.- Photodisc. Page 91: Stock Boston. Page 100: Grant Heilman Photography, Inc. Page 101: (c) Tim O'Hara/CORBIS. Page 102: Photo Researchers, Inc. Page 113: (T) (c) Beatrice Mategwa / Reuters / CORBIS All Rights Reserved; (B) AP Wide World Photos. Page 117: (T) Corbis/Bettmann; (B) The Stock Connection. Page 119: Galbe.com. Page 123: (T) Getty Images, Inc – Liaison; (B) Panos Pictures.

Chapter 4

Page 128: Courtesy Arizona Department of Health Services. Page 130: AP Wide World Photos. Page Page 131: (B/L) AP Wide World Photos; (T/R) New York, 2002 Photo by Martin Rowe, Lantern Books. Page 132: (L) AP Wide World Photos; (R) ©Miguel Fairbanks. Page 136: (c) Diego Lezama Orezzoli / CORBIS All Rights Reserved. Page 139: AP Wide World Photos. Page 141: Georgius Agricola De Re Metallica, Translated from the first latin edition of 1556 by Herbert Clark Hoover/Lou Henry Hoover, Pg. 337. Page 143: (T) (c) Jean Clottes and French Ministry of Culture and Communication, Regional Directon for Cultural Affairs - Rhone-Alps - Regional Department of Archaeology; (B) AP Wide World Photos. Page 144: (T) (c) Warren Morgan / CORBIS All Rights Reserved; (B) Arthur Lidov/National Geographic Society Image Collection. Page 145: AP Wide World Photos. Page 146: AP Wide World Photos. Page 151: AP Wide World Photos. Page 152: AP Wide World Photos. Page 155: (T) (c) China Photo / Reuters / CORBIS All Rights Reserved; (B) (c) Norbert Schaefer / CORBIS All Rights Reserved. Page 156: (T) (c) Najlah Feanny / CORBIS SABA All Rights Reserved.; (B) AP Wide World Photos. Page 159: AP Wide World Photos. Page 164: AP Wide World Photos. Page 165: AP Wide World Photos. Page 166: AP Wide World Photos

Chapter 5

Page 172: Cartoon by Ian Baker. Page 175: AP Wide World Photos. Page 176: (T) (c) Paul A. Souders / CORBIS All Rights Reserved; (B) (c) Gavriel Jecan / CORBIS All Rights Reserved. Page 178: (T) (c) Amet Jean Pierre / CORBIS Sygma All rights reserved; (B) Rischgitz / Getty Images, Inc. Page 179: (T) (c) Nathan Benn / CORBIS All Rights Reserved; (B) AP Wide World Photos. Page 180: AP Wide World Photos. Page 181: (c) Steve Azzara / CORBIS Sygma All rights reserved. Page 188: (T) AP Wide World Photos; (B) AP Wide World Photos. Page 189: AP Wide World Photos. Page 194: (c) Daniel Laine / CORBIS All Rights Reserved. Page 198: AP Wide World Photos. Page 201: (L) AP Wide World Photos; (R) AP Wide World Photos. Page 204: AP Wide World Photos. Page 205: Panoramic Images / Getty Images, Inc. Page 206: (c) Bob Krist / CORBIS All Rights Reserved. Page 209: AP Wide World Photos

Chapter 6

Page 212: (c) Antar Dayal / Illustration Works / Getty Images, Inc. Page 215: AP Wide World Photos. Page 216: AP Wide World Photos. Page 217: AP Wide World Photos. Page 219: AP Wide World Photos. Page 224: (T/L) (c) Alexander Natruskin / Retuers / Corbis; (T/R) (c) Sergio Moraes / Reuters / Corbis; (B/L) (c) Owen Franken / CORBIS All Rights Reserved; (B/R) (c) Matthew Polak / CORBIS All Rights Reserved. Page 225: (T) (c) Joseph Sohm / CORBIS All Rights Reserved; (B) AP Wide World Photos. Page 226: AP Wide World Photos. Page 227: AP Wide World Photos. Page 229: (T/L) (c) James Davis / Eye Ubiquitous / CORBIS All Rights Reserved; (B/L) Cristiano Mascaro / Getty Images, Inc.; (B/R) AP Wide World Photos. Page 230: AP Wide World Photos. Page 233: (c) Klaus Nather / Zefa / CORBIS All Rights Reserved. Page 234: AP Wide World Photos. Page 235: AP Wide World Photos. Page 236: Alex S. MacLean / Landslides. Page 238: AP Wide World Photos. Page 240: (T) AP Wide World Photos; (B) AP Wide World Photos. Page 241: (c) Gail Mooney / CORBIS All Rights Reserved. Page 242: (T) AP Wide

World Photos; (B) (c) Jason Hawkes/CORBIS. Page 244: (M/L) Pepys Library, Magdafene College, Cambridge; (M/R) AP Wide World Photos; (B/L/R) AP Wide World Photos. Page 245: AP Wide World Photos. Page 246: AP Wide World Photos.

Chapter 7

Page 250: (c) Claro Cortes IV / Reuters / CORBIS All Rights Reserved. Page 256: (T) AP Wide World Photos; (B) (c) Jeremy Horner / /CORBIS All Rights Reserved. Page 258: AP Wide World Photos. Page 262: (c) Yann Arthus-Bertrand / CORBIS All Rights Reserved. Page 264: (c) Lee Jae-Won / CORBIS All Rights Reserved. Page 267: (c) John Van Hasselt / CORBIS SYGMA. Page 270: (c) Luc Gnnago / Reuters / CORBIS. Page 272: (c) Peter Turnley / CORBIS All Rights Reserved. Page 275: AP Wide World Photos. Page 279: (c) Vince Streano / CORBIS All Rights Reserved. Page 287: S.M.A. #479 Del 20.05.1993. Page 289: (c) Steve Starr / CORBIS All Rights Reserved. Page 291: (c) Jagadeesh / Reuters / CORBIS All Rights Reserved. Page 294: (M/L) AP Wide World Photos; (M/R) Paul Nicklen / Getty Images, Inc.; (B) AP Wide World Photos.

Chapter 8

Page 300: (c) Tony Arruza / CORBIS All Rights Reserved. Page 303: (T) (c) Alain Le Garsmeur / CORBIS All Rights Reserved; (B) (c) Illustration by Mark McMahon; Photograph by Franklin McMahon / CORBIS All Rights Reserved. Page 305: AP Wide World Photos. Page 306: AP Wide World Photos. Page 307: AP Wide World Photos. Page 308: (T) (c) Michael S. Yamashita / CORBIS All Rights Reserved; (B) (c) Matthieu Paley / CORBIS All Rights Reserved. Page 310: (B/L) AP Wide World Photos; (B/R) AP Wide World Photos. Page 312: (c) Philip Gould / CORBIS All Rights Reserved. Page 313: (c) Yann Arthus-Bertrand / CORBIS All Rights Reserved. Page 317: (L) (c) Hulton-Deutsch Collection/CORBIS All Rights Reserved; (R) (c) Tom Bean / CORBIS All Rights Reserved. Page 318: AP Wide World Photos. Page 321: AP Wide World Photos. Page 322: (T) AP Wide World Photos; (B) AP Wide World Photos. Page 323: (c) Wendy Stone / CORBIS All Rights Reserved. Page 327: (c) Ted Spiegel / CORBIS All Rights Reserved. Page 328: AP Wide World Photos. Page 329: (T) (c) Jim Richardson / CORBIS All Rights Reserved; (B) AP Wide World Photos. Page 331: AP Wide World Photos. Page 333: (L) Agricultural Research Service – USDA; (R) AP Wide World Photos. Page 334: AP Wide World Photos. Page 336: Antoine Serra/Corbis Sygma. Page 339: AP Wide World Photos.

Chapter 9

Page 342: Beatrice Mategwa / Reuters / (c) CORBIS All Rights Reserved. Page 346: (B) AP Wide World Photos; (T) AP Wide World Photos. Page 347: (T) (c) Jason Hawkes / CORBIS All Rights Reserved; (B) AP Wide World Photos. Page 348: AP Wide World Photos. Page 354: AP Wide World Photos. Page 359: AP Wide World Photos. Page 360: AP Wide World Photos. Page 367: AP Wide World Photos. Page 368: AP Wide World Photos. Page 370: AP Wide World Photos. Page 371: AP Wide World Photos. Page 373: AP Wide World Photos. Page 380: AP Wide World Photos. Page 388: AP Wide World Photos.

Chapter 10

Page 392: Getty Images, Inc. Page 396: (c) Nik Wheeler / CORBIS All Rights Reserved. Page 399: (T) AP Wide World Photos; (B) (c) Historical Picture Archive/CORBIS. Page 400: (c) Historical Picture Archive/CORBIS. Page 401: (M/L) (c) Jose Fuste Raga / CORBIS All Rights Reserved; (M/R) (c) Yann Arthus-Bertrand / CORBIS All Rights Reserved; (B/L) AP Wide World Photos; (B/R) AP Wide World Photos. Page 405: (c) Arvind Garg / CORBIS All Rights Reserved. Page 407: AP Wide World Photos. Page 413: (c) Mike Finn-Kelcey / Reuters / CORBIS. Page 417: (T) AP Wide World Photos; (B) China Photos / Getty Images, Inc. Page 418: (L) (c) Wendy Stone / CORBIS All Rights Reserved; (R) (c) Catherine Karnow / CORBIS All Rights Reserved. Page 419: (T) Richard Nowitz / National Geographic Society / Getty Images Inc.; (B) AP Wide World Photos. Page420: (c) Tibor bogn-r / CORBIS All Rights Reserved. Page 422: AP Wide World Photos. Page 423: (c) Tibor Bogn-r / CORBIS All Rights Reserved.

Chapter 11

Page 426: (c) Royalty-Free / Corbis. All rights reserved. Page 429: AP Wide World Photos. Page 430: (c) Alan Schein Photography / CORBIS All Rights Reserved. Page 431: (c) Richard Berenholtz / CORBIS All Rights Reserved. Page 432: AP Wide World Photos. Page 433: AP Wide World Photos. Page 434: AP Wide World Photos. Page 435: AP Wide World Photos. Page 436: AP Wide World Photos. Page 437: AP Wide World Photos. Page 438: (T) AP Wide World Photos; (B) From CD Rom "Visual Paths Through Europe", part of a book published by Blackwell (2004), Cities of Europe. Book and CD-Rom are copyright of Yuri Kazepov, editor / Photographic credit: S. Hall, P. Lee and S. Sankey. Page 439: (T) AP Wide World Photos; (B) AP Wide World Photos. Page 440: AP Wide World Photos. Page 441: (T) (c) 2003 Artists Rights Society (ARS), New York / ADAGP, Paris / FLC; (B) AP Wide World Photos. Page 442: AP Wide World Photos. Page 443: (T) AP Wide World Photos; (B) AP Wide World Photos. Page 444: AP Wide World Photos. Page 445: AP Wide World Photos. Page 446: (T) (c) Liz Gilbert / Corbis Sygma; (B)) AP Wide World Photos. Page 447: (T) (c) Daniel Laine / CORBIS All Rights Reserved; (M) (c)James Marshall/Bettmann/CORBIS All Rights Reserved; (B) (c) William Campbell / Corbis Sygma. All Rights Reserved. Page 448: (L) AP Wide World Photos; (R) (c) Lynsey Addario / CORBIS All Rights Reserved. Page 449: (c) Janet Jarman / CORBIS All Rights Reserved. Page 450: AP Wide World Photos. Page 451: AP Wide World Photos. Page 452: AP Wide World Photos. Page 454: AP Wide World Photos. Page 455: AP Wide World Photos. Page 459: AP Wide World Photos. Page 460: STR/Agence France Presse/Getty Images. Page 461: (c) Laurent Zylberman / Corbis Sygma. Page 462: JEFF@JAGPPhotoInc.COM(c)2002

Chapter 12

Page 466: (c) Michael S. Yamashita / CORBIS All Rights Reserved. Page 469: (T/L) (c) Ashley Cooper / CORBIS All Rights Reserved; (T/R) (c) Charles O'Rear / CORBIS All Rights Reserved; (B/L) (c) Peter Turnley / CORBIS All Rights Reserved; (B/R) (c) PHOTRI / CORBIS SYGMA All Rights Reserved. Page 470: (T) (c) Brooks Kraft / CORBIS All Rights Reserved; (B) (c) Jean Miele / CORBIS All Rights Reserved. Page 474: (c) Georgina Bowater / CORBIS All Rights Reserved. Page 480: AP Wide World Photos. Page 486: AP Wide World Photos. Page 487: (M/L) AP Wide World Photos; (M/R) (c) Jehad Nga / CORBIS All Rights Reserved. Page 489: (c) Ed Kashi / CORBIS All Rights Reserved. Page 491: AP Wide World Photos.

Appendix

Page 494: AP Wide World Photos.

Index

A

Abacha, Sani, 486
Abbas, Mahmoud, 111
Absenteeism, 433
Absolute distance, 24
Accessibility, 26
  in spatial analysis, 22
  urban structure and, 428
Acid emissions, 160
Acid rain, 158–159, 160
  Asian Brown Cloud and, 491
  treaties on, 167
Activism
  anti-nuclear, 150–151
  environmental, 139–140, 166
  on genetically modified organisms, 336–337
  against globalization, 79–81
Acute hunger, 335
Adams, William, 168
Adaptation, 133
Addis Ababa, Ethiopia, 14–15
  slum housing in, 419
Adikiwe, Kate, 446
Administrative centers, 397, 398
Advanced Very High Resolution Radiometer (AVHRR) satellite imagery, 503
Advertising, 237–238
Advertising agencies, 292
Affluence
  environmental impact of, 132–133
  global inequality of, 75–76
  increasingly inequal distribution of, 12–13, 14–15
Afghanistan
  geopolitical significance of, 368–369
  new world order and, 367, 368–370
  Taliban in, 369–370
  war on terror and, 372–373
  women in, 201
Africa
  AIDS in, 104–105, 106
  colonialism in, 67–68
  corruption in, 486–487
  decolonization of, 362
  desertification in, 164–165, 334
  eco-migration from, 119
  European colonies in, 357–358, 359–360
  frontier regions in, 348
  fuelwood depletion in, 156, 257
  grasslands in, 164, 165
  Green Revolution and, 319
  Green Revolution in, 320
  HIV/AIDS in, 16–18, 488
  hunger in, 254
  informal sector in, 445
  international debt of, 269
  Islam in, 198
  kinship in, 190, 191–192
  language and ethnicity in, 200
  language extinction in, 192
  neocolonialism in, 68
  population in, 89
  pre-modern world-empires in, 47, 50
  refugees in, 112–113, 114–115
  rice production in, 327–328
  standard of living in, 75
  urbanization in, 412
  view of the U.S. in, 482
  women in, 206
African Americans
  hip-hop culture and, 180–181
  rural-to-urban migration of, 115–116
Afro-Asiatic language family, 200
Age of Discovery, 52–53
Age-sex pyramids, 92–99
  baby boom, 100
Agglomeration, geographical, 291–292
Agglomeration diseconomies, 277
  mature metropolises and, 420
Agglomeration effects, 274, 411
Aging population, 92, 96–98, 483
Agrarian lifestyle, 302
Agribusiness, 325
Agricola, Georg, 141
Agricultural density, 91
Agricultural industrialization, 316–317
Agricultural Origins and Dispersals (Sauer), 44
Agricultural revolutions, 309–317
  first, 310
  industrialization and, 316–317
  pre-modern, 44–45, 143–144
  second, 310–311
  third, 311, 316
Agriculture, 301–341
  agro-food system in, 324–325
  biotechnology in, 328–332, 335–337, 338, 475–476
  carrying capacity and, 258
  Columbian Exchange and, 148–152
  commercial, 304
  contract farming, 361
  cultivable land and, 257–258, 259
  definition of, 302
  domestication in, 143–144
  in the economic system, 321–322
  environmental impacts of, 164, 332–334
  environmental perception in, 214–215
  European expansion and, 146
  food regimes in, 325–327
  food surpluses in, 144–145
  global food system in, 334–339
  globalization of, 164
  global restructuring of, 317–327
  Green Revolution in, 318–320
  industrialization of, 316–317
  intensive subsistence, 307–308
  internal combustion engine in, 62–63
  invention of tools for, 145
  Latin American, 323–324
  organic, 330–331
  pastoralism, 308–309
  plantation, 50–51
  plantation system, 50–51
  political ecology and, 137
  population limits and, 119–120
  pre-modern, 44–45
  rural-urban boundaries and, 347
  shifting cultivation, 304–307
  slash-and-burn, 44, 305–306
  social change and, 327–328
  subsistence, 303
  sustainable, 261
  swidden, 44
  technology in, 327–332
  traditional geography of, 302–309
  urban, 337, 339
  urban development and, 397
  urbanization and, 144–145, 418
Agro-food system, 324–325
Agro-production system, 317
Ahmad Shah Durrani, 369
Ahya', 442
Airbus, 475
Air pollution, 453. *See also* Pollution
  Asian Brown Cloud, 491
Air travel
  future of, 475
  hubs in, 26
  time-space convergence and, 29
Albanians, 354–355

Al-Battani, 50
Al Biruni, 48
Alemán, Arnoldo, 486
Alexander the Great, 234
Al-Farghani, 50
Alfred P. Murrah Federal Office Building bombing, 371
Algeria, French rule in, 357–358
Ali, Muhammad, 180
Alien tort law, 79
Al Khoresm, 48
Al-Khwarazmi, 50
Allahabad, India, self-help housing in, 450
Al Quaeda, 18–19, 68, 370, 374, 479
  future of, 489–490
Altria, 281–282
American Century, 484
American Dream, 482
American Geographical Society, 37
Americanization, 207–208
American Society of Civil Engineers (ASCE), 433
Amsterdam, the arts in, 241
Analysis of data, 21–22
Anan, Koffi, 123, 378
Anasazi, environmental impact of, 145
Ancillary industries, 274
Antarctica, colonization in, 362, 365
Anthem, Arizona, as packaged community, 459, 460
Anti-clericalism, 195
Appadurai, Arjun, 72
Appalachian Regional Commission, 280
Aquaculture, 313–315
Arafat, Yasser, 111, 383
Architecture
  Beaux Arts, 440
  colonial, 405
  of European cities, 435–440
  International Style, 440
  Islamic, 442–443
  Modernist, 233, 440, 441
  new urbanism, 233, 236
  semiotics and, 228–229
Ardashev, Grigoriy, 413
Areas of settlement, 159
Arithmetic density, 89, 91
ASEAN (Association of South East Asian Nations), 7, 321, 322
Asia
  decolonization in, 364
  deindustrialization in, 267
  Green Revolution in, 318–319
  Internet in, 237
  population in, 89
  rice production in, 307–308
  urbanization in, 412, 414
  view of the U.S. in, 482
Asian Americans, birth rates among, 99
Asian Brown Cloud, 491
Asian Pacific Economic Cooperation, 166
Association of American Geographers, 37
Association of South East Asian Nations (ASEAN). *See* ASEAN (Association of South East Asian Nations)
Aswan High Dam, 159
Atlanta, GA
  inner-city poor neighborhoods in, 95
Australia
  gateway cities in, 402
  place marketing in, 241
  uranium mining in, 150–151
Autarky, 264
Automobile industry
  in China, 267
  future of, 473–475
  global assembly line in, 283, 286
Azimuthal projections, 496–497, 499
Aztec empire
  environmental impact of, 152
  as hearth area, 45
  smallpox in, 148

B

Baaker, James, 186
Baby boom, 96, 99
  aging of, 92, 96–98
  packaged communities for, 459
  population pyramids on, 100
Baby bust generations, 97–98
Babylon, 396
Back-office functions, 290–291
  splintering urbanism and, 422
Backward linkages, 274–275
Backwash effects, 276–277
Bacon, Francis, 141
Baghdad, Iraq, Al Hussein mosque in, 441
Bahrain, as offshore financial center, 293
Balance of payments, 290
Balfour Declaration, 235, 382
Balkans, refugees in, 113, 114. *See also* Eastern Europe
Baltimore, Maryland, place marketing in, 241, 242, 295
Bambaataa, Afrika, 180
Bangkok, Thailand
  air pollution in, 453
  as primate city, 410–411, 419
  sewage systems in, 451
  traffic in, 450, 451
Bangladesh
  eco-migration from, 119
  famine in, 335
  intensive subsistence agriculture in, 307
  women's self-help in, 207
Barak, Ehud, 383
Barbarism scenarios, 468, 469
Baron, Naomi, 173
Barter, 45
Bartholomew's Nordic projection, 497
Basque Homeland and Freedom movement, 379
Bazaars, 442
Beaux Arts style, 440
Beck, Ulrich, 19
Behavior, 225–226
  cognitive images and, 219–223
Belgium, industrialization in, 58, 59, 60
Bellinzona, Switzerland, 397, 399
Benneton, 287
Bergman, E.F., 22
Berkeley school of cultural geography, 176–177
Berlin, Germany
  in Golden Triangle, 59
  Wall, 347, 348, 366
Bhopal, India, Union Carbide accident, 453
Bid-rent, 428
B.I.G., 180
Bikini, nuclear testing on, 150–151
Bin Laden, Osama, 370
Biodiversity. *See* Biological diversity
Biological diversity, 130–131
  conventions protecting, 167
  domestication and, 144
Biomass energy, 153, 320
Bioprospecting, 167
Biorevolution, 329–332
Biotechnology, 58
  in agriculture, 324, 328–332, 335–337, 338
  definition of, 328
  future of, 475–476
  genetically modified organisms, 335–337, 338
Bioterrorism, 372
Bipolar world order, 377
Birmingham, England, municipal housing in, 438
Birth-control programs, 120, 122–124
  in India, 102
Birth rates, 99–101
Bismarck, Otto von, 68
Black Death, 129, 146, 147
Black Hebrew Israelism, 371
Black markets, 419–420, 444–445, 450
Blair, Tony, 373
Blau, Willem, 53
Blix, Hans, 373
Blue Revolution, 312–315

Body, as geographic scale, 8
Bolivia, coca growing in, 164
Bonn, Germany, skyline of, 436–438
Bono, 123
Boomburbs, 454
Borglum, Butzon, 247
Borlaug, Norman, 318
Bosnia-Herzegovina, 354–355
Boston, Massachusetts
    cognitive image of, 219, 220
    as gateway city, 402
    infrastructure problems in, 432
    place marketing in, 241
    reurbanization of, 421
Botswana, 18
Boundaries, 346–349
    formation of, 348–349
    frontier regions and, 347–348
Bowler, Ian, 311
Bra, Italy, 239
Brasilia, Brazil, 228–229, 440
Brazil
    Brasilia, 228–229, 440
    gay-pride parades, 204
    Green Revolution and, 319
    international debt of, 269
    landless movement in, 324
    regional development patterns in, 257
    rural-to-urban migration in, 118
    shifting cultivation in, 306
Bread riots, 323
Bride burning, 206–207
British Virgin Islands, as offshore financial center, 293
Brown, James, 180
Brundtland, Gro Harlem, 260
Brussels, Belgium, as world city, 411
Bryant Park Restoration Corp., 241
Bubonic plague, 129, 146, 147
Budapest, Hungary, social housing in, 439
Buddhism, 182, 185, 186, 188
    sacred spaces of, 229
Buddhism, 183, 184
Bukhara, 48
Burke, P., 95
Bush, George H.W., 366, 386, 387
Bush, George W.
    on Kyoto Protocol, 163
    Kyoto Protocol and, 490
    war on terror and, 372–374
Buttel, F., 332
Byzantium, urban origins in, 397

C

Cable television
    world distribution of, 197
Cabral, Pedro, 53
Cairo, Egypt, population of, 419
California
    diseconomies in, 277
    Proposition 13, 381
    separatist movements in, 379
Cambodia
    genocide in, 343, 366
    Vietnam War and, 365
Canada
    boundaries of, 346
    as core region, 75
    homelessness in, 435
    Internet in, 237
    Québecois movement, 190, 202–203, 379
    transnational migrants to, 110
Canal systems, 61–62
Cape Town, South Africa, as gateway city, 402
Capitalism
    definition of, 19
    fast vs. slow world, 77
    global, 377
    inauthentic landscapes and, 243
    jihad vs. McWorld and, 77–79
    merchant, 51, 399–400
    modernity and, 232
    terrorism and, 19
Carbon dioxide emissions, 135
    Kyoto Protocol on, 162–163
Caribbean countries
    colonialism in, 358
    religion in, 194–195
Carrying capacity, 258
Carson, Rachel L., 130, 140, 311, 333
Cartograms, 498–499, 502
Cartography, 52–53. *See also* Maps and mapping
Cascade diffusion. *See* Hierarchical diffusion patterns
Cash crops, 329–330
Cassini, Caesar-Francois, 54
Cassini, Dominique, 54
Cassini, Jacques, 54
Cassini, Jean Dominique, 54
Cassini triangulation surveys, 53, 54
Castro, Fidel, 364–365
Catholicism, 194–195
Cayman Islands, 292, 293
Censuses, 86–88
    limitations of, 86–88
Center for International Science Information Network (CIESN), 13
Central Andes region, cultural and political ecology in, 136–137
Central business districts (CBDs), 429–435
    European, 438
Central cities, 430
Centrality, 410
Central places, 405
Central place theory, 405
Centrifugal forces, 351–352
Centripetal forces, 351–352
Centro International de Mejorimiento de Maiz y Trigo (CIMMYT), 318
Change
    cultural, globalization and, 207–209
    cumulative, in places, 35–36
    general effects of, 36
    places in, 5
    unique outcomes of, 36
Chang Jiang, China, 9
Charlotte, N.C., as airline hub, 26
Chechnya, terrorism in, 370–371, 372–373
Chemical farming, 311, 321
Chernobyl incident, 151, 153
Cherokee Nation, forced migration of, 118
Chicago, Illinois
    immigrants in, 429–430
    as shock city, 403
Chicago Board of Trade, 303
Chicken industry, 301–302
Children
    as beggars, 393
    homeless, 435
    informal sector labor and, 445
    in labor, 255–256
    rights of, 378
China
    Asian Brown Cloud and, 491
    Buddhism in, 182, 185
    census in, 86
    coal use in, 155
    economic development in, 251, 266–267
    family planning in, 123
    films produced in, 196
    future of, 483–486
    geographic knowledge in pre-modern, 50
    hearth area in, 45
    intensive subsistence agriculture in, 307
    internal forced migration in, 118–119
    languages in, 193
    nuclear energy in, 153
    packaged landscapes in, 460–462
    place-making in, 9
    population distribution in, 89
    refugees in, 110
    regional development patterns in, 257
    SARS in, 18
    sewage systems in, 451
    Silk Road and, 47, 48
    urbanization in, 414, 416–417
    urban origins in, 397

Chinatowns, 205, 430, 431
Chloroflurocarbon (CFC) emissions, 167
Chorography, 49
Chorology, 49
Choropleth maps, 495
Christian Identity movement, 371
Christianity, 183–186
  fundamentalism in, 186, 189, 195
  Jerusalem in, 234–235
  pilgrimages in, 231
  terrorism and, 371
Christmas Atoll, nuclear tests on, 151
Chronic hunger, 335
Circle of poison, 333
Citadels, 442
Citibank, 290
Cities in a Globalizing World, 421
Citizen Law Enforcement and Analysis (CLEAR), 489
Citizenship, 349–350
CittaSlow movement, 236, 238–239
City-size distributions, 408–411
Classification from above, 30
Clean Development Mechanism, 163
Climate change
  disease spread and, 129–130
  Kyoto Protocol and, 162–163
  storms due to, 133–135
Clinton, George, 180, 213, 386, 387
Cloning, 329
Clothing industry, 284–286
Clovis point, 143, 144
Coal-powered technology, 57, 153
  environmental impact of, 155
Coded spaces, 225–232
  sacred, 229–232
  semiotics and, 226–229
Cogeneration, 159
Cognitive distance, 24
Cognitive images, 23–24, 219
  behavior and, 219–223
Cognitive space, 26
Cohorts, 92, 99
Cold War, 270, 271, 367, 468, 471
  cultural dissonance and, 488
  new world order and, 479
Collective memory, 5
Cologne, Germany, 397, 400
  architecture of, 437
Colombia, refugees from, 113
Colombo, Sri Lanka, as gateway city in, 402
Colón, Cristóbal, 52, 53
Colonial cities, 404–405
Colonialism, 356–360
  boundary formation and, 348
  decolonization and, 360–362
  definition of, 55
  diseases and, 147–148
  division of labor and, 63, 66–67
  environmental impact of, 145–152
  food regimes and, 325
  religion and, 182, 185, 186, 194–195
  resistance to, 67–68
Colonial mandate system, 360
Colonies, 429
Colonization. *See also* Imperialism
  definition of, 47
  New World, 54–55
  pre-modern, 47
Columbian Exchange, 147–148, 148–149, 183
Columbus, Christopher. *See* Colón, Cristóbal
Combined Heat and Power (CHP), 159
Commercial agriculture, 304
Commercial imperialism, 68
Commercialism, 227
Commission on Sustainable Development, 130
Commodification of nature, 141
Commodity chains, 69–72, 284, 286
Commodity concentration of exports, 268
Commonwealth of Independent States (CIS), 354
Communications systems
  cultural change and, 173–174
  information technology and, 476–478
  international communication flows and, 75, 76
  time-space convergence and, 28–29
Community
  Internet and, 20
  scale of, 8
  smart growth and, 454
Community art, 35
Comparative advantage, 63
Complementarity, 27
Computer-aided design (CAD), 287
Computer-aided manufacturing (CAM), 287
Computer-mediated communication (CMC), 173–174
Computers
  information technologies and, 476–478
  internationalization of finance and, 73
ConAgra, 330
Concentric zones, 430, 431
Confederations, 354
Conformal projections, 496–497, 499
Conglomerate corporations, 281–282
Congregation, 428–429
Connectivity, 26
Conservation, definition of, 139. *See also* Environment
Constantine I, 234
Consumer-driven commodity chains, 70
Consumerism
  environmental impact of, 132–133
  internal migration and, 116
  places as objects of, 237–240
  sexual identity and, 202
  shopping and, 227–228
  shopping behavior and, 220–221
Consumer markets
  global, 75
  homogenization of, 280, 282
Contagion diffusion, 30
Containerization, 73–74
Context, global, 43–83
Continuous partial attention (CPA), 174
Contour maps, 495
Contract farming, 361
Conventional worlds scenarios, 469
Convention on Biological Diversity, 167
Convention on the Prevention and Punishment of the Crime of Genocide, 343
Convention on the Rights of the Child, 378
Conversion, of land use, 159
Cook, James, 52
Core and periphery system
  fast vs. slow world in, 76–77
  globalization and, 75–81
  imperialism and, 356–366
  in the New World, 54–55
  regional development and, 277–280
  uneven economic development and, 252–257
  urbanization in, 414–415
Core-domain-sphere model, 30, 32
Core regions
  in 1900, 60
  aging of the population in, 96–97
  definition of, 55
  energy consumption in, 153–159
  globalization and, 75–81
  industrialization of, 55–61
  internal development of, 61–63
  Japan as, 60–61
  population growth in, 120–122
  U.S. as, 60, 61
Corporations, deterioration of society and, 471
Corruption, 486–487
Cortés, Hernán, 148
Cosmopolitanism, 240
Costa, Lucio, 228
Costa Rica
  ecotourism in, 295
  as offshore financial center, 293

Couclelis, H., 26
Counter-cultural movements, 233
Counterfeiting, 266–267
Counterurbanization, 420–421
Crawford, Margaret, 227–228
Creative destruction, 267, 277–279
  definition of, 278
Croatia, 354–355
  Serbian claims on, 31
Crop rotation, 304–307
Crowding, 433
Crude birth rate (CBR), 99–101
Crude death rates (CDR), 103–106
Crude density, 89, 91
Crusades, 235
Cuba
  East/West divide and, 364–365
  Hurricane Ivan in, 134–135
Cul-de-sacs, 442
Cultivable land, 259
  development and, 257–258
Cultivated land, 159, 164
Cultural complexes, 176–182
  definition of, 178, 182
Cultural dissonance, 488–489
Cultural ecology, 133, 136–137
Cultural forces, 379
Cultural geography, 173–211
  Berkeley school of, 176–177
  cultural complexes in, 176–182
  definition of, 175
  globalization and, 207–209
  landscape in, 225–226
  language in, 186–190
  religion, 182–186
  society and, 190–193
Cultural hearths, 188, 190
  of Islam, 199
Cultural identity, 488
Cultural innovation, 394
Cultural landscape, 176, 177
  humanistic approach to, 225
Cultural nationalism
  definition of, 193
  Islamic, 193–201
Cultural regions, 182
Cultural space, 25–26
Cultural systems, 183–193
  religion in, 183–186
Cultural traits, 177–178
Culture
  definition of, 174
  flows in, 72
  folk, 175
  as geographical process, 174–175
  global, 72, 208–209
  globalization and change in, 207–209
  hip-hop, 180–181
  identity and, 201–207
  marketing, 240–244
  material, 175
  popular, 175
  population distribution and, 89
  society and, 190–193
Culture industries, 237–238
Cumulative causation, 275–277, 404
Cyberspace, 232. *See also* Internet
Cycle of poverty, 433–434
Cyperspace, 20

D

Da Gama, Vasco, 53
DaimlerChrysler, 286
Dalai Lama, 186, 188
Dallas–Forth Worth, Texas
  population density in, 89, 90
Dams, 158
Daoud, Sardar Mohammad, 369
Dark Ages, 397, 471
Darwin, Charles, 57, 344
Data
  analysis of, 21–22
  gathering, 20–21
  visualization/representation of, 21, 500, 502
DDT, 333
Death and Life of Great American Cities, The (Jacobs), 471
Death rates, 85, 103–106
Debt forgiveness, 270
  economic development and, 123
  future of, 486–487
  by the G8, 3
  patterns of debt and, 268–270
Debt-for-nature swaps, 333–334
Debt trap, 270–271
Decentralization, 290–291
  mature metropolises and, 420
Decision-making capacity of urban settlement, 394
Decolonization, 360–362
Deep ecology, 140
Defensible space, 463
Defensive strongholds, 397, 398
Deforestation, 152, 159, 161
  definition of, 145
  in Europe, 146
  extent of, 161
  for hydroelectric power, 158
  medicines and, 167
  rates of, 164
Degenerative utopias, 243
Deindustrialization, 267, 277–279
  definition of, 278
  government intervention in, 279–280
  mature metropolises and, 420
De jure territories, 7, 348–349
  nested hierarchies of, 351
De la Blache, Vidal, 177
Del Cano, Juan Sebastián, 53
Del Webb, 459
Demand, elasticity of, 268, 270
Democracy
  Internet and, 237
  spread of, 12
Democratic Republic of Congo, 359
Democratic rule, 386
Demographic collapse, 148
Demographic transition theory, 106–108
  peripheral regions and, 415, 418–420
  urbanization and, 403
Demography, 86–88
  censuses and, 86–88
  definition of, 86
  GIS marketing applications using, 94–95
Deng Xiaoping, 266, 484
Denver, Colorado, telecommunications in, 23
Department of Reproductive Health and Research, 204
Dependency, 268
Dependency ratio, 99
Dependency theory, 273
Derelict landscapes, 223
Descansos, 176
Desertification, 164–165, 333–334
Developed regions, 252
Developing regions, 252–257
Developmentalism, 271, 273
Dhaka, Bangladesh, as primate city, 410
Dhavale, D., 456–457
Dialects, 186
Dias, Bartholomeu, 53
Diaspora, 183
Diffusion, spatial, 29–30
Digital divide, 237, 477–478
Diminishing returns, law of, 47
Direct foreign investment (DFI)
  in China, 267
Discrimination, 428–429
Disease
  future of, 487–488, 490
  geography and, 38
  globalization and spread of, 129–130
  global warming and, 163
  life expectancy and, 104–106
  in peripheral cities, 452–453
  in Spanish colonies, 147–148
Diseconomies, 277
Disneyfication, 243
Displaced persons, 109, 112–114
Distance, 24–25
  friction of, 25
  in spatial analysis, 22

Distance-decay function, 25
Distributed workforce, 478
Districts, in cognitive images, 219
Division of labor
colonization and, 63, 66–67
complementarity and, 27
gender, in agriculture, 306, 307
geographical, 260–264
globalization and, 72
international, 63, 66–67, 268, 280
Divorce, 434
Docklands, London, 241–243
Dome of the Rock, 235
Domesday Book, 177, 178
Domestication of plants/animals, 143–144, 305, 310
Dom Henrique (Portugal), 52–53
Domino theory, 363–366, 379
Dot maps, 495
Double cropping, 307–308
Doubling time, 101, 103
of city populations, 415
Drug trafficking, 76, 490
Dualism, 444, 445
Dubai, United Emirates, architecture of, 444
Dubrovnik, Croatia, architecture of, 437
DuPree, Rasheda, 24
Dust Bowl, 333
Duvalier, Jean-Claude, 486
Dwyer, Owen, 205
Dymaxion projection, 497–498, 501

E

Eade, J., 231
Earth as Modified by Human Action, The (Marsh), 138–139
Earth First!, 139
Earth Summits, 130, 131
Kyoto Protocol and, 162–163
on sustainable development, 260–261
Eastern Europe. *See also* Soviet Union, former
regional development patterns in, 257
urban structure in, 438–439
East/West divide, 363–366, 468
Ecclesiastical centers, 397, 399
Echo Boomers, 97–98
Eckert IV projection, 497
Ecofeminism, 140, 141
Ecological footprint, 260–261
Ecological imperialism, 148–152
Ecological model of land use, 429–430, 431
Ecological niches, 430, 431
Eco-migration, 119
Economic denationalization, 12
Economic development, 251–299
agriculture projects in, 323
Clean Develoment Mechanism and, 163
crude birth rates and, 100–101, 102
cumulative change in places from, 35–36
debt and, 268–270
definition of, 252
demographic transition theory and, 106–108
economic structure and, 260–264
future of, 478–479
geography and, 38
globalization and, 280–296
import substitution and, 51
international aid and, 270–271
international trade and, 264–265, 268–271
interpreting patterns of, 271–273
patterns of, 252–273
regional, 256–257, 273–280, 478–479
resources and, 257–260
stages of, 272
sustainable, 124, 260–261
tourism and, 292–296
unevenness in, 252–257
Economic integration
railroads in, 62
Economic structure, 260–264
geographical division of labor in, 260–264
primary activities in, 260
quaternary activities in, 260
secondary activities in, 260
tertiary activities in, 260
Economies of scale, 283
complementarity and, 27
Ecosystems
biodiversity and, 144
definition of, 144
Ecoterrorism, 139
Ecotourism, 295
Ecuador, ecotourism in, 295
Edge cities, 453
Edges, in cognitive images, 219
Education
crude birth rates and, 100–101
family-planning, 123–124
in index of human development, 254
urban poverty and, 433–434
Egypt
Aswan High Dam, 159
borders of, 349
population distribution in, 89, 90
urban origins in, 396–397
Ejidos, 324
Elasticity of demand, 268, 270
Electronic offices, 290–291
Electronics, 57
Elites, 478–479, 480
Ellis, Richard, 460
Elsea, D., 460–461
Email, 173–174
Embattled, the, 478–479, 480
Emerson, Ralph Waldo, 138
Emigration, 108
Emissions trading, 162–163
Enclaves, 429
End of History and the Last Man, The (Fukuyama), 366
Energy needs
development and, 257
environmental impact of, 152–159
future of, 473–478
Engels, Frederich, 120
Entertainment industry, 208
Environment, 129–171
affluence's effects on, 132–133
agriculture and, 332–334
ecological imperialism and, 148–152
eco-migration and, 119
fair trade movement and, 271, 274–275
free trade and, 81
globalization of, 166–168
globalization's effects on, 13, 15–16, 76
impact of ancient humans on, 142–145
Kyoto Protocol on, 162–163
land-use change and, 159–166
in peripheral cities, 452–453
politics on, 166–167
population limits and, 119–120
shifting cultivation and, 305–307
sustainability and, 167–168, 260–261, 490–491
threats to, 490
Environmental Defense Fund, 139
Environmental determinism, 57, 356
Environmental ethics, 140
Environmental justice, 140
Environmental perception, 214–215, 225–226. *See also* Places
Epidemics, 19
future of, 487–488, 490
globalization and, 129–130
life expectancy and, 104–106
virgin soil, 147–148
Equal-area projections, 497
Equidistant projections, 496, 499
Equivalent projections, 497
Erbil, 396

Eridu, 396
Erie Canal, 62
Erosion, 333–334
Essay on the Principle of Population, An (Malthus), 119–120
Estrada, Joseph, 486
Ethical Trading Initiative (ETI), 275
Ethics, 471
  environmental, 140
Ethiopia
  coffee in, 329
  eco-migration from, 119
Ethnicity
  in Africa, 200
  in censuses, 86–87
  censuses and, 86–88
  cultural dissonance and, 488–489
  definition of, 204
  identity and, 20
  nationalism and, 350–355
  neighboorhoods and, 429–430
  regionalism and, 379–380
  security issues and, 489–490
  social polarization and, 462–463
  space and, 204–205
  states and, 350
  terrorism and, 370–371, 372–373
Ethnocentrism, 37
  definition of, 57
Ethnoscapes, 72
Ethology, 217
Europe
  AIDS in, 105
  changing map of, 345–346
  imperialism by, 50–54, 145–152, 357–358
  industrialization in, 57–59
  population in, 89
  refugees in, 111, 114
  Renaissance in, 232
  urbanism in, 397–403
  urban structure in, 435–440
  view of the U.S. in, 483
European Dream, 482
European Union (EU), 7
  boundaries in, 346
  as core region, 75
  economic coordination efforts of, 322
  economic denationalization and, 12
  food safety in, 314
  future of, 481–483
  global integration zone, 456
  as supranational organization, 375–376
  Trans-European Network, 473–474
Exclusionary boundaries, 346
Existential imperative, 218
Expansion diffusion, 30, 31
Experience
  as object of consumption, 238–239
  in place-making, 219
Experiential consumption, 238–239
Experiential space, 25–26
Exploration
  geography and, 50–54
  imperialism and, 356, 361–362
Export-processing zones (EPZs), 288–289
Exports, peripheral region, 66–67. *See also* Trade
External areas, 50
External economies, 273
External edge cities, 453
Extinction
  language, 192
  Paleolithic period, 142–143
  through hunting, 143, 144

F

Fairtrade Labeling Organizations International (FLO), 274–275
Fair trade practices, 271
  movement for, 274–275
Faith-based terrorism, 371, 372
Family farms, 321, 322
Family-planning programs, 120, 122–124
Famines, 335
  eco-migration and, 119
  in Ethiopia, 15
Farm crisis, 328
Fast food, 236. *See also* McDonaldization
Fast world, 76
  Americanization and, 207–208
  international travel and, 239–240
  place and, 218–219
Federal states, 352–353
Federal Voting Rights Act, 388
Fertile Crescent, 45, 145
  urban origins in, 396–397
Fertility rates, 85, 99–101
  demographic transition theory and, 106–108
  family-planning programs and, 120, 122–124
Fertilizers
  biotechnology and, 328–331
  chemical farming and, 311
  environmental impact of, 332–333
  global use of, 316
  Green Revolution and, 318–320
Feudalism, 186, 311, 397–398, 399
Fieldwork, 20
Films, world production of, 196
Finance
  global office and, 289–292
  information technology in, 10
  internationalization of, 73, 280
  offshore financial centers, 292, 293
  splintering urbanism and, 422
  transnational network in, 377
  turnover time of capital in, 77
Finanscapes, 72
Finland, AIDS in, 106
Fire, environmental impact of, 142, 143
Fiscal squeeze, 430, 432, 433
  educational settings and, 433–434
Fisheries, 324
  aquaculture, 313–315
Fitzsimmons, Margaret, 133
Flexible production systems, 286–288
Floodplains, 310
Floods, sea-level changes and, 133, 135
Folk culture, 175
Food and Agriculture Organization (FAO), 323
Food chains, 138, 325
  integrated, 325–326
Food consumption patterns, 253–254
Food manufacturing, 311, 322
Food regimes, 325–327
Food security
  biotechnology and, 331, 476
  definition of, 335
  global food system and, 334, 335–337, 338
  shrimp fishing and, 314
Foot-and-mouth disease, 19
Forced migration, 109
  internal, 118–119
  internally displaced persons, 109, 112–114
  international, 111–115
  in Rwanda, 360
Ford, 286
Fordism, 286
Foreign direct investment (FDI), 73, 280
  splintering urbanism and, 422
Forests
  deforestation, 145, 146, 152, 158, 159, 161
  loss of, 159
Formal regions, 30
Forman, Murray, 181
Forward linkages, 274–275
Fossil fuels
  alternatives to, 159
  environmental impact of, 133, 152–159
  sustainable development and, 261
Framework Convention on Climate Change, 162–163

France
  Americanization of language in, 194
  canals in, 61
  cartography in, 53
  genres de vie in, 177
  Industrial Revolution in, 58, 59, 60
  interdependence of with Vietnam, 11
  language diversity in, 192
  nuclear tests by, 151
  resistance to globalization in, 79
  rural vs. urban politics in, 381
  technopoles in, 280
Francica, J., 95
Frank, André Gunder, 273
Free trade agreements
  environmental issues and, 81
  globalization and, 81
  transformationalist view of, 12
French Revolution, 367
Fresh fruit and vegetable food regime, 325–327
Friction of distance, 25
Frontier regions, 347–348
Frontiers, 346–349
Frostbelt, 116
Fuel cells, 159, 475
Fuelwood, 156
  countries using, 157
  economic development and, 257
Fujimori, Alberto, 486
Fukayama, Francis, 366
Fuller, Buckminster, 497–498, 501
Functional interdependence, 274
Functional regions, 30
Functional specialization, 409
Fundamentalism
  Christian, 186, 189, 195
  intensification of, 488
  terrorism and, 371
Future geographies, 467–493
  issues and threats in, 488–491
  mapping, 468–472
  regional, 478–488
  resources and technology and, 472–478
"Futurist Manifesto" (Marinetti), 232

G

Galactic metropolises, 453
Gambia, rice production in, 327–328
Gang culture, 180–181, 217
Garbage collection, 451
Garbage picking, 445
Gated communities, 455, 459, 462, 463
Gateway cities, 400–402
Gay marriage, 189
Gay-pride parades, 202, 204
Geldorf, Bob, 486
Gender
  agriculture and, 306, 307, 327–328
  definition of, 205
  division of labor by, 306, 307, 327–328
  employment and, 124
  environmental perception and, 214–215
  equality in and development, 255–256
  identity, 205–207
  kinship and, 192
  sexual identity and, 203
  technology access and, 307
Gender Empowerment Index, 255
General effects, 36
General Motors, 282, 286
Generation X, 97–98
Generation Y, 97–98
Generative functions of urban settlement, 395
Generica, 455, 459
Genetically modified organisms (GMOs), 335–337, 338
  bans on, 5, 6, 338
  future of, 475–476
Genetic engineering, 475–476
Genghis Khan, 368
Genoa, Italy
  antiglobalization riots in, 81
Genocide, 343–344
Genocide Watch, 343
Genre de vie, 177, 178
Gentrification, 430, 431, 432, 463
Geodemographic analysis, 92, 94–95
Geodemographic research, 503
Geographer, The (Vermeer), 53
Geographers, work of, 37–38
Geographical imagination
  definition of, 35
  developing, 35–36
Geographical path dependence, 273
Geographic information systems (GIS), 22, 500–504
  applications of, 500–503
  critiques of, 503
  geodemographic analysis with, 92, 94–95
  public participation, 95
Geographic knowledge
  Age of Discovery and, 52–53
  pre-modern world-empires and, 47, 49
Geographic literacy
  importance of, 4–10
  in the U.S., 4
Geographic scales, 7–8
  regional analysis and, 30–31
Geography
  careers in, 38
  exploration and, 50–54
  foundations of modern, 56–57
  future, 467–493
  globalization and, 19–20
  historical, 177
  humanistic approach to, 225–226
  importance of, 4–10
  importance of education in, 37
  politics of, 379–386
  power of, 36–38
  regional approach to, 49
  work of geographers and, 37–38
Geography (Strabo), 49
Geopolitics
  definition of, 344
  the state in, 344–346
  world order and, 349
Georeferencing, 94
Geostrategic theory, 362–363
Geothermal energy, 159
Germany
  age-sex pyramid for, 93
  Industrial Revolution in, 58, 59, 60
  refugees in, 110
Gerry, Edlbridge, 388
Gerrymandering, 387–388
G8 group
  antiglobalization protests and, 81
  debt forgiveness by, 3, 270, 486
  protests against, 166
Ghana
  international debt of, 269
Ghettos, 429
  race, 434
  social polarization and, 462–463
Ghuangzhou, China, as gateway city, 402
Gingrich, Newt, 88
GIS. *See* Geographic information systems (GIS)
Glacis militaire, 439
Glasnost, 354
Global change, 166
Global cities. *See* World cities
Global civil society, 378
Global context, 43–83
  contemporary globalization and, 69–81
  core and periphery in, 54–55
  industrialization and, 55–61
  new world geography in, 50–69
  periphery organization in, 63–69
  pre-modern world in, 44–50
Global economy, 478–488
Globalization
  of agriculture, 317–327
  Americanization and, 207–208
  causes and consequences of, 72–75
  of the clothing industry, 284–286
  consumer markets in, 75
  contemporary, 69–81

definition of, 10
disease and, 129–130
division of labor and, 72
economic development and, 280–296
of the environment, 166–168
environmental issues in, 13, 15–16, 145–152
fair trade movement and, 271, 274–275
flexible production systems and, 286–288
forces of, 317, 321–323
gender issues and, 124
geography and, 19–20
global assembly line and, 281–289
global office and, 289–292
health issues and, 16–18
the Internet in, 237
Islamic cities and, 443–444
jihad vs. McWorld and, 77–79
key issues in, 13–19
language and, 190
mobilization against, 79–81
perspectives on, 10, 12–13
place-making and, 232–237
security issues and, 18–19
splintering urbanism and, 421–422
technology in, 73, 75
transnational corporations in, 68
urban space and, 462–463
Globalized agriculture, 317–327
Global metropolitanism, 488–489
Global Positioning System (GPS), 23, 475
Global Report on Human Settlements 2001, 243, 462–463
Global scale, 8
Global sourcing, 283
Global warming, 13
Kyoto Protocol on, 162–163
Goals 2000: Educate America Act, 37
Godwin, William, 120
Gogol Bordello, 209
Golden Triangle, 59
Gonzaga, Vespasiano, 439
Goodman, David, 312
Gorbachev, Mikhail, 353–354, 372
Gore, Al, 386
Government
in agriculture, 322–323
corruption in, 486–487
in deindustrialization, 279–280
finance regulations and, 290
on genetically modified organisms, 338
Internet and, 237
supranational and world, 468
GPS. *See* Global Positioning System (GPS)
Graffiti, 217
Graham, Stephen, 421
Grameen Bank, 207
Grapes of Wrath, The (Steinbeck), 213–214
Grasslands, 159, 164–165
Great transitions scenarios, 469
Greek empire, 46, 47
geographic knowledge in, 49
urban design/planning in, 439
urban origins in, 397
Green Belt Movement, 130–131
Greenhouse gases, 135
Kyoto Protocol on, 162–163
Green Party, 167
Greenpeace, 139–140, 151
Green Revolution, 318–320
biorevolution compared with, 332
in Latin America, 324
Grenadines, banana industry in, 137
Griot, 180
Gross domestic product (GDP), 252
in developed vs. developing areas, 479
per capita, 252
in postindustrial countries, 261
Gross migration, 109
Gross national income (GNI), 252–253
as aid, 271
per capita, 253
Gross product, 394
Grouping, 30
Growth-control movements, 381
Growth poles, 279–280
Guangshou, China, urbanization in, 416, 417
Guatemala, refugees in, 113
Guest workers, 109–110
Guide to Geography (Ptolemy), 49
Gunnery, 50

H

Hajj, 230–231
Hall, Peter, 453, 454
Hamas, 383
Hanseatic League, 399–400
Harborplace, Baltimore, 241, 242
Harvey, David, 120, 240, 243
Hate movements, 367
Haussmann, Georges, 439–440
Havelaar, Max, 274
Hazardous waste, 167
Health care
deforestation and, 167
global spread of disease and, 16–18
in index of human development, 254
life expectancy and, 104–106
Health-care density, 91
Hearth areas, 44–46
cultural, 188, 190, 199
language, 188, 190
Heartland theory, 362–363, 366, 379
Hegemony, 68, 374, 479
Henry the Navigator, Prince, 52–53
Hersbrück, Germany, 36
Hierarchical diffusion patterns, 16–18, 30, 31
Highways, automated, 473, 475
Himba tribe minisystem, 45
Hinduism, 183, 184
sacred spaces in, 230
Hindu Kush, 368
Hinterlands, 50
Hip-hop culture, 180–181
Hispanics, birth rates among, 99
Hispaniola, disease and depopulation of, 147–148
Historical geography, 177
Historic districts, 243
HIV/AIDS
in Africa, 488
diffusion of, 3, 16–18, 30
in Ethiopia, 15
life expectancy and, 104–106
H.J. Heinz, 331
Hobbes, Thomas, 141
Ho Chi Minh City, Vietnam, 11
Holocaust, 343
Homeland security, geography and, 38
Homelessness, 434–435
Home work, 124
Homogenization
of consumer markets, 280, 282
cultural dissonance and, 488–489
Hondius, Jodocus, 53
Hong Kong
films produced in, 196
prosperity of, 266
Special Economic Zones and, 416–417
transnational migrants from, 110
urban gardens in, 339
Horizontal integration, 330
Housing
discrimination in, 434
social/public, 438
squatter settlements, 418–420, 447, 449–450
Human footprint, 13
Human geography
definition of, 4, 20
spatial analysis in, 22–30
studying, 20–30
tools for studying, 20–22
Humanistic approach, 225–226
Human rights issues, 378
sexual rights, 204
Human settlements scale, 8
Human trafficking, 490
Hunger, 254, 335

Hunters and gatherers, 142–143, 144, 303
  environmental impact of, 149, 152
Hunting and gathering, 303
Huntington, Ellsworth, 57
Hurricane Ivan, 134–135
Hurricane Katrina, 133–134
Hussein, Saddam, 372, 374
Hutus, 358–359
Hydraulic societies, 47
Hydrocarbons, 158–159
Hydroelectric power, 156, 158
Hydropower, 153
Hyperglobalists, 10, 12, 376

I

Ibn Sind, 48
IBP, Inc., 328
Ice Age, 143–144
Identity, 201–207
  cultural, 488
  ethnic, 20
  gender, 205–207
  places in, 5–6
  race, 205
  regional, 20
  sexual, 201–204
  territoriality and, 218
Ideoscapes, 72
Immigration, 108
  North American urban structure and, 429–430
Immigration and Naturalization Service, 86
Imperialism, 356–360
  commercial, 68
  core and periphery system and, 54–55
  decolonization and, 360–362
  definition of, 54
  ecological, 148–152
  environmental effects of, 145–152
  food regimes and, 325
  heartland theory in, 362–363
  new, 68–69
  New World, 50–54
  peripheral regions under, 67–68
  peripheral urbanization and, 404–405
Impermeability, 346–347
Imports, peripheral region, 66–67
Import substitution
  in China, 266–267
  cumulative causation and, 277
  definition of, 51, 270
Inca empire, 136
  environmental impact of, 152
Inclusionary boundaries, 346
Income
  global disparities in, 75–76
  inequality in, 479, 481
Index of human development, 254
India
  Asian Brown Cloud and, 491
  birth-control programs in, 102
  British rule in, 357, 359
  films produced in, 196
  future of, 483–484
  Green Revolution in, 318–319
  Hinduism in, 183
  languages in, 188, 190, 191
  Modi Revlon in, 208
  nuclear energy in, 153
  population distribution in, 89
  refugees in, 110
  sacred spaces in, 230
  self-help housing in, 450
  sewage systems in, 451
  Silk Road and, 47, 48
  software development in, 477
  urban origins in, 397
  women in, 206–207
Indo-European language family, 188, 190
Indonesia, population redistribution in, 118
Industrial cities, 404
Industrialization
  agricultural, 316–317
  in China, 266
  core region, 55–61
  demographic transition and, 108
  environmental impacts of, 152–166
  in Europe, 57–59
  geography in, 37–38
  in Japan, 60–61
  rural-to-urban migration and, 115–116
  in the U.S., 60, 61
Industrial resources, 258–260
Industrial Revolution
  agricultural revolution and, 311
  canal systems and, 61–62
  changes from, 35
  in Europe, 57–59
  fossil fuel use and, 152–153
  geographic scales and, 7
  imperialism in, 148–149
  origins of, 57
  urbanization and, 403–404
Infant mortality rates, 103–104, 105
  economic development and, 253
  in squatter settlements, 450
Inflation, 290, 322–323
Informal sector of the economy, 419–420, 444–445, 450, 452
Information economies, 280
Information highway, 20
Information Revolution, 478
Information technology, 476–478
  international finance and, 10
  internationalization of finance and, 73
  terrorism and, 490
Infrastructure
  cumulative causation and, 276
  deindustrialization and, 278–279
  North American urban, 432–433, 433
  in peripheral cities, 450–452
  splintering urbanism and, 421–422
  transferability and, 28
Initial advantage, 273
In-migration. *See* Immigration
Innovation, 5, 477
  cities in, 394
  cultural, 394
  world cities in, 421
Insiders, 218–219
  sense of place for, 33–35
Interdependence
  complementarity and, 27
  globalization and, 10–20
  perspectives on, 10, 12–13
  of places, 6–7
  transferability and, 27–29
  as two-way process, 8–10
Intergovernmental organizations (IGOs), 377
  in world cities, 411
Internal combustion engines, 57
  spatial reorganization and, 62–63
Internal edge cities, 453
Internal forced migration, 118–119
Internally displaced persons (IDPs), 109, 112–114
Internal migration, 108
Internal voluntary migration, 115–118
International affairs
  geography in, 37
International aid, 270–271
International Bank for Reconstruction and Development. *See* World Bank
International Bill of Human Rights, 378
International Conference on Population and Development (1994), 122
International Criminal Court (ICC), 378
International Crops Research Institute for the Semi-Arid Tropics, 320
International debt, 268–270
  debt forgiveness and, 3, 123
  future of, 486–487
International division of labor, 63, 66–67
  debt and, 268
  development and, 280
  global assembly line and, 281–289
International Energy Agency, 473
International Labor Organization (ILO), 255–256, 288

International law
  environmental, 167
  standardization and, 233
International migration, 108
International Monetary Fund (IMF), 166
  aid programs, 270–271
  EU in, 481
International nongovernmental organizations (INGOs), 377
  political influence of, 72
International organizations, 360
International regimes, 378
International Rescue Committee, 344
International Style, 440
International trade, 264–271
Internet
  backbone, visualization of, 502
  communication changed by, 173–174
  cyperspace and, 232
  digital divide and, 237, 477–478
  fast vs. slow world and, 77
  global connectivity via, 78
  growth of, 77
  human interaction and, 467
  NGO mobilization against globalization via, 79–80
  splintering urbanism and, 421–422
  terrorism and, 490
  time-space convergence and, 29
  vital records on, 87
Intersubjectivity, 34
Intertillage, 306
Intervening opportunity, 29
Intifada, 382–383
Intrafirm trade, 282
Invasion and succession, 430
Iran, refugees in, 110
Iraq
  Erbil, 396
  war on terror and, 372–374
Ireland, symbolic landscapes in, 32–33
Irish Republican Army (IRA), 370
Iroquois longhouses, 179
Irredentism, 31
Irrigation
  environmental impact of, 332–333
  environmental impacts of, 145, 152
  Green Revolution and, 319
  groundwater supplies and, 146
Islam, 183–186
  cultural nationalism and, 193–201
  definition of, 195
  Jerusalem in, 234–235
  number is adherents to, 194
  pilgrimages in, 230–231
  spread of, 198, 199
  terrorism based on, 371, 489–490
Islamic countries
  Algeria, 358
  geographic knowledge in pre-modern, 50
  jihad vs. McWorld and, 77–79
  population distribution in, 89
  pre-modern, 50
  urban structure in, 441–444
  view of the U.S. in, 479, 483
Islamism, 198, 201
Isoline maps, 495, 497
Isopleth maps, 495
Isotropic surfaces, 428
Israel
  forced migration and, 111, 115
  Jerusalem and, 234–235
  Palestinian conflict with, 382–385
Istanbul, Turkey
  cumulative change in, 35
  packaged landscapes in, 459
Italy
  Cassa del Mezzogiorno, 279
  electoral representation in, 386
  intersubjectivity in, 34
  regionalism in, 20
  Risorgimento, 32, 33
  slow city movement, 236, 238–239
  symbolic landscapes in, 32, 33
  zero population growth rate in, 99
I-35 Corridor, U.S., 456–457, 458
Iyer, Pico, 207–208

J

Jacobs, Jane, 471
Jami, 441–442
Japan
  as core region, 75
  deindustrialization in, 267
  energy needs in, 473
  import substitution in, 277
  industrialization in, 60–61
  Industrial Revolution in, 57
  Ministry of International Trade and Industry, 279
  place marketing in, 243
  salsa music in, 5, 6
  Tokaido megapolitan region, 456, 458
  urbanization in, 412
Jarvis, Howard, 381
Jerusalem
  conflict over, 234–235
  Palestinian-Israeli conflict and, 382
  as sacred space, 231, 232
Jewish nation, 349–350
Jihad, 77–79, 201
  definition of, 198
John XXIII (Pope), 195
Jones, John Paul, III, 205
Judaism, 183–186
  Black Hebrew Israelism and, 371
  Jerusalem in, 234–235
Judeo-Christian tradition
  imperialism and, 146
  view of man in, 141
  views of nature in, 135, 138
Justice
  environmental, 140
  spatial, 76, 253–254
Just-in-time production, 287
Jutland, Denmark, 20

K

Kansas City Stockyards, 327
Kant, Immanuel, 56
Karzai, Hamid, 370
Kasbahs, 442
Kenney, M., 332
Kenya
  civil servants in, 360–361
  cultural landscape in, 176
  Green Belt Movement, 130–131
  ostrich rearing project in, 323
  tourism and the environment in, 296
Kenyatta, Jomo, 68
Khiva, 48
Khmer Rouge, 343, 366
Khyber Pass, 368
Kin groups, 45
Kinship, 190–192
Kleptocracies, 486
Knowledge
  development and, 280–281
  as power, 19
Knowledge economies, 264
Koolhaas, Rem, 455
Kosovo, 354–355
Kruschev, Nikita, 372
Ktaz, Cindi, 214–215
Kuala Lumpur, Malaysia, information technology in, 477
Kurds
  ethnicity in, 379
  forced migration of, 111, 114
Kyoto Protocol, 162–163, 479, 490

L

Labeling, neighborhood, 433
Labor practices, 275
Labor supplies
  distributed workforce, 478
  global assembly line and, 283, 286
  underemployment and, 444
  urbanization and, 403
Labuan, as offshore financial center, 293
Lagos, Nigeria
  as primate city, 410
  as shock city, 446–447
Lake Baykal, Russia, 13, 16
Lambert Azimuthal Equal-Area projection, 497
Landmarks, 223, 224
  in cognitive images, 219

Land ownership patterns, 36
Land reform, 324
Landscape, 31–33, 223–232
  coded spaces, 225–232
  definition of, 223
  derelict, 223
  ordinary, 31–32, 223–229
  packaged, 455, 459, 460–461, 462
  postmodern, 240
  semiotics and, 226–229
  as text, 226
Land use
  agriculture and, 323–324
  environment and, 159–166
  urban, 428–429, 429
Lang, R.E., 456–457
Language, 173–174, 186–190
  definition of, 186
  ethnicity and, in Africa, 200
Language branches, 186, 188
Language families, 186, 188, 189
Language groups, 186, 188
Laos, Vietnam War and, 365
Large-scale maps, 495
Latin America
  agriculture in, 323–324
  population in, 89
  religion in, 194–195
  sewage systems in, 451
  view of the U.S. in, 482–483
Latitude, 22–23
Latvia, nationalism in, 354
Law
  environmental, 167
  geography in, 38
  standardization and, 233
Law of diminishing returns, 47
Lazarenko, Pavlo, 486
League of Nations, 360, 361
Lebanon, forced migration from, 111, 114
Le Corbusier, 440, 441
Lee Cooper jeans, manufacture of, 70–71
Le Heron, Richard, 330
Lemkin, Raphael, 343
Lenin, Vladimir, 352
Lesbian, gay, bisexual, and transgendered (LGBT) people, 201–203
Less developed countries (LDCs), 252
Lewis, Pierce, 453
Leyshon, Andrew, 377
Liberation theology, 194–195
Libya, borders of, 349
Life expectancy, 104–105
  economic development and, 253
  in peripheral cities, 452
Lifestyles, 238
Lifeworlds, 34, 218
Lincoln, Abraham, 380, 381
Literacy
  geographic, 4–10
  in index of human development, 254
Little Italys, 430, 431
Little Koreas, 430, 431
Live8 concerts, 486
Localization, 260
Localization economies, 273
Local scale, 8
  population and, 99, 100
Located charts, 495, 498
Location
  in spatial analysis, 22–24
  utility of, 25
Locational analysis, 94
Locational flexibility, 10
Location of public facilities, geography in, 37
Logical division, 30
Logistics zones, 422
London, England
  Covent Garden, 438
  financial center of, 422
  in Golden Triangle, 59
  place marketing in, 241–243
  population of, 400
  terrorist attacks in, 413
London Dumping Convention of 1972, 167
Longitude, 22, 23
Los Angeles, California
  census underrepresentation in, 87
  cognitive image of, 220, 221
Lourdes, France, 231, 232
Lovell, W. George, 147–148
Lynch, Kevin, 220

M

Maasai Mara National Reserve, 296
Maastricht Treaty, 376
Maathai, Wangari, 130, 131
Macedonia, 354–355
Machu Picchu, 152
Mackinder, Halford, 362–363, 366, 379
Madrid, Spain, terrorist attacks, 413
Magalhães, Fernando de, 53
Magellan, Ferdinand. *See* Magalhães, Fernando de
Malaria, 129
Mali, age-sex pyramid for, 92
Malthus, Thomas Robert, 119–120
Managua, Nicaragua, as primate city, 410
Man and Nature, or Physical Geography as Modified by Human Action (Marsh), 138–139
Mandela, Nelson, 193
Manila, Philippines
  air pollution in, 453
  packaged landscapes in, 459
  as primate city, 410
  slum housing in, 418
Manufacturing, decentralization of, 72
Manufacturing Belt, U.S., 60, 61, 62
Manufacturing value added (MVA), 261, 263
Mapping the Global Future, 478–488, 482–483
Map projections, 53, 495–500
Maps and mapping, 3, 495–504
  cartography, 52–53
  in data analysis, 22
  GIS and, 500–504
  Mercator projection, 366
  projections in, 53, 366, 495–500
  scales, 495
Maquiladoras, 288–289
Marcos, Ferdinand, 486
Marcuse, 462–463
Marginalized, the, 478–479, 480
Marinetti, Filippo, 232
Mariolle, Elaine, 214
Marketing
  geography in, 37–38
  GIS applications in, 94–95
  niche, 287
  place, 240–244
Marketing-driven commodity chains, 70
Markets, homogenization of, 280
Marks, Brian J., 312–315
Marley, Bob, 180
Marsh, George Perkins, 138–139
Marshall Plan, 270
Marvin, Simon, 421
Marx, Karl, 120, 194
Masai villages, 176
Masculinism, 57
Massachusetts, Proposition 2 1/2, 381
Mass communications, 232–233
Master-planned residential communities, 455, 459, 462, 463
Material culture, 240–244
Materialism
  international consumer markets and, 75
  shopping and, 227–228
  terrorism and, 19
Materials technologies, 476
Mature metropolises, 420–421
Mau Mau rebellion, 68
Mayan empire
  environmental impact of, 145
  epidemics in, 148
  as hearth area, 45
  urban origins, 397
McDonaldization, 207–208

McDonald's, 43, 80
  slow food movement and, 236, 238
McVeigh, Timothy, 371
McWorld, 77–79
Meaning, in place-making, 219
  semiotics and, 226–229
Meatpacking, 328
Mecca, 198
  pilgrimages to, 230–231
Mecca Cola, 79, 80
Mechanization, 311, 317
Media, global, 190, 197
  place-making and, 232–233
  in world cities, 411
Mediascapes, 72
Medical geography, 105–106
Medicare, 99
Medieval arras, 436
Medina, 198
Medinas, 443
Mediterranean transhumance routes, 309
Meek, James, 14–15
Megacities, 419–420
  splintering urbanism and, 422
Megapolitan regions, 454, 456–458
Meiji Restoration, 60–61
Meinig, Donald, 30, 32
Mendel, Gregor, 328
Mental maps. *See* Cognitive images
Mercator projection, 366, 496–497, 499
Merchant, Carolyn, 141
Merchant capitalism, 51, 399–400
Mergers and acquisitions, 281–282
Meridians, 23
Mesoamerica
  hearth areas in, 45
  urban origins in, 397
Mesopotamia
  environmental impacts in, 145
  urban origins in, 396, 397
  Ur in, 47
Mexico
  agriculture in, 323–324
  boundaries of, 346
  descansos in, 176
  epidemics in, 148
  Green Revolution in, 318, 319
  informal sector in, 445
  international debt of, 269
  local vs. national politics, 381
  maquiladoras, 288–289
  regional development patterns in, 256
  religion in, 194–195
  as salad bowl of North America, 360
  sewage systems in, 451
Mexico City
  air pollution in, 453
  as primate city, 410
  traffic in, 450
Microelectronics, 58
Middle America Main Streets, 223, 225
Middle cohort, 99
Middle East
  forced migration in, 111
  kinship in, 190–192
  urban origins in, 396
  view of the U.S. in, 483
  women in, 206
Migration, 108–119
  definition of, 108
  internal voluntary, 115–118
  international forced, 111–115
  international voluntary, 109–110
  North American urban structure and, 429–430
  reurbanization, 421
  rural-to-urban, 115–118
Milan, Italy
  metro system in, 26
  as world city, 411
Militarization of urban space, 488
Military expenditures, 254
Military intelligence, 503
Millenium Summit (2000), 122, 123, 124
Milosevic, Slobodan, 354–355, 486
Mining, environmental impact of, 150–151, 153, 155
Minisystems, 44–46
Minority groups
  birth rates among, 99
  congregation and segregation of, 428–429
  hip-hop culture and, 180–181
  rural-to-urban migration of, 115–116
Missionaries, 182, 185, 186, 188
Mitchell, Don, 225
Mitchell, Katharyne, 110
Mixed diffusion, 31
Mobility, population, 108–109
Mobilizing function of urban settlement, 394
Moderate Resolution imaging Spectroradiometer (MODIS), 21
Modernist architecture, 233
Modernity, 232
Modern movement, 440
Modification, of land use, 159
Modi Revlon, 208
Mollweide projection, 497, 499
Mombasa, Kenya, as gateway city, 402
Monarchy, 349
Money laundering, 76, 290
Monsanto Corporation, 336
Montenegro, 354–355
Moore, Michael, 331
Mormon culture region of the U.S., 30, 32, 182
Morrill, Richard, 25, 387
Mortality rates, 85, 103–106
  demographic transition theory and, 106–108
Mosques, 441–442
Mount Rushmore, 247
Muhammad, 195, 198
Multifiber Arrangement (MFA), 284, 286
Multinational states, 350
Municipal socialism, 438
Music
  hip-hop, 180–181
  world, 175, 208–209
Muslim, definition of, 195. *See also* Islam
Myrdal, Gunnar, 275–277

N

NAFTA (North American Free Trade Association), 7
  economic denationalization and, 12
  maquiladoras and, 288, 289
  political integration via, 375
  Québecois movement and, 203
Nagasaki, Japan, as gateway city, 402
Nairobi, Kenya
  as primate city, 410
  slum housing in, 418
Napoleon III, 439
NASA Moderate Resolution imaging Spectroradiometer, 21
NASDAQ (National Associated Automated Dealers Quotation System), 289
National Coalition for the Homeless, 435
National Commission on Terrorist Attacks Upon the United States, 374
National Council for Geographic Education, 37
National Geographic magazine, 4, 5
National Geographic Society, 37
National High System, 213–214
Nationalism, 351–355
  definition of, 350
  resurgence of, 20
  symbolic landscapes in, 32
  in the U.S., 481
National Law Center on Homelessness and Poverty, 435
National Marine Fisheries Service, 313
Nations, definition of, 349. *See also* States
Nation-states, 349–350
  hyperglobalist view of, 12

Native Americans
  environmental impact of, 149, 152
  forced migration of, 118
  longhouses, 179
  rites of passage, 179
  sacred spaces of, 229–230
NATO. *See* North Atlantic Treaty Organization (NATO)
Natural barriers as boundaries, 348
Natural decrease, 103, 104
Natural disasters
  cognitive images and, 222–223
  eco-migration and, 119
  urbanization and, 418
Natural gas, 153, 156
Natural increase, 103, 104
Natural resources
  development and, 257–260
  future of, 472–478
  industrial resources, 258–260
  materials technologies and, 476
  population distribution and, 89
  population limits and, 119–120
  primary economic activities and, 260, 262
Nature, 129–171. *See also* Environment
  commodification of, 141
  concept of, 130–141
  definition of, 131
  Neolithic view of, 144
  relationship between society and, 131–135
  science/technology and the concept of, 141
  social production of, 133
Nature Conservancy, 139
Nauru, as offshore financial center, 293
Navigation, 50
Nazi Holocaust, 343
Ndola, Zambia, self-help housing in, 450
Nearness principle, 25
Neighborhoods, 205
  decaying, 433–435
  European, stability of, 438
  immigrant, 429–430
  social polarization of, 462–463
Neocolonialism, 68–69, 361
Neo-Fordism, 286–288
Neoliberal policies
  definition of, 12
  deterioration of society and, 471
  sustainable development and, 124
Neolithic people, 143–144
Neo-Malthusians, 120
Nepal, child labor in, 256
Ness, Andrew, 460
Nestlé, 282, 288
NetGeneration, 97–98, 101
Netherlands
  industrialization in, 59
Net migration, 109
New England townscapes, 223, 225
Newly industrializing countries (NICs), 73, 261, 263
New Orleans, Louisiana
  Hurricane Katrina in, 133–134, 135
  postmodern landscape in, 240
Newton, Huey, 180
New urbanism, 233, 236
New World
  core and periphery in, 54–55
  ecological imperialism in, 148–152
  epidemics in, 147–148
  hearth areas, 45
  missions in, 182, 185, 186
New world order, 366–374, 479–488
  U.S. in, 479–481
New York City
  agglomeration effects in, 411
  as gateway city, 402
  global/local scales in, 8
  infrastructure problems in, 432
  interdependence of, 6–7
  place marketing in, 241
  population density in, 91
  World Trade Center attacks, 18–19
New Zealand
  agriculture in, 330–331
  uranium mining in, 150–151
NGOs. *See* Nongovernmental organizations (NGOs)
Niche marketing, 287
Nichols, Terry L., 371
NICs. *See* Newly industrializing countries (NICs)
Niemeyer, Oscar, 228–229
Niger, uranium mines in, 262
Niger-Congo language family, 200
Nigeria, regions in, 479
Nike, 286, 288
Nilo-Saharan language family, 200
NIMBYism, 381
Nitrogen oxides, 158–159
Nodes, in cognitive images, 219
Nomadism, 308–309
Nominal location, 22–23
Nongovernmental organizations (NGOs)
  environmental politics and, 167
  fair trade and, 275
  against globalization, 79–80
  in world cities, 411
Nonrenewable alternative energy sources, 159
Nontraditional agricultural exports (NTAEs), 324
North America
  energy consumption in, 257
  population in, 89
  pre-Columbian religions in, 186, 187
  settlement pattern of, 221–222
  urbanization in, 412
  urban structure in, 429–435
North American Free Trade Association (NAFTA). *See* NAFTA (North American Free Trade Association)
North Atlantic Treaty Organization (NATO), 365
North Korea, boundaries of, 346, 347
North/South divide, 356–360, 468
  definition of, 359
Northwest Territories Act of 1803, 350
Nuclear energy, 133, 152, 153
  countries using, 157
  uranium mining and, 150–151
Nuclear free zones, 5
Nuclear power, 57
Nutrition
  Columbian Exchange and, 148–149
  population density and, 91
  undernutrition, 335
Nutritional density, 91

O

Obasanjo, Olusegin, 123
Observation, of human geography, 20–21
Oceania, uranium mining in, 150–151
OECD. *See* Organization for Economic Cooperation and Development (OECD)
Offshore financial centers, 292, 293
Oil
  environmental impact of, 153, 156
  OPEC price increases on, 289–290
  price crises, 473
  production of, 257
Old-age cohort, 99
Old World, geography of, 47–50. *See also* Pre-modern world
100 mile cities, 454
On the Origin of Species (Darwin), 57
OPEC. *See* Organization of Petroleum Exporting Countries (OPEC)
Open door policy, 266, 484
Open space, 454
Operation Enduring Freedom, 370
Opportunity, intervening, 29
Optical character readers (OCRs), 290
Ordinary landscapes, 31–32, 223–229
Organic agriculture, 330–331
Organic theory of the state, 355–356
Organization for Economic Cooperation and Development (OECD), 75
  aid programs, 271
Organization of Petroleum Exporting Countries (OPEC)

in creation of fast world, 77
price increases by, 289–290
O'Riordan, Timothy, 168
Ottoman Empire, 235, 382
Silk Road and, 47, 48
Our Common Future, 260
Outermost edge-city complexes, 453
Outsiders, 218–219
sense of place and, 35
Outsourcing, 290–291
Overgrazing, 165
Overurbanization, 418–419
Ozone depletion, 76

P

Pacific Destiny, 484
Packaged landscapes, 455, 459, 460–461, 462
Pakistan
electoral representation in, 386
refugees in, 110
Paleolithic period, 142–143, 144
Palestine
forced migration from, 111, 115
Israeli conflict with, 382–385
Jerusalem and, 235
Palestinian Liberation Organization (PLO), 382–383
Panama, as offshore financial center, 293
Panama Canal, 63
Panama City, as gateway city, 402
Panregional influence, 411
Paris, France
as dominant city, 400
in Golden Triangle, 59
urban design/planning in, 439–440
Parochialism, 37
Parry, Martin, 332
Part-time work, 124
Pasteur, Louis, 328
Pastoralism, 308–309
Paths, in cognitive images, 219
Pearl River Delta, China, urbanization in, 414, 416–417
Pedicabs, 445
Peet, R., 48
Per capita consumption, 133
Perestroika, 354
Peripheral regions
in 1900, 60
aging of the population in, 97–98
definition of, 55
demographic transition and, 107–108
energy use in, 153, 156–157
environmental issues in, 167–168
future of, 486–487
globalization and, 75–81
imperialism and, 67–68
international division of labor and, 63, 66–67
Kyoto Protocol and, 163
megacities in, 415, 418–420
neocolonialism in, 68–69
in the New World, 54–55
organization of, 63–69
population growth in, 120–122
as Third World, 68–69
urbanization in, 404–405
Perot, H. Ross, 386, 387
Peru, refugees in, 113
Pesticides
biotechnology and, 328–331, 475–476
chemical farming and, 311
environmental impact of, 332–333
Green Revolution and, 318–320
pollution from, 130, 131
Peters, Arno, 500
Peters projection, 497, 500
Petrini, Carlo, 238
Philippines
electronic enterprise zones in, 291
films produced in, 196
informal sector in, 445
international debt of, 269
Overseas Contract Worker program, 109
Philips corporation, 281
Pilgrimages, 230–232
Pinchot, Gifford, 130, 139
Pink Gold Rush, 313
Pink spending, 202
Piracy, 266–267
Pittsburgh, PA, deindustrialization in, 279
Place-making, 8–10, 215–223
advertising in, 237–238
experience and meaning in, 219
globalization and, 232–237
images and behavior in, 219–223
race and, 205
territoriality and, 215–218
Places
definition of, 3
influence and meaning of, 5–6
insiders vs. outsiders in, 218–219
interdependence of, 6–7
interpreting, 213–249
love of, 223
marketing, 240–244, 244
in modern society, 232–246
as object of consumption, 237–240
Plantations, 50–51
Plazas, 435, 436
Plows, 145, 149
Plutonium, 150–151
Polarization
ethnicity and social, 462–463
socioeconomic, 477–478
Political ecology, 135, 136–137
Politics, 343–391
boundaries/frontiers in, 346–349
development of political geography and, 344–349
of geography, 379–386, 379–388
geopolitics, 344–346
global environmental, 166–167
INGOs in, 72
international/supranational organizations in, 374–378
nations in, 349–374
representation systems in, 386–388
states in, 344–346, 349–374
world order and, 349
Politics of geography vs. geography of politics, 379
Pollution, 135
from agriculture, 333–334
Asian Brown Cloud, 491
Lake Baykal, Russia, 13, 16
in peripheral cities, 452
from wood burning, 156
Pol Pot, 366
Polycentric metropolises, 453–463
packaged landscapes in, 455, 459, 462
Polyconic projection, 499
Popular culture, 175
Population, 85–127
age-sex pyramids, 92–99
aging of the, 92, 96–98, 483
birth/fertility rates and, 99–103
countries expecting declines in, 98
death/mortality rates and, 103–106
debates about, 119–120
demographic transition theory on, 106–108
demography and, 86–88
density and composition of, 89–92
distribution of, 88–89
doubling time, 101, 103
dynamics and processes in, 99–108
European expansionism and, 146, 147
factors in growth of, 85
movement/migration of, 108–119
policies/programs on, 120–124
shifting cultivation and, 306
sustainable development and, 124, 261
urbanization and, 412, 414
Population density, 89–92
agricultural, 91
crude, 89, 91
food-producing minisystems and, 45
health-care, 91
nutritional, 91
world, 88

Population policies, 120–124
Port au Prince, Haiti, as primate city, 410
Portsmouth, England, 244–245
Portugal, exploration by, 50, 52–53
Postindustrial economies, 261
  deindustrialization and, 279
Postmodernity, 236–237, 239
  food regimes and, 325–327
  landscapes of, 240
  packaged landscapes and, 455, 459, 462
Postmodern urban design, 459, 462
Poultry processing, 301–302
Poverty
  cycle of, 433–434
  in North American cities, 433–435
Pre-modern world, 44–50
  early empires, 46–47
  geography of, 47–50
  hearth areas, 44–46
Preservation, 139
Presidential elections, U.S., 386, 387
Price, Maria, 204–205
Primacy, 408–410
Primary activities, 260, 262
Primate cities, 409–410
Prime meridian, 22, 23
Privacy
  GIS marketing applications and, 95
  in Islamic cities, 441, 442, 443
Producer-driven commodity chains, 70
Producer services, 72
Product differentiation, 287
Projections, map, 53, 495–500
Property rights, 218
Proportional symbols, 495, 497
Proposition 2 1/2, 381
Proposition 13, 381
Propulsive industries, 279–280
Prostitution, 201
Protestant Reformation, 400
Protests, against globalization, 79–80
Proxemics, 217
Psychology, 214
Ptolemy, 49
Public participation geographic information systems (PPGIS), 95
Public policies
  on agriculture, 323–324
  on population, 120–124
Pudong, Shanghai, development zone, 422, 423
Pull factors, 109
Purchasing power parity (PPP), 252
Push factors, 109

Q

Quartering of urban space, 462–463
Quaternary activities, 260
Québecois movement, 190, 202–203, 379
Quiet Revolution, 203
Quiet supersonic aircraft technology (QSAT), 475
Qur'an, 195, 442

R

Race, 20
  censuses and, 86–88
  definition of, 205
  ghettos and, 434
  sense of place and, 205
Racialization, 205
Radioactive exposure, 150–151
Railroads
  economic integration and, 62
  geographic change from, 64–65
  high-speed, 473–474
  Industrial Revolution and, 58–59
  U.S., 62
Rain forests
  destruction of, 164 (*See also* Deforestation)
  shifting cultivation and, 304–307
Rangoon, Burma, as primate city, 410
Rank-size rule, 408–409
Rapid population growth
  age-sex pyramid for, 93
Rap music, 180
Ratzel, Friedrich, 56, 57, 344–345, 355–356
Ready Roundup, 336
Reapportionment, 387
Reciprocal social economy, 44
Redclift, Michael, 312
Redistribution of wealth, world-empires and, 46
Redistricting, 387
Redlining, 434
Refugees. *See* Forced migration
Regional analysis, 30–35
  landscape in, 31–33
  regionalization in, 30–31
  sense of place in, 33–35
Regional approach to geography, 49
Regional development, 273–280
  core-periphery patterns and, 277–280
  economic core creation in, 273–277
  patterns in, 256–257
Regional geography, 20
Regionalism, 379–380
  definition of, 31
  resurgence of, 20
Regionalization, 30–31
  globalization versus, 12
Regional movements, 379
Regional planning, 38
Regions
  definition of, 3, 20
  frontier, 347–348
  future of, 478–488
  identity and, 20
  urbanization in, 414–415
  world, 7
Relative distance, 24
Relative location, 23
Religion
  cultural regions and, 182
  cultural systems and, 183–186
  definition of, 183
  fundamentalism in, 186, 189, 195, 371, 488
  pilgrimages and, 230–232
  pre-Columbian, 186, 187
  resurgence of, 20
  sacred spaces and, 229–232
  terrorism based on, 371, 372
Religious right, 186, 189
Remote sensing, 20–21
Renaissance, 232, 400
  urban design/planning in, 439
Renewable energy sources, 152, 153
Renewing America's Infrastructure, 433
Representation of data, 21
Representative fractions, 495
Republican government, 349–350
Restaurants, as cultural sites, 239, 240
Reurbanization, 421
Reverse engineering, 266–267
Revlon, 208
Rice, Paul, 275
Rice production, 307–308
  Green Revolution in, 318–319
Rimini, Italy, tourism in, 295
Rio de Janeiro, Brazil, dualism in, 445
Risk society, 19, 490
Risk-taking, 222–223
Rites of passage, 178, 179
Ritter, Karl, 56, 57
Road accident memorials, 176
Robinson, Guy, 330
Robinson projection, 497, 500
Robotics, 58
Rockefeller Foundation, 318
Rock 'n' roll, 5
Roman empire, 46, 47
  geographic knowledge in, 49
  urban design/planning in, 439
  urban origins in, 397
Romanticism, 138
Rongelap, nuclear testing on, 150–151
Roosevelt, Theodore, 139
Rostow, W. W., 271
Route 66, 213–214
Royal Observatory, Greenwich, England, 22

Rural areas
rural-to-urban migration, 115–118
rural-urban boundaries and, 347
sectionalism and, 380–381, 386
Russia
AIDS in, 105
Chernobyl incident, 151, 153
environmental issues in, 13, 16
industrial resources in, 258
Lake Baykal, 13, 16
Rust, Paul, 14–15
Rustbelt, 116, 117
Rwanda, civil war in, 358–359

S

Sabbioneta, Italy, urban design of, 439
Sacred spaces, 229–232
Jerusalem, 234–235
Salinization, 152
Sallnow, M., 231
Salvador, Brazil, as gateway city, 402
Samarkand, 48
San Francisco, California
Chinatown, 205
wetlands conversion in, 165–166
Sanson, Nicolas, 53
Santeria, 194–195
San Xavier del Bac mission, 188
São Paulo, Brazil
as primate city, 410, 420
sewage systems in, 451
traffic in, 450
SARS (Severe Acute Respiratory Syndrome), 18
Satellite communications, 289
Saturnini, Paolo, 236
Sauer, Carl, 44, 133, 188, 225
on cultural landscape, 176, 177
Savannas, 164, 165
Savings, institutionalization of, 289
Savitch, H. V., 413
Sawici, D. S., 95
Scales, geographic, 7–8
Scales, map, 495
Science
concept of nature and, 141
deterioration of society and, 471
as power, 19
Scientific management, 286
Scientific revolution, 400
Scotland
birth/death rates in, 107, 108
St. Andrews, 397, 399
Sea level, changes in, 133, 135
Seattle, WA, reurbanization of, 421
Secondary activities, 260, 264
Sectionalism, 379–380
definition of, 31
Security issues. *See also* Food security
future of, 489–490
refugees, 111, 113–114
terrorism and, 18–19
Seed agriculture, 310
Segregation, 428–429
Seko, Mobutu Sésé, 486
Self, as geographic scale, 8
Self-actualization, 238
Self-determination, 379
Self-Employed Women's Association (SEWA), 207
Self-fulfilling prophecy, 434
Semiotics, 226–229, 227
definition of, 226
Semiperipheral regions
in 1900, 60
definition of, 55
gross national income in, 252–253
megacities in, 415, 418–420
in the New World, 55
Semple, Ellen Churchill, 57
Senegal, HIV/AIDS in, 18
Sense of place, 33–35
Separatism, 488–489
Québecois, 190, 202–203
Septic systems, 501
Serbia
Croatia claims by, 31
formation of, 354–355
Sewage systems, 451–452, 501
Sexual geographies, 201–204
Sexuality, 201
Shakur, Tupac, 180
Shan Chujie, 251
Shantytowns, 117–118, 418–420, 447, 449–450
Shifting cultivation, 304–307, 321
Shi'i, 198
Shipbuilding, 50
Shipping
containerization and, 73–74
steam-powered, 62, 66
Shock cities, 403, 444
Shopping behavior, 220–221
Shopping malls
behavior in, 220–221
semiotics of, 227–228
Short twentieth century, 468
Shrimp fishing, 312–315
Sickles, 145
Sierra Club, 139
Silent Spring (Carson), 130, 140, 311, 333
Silk Road, 47, 48, 50, 185
Siltation, 145
Singularity, 10
Sinopec Corp, 73
Site, definition of, 23
Situation, definition of, 23
Six-Day War (1967), 382
Skeptics, on globalization, 10, 12
Skylines, 436–438
Slash-and-burn cultivation, 44, 305–306
Slave trade, 194–195
Slovenia, 354–355
Slow city movement, 236, 238–239
Slow food movement, 236
Slow population growth, age-sex pyramid for, 93
Slow world, 76–77
Americanization and, 207–208
Slum housing, 418–420
Smallpox, 148
Small-scale maps, 495
Smart cars, 473, 475
Smart growth, 454–455
Smith, Neil, 133, 371, 481
Smog, 491
Snowbelt, 116
Social classes
gender systems and, 206–207
global social hierarchy and, 480
industrialization and, 59
urban polarization of, 462–463
Social construction of places, 5
Social Darwinism, 344–345
Social economy, reciprocal, 44
Socialism, public housing and, 438–439
Social organization, food-producing minisystems and, 45
Social polarization, 462–463
Social Security, 99
Social stratification, 12–13
Society
agriculture and changes to, 327–328
culture and, 190–193
definition of, 132
environmental impact of, 132–133
nature and, 131–133
postmodern, 236–237
Socioeconomic space, 25
Socio-spatial formations, 462–463
Soil degradation, 333–334
Solar energy, 58, 159
Somalia, humanitarian need in, 272
Sormolo, Hussein, 14–15
Soubirous, Bernadette, 231
South Africa
colonialism in, 361
internal forced migration in, 118–119
languages in, 193
on sexuality and human rights, 204
South America
colonialism in, 358
decolonization in, 363
deforestation in, 164

South Asia
eco-migration from, 119
hearth areas in, 45
hunger in, 254
pre-modern world-empires in, 47, 50
women in, 206–207
South Coast Plaza, Orange County, California, 227
South Korea
boundaries of, 346, 347
deindustrialization in, 267
nuclear energy in, 153
Sovereignty, 349–350
Soviet Union, former
Afghanistan and, 369
boundaries of, 347
breakup of, 354
Cold War and, 270, 271, 367
nationalism and, 351–355
regional development patterns in, 257
as superpower, 377
Space, 25–26
ethnicity and, 204–205
in modern society, 232–246
sexual identity and, 203–204
in spatial analysis, 22
Spain
Basque Homeland and Freedom movement, 379
colonies of, 147–148
descansos in, 176
exploration by, 50
industrialization in, 59
Industrial Revolution in, 36
urban system in, 408
Spatial analysis, 22–30
accessibility in, 26
distance in, 24–25
location in, 22–24
space in, 25–26
spatial interaction in, 27–30
Spatial change, 472–478
Spatial differentiation, 346
Spatial diffusion, 29–30, 31
Spatial interaction, 27–30
in spatial analysis, 22
Spatial justice, 76, 253–254
Spatial perspective
on population, 88–89
Spatial scales, 7–8
Special Economic Zones (SEZs), 416–417
Specialization
complementarity and, 27
food-producing minisystems and, 45
functional, 409
global office and, 291–292
labor pools and, 274–275
regional, 399–400
Splintering urbanism, 421–422
urban structure and, 427
Sprawl, 454–455, 460–461
Spread effects, 277
Squares, city, 435, 436
Squatter settlements, 418–420, 447, 449–450
Sri Lanka
clothing industry in, 284
Colombo as gateway city in, 402
refugees in, 114
terrorism in, 370
water supplies in, 451
St. Andrews, Scotland, 397, 399
St. Vincent, banana industry in, 137
Stalin, Josef, 352–353, 372
Standardization, 233
Stanton, Gregory, 343
Starbucks Coffee Co., 94
States
definition of, 7
federal, 352–353
geopolitical model of, 344–346
imperialism/colonialism and, 356–360
nationalism and, 350–355
in the New World, 55
Ratzel on, 56, 57, 344–345, 355–356
theories and practices of, 355–374
unitary, 352–353
Steamboats, 62
Steam engines, 57, 62
Steamship routes, 66
Steinbeck, John, 213
Stephenson, George, 62
Stone Age people, 142–143
Strabo, 49
Straight-line boundaries, 348
Strategic alliances, 288
Street vendors, 452
Subjective settings, 20
Subsidies, agricultural, 322
Subsistence agriculture, 303, 311
intensive, 307–308
Suburbanization, 116–117
definition of, 116
landscape and, 223, 225
North American, 429–430
sectionalism and, 380–381, 386
sprawl and, 454–455
Succession, invasion and, 430
Sudan
borders of, 349
environmental perception in, 214–215, 216
genocide in, 343
refugees from, 112–113, 114–115
Suez Canal, 63
Sulfur dioxide, 158–159
Sumatra, population redistribution to, 118
Sumer, city-states in, 396
Sunbelt, 116, 117
Sunna, 195
Sunni, 198
Supranational organizations, 7, 375–376
Suqs, 442
Surface mining, 155
Surveillance systems, 488, 489
Sustainable development, 124, 130, 490–491
definition of, 258
ecological footprint and, 260–261
environment and, 167–168
Swidden cultivation, 44, 305–306
Sydney, Australia
as gateway city, 402
place marketing in, 241
Symbolic landscapes, 32–33

T

Taipei, Taiwan, traffic in, 450
Taliban, 201, 369–370
Tax incentives, 288–289
Taxpayer revolts, 381
Taylor, P. J., 412
Technology
in agriculture, 327–332
concept of nature and, 141
definition of, 132
environmental impact of, 132–133
future of, 472–478
gender and access to, 307
globalization and, 73, 75
human interaction and, 467
industrialization and, 55–61, 58–59
industrial resources and, 258, 260
information, 476–478
materials, 476
Technology achievement index, 477
Technology systems, 55–56
Technopoles, 280, 422
Technoscapes, 72
Teen pregnancy, 434
Telegraphs, 66
Television
evangelism on, 186
international, 175
world distribution of, 197
Telework, 124
Terms of trade, 270
Territoriality, 215–218. *See also* States
boundaries and frontiers in, 346–349
power and, 345–346
urban structure and, 428–429
Territorial organization, 386

Territories
boundary formation and, 348
de jure, 7
Territories, definition of, 346
Terrorism
as attacks on globalization, 468
definition of, 367
eco-, 139
ethnicity and, 489–490
jihad vs. McWorld and, 78–79
new world order and, 366–374
refugee movements and, 109, 111, 113–114
regionalism and, 379, 380
security issues and, 18–19, 489–490
tourism and, 295–296
urban, 413
in world cities, 411, 413
Tertiary activities, 260, 264
Text, landscape as, 226
Textiles, 50–51
Text messaging, 174
Thailand, shrimp farms in, 313
Thatcher, Margaret, 243
Thematic maps, 495, 497
Third World, 68–69, 252. *See also* Peripheral regions
Thoreau, Henry David, 135, 138
Tibet, 188
Tibeto-Burmese language family, 188
Time-space convergence, 28–29
Times projection, 497
Timur, 369
Tobler, Waldo, 24–25
Tokaido megalopolis, Japan, 456, 458
Tokugawa regime, 60–61
Tokyo, Japan
terrorism in, 413
urban structure in, 427
Topographic maps, 495, 496
Topological space, 25, 26
Topophilia, 223
Topsoil, loss of, 333
Toronto, Canada, homelessness in, 435
Total fertility rate (TFR), 101–102
Toulouse, France, 11
Touré, 180
Tourism, 239–244
economic development and, 292–296
sustainable, 261
Township-and-range system, 350
Toyota, 282, 283, 286
Tractors, development of, 62–63, 317, 321
Trade
containerization and, 73–74
food-producing minisystems and, 45
international, 264–271
Silk Road in, 47, 48
Trade-off model of land use, 428
Trading blocs, 264
Traditional downtown centers, 453
Trail of Tears, 118
Transatlantic shipping, 66
Transcendentalism, 138
Trans-European Network (TEN), 473–474
Transfair USA, 275
Transferability, 27–29
Transformationalists, 10, 12–13
Transformative functions of urban settlement, 395
Transhumance, 308–309
Transnational corporations
in agribusiness, 325
commercial imperialism and, 68
definition of, 68
global assembly line and, 281, 282–286
globalization and, 10
global office and, 289–292
internationalization of finance and, 73
local differences and, 20
neo-Fordism, 286–288
Transnational crime, 489–490
Transnational migrants, 109–110
Transnational political integration, 375–376
Transportation systems
core region internal development and, 61–63
future of, 473–475
suburbanization and, 117
time-space convergence and, 28–29
Travel, international, 239–240
Tribal identity, 192–193
Tribes, 192–193
Trickle-down effects, 277
Trucks, development of, 63
Tsunamis, 119
Tübingen, Germany, piazzas in, 436
Tucson, Arizona
age-sex pyramid on, 101
San Xavier del Bac mission, 188
Tunis, North Africa, 208
Turkey, women in, 201
Turnover time, 77
Tutsi, 358–359
Typhoid, 129
Tyson Foods, Inc., 301–302
Tysons Corner, Virginia, 453, 454

U

Überlingen, Germany, 6
Uganda, HIV/AIDS in, 18
U.N. Food and Agricultural Organization, 164
Underclass, 434
Underdevelopment, 273
Underemployment, 444
Undernutrition, 335, 450
UNESCO (United Nations Economic, Social, and Cultural Organization), 243
World Natural Heritage sites, 16
UNICEF (United Nations International Children's Fund), 256, 393
Union Carbide, 453
Unique outcomes, 36
Unitary states, 352–353
United Kingdom
Afghanistan and, 369
birth/death rates in, 107, 108
canals in, 61–62
deindustrialization in, 278
Domesday Book, 177, 178
fair trade in, 275
historic places in, 243
imperialism by, 67–68
Industrial Revolution in, 36, 57–60
nuclear tests by, 151
place marketing in, 244–245
symbolic landscapes in, 34
telegraph network in, 66
United Nations
Environment Program, 167, 491
International Bill of Human Rights, 378
Law of the Sea, 167
Millenium Summit (2000), 122, 123, 124
Palestinian-Israeli conflict and, 382
political integration via, 375
population policy conferences, 122
on urbanization, 412
United Nations Center for Human Settlements (UNCHS), 233, 243, 337, 393, 394
splintering urbanism and, 421
United Nations Development Program, 75–76
Gender Empowerment Index, 255
on genetic engineering, 475
index of human development, 254
Millennium Development Goals, 124
United Nations International Children's Fund (UNICEF). *See* UNICEF
United States
age-sex pyramid for, 92–93, 99
aid programs, 271
AIDS in, 105
baby boom in, 96–97
boundaries of, 346
canals in, 62
census in, 86–88
chicken industry in, 301–302
cognitive images of, 221–222

United States (*cont.*)
  as core region, 75
  cultural nationalism and, 193–201
  cultural regions, 182
  deindustrialization in, 278
  demographic center of, 116
  domino theory and, 364–365
  Economic Development Administration, 279
  electoral systems in, 386, 387–388
  environmental views in, 135, 138–140
  farm crisis in, 328
  geographic education in, 37
  geography knowledge in, 4
  hegemony of, 68, 374, 479
  homelessness in, 434–435
  Hurrican Katrina in, 133–134
  income inequality in, 479
  industrial resources in, 258
  Industrial Revolution in, 57
  internal forced migration in, 118
  internal voluntary migration in, 115–117
  Kyoto Protocol and, 163
  Manufacturing Belt, 60, 61, 62
  megapolitan regions in, 456–457, 458
  new imperialism by, 68–69
  in new world order, 479–481
  perceptions of, 4
  population density in, 89, 90
  railroads in, 64–65
  regional development patterns in, 257
  resistance to globalization in, 79
  reurbanization in, 421
  Route 66, 213–214
  sectionalism in, 379–380
  shopping in, 227
  shrimp fishing in, 312–315
  sprawl in, 454–455
  as superpower, 366–367, 377
  township-and-range system in, 350
  trade deficits, 290
  urban system in, 408, 409
Universal Zulu Nation, 180
Universities, roles of, 471
University centers, 397, 398
Ur, Mesopotamian city of, 47, 396
Uranium mines, 150–151, 262
Urban agriculture, 337, 339
Urban design and planning
  European, 439–440
  packaged landscapes and, 455, 459, 462
  postmodern, 459, 462
  smart growth and, 454–455
Urban ecology, 396
Urban forms, 395–396
Urbanism
  definition of, 396
  new, 233, 236
  splintering, 421–422
Urbanization, 393–425
  agriculture in, 144–145
  competition between cities and, 381, 386
  current, 412, 414
  early world-empires and, 46–47
  environmental impacts of, 152–166
  mature metropolises and, 420–421
  megacities, 419–420
  origins of, 396–405
  overurbanization, 418–419
  place and, 218–219
  rates of growth in, 394, 395
  rural-to-urban migration and, 115–116
  rural-urban boundaries and, 347
  sectionalism and, 380–381, 386
  shantytowns and, 117–118
  splintering urbanism, 421–422
  terrorism and, 413
  urban geography and, 394–396
  urban structure and, 427–465
  urban systems and, 405–421
  world cities in, 411–412
Urban planning
  geography and, 38
Urban structure, 427–465
  European, 435–440
  Islamic, 441–444
  land use and, 428
  militarization in, 488
  North American, 429–435
  in peripheral regions, 444–453
  polycentric metropolises, 453–463
  spatial organization in, 428–429
  traditional patterns of, 429–453
Urban systems, 405–421
  definition of, 395–396
Uruk, 396
U.S. Endangered Species Act, 81
U.S. Environmental Protection Agency (EPA), 81
U.S. Land Ordinance of 1785, 350
U.S. National Intelligence Council, 478–488, 482–483, 489
U.S. Postal Service, 290
Utility, of locations, 25

V

Vallon-pont-d'arc, France, cave paintings, 142, 143
Van Kempen, Ronald, 462–463
Vermeer, Johannes, 53
Vernacular landscapes. *See* Ordinary landscapes
Vertical disintegration, 287–288
Vertical integration, 330
Vietnam, interdependence of with France, 11
Vietnam War, 365–366
Ville Radieuse, La, 440, 441
Virgin soil epidemics, 147–148
Visual consumption, 238–239
Visual culture, 240–244
Visualization of data, 21
  maps in, 500, 502
Vital records, 86
Volkswagen, 286
Voluntary migration, 109–110
Von Humboldt, Alexander, 56–57
Voodoo, 194–195
Vulgaria, 223, 225, 226

W

Wachusett Reservoir, Massachusetts, planning map, 503
Walden (Thoreau), 135
Wallerstein, Immanuel, 46, 273
Wal-Mart, 70, 282
War on terror, 372–374
  new world order and, 367, 370–374
  refugees due to, 109, 111, 113–114
Washington, D. C.
  cognitive image of, 24
  poverty areas in, 434
  situation of, 23
  symbolic landscape in, 32
Water power technology, 51, 57, 153
Water supplies
  Green Revolution in, 320
  irrigation and, 146
  in North American cities, 432
  in peripheral cities, 451–452
  as supranational organization, 375
  wetlands and, 165–166
Wealth
  global inequality of, 75–76
  increasingly inequal distribution of, 12–13, 14–15, 479, 481
  world-empires and redistribution of, 46
Weapons trafficking, 490
Weather changes, 133–135. *See also* Climate change
West Africa, cocoa production in, 268, 270
West African Rice Development Association, 327–328
Westernization
  cultural nationalism and, 193–201
  jihad vs. McWorld and, 78–79
West Nile Virus (WNV), 129–130
Wetlands, 159, 165–166
Wheat and livestock food regime, 325
Whiteness, 205

Whitworth, Courtney, 204–205
WHO. *See* World Health Organization (WHO)
Wiethaler, Cornelia, 6
Wind energy, 159
Windows of locational opportunity, 277
Women
  of childbearing age, 92
  ecofeminism and, 140
  education of
    birth rates and, 100–101, 123–124
  equality of and development, 255–256
  in Islam, 201
  in Islamic urban design, 442
  in pastoralism, 309
  rap music on, 181
  in South Asia, 206–207
  status of
    birth rates and, 123–124
Wood burning, 156
  countries using, 157
  economic development and, 257
World Bank
  in agricultural development, 323
  aid programs, 270–271
  EU in, 481
  on knowledge and development, 281
  on Vietnam, 11
  on water supplies, 451
World cities, 291–292, 404, 411–412
  splintering urbanism and, 421–422
  terrorism in, 413
World Commission on Environment and Development, 260
World Council of Churches, 497, 500
World Development Reports, 281
World Economic Forum (2005), 73
World-empires
  colonization by, 47
  geographic knowledge and, 49
  pre-modern, 46–47
  urban origins in, 397
World Health Organization (WHO), 86
  on sexual rights, 204
  on urban pollution, 453
World Heritage Lists, 243
World music, 5, 175, 208–209
World Natural Heritage sites, 16
World regions, 7
World Summit on Sustainable Development (2002), 491
World-system, 43
  core and periphery in, 54–55
  definition of, 46
  dependency in, 273
  New World in, 50–69
World Trade Center attacks, 18–19, 413
  economic impact of, 472
  globalization and, 468
  jihad vs. McWorld and, 78–79
  new world order and, 367, 370–374
  refugee movements and, 109
  tourism and, 295–296
World Trade Organization (WTO)
  on biodiversity, 167
  China in, 266
  economic coordination efforts of, 322
  environmental standards and, 81
  on genetically modified organisms, 336, 337
  protests against, 80–81, 166
World Urbanization Prospects: The 2003 Revision, 412
World War I, Japanese industrialization and, 61
World War II
  baby boom and, 96
  National Highway System and, 213
World Watch Institute, 139
World Wide Web, 77. *See also* Internet
World Wildlife Fund, 334
WTO. *See* World Trade Organization (WTO)

Y

Yi-Fu Tuan, 229
Yokes, in agriculture, 145
Yosemite National Park, 296
Youth cohort, 99
Youth culture, 180–181
Yugoslavia, 354–355, 380

Z

Zahir, Mohammad, 369
Zambia
  HIV/AIDS in, 18
  self-help housing in, 450
Zero population growth (ZPG), 122
  age-sex pyramid for, 93
Zimbabwe, HIV/AIDS in, 18
Zionism, 382
Zoning, 454
Zoonotic diseases, 490
Zug, Switzerland, 14–15

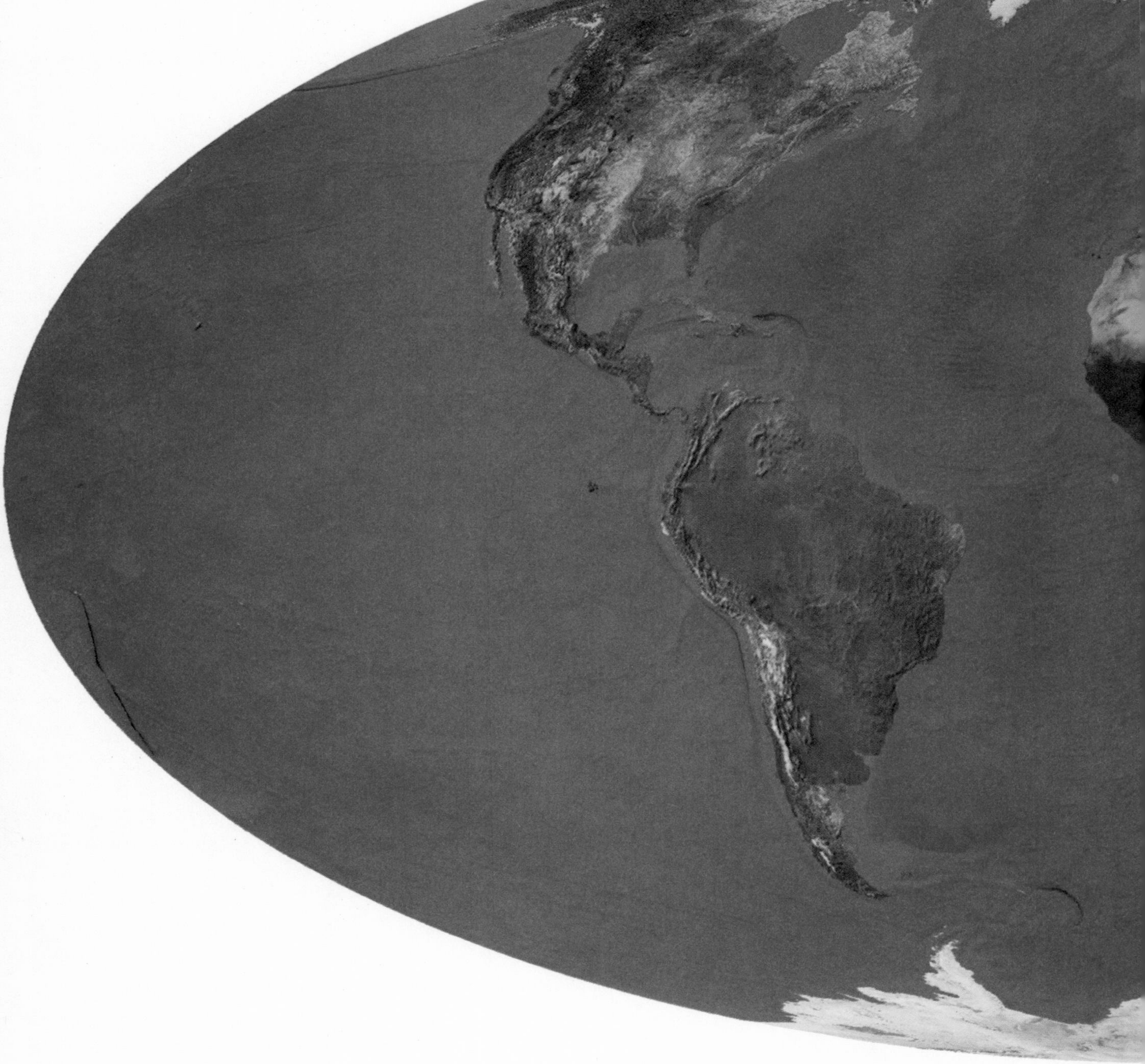

World Topography

World States See page 497 of the Appendix for an explanation of this projection